CW00384506

Numerical Techniques for Microwave and Millimeter-Wave Passive Structures

Numerical Techniques for Microwave and Millimeter-Wave Passive Structures

Edited by

TATSUO ITOH

Department of Electrical and Computer Engineering
The University of Texas at Austin
Austin, Texas

A Wiley-Interscience Publication

JOHN WILEY & SONS

New York Chichester Brisbane Toronto Singapore

Library of Congress Cataloging in Publication Data:

Tatsuo Itoh

Numerical techniques for microwave and millimeter-wave passive structures / edited by Tatsuo Itoh.

p. cm.

"A Wiley-Interscience publication."

Includes bibliographies and index.

ISBN 0-471-62563-9

1. Microwave devices—Design and construction—Data processing.
2. Numerical calculations. I. Itoh, Tatsuo.

TK7876.N86 1988

621.381′3—dc19

88-28620
CIP

Printed in the United States of America

10 9 8 7 6 5 4 3 2

CONTRIBUTORS

M. D. Abouzahra
Lincoln Laboratory
Massachusetts Institute of Technology
Lexington, Massachusetts 02173

J. B. Davies
Department of Electronic and Electrical Engineering
University College
London WC1E 7JE England

K. C. Gupta
Department of Electrical and Computer Engineering
University of Colorado
Boulder, Colorado 80309

Wolfgang J. R. Hoefer
Laboratory for Electromagnetics and Microwaves
Department of Electrical Engineering
University of Ottawa
Ottawa, Ontario K1N 6N5 Canada

Tatsuo Itoh
Department of Electrical and Computer Engineering
The University of Texas at Austin
Austin, Texas 78712

Juan R. Mosig
Laboratoire d'Electromagnétisme et d'Acoustique (LEMA)
Départment d'Électricité—Écublens
École Polytechnique Fédérale de Lausanne
Écublens, CH-1015 Lausanne, Switzerland

Wilfred Pascher
Allgemeine und Theoretische Elektrotechnik
FernUniversität
D5860 Iserlohn, Federal Republic of Germany

Reinhold Pregla
Allgemeine und Theoretische Electrotechnik
FernUniversität
D5860 Iserlohn, Federal Republic of Germany

Y. C. Shih
Microwave Products Division
Hughes Aircraft Corporation
Torrance, California

R. Sorrentino
Department of Electronic Engineering
Universita di Roma Tor Vergata
Rome 00173 Italy

Tomoki Uwano
Wireless Research Laboratory
Matsushita Electric Industrial Company
Osaka 571 Japan

Ingo Wolff
Department of Electrical Engineering and
Sonderforschungsbereich 254
Universität Duisberg
D4100 Duisberg, Federal Republic of Germany

PREFACE

Recent advances in millimeter-wave integrated circuits, particularly in the monolithic form, have increased the necessity of accurate computer-aided design (CAD). Unlike hybrid microwave integrated circuits at low frequencies, it is extremely difficult and essentially impossible to adjust the circuit characteristics of monolithic circuits once they are fabricated. Therefore, an accurate CAD program is essential for design of these circuits.

The starting point for the development of CAD programs is an accurate characterization of the passive and active structures involved in the circuits. Although most CAD programs are based on curve-fitting formulas and look-up tables and not on accurate numerical characterization, the latter can be used if it is fast enough. In addition, it can be used to generate look-up tables and to check the accuracy of empirical formulas.

In line with significant advances in the capabilities of computers, numerical methods have also been advanced in the last several years. Some methods have been made more efficient and some that originated in a different discipline such as engineering mechanics have found applications in electromagnetic wave problems.

This book is intended for graduate students and working engineers who want to acquire a working knowledge of numerical methods for passive structures. Such a talent is believed to be one of the most important assets for the next generation engineers. The book is written by a number of internationally recognized researchers on each subject. However, they recognize the fact that many novices trying to learn these difficult subjects have a hard time in catching up with the latest developments. For this reason, each chapter is written as comprehensively as the book format allows. Typically, each chapter starts with a brief historical background and a description of the method, followed by detailed formulations for practical examples. Computer program descriptions are included where appropriate.

The book starts with an Introduction and Overview to give the reader a quick guided tour of most of the numerical methods that are treated in detail in subsequent chapters. Although the contents of each chapter are comprehensive enough, it is by no means intended for the reader to acquire all the necessary tools to solve his own problems. However, it is believed that this book provides sufficient grounding for those attempting to plunge into numerical endeavors for passive structures in microwave and millimeter-wave circuits.

TATSUO ITOH

CONTENTS

Chapter 1 Introduction and Overview 1

Tatsuo Itoh

1. Finite Difference Method 2
2. Finite Element Method 4
3. TLM Method 6
4. Integral Equation Method 8
5. Moment Methods and Galerkin's Method 9
6. Mode-Matching Method 12
7. Transverse Resonance Technique 14
8. Method of Lines 16
9. Generalized Scattering Matrix Method 19
10. Spectral Domain Method 22
11. Equivalent Waveguide Method 25
12. Planar Circuit Model 26
13. Conclusion 29

References 30

Chapter 2 The Finite Element Method 33

J. B. Davies

1. Introduction 33
2. The Method of Weighted Residuals 35
3. The Variational Method 40
4. Using a Variational Expression 52
5. The Finite Element Method 59
6. Integral Formulation of Problems 80
7. Antennas and Scattering from Conductors 86
8. Waveguides—Hollow, Dielectric, and Optical 95
9. Finite Differences in Space and Time 102
10. Matrix Computations 108
11. A Finite Element Computer Program for Microstrips 123

References 131

Chapter 3 Integral Equation Technique 133

Juan R. Mosig

1. Introduction 133
2. Conventions and Notations 135

3. Integral Equations for Layered Medium Problems 137
4. The Fields of an Elementary Source in a Layered Medium 146
5. Two Practical Substrates 152
6. The Surface Waves 155
7. Zero, Low, and High Frequencies 161
8. Numerical Evaluation of Sommerfeld Integrals 166
9. Results for the Potentials 176
10. The Method of Moments 184
11. Excitation and Multiport Analysis 193
12. Practical Applications 198
13. Concluding Remarks 209
References 211

Chapter 4 Planar Circuit Analysis **214**

K. C. Gupta and M. D. Abouzahra

1. Introduction 214
2. Planar Circuit Analysis 216
3. Green's Function Approach 221
4. Impedance Green's Functions 224
5. Contour Integral Approach 247
6. Analysis of Planar Components of Composite Configurations 256
7. Planar Circuits with Anisotropic Spacing Media 263
8. Applications of the Planar Circuit Concept 270
9. Summary 311
Appendix A. Impedance Matrices for Planar Segments with Regular Shapes 313
Appendix B. Consistency of the Mode-Matching Approach with the Green's Function Approach 326
References 328

Chapter 5 Spectral Domain Approach **334**

Tomoki Uwaro and Tatsuo Itoh

1. Introduction 334
2. General Approach for Shielded Microstrip Lines 335
3. The Immittance Approach 345
4. Formulations for Slotlines, Finlines, and Coplanar Waveguides 351
5. Numerical Computation 355
Appendix A. Some Mathematical Identities 359
Appendix B. Derivation of Eq. (21) 360
Appendix C. Derivation of Eqs. (32), (33), and (34) 364
Appendix D. Derivation of the Poynting Power Flow Formulation in the Immittance Approach 366
Appendix E. Computer Program Example 370
References 380

Chapter 6 The Method of Lines **381**

Reinhold Pregla and Wilfrid Pascher

Preface 1
1. Introduction 1
2. The Full Wave Analysis of Planar Waveguide Structures by the Method of Lines 383
3. Extensions 400
4. Numerical Results 423
5. About the Nature of the Method of Lines 430

Appendix A. Determination of the Eigenvalues and Eigenvectors of P 434
Appendix B. Calculation of the Matrices $\boldsymbol{\delta}$ 438
Appendix C. The Component of ε_h at an Abrupt Transition 440
Appendix D. Eigenvalues and Eigenvectors for Periodic Boundary Conditions 441
Appendix E. Calculation of Characteristic Impedance 442
References 444

Chapter 7 The Waveguide Model for the Analysis of Microstrip Discontinuities **447**

Ingo Wolff

1. Introduction 447
2. The Waveguide Model of the Microstrip Line 449
3. Mathematical Analysis of Microstrip Discontinuities 466
4. Convergence and Numerical Results 481
5. Summary 489

References 491

Chapter 8 The Transmission Line Matrix (TLM) Method **496**

Wolfgang J. R. Hoefer

1. Introduction 496
2. Historical Background 498
3. Huygens's Principle and Its Discretization 498
4. The Two-Dimensional TLM Method 500
5. The Three-Dimensional TLM Method 532
6. Errors and Their Correction 548
7. Variations of the TLM Method 552
8. Applications of the TLM Method 568
9. Discussion and Conclusion 571

Appendix. A Two-Dimensional Inhomogeneous TLM Program for the Personal Computer 573
References 587

Chapter 9 The Mode-Matching Method **592**

Y. C. Shih

1. Introduction 592
2. Formulation 593
3. Relative Convergence Problem 603
4. Numerical Examples 604
5. Conclusion 613

Appendix A. Scattering Parameters for Bifurcated Waveguides 614
Appendix B. Scattering Parameters for Step Discontinuity 616
Appendix C. Comparison of Different Formulations 618
References 619

Chapter 10 Generalized Scattering Matrix Technique **622**

Tatsuo Itoh

1. Introduction 622
2. Definition of Generalized Scattering Matrix 623
3. Simple Use of the Generalized Scattering Matrix 624
4. Examples for Cascaded Junctions 627
5. Conclusion 631

Appendix. Computer Program Description 631
References 636

Chapter 11 Transverse Resonance Technique **637**

R. Sorrentino

1. Introduction 637
2. Inhomogeneous Waveguides Uniform Along a Transverse Coordinate 643
3. Conventional Transverse Resonance Technique for Transversely Discontinuous Waveguides 649
4. Generalized Transverse Resonance Technique for Transversely Discontinuous Inhomogeneous Waveguides 654
5. Analysis of Discontinuities and Junctions by the Generalized Transverse Resonance Technique 664
6. Examples of Computer Programs 678

Appendix. Field Expansion in Waveguides 690
References 693

Index **697**

1

Introduction and Overview*

Tatsuo Itoh

Department of Electrical and Computer Engineering
The University of Texas at Austin
Austin, Texas

Numerical characterizations and modelings of guided-wave passive components have been an important research topic in the past two decades. The necessity of such activities has become increasingly obvious in recent years. This is due to increased research and development in millimeter-wave integrated circuits and monolithic integrated circuits. It is no longer economical, or in many cases even feasible, to tune the circuits once they are fabricated. Therefore, extremely accurate characterization methods are needed to model the structures.

Because most structures used in today's printed and planar integrated circuits are not amenable to closed-form analytical expressions, the numerical methods needed for characterizations are in fact a necessary evil. Circuit designers would like to use CAD packages, which in most cases consist of curve-fitting or empirical formulas. However, the validity of these formulas must be supported by accurate characterizations. In addition, any numerical methods for characterizations need to be as efficient and economical as possible in both CPU time and temporary storage requirements, although recent rapid advances in computers impose less severe restrictions on the efficiency and economy of the method. Another aspect important in the development of numerical methods has been the versatility of the method. In reality, however, numerical methods are chosen on the basis of trade-offs between accuracy, speed, storage requirement, versatility, etc., and are often structure-dependent. Since the advent of microwave integrated circuits, a number of methods have been invented and the somewhat more classical methods have been refined for these modern structures.

When a specific structure is analyzed, one has to make a choice as to which method is best suited for the structure. Obviously, the choice is not

*Modified from an article that originally appeared in *Annales des Telecommunications*, vol. 41, pp. 449–462, Sept.–Oct. 1986.

unique. Therefore, the user must make a critical assessment for each candidate method. In this chapter, we list a number of numerical methods and present generally accepted appraisals for them. It is not possible to make an exhaustive list of all available methods. Rather, the most representative ones are reviewed. Although the assessment will be aimed at the characterization of three-dimensional passive structures, many methods are also effective for two-dimensional problems. Furthermore, two-dimensional methods can be used in an integral part of the composite characterization program for three-dimensional structures.

1. FINITE DIFFERENCE METHOD

The finite difference method [1] is best illustrated by means of a problem characterized by the two-dimensional Laplace equation

$$\frac{\partial^2 \phi}{\partial x^2} + \frac{\partial^2 \phi}{\partial y^2} = 0 \tag{1}$$

An extension to a three-dimensional problem is more complicated but is straightforward. The region of interest is divided into mesh points separated by the distance h. Instead of solving (1) directly, the method entails the solution of its discretized version. Let us take the coordinate origin at point A in Fig. 1. The potentials ϕ_B, ϕ_C, ϕ_D, and ϕ_E, at points B, C, D, and E can be expressed in terms of Taylor expansions:

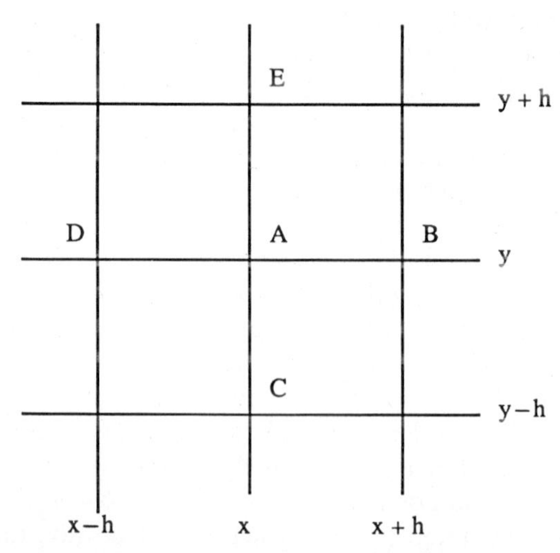

Fig. 1 A typical mesh for a two-dimensional finite difference method.

$$\phi_B = \phi_A + h\left(\frac{\partial\phi}{\partial x}\right)_A + \frac{h^2}{2!}\left(\frac{\partial^2\phi}{\partial x^2}\right)_A + \frac{h^3}{3!}\left(\frac{\partial^3\phi}{\partial x^3}\right)_A + O(h^4)$$

$$\phi_D = \phi_A - h\left(\frac{\partial\phi}{\partial x}\right)_A + \frac{h^2}{2!}\left(\frac{\partial^2\phi}{\partial x^2}\right)_A - \frac{h^3}{3!}\left(\frac{\partial^3\phi}{\partial x^3}\right)_A + O(h^4)$$

$$\phi_E = \phi_A + h\left(\frac{\partial\phi}{\partial y}\right)_A + \frac{h^2}{2!}\left(\frac{\partial^2\phi}{\partial y^2}\right)_A + \frac{h^3}{3!}\left(\frac{\partial^3\phi}{\partial y^3}\right)_A + O(h^4)$$

$$\phi_C = \phi_A - h\left(\frac{\partial\phi}{\partial y}\right)_A + \frac{h^2}{2!}\left(\frac{\partial^2\phi}{\partial y^2}\right)_A - \frac{h^3}{3!}\left(\frac{\partial^3\phi}{\partial y^3}\right)_A + O(h^4)$$

where the subscript A indicates the quantities evaluated at A and $O(h^4)$ is a quantity of the order of h^4. Adding these equations, we obtain

$$\phi_B + \phi_C + \phi_D + \phi_E = 4\phi_A + h^2\left(\frac{\partial^2\phi}{\partial x^2} + \frac{\partial^2\phi}{\partial y^2}\right)_A + O(h^4)$$

The second term on the right-hand side vanishes since ϕ is required to satisfy (1) everywhere. Hence,

$$\tfrac{1}{4}(\phi_B + \phi_C + \phi_D + \phi_E) = \phi_A \tag{2}$$

is a good approximation of (1) as long as h is small enough to neglect $O(h^4)$ terms. Somewhat different equations are used if point A is located on the boundary between two media [2]. At the boundary point, ϕ itself, its derivative in the form of a finite difference, or a combination of the two is specified [2].

All of these procedures are repeated at each mesh point. The result is a matrix equation

$$\boldsymbol{M\phi} = \mathbf{B} \tag{3}$$

The right-hand-side vector **B** contains information given by the boundary points. It is readily seen from (2) that the coefficient matrix $\boldsymbol{M}$ contains a large number of zero elements and only the diagonal and nearby elements are filled. For this reason, in most cases, (3) is solved not by matrix inversion but by an interactive method. A certain scheme called the successive overrelaxation method is employed to accelerate convergence of the solution [2].

This method is well known to be the least analytical. The mathematical preprocessing is minimal, and the method can be applied to a wide range of structures including those with odd shapes. A price one has to pay is numerical inefficiency. Certain precautions have to be taken into account when the method is used for an open-region problem in which the region is truncated to a finite size. Also, the method requires that mesh points lie on the boundary.

2. FINITE ELEMENT METHOD

The finite element method [3–7] is somewhat similar to the finite difference method. However, it has variational features in the algorithm and contains several flexible features. Details are presented in Chapter 2 of this book.

In the finite element method, instead of the partial differential equations with boundary conditions, corresponding functionals are set up and variational expressions are applied to each of the small areas or volumes subdividing the region of interest. Usually, these small segments are polygons such as triangles and rectangles for two-dimensional problems and tetrahedral elements for three-dimensional problems. Because of this type of discretization, hardly any restrictions can be imposed on the shape of the structure.

The essence of this method is illustrated below for a problem of Laplace equation (1) in a two-dimensional region in Fig. 2. The solution of (1) subject to the boundary condition is equivalent to minimizing the functional

$$
\begin{aligned}
I(\phi) = \langle \phi, \nabla^2 \phi \rangle &= \iint_S \phi \left(\frac{\partial^2 \phi}{\partial x^2} + \frac{\partial^2 \phi}{\partial y^2} \right) dx\, dy \\
&= - \iint_S \left[\left(\frac{\partial \phi}{\partial x} \right)^2 + \left(\frac{\partial \phi}{\partial y} \right)^2 \right] dx\, dy \qquad (4)
\end{aligned}
$$

This integral is carried out as a collection of the contributions from all of the small polygonal (triangular in this example) areas. In each polygon ϕ is approximated as a polynomial of the variables x and y.

$$
\phi = a + a_x x + a_y y
$$

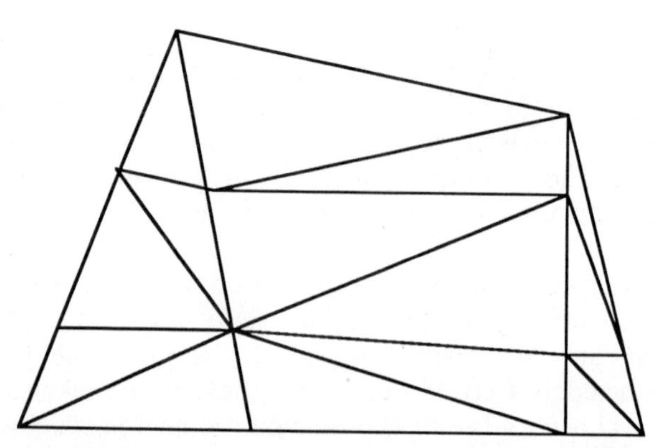

Fig. 2 Typical subdivision of a cross section in a two-dimensional finite element analysis.

The coefficients a, a_x, and a_y can be expressed in terms of the values of ϕ at each vertex of the triangle:

$$\phi_p = a + a_x x_p + a_y y_p\,, \qquad p = i,\, j,\, k$$

where the subscript $p = i, j, k$ identifies three vertices. Since only a_x and a_y are required for evaluation of (4), they are written as

$$\begin{bmatrix} a_x \\ a_y \end{bmatrix} = \mathbf{A} \begin{bmatrix} \phi_i \\ \phi_j \\ \phi_k \end{bmatrix}$$

The value of $I(\phi)$ for one polygon is

$$I_{ijk}(\phi) = [\phi_i, \phi_j, \phi_k]\mathbf{A}^{\mathrm{t}}\mathbf{A} \begin{bmatrix} \phi_i \\ \phi_j \\ \phi_k \end{bmatrix} |\Delta S| \tag{5}$$

where the superscript t indicates transpose and $|\Delta S|$ is the area of this polygon given by

$$\Delta S = \frac{1}{2} \begin{vmatrix} 1 & x_i & y_i \\ 1 & x_j & y_j \\ 1 & x_k & y_k \end{vmatrix}$$

To minimize $I_{ijk}(\phi)$, the Rayleigh–Ritz technique is used.

$$\frac{I_{ijk}}{\partial \phi_i} = \frac{I_{ijk}}{\partial \phi_j} = \frac{I_{ijk}}{\partial \phi_k} = 0 \tag{6}$$

Use of (5) in (6) results in

$$\mathbf{A}^{\mathrm{t}}\mathbf{A} \begin{bmatrix} \phi_i \\ \phi_j \\ \phi_k \end{bmatrix} = 0$$

When this process is applied to all of the polygons in S, one obtains

$$\mathbf{Z} \begin{bmatrix} \phi_1 \\ \phi_2 \\ \cdot \\ \cdot \\ \cdot \\ \phi_N \end{bmatrix} = 0 \tag{7}$$

Since some of the ϕ_i's located on the boundaries are known, (7) can be solved for potentials of all interior points. Algorithms for wave equations in two and three dimensions have been worked out extensively [8]. One of the problems of the finite element methods is the existence of the so-called spurious zeros. Such zeros correspond to unphysical field structures. The exact cause of this phenomenon is not yet resolved. Several schemes are available to reduce or eliminate these zeros. Typically, they are based on the variational expression that contains an additional constraint $\nabla \cdot \mathbf{H} = 0$ [9].

A certain precaution needs to be exercised when the finite element method is applied to an open-region problem such as a dielectric waveguide circuit. In many cases, the region to which the method is applied is truncated at a finite extent. In some situations—for instance, the waveguide near cutoff—such truncations are not straightforward because the field decays very slowly [6].

Recently, the boundary element method has been proposed [10, 11]. This is a combination of the boundary integral equation and a discretization technique similar to the finite element algorithm as applied to the boundary. Essentially, the wave equation for the volume is converted to the surface integral equation by way of the Green's identity. The surface integrals are discretized into N segments (elements), and their evaluation in each element is performed after the field quantities are approximated by polynomials.

One of the advantages of this method lies in the reduction in number of storage locations and CPU time resulting from the reduction in the number of dimensions.

3. TLM METHOD

In the TLM method [12, 13], the field problem is converted to a three-dimensional equivalent network problem. This method is essentially for simulation of the wave propagation phenomena in the time domain rather than for characterization of the structure. As such, it is very versatile. In the generic form of the three-dimensional TLM method, the space is discretized into a three-dimensional lattice with a period Δl. Six field components are represented by a hybrid TLM cell as shown in Fig. 3. Boundaries corresponding to the electric wall and the magnetic wall are represented by short-circuiting shunt nodes and open-circuiting shunt nodes on the boundary. Magnetic and dielectric materials can be introduced by adding short-circuited series stubs of length $\Delta l/2$ at the series nodes (magnetic field components) and open-circuited $\Delta l/2$ stubs at the shunt nodes (electric field components). The losses can be represented by resistively loading the shunt nodes. After the time-domain response is obtained, the frequency response is found by the Fourier transform.

There are several precautions to be exercised. Due to the introduction of periodic lattice structures, a typical passband–stopband phenomenon appears in the frequency-domain data. The frequency range must be below the upper bound of the lowest passband and is determined by the mesh size Δl.

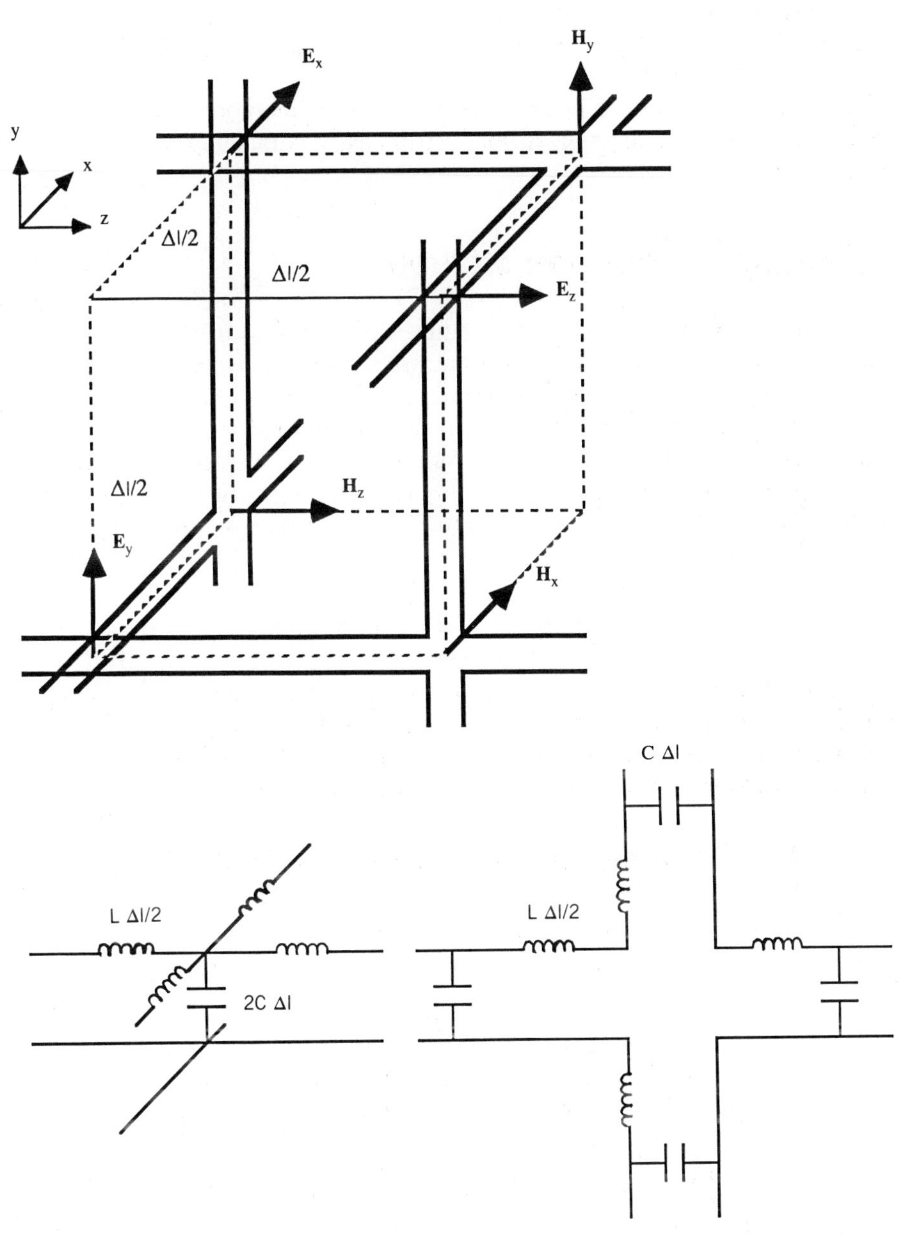

Fig. 3 A hybrid TLM cell.

There are a number of sources of error. Several remedial procedures have also been reported. In a 1985 article, Hoefer reviews the TLM method extensively [14]. A chapter written by Hoefer appears in this book (Chapter 8).

The structures that can be analyzed by the TLM method are quite varied. A typical problem is a shielded microstrip cavity containing a step discontinuity [13].

4. INTEGRAL EQUATION METHOD

The field in a three-dimensional structure can be found from the unknown quantities over a certain boundary that are solved for by this method [15]. Moderate to extensive analytical preprocessing is often required.

A typical integral equation for a three-dimensional passive component can be derived formally in the following manner. Consider the microstrip resonator problem of Fig. 4. From the superposition principle of a linear system, the total electric field tangential to the surface of the substrate is given by

$$\mathbf{E}^i(\mathbf{r}) + \int_S \mathbf{Z}(\mathbf{r}, \mathbf{r}')\mathbf{J}(\mathbf{r}')\, ds = \mathbf{E}_t(\mathbf{r}) \tag{8}$$

where $\mathbf{J}$ is the vector density on the microstrip surface S and $\mathbf{Z}$ is the dyadic Green's function. The incident field $\mathbf{E}^i$ vanishes if the formulation is for an eigenvalue problem. The integral equation is derived from the recognition that the total electric field $\mathbf{E}_t$ must be zero on the strip S. Hence,

$$\int_S \mathbf{Z}(\mathbf{r}, \mathbf{r}')\mathbf{J}(\mathbf{r}')\, ds = -\mathbf{E}^i(\mathbf{r}), \qquad \mathbf{r} \in S \tag{9}$$

Obviously, a homogeneous equation is found for an eigenvalue problem for which $\mathbf{E}^i = 0$.

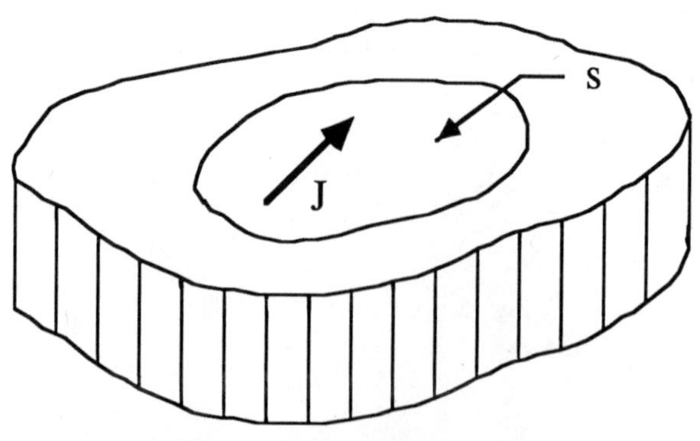

Fig. 4 Microstrip resonator.

Derivation of $\mathbf{Z}(\mathbf{r}, \mathbf{r}')$ is obviously an important and often difficult task. One possible way is to first introduce a two-dimensional Fourier transform with respect to two directions parallel to the substance surface, transform the Helmholtz equation to a one-dimensional ordinary differential equation with respect to the vertical direction, and find the solution. The Green's function is then found from the two-dimensional inverse Fourier transform.

The integral equation itself is transformed to a set of linear simultaneous equations for numerical inversion. The transformation is done by one of several methods such as the moment method [16]. In some cases, a variational expression derived from the integral equation is sufficient for the solution [17]. For instance, in his now historic paper, Yamashita solved a quasi-TEM microstrip line problem by a variational expression in the spectral (Fourier transform) domain. Under the quasi-TEM approximation, the capacitance per unit length of the microstrip line must be computed from the charge distribution on the strip of width $2w$. The integral equation for the unknown charge distribution is

$$\int_{-w}^{w} G(x - x'; y = y' = d)\rho(x')\, dx' = V\,, \qquad |x| < w \tag{10}$$

Instead of solving this equation, the variational expression is used for finding the line capacitance by

$$\frac{1}{C} = \frac{\int_{-w}^{w} \rho(x) \int_{-w}^{w} G(x - x'; y = y' = d)\rho(x')\, dx'\, dx}{\left[\int_{-w}^{w} \rho(x)\, dx\right]^2} \tag{11}$$

Although this can be used directly, Yamashita evaluated a Fourier-transformed version of (11) for more express processing of numerical calculations.

In the variational method, the first-order error in the choice of an approximate $\rho(x)$ results in a second-order (quadratic) error in C. It should be noted, however, that the above statement is not a guarantee for an accurate solution for C. The magnitude of the error in C is directly related to the choice of an approximate $\rho(x)$. It is important to select $\rho(x)$ as closely as possible to the true but unknown charge distribution.

The integral equation method is extensively discussed in Chapter 3 in which the method of moments discussed below is also treated.

5. MOMENT METHODS AND GALERKIN'S METHOD

These are popular means for discretizing a continuous operator equation such as an integral equation [16]. In the narrowest sense, the moment method employs step functions as the basis functions and delta functions as the testing functions. However, choices for basis and testing functions can be much more flexible. The basis and testing functions are identical in the Galerkin's method, and the resulting solutions are known to be variational [18].

The formal procedure of the moment method is now presented. Assume that the integral equation given is

$$\int_D G(\mathbf{r}, \mathbf{r}')f(\mathbf{r}')\, dr' = \rho(\mathbf{r}), \qquad \mathbf{r} \in D \tag{12}$$

where G is the Green's function and ρ is the known "excitation" term. The first step in the moment method is to expand the unknown function f in terms of a linear combination of known basis functions $\phi_n(\mathbf{r})$ with $n = 1, 2, \ldots, N$.

$$f(\mathbf{r}) = \sum_{n=1}^{N} c_n \phi_n(\mathbf{r}) \tag{13}$$

where c_n is the unknown coefficient to be determined. When (13) is substituted into (12), we obtain

$$\sum_{n=1}^{N} c_n \int_D G(\mathbf{r}, \mathbf{r}')\phi_n(\mathbf{r}')\, dr' = \rho(\mathbf{r}) \tag{14}$$

The second step is to take inner products of (14) with testing functions $\chi_m(r)$, $m = 1, 2, \ldots, N$. The results are

$$\sum_{n=1}^{N} K_{mn} c_n = b_m \tag{15}$$

where

$$K_{mn} = \left\langle \chi_m(\mathbf{r}), \int_D G(\mathbf{r}, \mathbf{r}')\phi_m(\mathbf{r}')\, dr' \right\rangle \tag{16}$$

$$b_m = \langle \chi_m(\mathbf{r}), \rho(\mathbf{r}) \rangle \tag{17}$$

The symbol $\langle \ \rangle$ indicates the inner product and is typically an integral with respect to r over the region D. It is clear that (15) is a set of linear equations of size $N \times N$.

There are several choices available for the basis functions $\phi_n(\mathbf{r})$ and the testing functions $\chi_n(\mathbf{r})$. One of the simplest is the choice in the so-called point-matching method. In this method, the following selection is made:

$$\phi_n(\mathbf{r}) = U(\mathbf{r}_n) = \begin{cases} 1 & \mathbf{r} \equiv [\mathbf{r}_n - \mathbf{\Delta}/2, \mathbf{r}_n + \mathbf{\Delta}/2] \\ 0 & \text{otherwise} \end{cases}$$

$$\chi_n(\mathbf{r}) = \delta(\mathbf{r} - \mathbf{r}_n)$$

where U is the unit pulse function, which is zero outside the narrow range of $\mathbf{\Delta}$ around a discretized point $\mathbf{r}_n$ in the domain, and δ is the Dirac delta function. It is clear now that, if $|\mathbf{\Delta}|$ is small enough,

$$K_{mn} = G(\mathbf{r}_m, \mathbf{r}_n)|\mathbf{\Delta}| \tag{18}$$

$$b_m = \rho(\mathbf{r}_m) \tag{19}$$

Due to the choice of the functions, no integral operations are needed. Hence, the analytical preprocessing is extremely simple. The price one has to pay for this simplicity is the large matrix size N for accurate solutions. The method is quite structure-independent and can be applied to a large class of odd-shaped geometries.

There are several improved versions of the point-matching method. For instance, higher-order functions or piecewise sinusoidal functions can be used for basis functions [19].

Another popular method is the Galerkin's method, which essentially results in the same procedures as the Rayleigh–Ritz method. In Galerkin's method, the basis functions and the testing functions are identical and are defined over the entire range.

$$\chi_n(\mathbf{r}) = \phi_n(\mathbf{r}) , \qquad \mathbf{r} \in D$$

In this case, the matrix element k_{mn} and the vector element b_m become

$$K_{mn} = \int_D \phi_m(\mathbf{r})\, dr \int_D G(\mathbf{r}, \mathbf{r}')\phi_n(\mathbf{r}')\, dr' \tag{20}$$

$$b_m = \int_D \phi_m(\mathbf{r})\rho(\mathbf{r})\, dr \tag{21}$$

It is known that the "results" from the Galerkin's method with a real operator are variational. The "results" could be the eigenvalues for $\rho(\mathbf{r}) = 0$ and could be some scalar product quantities such as $\langle \rho(\mathbf{r}) f(\mathbf{r}) \rangle$. A problem of this method is that double integrals have to be evaluated for each matrix element. However, the size N of the matrix can be substantially smaller than the one required in the point-matching method. In many cases, $N = 1$ can result in a reasonably accurate solution as long as a good choice is made for the basis function. The Galerkin's method is more flexible than the straightforward variational method in the derivation of a variational quantity. It is possible to improve the accuracy of the approximation of the unknown function $f(\mathbf{r})$ simply by increasing the matrix size N. However, proper choice of the basis functions is still important. If they are substantially different from the correct solution, convergence of the solution is poor and a large matrix is required.

6. MODE-MATCHING METHOD

This method is typically applied to the problem of scattering into waveguiding structures on both sides of the discontinuity such as the one in Fig. 5. The fields on both sides of the discontinuity are expanded in terms of the modes in the respective regions with unknown coefficients [20]. To illustrate the method, let us choose a simple step discontinuity with a TE_{n0} excitation. The first step is to expand the E_y and the H_x in terms of modal functions $\phi_{an}(x)$ and $\phi_{bn}(x)$, $n = 1, 2, \ldots$. Next, continuity of E_y and H_x is applied at the discontinuity $z = 0$.

$$E_y: \sum_{n=1}^{\infty} (A_n^+ + A_n^-)\phi_{an} = \begin{cases} \sum_{n=1}^{\infty} (B_n^+ + B_n^-)\phi_{bn} & 0 < |x| < b \\ 0 & b < |x| < a \end{cases} \tag{22}$$

$$H_x: \sum_{n=1}^{\infty} (A_n^+ - A_n^-)Y_{an}\phi_{an} = \sum_{n=1}^{\infty} (B_n^+ - B_n^-)Y_{bn}\phi_{bn} \qquad 0 < |x| \le b \tag{23}$$

where the superscripts + and − indicate the amplitude of the modal wave

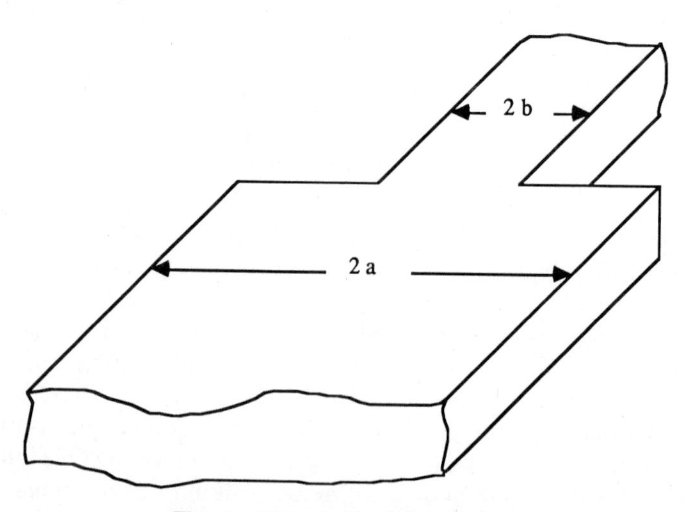

Fig. 5 Waveguide discontinuity.

propagating in the positive and negative directions of z respectively. Hence, A_n and B_n are the incident terms, and usually only one of them with a designated n is nonzero for a single-mode excitation. The next step is to eliminate the x dependence in (22) and (23). If we use the orthogonality of $\phi_{bn}(x)$ in the region $0 < |x| < b$, we find

$$\sum_{n=1}^{\infty} H_{nm}(A_n^+ + A_n^-) = B_m^+ + B_m^- \tag{24}$$

$$\sum_{n=1}^{\infty} H_{nm} Y_{an}(A_n^+ - A_n^-) = Y_{bm}(B_m^+ - B_m^-) \tag{25}$$

On the other hand, the orthogonality of $\phi_{an}(x)$ for $0 < |x| < a$ can only be used for (22) because (23) is defined only for $0 < |x| < b$.

$$A_m^+ + A_m^- = \sum_{n=1}^{\infty} H_{mn}(B_n^+ + B_n^-) \tag{26}$$

where

$$H_{mn} = \int_0^b \phi_{am}(x)\phi_{bn}(x)\, dx$$

From this point, several approaches are possible. One way is to eliminate the unknown B_m from (24) and (25). Then

$$\sum_{n=1}^{\infty} (Y_{an} + Y_{bm})H_{nm}A_n^- = \sum_{n=1}^{\infty} (Y_{an} - Y_{bm})H_{nm}A_n^+ + 2Y_{nm}B_n^- , \quad m = 1, 2, \ldots \tag{27}$$

This is a set of linear simultaneous equations for the unknown A_n and is called a formulation of the first kind. In the solution process, the matrix size must be truncated to a finite size so that $n, m = 1, 2, \ldots, N$.

There are several alternative formulations in addition to (27). All of them are theoretically equivalent. They may be different numerically, however [21]. The mode matching method is extensively discussed in Chapter 9.

Mode matching is often applied to find the guided mode in a waveguide with a complicated cross-sectional structure. Strictly speaking, this application, however, should be called the field-matching method. Let us consider finding the guided mode in a shielded microstrip line with a thick center conductor as shown in Fig. 6. In this method, the fields in the subdivided regions in the cross section are expanded in terms of appropriate orthogonal sets with a common but unknown propagation constant. Some of the boundary conditions are satisfied by individual terms in the expansions. For instance, by expanding the fields into sinusoidal series, the boundary

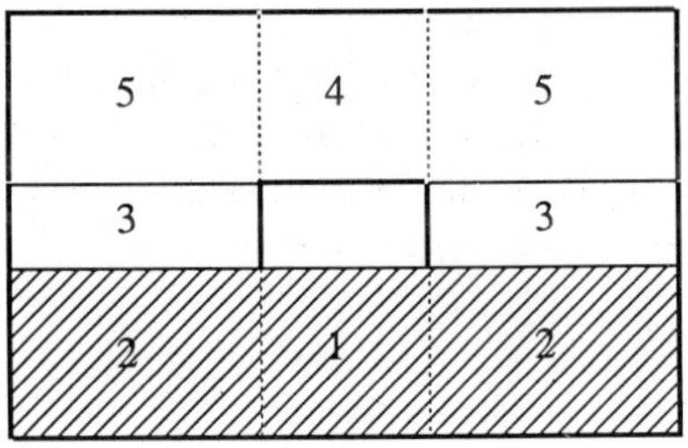

Fig. 6 Cross section of a shielded microstrip line with finitely thick strips.

conditions on the metal conductors can be satisfied. The continuity conditions of the tangential electric and magnetic fields are now imposed along each interface. After the orthogonality of the expansion functions is used, we obtain linear simultaneous homogeneous equations for unknown expansion coefficients in each region. We look for a value of the propagation constant that makes the determinant of this system of equations zero [22].

7. TRANSVERSE RESONANCE TECHNIQUE

This technique is somewhat similar to the mode-matching method and is suited for characterization of the discontinuity in a planar waveguide structure. The method is illustrated by way of a finline discontinuity shown in Fig. 7.

First, two shorting end plates are placed in the waveguide case at such distances from the discontinuity that all the higher-order modes excited at the discontinuity are negligible. Only the dominant modes can propagate in the two finline sections. The objective of the analysis is to find the resonant frequency from which one can extract information on the discontinuity [23].

Due to the symmetry of the bilateral finline configuration, only one half of the cross section $-a_1 < x < a_2$ in Fig. 7 is considered.

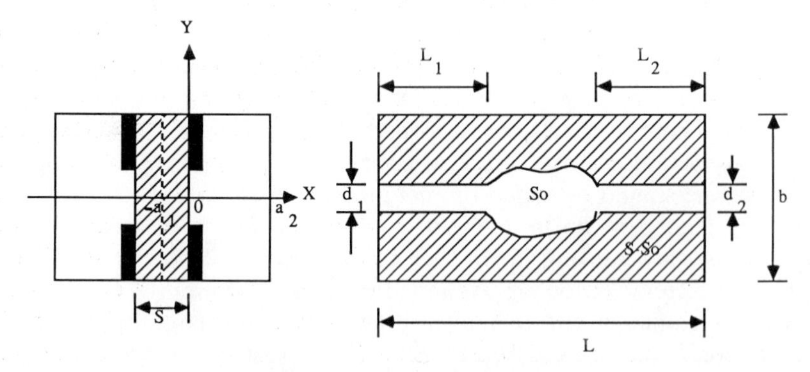

Fig. 7 Finline discontinuity.

The electromagnetic field in the dielectric region (region 1; $-a_1 \leq x \leq 0$) and in the air region (region 2, $0 \leq x \leq a_2$) can be expanded in terms of TE and TM modes of a rectangular waveguide with inner dimensions L and b. We obtain the following expressions for the transverse **E**- and **H**-field components in the two regions:

Dielectric Region: $-a_1 \leq x \leq 0$

$$\mathbf{E}_{t1} = \sum_{mn} A'_{mn} \cos k'_{mn}(x + a_1)\mathbf{x} \times \nabla_t \psi_{mn} + \frac{1}{j\omega\varepsilon_0\varepsilon_r} \sum_{mn} B'_{mn} k'_{mn} \cos k'_{mn}(x + a_1) \nabla_t \phi_{mn}$$

$$\mathbf{H}_{t1} = \frac{-1}{j\omega\mu_0} \sum A'_{mn} k'_{mn} \sin k'_{mn}(x + a_1) \nabla_t \psi_{mn} + \sum B'_{mn} \sin k'_{mn}(x + a_1) \nabla_t \phi_{mn} \times \hat{\mathbf{x}} \tag{28}$$

Air Region: $0 \leq x \leq a_2$

$$\mathbf{E}_{t2} = \sum_{mn} A_{mn} \sin k_{mn}(x - a_2)\hat{\mathbf{x}} \times \nabla_t \psi_{mn} - \frac{1}{j\omega\varepsilon_0} \sum_{mn} B_{mn} k_{mn} \sin k_{mn}(x - a_2) \nabla_t \phi_{mn}$$

$$\mathbf{H}_{t2} = \frac{1}{j\omega\mu_0} \sum_{mn} A_{mn} k_{mn} \cos k_{mn}(x - a_2) \nabla_t \psi_{mn} + \sum_{mn} B_{mn} \cos k_{mn}(x - a_2) \nabla_t \phi_{mn} \times \hat{\mathbf{x}} \tag{29}$$

where

$$\psi_{mn} = P_{mn} \cos \frac{m\pi z}{l} \cos \frac{n\pi y}{b}$$

$$\phi_{mn} = P_{mn} \sin \frac{m\pi z}{l} \sin \frac{n\pi y}{b}$$

$$P_{mn} = \left(\frac{\delta_m \delta_n}{b}\right)^{1/2} \frac{1}{\gamma_{mn}} \qquad \delta_i = \begin{cases} 1 & i = 0 \\ 2 & i \neq 0 \end{cases}$$

$$\gamma_{mn}^2 = \left(\frac{m\pi}{l}\right)^2 + \left(\frac{n\pi}{b}\right)^2$$

$$k_{mn}^2 = k_0^2 - \gamma_{mn}^2 \qquad k'^{\,2}_{mn} = k_0^3 \varepsilon_r - \gamma_{mn}^2$$

$$k_0^2 = \omega^2 \mu_0 \varepsilon_0$$

when ψ_{mn} and ϕ_{mn} are the TE and TM scalar potentials. Note that (28) and

(29) already satisfy the boundary conditions at $x = -a_1$ and a_2. The boundary conditions at $x = 0$ are

$$\mathbf{E}_{t1} = \mathbf{E}_{t2} = \begin{cases} \mathbf{E}_{t0} & \text{on } S_0 \\ 0 & \text{on } S - S_0 \end{cases} \tag{30}$$

$$\mathbf{H}_{t1} = \mathbf{H}_{t2} = \mathbf{H}_{t0} \qquad \text{on } S_0 \tag{31}$$

where $\mathbf{E}_{t0}$ and $\mathbf{H}_{t0}$ are unknown functions of z, y.

From this point, we could proceed in a manner similar to the one in the mode-matching method in Section 6. However, we will take a different approach. $\mathbf{E}_{t0}$ and $\mathbf{H}_{t0}$ are expanded in terms of a set of orthonormal vector functions $\mathbf{e}_v$ and $\mathbf{h}_\mu$ defined on the aperture S_0.

$$\mathbf{E}_{t0} = \sum V_v \mathbf{e}_v \tag{32}$$

$$\mathbf{H}_{t0} = \sum I_\mu \mathbf{h}_\mu \tag{33}$$

Substituting (28), (29), (32), and (33) into (30) and (31) and using the orthogonal properties of ψ_{mn}, ϕ_{mn}, $\mathbf{e}_v$, and $\mathbf{h}_\mu$, we obtain homogeneous equations. We eliminate A_{mn}, A'_{mn}, B_{mn}, B'_{mn}, and I_μ and obtain the homogeneous equation

$$\sum k_{\mu\nu} V_\nu = 0 \tag{34}$$

when $k_{\mu\nu}$ contains summations over m and n. The nontrivial solution of (34) results in the resonant frequency of the structure.

As demonstrated above, this technique is useful when the discontinuity is located only over a plane including the guide axis, that is, when the discontinuity does not involve a change in height. A more detailed discussion appears in Chapter 11.

8. METHOD OF LINES

In this method, two of the three dimensions are discretized for numerical processing while the analytical expressions are sought in the remaining dimension. The essential feature of this method is first explained by way of a simple two-dimensional problem of finding the propagation constant of a microstrip line in Fig. 8 [24]. First, the x direction is discretized by a family of N straight lines parallel to the y axis separated by h. When the partial derivative with respect to the x coordinate is replaced with the difference formula, the two scalar potentials ψ^e and ψ^h necessary for describing the hybrid field satisfy

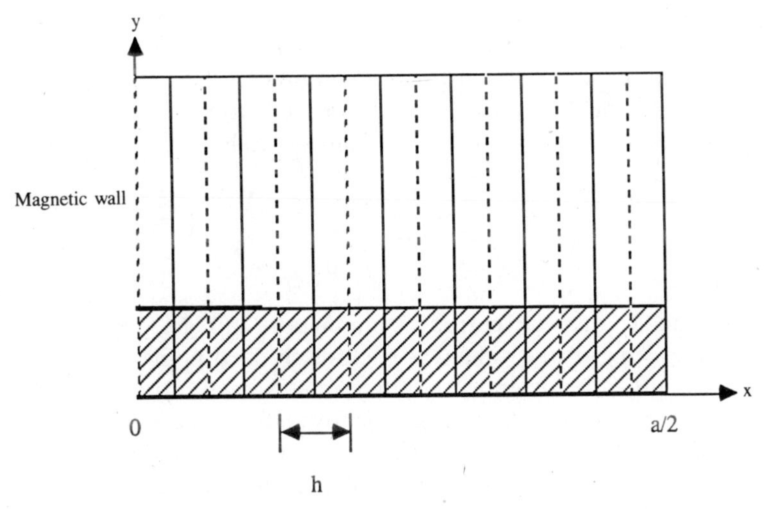

Fig. 8 One half of the cross section of a microstrip line for the methods-of-lines procedure.

$$\frac{d^2\psi_i}{dy^2} + \frac{1}{h^2}[\psi_{i-1}(y) - 2\psi_i(y) + \psi_{i+1}(y)] + (k^2 - \beta^2)\psi_i(y) = 0\,, \qquad i = 1, 2, \ldots, N \tag{35}$$

or in matrix form,

$$h^2 \frac{d\boldsymbol{\psi}}{dy^2} - [\mathbf{P} - h^2(k^2 - \beta^2)\mathbf{I}]\boldsymbol{\psi} = 0 \tag{36}$$

where **I** is the identity matrix and **P** is a tridiagonal matrix determined by the lateral boundary conditions at $x = 0$ and $a/2$. The discretization lines for ψ^e and ψ^h are shifted by half the discretization distance, $h/2$, so that the lateral boundary conditions are easily implemented. The essential feature of the method lies in the diagonalization of (36) so that the equation for the potential can be solved independently for each discretization i. This is accomplished by the transformation

$$\boldsymbol{T}^t\boldsymbol{\psi} = \mathbf{U} \tag{37}$$

where $\boldsymbol{T}^t$ denotes the transpose of $\boldsymbol{T}$, which is an orthogonal matrix and is determined by the lateral boundary conditions. The uncoupled equations take the form

$$h^2 \frac{d^2U_i}{dy^2} - [\lambda_i - h^2(k^2 - \beta^2)]U_i = 0 \qquad i = 1, 2, \ldots, N \tag{38}$$

when λ_i is the eigenvalue of **P**. The equations of this form for the two scalar potentials are solved for each homogeneous region. Then we impose the boundary conditions at the substrate–air interface.

Finally, the condition that the tangential electric fields on the strip be zero is imposed in the original domain, and the following matrix equation is derived.

$$R\begin{bmatrix} \mathbf{J}_x \\ \mathbf{J}_z \end{bmatrix} = \begin{bmatrix} \mathbf{0} \\ \mathbf{0} \end{bmatrix} \tag{39}$$

where $\mathbf{J}_x$ and $\mathbf{J}_z$ are the current components and are vectors with the elements consisting of the values at each discretized point.

The method can be extended to three-dimensional problems such as the microstrip resonator [25] (see Fig. 9). Instead of the central difference formula, the forward difference formula is used for the first derivative of the potentials ψ with respect to the x variable. In matrix notation,

$$h_x \frac{\partial \psi}{\partial x} \rightarrow \boldsymbol{D}_x \boldsymbol{\psi}$$

The difference matrix $\boldsymbol{D}_x$ is bidiagonal and is dependent on the lateral boundary conditions. Once again, the discretized Helmholtz equations for the two potentials ψ^e and ψ^h are treated. By way of the orthogonal transformation matrices, the difference matrix equations are transformed to diagonal forms. From application of the interface conditions, the matrix relation between the electric field and the current is obtained in the transformed domain. This relation is transformed back to the original domain, and the final boundary condition on the strip is imposed. From

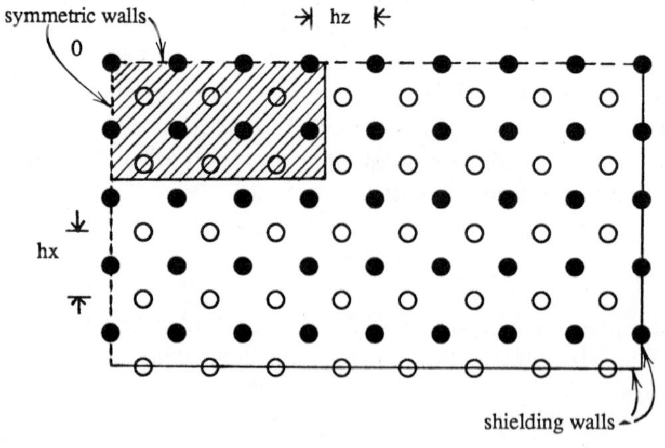

Fig. 9 One quarter of the top view of a microstrip resonator for the method-of-lines procedure.

nontriviality of the solution, the eigenvalue equation of the resonant frequency is obtained.

The method of lines has been applied to a number of practical but analytically complex structures. Examples include a triangular microstrip resonator and a periodic microstrip structure. For more thorough treatment and recent development, the readers are referred to Chapter 6.

9. GENERALIZED SCATTERING MATRIX METHOD

Although this has been developed for analyzing complicated discontinuity problems, it can be used for the characterization of cascaded discontinuities often seen in passive components such as the **E**-plane filter [26]. The generalized scattering matrix combines the mutual interaction of two discontinuities via the dominant and higher-order modes. This method has to be used with other techniques such as the mode-matching method, which characterizes a single discontinuity.

Let us illustrate the method by means of the cascaded discontinuity in Fig. 10 [27]. The first step is to characterize all the discontinuities involved in the microwave circuit. This characterization is expressed in terms of the generalized scattering matrix, which is closely related to the scattering matrix used in microwave network theory but differs in that the higher-order modes are included in addition to the dominant mode. Hence, the generalized scattering matrix is in general of infinite order. Consider that junction 1 is excited with the pth mode with unit amplitude from the left. If the complex amplitude of the nth mode of the reflected wave to the left is A_n, the (n, p) entry of the generalized scattering matrix $S^{11}(n, p)$ is A_n. Similarly, if the amplitude of the mth mode transmitted to the right is B_m, $S^{21}(m, p)$ is B_m. The generalized scattering matrix $\boldsymbol{S}_1$ of junction 1 is

$$\boldsymbol{S}_1 = \begin{bmatrix} \boldsymbol{S}^{11} & \boldsymbol{S}^{12} \\ \boldsymbol{S}^{21} & \boldsymbol{S}^{22} \end{bmatrix} \tag{40}$$

Similarly, for junction 2, the generalized scattering matrix $\boldsymbol{S}_2$ is

$$\boldsymbol{S}_2 = \begin{bmatrix} \boldsymbol{S}^{33} & \boldsymbol{S}^{34} \\ \boldsymbol{S}^{43} & \boldsymbol{S}^{44} \end{bmatrix} \tag{41}$$

Obviously, we need to find all the scattering matrix elements by some means such as the mode-matching technique before proceeding further.

The next step is to combine $\boldsymbol{S}_1$ and $\boldsymbol{S}_2$ to find the composite matrix

$$\boldsymbol{S} = \begin{bmatrix} \boldsymbol{S}^{AA} & \boldsymbol{S}^{AC} \\ \boldsymbol{S}^{CA} & \boldsymbol{S}^{CC} \end{bmatrix} \tag{42}$$

of the cascaded junctions. It turns out that

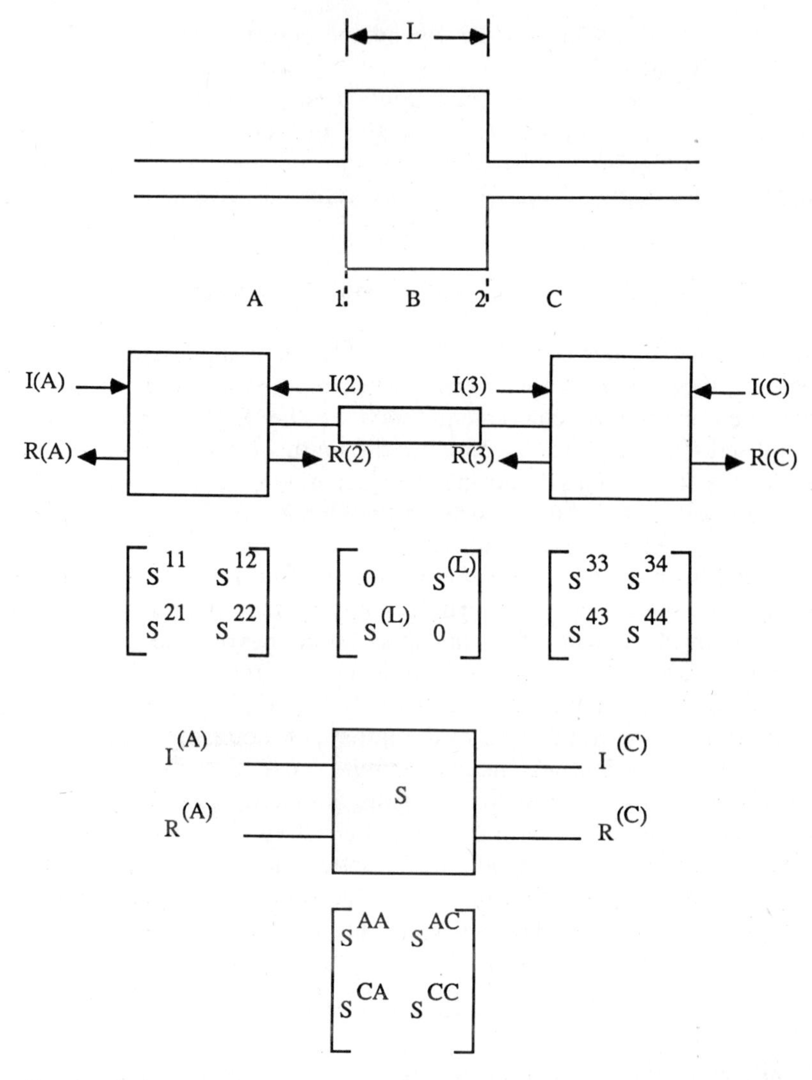

Fig. 10 Analysis procedure of a cascaded junction by the generalized scattering matrix analysis.

$$S^{AA} = S^{11} + S^{12}S^{(L)}U_2S^{33}S^{(L)}S^{21} \tag{43a}$$

$$S^{AC} = S^{12}S^{(L)}U_2S^{34} \tag{43b}$$

$$S^{CA} = S^{43}S^{(L)}U_1S^{21} \tag{43c}$$

$$S^{CC} = S^{44} + S^{43}S^{(L)}U_1S^{22}S^{(L)}S^{34} \tag{43d}$$

where

$$U_1 = (I - S^{22}S^{(L)}S^{33}S^{(L)})^{-1}, \qquad U_2 = (I - S^{33}S^{(L)}S^{22}S^{(L)})^{-1}$$

and $S^{(L)}$ is the transmission matrix for the waveguide between the two junctions.

$$S^{(L)} = \begin{bmatrix} e^{-\gamma_1 L} & & & 0 \\ & e^{-\gamma_2 L} & & \\ & & \cdot & \\ & & & \cdot \\ & & & & \cdot \\ 0 & & & \end{bmatrix} \tag{44}$$

I is the identity matrix and γ_n is the propagation constant of the nth mode.

The method can be applied to a complicated discontinuity by decomposing it to several discontinuities of less complicated geometry for which the solution is available. This application is illustrated by the offset discontinuity shown in Fig. 11*a* [27]. To this end, an auxiliary structure is introduced in Fig. 11*b*. Note that the original offest discontinuity can be recovered by setting δ to zero after all the formulations are carried out. Therefore, the generalized scattering matrices for junctions $J1$ and $J2$ are first obtained. They can be combined to find the composite matrix by way of (43) except that $S^{(L)} = I$ when $\delta \rightarrow 0$.

Note that in the above formulations, all the interactions via higher-order modes are included in addition to the dominant mode contributions. All the

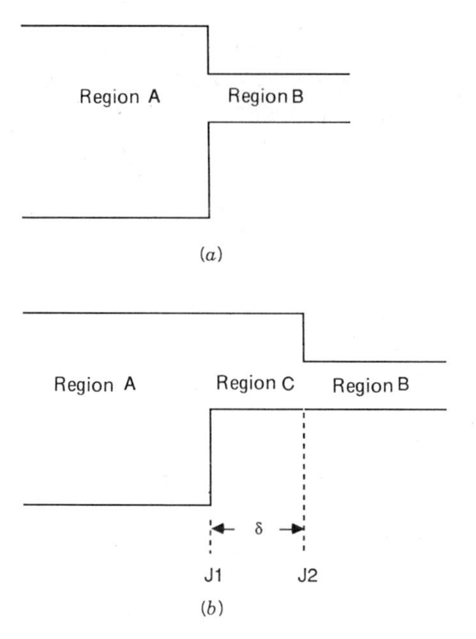

Fig. 11 Offset microstrip step discontinuity and an auxiliary structure for analysis.

matrices are of infinite order. In practice, however, they must be truncated to a finite size. It turns out that small (e.g., 2×2 and 3×3) matrices provide excellent results even when $\delta = 0$ [27]. Chapter 10 deals with the generalized scattering matrix technique in more detail.

10. SPECTRAL DOMAIN METHOD

This is a Fourier-transformed version of the integral equation method applied to microstrips or other printed line structures. It is one of the most preferred methods in recent years. The method is known to be efficient but is restricted in general to well-shaped structures that involve infinitely thin conductors. The method is illustrated by means of a microstrip resonator in Fig. 12 [28].

It is known that the hybrid fields in the structure can be found from two scalar potentials ϕ and ψ associated with the E_y and H_y fields in both the substrate and air regions. When all the field components are Fourier transformed in both the x and z directions with the transform variables α and β, the Helmholtz equations to be satisfied with ϕ and ψ and all the field

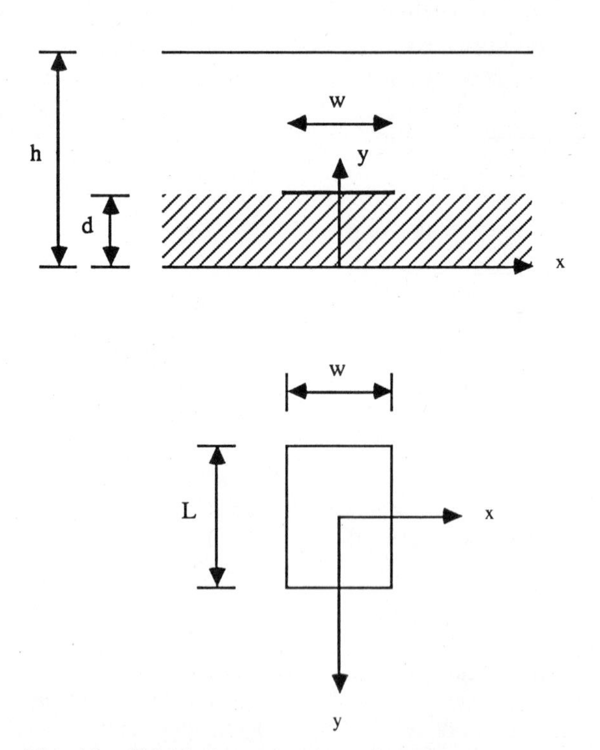

Fig. 12 Shielded rectangular microstrip resonator.

components are now reduced to one-dimensional ordinary differential equations for y only. Appropriate solutions to these equations are now found so that the boundary conditions at the bottom and top conducting planes are satisfied. This implies that $E_x = E_z = 0$ there and hence $\tilde{E}_x = \tilde{E}_z = 0$, where the latter symbols with tildes ($\tilde{E}_z$, etc.) denote Fourier-transformed quantities.

Next, the boundary conditions at the interface $y = d$ are applied in the Fourier-transform domain (spectral domain). Notice that in the space domain they are

$$E_{x1} = E_{x2} \qquad E_{z1} = E_{z2} \tag{45}$$

$$E_{x1} = \begin{cases} 0 \\ f(x) \end{cases} \qquad E_{z1} = \begin{cases} 0 \\ g(x) \end{cases} \qquad \begin{matrix} \text{on strip} \\ \text{outside} \end{matrix} \tag{46}$$

$$H_{x1} - H_{x2} = \begin{cases} J_z \\ 0 \end{cases} \qquad H_{z1} - H_{z2} = \begin{cases} J_x \\ 0 \end{cases} \qquad \begin{matrix} \text{on strip} \\ \text{outside} \end{matrix} \tag{47}$$

These conditions are Fourier-transformed with respect to the x and z directions. All the field components expressed in the spectral domain are substituted into the boundary conditions in the spectral domain. When all the unknown coefficients of the field expressions are eliminated, the following coupled algebraic equations are obtained.

$$\tilde{E}_x(\alpha, \beta) = \tilde{G}_{xx}(\alpha, \beta, k)\tilde{J}_x(\alpha, \beta) + \tilde{G}_{xz}(\alpha, \beta, k)\tilde{J}_z(\alpha, \beta) \tag{48a}$$

$$\tilde{E}_z(\alpha, \beta) = \tilde{G}_{zx}(\alpha, \beta, k)\tilde{J}_x(\alpha, \beta) + \tilde{G}_{zz}(\alpha, \beta, k)\tilde{J}_z(\alpha, \beta) \tag{48b}$$

The above equations correspond to the coupled homogeneous integral equations obtainable in the space domain.

$$0 = \int_{\text{strip}} G_{xx}(x, x', z, z', k)J_x(x', z')\, dx'\, dz' + \int_{\text{strip}} G_{xz}(x, x', z, z', k)J_z(x', z')\, dx'\, dz' \tag{49a}$$

$$0 = \int_{\text{strip}} G_{zx}(x, x', z, z', k)J_x(x', z')\, dx'\, dz' + \int_{\text{strip}} G_{zz}(x, x', z, z', k)J_z(x', z')\, dx'\, dz' \qquad (x, z) \equiv \text{strip} \tag{49b}$$

Also notice that eqs. (48) contain four unknowns E_x, E_z, E_x, and J_z. In the solution process, however, E_x and J_z are eliminated and (48) can be solved only for J_x and J_z.

The solution of (48) is undertaken by means of the Galerkin procedure.

To this end, J_x and J_z are first expanded in terms of known basis functions.

$$\tilde{J}_x(\alpha, \beta) = \sum_{m=1}^{M} a_m \tilde{J}_{xm}(\alpha, \beta) \tag{50a}$$

$$\tilde{J}_z(\alpha, \beta) = \sum_{n=1}^{N} b_n \tilde{J}_{zn}(\alpha, \beta) \tag{50b}$$

It is important to select $\tilde{J}_{xm}$ and $\tilde{J}_{zn}$ such that their inverse transforms $J_{xm}(x, z)$ and $J_{zn}(x, z)$ are nonzero only on the strip. Furthermore, they should incorporate appropriate edge conditions for faster convergence of the solution [29].

Expressions (50) are substituted into (48), and inner products with each of J_{xm} and J_{zn} are formed. The results are the following linear simultaneous equations.

$$\sum_{m=1}^{M} K_{pm}^{xx}(k)a_m + \sum_{n=1}^{N} K_{pn}^{xz}(k)b_n = 0\,, \qquad p = 1, 2, \ldots, M \tag{51a}$$

$$\sum_{m=1}^{M} K_{qm}^{zx}(k)a_m + \sum_{n=1}^{N} K_{qn}^{zz}(k)b_n = 0\,, \qquad q = 1, 2, \ldots, N \tag{51b}$$

The right-hand sides become zero by virtue of Parseval's relation. Equation (51) is solved for the unknowns a_m and b_n. To have a meaningful solution, the determinant of the coefficients matrix must be zero. From this requirement, the resonant frequency ($k = \omega\sqrt{\varepsilon_0 \mu_0}$) is obtained. All the field components are obtained from a_m and b_n.

The method has been applied to two-dimensional waveguide problems [30] and has been extended to discontinuity problems [31].

The derivation of (48) is often involved although straightforward. This is particularly true if one deals with multilayered structures or structures with conductors at several interfaces. The process can be significantly simplified by means of the immittance approach based on the coordinate transformation and the equivalent transmission lines [32]. The result for the structure in Fig. 12 is

$$G_{xx} = N_x^2 Z_{11}^e + N_z^2 Z_{11}^h \tag{52a}$$

$$G_{zx} = G_{xz} = N_x N_z(-Z_{11}^e + Z_{11}^h) \tag{52b}$$

$$G_{zz} = N_z^2 Z_{11}^e + N_x^2 Z_{11}^h \tag{52c}$$

$$N_x = \frac{\alpha}{\sqrt{d^2 + \beta^2}}\,, \qquad N_2 = \frac{\beta}{\sqrt{\alpha^2 + \beta^2}} \tag{53}$$

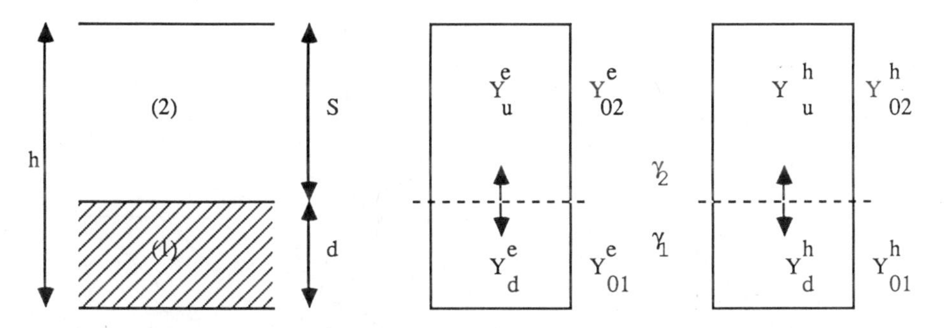

Fig. 13 Equivalent transmission lines for derivations of Green's functions in the spectral domain. γ_1 and γ_2 are propagation constants.

$$Z_{11} = \frac{1}{Y_u^e + Y_d^e}, \qquad Z_{11} = \frac{1}{Y_u^h + Y_d^h} \tag{54}$$

Y_u^e and Y_d^e are input admittance looking upward and downward respectively at the air–dielectric interface for the TM transmission line in Fig. 13. Y_u^h and Y_d^h are similarly defined. It is clear that G_{xx}, etc., are now written down almost by inspection of the structure. Chapter 5 deals with details of the spectral domain method.

11. EQUIVALENT WAVEGUIDE MODEL

This is not the numerical method but a formalism used for analysis of microstrip discontinuity problems. After the microstrip problem is converted to the equivalent waveguide model, one of the suitable numerical methods is used for characterizing the discontinuities.

Originally, this technique was introduced by Oliner for stripline structures [33]. Its application to microstrip line structures was introduced by the research group headed by I. Wolff [34, 35]. Details will be presented in Chapter 7. Let us illustrate this technique by means of the microstrip step discontinuity shown in Fig. 14. Its essential feature is the identification of a hypothetical waveguide representing the microstrip line. For instance, the microstrip discontinuity is replaced with the equivalent waveguide structure. The equivalent waveguide has the same height as the substrate thickness, two perfectly conducting top and bottom walls, and the two magnetic side walls. It is filled with a hypothetical medium with the effective dielectric constant, and its width is equal to the effective width. The effective dielectric constants of regions A and B are given by

$$\varepsilon_A = (\beta_A / k)^2 \tag{55a}$$

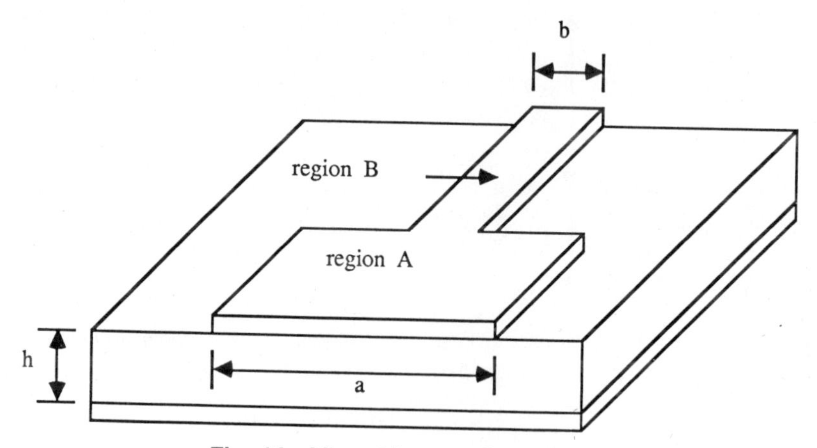

Fig. 14 Microstrip step discontinuity.

$$\varepsilon_B = (\beta_B/k)^2 \tag{55b}$$

and the effective widths are

$$a = \frac{120\pi}{\sqrt{\varepsilon_A}}\left(\frac{h}{Z_{01}}\right) \tag{56a}$$

$$b = \frac{120\pi}{\sqrt{\varepsilon_B}}\left(\frac{h}{Z_{02}}\right) \tag{56b}$$

In these equations β_A and β_B are the phase constants and Z_{01} and Z_{02} are the characteristic impedances of the dominant microstrip mode in the respective regions. These four quantities must be calculated by a standard technique such as the spectral domain method. Naturally, they are functions of frequency.

Once the equivalent waveguide structure is obtained for each microstrip section, the discontinuity problem is transformed to that of the closed waveguide configuration. A number of techniques are available, including the mode-matching method, to characterize such a discontinuity.

The method is inherently limited to the case where the surface wave excitation and radiation phenomena at the discontinuity are negligible. Reasonably accurate data have been obtained as long as the frequency is relatively low.

12. PLANAR CIRCUIT MODEL

This is also a formalism for analysis of planar passive components. The eigenmode expansion and the integral equation are often used for this model. In addition, the so-called segmentation and desegmentation techniques are powerful supplements for the planar circuit approach.

The concept of planar circuits was introduced by Okoshi and Miyoshi [36]. A planar circuit is defined as a microwave structure in which one of the three dimensions, say z, is much smaller than the wavelength whereas the remaining two are comparable to the wavelength. Hence it is possible to assume that the field is invariant in the z direction ($\partial/\partial z \equiv 0$). One then needs to deal with a two-dimensional Helmholtz equation. When the magnetic side wall of Fig. 15 is assumed, except for the ith port where a transmission line of width W_i is connected, the equation for E_z and the boundary conditions are

$$\nabla_t^2 E_z + k^2 E_z = 0 \tag{57}$$

$$\tilde{\mathbf{n}} \cdot \nabla E_z = \begin{cases} 0 & \text{on magnetic wall} \\ -j\omega\mu \mathbf{J}_s \cdot \hat{\mathbf{n}} & \text{on } W_i \end{cases} \tag{58}$$

where $\mathbf{J}_s$ is the surface current density. The solution to this problem can be found once the Green's function is available.

$$\nabla^2 G + k^2 G = -\delta(\mathbf{r} - \mathbf{r}') \tag{59}$$

One of the methods is the direct solution of the integral equation derived from Green's theorem. Green's function is

$$G = -\frac{j}{4} H_0^{(2)}(kr)\,, \qquad r = |\mathbf{r} - \mathbf{r}'| \tag{60}$$

where $H_0^{(2)}$ is the zeroth-order Hankel function of the second kind. Then

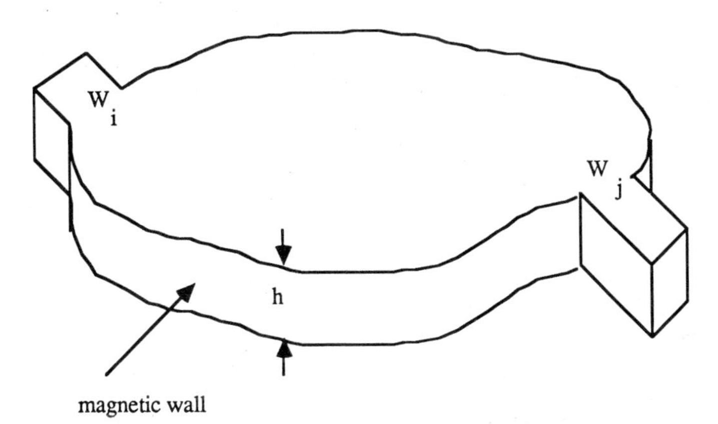

Fig. 15 Planar circuit model.

$$v(\mathbf{r}) = \frac{1}{4j} \int_C k \cos\theta\, H_1^{(2)}(kr) v(\mathbf{r}')\, dr - \frac{\omega\mu}{4} \sum \int_{w_i} H_0^{(2)}(kr) \hat{\mathbf{n}} \cdot \mathbf{J}_s(\mathbf{r}')\, dr \tag{61}$$

where $v = hE_z$ and θ is the angle between the normals at $\mathbf{r}'$ and $\mathbf{r}' - \mathbf{r}$. Discretization of (61) provides an impedance relation between the terminal voltage and the current at each port.

The above method is applicable to a structure with an arbitrary shape. However, when the shape of the circuit is more regular, say, rectangular or circular, another method is more convenient and informative. In this second method, Green's function with the boundary condition $\mathbf{n} \cdot \nabla G = 0$ is expanded in terms of its eigenfunctions.

$$G(\mathbf{r}, \mathbf{r}') = j\omega\mu \sum_{\nu=0}^{\infty} \frac{\phi_\nu(\mathbf{r})\phi_\nu(\mathbf{r}')}{k_\nu^2 - \kappa^2} \tag{62}$$

From this expression, the impedance relationship between the terminal voltage and current can be found.

In general, the voltage v and the injected current j at the ith port can be written in terms of Fourier expansions [37].

$$v = \sum_{m=0}^{\infty} v_i^{(m)} \sqrt{\delta_m} \cos \frac{m\pi l}{w_i} \tag{63}$$

$$j = \sum_{n=0}^{\infty} I_i^{(n)} \sqrt{\delta_n} \cos \frac{n\pi l}{w_i} \tag{64}$$

$$\delta_m = \begin{cases} 1 & m = 0 \\ 2 & m \neq 0 \end{cases}$$

where l is the coordinate along the ith port $(0 \le l \le w_i)$. The generalized impedance matrix is defined as

$$V_i^{(m)} = \sum_{n=0}^{\infty} \sum_{j=1}^{N} Z_{ij}^{(mn)} I_j^{(n)} \tag{65}$$

$$Z_{ij}^{(mn)} = \frac{j\omega\mu h \sqrt{\delta_m \delta_n}}{w_i w_j} \sum_{\nu=0}^{\infty} \frac{g_{\nu i}^{(m)} g_{\nu j}^{(n)}}{k_\nu^2 - k^2} \tag{66}$$

$$g_{\nu i}^{(m)} = \int_{w_i} \phi_\nu \cos \frac{m\pi}{w_i}\, d \tag{67}$$

$Z_{ij}^{(mn)}$ gives the mth-order voltage in the ith port when a unit nth-order current is injected at the jth port with all other currents zero.

The planar circuit approach can be extended by introducing segmentation of planar elements. It is recognized that Green's function and eigen-

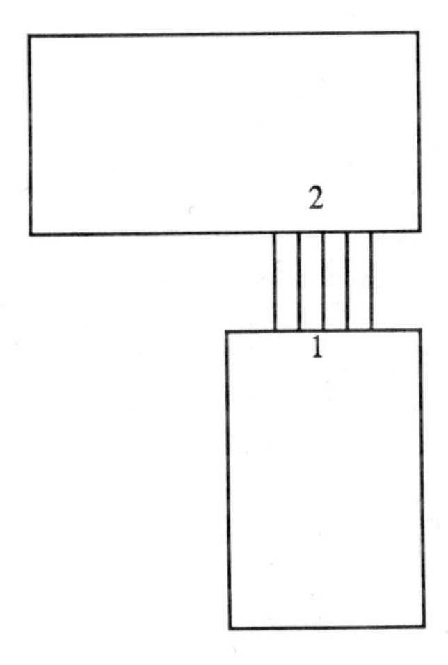

Fig. 16 Segmentation procedure.

functions are known for only a limited number of structural shapes. However, a more complicated shape can be segmented into elementary shapes for which the impedance matrix can be calculated by the method described above. The common ports are connected to form the original circuit (Fig. 16). The same physical port 1 (the one only in Fig. 16) can be decomposed to infinite electrical ports corresponding to the Fourier expansion in (63) and (64), which are truncated to a finite size.

In the so-called segmentation method [38] the interconnection is discretized into a finite number of physical ports after the voltage and current along the interconnection are approximated by step functions.

Another extension of the planar circuit approach is the desegmentation method [39]. This technique is applicable to geometry that can be analyzed easily by either the original planar circuit method or the segmentation method if a simple element is added. Details of the planar circuit model are presented in Chapter 4.

13. CONCLUSION

This chapter has provided brief descriptions for representative numerical techniques useful for millimeter-wave passive structures. Many of these methods are discussed in greater detail in the subsequent chapters in this book. As illustrated here, each method has advantages and disadvantages. For instance, although the finite element method requires considerable computation time and many memory locations, it is a versatile technique. On the other hand, the spectral domain technique is numerically rather efficient, but its range of applicability is limited. Various aspects of numerical methods are compared in Table 1. The evaluation is not quantitative but qualitative. There is no clear-cut boundary assigned between "moderate"

Table 1 Comparison of Numerical Methods

Method	Storage Requirement	CPU Time	Generality	Preprocessing
Finite difference	L	L	VG	Nil
Finite element	L	ML	VG	S
Boundary element	M	M	VG	S
Transmission line matrix	ML	ML	VG	S
Integral equation	SM	SM	G	M
Mode matching	M	SM	G	M
Transverse resonance	SM	SM	Ma	M
Method of lines	M	S	G	L
Spectral domain	S	S	Ma	L

L = large, M = moderate, S = small, VG = very good, G = good, Ma = marginal.

and "large." Additionally, considerable variation exists in each aspect for a particular method. In many cases, an experienced researcher can accelerate numerical processing by a number of techniques and skills. Therefore, Table 1 only serves as a rough guide for comparison.

Finally, the steady improvement of personal computers affords an additional opportunity for numerical analysis tasks. Some simple characterizations can be done directly with a modern personal computer. In addition, these machines can be used for database or look-up tables. The data can be generated by a larger machine or in some cases by the personal computer itself.

REFERENCES

1. G. Mur, "Finite difference method for the solution of electromagnetic waveguide discontinuity problem," *IEEE Trans. Microwave Theory Tech.*, vol. MTT-22, pp. 54–57, Jan. 1974.
2. H.E. Green, "The numerical solution of some important transmission-line problems," *IEEE Trans. Microwave Theory Tech.*, vol. MTT-13, pp. 676–692, Sept. 1965.
3. P. Silvester, *Finite Elements for Electrical Engineers*, Cambridge University Press, New York, 1983.
4. P. Daly, "Hybrid-mode analysis of microstrip by finite element method," *IEEE Trans. Microwave Theory Tech.*, vol. MTT-19, pp. 19–25, Jan. 1971.
5. A.F. Thomson and A. Gopinath, "Calculation of microstrip discontinuity inductances," *IEEE Trans. Microwave Theory Tech.*, vol. MTT-23, pp. 648-655, Aug. 1975.

6. B. M. A. Rahman and J. B. Davie, "Finite element analysis of optical and microwave waveguide problems," *IEEE Trans. Microwave Theory Tech.*, vol. MTT-32, pp. 20–28, Jan. 1984.
7. P. Silvester, "Finite element analysis of planar microwave networks," *IEEE Trans. Microwave Theory Tech.*, vol. MTT-21, pp. 104–108, Feb. 1973.
8. R. L. Ferrari, "Finite element analysis of three-dimensional electromagnetic devices," *15th Eur. Microwave Conf. Dig.*, pp. 1064–1069, Sept. 1985.
9. B. M. A. Rahman and J. B. Davies, "Penalty function improvement of waveguide solution by finite elements," *IEEE Trans. Microwave Theory Tech.*, vol. MTT-32, pp. 922–928, Aug. 1984.
10. C. A. Brebbia, *The Boundary Element Method for Engineers*, Pentech Press, London, 1978.
11. S. Kagami and I. Fukai, "Application of boundary element method to electromagnetic field problems," *IEEE Trans. Microwave Theory Tech.*, vol. MTT-32, pp. 455–461, Apr. 1984.
12. W. J. R. Hoefer and A. Ros, "Fin line parameters calculated with the TLM-method," *IEEE MTT-S Int. Microwave Symp. Dig.*, pp. 341–343, Apr.–May, 1979.
13. S. Akhtarzad and P. B. Johns, "Three-dimensional transmission-line matrix computer analysis of microstrip resonators," *IEEE Trans. Microwave Theory Tech.*, vol. MTT-23, pp. 990–997, Dec. 1975.
14. W. J. R. Hoefer, "The transmission-line matrix method—theory and applications," *IEEE Trans. Microwave Theory Tech.*, vol. MTT-33, pp. 882–893, Oct. 1985.
15. W. C. Chew and J. A. Kong, "Resonance of the axial-symmetric modes in microstrip disk resonators," *J. Math. Phys.*, vol. 21, pp. 582–591, Mar. 1980.
16. R. F. Harrington, *Field Computation by Moment Methods*, Macmillan, New York, 1968.
17. E. Yamashita and R. Mittra, "Variational method for the analysis of microstriplines," *IEEE Trans. Microwave Theory Tech.*, vol. MTT-16, pp. 251–256, Aug. 1968.
18. D. S. Jones, *The Theory of Electromagnetism*, Pergamon, New York, 1964.
19. R. W. Jackson and D. M. Pozer, "Full-wave analysis of microstrip open-end and gap discontinuities," *IEEE Trans. Microwave Theory Tech.*, vol. MTT-33, pp. 1036–1042, Oct. 1985.
20. Y. C. Shih and K. G. Gray, "Convergency of numerical solutions of step-type waveguide discontinuity problems by modal analysis," *IEEE MTT-S Int. Microwave Symp. Dig.*, pp, 233–235, May 1983.
21. T. S. Chu, T. Itoh, and Y.-C. Shih, "Comparative study of mode-matching formulations for microstrip discontinuity problems," *IEEE Trans. Microwave Theory Tech.*, vol. MTT-33, pp. 1018–1023, Oct. 1985.
22. G. Kowalski and R. Pregla, "Dispersion characteristics of shielded microstrips with finite thickness," *Arch. Elektron. Ubertragungstech.*, vol. 25, pp. 193–196, Apr. 1971.
23. R. Sorrentino and T. Itoh, "Transverse resonance analysis of finline discontinuities," *IEEE Trans. Microwave Theory Tech.*, vol. MTT-32, pp. 1633–1638, Dec. 1984.

24. U. Schulz and R. Pregla, "A new technique for the analysis of the dispersion characteristics of planar waveguides and its application to microstrips with tuning septums," *Radio Sci.*, vol. 16, pp. 1173–1178, Nov.–Dec. 1981.
25. S. B. Worm and R. Pregla, "Hybrid-mode analysis of arbitrarily shaped planar microwave structures by the method of lines," *IEEE Trans. Microwave Theory Tech.*, vol. MTT-32, pp. 191–196, Feb. 1984.
26. Y.-C. Shih, T. Itoh, and L. Q. Bui, "Computer-aided design of millimeter-wave E-plane filters," *IEEE Trans. Microwave Theory Tech.*, vol. MTT-31, pp. 135–142, Feb. 1983.
27. T. S. Chu and T. Itoh, "Analysis of cascaded and offest microstrip step discontinuities by the generalized scattering matrix technique," *IEEE Trans. Microwave Theory Tech.*, vol. MTT-34, pp. 280–284, Feb. 1986.
28. T. Itoh, "Analysis of microstrip resonators," *IEEE Trans. Microwave Theory Tech.*, vol. MTT-22, pp. 946–952, Nov. 1974.
29. R. H. Jansen, "Unified user-oriented computation of shielded, covered and open planar microwave and millimeter-wave transmission-line characteristics," *IEE J. Microwaves, Opt. Acoust.*, vol. 3, pp. 14–22, Jan. 1979.
30. L. P. Schmidt, T. Itoh, and H. Hofmann, "Characteristics of unilateral fin-line structures and arbitrarily located slots," *IEEE Trans. Microwave Theory Tech.*, vol. MTT-29, pp. 352–355, Apr. 1981.
31. J. Boukamp and R. H. Jansen, "The high-frequency behavior of microstrip open ends in microwave integrated circuits including energy leakage," *14th Eur. Microwave Conf. Dig.*, pp. 142–147, Sept. 1984.
32. T. Itoh, "Spectral domain immittance approach for dispersion characteristics of generalized printed transmission lines," *IEEE Trans. Microwave Theory Tech.*, vol. MTT-28, pp. 733–736, July 1980.
33. A. A. Oliner, "Equivalent circuits for discontinuity in balances strip transmission line," *IRE Trans. Microwave Theory Tech.*, vol. MTT-3, pp. 134–143, Mar. 1955.
34. I. Wolff and N. Knoppik, "Rectangular and circular microstrip disk capacitors and resonators," *IEEE Trans. Microwave Theory Tech.*, vol. MTT-22, pp. 857–864, Oct. 1974.
35. G. Kompa, "Frequency dependent behavior of microstrip offset junction," *Electron. Lett.*, vol. 11, pp. 537–538, Oct. 1975.
36. T. Okoshi and T. Miyoshi, "The planar circuit—an approach to microwave integrated circuitry," *IEEE Trans. Microwave Theory Tech.*, vol. MTT-20, pp. 245–252, Apr. 1972.
37. R. Sorrentino, "Planar circuits, waveguide models, and segmentation method," *IEEE Trans. Microwave Theory Tech.*, vol. MTT-33, pp. 1057–1066, Oct. 1985.
38. T. Okoshi, Y. Uehara, and T. Takeuchi, "The segmentation method—an approach to the analysis of microwave planar circuits," *IEEE Trans. Microwave Theory Tech.*, vol. MTT-24, pp. 662–668, Oct. 1976.
39. P. C. Sharma and K. C. Gupta, "Desegmentation method for analysis of two-dimensional microwave circuits," *IEEE Trans. Microwave Theory Tech.*, vol. MTT-29, pp. 1094–1098, Oct. 1981.

2

The Finite Element Method

J. B. Davies
Department of Electronic and Electrical Engineering
University College
London, England

1. INTRODUCTION

This chapter is concerned with currently used techniques for general electromagnetic field computation and applications to microwave and optical components, antennas, and scattering from craft. Concentration is on *finite elements*, a method that is well developed and has perhaps the widest scope or versatility in dealing with a vast range of components, geometries, and material distributions.

Our strategy is to first introduce the foundation method of *weighted residuals*, an approach that is both straightforward and general. This leads to the *variational method*, upon which we choose to base the *finite element* method, described in Section 5, which is the core of the chapter. The closely related method of *finite differences* is considered in Section 9, and throughout the text an attempt is made to relate the different available options and discuss their advantages and disadvantages. In Section 6, which considers problems reduced from three dimensions to two or one, it is hoped that the reader can appreciate the costs and limitations of using an integral equation approach, whether it be for the solution of integrated optical guides or radar cross sections.

In Sections 7 and 8, we consider applications of antennas and scattering and of various uniform (in z) waveguides. A complete finite element computer program is presented in the final section. This is both to illustrate the methods discussed earlier (especially Sections 3–5) and to provide the essence of a useful program for the quasi-static solution of microwave planar guides, in particular of microstrips.

For components or problems needing an electromagnetic description, explicit or closed-form expressions for the fields are always to be preferred when possible. This chapter is concerned with the large range of components and structures where such closed-form solutions are *not* adequate and so some form of computer analysis is necessary.

Thankfully, many real situations involving fields—such as those involving beams of light or radio waves traveling in straight lines—can be solved easily (if approximately). The plane wave

$$E_x = E_0 \exp[j(\omega t - kz)] \tag{1}$$

or, if we wish to be smarter, the Gaussian beam [1]

$$E_x = E_0 \exp[j(\omega t - kz)] \frac{w(0)}{w(z)} \exp\left(-\frac{x^2 + y^2}{w(z)}\right) \tag{2}$$

(plus some phase terms) are good enough representations of physical fields (or solutions to Maxwell's equations) to meet our needs for free-space radiation. They will adequately represent the fields for (a) VHF or microwave signals between transmitter and receiving antennas, (b) individual beams of light through the lens system of a camera or microscope, and (c) transmitted and returned signals of a microwave radar.

In cases (a) and (c), the so-called near fields (and indeed the behavior) around the antennas are likely to be too complicated for simple analysis; in case (c), the important problem of finding the scattering or monostatic radar cross section of any object that returns the signal will similarly be complicated.

Another area where computer methods are necessary for proper analysis and thus understanding is that of hollow metal waveguides. After Maxwell introduced his equations, it was not long before Rayleigh used Bessel functions to give the analysis (and so the prediction) of radio wave propagation along a hollow circular conducting guide. The fields (and therefore the cutoff wavelength, phase velocity, etc.) can also be written down for rectangular and elliptical cross sections, but that is effectively the end of the list of cross section shapes that can be analyzed. Dozens, probably hundreds, of different nonstandard cross sections have been needed by microwave engineers, who have to work in ignorance of the details of the guide they are using or somehow find a computer solution for their guide. Ridged waveguides are probably the most common such "mathematically inconvenient" waveguides, having a potentially wide bandwidth (of mono-mode operation).

As for optical guides, stepped-index fiber and simple film (planar) guides are effectively the only structures that can be analyzed without recourse to the computer, so that all integrated-optics types of guides and most current optical fibers require computer methods.

Finally, when it comes to semiconductor devices, component designers are similarly limited in the structures they can have where the fields (with time-dependent voltages and currents) are in closed form. Solutions to the coupled transport and electric field equations can be well approximated in one dimension, but again a more realistic analysis requires some computer methods. Although semiconductors will not be referred to again in this chapter, the methods we discuss are indeed used directly in semiconductor analysis.

1.1. Analysis, Synthesis, Design, and Optimization

Emphasis throughout the text is on analysis—that is, given a specific component or structure, find its behavior. However, just as with circuit analysis (or structural analysis of bridges or dams), the usual objective is to gain a capability in *analysis for design*.

Once an analysis capability is available and trusted (by comparison with experiment), one design algorithm is:

1. Think of a possible structure and/or design.
2. Analyze it.
3. *If* performance is adequate for specification or application
   then exit with design that is adequate (it may, of course, not be the best or cheapest, but . . .)
   else vary some parameter or add a new parameter and return to 2.

A general algorithm is outlined in the equivalence

$$\text{Analysis} + \text{optimization} = \text{synthesis}$$

For instance, suppose you could find the scattering matrix, or complex reflection coefficient, of a given filter at any frequency. You could evaluate the magnitude of the reflection coefficient and numerically integrate over some specified range of frequencies. Using this as a figure of merit, you could combine it with an optimization routine (such as those available "off the shelf" in any decent-sized computer), and this would allow you to design (synthesize) a broadband filter with the best Voltage Standing Wave Ratio (VSWR) over a specified frequency range. The "best" would depend on your definition of "figure of merit" and on how many free parameters you allowed yourself.

2. THE METHOD OF WEIGHTED RESIDUALS

This is a very general scheme for projecting Maxwell's equations into a form suitable for numerical solution by standard matrix methods. We will use it as the base from which we derive all of the methods considered here.

To ease the description, we assume all media to be linear and loss-free; we also assume that permeability and permittivity are time-independent and scalar (the materials are isotropic). With perhaps one exception, all space between conducting media will be taken to be uniform. All these restrictions can be removed, but at a cost in terms of complication.

Precise knowledge of an electric or magnetic field over some region of space and time generally needs an infinite amount of data. This cannot be stored on a digital computer with finite word length and finite storage capacity; even less can a precise solution to such a problem be achieved in a finite time of computing. *Some approximation must be made*. For problems with linear media (i.e., linear relations between **E** and **D**, **H** and **B**), this approximation invariably results in a matrix. To the mathematician, this is a matter of taking a *projection* from an infinite-dimensional Hilbert space onto a finite Euclidean space.

Although there are many ways in which this projection can be made, most of them can be described as special cases of a general method or procedure. One *can* describe the methods of (say) point matching or finite differences directly, but it surely helps to unify the many methods. A unified approach can be made via:

The Galerkin method [1]
The method of moments [2, 3]
The method of weighted residuals [4, 5]

The differences between these three methods are rather fine, but we will take the approach and language of *weighted residuals*.

We start with a "deterministic" problem:

$$Lu = v \tag{3}$$

where v represents some known excitation and u is the unknown and wanted field. L is a linear operator involving differentiation or integration, possibly both.

For illustration we can take the simple electrostatic problem of a capacitor where the unknowns are either the static potential ϕ or the surface charge density q. We know that the potential at distance R from a unit point charge is $1/4\pi\varepsilon_0 R$. (To give it a name, this is the free-space electrostatic Green's function.) By integrating over all the conducting surface, we have the potential at any point $\mathbf{r}$:

$$\phi(\mathbf{r}) = \int\int \frac{q(s)\,ds}{4\pi\varepsilon_0|\mathbf{r}-\mathbf{s}|} \tag{4}$$

The problem is specified when the capacitor plates (with surfaces S_1 and S_2) are allocated voltages such as

$$\phi(\mathbf{r}) = \begin{cases} 1 & \text{for } \mathbf{r} \text{ on } S_1 \\ 0 & \text{for } \mathbf{r} \text{ on } S_2 \end{cases} \tag{5}$$

This is our excitation v, for eq. (3), while our unknown u is the charge distribution over S_1 and S_2 and L is the integral operator [from eq. (4)] operating over the domain of suitable functions on the conductors S_1 and S_2.

Suppose we approximate the unknown u by

$$u = \sum_{i=1}^{N} u_i b_i(\mathbf{s}) \tag{6}$$

where $b_1, b_2, b_3, \ldots$ are some known *basis functions*, usually forming a complete set over the conducting surface. The problem is now to choose the (Fourier-like) coefficients $u_1, u_2, \ldots, u_N$ to approximate as well as possible the unknown solution u to eq. (3). In general it will be impossible to satisfy (3) precisely, so how are we to define "approximate as well as possible"? We do not know the exact solution for u and so cannot even discuss any error in u. We can, however, define what is called the *error residual*:

$$R(\mathbf{s}) = Lu - v = L \sum_{1}^{N} u_i b_i(\mathbf{s}) - v(\mathbf{s}) \tag{7}$$

which clearly is zero when and only when we have a precise solution to eq. (3). Now we have a realistic objective: we can try to make the residual $R(\mathbf{s})$ "small."

We now choose another set of test or *weight functions* $w_1, w_2, w_3, \ldots$, complete over the range of $\mathbf{s}$; we introduce an "inner product" (scalar product) formally written as $\langle x(\mathbf{s}), y(\mathbf{s}) \rangle$ and, rather than ask the impossible [that $R(\mathbf{s}) = 0$ for all $\mathbf{s}$], we insist that $R(\mathbf{s})$ be orthogonal to each of the functions $w_1, \ldots, w_N$. This results in the following N equations with N unknown u's:

$$\left\langle \left\{ L \sum_{i=1}^{N} u_i b_i(\mathbf{s}) - v(\mathbf{s}) \right\}, \{w_j(\mathbf{s})\} \right\rangle = 0 \qquad \text{for } j = 1, 2, \ldots, N \tag{8}$$

In the above equation, $v(\mathbf{s})$, $b_i(\mathbf{s})$, and $w_j(\mathbf{s})$ are known functions of $\mathbf{s}$; L is a known operator; u_i are the unknown and wanted scalars, which when put into eq. (6) give our approximate solution.

Because of the linearity of eq. (8), it can be put into matrix form:

$$[\boldsymbol{L}][\boldsymbol{u}] = [\boldsymbol{v}] \tag{9}$$

and we are now using a compact matrix notation (just as in quantum matrix mechanics), where

$[\boldsymbol{u}]$ denotes the unknown column vector containing elements $u_1, u_2, \ldots, u_N$.

$[\boldsymbol{v}]$ denotes the known column vector with elements $\langle v(\mathbf{s}), w_j(\mathbf{s}) \rangle$.

$[\boldsymbol{L}]$ is the known $N \times N$ matrix with (i, j)th element $\langle Lb_i(\mathbf{s}), w_j(\mathbf{s}) \rangle$.

Our equation $Lu = v$ has therefore been "projected" by approximation into the matrix equation $[\boldsymbol{L}][\boldsymbol{u}] = [\boldsymbol{v}]$. This can be solved routinely—more will be said later about this aspect.

The above procedure is called the weighted residuals method, the method of moments, or the generalized Galerkin method. Before getting into specific details about the choice of basis functions $\{b_i\}$ and weighting functions $\{w_i\}$, there are three particular avenues along which we can continue regarding the choice of $\{w_i\}$:

First,

$$w_i(\mathbf{s}) = b_i(\mathbf{s}) \tag{10}$$

gives the straight Galerkin method; this often results in a formulation identical to that of the variational method, a central method that will be studied later.

Second, $\{w_i\}$ and $\{b_i\}$ can be chosen as quite different and effectively unrelated sets, such as polynomials and sinusoids!

Third, there is the special choice

$$w_i(\mathbf{s}) = Lb_i(\mathbf{s}) \tag{11}$$

which gives what is called the *least squares residual* method. It is so called because it corresponds to the minimization of $\langle R(\mathbf{s}), R(\mathbf{s})\rangle$. This seems the surest and most explicit way of minimizing the error residual, though it is not the most commonly used method.

EXERCISE

Prove that, indeed, choosing the set of elements $\{u_i\}$ to minimize $\langle R(\mathbf{s}), R(\mathbf{s})\rangle$ does give eq. (11). Minimization can be achieved by differentiating with respect to each element u_i and putting the result equal to zero. Note that inner products are usually arranged so that $\langle a, b\rangle = \langle b, a\rangle$. For our purposes, this follows immediately when the inner product is taken as the integral over $\mathbf{s}$ of the scalar product $\mathbf{a} \cdot \mathbf{b}$.

2.1. Point Matching

So far in this section, we have outlined the weighted-residual way of reducing a differential or integral equation, in the form of eq. (3), into matrix form (9). We now consider particular choices of basis and weighting functions. Unquestionably, the simplest is the method of *point matching*. Here, one chooses that the error residual of (7) be forced to vanish at N selected points. This corresponds to using Dirac delta functions as weighting functions:

$$w_j(\mathbf{s}) = \delta(\mathbf{s} - \mathbf{s}_j) \tag{12}$$

which, when substituted into the inner products following eq. (9), gives

$$[v_j] = v(\mathbf{s}_j) \qquad \text{and} \qquad [L_{ij}] = Lb_i(\mathbf{s})|_{\mathbf{s}=\mathbf{s}_j} \tag{13}$$

Integration with a Dirac function in the integrand is so simple (using the "sifting" property) that the rather imposing inner products involving integrations give simple evaluations at specific points; point matching therefore makes for the ultimate simplification of testing or weighting functions. In terms of the earlier electrostatic example, we now look for a charge distribution over the surface of the capacitor plates that gives the correct potential at N points selected on the same surface.

If point matching is our choice of weighting functions, there remains the selection of basis functions. The literature is divided in its use of names here, but we follow Harrington's usage [2] with "the method of subsections." Here, the region or domain of $\mathbf{s}$ is divided into subregions or subsections, and each basis function $b_i(\mathbf{s})$ has nonzero value only on one subsection. This simplifies the inner product integrations. Some simple forms of $b_i(\mathbf{s})$ are:

(i) $\delta(\mathbf{s} - \mathbf{s}_i)$, where the $\{\mathbf{s}_i\}$ would generally be different points than the $\{\mathbf{s}_j\}$ of the weighting functions (in this case the term subsections is redundant, a preliminary division into subsections being unnecessary).

(ii) $b_i(\mathbf{s}) = 1$; this is a "pulse" function defined over each subsection.

(iii) Sets of polynomial or sinusoidal functions.

To continue with the two-conductor electrostatic problem, consider (ii) above: numerical solution by point matching with the use of subsection basis pulse functions. The two conductors are divided into subsections $i = 1, 2, \ldots, N$, over each of which we have a presumed uniform surface charge density u_i. There will be N "testing points" $\mathbf{s}_j$, which are conveniently placed one on each subsection, at which the stipulated voltage has to be, say, 1 at points $1, 2, \ldots, N/2$ and 0 at points $(N/2) + 1, \ldots, N$. This defines our column vectors $[\boldsymbol{u}]$ and $[\boldsymbol{v}]$ of eq. (9), with u_i as the unknowns.

As eq. (4) gives the operator L of the problem, the matrix $[\boldsymbol{L}]$ has the (i, j) element

$$Lb_i(\mathbf{s})|_{\mathbf{s}=\mathbf{s}_j} = \int\int_{S_i} \frac{ds}{4\pi\varepsilon_0|\mathbf{s} - \mathbf{s}_j|} \tag{14}$$

where S_i denotes integration being over the ith subsection surface. Because the electric charge is presumed uniform over each subsection, these matrix elements of $[\boldsymbol{L}]$ are purely geometry-dependent (rather than solution-dependent). For certain geometries of subsection, closed-form (explicit, analytic) expressions are available for the integrals of eq. (14), including the one for the potential at $\mathbf{s}_j$ due to uniform charge on the *same* subsection. For other geometries, approximate forms have to be used, including numerical integration.

It was mentioned above that eq. (10), if weighting and basis functions are chosen to be the same, is often equivalent to the variational method; we now consider this method.

3. THE VARIATIONAL METHOD

Our main rationale for including variational methods in this chapter is that whenever possible we choose to base finite elements on the variational approach (and recommend that others do also). Although finite elements can be established without this variational prelude, and the procedure is briefly considered, variational methods have a value in their own right.

A variational method is often an approach where just one parameter is the required answer; for example, if one just wants a resonant frequency, reflection coefficient, radar cross section, phase shift, or, for the Schrödinger equation, just an energy level or reflection coefficient, etc. The variational approach has two major benefits:

1. Numerical methods can often be set up rather efficiently.
2. Perturbation theory and applications can be very quickly and methodically set up.

Just for openers, we will give a simple illustration, with explanations later!

Suppose we want to find the cutoff frequency of the dominant TE_0 mode in a circular hollow conducting waveguide; that is, to find the smallest nonzero k that gives a solution to

$$\nabla^2 H_z + k^2 H_z = 0 \tag{15}$$

subject to satisfying

$$\frac{\partial H_z}{\partial r} = 0 \tag{16}$$

on the circle $r = a$.

(Of course we know that the desired mode is the TE_{01} mode with $k =$

3.832 . . . $/a$, coming from the equation $J_1(ka) = 0$; we can use this to check our forthcoming answer!)

Later on in the section we will derive the formula

$$k^2 \approx \frac{\iint (\nabla\phi)^2 \, dS}{\iint \phi^2 \, dS} \tag{17}$$

We try ϕ as a third-degree polynomial, with coefficients such that $d\phi/dr = r(r - a)$ and $\int\int \phi \, dS = 0$. (The good reasons for this choice will not be discussed.)

This gives $\phi = 7a^3 - 30ar^2 + 20r^3$, and substituting into eq. (17) gives us $ka = 3.839$, compared with the exact result 3.832. So we have obtained a result with 0.2% error without using a Bessel function!

Three questions can be put about the variational method:

1. How does one obtain a variational expression?
2. What *is* a variational expression?
3. How does one use a variational expression?

Questions 1 and 2 will be considered in Sections 3.1 and 3.2, and question 3 in Section 4.

3.1. How to Obtain a Variational Expression

We do not give a very detailed answer to question 1, but briefly, four ways to obtain a variational expression are:

1. Start from an energy-type expression—it might do! In classical mechanics we would use the Lagrangian function $T - U$, where T and U are the kinetic and potential energies, respectively. In electromagnetics, stored electric and/or stored magnetic energy [illustrated later in eq. (161)], and sometimes power flow, can be used.
2. Start from the equations you want to solve, multiply by a $\delta\phi$ term [as we will see later in eq. (24)], and try tricks with integration by parts.
3. Write down a quadratic or Hermitian form in *all* the fields involved, and see if that will do.
4. Find someone who has tackled the same class of problem, and use *their* variational expression!

Without question, the fourth is the fastest way to find a suitable variational expression.

The reference list at the end of the chapter includes a selection both of papers to provide a source of variational formulas for microwave and optical purposes [6, 7, 12, 18] and of texts with general developments of variational methods [1, 8–10].

3.2. What Is a Variational Expression?

Now to answer question 2. Generally, a variational expression is of the form:

Parameter = SV {expression involving fields, potentials, etc.}

where SV is short for "stationary value of." Sometimes instead of SV we have "min" for "minimum value of." Sometimes—and this is important—the parameter is a figure of special interest, such as resonant frequency, radiation resistance, characteristic impedance, or cutoff frequency.

3.2.1. *Fermat's Principle—a Simple Variational Form*

Fermat's principle illustrates perhaps the simplest variational expression, which can be expressed roughly as:

> A ray of light takes less time along the actual path between two points than it would along any other conceivable path.

(See Fig. 1.) This principle can be used to set up all the rules of geometrical optics.

Given the finite velocity of light, a straight line is the quickest path from the lamp to the eye, and so, according to Fermat's principle, it is the physical path taken. But how can we interpret Fermat's principle for a varying refractive index, such as light (or microwaves) cutting its way through the atmosphere? Generally it can be expressed as

$$T = \min \int_{P_1}^{P_2} \frac{ds}{c(s)} \tag{18}$$

Here, s is the distance measured along some path joining points P_1 and P_2, with ds an "elementary distance," and $c(s)$ denotes the local velocity of light at point s.

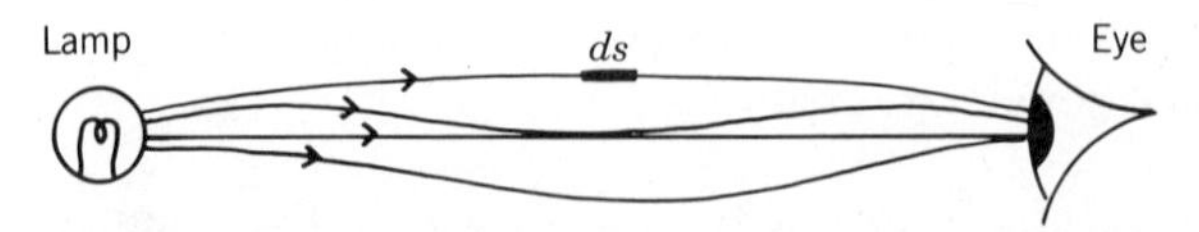

Fig. 1 Conceivable light paths between two points.

The right-hand side of eq. (18) is not an ordinary function (which is usually explicit); it is called a *functional*, and in this case its value depends on the particular path chosen.

Note some features of this formulation, common to any variational approach:

1. Consider any possible or "admissible" path.
2. Find the associated value of the right-hand side.
3. Suppose now the path is shifted slightly.
4. Find the new value of the right-hand side.

Note that, for our expression to be variational, it is *necessary* that there is (to first order) no change in the value of the right-hand side for the physically correct path.

To illustrate, let us *derive* Snell's law of refraction, assuming Fermat's principle is true. We consider at what location light will cross an air/glass interface, as depicted in Fig. 2. Since Fermat tells us immediately that in uniform medium, light travels in straight lines, we need only consider the above possible routes from the source to the sink and ask, does it go via A, or B, or $C, \ldots$? In Fig. 3, we plot, over the range of $A, B, \ldots$, the resulting value of T from eq. (18), the "time of (f)light"!

Suppose, as suggested above, we take any path and shift it slightly. *Only*

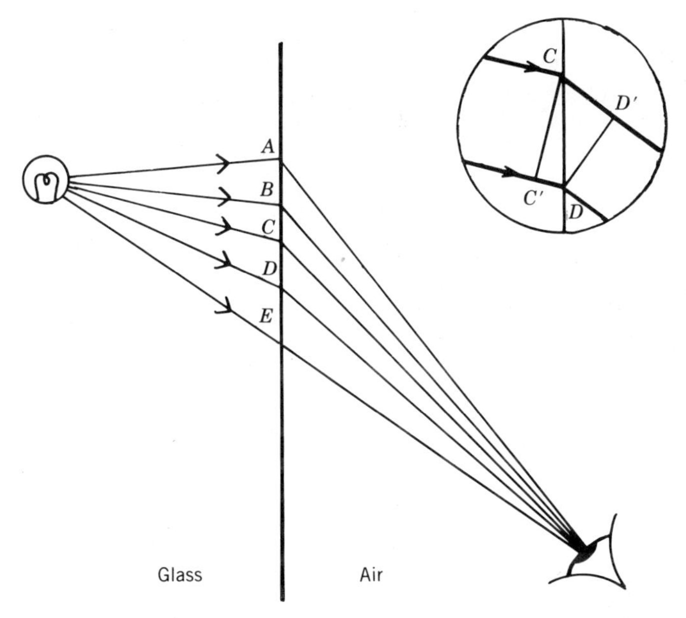

Fig. 2 Possible light paths between points in glass and air.

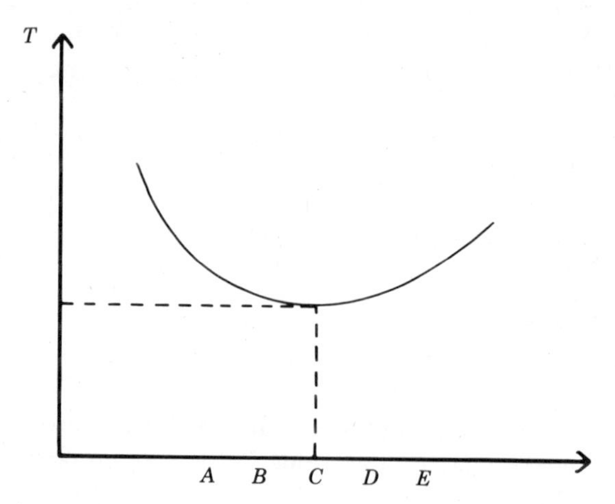

Fig. 3 Time, by eq. (18), as a function of location where light travels from glass to air.

at the minimum value of T will T not be changed to first order; according to Fermat, this corresponds to the physically correct path. Suppose, from Fig. 3, that C is the point with minimum T and that D is a very nearby point. The inset to Fig. 2 shows the two possible light paths from source to sink, immediately around C and D. Suppose that C–C' is a phase front through C (that is, C and C' are equidistant from the source) and D–D' is a phase front through D. If Fermat is correct and C is the route taken by a physical ray of light, then the time for light to travel from C' to D must equal the time for it to travel from C to D'. If η_1, η_2 are the two refractive indices and c is the free-space velocity of light, then we have these two times to be

$$C'D/(c/\eta_1) = CD'/(c/\eta_2)$$

and so, with θ_1, θ_2 as the ray angles to the nomal

$$CD\,\frac{\sin\theta_1}{c/\eta_1} = CD\,\frac{\sin\theta_2}{c/\eta_2}$$

which is Snell's law

$$\frac{\sin\theta_1}{c/\eta_1} = \frac{\sin\theta_2}{c/\eta_2}$$

We have therefore deduced Snell's law of refraction from Fermat's principle; in a way, they are equivalent formulations of the simple laws of refraction.

3.2.2. *Basic Proof of a Variational Form*

Now we consider another variational form, a simple electromagnetic field problem where a minimization occurs. It is closely related to the static field problem, which *is* important for the low-frequency (quasi-TEM mode) behavior of microstrips and coplanar guides.

In a steady current flow situation, when a voltage is applied between two conductors immersed in a resistive medium (perhaps with inhomogeneous resistivity), one can ask, How do the currents flow? How do they distribute themselves? One answer (related to Thompson's theorem) is that the potential distribution and current flow are such as to minimize dissipated (ohmic) power. To put this answer into equation form, consider the two-dimensional flow of steady current in a (possibly nonuniform) resistive sheet. To find an expression for the dissipated power, consider first a small elementary rectangle of dimensions L by W and of resistivity R ohms per square. This means that it will have a resistance RL/W from end to end and so will dissipate power:

$$\frac{(V_1 - V_2)^2}{RL/W} = \left(\frac{V_1 - V_2}{L}\right)^2 \frac{1}{R} LW$$

Therefore heat loss per elementary area dS (= LW) is (with electric field E and potential ϕ)

$$E^2 \frac{1}{R} dS = \frac{1}{R} (\nabla\phi)^2 dS$$

If we want to use this "extremum of power" as a starting point for a variational expression, it will be

$$J(\phi)) = \int\int \frac{1}{R} (\nabla\phi)^2 dS \tag{19}$$

Apparently, if we substitute into (19) all sorts of potential distributions $\phi(x, y)$, the smallest value for J will occur when (and only when) we substitute the physically *correct* $\phi(x, y)$. We will, of course, get the correct value of dissipated heat.

Specifically we are proposing that (19) is a variational expression for steady current flow. But how do we prove it? Our proof follows the four steps outlined after eq. (18).

Consider a particular potential distribution $\phi(x, y)$, which must be an admissible function. With this problem, for a function to be admissible, ϕ can be any "well-behaved function" of x and y providing ϕ takes on the physically assigned values on the boundaries. (An admissible function roughly corresponds to an admissible path with Fermat, which must be continuous and go from the source to the sink. To consider other paths or functions just yields nonsense!) Suppose now that ϕ changes to $\phi + \delta\phi$, which must again be an admissible function. $\delta\phi$ must therefore vanish on the boundary. We take $\delta\phi$ to be small but arbitrary. Therefore,

$$J(\phi + \delta\phi) = \int\int \nabla(\phi + \delta\phi) \cdot \nabla(\phi + \delta\phi)\, dS$$

$$= \int\int (\nabla\phi)^2\, dS + 2\int\int \nabla\phi \cdot \nabla\delta\phi\, dS + \int\int (\nabla\delta\phi)^2\, dS \tag{20}$$

In eq. (20), the factor $1/R$ of eq. (19) has been dropped, resistivity being taken as constant. Now, as in step 4 after eq. (18), we are going to insist that this expression varies only to second order with $\delta\phi$, from the original J value of

$$J(\phi) = \int\int (\nabla\phi)^2\, dS \tag{21}$$

Subtracting (21) from (20) and ignoring the second-order term with $(\nabla\delta\phi)^2$, we have what is called "the first variation in J."

By definition,

$$\delta J \equiv J(\phi + \delta\phi) - J(\phi) \tag{22}$$

which for this case is approximately

$$\delta J = 2\int\int \nabla\phi \cdot \nabla\delta\phi\, dS \tag{23}$$

Green's theorem (one of the many, and this is a standard trick in this part of variational approaches—effectively we are integrating by parts) gives us

$$\int\int \{\nabla u \cdot \nabla v + v\, \nabla^2 u\}\, dS = \int v\, \nabla u\, dc$$

so that eq. (23) can be rewritten as

$$\delta J = -2\int\int \delta\phi\, \nabla^2\phi\, dS + 2\int \delta\phi\, \nabla\phi\, dc \tag{24}$$

The last term is in fact zero, because $\delta\phi = 0$ around the boundary [as noted before eq. (20)]. For the first variation δJ to vanish for *any* small $\delta\phi$, eq. (24) tells us that everywhere we must have

$$\nabla^2\phi = 0 \tag{25}$$

So ϕ must be the correct solution to the steady current problem!

Note 1. The crucial step, from the integral of (24) vanishing to the integrand vanishing in (25), relies on the stationarity being true for *any* small change in $\delta\phi$. It corresponds to Fermat's principle referring to *any* conceivable path.

Note 2. So far, we have only proved the *stationariness* of J for the physically correct ϕ. That the expression is in fact *minimized* can be seen directly from eq. (20). The second term of the equation is δJ, and we have found the consequence of it vanishing. The third term is the integral of an expression that cannot be negative. $\delta J = 0$ must therefore give a minimum value to $J(\phi)$.

Note 3. Equation (25) would be described as the Euler equation, or, better still, the Euler–Lagrange equation corresponding to the original variational expression, eq. (19).

Note 4. The variational form (19) will be used as the basic equation in formulating the complete computer program presented in Section 11.

The above variational form started from eq. (19), on the physical basis of Thompson's theorem, and led to eq. (25), just as Fermat's principle led to Snell's law.

3.2.3. A Variational Form for Waveguides

Now we will try another variational expression. The philosophy or strategy of our approach is to take a proposed variational form and find its Euler equation, that is, find what equations have to be satisfied to make the functional J stationary.

If the Euler equations correspond to a physical problem, then we have a useful variational form; if not, we've wasted a few minutes of our lives. The original equation is still, arguably, a variational expression—it just has no interest for us.

Let us consider a plane surface S enclosed by a contour C.

Letting

$$J(\phi) = \frac{\iint (\nabla\phi)^2 \, dS}{\iint \phi^2 \, dS} \tag{26}$$

then

$$J(\phi) \iint \phi^2 \, dS = \iint (\nabla\phi)^2 \, dS \tag{27}$$

and so

$$J(\phi + \delta\phi) \iint (\phi + \delta\phi^2)\, dS = \iint (\nabla\phi + \nabla\delta\phi)^2\, dS \tag{28}$$

where $\delta\phi$ is our small variation in the admissible function. Subtracting eq. (27) from (28) (and ignoring any terms with degree 2 in $\delta\phi$) gives the equation for δJ:

$$\delta J \iint \phi^2\, dS + J \cdot 2 \iint \phi\delta\phi\, dS = 2 \iint \nabla\phi \cdot \nabla\delta\phi\, dS \tag{29}$$

where now J simply denotes $J(\phi)$ and $\delta J = J(\phi + \delta\phi) - J(\phi)$.

Again we wield Green's theorem to express the right-hand side of (29) as

$$2 \int \delta\phi \cdot \frac{\partial\phi}{\partial n}\, dc - \iint \delta\phi \cdot \nabla^2\phi\, dS$$

For stationariness, $\delta J = 0$, and we must therefore have

$$J \iint \phi\, \delta\phi = \int \delta\phi \cdot \frac{\partial\phi}{\partial n}\, dc - \iint \delta\phi \cdot \nabla^2\phi\, dS \tag{30}$$

or, to rearrange it,

$$\iint \delta\phi\{\nabla^2\phi + J\phi\}\, dS = \int \delta\phi \cdot \frac{\partial\phi}{\partial n}\, dc \tag{31}$$

We have not yet discussed what functions are admissible. First they must be sufficiently differentiable (this can be checked by seeing the consequence of any violation). But in addition:

1. Suppose that for admissibility ϕ has to be zero on the boundary C. Then $\delta\phi = 0$ on C and the right-hand side of (31) is zero; our Euler equation must therefore be

$$\nabla^2\phi + J\phi = 0 \tag{32}$$

which, of course, corresponds to the Helmholtz equation

$$\nabla^2\phi + k^2\phi = 0 \tag{33}$$

for a waveguide with cutoff wavelength $\lambda_c = 2\pi/k$, and we identify $J = k^2$. Because we insisted on $\phi = 0$, the modes we obtain for the waveguide must be TM (**E**) rather than TE (**H**).

2. Suppose now, for admissibility, ϕ has to satisfy

$$\frac{\partial\phi}{\partial n} = 0 \tag{34}$$

on the boundary C. Again the right-hand side of eq. (31) is precisely zero, and everything in the previous paragraph holds except that eq. (34) is the boundary condition to give us TE (**H**) modes.

We can therefore write

$$k^2 = J(\phi) = \text{SV} \frac{\iint (\nabla\phi)^2 \, dS}{\iint \phi^2 \, dS} \tag{35}$$

as a variational expression related to the problem of modes in a hollow conducting waveguide of cross section S. By insisting on either ϕ or its normal derivative being zero on C, we obtain corresponding TM or TE modes.

In fact, the above expression is more than just stationary—it is minimal. *Any* admissible ϕ substituted into (35) will give a result *above* the (lowest or dominant mode) cutoff wavenumber. For modes above the dominant, eq. (35) is still stationary—it just ceases to be minimal.

EXERCISE

As a variation on the above theme involving a resistive medium, find the Euler equation corresponding to the variational expression

$$J(\phi) = \iint f(x, y)(\nabla\phi)^2 \, dS \tag{36}$$

where $f(x, y)$ is some given real scalar function. The result is useful even at microwave frequencies.

To save time searching through the almost unlimited types of Green's theorems, the following equation may be useful:

$$\iint_S p \, \nabla q \cdot \nabla r \, dS = -\iint_S q \nabla \cdot (p \, \nabla r) \, dS + \int_C pq \frac{\partial r}{\partial n} \, dc \tag{37}$$

As an aside, we mention how variational expressions are regularly used in other topics.

1. If eq. (19) is applied to a capacitor rather than to the earlier resistor, we obtain instead of heat, the more useful parameter of capacitance. This is used by Collin [8] to obtain various approximations to characteristic impedances of TEM structures in his classic text.

2. Equation (35) will work equally well for the resonant frequency of an acoustic container (ϕ is then the instantaneous pressure). Integration would

clearly be over three dimensions in general. The problem of a vibrating string is just the same equation, but in one dimension.

3. The equation

$$E = \mathrm{SV} \frac{\iiint \psi^* \mathbf{H} \psi \, dv}{\iiint \psi^* \psi \, dv} \tag{38}$$

will give the quantum-mechanical energy levels, solutions to the time-independent Schrödinger equation (where ψ is the wave function and $\mathbf{H} \equiv [-(\hbar^2/2m) \cdot \nabla^2 + V]$).

4. The equation

$$\omega^2 = \mathrm{SV} \frac{\iiint (\nabla \times \mathbf{H})^* \|\varepsilon\|^{-1} (\nabla \times \mathbf{H}) \, dv}{\iiint \mathbf{H}^* \|\mu\| \mathbf{H} \, dv} \tag{39}$$

gives the resonant frequency of an electromagnetic resonator that can include any distribution of dielectric or magnetic material, including anisotropic material. One can then deal with uniaxial dielectrics, such as $LiNbO_3$, which is regularly used in optical components, and even gyrotropic materials, such as magnetized ferrites, which are regularly used in microwave and UHF components.

3.2.4. *Natural Boundary Conditions*

Equation (26) started as the variational expression

$$J(\phi) = \frac{\iint (\nabla \phi)^2 \, dS}{\iint \phi^2 \, dS} \tag{40}$$

and we found that for suitable admissible functions (either $\phi = 0$ or $\partial \phi / \partial n = 0$ on the boundary) the corresponding Euler equation was the Helmholtz equation

$$\nabla^2 \phi + k^2 \phi = 0 \tag{41}$$

That is, for $J(\phi)$ to be stationary, or $\delta J = 0$, our function must satisfy (41).

Suppose now we relax our conditions on admissible functions—that we no longer restrict the functions on the boundary. This takes us back to eq. (31), which is

$$\iint_S \delta\phi\{\nabla^2\phi + J\phi\}\, dS = \int_C \delta\phi \frac{\partial\phi}{\partial n}\, dc \tag{42}$$

Stationarity of eq. (40) for *any* small $\delta\phi$ forced us to the Euler equation, (41); that was the only way that (42) could be satisfied for all $\delta\phi$. But the right-hand side of eq. (42) is rather like the left-hand side, in that both have $\delta\phi$ "in isolation." Hence, just as for the $\{\nabla^2\phi + J\phi\}$ term, so for the $\partial\phi/\partial n$ term we can argue that for eq. (42) to be true for any $\delta\phi$ we must have

$$\frac{\partial\phi}{\partial n} = 0 \tag{43}$$

These are the natural boundary conditions relating to the variational expression, eq. (40). They are rather like "Euler equations at the boundary."

$$\text{Insistence that } \delta J = 0 \Rightarrow \begin{cases} \text{Euler equations are satisfied} \\ \qquad\qquad \text{and} \\ \text{Natural boundary conditions} \\ \text{are also satisfied} \end{cases}$$

On our first dealing with eq. (14) we took a shortcut by *assuming* certain boundary conditions to be satisfied and *then* applying stationarity arguments. If, however, we do not restrict the trial function, then we find that the boundary condition $\partial\phi/\partial n = 0$ is naturally implied.

Note 1. We say implied; the boundary conditions are *not exactly* satisfied, any more than (generally speaking) the Euler equations are *exactly* satisfied by any trial function we choose to use. This misconception does appear in the literature!

Note 2. If boundary conditions have to be satisfied for an expression to be stationary, they are termed *principal* or *essential boundary conditions—in contrast to the natural boundary conditions.*

The importance of natural boundary conditions lies in the fact that it is often difficult, sometimes impossible, to arrange for the essential conditions to be satisfied. When this applies we can find the natural conditions, and (rather like the Euler equations) if they correspond to our physical problem we are in business.

Sometimes these natural boundary conditions are not the ones we want—they do not correspond to our physical problem. Then we can try adding a term or two to our variational expression. For example, using eq.

(40) as our variational expression, stationarity implied that eq. (42) had to be satisfied for any $\delta\phi$. Now instead of eq. (40) we try an extra term:

$$J(\phi) = \frac{\iint (\nabla\phi)^2 \, dS - F \int (\phi \, \partial\phi/\partial n) \, dc}{\iint \phi^2 \, dS} \tag{44}$$

where F is a constant factor, to be chosen later. By our standard variational-proving procedure, we find that the following equation must be satisfied:

$$2 \iint \delta\phi(\nabla^2\phi + J\phi) \, dS = (2 - F) \int \delta\phi \, \frac{\partial\phi}{\partial n} \, dc - F \int \phi \, \frac{\partial(\delta\phi)}{\partial n} \, dc \tag{45}$$

Naturally, choosing $F = 0$ gives the original result, with the left-hand side giving our Euler equation, eq. (41), and the right-hand side our natural boundary condition, eq. (43). However, from eq. (45) we see that choosing $F = 2$ will give us a new natural boundary condition:

$$\phi = 0 \tag{46}$$

Note that if it is convenient to choose trial functions that indeed satisfy the essential (i.e., physical) boundary conditions, then the concept of natural boundary conditions is irrelevant and effectively undefinable. When it is not convenient to so choose them, it is then necessary to ensure that the natural boundary conditions are the ones of our physical problem.

4. USING A VARIATIONAL EXPRESSION

After eq. (17) we posed three questions, and we have answered the first two—what is, and how do we obtain, a variational expression? Now to use it! There are perhaps three ways of using it:

(i) By directly substituting into the expression a specific (hopefully good) trial function.

(ii) By using it to obtain a perturbation formula.

(iii) By substituting a set of basis functions as described in Section 2 in discussing weighted residuals, and methodically choosing good coefficients $\{u_i\}$ in an expansion such as eq. (6). This is the scheme of our computational approach and will be elaborated after we consider (i) and (ii).

(i) The example after eq. (17) illustrated this approach. It is relevant *only* when the variational expression gives a parameter in which we are interested. If we indeed have such a form (e.g., for cutoff frequency or radar cross section), then there is nothing to stop us from substituting any eligible trial function. Because of the stationary nature of the variational expression, any reasonably good trial function will give a very good numerical result for our wanted parameter.

This was the rationale behind our success following eq. (17).

(ii) *Perturbation theory* is any theory that gives an (approximate) answer to a problem that is a small perturbation from some other problem to which we have the answer. As an example, take as our base problem that of the dominant TE_{01} rectangular waveguide mode. Suppose, to quantify manufacturing tolerances, we wanted to know the effect on cutoff frequency of (1) the angles not being quite 90° or (2) some of the corners being rounded or filleted. Perturbation formulas would be ideal here, giving us the cutoff frequency explicitly (for either small changes of angle from 90° or small radius of corners).

Perturbation formulas are derived:

(a) By ad hoc methods (such as by obtaining an answer as a Taylor series in terms of the small perturbation parameter—this can be tortuous!).

(b) Directly from a variational expression. If there is such an expression available, it is undoubtedly the quicker and more powerful approach.

A perturbation formula will automatically emerge if, into a variational expression valid for the perturbed problem, we substitute explicitly as a trial function the known solution to the base (original) problem.

The known solution must, of course, be an admissible function, and *either* the known solution is an essential boundary condition *or* (more usually) the natural boundary conditions corresponds to our physical problem.

To solve the example mentioned above, we will look for the cutoff frequency of the TE_{01} mode in the waveguide shown in Fig. 4. We use a variational expression valid for the problem—eq. (40) for a TE mode. For our deformed rectangular guide, it is therefore a matter of substituting

$$(H_z =)\ \phi = \cos(\pi x/a) \tag{47}$$

into

$$k^2 \approx \frac{\displaystyle\iint_{S'} (\nabla\phi)^2\, dS}{\displaystyle\iint_{S'} \phi^2\, dS} \tag{48}$$

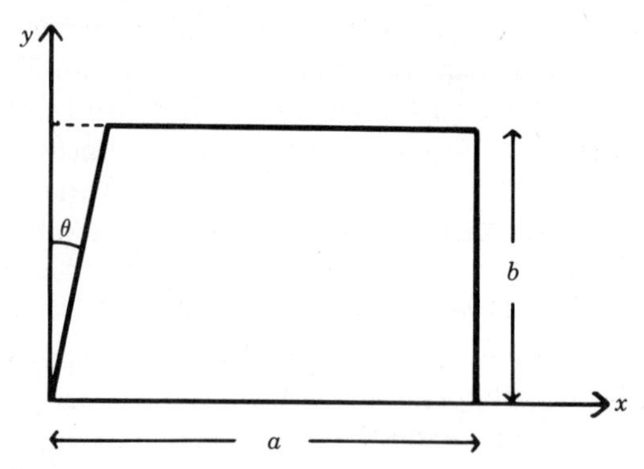

Fig. 4 Cross section of slightly deformed rectangular waveguide.

Here S' denotes the cross section of the deformed rectangle. Let S denote the original rectangle and $s = S - S'$ the small triangular difference.

Integration over the original rectangle gives

$$k_0^2 \approx \frac{\iint_S (\nabla\phi)^2 \, dS}{\iint_S \phi^2 \, dS} \tag{49}$$

which must have the value $k_0^2 = (\pi/a)^2$.

Subtracting eq. (49) from (48) gives the perturbation:

$$k^2 - k_0^2 = \frac{\iint_{S'} (\nabla\phi)^2 \, dS}{\iint_{S'} \phi^2 \, dS} - \frac{\iint_S (\nabla\phi)^2 \, dS}{\iint_S \phi^2 \, dS}$$

$$= k_0^2 \left\{ \frac{\iint_{S'} (\nabla\phi)^2 \, dS}{\iint_{S'} \phi^2 \, dS} \frac{\iint_S \phi^2 \, dS}{\iint_S (\nabla\phi)^2 \, dS} - 1 \right\}$$

$$= k_0^2 \left\{ \left(1 - \frac{\iint_s (\nabla\phi)^2 \, dS}{\iint_S (\nabla\phi)^2 \, dS} \right) \left(1 - \frac{\iint_s \phi^2 \, dS}{\iint_S \phi^2 \, dS} \right)^{-1} - 1 \right\}$$

$$= k_0^2 \left\{ \frac{\iint_s \phi^2 \, dS}{\iint_S \phi^2 \, dS} - \frac{\iint_s (\nabla\phi)^2 \, dS}{\iint_S (\nabla\phi)^2 \, dS} \right\} \tag{50}$$

In the above, only first-order terms have been retained, and the binomial theorem has been used. We have, from eq. (47),

$$\phi = \cos(\pi x / a)$$

Therefore,

$$|\nabla\phi| = -\frac{\pi}{a} \sin\frac{\pi x}{a} \qquad \text{and} \qquad (\nabla\phi)^2 = \left(\frac{\pi}{a}\right)^2 \sin^2\frac{\pi x}{a}$$

Evaluating first the integrals in the numerators of eq. (50), we note that $(\nabla\phi)^2$ is zero over the thin triangle s. Equation (50) therefore becomes

$$k^2 - k_0^2 = k_0^2 \frac{\iint_s \phi^2 \, dS}{\iint_S \phi^2 \, dS} = \frac{k_0^2(\frac{1}{2}\theta)(b^2)}{\int_S \cos^2(\pi x/a) \, dx \, dy} = k_0^2 \frac{\frac{1}{2}\theta b^2}{\frac{1}{2}ab} = k_0^2 \theta \frac{b}{a}$$

or

$$\frac{k - k_0}{k_0} \approx \frac{1}{2} \theta \frac{b}{a} \tag{51}$$

This is the perturbation formula we sought.

Of course, by a Taylor expansion, the *exact* expression for the perturbation must be

$$k - k_0 = c_1\theta + c_2\theta^2 + c_3\theta^3 + \cdots$$

Our perturbation formula, eq. (51), must give *exactly* the first term of this Taylor series.

4.1. The Rayleigh–Ritz Method

At the beginning of this section, we listed three ways of using a variational expression; this is the last of the three, and a way we will exploit.

Suppose we have a variational expression—and to make the discussion easier we will assume it is a minimum—namely,

$$J_0 = \min_{\phi} [J(\phi)] \tag{52}$$

that gives us what we want as an Euler equation. We could get an idea of ϕ by substituting various trial functions $\phi_1, \phi_2, \phi_3, \ldots$ (which must clearly be admissible) into eq. (52) and picking out the smallest.

It might be more systematic to try a combination of these basis functions [like our eq. (6) of the weighted residuals] and pick out the best combination. The best here will be the smallest and will also be stationary with respect to small changes. Trying eq. (6) as our trial function yields

$$\phi = u_1\phi_1 + u_2\phi_2 + u_3\phi_3 + \cdots + u_N\phi_N \tag{53}$$

where $\{u_i\}$ (the set $u_1, u_2, \ldots, u_N$) are N arbitrary coefficients (which in this context we call Ritz parameters). We presume that any choice of $\{u_i\}$ will give via eq. (53) an admissible function ϕ. Substituting into eq. (52) gives us

$$J_0 = \min [J(u_1\phi_1 + \cdots + u_N\phi_N)] \tag{54}$$

where for specified basis functions ϕ_i, J must be a function of the coefficients $\{u_i\}$. To find the best combination of $\{u_i\}$, we differentiate (54) with respect to one of the u's, say u_i, to give $\partial J/\partial u_i$.

If we have the best (minimizing) combination, then certainly

$$\frac{\partial J}{\partial u_i} = 0 \tag{55}$$

This is true for all u_i, and so there will be N such equations (55) (with $i = 1, 2, \ldots, N$) in N unknowns $(u_1, u_2, \ldots, u_N)$.

To illustrate the Rayleigh–Ritz method, we consider the vibrating string problem, where we look for the resonant frequencies of a uniform string stretched between two points, as in a violin or guitar. The problem is merely a one-dimensional version of the two-dimensional waveguide problem discussed earlier for TM modes. It just makes the algebra easier!

The vibrating string problem is to find solutions to

$$\frac{d^2y}{dx^2} + k^2y = 0 \tag{56}$$

subject to

$$y(-1) = y(+1) = 0 \tag{57}$$

We already know the answers to be:

y	k^2
$y_1 = \cos(\pi x/2)$	$(\pi/2)^2$
$y_2 = \sin(2\pi x/2)$	$(2\pi/2)^2$
$y_3 = \cos(3\pi x/2)$	$(3\pi/2)^2$
etc.	

Just as for TM modes in a waveguide [eq. (35)], we obtain for the vibrating string the variational form

$$k^2 = \text{SV } J(y) = \text{SV } \frac{\int_{-1}^{1} (dy/dx)^2 \, dx}{\int_{-1}^{1} y^2 \, dx} \tag{58}$$

Knowing the answer for the lowest resonance, or by carefully observing a vibrating string, we could try our luck and think of a single trial function. It must satisfy (57). If the simplest function is a polynomial, then the simplest function satisfying (57) is

$$y = (x-1)(x+1) = x^2 - 1 \tag{59}$$

Substitution into (58) gives

$$k^2 \approx J(x^2 - 1) = \frac{\int_{-1}^{1} (2x)^2 \, dx}{\int_{-1}^{1} (x^2 - 1) \, dx}$$

$$= \frac{\int_{-1}^{1} 4x^2 \, dx}{\int_{-1}^{1} (x^4 - 2x^2 + 1) \, dx} = \frac{4/3}{(1/5 - 2/3 + 1)}$$

$$= 2.5 \tag{60}$$

This compares with the exact answer, 2.4674 . . . .

Accurate as it is, we now go to Rayleigh–Ritz. Taking the above guess as the first basis function ϕ_1, the next eligible polynomial would be $x(x-1)(x+1)$. Substituting it alone into (58) gives

$$J(x^3 - x) = 10.5$$

Not a very impressive approximation to 2.4674! It is, in fact, a reasonable approximation to $\pi^2 = 9.87$, the lowest resonance with an odd x dependence. (The whole problem is symmetrical about $x = 0$, and so solutions could be separately sought with odd and even x dependence.)

A more useful ϕ_2 would be $(x^2 - 1)(x^2 + 1)$, the next even polynomial in x that vanishes at $x = -1$ and $x = +1$. We therefore try

$$y = \phi = u_1\phi_1 + u_2\phi_2 \quad \text{with } \phi_1 = x^2 - 1,\ \phi_2 = x^4 - 1 \tag{61}$$

This results in

$$J(u_1\phi_1 + u_2\phi_2) = \frac{105u_1^2 + 2u_1u_2 \cdot 126 + u_2^2 \cdot 180}{42u_1^2 + 2u_1u_2 \cdot 48 + u_2^2 \cdot 56} \tag{62}$$

Altering the ratio u_1/u_2 alters the "mixture" of ϕ_1 and ϕ_2 in our trial function, and consequently the value of J given by (62). To get the best possible approximation with just ϕ_1 and ϕ_2, we want the stationary (possibly minimum) value of J. Putting $u_1 = 1$ and differentiating J with respect to u_2 gives

$$\frac{dJ(\phi_1 + u_2\phi_2)}{du_2} = 0 \qquad \text{when } u_2 = \frac{-70 \pm \sqrt{2128}}{132} \tag{63}$$

The two roots are $u_2 = -0.18808315$ and $u_2 = -0.8797746$ with the associated values

$$J = 2.467438 \qquad \text{and} \qquad J = 25.53255$$

The first value is a much better approximation (than our previous 2.5) to the exact $(\frac{1}{2}\pi)^2 \approx 2.467401$. The second value is in fact an approximation to the second resonance with an even x dependence, namely, $(1\frac{1}{2}\pi)^2 \approx 22.21$. In Fig. 5, the two resulting ϕ's are plotted, as given by eq. (61) with $u_1 = 1$, and u_2 given by eq. (63). These resulting solutions can be seen as very reasonable approximations to (a negative constant times) $\cos\{\frac{1}{2}\pi x\}$ and $\cos\{1\frac{1}{2}\pi x\}$.

If we wanted to continue expanding our polynomial basis functions in this Rayleigh–Ritz method, the next sensible basis function might be

$$\phi_3 = x^6 - 1 \tag{64}$$

The resulting J would be like eq. (62), but now as a ratio of quadratic forms in u_1, u_2, and u_3 (rather than in just u_1 and u_2). Putting $u_1 = 1$ and differentiating J with respect to u_2 and u_3 and putting each to zero would give

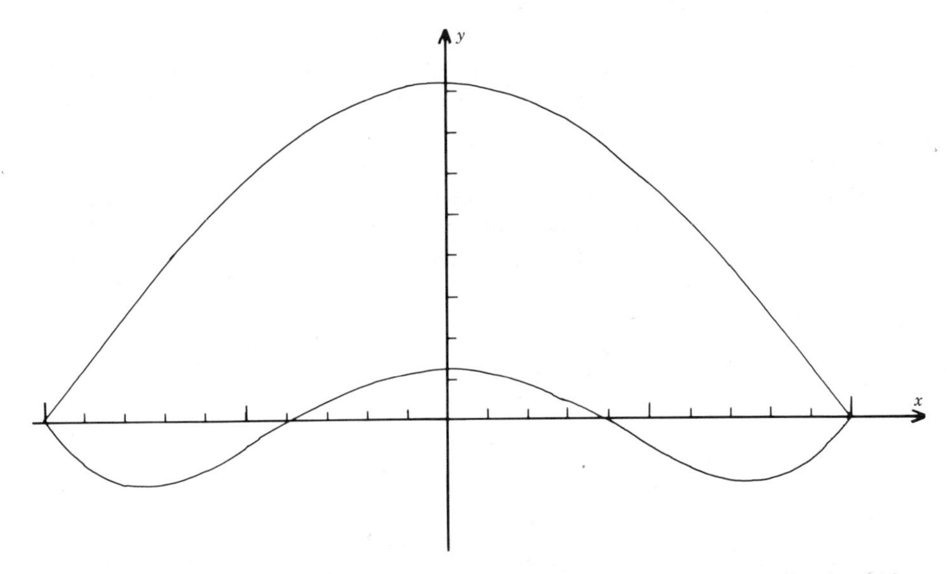

Fig. 5 Rayleigh–Ritz results for the first two "even" solutions to the vibrating string problem of eqs. (56) and (57).

1. A third approximation to $(\frac{1}{2}\pi)^2$, better than 2.5 or 2.467438.
2. A second approximation to $(1\frac{1}{2}\pi)^2$, better than 25.53.
3. A first approximation to $(2\frac{1}{2}\pi)^2$.

5. THE FINITE ELEMENT METHOD

Basis and weighting functions were introduced in the weighted residuals method of Section 2. For the solution to converge correctly, these functions must clearly be "complete"; loosely speaking, at any location s, the finite sum

$$u = \sum_{i=1}^{N} u_i b_i(s) \tag{65}$$

should approach the physical field as N approaches infinity.

Similarly with the Rayleigh–Ritz method; in this case only one series is needed, but it must equally be complete.

The finite element method is essentially either of these two methods where the domain of interest is *first* divided into subdomains or elements, and different expansions like (65) are used over each element. Usually they have the same *form* over all elements (such as polynomials or sinusoids), but they have different coefficient values $\{u_i\}$. To be admissible functions, they must satisfy some conditions between elements; usually we will insist on continuity, but otherwise there is no need for the elements to be of equal size or proportions or (with care) even to have the same degree of polynomial, sine, etc.

The simplest example in one dimension would surely be the piecewise continuous linear function or, more elaborately, the piecewise quadratic function, as shown in Fig. 6. These are clearly much like the basis functions of eqs. (59) and (61), except for their piecewise rather than global definitions.

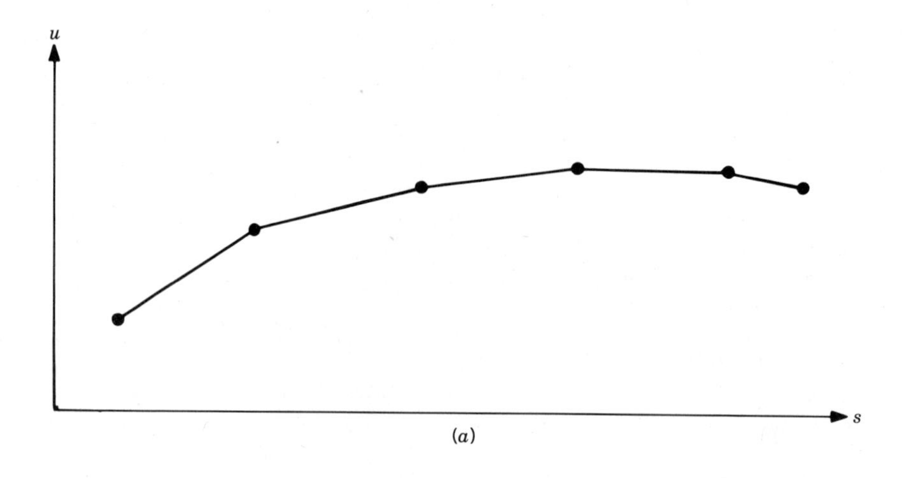

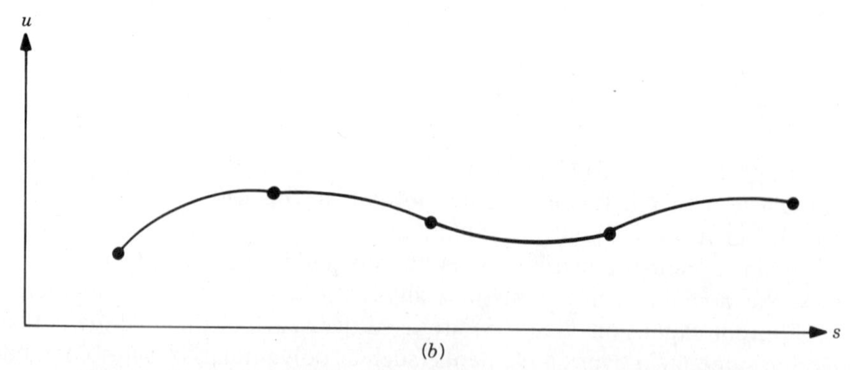

Fig. 6 Piecewise linear (*a*) or quadratic (*b*) continuous functions.

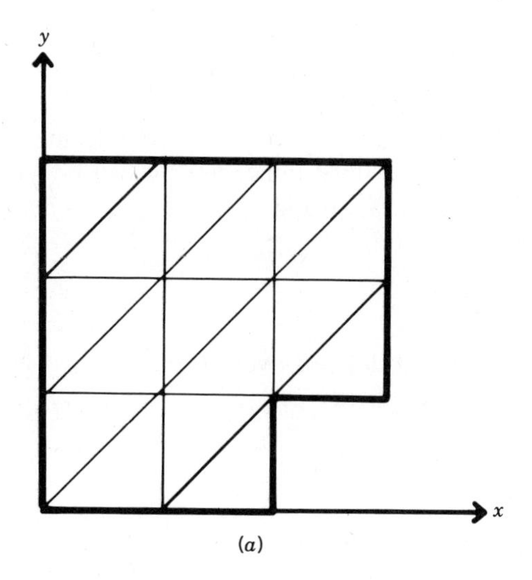

(a)

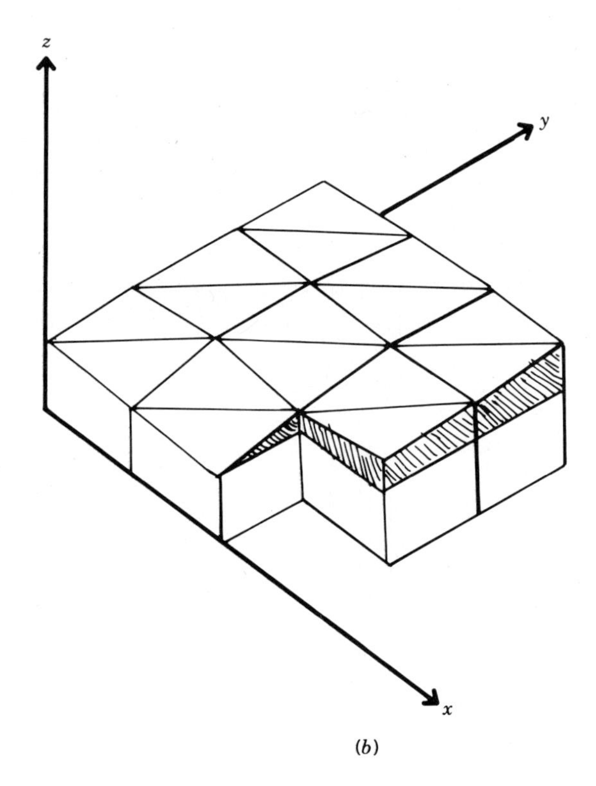

(b)

Fig. 7 (*a*) Example of triangular elements over an L-shaped region. (*b*) Display of potential U as a function of x and y.

In two dimensions the elements are often triangles or rectangles.

The simplest use of triangular elements is by "first-order" elements, where a first-degree polynomial $(a + bx + cy)$ is used over each element. This function will be continuous across adjacent triangles, so that if the potential is interpreted like a third dimension, it can be viewed as a surface with many triangular facets, as in Fig. 7.

5.1. Nodal Values

Figure 7*b* shows a potential defined over 16 first-order triangular elements. It could then be defined explicitly as

$$U(x, y) = a_i + b_i x + c_i y \tag{66}$$

with a different set of coefficients $\{a_i, b_i, c_i\}$ on each triangle. To ensure admissibility of our function U, we would have to arrange continuity at all triangle edges (it would be adequate to arrange for continuity at the 15 vertices). However, a much smarter way is to define *not* in terms of different coefficients $\{a, b, c\}$ over each triangle but in terms of the *nodal values*. These are the (unknown but wanted) values of our field or potential at the element vertices. Luckily the triangle has three vertices—just the number of coefficients per triangle, so all is well!

Suppose the typical triangle has vertices (x_1, y_1), (x_2, y_2), and (x_3, y_3) with nodal values U_1, U_2, and U_3. Substituting into eq. (66) at each vertex gives

$$\begin{bmatrix} U_1 \\ U_2 \\ U_3 \end{bmatrix} = \begin{bmatrix} 1 & x_1 & y_1 \\ 1 & x_2 & y_2 \\ 1 & x_3 & y_3 \end{bmatrix} \begin{bmatrix} a \\ b \\ c \end{bmatrix} \tag{67}$$

Rearranging (67) as

$$\begin{bmatrix} a \\ b \\ c \end{bmatrix} = \begin{bmatrix} 1 & x_1 & y_1 \\ 1 & x_2 & y_2 \\ 1 & x_3 & y_3 \end{bmatrix}^{-1} \begin{bmatrix} U_1 \\ U_2 \\ U_3 \end{bmatrix} \tag{68}$$

and putting (66) in a scalar product form of vectors gives

$$U(x, y) = [1 \quad x \quad y] \begin{bmatrix} 1 & x_1 & y_1 \\ 1 & x_2 & y_2 \\ 1 & x_3 & y_3 \end{bmatrix}^{-1} \begin{bmatrix} U_1 \\ U_2 \\ U_3 \end{bmatrix} \tag{69}$$

Note. The notation of eq. (69) is typical of the very structured notation used in finite elements. $\boldsymbol{U}$ is given by the product of three matrices, and manipulation of $\boldsymbol{U}(x, y)$ can be made quite "orderly." The first matrix is a list of *shape functions* (to use the jargon term of finite elements). The third matrix is a list of nodal values. The middle matrix represents the unique transformation between them over the triangle. We will return to this later.

If we combine the first two matrices, we obtain

$$\boldsymbol{U}(x, y) = [A_1 \quad A_2 \quad A_3]\begin{bmatrix} U_1 \\ U_2 \\ U_3 \end{bmatrix} \tag{70}$$

where

$$A_1(x, y) = \{(x_2y_3 - x_3y_2) + (y_2 - y_3)x + (x_3 - x_2)y\}/\det \tag{71}$$

where det is the determinant value of the 3×3 matrix in eq. (67) and A_2, A_3 are given by cyclic exchange of $1 \rightarrow 2 \rightarrow 3$ in (71).

The point of these last two equations is that (71) is an *interpolation polynomial*. Its distinguishing feature is that it has value 1 at a single node (vertex) but value 0 at other nodes. So A_1 takes on value 1 at node 1 (only), and similarly A_2 takes on the value 1 at node 2, and A_3, at node 3.

The three nodal values U_1, U_2, and U_3 of eq. (70), or the "ensemble" of values (like the 15 of Fig. 7*b*) become the basis coefficients $\{u_i\}$ of our weighted residual [eq. (6)] or Rayleigh–Ritz [eq. (53)].

At this point we can conveniently refer to Silvester and Ferrari's text [11]. They dive straight into finite elements (with no reference to variational or weighted-residual methods), and on pages 7–25 develop fairly detailed "bookkeeping" to present an excellent little complete finite element program package in Fortran. It involves a mere 250 lines of executable statements and is professionally structured and documented. Using first-order polynomial functions, their program solves fairly arbitrarily shaped structures for the Laplace (or more general Poisson) equation, with different realistic boundary conditions.

5.2. Shape Functions for Triangles

We obtained eqs. (70) and (71) by inversion of the transforming matrix in (69). But when considering more nodes with more complicated functions, it is advantageous to work directly, by considering these interpolation polynomials. The representation of $\boldsymbol{U}(x, y)$ by eqs. (70) and (71) is called a shape function. We used interpolation functions, which in fact were normalized with respect to their value (they assumed values from 0 to 1 as x and y ranged over the triangle). A more important normalization can be done with respect to the x and y coordinates; then the same function can be applied to any kind of triangle.

Shape functions can be considered for one, two, or three dimensions; we will study only those for two dimensions. The most common elements are triangular or rectangular. We will study only triangles, they have a unique shape for elements using polynomial shape functions.

5.2.1. *Complete Polynomials*

$a + bx + cy$ is a *complete polynomial* of first degree.

$a + bx + cy + dx^2 + exy + fy^2$ is a complete polynomial of second degree.

Omitting any term from either of the above expressions would yield an *incomplete polynomial.* It is desirable, though not paramount, to use complete polynomials. One feature of them is that they are invariant with respect to coordinate rotation; they are isotropic—there is no bias or preferred direction.

As has already been illustrated, the first-degree polynomial involves three coefficients and so can be expressed in terms of three nodal values at the triangle vertices. The second-degree polynomial needs six coefficients and can similarly be expressed in terms of values of six nodes, located at the vertices and midpoints of the sides, as in Fig. 8.

We see from Fig. 9, which identifies Pascal's triangle in x and y with the geometric layout of nodes and triangular elements, that a complete polynomial of any degree is precisely and uniquely defined in terms of nodal values at the nodes uniformly distributed over the triangular element—three for our first order, six for the second order, and so on. Before elaborating on

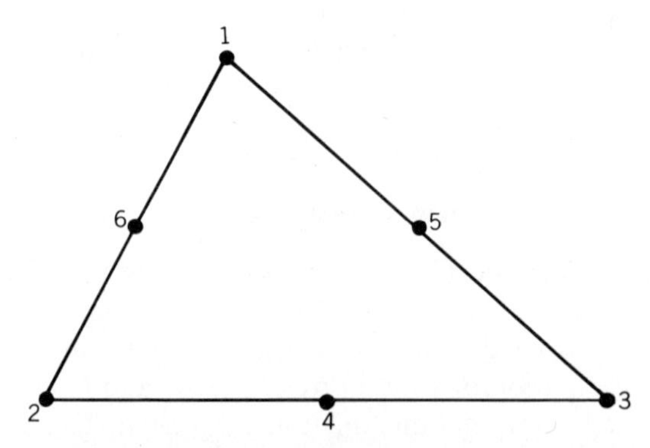

Fig. 8 Nodes for a second-degree complete polynomial on a triangular element.

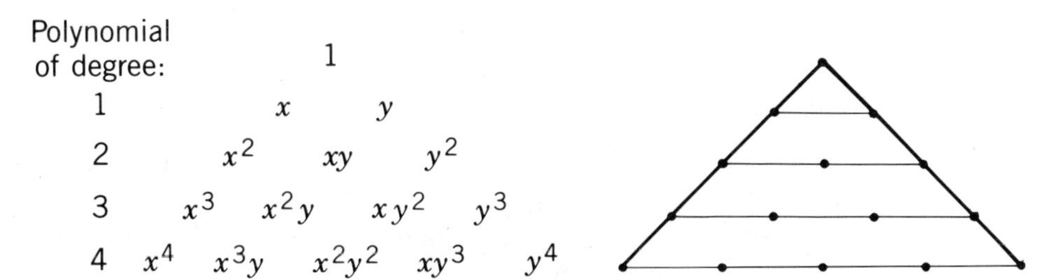

Fig. 9 Pattern of nodes on a single element compared with Pascal's triangle, for complete polynomials of increasing degrees.

some needed interpolating polynomials, we introduce local coordinates, so as to normalize all triangles to effectively one prototype.

5.2.2. *Local (Area) Coordinates*

Consider the typical point P somewhere in the triangle of Fig. 10 with vertices v_i, v_2, and v_3. Denote by L_1 the ratio

$$L_1 = \frac{\text{area of triangle } Pv_2v_3}{\text{area of triangle } v_1v_2v_3} \tag{72}$$

L_1 is proportional to the perpendicular distance of P from side v_2v_3; more notable is the fact that it takes on value 1 at v_1, 0 at v_2, and 0 at v_3; it is therefore the unique interpolating first-degree polynomial for node 1. L_2 and L_3 are similarly defined. It then follows immediately from the area definition that

$$L_1 + L_2 + L_3 = 1 \tag{73}$$

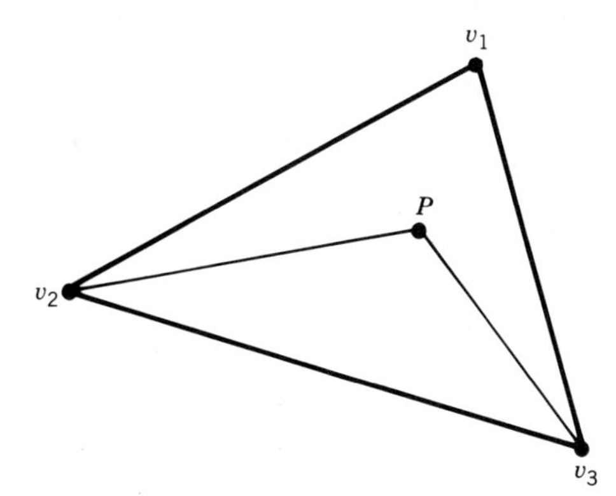

Fig. 10 Triangular element.

Because of their linearity, L_1, L_2, and L_3 are "equivalent" or constitute a substitute for the usual x, y coordinates. For the triangle, they are more natural, having no bias to any vertex. They are called *local coordinates* or *area coordinates*.

To obtain first-degree shape functions, L_1, L_2, and L_3 are simply substituted for A_1, A_2, and A_3 of eqs. (70) and (71).

Equation (72) also provides the formal transformation/relation between L_1 and $1, x, y$, with similar versions for L_2 and L_3.

To generalize now to higher-order elements, consider first the problem of interpolating a higher degree, just in one dimension. Specifically, how do we choose a polynomial in x, of degree N, that has value 1 at node x_0 but vanishes on all other nodes $x_1, x_2, \ldots, x_N$?

By examination, the polynomial

$$P(x) = \frac{(x - x_1)\cdots(x - x_N)}{(x_0 - x_1)\cdots(x_0 - x_N)} \tag{74}$$

has just the properties we need. It is known as the *Lagrange interpolation polynomial* of degree N and applies with the nodes $x_0, x_1, \ldots, x_N$ in any order (see Fig. 11, for example).

Having defined these neatly normalized area coordinates and interpolation polynomials, we now stitch them together to give, for the three-node triangle of Fig. 12, first-order shape functions:

$$N_1 = L_1 \qquad N_2 = L_2 \qquad N_3 = L_3 \tag{75}$$

For the six-node triangle, we have second-order shape functions

$$N_1 = (2L_1 - 1)L_1 \tag{76}$$

and similarly for N_2 and N_3 at the other two vertices. For the midpoints we have

$$N_4 = 4L_1L_2 \tag{77}$$

and similarly for N_5 and N_6 at the other two midpoints.

Third- and higher-order elements can similarly be realized.

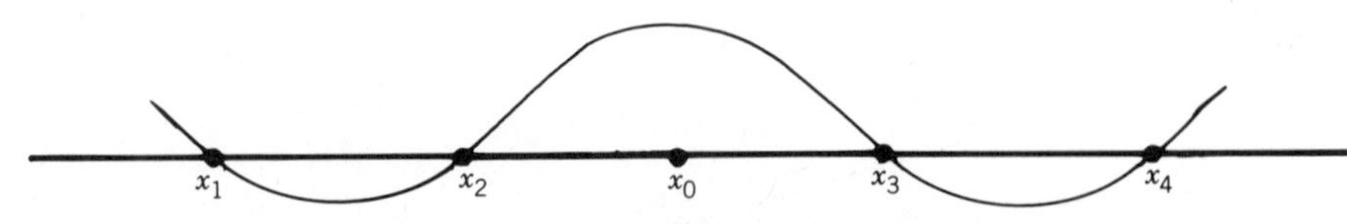

Fig. 11 Lagrange interpolation polynomial of degree 4.

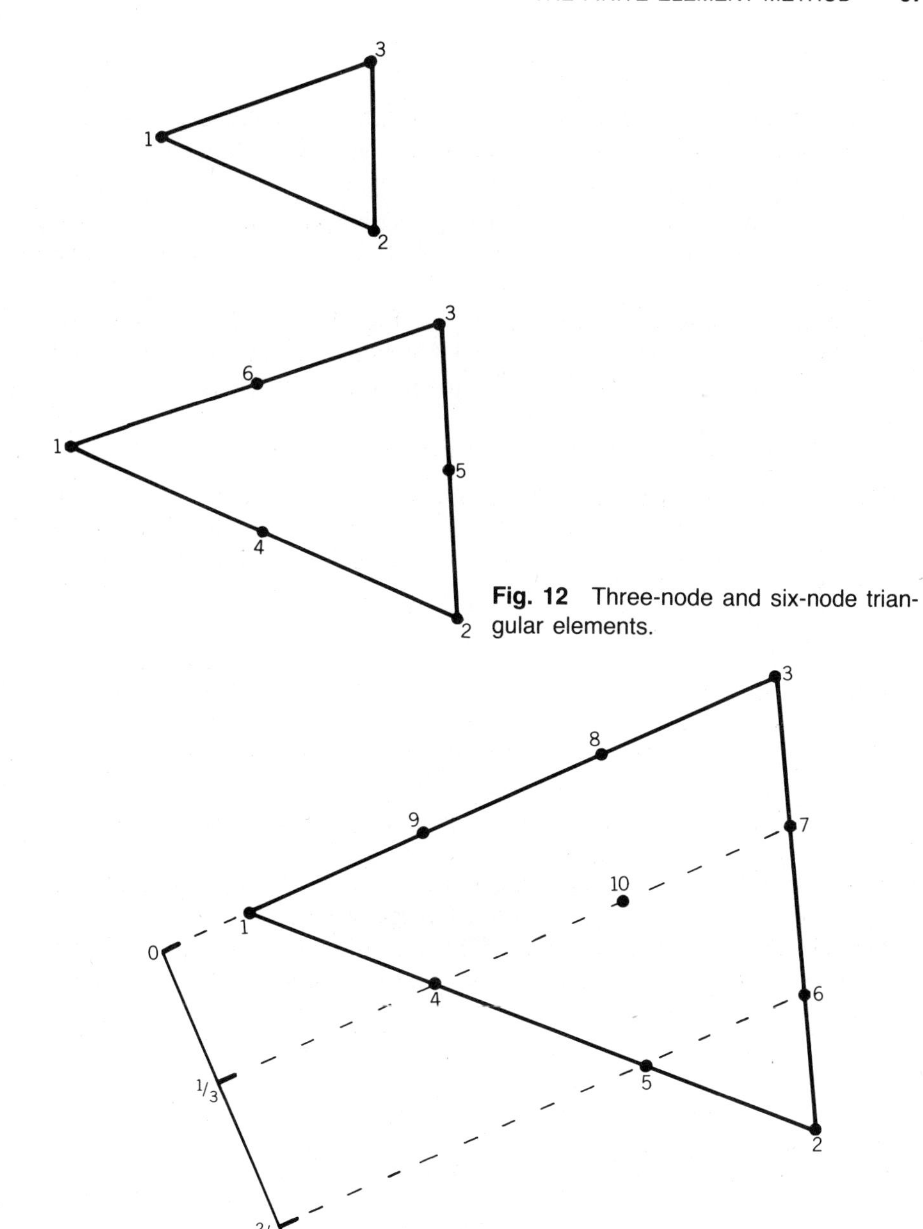

Fig. 12 Three-node and six-node triangular elements.

Fig. 13 Nodes for third-order element, with typical local coordinate.

EXERCISES

1. Check that eqs. (75)–(77) are indeed what they claim to be: two-dimensional interpolation polynomials that have value 1 at the associated node and value 0 at all other nodes.
2. Obtain in a similar manner shape functions for the third-order triangle, say of the representative N_1, N_4, and N_{10}. Figure 13 may assist here.
3. At the beginning of the section, in Fig. 5, we sketched possible piecewise linear or quadratic trial functions. Sketch the normalized shape functions over a single "element" for the two cases.

5.2.3. *Continuity of Functions from Element to Element*

So far, we have considered our trial functions as being defined over each single element as a linear combination of shape functions—defined via the nodal values. With the third-order element from exercise 2 above, explicitly,

$$U = U_1N_1 + U_2N_2 + \cdots + U_{10}N_{10} \tag{78}$$

Considering adjacent elements, there are then the important questions:

Can nodes on common boundaries be indeed "common" to both elements?

Will the resulting trial functions be continuous across the boundaries?

For the first-order shape functions (as in Fig. 7), it is obviously all right. We will consider the more complicated third-order situation.

Suppose that U is defined by eq. (78) over triangle A of Fig. 14. What value does U take along the common boundary? By our use of interpolation polynomials in two dimensions, we know that of the 10 shape functions

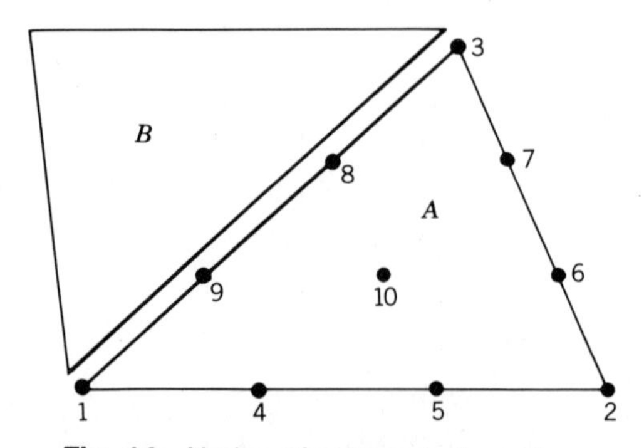

Fig. 14 Nodes of adjacent elements.

Fig. 15 Interpolation polynomials along common boundary.

N_1–N_{10}, all but N_1, N_9, N_8, and N_3 vanish at every point along that boundary. N_1, N_9, N_8, and N_3 are defined (via Exercise 2) in terms of L_1, L_2, and L_3. But along our special boundary, $L_2 = 0$, and L_1 and L_3 are linearly related to the coordinate (call it c) along the boundary. It then follows that along the boundary, N_1, N_9, N_8, and N_3 must be the unique interpolation polynomials in c, as shown in Fig. 15.

If we went through the same argument with the adjacent triangle B, we would find that its shape functions have the same property, that is, all ten vanish along the common boundary except for four that are the interpolation polynomials in c. It therefore follows that if we share nodes between adjacent elements the resulting function represented must be continuous across the boundary.

5.3. Global Coordinates and Local Coordinates

In Fig. 16 we see a typical triangular element, where L_1, L_2, and L_3 are the (area) local coordinates and (x, y) are the global coordinates, the usual Cartesian coordinates. They are related (using the interpolation polynomial property of the L's) by

$$\begin{aligned} x &= x_1 L_1 + x_2 L_2 + x_3 L_3 \\ y &= y_1 L_1 + y_2 L_2 + y_3 L_3 \end{aligned} \tag{79}$$

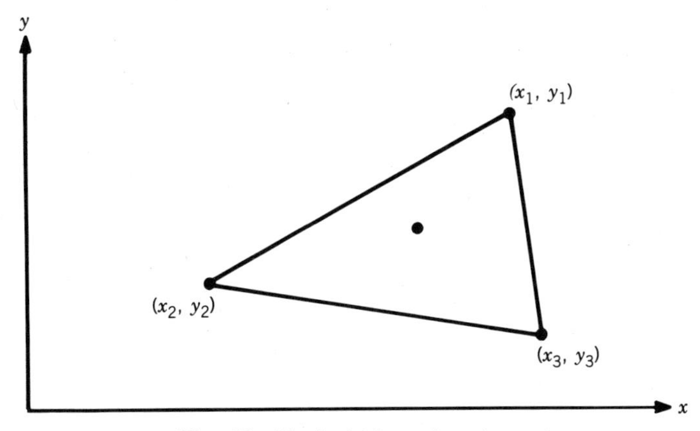

Fig. 16 Typical triangular element.

Suppose that, for the solution of the waveguide problem via eq. (26), we needed to evaluate

$$\iint U^2 \, dx \, dy \tag{80}$$

Suppose we evaluate (80) over the typical triangle of Fig. 16 with third-order shape functions. Using the matrix-style notation of eq. (70) we have

$$U = [N_1 \quad N_2 \quad \cdots \quad N_{10}] \begin{bmatrix} U_1 \\ U_2 \\ \cdot \\ \cdot \\ \cdot \\ U_{10} \end{bmatrix} \tag{81}$$

and so

$$\iint U^2 \, dx \, dy = \iint [U_1 \quad U_2 \quad \cdots \quad U_{10}]$$

$$\times \begin{bmatrix} N_1 \\ N_2 \\ \cdot \\ \cdot \\ \cdot \\ N_{10} \end{bmatrix} [N_1 \quad N_2 \quad \cdots \quad N_{10}] \begin{bmatrix} U_1 \\ U_2 \\ \cdot \\ \cdot \\ \cdot \\ U_{10} \end{bmatrix} dx \, dy \tag{82}$$

$$= [U_1 \quad U_2 \quad \cdots \quad U_{10}] \begin{bmatrix} & & \\ & \text{"matrix"} & \\ & & \end{bmatrix} \begin{bmatrix} U_1 \\ U_2 \\ \cdot \\ \cdot \\ \cdot \\ U_{10} \end{bmatrix} \tag{83}$$

where the 10×10 "matrix" has the (i, j)th element

$$\iint N_i N_j \, dx \, dy \tag{84}$$

Note that this integral involves our normalized shape functions; only the $dx\,dy$ needs transforming to local coordinates, via eq. (79). All the N's are polynomials in L_1, L_2, and L_3, so that the integrands of eq. (84) are linear combinations of $L_1^m L_2^n L_3^p$ over the triangle. These are straightforward, and to quote one typical integration formula:

$$\iint L_1^m L_2^n L_3^p \, dA = \frac{m!n!p!}{(m+n+p+2)!} (2A) \tag{85}$$

In this style, integrals like eq. (80) can be evaluated as matrix forms like (83), and the detailed integrations of shape functions can be done once and for all.

If we need more complicated expressions for our variational form (or Rayleigh–Ritz forms), like the other one for waveguides

$$\iint (\nabla\phi)^2 \, dx \, dy$$

again, the differentiation with respect to x and y is transformed by eq. (79) to differentiation with respect to local coordinates.

Most standard differentiations and integrations that arise (such as div, grad, and curl in the common coordinate systems) have been tabulated and are available in the literature, clearly using normalized elements.

5.4. Assembly of Element Matrices

For every term in the variational expression, each element will have a term like:

$$[U_1 \quad U_2 \quad \cdots \quad U_{10}] \begin{bmatrix} \\ \text{matrix} \\ \\ \end{bmatrix} \begin{bmatrix} U_1 \\ U_2 \\ \cdot \\ \cdot \\ \cdot \\ U_{10} \end{bmatrix} \tag{86}$$

By the nature of an integral, since it is the sum of integrals over each element, the *global matrix* is simply an assembly of the various element matrices like eq. (86). The only precaution is that many nodes are common to two or more elements.

Suppose that the total domain for analysis consisted of three triangles. Each triangle would result in a 10×10 matrix like (86)—say matrices $\boldsymbol{A}_{i,j}$, $\boldsymbol{B}_{i,j}$, and $\boldsymbol{C}_{i,j}$ where i, j are the "local numberings" of nodes, from 1 to 10. Renumbering of the nodes in "global numbers" from 1 to 22 is shown in Fig. 17.

The required global matrix $\boldsymbol{G}_{p,q}$ comes from adding the element matrices $\boldsymbol{A}$, $\boldsymbol{B}$, and $\boldsymbol{C}$, but for $\boldsymbol{G}$ using the global numbering, so that, for example,

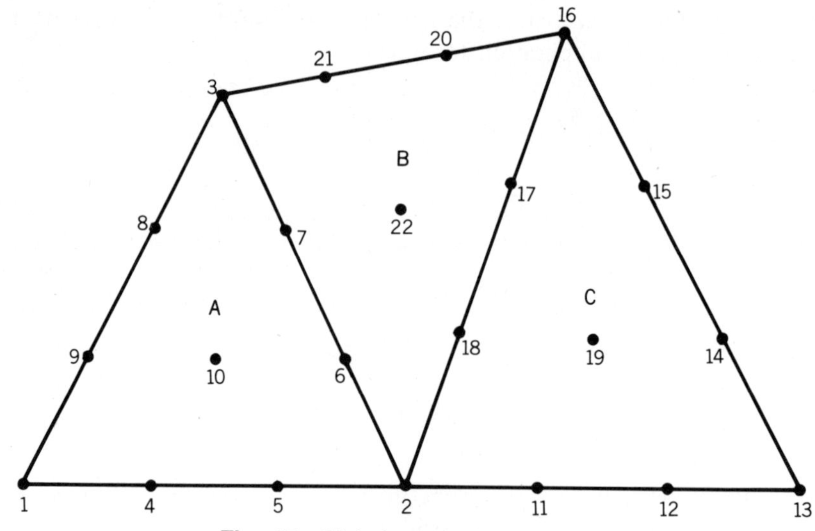

Fig. 17 Global numbering of nodes.

$$G_{1,1} = A_{1,1} \qquad G_{4,9} = A_{4,9} \qquad G_{10,7} = A_{10,7}$$

but

$$G_{3,3} = A_{3,3} + \text{(say) } B_{1,1} \qquad \text{and} \qquad G_{3,7} = A_{3,7} + B_{1,4}$$

Also,

$$G_{2,2} = A_{2,2} + B_{2,2} + C_{2,2}$$

The rule is (for our example) that $G_{i,j}$ consists of a contribution from only one element *unless both* of the indices i and j are from the set of numbers 3, 7, 6, 2, 18, 17, or 16, that is, along a shared boundary.

5.5. Sparsity of Global Matrices

The most important detail of the final matrix, or matrices, is that any matrix element corresponding to nodes that do *not* lie in any one element will be zero. For example,

$$G_{1,22} = G_{8,21} = G_{5,14} = 0$$

Specifically, the first row will include $G_{1,1}$, $G_{1,2}$, . . . , up to $G_{1,10}$, but thereafter $G_{1,11}$ to $G_{1,22}$ will all be zero. And similarly for the first column.

This sparsity is a rather crucial aspect of finite elements and leads to the use of special algorithms for the matrix solution. (Some are considered in

Section 10.) Special storage schemes are used, and sometimes elements are recalculated whenever needed, rather than being stored, to allow for the solution of very large order matrices.

EXERCISES

1. How many of the 484 elements in our 22×22 global matrix will inevitably be zero?
2. A square-area problem is divided like a chessboard, with each small square divided by a diagonal into two first-order triangular elements. Find (a) the maximum number of nonzero elements in any row or column and hence (b) the approximate *density* of the global matrices. Density is defined below.
3. What would be the answers to Exercise 2 if there were 16×16 rather than 8×8 squares?

Definition

$$\text{Density of a matrix} = 1 - \text{sparsity of matrix}$$
$$= \frac{\text{number of nonzero elements}}{\text{number of elements}}$$

5.6. Infinite Elements

Many waveguiding structures have (in principle) fields extending away to infinity; microwave image guides, optical fibers, and optical channel guides are examples. Any mathematical modeling must strive to be realistic.

A common strategy is to model the structure by an enclosing metallic box. This may be realistic, or it may not. If indeed the fields are significant to a "great distance" (many wavelengths) from the main guide, such as just above cutoff, this model will introduce an error or artifact. But if, say, we need to use finite elements for 5 or 50 wavelengths (especially in each of two or three dimensions) to adequately represent the fields, we soon run out of computer storage.

One ad hoc method (which is typical of a procedure to check for adequately good modeling) is to solve with a small enclosing metal box, then double the size of the box and solve, and double again, and so on. If and when the solution settles down, we may presume that the error introduced is adequately small.

A more accurate way is to make use of *infinite elements* [12]. Consider a dielectric waveguide consisting of a dielectric rod of uniform rectangular cross section surrounded by air. Cover the rectangle with first-order triangular elements, as in Fig. 18 (where, for simplicity, only a few elements are shown).

Over the indicated right-hand triangle, the field would be represented by

$$U_1N_1 + U_2N_2 + U_3N_3 \tag{87}$$

using the usual area coordinates for N_1, N_2, N_3. Suppose we define an "infinite element" as being the infinite strip extending to the right from our triangle, and represent the field by

$$\{U_3(y/b) + U_2[(b-y)/b]\}\exp[-A(x-a)] \tag{88}$$

where a and b are the x and y dimensions of the triangle.

The representations in (87) and (88) are continuous across their boundary, and the exponential factor satisfies the necessary "radiation condition" (see Jones [1]).

Similar infinite elements could be used, with, say,

$$\exp[-B(y-b)], \qquad \exp[C(x+a)], \qquad \text{and} \qquad \exp[D(y+b)]$$

as the factors above, to the left of, and below our finite elements.

More elaborate infinite elements can be used, with higher degrees parallel to the adjacent finite element and more fancy functions extending to infinity [13]. With the above simple elements, the positive constants $A, B, \ldots$ would be guessed at.

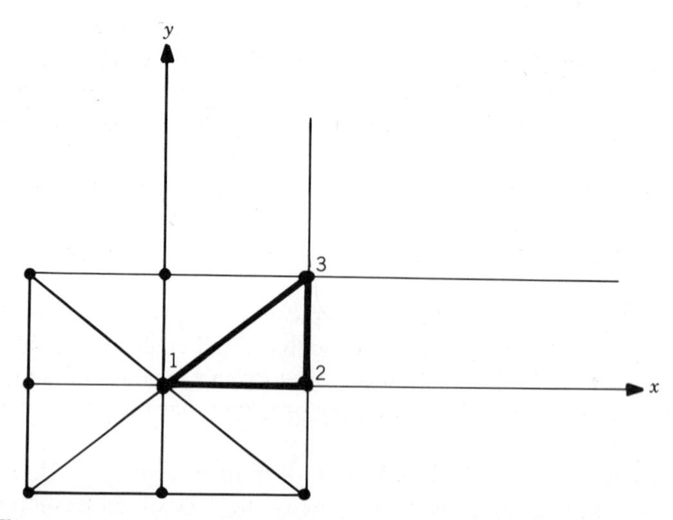

Fig. 18 Region of finite elements with adjacent infinite element.

EXERCISES

1. Think of a suitable expansion to use at the top-right quadrant of Fig. 18, which extends away to infinity and is continuous with the two necessary infinite elements.
2. Suppose you computed some answers with $A = 0$, then with 0.1, 0.2, ... How could you get an idea of the "best" A value? Could advantage be taken of a variational property?

5.7. Application of Boundary Conditions to the Global Matrix

In all the above finite element discussion, boundary conditions have been ignored (except for dealing with the more difficult condition at infinity!). The three types that commonly arise in high-frequency electromagnetics are

$$\text{Homogeneous Dirichlet} \qquad \phi = 0 \tag{89}$$

$$\text{Inhomogeneous Dirichlet} \qquad \phi = \text{prescribed value} \tag{90}$$

$$\text{Homogeneous Neumann} \qquad d\phi/dn = 0 \tag{91}$$

Different ones of these three may apply around the boundary, but this causes no difficulty. To illustrate a mixture of such boundary conditions, we consider the TEM mode analysis of a stripline (see Fig. 19).

The cross section shows a rectangular conducting strip mounted centrally within a rectangular hollow conductor. It is well known [8] that at any frequency the **E** and **H** fields of the TEM mode are simply $\exp\{j(\omega t - kz)\} \times$ the electrostatic or magnetostatic fields associated with the cross section. Moreover, **E** and **H** are related by $\mathbf{E} = Z_0 \cdot \mathbf{u}_z \times \mathbf{H}$, and so it is sufficient to know either the associated electrostatic or magnetostatic fields. We therefore consider here the electrostatic problem where the potential is physically required to be (say) 0 on the surrounding metal box and 1 on the central conducting strip.

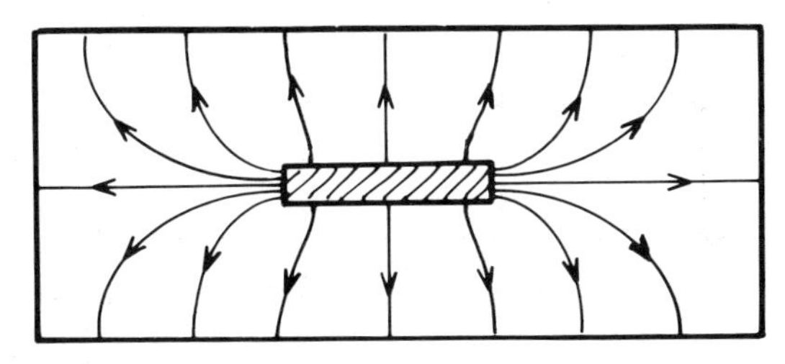

Fig. 19 Cross section of stripline, modeled to illustrate various boundary conditions.

In solving this by finite elements, we will use the variational form eq. (19) (but omitting the term $1/R$):

$$J(\phi) = \int\int (\nabla\phi)^2 \, dS \tag{92}$$

which we proved to have eq. (25) as its associated Euler equation. Eq. (25) is, of course, the Laplace equation of electrostatics. This proof [proceeding from (19) to (25)] relied on admissible potential functions taking on their physically correct values at the boundaries, which were conductors.

If we were doing a computer analysis of the problem, a smart way to economize would be to take advantage of the two planes of symmetry, as shown in Fig. 19; along these symmetry planes the physical boundary conditions would be precisely eq. (91). (We purposely use here the term *physical* boundary conditions; this is to avoid confusion with "essential" or "natural," which are at the discretion of the person choosing the variational expression!)

Suppose that first-order elements were spread over the domain, as shown in Fig. 20 and that the global matrix had been assembled in terms of the 18 nodes.

The Dirichlet-type boundary conditions are easily dealt with; we merely set nodal values U to

$$\begin{array}{ll} 0.0 & \text{at nodes } 1, 2, 3, 4, 5, 10, 15, 18 \\ 1.0 & \text{at nodes } 11, 12, 13, 16 \end{array} \tag{93}$$

The Neumann conditions, eq. (91), are more difficult to apply. As generally with any boundary conditions, we can:

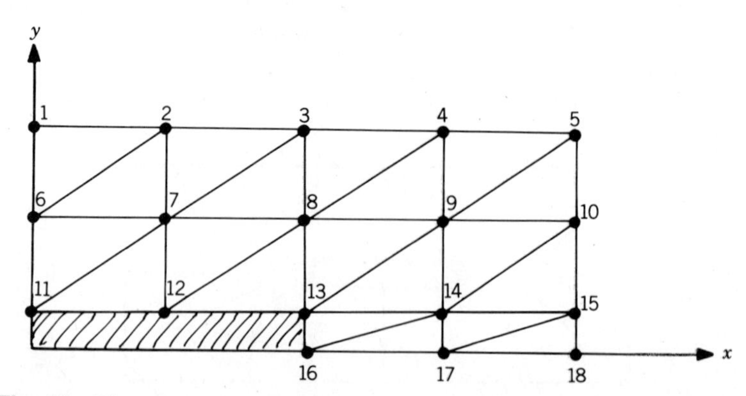

Fig. 20 Elements and nodes for one quarter of the problem in Fig. 19.

either make sure that they are natural boundary conditions corresponding to our chosen variational expression (we will not pursue this possibility);

or somehow "force" them [as we have just dealt with the Dirichlet conditions, (89) and (90), by means of (93)];

or, when the boundary condition is due to a symmetry, use the special trick of considering the whole structure, but when the complete finite-element procedure is assembled we *then* identify (make equal) corresponding nodal values.

For our problem, in order to illustrate the procedure, we choose to indeed force the boundary conditions, and we note that at node 6, eq. (91) would be satisfied if we arranged $U_7 = U_6$. All the Neumann conditions would therefore be satisfied if we made

$$U_2 = U_1 \quad U_7 = U_6 \quad U_{12} = U_{11} \quad U_{13} = U_{16} \quad U_{14} = U_{17} \quad U_{15} = U_{18} \tag{94}$$

Because of the earlier Dirichlet conditions, we have already fixed values for U_1, U_{11}, U_{16} and U_{18} which therefore apply to U_2, U_{12}, U_{13}, U_{15}. It is therefore only necessary to constrain

$$U_7 = U_6 \qquad \text{and} \qquad U_{14} = U_{17} \tag{95}$$

The only "slack" or free variables left in the system are therefore the four:

$$U_7(= U_6) \qquad U_8 \qquad U_9 \qquad U_{14}(= U_{17}) \tag{96}$$

5.8. From the Global Matrix to the Equations for Solution

Continuing with our stripline problem, we started with a global matrix involving 18 nodal values. After applying boundary conditions there are 12 prescribed nodal values [eq. (93)] and two "identities" [eq. (95)], leaving just four free variables [eq. (96)]. Suppose the global matrix came from applying the variational (92), namely,

$$J(\phi) = \int\int (\nabla\phi)^2 \, dS \tag{97}$$

In terms of our 18 nodal values, (97) becomes

$$J = [U_1 \quad U_2 \quad \cdots \quad U_{18}] \begin{bmatrix} G_{1,1} & G_{1,2} & \cdots & G_{1,18} \\ G_{2,1} & G_{2,2} & \cdots & G_{2,18} \\ \cdots & \cdots & \cdots & \cdots \\ G_{18,1} & & \cdots & G_{18,18} \end{bmatrix} \begin{bmatrix} U_1 \\ U_2 \\ \cdot \\ \cdot \\ \cdot \\ U_{18} \end{bmatrix} \tag{98}$$

or in summation notation,

$$J = \sum_i \sum_j U_i G_{i,j} U_j \tag{98'}$$

We now "shrink" the system because of the duplication in $U_6 = U_7$ and $U_{14} = U_{17}$. Specifically, we redefine

$$G_{7,i} = G_{6,i} + G_{7,i} \qquad \text{for all } i \neq 6 \text{ or } 7$$

then

$$G_{i,7} = G_{7,i} \qquad \text{for all } i$$

and then

$$G_{7,7} = G_{6,6} + 2G_{6,7} + G_{7,7}$$

This "merges" 6 and 7.

Then we similarly merge 14 with 17 and remove rows and columns 6 and 17 from (98).

The notation now becomes rather messy, but to complete the story, and using p and f to denote prescribed and free, we rewrite the reduced (98) as

$$J = [\boldsymbol{U}_f^t \quad \boldsymbol{U}_p^t] \begin{bmatrix} G_{ff} & G_{fp} \\ G_{pf} & G_{pp} \end{bmatrix} \begin{bmatrix} \boldsymbol{U}_f \\ \boldsymbol{U}_p \end{bmatrix} \tag{99}$$

where the matrices have been "partitioned" into blocks, so that $\boldsymbol{U}_f$ denotes the column vector containing the *free* nodal values and $\boldsymbol{U}_p$ denotes the column vector containing the *prescribed* nodal values (and $\boldsymbol{U}_f^t$, $\boldsymbol{U}_p^t$ denote their transposes). Specifically:

$$\boldsymbol{U}_f^t = [U_7, U_8, U_9, U_{14}] \tag{100}$$

and

$$\begin{aligned} \boldsymbol{U}_p^t &= [U_1, U_2, U_3, U_4, U_5, U_{10}, U_{11}, U_{12}, U_{13}, U_{15}, U_{16}, U_{18}] \\ &= [0, \ 0, \ 0, \ 0, \ 0, \ 0, \ 1, \ 1, \ 1, \ 0, \ 1, \ 0] \end{aligned} \tag{101}$$

and G_{ff} and so on, contain the corresponding elements from (98), so that G_{ff} is a 4×4 matrix, and G_{fp}, G_{pf}, and G_{pp} are 4×12, 12×4, and 12×12 matrices, respectively.

Now we are in a position to apply the Rayleigh–Ritz procedure; we need to differentiate J of eq. (99) with respect to each of the four remaining free nodes, and put the result equal to zero. From eq. (99) we know that J is a quadratic form (or second-degree polynomial) in U_7, U_8, U_9, and U_{14}, so that $\partial J/\partial U_i$ for $i = 7, 8, 9$, and 14 gives four equations, linear in U_7, U_8, U_9, and U_{14}, with a purely temporary notation:

$$\frac{\partial J}{\partial U_7} = [G_{7f}][U_f] + [G_{7p}][U_p] = 0$$

which combine to

$$[G_{ff}][U_f] = -[G_{fp}][U_p] \tag{102}$$

This is the final equation that when solved for U_f will give our sought-after solution!

$\boldsymbol{U}_p$ is the known vector from (101), $\boldsymbol{U}_f$ is the unknown and wanted vector from (100). $\boldsymbol{G}_{ff}$ and $\boldsymbol{G}_{fp}$ are (4×4 and 4×12, respectively) sets of appropriately assembled elements from our original global matrix, shrunk and then partitioned from (98) to (99).

Note 1. To ease illustration, the stripline example, as we modeled it in Fig. 20, contained 12 prescribed and only four free nodal values. Any realistic problem has far more free than prescribed values, for the same reason that jigsaw puzzles have more interior than edge pieces!

Note 2. Having solved (102), we would generally have the resulting electric potential, field, or whatever, from which *all* important parameters could be derived—for example, impedance, capacitance, phase velocity, and radiation pattern in the case of an antenna.

Note 3. The above strategy, from eq. (92) to eq. (102), is used in the complete computer program presented in Section 11.

EXERCISE

In the section above, from eq. (96), having obtained the global matrix we applied the Rayleigh–Ritz procedure to (21) or (97), to end up with eq. (102). Go through the same procedure for the waveguide problem, from, say, the variational form (35), and show that in place of (102) the result must have the form:

$$[\boldsymbol{A}][\boldsymbol{U}] = k^2[\boldsymbol{B}][\boldsymbol{U}] \tag{103}$$

We have almost stumbled across this equation already; eq. (62) is, after differentiation, just a particular case of eq. (103)!

Note that, in this exercise, boundary conditions are far easier than in the stripline problem, being always "homogeneous" [zero values are assigned, as in eqs. (89) and (91)]. This means that we do not have to retain any "prescribed" nodal values in $\boldsymbol{U}_p$ (previously some elements of $\boldsymbol{U}_p$ had to be nonzero to give a realistic electrostatic situation).

For TM modes, $\phi = 0$ around the entire boundary, eqs. (99)–(102) have $\boldsymbol{U}_p = 0$, and we can simply delete all corresponding rows and columns from the global matrix (or equivalently ignore the matrices like $\boldsymbol{G}_{pp}$, $\boldsymbol{G}_{pf}$, and $\boldsymbol{G}_{fp}$).

For TE modes we can *either* take advantage of natural boundary conditions *or* apply the earlier technique for Neumann boundary conditions [after eq. (93)], which results in equating pairs of nodal values and shrinking the matrices [as after (98)].

6. INTEGRAL FORMULATION OF PROBLEMS

In Section 2 an electrostatic problem was put as [eq. (4)]

$$\phi(\mathbf{r}) = \int\int \frac{q(s)\, ds}{4\pi\varepsilon_0 |\mathbf{r} - \mathbf{s}|} \tag{104}$$

If we know the physical charge distribution $q(\mathbf{s})$ over the conductors, then by linearity and Coulomb's law, (104) would give the potential anywhere, including the conductors. Its use was not pursued in detail, but we do so now for the stripline structure already tackled after eq. (91).

In the earlier finite element approach, we took $\phi(x, y)$ as the unknown; now we take $q(\mathbf{s})$ as the unknown charge distribution over the conductor surfaces, $\mathbf{s}$ measuring distance around the cross section. (See Fig. 21.)

Because of uniformity along the stripline structure, $q(\mathbf{s})$ must now be a line charge density, in coulombs per meter. Again, because it is effectively a two-dimensional problem, (104) needs to be revised to

$$\phi(\mathbf{r}) = \int\int \frac{q(\mathbf{s}) \ln (|\mathbf{r} - \mathbf{s}|/R_0)\, ds}{2\pi\varepsilon_0} \tag{105}$$

due to the different potential distribution around a line charge compared with a point charge. An arbitrary dimension R_0 is necessary; mathematically the natural logarithm function gives a correct potential, with its radial

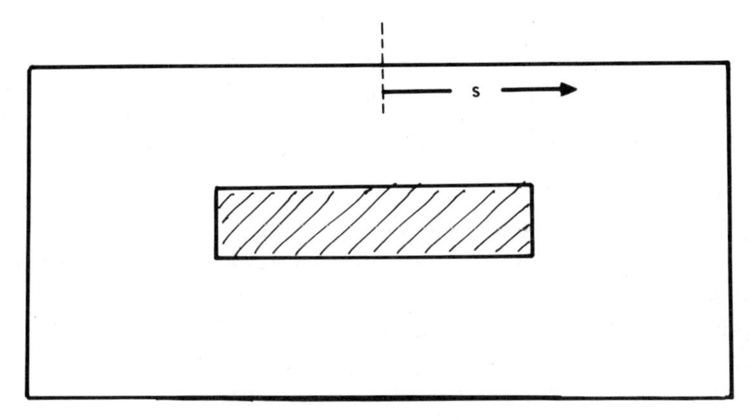

Fig. 21 Cross section of stripline.

derivative the physically correct **E** field, but R_0 is needed to establish a particular potential, an arbitrary constant of integration. Alternatively one can argue that the ln function must be of a dimensionless quantity—its series expansion form would be meaningless in terms of meters!

The charge distribution $q(\mathbf{s})$ is our unknown; the potential or voltage distribution $v(\mathbf{s})$ is known. It seems appropriate to apply the weighted-residual method described in Section 2.

Each of the conductor surfaces must be divided into subsections or one-dimensional finite elements. Much of Section 5's approach is appropriate, except that it is easier in one dimension. Questions of complete polynomials do not apply: there is only one independent variable. Suppose we divide the conductors into a total of N elements, such as the 32 in Fig. 22 (we postpone taking advantage of symmetry).

Our basis functions [eq. (8)] will be of finite element style: zero everywhere except over a single element. Over each element they may be

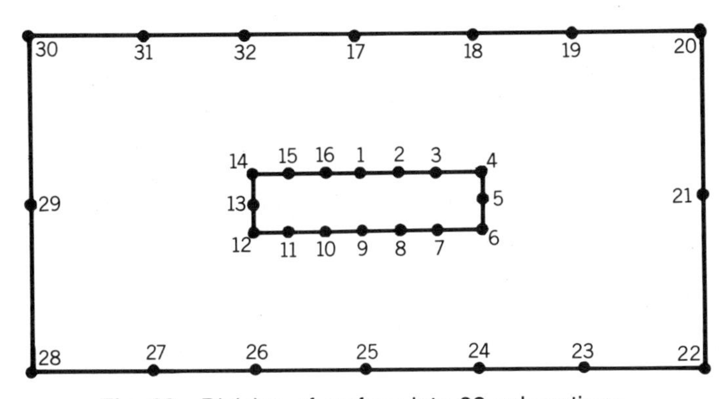

Fig. 22 Division of surface into 32 subsections.

(i) Constant
(ii) Polynomial (interpolation would be very appropriate)
(iii) A single Dirac delta function
(iv) Any other especially appropriate function

An example of a special function would be one including the known form of singularity of charge density at the four corners, namely, $r^{-1/3}$. This would avoid having our shape functions trying to approximate an unbounded function!

Another example will arise with the wire antenna, where sinusoids are worth considering.

The decision has to be taken whether to (1) use the Galerkin method, where

$$w_i(\mathbf{s}) = b_i(\mathbf{s}) \tag{106}$$

(2) choose $w_i(\mathbf{s})$ differently, from the choice (i)–(iv), above, or (3) use a variational approach.

Before making a decision, it is worth checking the consequences of any choice, especially the form of the resulting equations. Repeating eq. (8) for the equation resulting from making the weighting functions orthogonal to the residual error gives

$$\left\langle \left\{ L \sum_{i=1}^{N} u_i b_i(\mathbf{s}) - v(\mathbf{s}) \right\}, \{w_j(\mathbf{s})\} \right\rangle = 0 ; \qquad j = 1, 2, \ldots, N \tag{107}$$

where the linear operator L is

$$L = \int\int \frac{\ln(|\mathbf{s} - \mathbf{s}'|/R_0)}{2\pi\varepsilon_0}\, ds \tag{108}$$

The inner product $\langle\ ,\ \rangle$ in (107) involves an integral over the two conductors; L also involves an integral over the two conductors. One integral comes from the distribution of charge, and the other from the requirement of the voltage being correct everywhere.

We have three important consequences of this new integral equation approach, contrasting with the finite elements of Section 5, which was based on the variational expression

$$J(\phi) = \int\int (\nabla\phi)^2\, dS \tag{109}$$

The consequences are relevant for any such integral equation approach:

1. The resulting matrix is totally dense; we have generally no zero elements. (Physically, every source point affects every observation point.)
2. Each component of the matrix is *liable* to involve double integrals (integration over all source points and over all observation points).
3. As recompense for the above disadvantages, there will surely be a need for fewer elements, nodes, or unknowns (again, like the edge pieces of a jigsaw puzzle), and so the resulting matrix order will be smaller.

To be quantitative about some of these points, for the dense (full) matrix of order N:

1. Computing time for solution of the resulting matrix equation is invariably proportional to N^3.
2. The number of elements is N^2 (perhaps effectively $N^2/2$ if the matrix is symmetrical), with a proportional effect on computer storage, and on computing time required to evaluate the elements. This time may be quite significant if double integrals are being evaluated numerically.

To return then to decisions, there are at least 16 possibilities!

Weighted residuals can be used with *each* $b_i(\mathbf{s})$ and $w_i(\mathbf{s})$ selected from the four possibilities (i)–(iv). These will include the four choices of Galerkin (namely, using the same basis and weighting functions) from (i) to (iv).

We will not explore all 16 but first try Galerkin (i). In finite element nomenclature, all the shape functions $b_i(\mathbf{s})$ of eq. (107) are 1.0. Suppose our nodal values are

$$U_i \qquad \text{for } i = 1, N$$

to represent the uniform charge density on the ith element. The same shape function for $w_j(\mathbf{s})$ effectively requires that the *mean* voltage along the element be the required value for the particular conductor. For notation's sake, we take V_i (for $i = 1, N$) as the required voltages, arising from

$$V_i = \frac{\int_{i\text{th element}} v(\mathbf{s}) w_i(\mathbf{s})\, ds}{\int_{i\text{th element}} ds} \tag{110}$$

using the notation from eq. (107). That equation now becomes

$$\int \left[\int \sum_{i=1}^{N} U_i \frac{\ln\{|\mathbf{s}_i - \mathbf{s}_j|/R_0\}}{2\pi\varepsilon_0}\, ds_i - v(\mathbf{s}_j) \right] ds_j = 0$$
$$\text{for } j = 1, \ldots, N \tag{111}$$

where the outer integral $\int \{\ldots\} ds_j$ is over the jth element, and the inner integral $\int \{\ldots\} ds_i$ is over the ith element.

Rearranging (111) into matrix form, we have

$$[G][U] = [V] \quad \text{or} \quad \sum_{i=1}^{N} G_{j,i} U_i = V_j \qquad j = 1, \ldots, N \tag{112}$$

where

$$G_{j,i} = \int \int \ln \left(\frac{|\mathbf{s}_i - \mathbf{s}_j|}{R_0} \right) ds_i \, ds_j \tag{113}$$

and

$$V_j = \int v(\mathbf{s}_j) \, ds_j \tag{114}$$

The integrals in (114) are trivial, as the voltage $v(\mathbf{s})$ must be constant along any element. The double integrals of (113) must be evaluated for all pairs of elements; in general they may possibly be integrated in closed form, but commonly they would be numerically integrated (there are general-purpose routines for doing this).

The only complications that arise are when $i = j$, or i and j denote adjacent elements. The integrand of (113) now includes a singularity, but it is integrable; it has to be integrable because of the physics it represents, concerning voltage along an element including or up to an element of charge distribution.

Taking advantage of symmetry is straightforward, and as before, it reduces the order of matrix $\boldsymbol{G}$ and vectors $\boldsymbol{U}$ and $\boldsymbol{V}$ by a factor of 4; benefits for this dense matrix $\boldsymbol{G}$ are especially important, reducing computer storage by 16-fold and computer time by from a factor of 4 (for the evaluation of the elements $G_{i,j}$) to a factor of 64 [for the solution of eq. (112)]. Silvester and Ferrari [11] deal with this symmetry (and indeed this problem) and (on page 106 of that reference) give the fourfold enlargement of Green's function; in place of the single function $\ln |s_i - s_j|$ of eq. (113) that gives the voltage at $\mathbf{s}_j$ due to a line charge at $\mathbf{s}_i$, one obtains the fourfold Green's function as the voltage at $\mathbf{s}_j$ due to the sum of the charge at $\mathbf{s}_i$ plus charges at the three "images." By using symmetry, then, each matrix element involves about four times as much arithmetic, but there are one-sixteenth as many elements to evaluate.

6.1. From Two Dimensions to One?

The most important distinction between the formulations of this and Section 5 is that here the unknown is $q(\mathbf{s})$ around the one-dimensional conducting surface; the earlier formulation was in terms of $\phi(x, y)$ as the unknown function over the two-dimensional cross section. When the details are sorted out, it leads to the three consequences listed after eq. (109).

One other distinction—that a *variational* approach was used before [in terms of $\phi(x, y)$] but not in the above [in terms of $q(\mathbf{s})$]—is incidental and does *not* affect the important differences.

To emphasize this distinction, a variational formulation will now be outlined for the integral equation approach. Consider

$$J(q(\mathbf{s})) = 2\int V(\mathbf{s})q(\mathbf{s})\,ds + \frac{1}{2\pi\varepsilon}\int\int q(\mathbf{s}_i)q(\mathbf{s}_j)\left\{\ln\frac{|\mathbf{s}_i - \mathbf{s}_j|}{R_0}\right\}ds_i\,ds_j \tag{115}$$

EXERCISES

1. Confirm that eq. (115) is variational, with eq. (105) as the corresponding Euler equation. Earlier dealings with proving variational forms always needed a Green's theorem, for integration by parts. Now we only need the symmetry property of Green's function, so that $\mathbf{s}_i$ and $\mathbf{s}_j$ can be simply exchanged when necessary.
2. Using (115) as the starting point for a finite element solution, show that if Q_i are the nodal values of charge distribution $q(\mathbf{s})$, the resulting equation for solution must be *of the form* (i.e., skip all possible details, but the approach can include any shape functions, etc., without particular attention):

$$[\boldsymbol{G}_{i,j}][\boldsymbol{Q}_i] = [\boldsymbol{W}_j] \tag{116}$$

where $[\boldsymbol{Q}_i]$ and $[\boldsymbol{W}_j]$ are column vectors, $[\boldsymbol{G}_{i,j}]$ and $[\boldsymbol{W}_j]$ being known, and $[\boldsymbol{G}_{i,j}]$ is a symmetric matrix. This symmetry (that $\boldsymbol{G}_{i,j} = \boldsymbol{G}_{j,i}$) follows from the variational or Rayleigh–Ritz approach and is of practical importance, halving computer storage and computing time.

Equation (116) is, of course, just like (101) in form, with the earlier equation being in terms of potential.

6.2. Comments on the Integral Equation Approach

1. It reduces the number of dimensions (here from two dimensions to one) over which basis and weighting functions need to span.
2. Whether approached by direct Galerkin [resulting in eq. (112)] or by a variational form [resulting in (116)], we end up with an explicit (approximate) charge distribution $q(\mathbf{s})$. From this, the potential anywhere can be found [via (104)], as can more global parameters like capacitance, which gives the characteristic impedance of the TEM-mode stripline transmission line.

3. The variational approach can be advantageous if it uses a physically useful parameter (like capacitance) as its stationary form. For any approximation to fields, potentials, and so on, a stationary, rather than a nonstationary, expression for (say) capacitance can clearly by expected to give superior accuracy.
4. Comparing the variational approaches from (115) and (97) [resulting in (116) and (102)], it happens that one gives an upper bound and the other a lower bound to capacitance and therefore to Z_c. This was used in the early text of Collin [8]. More recently, it has been used, together with duality and complementary principles, to obtain many useful finite element solutions to TEM and quasi-TEM modes of microstrips and similar guides [14].
5. Another obvious choice from the 16 possibilities of weighted residuals mentioned before (110) would be point matching, using either (or both) point charges or/and points at which $v(\mathbf{s})$ is forced to be correct. The usual double integration [mentioned in comment 2 after eq. (109)] would then involve one, or even no, integration! Point charges could also be implemented into the variational approach [say, into eq. (115)]—*un embarras de choix*. . . .
6. Finally we could describe the methods of this section as the "boundary element" method. This descriptive term is used for the solution by finite elements of a problem that has been first reduced to dealing with the boundary rather than the whole region of the problem. The term has crept into the literature, and, while being a convenient name, it is unfortunately used to suggest that it is an *alternative* method to finite elements, when of course it *is* a finite element method!

As has been emphasized in this section, one may choose to formulate the problem (in terms of unknowns) *either* over the basic three dimensions *or* over boundaries of fewer dimensions. In either case, one can (try to) apply the range of techniques of weighted residuals, which includes finite elements. In either case also, one can (try to) use a variational formulation.

7. ANTENNAS AND SCATTERING FROM CONDUCTORS

In this section, we consider the important areas of electromagnetic *radiation* and *scattering* from conducting bodies. The radiation aspects come from considering an antenna, where excitation of currents or voltages within or around the conductor cause radiation. The scattering aspects include the classic radar terms "effective scattering area" or "radar cross section." Here we suppose that a plane wave is incident on a conducting body (such as an aircraft); the problem is to find the scattered fields. Fields scattered back in the direction toward the incident wave give rise to the radar cross section σ, defined [1, 15] as the ratio of power reflected to incident power density and involved in the so-called radar equation:

$$\frac{P_{\mathrm{rec}}}{P_{\mathrm{trans}}} = \frac{G^2\lambda^2\sigma}{(4\pi)^3R^4} \tag{117}$$

As we shall see, this scattering problem is very similar to that of the antenna mounted on (or consisting of) the same conducting body. The only difference concerns the excitation.

Most of the principles apply equally to two- and three-dimensional problems, but we shall start with two dimensions, it being so much easier.

7.1. Two-Dimensional Scattering or Radiation

In the integral formulation of the last section, eq. (105) used Green's function $\ln(|\mathbf{r}|/R_0)/2\pi\varepsilon_0$.

This gives the effect (of potential, at radius $\mathbf{r}$) due to a unit line charge. The associated electric field would be

$$E_r = 1/2\pi r\varepsilon_0 \tag{118}$$

If we move from electrostatics to magnetostatics, we similarly know that a steady current of 1 A along an infinite conducting wire would give, by Ampère's law, an associated magnetic field:

$$H_\theta = 1/2\pi r \tag{119}$$

If the driving current is $\exp(j\omega t)$, at $r = 0$ along the z axis, then by symmetry there must be a cylindrically radiating wave, with H_θ and E_r. Now, besides a magnetic field there is an associated electric field given by

$$E_z = -\frac{\omega\mu}{4} H_0^{(2)}(kr) \tag{120}$$

where $k = \omega\sqrt{\mu\varepsilon}$ and $H_0^{(2)}$ is a Hankel form of Bessel function [see paragraphs before and after eq. (131)].

Equation (120) is our new Green's function for the problem, like the electrostatic one of (118).

Consider a uniform TM plane wave incident on a uniform conducting cylinder of any cross section, as in Fig. 23. TM in this context means that the magnetic field is transverse to the cylindrical axis of the conductor. Using r, z, θ coordinates, by symmetry, *all* electric fields must be parallel to the incident E_z and the cylindrical conductor. *All* magnetic fields must lie in the $r\theta$ plane. All fields can be obtained from E_z, and so it is economical to work in terms of that one component; this conveniently means that we have a "scalar field problem."

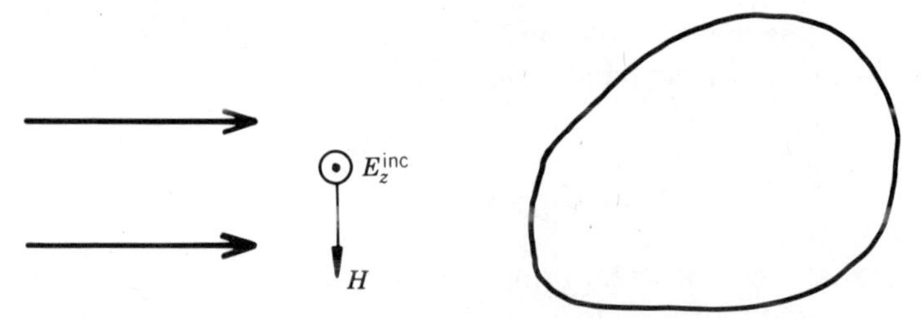

Fig. 23 Plane wave incident on a conducting cylinder.

Currents must be set up on the conductor that will cause a scattered wave to radiate. Let $J_z(\mathbf{s}')$ be the axial (and only) component of surface current density on the surface at $\mathbf{s}'$. The electric field at $\mathbf{s}$ caused by $J_z(\mathbf{s}')$ must be

$$E_z(\mathbf{s}) = -\frac{\omega\mu}{4}\int_S J_z(\mathbf{s}')H_0^{(2)}(k|\mathbf{s}-\mathbf{s}'|)\,ds' \tag{121}$$

These scattered fields must be such that on the conducting surface

$$E_z^{\text{inc}} + E_z^{\text{scat}} = 0 \tag{122}$$

Applying (122) to all points on the surface S and using (121) gives

$$E_z^{\text{inc}}(\mathbf{s}) = \frac{\omega\mu}{4}\int_S J_z(\mathbf{s}')H_0^{(2)}(k|\mathbf{s}-\mathbf{s}'|)\,ds' \tag{123}$$

which is an integral equation exactly like our earlier static eq. (105):

$$\phi(\mathbf{s}) = \frac{1}{2\pi\varepsilon}\int_S q(\mathbf{s}')\ln\left(\frac{|\mathbf{s}-\mathbf{s}'|}{R_0}\right)ds' \tag{124}$$

Each is an integral equation of the first kind:

$$\text{Excitation}(\mathbf{s}) = \int_V \text{source}(\mathbf{s}')\cdot\text{Green's function }(|\mathbf{s}-\mathbf{s}'|)\,dv \tag{125}$$

So far in this section, we have considered the incident wave to be TM, that is, a plane wave with its polarization (E field) parallel to the conducting cylinder. In practice, an incident wave can be of any polarization but could be decomposed into the two orthogonal linear polarizations; these would conveniently be a TM wave (with E parallel to) and a TE (with E at right angles to) the cylindrical axis. Turning, then, to this other polarization, we consider an incident TE wave (with E in the $r\theta$ plane and H along the z axis). Now an equation corresponding to (123) is

$$H_z^{\mathrm{inc}}(\mathbf{s}) = -J_t(\mathbf{s}) + \frac{j\mathbf{u}_z}{4} \cdot \mathrm{curl} \int J_t(\mathbf{s}')H_0^{(2)}(k|\mathbf{s}-\mathbf{s}'|)\, d\mathbf{s}' \qquad (126)$$

This is an integral equation of the second kind, having the unknown function (J_t in this case) outside as well as inside the integral sign.

The additional term arises because for the TM polarization the boundary condition is

$$E_z^{\mathrm{inc}} + E_z^{\mathrm{scat}} = 0 \qquad (127)$$

whereas the TE polarization needs

$$H_z^{\mathrm{inc}} + H_z^{\mathrm{scat}} = J_t(\mathbf{s}) \qquad (128)$$

the "scattered" fields in each case being given by the integral term, from (123).

In principle, solution of either of the integral equations [(123) or (126)] is straightforward by the weighted-residual method, including finite elements. Clearly we divide the surface into elements and choose shape or basis functions to represent $J_t(\mathbf{s})$. For any choice, $J_t(\mathbf{s})$ will be represented linearly in terms of the nodal values U_1 to U_n, say, and the resulting matrix equation for solution will be the familiar

$$[G][U] = [V] \qquad (129)$$

7.2. Antennas or Scatterers

The waves just considered were excited by arranging an incident plane wave. *Any* form of incident wave would do; via the boundary conditions, the "scattered" fields are *forced* to exist.

An alternative excitation is, as in a transmitting antenna, to force some particular surface currents to flow, from within the conductor. For the TM polarization, this means that in (122) and (123)

$$E_z^{\mathrm{inc}} = 0 \qquad (130)$$

The current distribution $J_z(\mathbf{s}')$ can be split into prescribed and free parts, just like the potentials from (98)–(101), and the resulting equation would be like (102), in fact, (129).

The scattering and antenna problems are virtually the same, in needing to set up the integral equations (123) or (126). Their differences are in the details; for example, a slot excitation might force a discontinuity in $J_z(\mathbf{s}')$ and allow a nonzero longitudinal voltage or E_z to exist at (some of) the surface. Our same integral equations will still be the starting point—the boundary conditions needing appropriate attention.

7.3. The Interior Problem

The foregoing scattering problem considered a plane wave incident on a uniform conducting cylinder of arbitrary cross section, presumed solid. A hollow cylinder would behave the same if its thickness were many skin depths. Suppose it was 5 skin depths thick; the incident and scattered waves of Section 7.1 would be very loosely coupled to the interior, by about 40 dB. For an incident TM polarization, this would *have* to set up a TM waveguide mode inside the hollow tube, effectively at cutoff (when $\beta = 0$ or guide wavelength = infinity) because the structure and excitation is z-independent.

Having used an integral equation approach for the scattering, exterior problem, we now ask whether a similar approach can be taken for the interior problem, the hollow conducting waveguide.

The "exterior" integral equation (123) was based on (120), the Green's function for the E_z field at r due to a unit time-harmonic current filament at $r = 0$. With implied time dependence $\exp(j\omega t)$, $H_0^{(2)}(kr)$ gives a uniform *outward* radiating wave (appropriate for the scattered waves). $H_0^{(1)}(kr)$ similarly gives an *inward* radiating wave.

The appropriate choice of Green's function for the hollow conducting waveguide, from which there is zero net radial radiation, is

$$Y_0(kr) = \tfrac{1}{2}[H_0^{(2)}(kr) - H_0^{(1)}(kr)]j \tag{131}$$

This is the Bessel function corresponding to a radial standing wave with a singularity appropriate to a line source at $r = 0$.

[For time-harmonic waves in $r\theta z$ coordinates, $J_0(kr)$ is the other standing-wave Bessel function, obtained by changing the $-$ to $+$ in (131); J_0 and Y_0 are related to $H_0^{(1)}$ and $H_0^{(2)}$ just as $\sin x$ and $\cos x$ are related to $\exp(+jx)$ and $\exp(-jx)$ in the easier xyz coordinate system.]

An integral representation of E_z within the hollow guide can now be set up and applied to give the boundary condition of E_z vanishing on the surface:

$$\int J_z(\mathbf{s}')Y_0(k|\mathbf{s} - \mathbf{s}'|)\, ds' = 0 \qquad \text{for all } \mathbf{s} \text{ on } S \tag{132}$$

This is a total, adequate formulation of the hollow waveguide problem, but it is different in nature from (123) and (126), being an eigenvalue problem. *Only* for correct (eigen-) values of k can a solution of (132) be obtained. These are the discrete cutoff wavenumbers of TM modes in the hollow guide.

A finite element solution of (132) could be tackled almost exactly like the integral equation version of the stripline problem from (104) to (114). Decisions about shape functions and so on would be the same [as from (105) to (107)], and we would again have a singular Green's function with its associated delicate treatment. The earlier $\phi(\mathbf{s})$ and $q(\mathbf{s}')$ become $E_z(\mathbf{s})$ and $J_z(\mathbf{s}')$. The big difference is in the change from deterministic to eigenvalue form. Whatever choices are made about shape functions, Galerkin, and so on, the resulting matrix equation is

$$[G][U]=[\mathbf{0}] \tag{133}$$

that is, like (129) but without an excitation or "forcing function." For a nontrivial solution to (133) we must have

$$\det[G(k)]=0 \tag{134}$$

The matrix elements of $[G]$ clearly depend [via the Green's function (131)] on k, and we now seek the particular values of k that allow (134) to be satisfied.

Numerically, the bad news about (134) is that there is generally *no automatic*, *guaranteed algorithm* for finding solutions; this does not stop us from indeed finding solutions—they may emerge without difficulty—but generally it needs *some* human intervention.

7.4. Three-Dimensional Scatterers and Antennas

We have introduced the integral equations of scattering in two dimensions, with its easier scalar form and Green's functions, and have outlined a finite element/weighted residual method of solving the equations.

Going to three dimensions is largely analytical, in setting up the more complicated integral equations. The finite element approach remains basically as before; the more complicated problem makes for greater detail and finer judgment.

In three dimensions, our first decision is the choice of integral equation to use—the electric field integral equation (EFIE) or the magnetic field integral equation (MFIE).

Equation (121) gave the time-harmonic electric field E_z in terms of J_z for the effectively two-dimensional situation. Equation (107) similarly gave the potential ϕ in terms of q. In genuine 3-D, the fields come from

$$\mathbf{H}=\operatorname{curl}(\mathbf{A}) \tag{135}$$

and

$$\mathbf{E}=-j\omega\mathbf{A}-\operatorname{grad}\phi \tag{136}$$

where the vector and scalar potentials are given by

$$\mathbf{A} = \mu \int \mathbf{J}(\mathbf{s}') \frac{\exp(-jk|\mathbf{s}-\mathbf{s}'|)}{4\pi|\mathbf{s}-\mathbf{s}'|}\, ds' \tag{137}$$

and

$$\phi = \frac{1}{\varepsilon} \int \rho(\mathbf{s}') \frac{\exp(-jk|\mathbf{s}-\mathbf{s}'|)}{4\pi|\mathbf{s}-\mathbf{s}'|}\, ds' \tag{138}$$

The charge and current densities of (137) and (138) are related by charge conservation to

$$\operatorname{div}(\mathbf{J}) - j\omega\rho = 0 \tag{139}$$

Equations (135)–(138) are, indeed, three-dimensional versions of (121).

Skipping the detailed derivation [e.g., substituting (139) into (138), then (137) into (135) and (136), and then into a three-dimensional vector form of (122) gives (141)], general forms of MFIE and EFIE [like the two-dimensional (123) and (126)] are (Jones [1], pp. 463–478; Moore and Pizer [15], pp. 29 and 56)

$$\mathbf{n} \times \mathbf{H}^{\text{inc}}(\mathbf{s}) = \frac{1}{2} J(\mathbf{s}) - \mathbf{n} \times \iint \mathbf{J}(\mathbf{s}') \times \operatorname{grad}\left\{ \frac{\exp(-jk|\mathbf{s}-\mathbf{s}'|)}{4\pi|\mathbf{s}-\mathbf{s}'|} \right\} dS' \tag{140}$$

$$-j\omega\varepsilon\mathbf{n}(\mathbf{s}) \times \mathbf{E}^{\text{inc}}(\mathbf{s}) = \mathbf{n}(\mathbf{s}) \times \left[k^2 \iint \mathbf{J}(\mathbf{s}') \left\{ \frac{\exp(-jk|\mathbf{s}-\mathbf{s}'|)}{4\pi|\mathbf{s}-\mathbf{s}'|} \right\} dS' + \operatorname{grad} \iint \operatorname{div}\left\{ \mathbf{J}(\mathbf{s}') \frac{\exp(-jk|\mathbf{s}-\mathbf{s}'|)}{4\pi|\mathbf{s}-\mathbf{s}'|} \right\} dS' \right] \tag{141}$$

These are the integral equations for dealing with scattering (or radiation) from arbitrarily shaped conducting surfaces, just being complicated 3-D versions of the two-dimensional equations (120)–(130). The MFIE is strictly valid only for closed surfaces; this arises as the fields are required to vanish *within* an enclosed solid. From the above integral equations, finite elements would proceed in an orthodox manner, with basis and weighting functions being required over the conducting surface.

7.5. Uniqueness

Uniqueness theorems are part of the day-to-day diet of mathematicians. Fortunately they are unnecessarily fussy for many practical problems, one reason being that they often result in simply reminding us that we must set up the problem properly in the first place. [For example, our earlier stripline problem would *not* have a unique field solution if we failed to specify $v(\mathbf{s})$ adequately; with care, these details should always be physically obvious.]

Having obtained a solution to either of the integral equations, (140) and (141), we then ask the question: Is my solution indeed unique? For if not, how do we know we have the correct solution—for physically there is only one solution!

For the EFIE, it turns out that we do *not* have a unique solution if the wavenumber k is a value corresponding to a cavity resonance of the hollow cavity coincident with our scattering volume. Physically, the situation is like our interior problem introduced between eqs. (130) and (131), where a thin cavity wall allows coupling between the interior and exterior regions. If the skin depth were precisely zero, there would be *no* coupling. Numerically, this uniqueness failure would be irrelevant if our computer dealt precisely with its numbers. But the finite precision, or roundoff error, means that the interior and exterior problems are inevitably coupled.

Jones [1] illustrates this with comparable results from an acoustic problem (Fig. 24).

If resonances of the cavity cause "breakup" of the EFIE, what about the MFIE? Again, all is well *unless* k corresponds to resonances of a cavity of the same shape with a "magnetic wall" surface.

These troubles have been tackled (with inevitable complications), and many results have been obtained with our two integral equations, but for

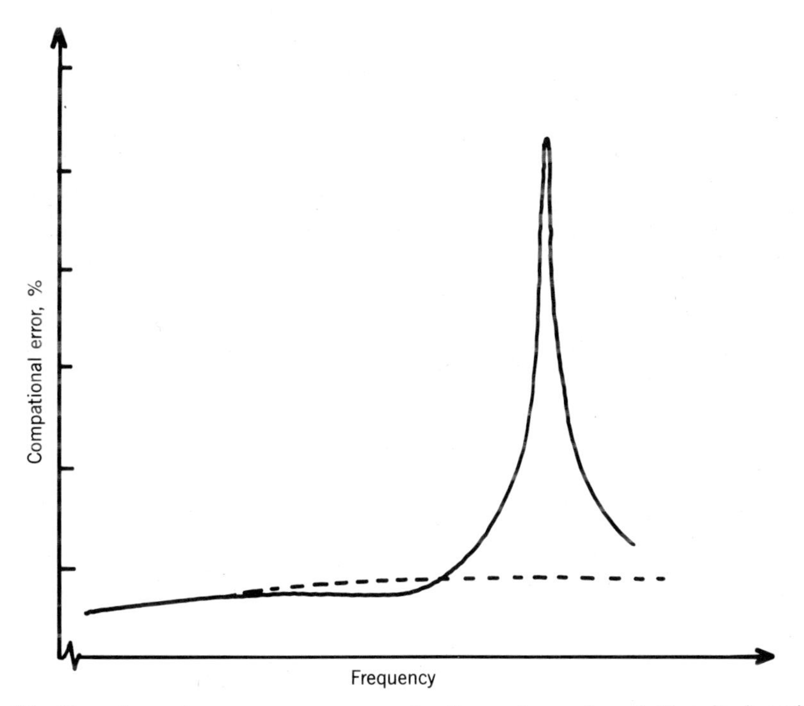

Fig. 24 Error in surface pressure on a vibrating sphere. (——) Results ignoring internal resonance; (– – –) results correcting for internal resonance.

versatile ability to deal with a wide range of conducting shapes, a more popular avenue has been that of wire-grid modeling.

7.6. Wire-Grid Modeling

When the scatterer (or antenna) is composed of conducting sheets, or is effectively solid, it turns out that it is *a good procedure* to model the conducting sheets by a mesh of filamentary conducting wires. One good reason lies in the experimental verification that a wire grid *can* model a conducting sheet well, provided that the mesh spacing is small compared with the wavelength. The reassurance is clearly limited; in theory and in the laboratory one has to ask, Precisely what diameter wire is used? How important is conductivity? What happens near a sharp conducting corner? One must *always* look out for tests of result quality.

7.6.1. The Thin-Wire Approximation

If we consider conductors to be *only* in the form of thin wires, all integrals over $\mathbf{s}'$ simplify. Providing the wire diameter is very small compared with free-space wavelength, the surface densities $\mathbf{J}(\mathbf{s}')$ and $\rho(\mathbf{s}')$ are replaced by line densities:

$$I(l) = \text{area}\, \mathbf{J}(\mathbf{s}')$$

and (142)

$$q(l) = \text{area}\, \rho(\mathbf{s}')$$

where l denotes distance along the wire.

Boundary conditions also simplify; only the component of $\mathbf{E}$ parallel to the wire needs to vanish at the wire.

Earlier the full vector field forms (135)–(139) led to the full vector integral equations (140) (MFIE) and (141) (EFIE). Only the EFIE formulation remains valid for thin wires, and this leads to Pocklington's 1897 integral equation [11, 15];

$$\hat{\mathbf{l}} \cdot \mathbf{E}^{\text{inc}}(\mathbf{l}) = j \int \left\{ \omega\mu \hat{\mathbf{l}} \cdot \mathbf{I}(\mathbf{l}') - \frac{I(\mathbf{l}')}{\omega\varepsilon} \frac{\partial^2}{\partial l \partial l'} \right\} \frac{\exp(-jk|\mathbf{l} - \mathbf{l}'|)}{4\pi|\mathbf{l} - \mathbf{l}'|}\, dl' \quad (143)$$

where $\hat{\mathbf{l}}$ and $\hat{\mathbf{l}}'$ denote unit vectors along the wire at $\mathbf{l}$ and $\mathbf{l}'$.

Having shrunk our three-dimensional problem into a one-dimensional domain (from the point of our unknown being in one dimension), we now have the familiar weighted residual/finite element choice of basis and weighting functions but only in one dimension along the wires!

Moore and Pizer [15] have used nine different computer program packages in obtaining results used in their textbook, all of them based on wire-grid modeling. Nearly all of them use Dirac delta functions for the weighting (testing) functions to simplify integration. This is certainly more physically reasonable than using delta functions for the current sources.

One bad feature of this choice is that a symmetric matrix occurs only with a Gelerkin choice, that is, when basis and weighting functions are the same.

Moore and Pizer [15] devote many pages to the choice of basis and weighting functions and quote the experiences of many workers. Graduation from delta functions to pulse functions (piecewise constant) and to sinusoids generally improves convergence, especially if both basis and weighting functions are given "good treatment."

One other advantage of Galerkin, besides the numerical advantage of symmetry of $[\mathbf{Z}]$, has more to do with the physics; the resulting fields fundamentally satisfy reciprocity. (It must surely be embarrassing, after spending a fortune on computing, to find that an antenna has a different radiation pattern as a transmitter than as a receiver!)

Whatever basis and weighting functions are chosen (decisions on elements and shape functions), weighted residuals appled to Pocklington's equation, eq. (143), must result in the familiar matrix equation

$$[\mathbf{Z}][\mathbf{I}] = [\mathbf{V}] \tag{144}$$

$[\mathbf{I}]$ contains the nodal values for the currents along the wires, $[\mathbf{V}]$ contains voltage differences from node to node, and $[\mathbf{Z}]$ contains the inevitable (generalized, compared with circuit theory) impedance matrix. This makes for a very versatile formulation, as most physical boundary conditions can be modeled reasonably. A voltage difference across one or more gaps, or current feed into a section of conductor, can be specified by appropriate matrix manipulation. Junctions of wires are straightforwardly dealt with (by applying Kirchhoff's law of current continuity, and forcing constraints between adjacent currents). Finite conductivity of the wires is dealt with by simply adding a term to eq. (143).

8. WAVEGUIDES—HOLLOW, DIELECTRIC, AND OPTICAL

The finite element procedure was introduced in Section 5. We now exploit it for successively more complicated waveguides.

8.1. Hollow Conducting Waveguides

Rectangular and circular waveguides are the popular versions of hollow guides. Other shapes are needed, for example, to achieve special field distributions for antenna feeds or for industrial drying.

Variational expressions for the cutoff wavenumber k are ready from (35):

$$k^2 = J(\phi) = \text{SV} \frac{\iint (\nabla\phi)^2 \, dS}{\iint \phi^2 \, dS} \tag{145}$$

If we apply essential boundary conditions,

$$\phi = 0 \tag{146}$$

we will obtain TM modes.

TE modes will emerge either by restricting trial functions to satisfying boundary conditions,

$$\frac{\partial \phi}{\partial n} = 0 \tag{147}$$

or by leaving them free and relying on (147) being the natural boundary conditions for (145).

From any resulting ϕ we can obtain the transverse dependence of all fields [8], by using

TM(E) Modes

$$\begin{aligned} &E_z = \phi_e \qquad H_z = 0 \\ &E_t = -\frac{j\beta}{k^2} \nabla_t(\phi_e) \qquad H_t = -\frac{j\sqrt{k^2+\beta^2}}{k^2 Z} \mathbf{u}_z \times \nabla_t(\phi_e) \end{aligned} \tag{148}$$

TE(M) Modes

$$\begin{aligned} &H_z = \phi_m \qquad E_z = 0 \\ &H_t = -\frac{j\beta}{k^2} \nabla_t(\phi_m) \qquad E_t = +\frac{j\sqrt{k^2+\beta^2}}{k^2} Z\mathbf{u}_z \times \nabla_t(\phi_m) \end{aligned} \tag{149}$$

and the total power flow P from

$$\left.\begin{matrix} PZ & (\text{for TM}) \\ P/Z & (\text{for TE}) \end{matrix}\right\} = \frac{\beta\sqrt{k^2+\beta^2}}{2k^2} \iint \phi^2 \, dS$$

where

$$Z = \sqrt{\mu/\varepsilon}$$

$$\nabla_t \equiv \nabla - \mathbf{u}_z \frac{\partial}{\partial z} = \mathbf{u}_x \frac{\partial}{\partial x} + \mathbf{u}_y \frac{\partial}{\partial y}$$

As given in the exercise resulting in eq. (103), the matrix equation emerging from a finite element choice of basis functions for ϕ in the Rayleigh–Ritz application to eq. (145) must be of the form

$$[\boldsymbol{A}][\boldsymbol{U}] = k^2[\boldsymbol{B}][\boldsymbol{U}] \tag{150}$$

There is a fair heritage of computer programs for this classical waveguide problem [16, 17]. One advanced finite element program [17] uses triangular elements with complete polynomials up to degree 6. It is professionally documented and is a reasonably "friendly" package. Its basic failing, these many years later, is that it uses an old and inefficient matrix-solving routine for eq. (150), but that can be updated straightforwardly.

8.2. Dielectric and Optical Waveguides

We now move to more complicated structures. Because of the extra complexity, we will deal more with the principles and less with the fine details.

Research interest has grown in this field, especially since 1980 for integrated optical structures. Besides the well-known optical fiber, typical cross sections of guides are as shown in Fig. 25.

Optical fibers have been fairly thoroughly studied, mainly without finite elements, because of their axial symmetry and the availability of Bessel functions.

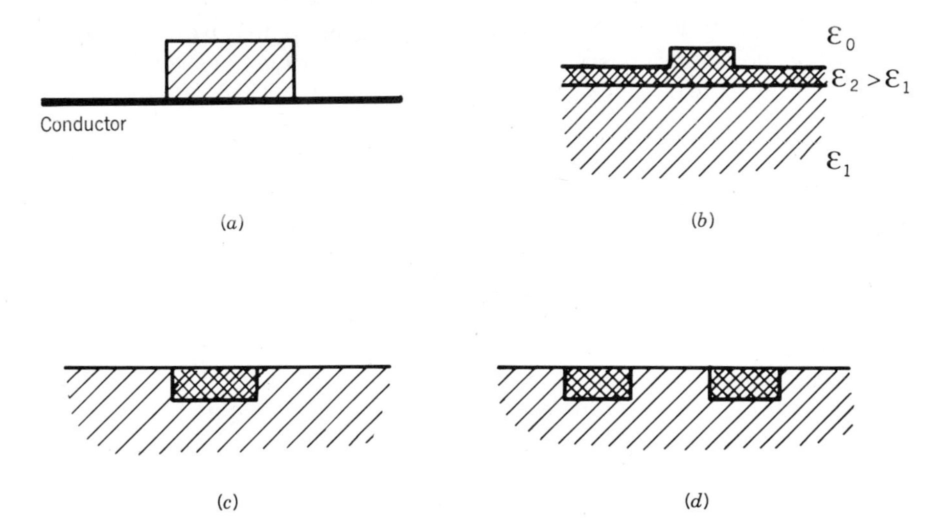

Fig. 25 Cross section of (*a*) microwave image guide, (*b*) optical rib guide, (*c*) optical channel guide, and (*d*) channel directional coupler.

Optical planar guides [as in (b)–(d) in Fig. 25] provide the biggest challenge of these structures because of the very fine gradations (of the order of 0.01) in refractive index profile. Another challenge comes because of the common use of anisotropic substrates (e.g., $LiNbO_3$). The proliferation of different two-dimensional profiles used in optical guides makes the finite element method very convenient for their study.

8.2.1. Choice of Field Variables

In electrostatic problems, only one scalar function is required—the potential ϕ, as used in Sections 2, 3, and 5. This remains true even with inhomogeneous media (see the exercise in Section 3.2.3), where the variational form

$$\iint \varepsilon(\nabla\phi)^2\, dS \tag{151}$$

has as Euler equation

$$\nabla\cdot(\varepsilon\nabla\phi) = 0 \tag{152}$$

Importantly, this means that finite elements based on (151) can deal, *with equal ease*, with electrostatic problems involving *arbitrary* transverse variation of dielectric constant. Note that this flexibility does not apply to the integral equation version of the problem (Section 6).

In the hollow-waveguide problems, again it is adequate to work in terms of one scalar function—the E_z or H_z separately for the TM and TE, as in (148) and (149). Unfortunately, the flexible extension to inhomogeneous dielectric constant does not apply. This is because (in contrast to the electrostatic situation) the problem is fundamentally more involved; specifically, it is no longer possible to find the fields from any single scalar function; at least two scalars are needed. The modes are not TE or TM, but some combination (and so are called *hybrid* modes).

One procedure is to take the total fields to be indeed a straight sum of the fields derived from the TM and TE expressions. Specifically, consider

$$\mathbf{E}(x, y, z, t) = \mathbf{e}(x, y)\exp[j(\omega t - \beta z)] \tag{153}$$

$$\mathbf{H}(x, y, z, t) = \mathbf{h}(x, y)\exp[j(\omega t - \beta z)] \tag{154}$$

All field components can now be obtained from the addition of those given in (148) and (149), with separate ϕ_e and ϕ_m denoting the $e_z(x, y)$ and $h_z(x, y)$ parts of the hybrid modes. We now have

$$e_z = \phi_e \tag{155}$$

$$h_z = \phi_m \tag{156}$$

$$\mathbf{e}_t = -\frac{j\beta}{k^2}\nabla_t(\phi_e) + \frac{j\sqrt{k^2+\beta^2}}{k^2} Z\cdot\mathbf{u}_z \times \nabla_t(\phi_m) \tag{157}$$

$$\mathbf{h}_t = -\frac{j\beta}{k^2}\nabla_t(\phi_m) - \frac{j\sqrt{k^2+\beta^2}}{k^2 Z} \mathbf{u}_z \times \nabla_t(\phi_e) \quad . \tag{158}$$

A finite element solution along these lines would be:

1. Take a variational form (for ω or β) in terms of $e_z(x, y)$ and $h_z(x, y)$.
2. Use normal two-dimensional representations separately for $e_z(x, y)$ and $h_z(x, y)$ (with separate nodal values for e_z and h_z, but using the same elements and probably the same shape functions).
3. Apply Rayleigh–Ritz to again obtain the standard matrix eigenvalue equation, (150), for solution.

A more versatile and robust approach has been used and described [12, 19]. The same references (and [6]) give a variational form for the frequency at which (154) applies (thus giving all modes):

$$\omega^2 = \frac{\iint (\text{curl}\,\mathbf{H})^* \|\varepsilon\|^{-1} (\text{curl}\,\mathbf{H})\, dS}{\iint \mathbf{H}^* \|\mu\| \mathbf{H}\, dS} \tag{159}$$

Just as with the ε of the electrostatic problem giving (151), so (159) allows ε or μ to vary arbitrarily across the cross section, which is clearly useful for the cross sections shown in Fig. 25. Moreover, (159) allows ε or μ to be arbitrarily tensor (though Hermitian, that is, loss-free), thus allowing for analysis of crystalline materials such as $LiNbO_3$. Taking advantage of varying ε or μ does not increase the matrix order [of (150)]; it merely complicates evaluation of the matrix elements.

Variational expressions have been used (for ω or β) in terms of:

1. E_z and H_z as described above (requiring 2 unknowns)
2. $\mathbf{H}$ as in eq. (159) (requiring 3 unknowns)
3. $\mathbf{E}$ (requiring 3 unknowns)
4. $\mathbf{H}$ and $\mathbf{E}$ (requiring 6 unknowns)
5. $\mathbf{H}_t$ and/or $\mathbf{E}_t$ (requiring 4 or 2 unknowns)

A formula for type 3 would be the dual of (159),

$$\omega^2 = \frac{\iint (\text{curl}\,\mathbf{E})^* \|\mu\|^{-1} (\text{curl}\,\mathbf{E})\, dS}{\iint \mathbf{E}^* \|\varepsilon\| \mathbf{E}\, dS} \tag{160}$$

Just to give a hint of where formulas like these come from, consider the sum

$$
\begin{aligned}
J &= \text{mean stored (electric energy} - \text{magnetic energy)} \\
&= \tfrac{1}{2} \int\int (\mathbf{D} \cdot \mathbf{E}^* - \mathbf{B} \cdot \mathbf{H}^*) \, dS \\
&= \tfrac{1}{2} \int\int (\mathbf{E} \cdot \varepsilon \mathbf{E}^* - \mathbf{H} \cdot \mu \cdot \mathbf{H}^*) \, dS
\end{aligned}
\tag{161}
$$

Substituting for **E** by use of the appropriate Maxwell curl equation gives an expression solely in terms of **H**:

$$
J = \int\int \left\{ \mathbf{H} \cdot \mu \cdot \mathbf{H}^* - \frac{(\nabla \times \mathbf{H}) \cdot (\nabla \times \mathbf{H}^*)}{\omega^2 \cdot \varepsilon} \right\} dS \tag{162}
$$

Stationarity of this formula, and taking care if tensors ε and μ are to be dealt with, gives (159).

Now we briefly consider the possibilities 1–5 [listed preceding (160)] for the finite element solution of microwave and optical guides.

The distinguishing property of all the guides in Fig. 25 is their need for a transversely varying dielectric constant; indeed, this variation is needed to provide guiding properties! Because, at these inhomogeneous boundaries, tangential fields **E** and **H** and normal fluxes **D** and **B** are continuous, it follows that across a typical cross section

$$E_z,\ H_x,\ H_y, \text{ and } H_z \text{ are continuous}$$

but

$$E_x,\ E_y,\ D_x, \text{ and } D_y \text{ are not } \textit{generally} \text{ continuous}$$

Variational forms 1 and 2, and the $\mathbf{H}_t$ form of 5, from above, are the only ones that allow us to use functions that are continuous everywhere. Any use of vector **E** would involve discontinuous functions, with associated complications.

To illustrate results that have been obtained [19] using eq. (159), Fig. 26*a* gives dispersion curves and Fig. 26*b* shows a field plot down the symmetry plane for an optical $LiNbO_3$ channel guide. The substrate is uniaxial, that is, it has a diagonal tensor permittivity with elements, say, ε_1, ε_2, ε_2. Here it is assumed that the optical axis is in the transverse x direction. The substrate refractive index values are 2.20, 2.29, 2.29, while values in the 1 μm $\times$ 5 μm channel are 2.222, 2.3129, 2.3129.

Dispersion is given here in terms of effective refractive index [20], the

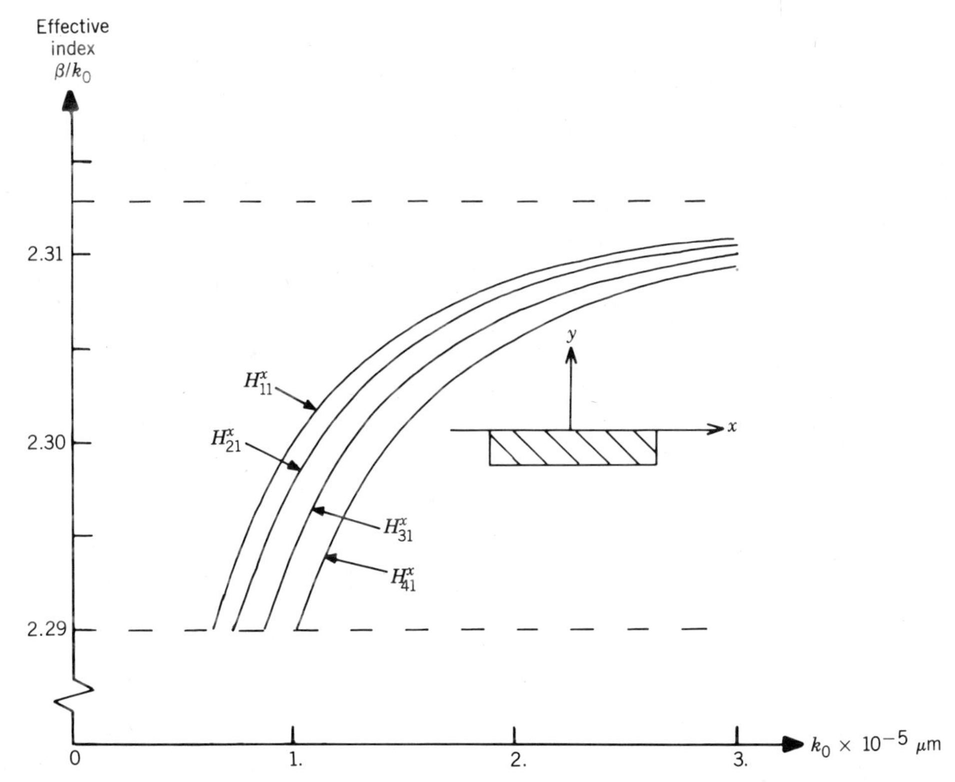

Fig. 26a Channel waveguide of lithium niobate: dispersion of four nodes.

index value of infinite material for which a plane TEM wave would have the same phase propagation constant. A convenient feature of effective refractive index is that for a propagating mode it must lie between the smallest and the largest values of refractive index of any materials present.

The results given above for the $LiNbO_3$ channel guide were obtained [19] using infinite elements as described in Section 5.6. This had the advantage of modeling the fields H_x, H_y, and H_z by continuous functions over the whole, unbounded transverse plane, and without increasing the matrix order. The results were obtained with a modest matrix order of 1120—about 370 nodal values each of H_x, H_y, and H_z.

The approach of (159) in this section has been based on an "exact-in-the-limit" approach, using a full 3-vector for **H**. An alternative attack is based on the so-called scalar field approximation. This approach is common in the optical literature [20, 21] and has been pursued using finite elements [21]. When applicable, it has the clear advantage of using a smaller matrix order (by a factor of 3). It also avoids a problem of spurious modes [19]. On the

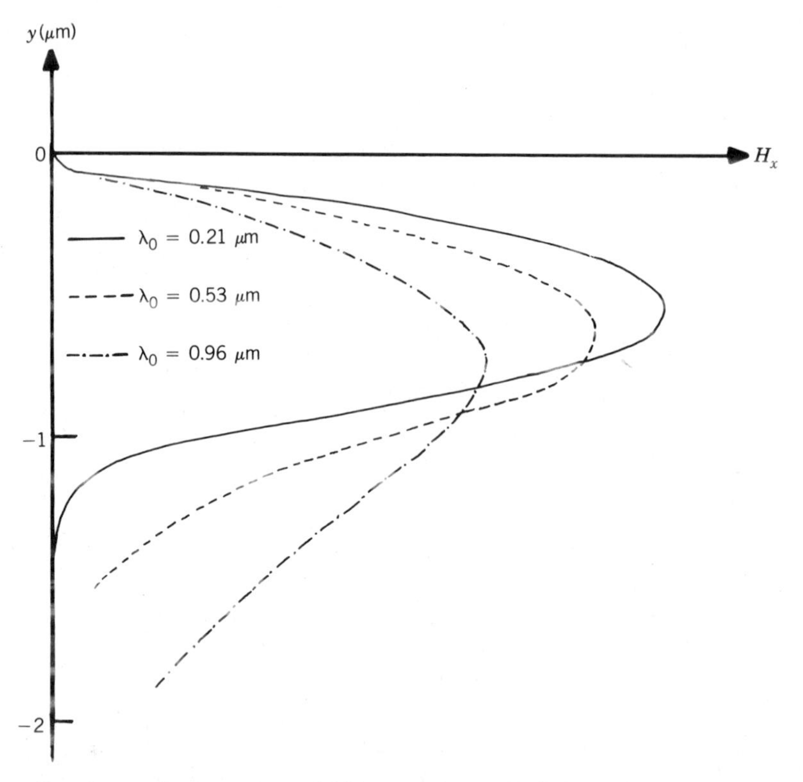

Fig. 26b Channel waveguide of lithium niobate: H_x field profile along the y-axis symmetry line for the dominant H^x_{11} mode. After [19] © 1984 IEEE.

other hand, it is more difficult to treat anisotropic dielectric; its a priori approximation can break down (e.g., when the mode lacks any TE-like or TM-like nature) and at the least needs the "exact" approach to assess its accuracy.

Much work is still progressing in this direction of finite element solution of optical and microwave structures. For current and more detailed information, it is best to refer to recent journals (e.g., *IEEE Transactions on Microwave Theory and Techniques and IEEE Journal of Light Wave Technology*).

9. FINITE DIFFERENCES IN SPACE AND TIME

This section will present the very briefest account of the use of finite differences, in the context of this work. Finite differences provide an alternative to finite elements for the solution of partial differential equations in space. But more strongly, they provide the most common way (where finite elements are not especially appropriate) of dealing numerically with transient time dependence.

We will be interested in using finite differences along (some of) the x, y, z, and t axes. In all cases, we consider a function to be "sampled" at (usually regular) intervals along the variable x, y, z, or t.

In one dimension, considering, say, a real function $f(s)$, a derivative is commonly defined by

$$f'(a) = \lim_{h\to 0}\left\{\frac{f(a+h)-f(a)}{h}\right\} \tag{163}$$

This suggests that any required derivative may be reasonably approximated by

$$f'(a) \approx \left\{\frac{f(a+h)-f(a)}{h}\right\} \tag{164}$$

An equally eligible approximation might be

$$f'(a) \approx \left\{\frac{f(a)-f(a-h)}{h}\right\} \tag{165}$$

Equations (164) and (165) are called, respectively, a *forward difference formula* and a *backward difference formula*.

A *central difference formula* is

$$f'(a) \approx \left\{\frac{f(a+h)-f(a-h)}{2h}\right\} \tag{166}$$

The errors (and justification) of the above formulas can be obtained [1] directly from Taylor's theorem, and are proportional to:

h for the forward and backward formulas

and

h^2 for the central formula

The better accuracy of the central formula can be visualized from Fig. 27, where it is clear which of the three chords is nearest to being parallel to the tangent.

A central difference formula for the *second* derivative follows from successive application of the first derivative formula:

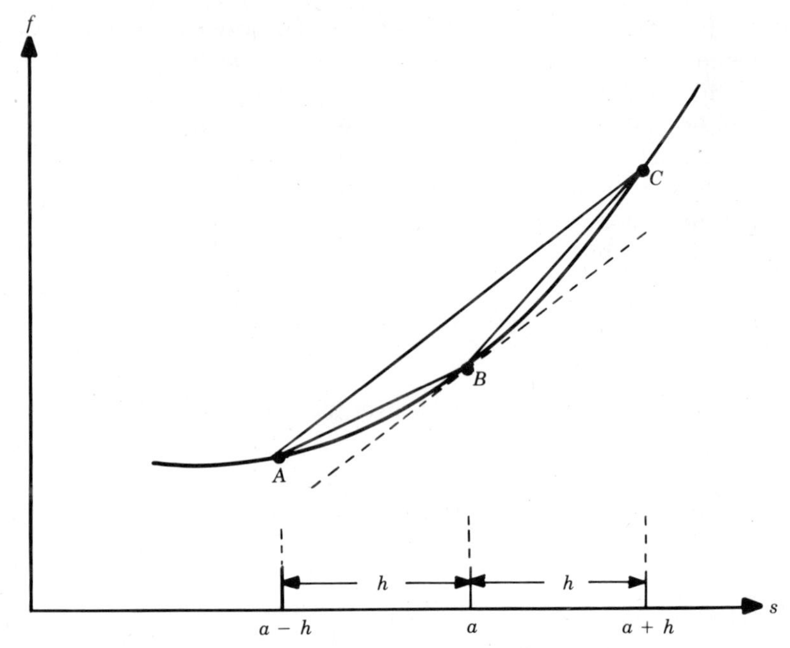

Fig. 27 Chords *B–C*, *A–B*, *A–C* corresponding to difference formulas (164)–(166).

$$
\begin{aligned}
f''(a) &\approx \frac{f'(a+0.5h) - f'(a-0.5h)}{h} \\
&\approx \frac{[f(a+h) - f(a)]/h - [f(a) - f(a-h)]/h}{h} \\
&\approx \frac{f(a+h) - 2f(a) + f(a-h)}{h^2}
\end{aligned}
\tag{167}
$$

The error in (167) is again proportional to h^2.

9.1. Finite Differences in Space

Finite differences can be used as an alternative to finite elements in virtually all the work described in this chapter. Going back to our unified approach using weighted residuals, it is rather like using regularly spaced Dirac delta functions—just as in digital filters, working with a sampled data stream.

Finite difference methods can be set up various ways; one elegant way is to use a variational expression. We will illustrate with an example.

Consider a hollow conducting waveguide of arbitrary cross section. Instead of the earlier variational form, eq. (145), it is slightly more convenient (allowing use of the more accurate central differences) to use another variational form:

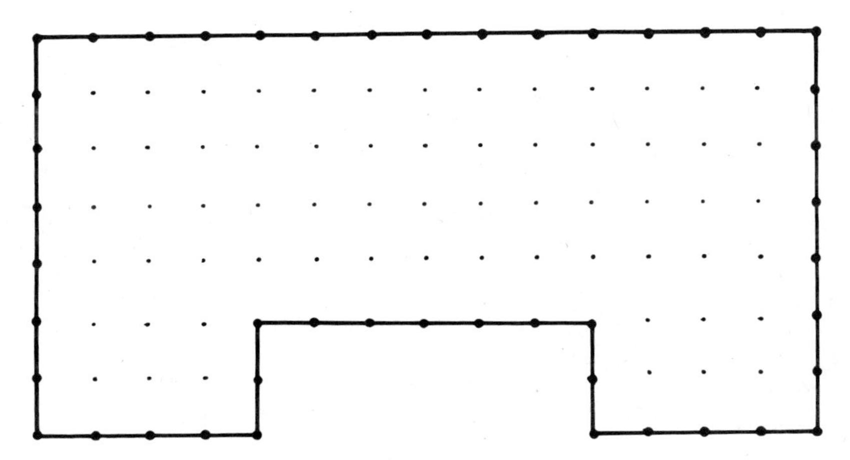

Fig. 28 Finite difference mesh of nodes.

$$k^2 = \frac{\iint \phi \nabla^2 \phi \, dS}{\iint \phi^2 \, dS} \tag{168}$$

To approximate (168), we superimpose a regular square mesh over the guide cross section and consider values of the usual ϕ only at the nodes. (See Fig. 28.)

To evaluate the integrals of (168), the denominator will clearly be

$$\iint \phi^2 \, dS = h^2 \sum_i \phi_i^2 \tag{169}$$

with summation over all nodes and with special adaption near boundaries. To approximate the numerator, referring to Fig. 29, (167) will give

$$\frac{\partial^2 \phi}{\partial x^2} \approx \frac{\phi_E + \phi_W - 2\phi_O}{h^2} \tag{170}$$

and with a similar expression in the y direction we have the classic *five-point formula*

$$\left(\frac{\partial^2}{\partial x^2} + \frac{\partial^2}{\partial y^2}\right)\phi \approx \frac{\phi_N + \phi_S + \phi_E + \phi_W - 4\phi_O}{h^2} \tag{171}$$

Applying (169) and (171) to (168) must give

$$[\boldsymbol{U}]^t[\boldsymbol{A}][\boldsymbol{U}] = k^2[\boldsymbol{U}]^t[\boldsymbol{U}] \tag{172}$$

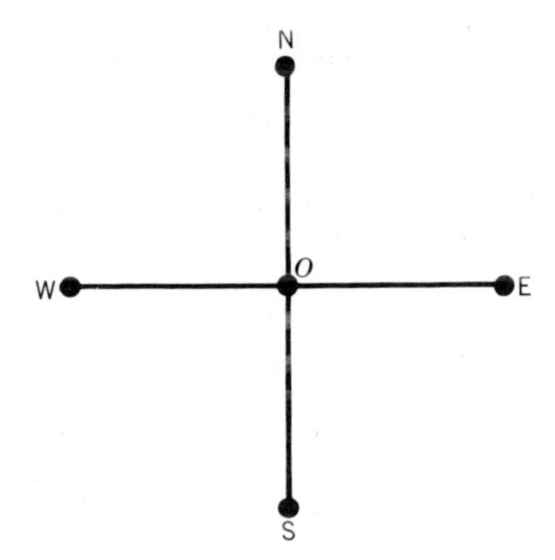

Fig. 29 Typical five-point star of neighboring nodes.

which, after the usual Rayleigh–Ritz procedure [applying eq. (55)] gives

$$[A][U] = k^2[U] \tag{173}$$

So the method of finite differences gives the same matrix for solution as finite elements [(150)] *except* that (173) is slightly simpler in having $[B] = [I]$, the unit matrix.

In identical style, finite differences could be applied to our stripline/electrostatic problem of Section 5.

Finite differences have been, and will continue to be, used alongside finite elements. A comparison of the methods is somewhat blurred, but some of the pros and cons can be listed.

Finite differences have the following advantages:

1. They can be easier to set up. Unlike the finite element approach, they do not need a Galerkin, a weighted-residual, or even a variational approach. The finite difference formulas (164)–(167) and (171) can be substituted *directly* into the differential equations of the problem.
2. Following from the above, finite elements can involve more programming, especially if you are working with vector operators in coordinates not already programmed and documented.
3. They can give slightly simpler matrices for solution: for example, (173) can sometimes be solved in half as much time as (150) and/or need half as much matrix storage.

Finite elements have the following advantages:

1. Fitting the elements to odd-shaped boundaries is simpler than with finite differences. This is clear when triangular elements are being

used, and so-called curved elements are also routinely used to give a better fit.

2. Altering the density of elements or order of elements (matching according to regions of rapid field variation) is easier than with finite differences.
3. The field or potential is defined explicitly everywhere. This makes for easier manipulation such as when evaluating spatial derivatives to give related fields, impedance, and so on.

9.2. Time Domain Solutions; Finite Differences in Time

All time dependence so far has been of the monochromatic, single-frequency type; when there has been any time dependence, it has been $\exp(j\omega t)$. Now we consider the transient case, with the possibility of arbitrary time variation. Whether discussing waveguides, antennas, or whatever, we consider again a sample data stream, where fields are considered only at the times

$$t = t_0, t_1, t_2, \ldots$$

Considering any or all of the fields (**E**, **H**, **J**, etc.) at these specific times, we can go back to Maxwell's equations or derived formulas with their general time-dependent forms, such as

$$\operatorname{curl} \mathbf{E} + \mu \frac{\partial \mathbf{H}}{\partial t} = 0 \qquad \operatorname{curl} \mathbf{H} - \varepsilon \frac{\partial \mathbf{E}}{\partial t} = \mathbf{J} \tag{174}$$

An integral equation for wire-grid modeling in the time domain follows from [15]:

$$\mathbf{E}(\mathbf{r}, t) = -\frac{\partial \mathbf{A}}{\partial t}(\mathbf{r}, t) - \nabla\phi(\mathbf{r}, t) \tag{175}$$

where

$$\mathbf{A}(\mathbf{r}, t) = \frac{\mu}{4\pi} \int \mathbf{I}\left(l', t - \frac{R}{c}\right) dl' \tag{176}$$

$$\phi(\mathbf{r}, \mathrm{t}) = \frac{1}{4\pi\varepsilon} \int q\left(l', t - \frac{R}{c}\right) dl' \tag{177}$$

$$\frac{\partial q(l, t)}{\partial t} = -\frac{\partial I(l, t)}{\partial l} \tag{178}$$

where l and l' measure distance along the wires and $R = |\mathbf{r}' - \mathbf{r}|$.

Replacing time derivatives in the above, by using some of the finite difference formulas (164)–(167), must result in a linear relation (at its most general, involving spatial integrals and/or differentiation) between fields at times t_{i-1}, t_i, and t_{i+1}.

If, in addition, finite elements (weighted residuals, etc.) are applied in the space domain, we must end up with

$$[\boldsymbol{A}][\boldsymbol{F}(t_{i-1})] + [\boldsymbol{B}][\boldsymbol{F}(t_i)] + [\boldsymbol{C}][\boldsymbol{F}(t_{i+1})] = [\boldsymbol{0}] \tag{179}$$

where the $\boldsymbol{F}$'s are column vectors of the nodal values at times t_{i-1}, t_i, and t_{i+1}.

Solution of (179) needs some care, but it means that, knowing the fields at $t = t_0$ and $t = t_1$, we can "time-step" to derive the fields at $t = t_2$. Application again gives t_3, and so on. (Care is required because some formulas are conditionally convergent, needing careful choice of time and/or space intervals; some formulas are divergent, and some are unconditionally convergent.)

What, then, are the advantages and disadvantages of numerical time-domain solutions compared with single-frequency, time-harmonic solutions? Given the fundamental Fourier transform equivalence between time and frequency, and especially the availability of fast Fourier transforms [22]:

1. There will rarely be any point in obtaining time-domain solutions if answers are needed only at one frequency, or even two. However, if results are needed over a wide frequency range and/or the "signal" is defined as a time waveform, then time-domain solutions can be the most cost effective.
2. There is a tendency for time domain solutions to run away with CPU time, and so they are not lightly used in the antenna business, where 3-space dimensions are already involved.

10. MATRIX COMPUTATIONS

Before discussing any methods of solving matrix equations, we consider the rather fundamental matrix property of *condition number*. One could well argue that *anyone using matrix computations*, no matter how ignorant they may (happily) be of the detailed computing, should be aware of a matrix condition.

10.1. Condition of a Matrix

Multiplying or dividing two floating point numbers gives an error of the order of the last preserved bit. If, say, two numbers are held to eight decimal digits, the resulting product (or quotient) will effectively have its least significant bit truncated and therefore have a relative uncertainty of around $\pm 10^{-8}$.

By contrast, with matrices, multiplying (evaluating $\boldsymbol{y} = \boldsymbol{Ax}$) or dividing (solving $\boldsymbol{Ax} = \boldsymbol{y}$ for $\boldsymbol{x}$) can lose all significant figures!

10.1.1. Vector and Matrix Norms

To introduce the idea of "length" into vectors and matrices, we have to consider *norms*.

Vector Norm. If $\boldsymbol{x}^{\mathrm{t}} = [x_1, x_2, \ldots, x_n]$ is a real or complex vector, a general norm is denoted by $\|\boldsymbol{x}\|_N$ and is defined by

$$\|\boldsymbol{x}\|_N = \left\{ \sum_{i=1}^{n} |x_i|^N \right\}^{1/N} \tag{180}$$

So the usual Euclidian norm, or length, is

$$\|\boldsymbol{x}\|_2 = \sqrt{|x_1|^2 + \cdots + |x_n|^2} \tag{181}$$

Other norms are also used, for example, $\|\boldsymbol{x}\|_1$ and $\|\boldsymbol{x}\|_\infty$, the latter corresponding to the $|\boldsymbol{x}_i|$ greatest in magnitude.

Matrix Norm. If $\boldsymbol{A}$ is an $n \times n$ real or complex matrix, we denote its norm $\|\boldsymbol{A}\|$, defined by

$$\|\boldsymbol{A}\| = \max_{x \neq 0} \frac{\|\boldsymbol{Ax}\|}{\|\boldsymbol{x}\|} \tag{182}$$

According to our choice of N, in defining the vector norms by (180), we have corresponding $\|\boldsymbol{Ax}\|_1$, $\|\boldsymbol{Ax}\|_2$, $\|\boldsymbol{Ax}\|_\infty$, the Euclidian $N = 2$ being the most common. Note that (ignoring the question, How do we find its value?), for given $\boldsymbol{A}$ and N, $\boldsymbol{A}$ has some specific numerical value greater than or equal to zero.

10.2. Condition of a Linear System $Ax = y$

This is an example of Hadamard's general concept of a "well-posed problem," which is roughly one where the result is not *too* sensitive to small changes in the problem specification. Another example arises in the concept of the supergain antenna.

Definition. The problem of finding $\boldsymbol{x}$, satisfying $\boldsymbol{Ax} = \boldsymbol{y}$, is *well-posed* or *well-conditioned* if:

(i) A unique $\boldsymbol{x}$ satisfies $\boldsymbol{Ax} = \boldsymbol{y}$, and
(ii) Small changes in either $\boldsymbol{A}$ or $\boldsymbol{y}$ result in small changes in $\boldsymbol{x}$.

For a quantitative measure of how well conditioned, we need something like $\delta\boldsymbol{x}/\delta\boldsymbol{y}$ or $\delta\boldsymbol{x}/\delta\boldsymbol{A}$! *Suppose* $\boldsymbol{A}$ is fixed but $\boldsymbol{y}$ changes slightly to $\boldsymbol{y}+\delta\boldsymbol{y}$, with the associated $\boldsymbol{x}$ changing to $\boldsymbol{x}+\delta\boldsymbol{x}$. We then have

$$\boldsymbol{A}\boldsymbol{x}=\boldsymbol{y} \tag{183}$$

and so

$$\boldsymbol{A}(\boldsymbol{x}+\delta\boldsymbol{x})=\boldsymbol{y}+\delta\boldsymbol{y}$$

Subtracting gives

$$\boldsymbol{A}\,\delta\boldsymbol{x}=\delta\boldsymbol{y} \qquad \text{or} \qquad \delta\boldsymbol{x}=\boldsymbol{A}^{-1}\,\delta\boldsymbol{y} \tag{184}$$

From our definition (182), we must have, for any $\boldsymbol{A}$ and z,

$$\frac{\|\boldsymbol{A}z\|}{\|z\|}\le\|\boldsymbol{A}\| \qquad \text{and so} \qquad \|\boldsymbol{A}z\|\le\|\boldsymbol{A}\|\cdot\|z\| \tag{185}$$

Taking the norm of both sides of (184) and using inequality (185) gives

$$\|\delta\boldsymbol{x}\|=\|\boldsymbol{A}^{-1}\,\delta\boldsymbol{y}\|\le\|\boldsymbol{A}^{-1}\|\cdot\|\delta\boldsymbol{y}\| \tag{186}$$

Taking the norm of (183) and using inequality (185) gives

$$\|\boldsymbol{y}\|=\|\boldsymbol{A}\boldsymbol{x}\|\le\|\boldsymbol{A}\|\cdot\|\boldsymbol{x}\| \tag{187}$$

Finally, multiplying corresponding sides of (186) and (187) and dividing by $\|\boldsymbol{x}\|\cdot\|\boldsymbol{y}\|$ gives our fundamental result:

$$\frac{\|\delta\boldsymbol{x}\|}{\|\boldsymbol{x}\|}\le\|\boldsymbol{A}\|\cdot\|\boldsymbol{A}^{-1}\|\,\frac{\|\delta\boldsymbol{y}\|}{\|\boldsymbol{y}\|} \tag{188}$$

For any square matrix $\boldsymbol{A}$ we introduce its *condition number* and define

$$\text{cond}\,(\boldsymbol{A})=\|\boldsymbol{A}\|\cdot\|\boldsymbol{A}^{-1}\| \tag{189}$$

We note that a "good" condition number is small, near to 1.

10.1.2. Relevance of the Condition Number

The condition number $\|\delta\boldsymbol{y}\|/\|\boldsymbol{y}\|$ can be interpreted as a measure of relative uncertainty in the vector $\boldsymbol{y}$. Similarly, $\|\delta\boldsymbol{x}\|/\|\boldsymbol{x}\|$ is the associated relative uncertainty in the vector $\boldsymbol{x}$.

From eqs. (188) and (189), cond $(\boldsymbol{A})$ gives an upper-bound (worst-case) factor of degradation of precision between $\boldsymbol{y}$ and $\boldsymbol{x} = \boldsymbol{A}\boldsymbol{y}$. Note that if we reversed $\boldsymbol{A}$ and $\boldsymbol{A}^{-1}$ throughout the theory from (183), eqs. (188) and (189) would remain exactly the same. These two equations therefore give the important result that:

If $\boldsymbol{A}$ denotes the *precise* transformation $\boldsymbol{y} = \boldsymbol{A}\boldsymbol{x}$ and $\delta\boldsymbol{x}, \delta\boldsymbol{y}$ are small related changes in $\boldsymbol{x}$ and $\boldsymbol{y}$, the ratio

$$\frac{\|\delta\boldsymbol{x}\| / \|\boldsymbol{x}\|}{\|\delta\boldsymbol{y}\| / \|\boldsymbol{y}\|}$$

must lie between $1/\text{cond}(\boldsymbol{A})$ and $\text{cond}(\boldsymbol{A})$

10.2.2. Numerical Example

Here is a numerical example using integers for total precision. Suppose:

$$\boldsymbol{A} = \begin{bmatrix} 100 & 99 \\ 99 & 98 \end{bmatrix}$$

We then have:

$$\boldsymbol{A}\begin{bmatrix} 1000 \\ -1000 \end{bmatrix} = \begin{bmatrix} 1000 \\ 1000 \end{bmatrix}$$

Shifting $\boldsymbol{x}$ slightly gives

$$\boldsymbol{A}\begin{bmatrix} 1001 \\ -999 \end{bmatrix} = \begin{bmatrix} 1199 \\ 1197 \end{bmatrix}$$

Alternatively, shifting $\boldsymbol{y}$ slightly gives

$$\boldsymbol{A}\begin{bmatrix} 803 \\ -801 \end{bmatrix} = \begin{bmatrix} 1001 \\ 999 \end{bmatrix}$$

So a small change in $\boldsymbol{y}$ can cause a big change in $\boldsymbol{x}$, or vice versa. We have this clear moral, concerning any matrix multiplication or (effectively) inversion:

For a given $\boldsymbol{A}$, either multiplying $\boldsymbol{A}\boldsymbol{x}$ or "dividing" $(\boldsymbol{A}^{-1}\boldsymbol{y})$ can be catastrophic, the degree of catastrophe depending on cond $(\boldsymbol{A})$ and on the direction of change in $\boldsymbol{x}$ or $\boldsymbol{y}$.

In the above example, cond $(\boldsymbol{A})$ is about 4000.

10.3. Matrix Computations

After the above cautionary account of matrix condition, we now consider methods of solving matrix equations.

All the field computing methods studied in this chapter have resulted in

$$\boldsymbol{A}\boldsymbol{x} = \boldsymbol{y} \tag{190}$$

or

$$\boldsymbol{A}\boldsymbol{x} = \boldsymbol{0}\,, \qquad \text{requiring } \det(\boldsymbol{A}) = 0 \tag{191}$$

[we will consider this a special case of (190)] or

$$\boldsymbol{A}\boldsymbol{x} = k^2\boldsymbol{B}\boldsymbol{x} \tag{192}$$

where $\boldsymbol{A}$ (and $\boldsymbol{B}$) are known $n \times n$ matrices and $\boldsymbol{x}$ and k^2 are unknown.

Usually $\boldsymbol{B}$ (and sometimes $\boldsymbol{A}$) is positive definite, meaning that $\boldsymbol{x}^t\boldsymbol{B}\boldsymbol{x} > 0$ for all $\boldsymbol{x}$; this typically follows because it comes from terms such as

$$\int\int \mathbf{H}^*\mathbf{H}\, dS$$

$\boldsymbol{A}$ is sometimes complex, but numerically the difference is straightforward, so we will consider $\boldsymbol{A}$ to be real.

As an example of (192), the vibrating string of Section 4 gave (62), which, after applying (55) gives

$$\begin{bmatrix} 105 & 126 \\ 126 & 180 \end{bmatrix}\begin{bmatrix} u_1 \\ u_2 \end{bmatrix} = k^2\begin{bmatrix} 42 & 48 \\ 48 & 56 \end{bmatrix}\begin{bmatrix} u_1 \\ u_2 \end{bmatrix} \tag{193}$$

with eigenvalues and vectors given after eq. (63).

10.3.1. Types (Sparsity Patterns) of Matrices A and B

The main categories are:

1. Full (dense)
2. Band (sparse)
3. Variable band (sparse)
4. Arbitrarily sparse (sparse)

where all of the above have their nonzero elements stored, and, finally,

5. Any pattern, but where the elements are *not stored*; that is, elements are "generated" or calculated each time they are needed in the solving algorithm.

Methods of solution are basically:

Direct, where the solution emerges in a finite number of calculations (if we temporarily ignore roundoff error due to finite word length).

Indirect, or iterative, where a step-by-step procedure converges toward the correct solution.

Indirect methods can be specially suited to sparse matrices (especially when the order is large) because they can often be implemented without the need to store the entire matrix $\boldsymbol{A}$ (or intermediate forms of matrices) in high-speed RAM storage.

All the common direct routines are available in software libraries and in books and journals, commonly in Fortran, Pascal, or Basic [3, 22, 23].

More detailed accounts of matrix methods have also been published [1, 3, 22–24].

10.4. Direct Methods for Solving $\boldsymbol{Ax} = \boldsymbol{y}$

The classic solution method of (190) is the Gauss method. Given the system

$$\begin{bmatrix} 1 & 4 & 7 \\ 2 & 5 & 8 \\ 3 & 6 & 11 \end{bmatrix} \begin{bmatrix} X_1 \\ X_2 \\ X_3 \end{bmatrix} = \begin{bmatrix} 1 \\ 1 \\ 1 \end{bmatrix} \tag{194}$$

we subtract 2 times the first row from the second row, and then we subtract 3 times the first row from the third row, to give

$$\begin{bmatrix} 1 & 4 & 7 \\ 0 & -3 & -6 \\ 0 & -6 & -10 \end{bmatrix} \begin{bmatrix} X_1 \\ X_2 \\ X_3 \end{bmatrix} = \begin{bmatrix} 1 \\ -1 \\ -2 \end{bmatrix} \tag{195}$$

and then subtracting 2 times the second row from the third row gives

$$\begin{bmatrix} 1 & 4 & 7 \\ 0 & -3 & -6 \\ 0 & 0 & 2 \end{bmatrix} \begin{bmatrix} X_1 \\ X_2 \\ X_3 \end{bmatrix} = \begin{bmatrix} 1 \\ -1 \\ 0 \end{bmatrix} \tag{196}$$

The steps from (194) to (196) are termed *triangulation* or *forward elimination*. The triangular form of the left-hand matrix of (196) is crucial; it allows the next steps.

The third row immediately gives

$$\underline{X_3 = 0} \tag{197a}$$

and substitution into row 2 gives

$$(-3)X_2 + 0 = -1\,, \qquad \text{so} \qquad \underline{X_2 = \tfrac{1}{3}} \tag{197b}$$

and then substitution into row 1 gives

$$X_1 + 4(\tfrac{1}{3}) + 0 = 1\,, \qquad \text{so} \qquad \underline{X_1 = -\tfrac{1}{3}} \tag{197c}$$

The steps through eqs. (197) are termed *back-substitution*. We now ignore a complication of "pivoting" [1, 22–24].

Important points about this algorithm are listed below.

1. Computing time is proportional to n^3. This means that doubling the number of finite element nodes in an integral equation solution (which gives a full matrix) will increase CPU time by up to 8 times!

2. The determinant comes immediately as the product of the diagonal elements of (196).

3. Algorithms that take advantage of the special band and variable band are very straightforward [3, 23, 24], just changing the limits of the DO or FOR loops, and some bookkeeping. For example, in a matrix of "semi-bandwidth" 4, the first column has nonzero elements only in the first four rows, as in Fig. 30. Then only those four numbers need storing, and only the three elements below the diagonal need to be eliminated in the first column.

4. Oddly, it turns out that, in our context, one should *never* find the inverse matrix $\boldsymbol{A}$ in order to solve $\boldsymbol{Ax} = \boldsymbol{y}$ for $\boldsymbol{x}$. Even if it needs doing for a number of different right-hand-side vectors $\boldsymbol{y}$, it is better to keep a record of the triangular form of (196) and back-substitute as necessary.

5. Other methods very similar to Gauss are due to Crout and Choleski (and, interestingly, the original Crout paper was in an electrical engineering journal). The latter is (only) for use with symmetric matrices. Its advantage is that time *and* storage are half that of the orthodox Gauss. For these

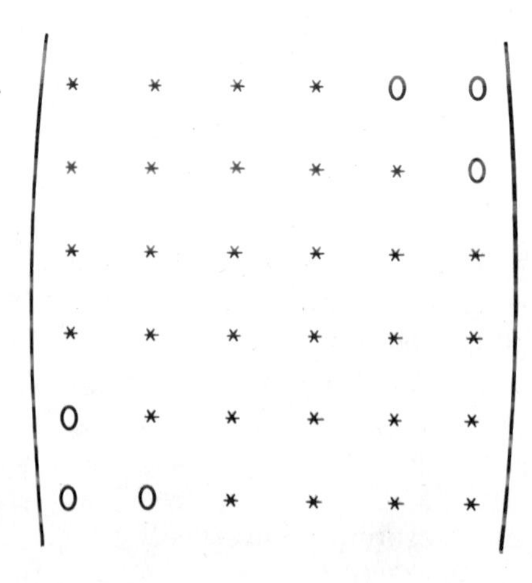

Fig. 30 Zeros and nonzeros in band matrix of semibandwidth 4.

reasons, the Choleski algorithm is chosen for use in the complete computer program presented in Section 11.

6. There is a drastic variant of Gauss's method that is ideal for finite element work [24]. It accepts a matrix $\boldsymbol{A}$ of *arbitrary* sparsity; it basically starts with an empty matrix, inserts just the nonzero elements, and cleverly applies Gauss without causing too much "fill-in"—without spreading its arithmetic too much outside the original pattern of nonzeros.

7. Another variant of the Gauss algorithm has been especially developed for finite element work. In the frontal method, elimination takes place in a carefully controlled manner, with intermediate results being kept in backing storage. Again, the method is well studied and documented [3, 23].

10.5. Iterative Methods for Solving $\boldsymbol{Ax} = \boldsymbol{y}$

We will outline two methods: (1) the conjugate gradient algorithm and (2) the Jacobi (simultaneous displacement) method with the closely related Gauss–Seidel (successive displacement) algorithm.

10.5.1. The Conjugate Gradient Algorithm

Although known and used for decades, the conjugate gradient method has come, in the 1980s, to be adopted as one of the most popular iterative algorithms for solving $\boldsymbol{Ax} = \boldsymbol{y}$. Its rationale starts as a matrix version of our weighted-residual approach where, as in eq. (7), we introduced and made small an error residual $R = Lu - v$.

The equation to be solved for x,

$$\boldsymbol{Ax} = \boldsymbol{y} \tag{198}$$

can be recast as finding x to minimize the error residual, a column vector $\boldsymbol{r}$ defined as a function of $\boldsymbol{x}$ by

$$\boldsymbol{r} = \boldsymbol{Ax} - \boldsymbol{y} \tag{199}$$

For an iterative method, this means, from a given value of x_i, finding a "better" value x_{i+1}, with a "smaller" residual $\boldsymbol{r}$.

The norm $\|\boldsymbol{r}\|$ of this residual vector is a convenient measure of $\boldsymbol{r}$ to be minimized, and the conventional steepest-gradient method of optimization [1, 22] would seem surely the best way of proceeding from one step to the next, toward the bottom of the valley. Unfortunately, in numerical work, the valleys are rarely isotropic in the parameters being varied (all the elements of x), and the steepest-descent algorithm is too myopic to notice!

The conjugate gradient method instead evaluates the gradient (with respect to $\boldsymbol{x}$) of $\frac{1}{2}\|\boldsymbol{r}(\boldsymbol{x})\|^2$ at a point $\boldsymbol{x}$:

$$\nabla(\tfrac{1}{2}\|\boldsymbol{r}(\boldsymbol{x})\|^2)$$

and then minimizes $\|\boldsymbol{r}\|$ along a particular line $(\boldsymbol{x} + v\,\delta\boldsymbol{x})$; that is, it finds the value of v that minimizes $\|\boldsymbol{r}(\boldsymbol{x} + v\,\delta\boldsymbol{x})\|$.

It transpires [1, 22] that

$$\nabla(\tfrac{1}{2}\|\boldsymbol{r}(\boldsymbol{x})\|^2) = \boldsymbol{A}^{\mathrm{t}}(\boldsymbol{A}\boldsymbol{x} - \boldsymbol{y}) \tag{200}$$

and

$$v = \frac{-\delta\boldsymbol{x}\nabla(\tfrac{1}{2}\|\boldsymbol{r}\|^2)}{\|\boldsymbol{A}\,\delta\boldsymbol{x}\|^2} \tag{201}$$

An immediate feature of expressions (200) and (201) is that reference to matrix $\boldsymbol{A}$ is only via simple matrix products; for given values of the matrices $\boldsymbol{A}$, $\boldsymbol{y}$, $\boldsymbol{x}_i$, and $\delta\boldsymbol{x}_i$, we need only form $\boldsymbol{A}$ times a vector ($\boldsymbol{A}\boldsymbol{x}$ or $\boldsymbol{A}\,\delta\boldsymbol{x}_i$) and $\boldsymbol{A}^{\mathrm{t}}$ times a vector ($\boldsymbol{A}\boldsymbol{x}_i - \boldsymbol{y}$). These can be formed from a given sparse $\boldsymbol{A}$ without unnecessary multiplication (of nonzeros) or storage. In Reference 22 and many commercial packages, the user has to arrange these scalar products, while guidance through the overall strategy is provided by the package.

More robust versions of the algorithm see to a preliminary preconditioning of the matrix $\boldsymbol{A}$ to alleviate the problem that the condition number of (200) is the square of the condition number of $\boldsymbol{A}$—a serious problem if $\boldsymbol{A}$ is not safely positive-definite. This leads to the popular PCCG (preconditioned conjugate gradient) algorithm, as a complete package, for which reference must be made to the literature and to commercial packages.

10.5.2. Jacobi and Gauss–Seidel

Two algorithms that are simple to implement are the closely related Jacobi (simultaneous displacement) and Gauss–Seidel (successive displacement, or relaxation) algorithms.

Suppose the set of equations for solution are

$$\begin{aligned} a_{11}x_1 + a_{12}x_2 + a_{13}x_3 &= y_1 \\ a_{21}x_1 + a_{22}x_2 + a_{23}x_3 &= y_2 \\ a_{31}x_1 + a_{32}x_2 + a_{33}x_3 &= y_3 \end{aligned} \tag{202}$$

This can be reorganized to

$$x_1 = (y_1 - a_{12}x_2 - a_{13}x_3)/a_{11} \tag{203a}$$

$$x_2 = (y_2 - a_{23}x_3 - a_{21}x_1)/a_{22} \tag{203b}$$

$$x_3 = (y_3 - a_{31}x_1 - a_{32}x_2)/a_{33} \tag{203c}$$

Suppose we had the vector $\boldsymbol{x}^{(0)} = [x_1, x_2, x_3]^{(0)}$ and substituted it into the right-hand side of eqs. (203) to yield on the left-hand side the new vector $\boldsymbol{x}^{(1)} = [x_1, x_2, x_3]^{(1)}$. Successive substitutions will give the sequence of vectors

$$\boldsymbol{x}^{(0)}, \boldsymbol{x}^{(1)}, \boldsymbol{x}^{(2)}, \boldsymbol{x}^{(3)}, \ldots$$

Because eqs. (203) are merely a rearrangement of the equations for solution, the "correct" solution substituted into (203) must be self-consistent—it must yield itself! The sequence will:

either converge to the correct solution

or diverge.

This is the Jacobi or simultaneous displacement iterative scheme.

Note that when eq. (203b) is applied, a new value of x_1 will be available from (203a), which *could* be used instead of the previous x_1 value. And similarly for x_1 and x_2 when applying (203c). This is the Gauss–Seidel or successive displacement iterative scheme, illustrated here with an example to show that the computer program is barely more complicated than writing down the equations.

```
05 REM example of successive displacement
10 LET X1 = 0
20 LET X2 = 0
30 LET X3 = 0                               ;Equations being solved are:
40 FOR I=1 TO 10
50     LET X1=(4 + X2 - X3)/4               ;  4.x1  -   x2  +    x3  = 4
60     LET X2=(9-2*X3 - X1)/6               ;    x1  +6.x2  + 2.x3  = 9
70     LET X3=(2 + X1+2*X2)/5               ;   -x1  -2.x2  + 5.x3  = 2
80     PRINT X1,X2,X3
90 NEXT
OK
RUN
 1              1.333333       1.133334
 1.05           .9472222       .9888889
 .9895833       1.00544        1.000093
 1.001337       .9997463       1.000166
 .9998951       .9999621       .9999639
 .9999996       1.000012       1.000005
 1.000002       .9999981       .9999996
 .9999996       1              1
 1              1              1
 1              1              1
OK
C
C   GAUSS-SEIDEL TO SOLVE LAPLACE BETWEEN SQUARE INNER & OUTER.
C
      DIMENSION Z(11,11)
      DATA Z/121*0./
      DO 1 I=4,8
        DO 2 J=4,8
          Z(I,J)=1.
```

```
    2   CONTINUE
    1 CONTINUE
C
      DO 3 N=1,30
        DO 4 I=2,10
          DO 5 J=2,10
      IF(Z(I,J).LT.1.)Z(I,J)=.25*(Z(I-1,J)+Z(I+1,J)+Z(I,J-1)+Z(I,J+1))
    5     CONTINUE
    4   CONTINUE
C
      WRITE(6,6)N,(Z(3,J),J=1,11)
    6 FORMAT(1X,I2,11F7.4)
    3 CONTINUE
C
      WRITE(6,7)Z
    7 FORMAT(14H FINAL RESULT=,//(//1X,11F7.4))
      STOP
      END
```

```
 1 0.0000 0.0000 0.0000 0.2500 0.3125 0.3281 0.3320 0.3330 0.0833 0.0208 0.0000
 2 0.0000 0.0000 0.1250 0.3750 0.4492 0.4717 0.4785 0.4181 0.1895 0.0699 0.0000
 3 0.0000 0.0469 0.2109 0.4473 0.5225 0.5472 0.5399 0.4737 0.2579 0.1098 0.0000
 4 0.0000 0.0908 0.2690 0.4922 0.5653 0.5865 0.5742 0.5082 0.3016 0.1369 0.0000
 5 0.0000 0.1229 0.3076 0.5205 0.5902 0.6084 0.5944 0.5299 0.3293 0.1542 0.0000
 6 0.0000 0.1446 0.3326 0.5381 0.6047 0.6211 0.6067 0.5435 0.3465 0.1650 0.0000
 7 0.0000 0.1586 0.3484 0.5488 0.6133 0.6288 0.6143 0.5519 0.3572 0.1716 0.0000
 8 0.0000 0.1675 0.3582 0.5553 0.6184 0.6334 0.6190 0.5572 0.3637 0.1756 0.0000
 9 0.0000 0.1729 0.3641 0.5592 0.6215 0.6362 0.6218 0.5604 0.3676 0.1780 0.0000
10 0.0000 0.1763 0.3678 0.5616 0.6234 0.6379 0.6236 0.5624 0.3700 0.1794 0.0000
11 0.0000 0.1783 0.3700 0.5631 0.6245 0.6390 0.6247 0.5636 0.3713 0.1802 0.0000
12 0.0000 0.1795 0.3713 0.5639 0.6253 0.6396 0.6254 0.5643 0.3722 0.1807 0.0000
13 0.0000 0.1802 0.3721 0.5645 0.6257 0.6400 0.6258 0.5647 0.3727 0.1810 0.0000
14 0.0000 0.1807 0.3726 0.5648 0.6259 0.6403 0.6260 0.5649 0.3729 0.1812 0.0000
15 0.0000 0.1810 0.3729 0.5650 0.6261 0.6404 0.6261 0.5651 0.3731 0.1813 0.0000
16 0.0000 0.1811 0.3731 0.5651 0.6262 0.6405 0.6262 0.5652 0.3732 0.1813 0.0000
17 0.0000 0.1812 0.3732 0.5652 0.6263 0.6406 0.6263 0.5652 0.3733 0.1813 0.0000
18 0.0000 0.1813 0.3732 0.5652 0.6263 0.6406 0.6263 0.5653 0.3733 0.1814 0.0000
19 0.0000 0.1813 0.3733 0.5653 0.6263 0.6406 0.6263 0.5653 0.3733 0.1814 0.0000
20 0.0000 0.1814 0.3733 0.5653 0.6263 0.6406 0.6263 0.5653 0.3733 0.1814 0.0000
21 0.0000 0.1814 0.3733 0.5653 0.6263 0.6406 0.6263 0.5653 0.3733 0.1814 0.0000
22 0.0000 0.1814 0.3733 0.5653 0.6263 0.6406 0.6263 0.5653 0.3733 0.1814 0.0000
23 0.0000 0.1814 0.3733 0.5653 0.6263 0.6406 0.6263 0.5653 0.3734 0.1814 0.0000
24 0.0000 0.1814 0.3733 0.5653 0.6263 0.6406 0.6263 0.5653 0.3734 0.1814 0.0000
25 0.0000 0.1814 0.3734 0.5653 0.6263 0.6406 0.6263 0.5653 0.3734 0.1814 0.0000
26 0.0000 0.1814 0.3734 0.5653 0.6263 0.6406 0.6263 0.5653 0.3734 0.1814 0.0000
27 0.0000 0.1814 0.3734 0.5653 0.6263 0.6406 0.6263 0.5653 0.3734 0.1814 0.0000
28 0.0000 0.1814 0.3734 0.5653 0.6263 0.6406 0.6263 0.5653 0.3734 0.1814 0.0000
29 0.0000 0.1814 0.3734 0.5653 0.6263 0.6406 0.6263 0.5653 0.3734 0.1814 0.0000
30 0.0000 0.1814 0.3734 0.5653 0.6263 0.6406 0.6263 0.5653 0.3734 0.1814 0.0000

FINAL RESULT=
 0.0000 0.0000 0.0000 0.0000 0.0000 0.0000 0.0000 0.0000 0.0000 0.0000 0.0000

 0.0000 0.0907 0.1814 0.2615 0.2994 0.3099 0.2994 0.2615 0.1814 0.0907 0.0000

 0.0000 0.1814 0.3734 0.5653 0.6263 0.6406 0.6263 0.5653 0.3734 0.1814 0.0000

 0.0000 0.2615 0.5653 1.0000 1.0000 1.0000 1.0000 1.0000 0.5653 0.2615 0.0000

 0.0000 0.2994 0.6263 1.0000 1.0000 1.0000 1.0000 1.0000 0.6263 0.2994 0.0000

 0.0000 0.3099 0.6406 1.0000 1.0000 1.0000 1.0000 1.0000 0.6406 0.3099 0.0000

 0.0000 0.2994 0.6263 1.0000 1.0000 1.0000 1.0000 1.0000 0.6263 0.2994 0.0000

 0.0000 0.2615 0.5653 1.0000 1.0000 1.0000 1.0000 1.0000 0.5653 0.2615 0.0000

 0.0000 0.1814 0.3734 0.5653 0.6263 0.6406 0.6263 0.5653 0.3734 0.1814 0.0000

 0.0000 0.0907 0.1814 0.2615 0.2994 0.3099 0.2994 0.2615 0.1814 0.0907 0.0000

 0.0000 0.0000 0.0000 0.0000 0.0000 0.0000 0.0000 0.0000 0.0000 0.0000 0.0000
STOP
```

(The only reason for using Basic is that this text was written at a PC keyboard.)

Whether the algorithm converges or not depends on the matrix $\boldsymbol{A}$ and (surprisingly) not on the right-hand-side vector $\boldsymbol{y}$ of (190). Convergence does not even depend on the starting value of the vector, which only affects the necessary number of iterations.

We will skip over any formal proof of convergence [1]. But to give the sharp criteria for convergence, first we split $\boldsymbol{A}$ as

$$\boldsymbol{A} = \boldsymbol{L} + \boldsymbol{D} + \boldsymbol{U}$$

where $\boldsymbol{L}$, $\boldsymbol{D}$, and $\boldsymbol{U}$ are the lower, diagonal, and upper triangular parts of $\boldsymbol{A}$. Then the schemes converge if and only if all the eigenvalues of the matrix

$$-\boldsymbol{D}^{-1}(\boldsymbol{U} + \boldsymbol{L}) \qquad \text{for simultaneous displacement}$$

or

$$-(\boldsymbol{D} + \boldsymbol{L})^{-1}\boldsymbol{U} \qquad \text{for successive displacement}$$

lie within the unit circle.

For applications, it is simpler to use some *sufficient* conditions, when possible, such as:

1. If $\boldsymbol{A}$ is symmetric and positive definite, then successive displacement converges.
2. If, in addition, $A_{i,j} < 0$ for all $i = j$, then simultaneous displacement also converges.

Condition 1 is commonly satisfied; for example, the stripline problem that started with the variational form (97) gave the global matrix (98) and then (104) for solution. Clearly the matrix $\boldsymbol{A}$ is positive definite [see the definition following eq. (92)], the stored energy indeed being positive for *any* nodal values.

Usually, the successive method converges in about half the computer time of the simultaneous method, but strictly there are matrices where one method converges and the other does not, and vice versa.

10.5.3. Application of Successive or Simultaneous Displacements to Finite Element or Finite Difference Equations

Suppose we tackle a problem like the stripline as in Section 5 but to make the description easier we assume square inner and outer conductors. We will also ignore (not take the usual advantage of) the symmetry.

Skipping over variational and finite element methods, we go for the simplest finite difference formulation and superimpose a square mesh over the cross section of Fig. 31. We consider the unknown nodal values of ϕ at the interior nodes and apply the finite difference equation (171),

$$\nabla^2\phi \approx (\phi_N + \phi_S + \phi_E + \phi_W - 4\phi_O)/h^2 \tag{204}$$

directly to the Laplace equation

$$\nabla^2\phi = 0 \tag{205}$$

to give

$$\phi_N + \phi_S + \phi_E + \phi_W - 4\phi_O = 0 \tag{206}$$

Applying eq. (206) to point 1 of the mesh gives

$$0 + \phi_{10} + \phi_2 + 0 - 4\phi_1 = 0 \tag{207}$$

and applying it to point 2 gives

$$0 + \phi_{11} + \phi_3 + \phi_1 - 4\phi_2 = 0 \tag{208}$$

A typical interior point 11 gives

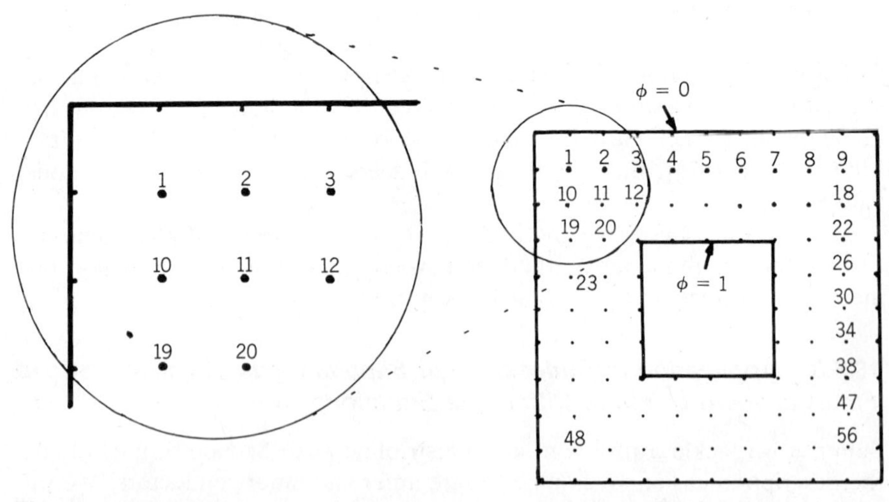

Fig. 31 Finite difference mesh for solution of the electrostatic field in square coaxial line.

$$\phi_2 + \phi_{20} + \phi_{12} + \phi_{10} - 4\phi_{11} = 0 \tag{209}$$

and point 13, near the inner conductor, will give

$$\phi_4 + 1 + \phi_{14} + \phi_{12} - 4\phi_{13} = 0 \tag{210}$$

In this way, we can assemble all 56 equations from the 56 mesh points of Fig. 31 in terms of the 56 unknown ϕ's.

The resulting 56 equations must be expressible as

$$\boldsymbol{A}\boldsymbol{x} = \boldsymbol{y} \tag{211a}$$

or

$$\begin{bmatrix} -4 & 1 & & \cdots & 1 & & & & \\ 1 & -4 & 1 & \cdots & & 1 & & & \\ & 1 & -4 & \cdots & & & 1 & & \\ & & & \cdot & & & & & \\ & & & & \cdot & & & & \\ & & & & & \cdot & & & \\ & & & & & & & & \\ & & & & & & & -4 & 1 \\ & & & & & & & 1 & 4 \end{bmatrix} \begin{bmatrix} \phi_1 \\ \phi_2 \\ \phi_3 \\ \cdot \\ \cdot \\ \cdot \\ \phi_{13} \\ \cdot \\ \cdot \\ \cdot \\ \phi_{55} \\ \phi_{56} \end{bmatrix} = \begin{bmatrix} 0 \\ 0 \\ 0 \\ \cdot \\ \cdot \\ \cdot \\ -1 \\ \cdot \\ \cdot \\ \cdot \\ 0 \\ 0 \end{bmatrix} \tag{211b}$$

The unknown vector $\boldsymbol{x}$ of (211a) is simply $(\phi_1, \phi_2, \ldots, \phi_{56})^t$.

The right-hand-side vector $\boldsymbol{y}$ of eqs. (211) consists of zeros *except* for the -1's coming from equations like (210) corresponding to points next to the inner conductor with potential 1; namely, from points 12–17, 20, 21, . . . , 41–45.

The 56×56 matrix $\boldsymbol{A}$ has mostly zero elements, except for -4 on the diagonal, and either two, three, or four $+1$'s somewhere else on each matrix row, the number and distribution depending on the geometric node location.

One has to be careful not to confuse the row-and-column numbers of the two-dimensional array $\boldsymbol{A}$ with the x and y coordinates of the physical problem. Each number 1–56 of the mesh points in Fig. 31 corresponds precisely to the row number of matrix $\boldsymbol{A}$ and the row number of column vector $\boldsymbol{x}$.

Equations (211) are standard, as given in (190), and *could* be solved with a standard library package, with routine Gauss, or better still with a band-matrix version of Gauss. Even better would be the Gauss–Seidel or successive displacement. As eqs. (202) and (203) showed, this consists of taking the original simultaneous equations, putting all diagonal matrix terms on the left-hand side as in (203), and effectively putting them in a FOR or DO loop (after initializing the "starting vector").

Equation (206) is the typical equation, and putting the diagonal term on the left-hand side gives

$$\phi_O = (\phi_N + \phi_S + \phi_E + \phi_W)/4 \tag{212}$$

Fifty-six lines of Fortran (or Pascal, or Basic, or, . . .) could be written from (212), just like the two lines of Basic following eqs. (203), using the necessary coefficients from a stored 56×56 matrix (or fewer lines, using subscripted variables). But the elements of $\boldsymbol{A}$ are all either 0, +1, or −4 and are easily generated during the algorithm, rather than actually being stored in an array. This simplifies the computer program, and instead of $\boldsymbol{A}$, the only array needed holds the current value of vector elements

$$\boldsymbol{x}^{t} = (\phi_1, \phi_2, \ldots, \phi_{56}) .$$

The program is simplified further (at the risk of confusing the one-dimensional vector x with a two-dimensional array) by keeping x in a two-dimensional array $Z(11, 11)$ to be identified spatially with the two-dimensional Cartesian coordinates of the physical problem (Fig. 30).

The resulting program for solution by successive displacements needs only two assignment statements. The first of these,

```
Z(I,J)=1.
```

corresponds to statements 10, 20, and 30 of the earlier Basic program (Section 10.5.2), setting the vector $\boldsymbol{x}^{(0)}$ to an initial value. (A good guess would speed convergence, but is it worth the trouble?) The second,

```
IF(Z(I,J).LT.1.)Z(I,J)=etc.
```

and its immediate two enclosing loops, corresponds to statements 50, 60, and 70 of the earlier Basic program, applying the iteration once.

The outer loop,

```
DO 3 N=1,25

3 CONTINUE
```

applies the iteration 25 times; as results show, in fact, 20 iterations are adequate for convergence to the accuracy of printout. The intermediate printout gives, at each iteration, a sample of potentials along a line of nodes. This is followed by FINAL RESULT, giving the final potentials over the whole cross section.

11. A FINITE ELEMENT COMPUTER PROGRAM FOR MICROSTRIPS

To illustrate the finite elements methods discussed, and to give the reader the chance to expand and experiment, a complete program is now presented. Finite element programs given in the literature [3, 11, 17] are usually versatile in dealing with arbitrary shapes, where the user is entitled to give any reasonable shape via a data string defining the element vertices. By contrast, we present a *particular* problem, but one that is common in microwaves, and with a program strategy that can easily be developed to deal with more complicated planar structures, such as (single or coupled) coplanar waveguides or asymmetric couplers.

The problem chosen is that of a microstrip in an enclosed box, effectively like that discussed in Section 5.7, using the same first-degree triangular elements and variational form

$$J = \int\int \frac{\varepsilon}{\varepsilon_0} (\nabla\phi)^2 \, dS \tag{213}$$

Facilities are included for a suspended substrate, and variable mesh divisions can be employed (three regions vertically and two horizontally) so that the user can choose finer elements in regions of higher field values. The strategy used in the program (in dealing with a global matrix, which is then reduced) is precisely that of Section 5, proceeding from eq. (92) to the final matrix equation for solution, eq. (102).

As well as evaluating the field (or voltage) distribution for the given microstrip structure, the total capacitance is calculated, being indeed J of eq. (213). This can be used, as already mentioned and more generally demonstrated by Daly [14], by evaluating C with the dielectric included and then evaluating C_0 by repeating with the dielectric effectively removed, that is, with $\varepsilon = \varepsilon_0$ everywhere. The so-called quasi-static solution for a microstrip then follows, via

$$\varepsilon_{\text{eff}} = C/C_0$$

and phase velocity

$$v_p = c/\sqrt{\varepsilon_{\text{eff}}}$$

Used in this way, the capacitance does not need to be "normalized."

Indeed, dimensions can be in micrometers, millimeters, or meters, as the user chooses.

Having introduced the program, and having stressed the teaching role of the program, a few associated reservations and precautions should be emphasized. First, the program is written for (human) readability rather than efficiency. The strategy used is *not* the most efficient (in computer storage or CPU time) but was chosen with the aim of clarity. A simple (but classic) dense matrix algorithm is used; more efficient (and complicated) algorithms could be used instead [3, 23]. The simplest possible program structure (with a main program that only CALLs the constituent sections) is used here. For the same reason, the program is as compact as possible, with no error checks.

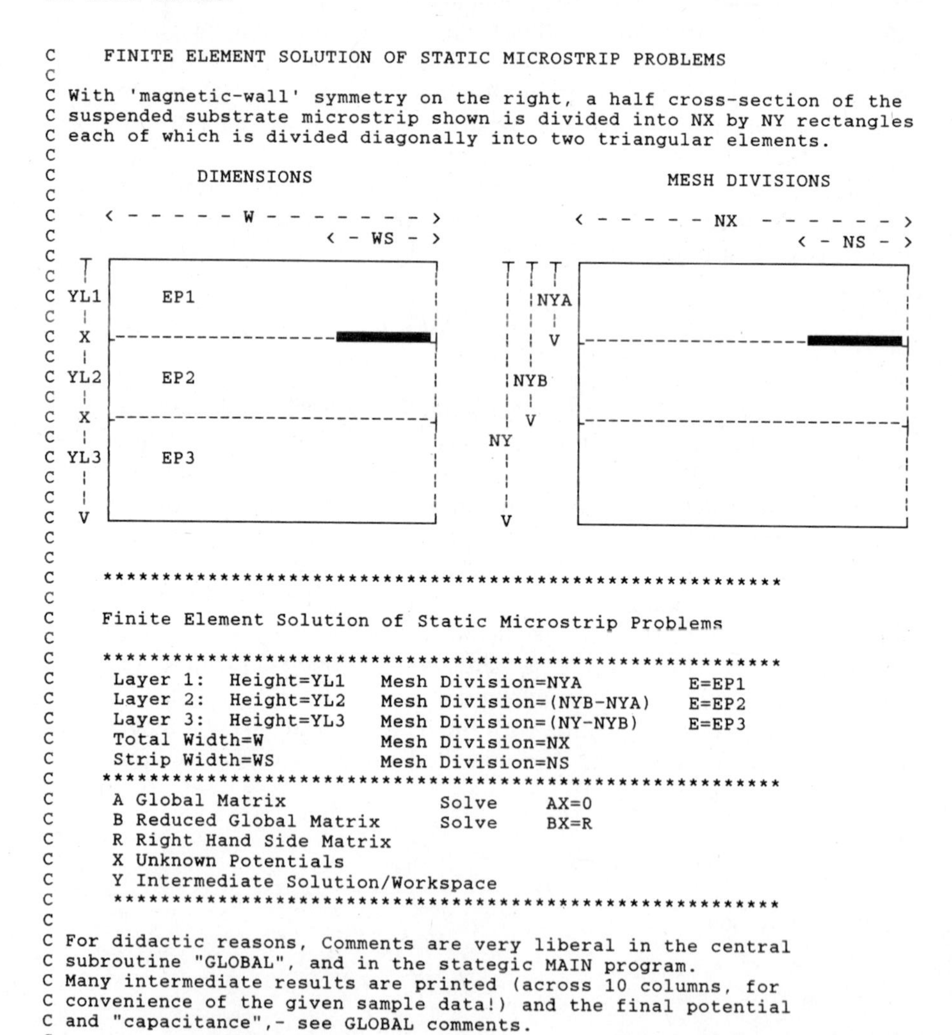

```
C     FINITE ELEMENT SOLUTION OF STATIC MICROSTRIP PROBLEMS
C
C With 'magnetic-wall' symmetry on the right, a half cross-section of the
C suspended substrate microstrip shown is divided into NX by NY rectangles
C each of which is divided diagonally into two triangular elements.
C
C               DIMENSIONS                                  MESH DIVISIONS
C
C     < - - - - - W - - - - - - - >               < - - - - - NX  - - - - - - >
C                          < - WS - >                                < - NS - >
C   T  ______________________________         T  T  T  ______________________________
C   |                                |        |  |  |  |                              |
C YL1     EP1                        |        |  |NYA  |                              |
C   |                                |        |  |  |  |                              |
C   X  |-------------------█████████|        |  |  V  |--------------------█████████|
C   |                                |        |  |     |                              |
C YL2     EP2                        |        |NYB     |                              |
C   |                                |        |  |     |                              |
C   X  |-----------------------------|        |  V     |------------------------------|
C   |                                |        NY       |                              |
C YL3     EP3                        |        |        |                              |
C   |                                |        |        |                              |
C   |                                |        |        |                              |
C   V  |_____________________________|        V        |______________________________|
C
C
C     ************************************************************
C
C     Finite Element Solution of Static Microstrip Problems
C
C     ************************************************************
C      Layer 1:  Height=YL1    Mesh Division=NYA          E=EP1
C      Layer 2:  Height=YL2    Mesh Division=(NYB-NYA)    E=EP2
C      Layer 3:  Height=YL3    Mesh Division=(NY-NYB)     E=EP3
C      Total Width=W           Mesh Division=NX
C      Strip Width=WS          Mesh Division=NS
C     ************************************************************
C      A Global Matrix                  Solve    AX=0
C      B Reduced Global Matrix          Solve    BX=R
C      R Right Hand Side Matrix
C      X Unknown Potentials
C      Y Intermediate Solution/Workspace
C     ************************************************************
C
C For didactic reasons, Comments are very liberal in the central
C subroutine "GLOBAL", and in the stategic MAIN program.
C Many intermediate results are printed (across 10 columns, for
C convenience of the given sample data!) and the final potential
C and "capacitance",- see GLOBAL comments.
C
```

```
C For ease, identical DIMENSION and COMMON is used in all
C subprograms:
C
      DIMENSION A(100,100),B(100,100),Y(100)
      DIMENSION KA(100),KB(100),KN(100),X(100),R(100)
      COMMON/D1/A,B,Y,NX,NYA,NYB,NY,NS,NORDER,MORDER
      COMMON/D2/KA,KB,KN,X,R,NX1,NY1,NS1
      COMMON/D3/DX1,DX2,DY1,DY2,DY3,EP1,EP2,EP3
C
C Collect input data:
C
      CALL INPUT
C
C Set boundary potentials and strip:
C
      CALL BOUND
C
C Assemble global matrix A:
C
      CALL GLOBAL
C
C Reorder and reduce matrix A to B:
C
      CALL REDUCE
C
C Decompose B into LU matrices:
C
      CALL CHOLES
C
C Solve equations and write solutions:
C
      CALL SOLVE
C
      STOP
      END
C-------------------------------------------------------------
      SUBROUTINE INPUT
      DIMENSION A(100,100),B(100,100),Y(100)
      DIMENSION KA(100),KB(100),KN(100),X(100),R(100)
      COMMON/D1/A,B,Y,NX,NYA,NYB,NY,NS,NORDER,MORDER
      COMMON/D2/KA,KB,KN,X,R,NX1,NY1,NS1
      COMMON/D3/DX1,DX2,DY1,DY2,DY3,EP1,EP2,EP3
C
      WRITE(*,3)
    3 FORMAT(' GIVE HEIGHTS YL1,YL2,YL3:'/)
      READ(*,4)YL1,YL2,YL3
    4 FORMAT(E20.10)
C
      WRITE(*,5)
    5 FORMAT(' GIVE DIELECTRIC CONSTANTS EP1,EP2,EP3:'/)
      READ(*,6)EP1,EP2,EP3
    6 FORMAT(E20.10)
C
      WRITE(*,7)
    7 FORMAT(' GIVE TOTAL AND STRIP WIDTHS:'/)
      READ(*,8)W,WS
    8 FORMAT(E20.10)
C
      WRITE(*,1)
    1 FORMAT(' GIVE MESH DIVISION NUMBERS NYA,NYB,NY,NX,NS:'/)
      READ(*,2)NYA,NYB,NY,NX,NS
    2 FORMAT(I3)
C
```

```
      WRITE(6,201)YL1,YL2,YL3
  201 FORMAT(' YL1=',F8.4,' YL2=',F8.4,' YL3=',F8.4)
      WRITE(6,203)EP1,EP2,EP3
  203 FORMAT(' EP1=',F8.4,' EP2=',F8.4,' EP3=',F8.4)
      WRITE(6,202)W,WS
  202 FORMAT(' W=',F8.4,' WS=',F8.4)
      WRITE(6,204)NYA,NYB,NY
  204 FORMAT(' NYA=',I2,' NYB=',I2,' NY=',I2)
      WRITE(6,205)NX,NS
  205 FORMAT(' NX=',I2,' NS=',I2)
      DY1 = YL1/FLOAT(NYA)
      DY2 = YL2/FLOAT(NYB-NYA)
      DY3 = YL3/FLOAT(NY -NYB)
      DX1 = (W-WS)/FLOAT(NX-NS)
      DX2 = WS/FLOAT(NS)
      RETURN
      END
C-------------------------------------------------------------
      SUBROUTINE BOUND
C
C This sets the arrays KA,KB,KN to re-order node numbers:
C KA Original Nodes, Unknown(=1), Known zero(=0), Known nonzero(=2)
C KB Stores original node no for reduced array
C KN Stores original node no for known but nonzero nodes
      DIMENSION A(100,100),B(100,100),Y(100)
      DIMENSION KA(100),KB(100),KN(100),X(100),R(100)
      COMMON/D1/A,B,Y,NX,NYA,NYB,NY,NS,NORDER,MORDER
      COMMON/D2/KA,KB,KN,X,R,NX1,NY1,NS1
      COMMON/D3/DX1,DX2,DY1,DY2,DY3,EP1,EP2,EP3
C
      NX1 = NX+1
      NY1 = NY+1
      NS1 = NS+1
C
C Initialization
C
      MORDER = NX1*NY1
      DO 10 I = 1,MORDER
         KA(I) = 1
   10 CONTINUE
C
C Top and bottom conductors
C
      DO 20 I = 1,NX1
         KA(I) = 0
         KA(NY*NX1+I) - 0
   20 CONTINUE
C
C Left side conductor
C
      DO 30 I = 1,NY1
         KA((I-1)*NX1+1) = 0
   30 CONTINUE
C
C Metal strip
C
      DO 40 I = 1,NS1
         IJ = NYA*NX1+NX-NS+I
         KA(IJ) = 2
   40 CONTINUE
C
C Evaluate KB holds unknown nodes only
C
      ICOUNT = 0
```

```
      DO 50 I = 1,MORDER
         IF(KA(I).NE.1) GOTO 50
         ICOUNT = ICOUNT+1
         KB(ICOUNT) = I
   50 CONTINUE
      NORDER = ICOUNT
C
C Calculate KN, known nonzero potential nodes
C
      ICN = 0
      DO 60 I = 1,MORDER
         IF(KA(I).NE.2) GOTO 60
         ICN = ICN+1
         KN(ICN) = I
   60 CONTINUE
      WRITE(6,101)(KA(I),I = 1,MORDER)
  101 FORMAT(' ALL NODES='/(1X,10I4))
      WRITE(6,102)(KB(I),I = 1,NORDER)
  102 FORMAT(//////' UNKNOWN NODES='/(1X,10I4))
      WRITE(6,103)(KN(I),I = 1,ICN)
  103 FORMAT(' KNOWN POT NODES='/(1X,10I4))
      RETURN
      END
C-----------------------------------------------------------------
      SUBROUTINE GLOBAL
      DIMENSION A(100,100),B(100,100),Y(100)
      DIMENSION KA(100),KB(100),KN(100),X(100),R(100)
      COMMON/D1/A,B,Y,NX,NYA,NYB,NY,NS,NORDER,MORDER
      COMMON/D2/KA,KB,KN,X,R,NX1,NY1,NS1
      COMMON/D3/DX1,DX2,DY1,DY2,DY3,EP1,EP2,EP3
C
C Set matrix order (=number of nodes) as:
C
      MORDER = (NX + 1)*(NY + 1)
C
C Set global matrix to zero, ready for cumulative summing of the
C variational integral (which is 2*energy = C.V.V and capacitance)
C viz. cross-section integral of (epsilon.(grad(phi))squared).dS
C
      DO 6 I = 1,MORDER
         DO 5 J = 1,MORDER
            A(I,J) = 0.
    5    CONTINUE
    6 CONTINUE
C
C The following 2 DO-loops  (DO 1 I=... and DO 2 J=...)
C scan over the whole range of nodes to the right of X=0
C and below Y=0.
C At each of these nodes,the overall global matrix has its
C element values accumulated.
C Relevant values from, first the upper left triangle,
C and then the lower right triangle, as in the diagram below.
C Within the outer loop, variations in epsilon and vertical
C element dimensions are updated, so that all elements are
C dealt with in the one inner loop.
C Run down rows of nodes, viz. with y:
C
      DO 1 I = 1,NY
C
C Set epsilon according to layer:
C
         E = EP1
         IF((I.GT.NYA).AND.(I.LE.NYB)) E = EP2
```

```
         IF(I.GT.NYB) E = EP3
C
C Set vertical element dimension:
C
         DY = DY1
         IF(I.GT.NYA) DY = DY2
         IF(I.GT.NYB) DY = DY3
C
C Run along nodes, viz. with X                          3--2
C Evaluate global node numbers K1, K2, K3 ,K4          | /|
C Corresponding to:                                    |/ |
C (I,J), (I,J-1), (I-1,J-1) AND (I-1,J)                4--1
C
         DO 2 J = 1,NX
            DX = DX1
            IF(J.GT.(NX-NS)) DX = DX2
C
C Set the often-used constants,- in fact the only values
C of matrix elements (per finite element) that arise !
C
            DXBYDY =  E*DX/DY
            DYBYDX =  E*DY/DX
            DXYD =    DXBYDY + DYBYDX
            K1 = (NX + 1)*I + J + 1
            K2 = K1 - (NX + 1)
            K3 = K2 - 1
            K4 = K1 - 1
C
C Accumulate in global matrix contribution from
C upper left triangle, with vertices K2, K3, K4.
C
            A(K4,K4) = A(K4,K4) + DXBYDY
            A(K3,K3) = A(K3,K3) + DXYD
            A(K2,K2) = A(K2,K2) + DYBYDX
            A(K3,K2) = A(K3,K2) - DYBYDX
            A(K3,K4) = A(K3,K4) - DXBYDY
            A(K2,K3) = A(K2,K3) - DYBYDX
            A(K4,K3) = A(K4,K3) - DXBYDY
C
C Accumulate in global matrix, contribution from
C lower right triangle, with vertices K4, K1, K2.
C
            A(K4,K4) = A(K4,K4) + DYBYDX
            A(K1,K1) = A(K1,K1) + DXYD
            A(K2,K2) = A(K2,K2) + DXBYDY
            A(K4,K1) = A(K4,K1) - DYBYDX
            A(K2,K1) = A(K2,K1) - DXBYDY
            A(K1,K4) = A(K1,K4) - DYBYDX
            A(K1,K2) = A(K1,K2) - DXBYDY
    2    CONTINUE
    1 CONTINUE
      RETURN
      END
C-----------------------------------------------------------------
      SUBROUTINE REDUCE
      DIMENSION A(100,100),B(100,100),Y(100)
      DIMENSION KA(100),KB(100),KN(100),X(100),R(100)
      COMMON/D1/A,B,Y,NX,NYA,NYB,NY,NS,NORDER,MORDER
      COMMON/D2/KA,KB,KN,X,R,NX1,NY1,NS1
      COMMON/D3/DX1,DX2,DY1,DY2,DY3,EP1,EP2,EP3
      WRITE(6,24)(A(I,I),I = 1,MORDER)
   24 FORMAT(' MATRIX A DIAG='/(1X,10F7.3))
      DO 50 I = 1,NORDER
```

```
         DO 51 J = 1,NORDER
            IN = KB(I)
            JN = KB(J)
            B(I,J) = A(IN,JN)
   51    CONTINUE
   50 CONTINUE
      WRITE(6,152)(B(I,I),I = 1,NORDER)
  152 FORMAT(' MATRIX B DIAG='/(1X,10F7.3))
C
C Generate R from known potentials
C
      DO 60 I = 1,NORDER
         IX = KB(I)
         R(I) = 0.0
         DO 61 J = 1,NS1
            IY = KN(J)
            R(I) = R(I)-A(IX,IY)
   61    CONTINUE
   60 CONTINUE
      WRITE(6,153)(R(I),I = 1,NORDER)
  153 FORMAT(' RHS ARRAY='/(1X,10F7.3))
      RETURN
      END
C-------------------------------------------------------------
      SUBROUTINE CHOLES
C
C This subroutine performs the Cholesky decomposition of the
C matrix B. Its upper triangular factor is returned overwriting B.
C B is decomposed to LU and only U is written on B.
C NORDER = Matrix order
C
      DIMENSION A(100,100),B(100,100),Y(100)
      DIMENSION KA(100),KB(100),KN(100),X(100),R(100)
      COMMON/D1/A,B,Y,NX,NYA,NYB,NY,NS,NORDER,MORDER
      COMMON/D2/KA,KB,KN,X,R,NX1,NY1,NS1
      COMMON/D3/DX1,DX2,DY1,DY2,DY3,EP1,EP2,EP3
C
      ZERO = 0.0
      DO 50 K = 1,NORDER
         IF(B(K,K).LE.ZERO) GO TO 70
         B(K,K) = SQRT(B(K,K))
         KP1 = K+1
         IF (KP1.GT.NORDER) GO TO 50
         DO 10 I = KP1,NORDER
            B(K,I) = B(K,I)/B(K,K)
   10    CONTINUE
         DO 30 I = KP1,NORDER
            DO 20 J = I,NORDER
               B(I,J) = B(I,J)-B(K,I)*B(K,J)
   20       CONTINUE
   30    CONTINUE
   50 CONTINUE
      RETURN
   70 WRITE (6,80)
   80 FORMAT (' ERROR RETURN FROM CHOLESKY DECOMPOSITION')
      STOP
      END
C-------------------------------------------------------------
      SUBROUTINE SOLVE
      DIMENSION A(100,100),B(100,100),Y(100)
      DIMENSION KA(100),KB(100),KN(100),X(100),R(100)
      COMMON/D1/A,B,Y,NX,NYA,NYB,NY,NS,NORDER,MORDER
      COMMON/D2/KA,KB,KN,X,R,NX1,NY1,NS1
      COMMON/D3/DX1,DX2,DY1,DY2,DY3,EP1,EP2,EP3
```

```
C
C First perform forward substitution to solve for Y
C Solve  LY = R
C
      Y(1) = R(1)/B(1,1)
      DO 300 K = 2,NORDER
         SUM = R(K)
         KM1 = K-1
         DO 400 J = 1,KM1
            SUM = SUM-B(J,K)*Y(J)
  400    CONTINUE
         Y(K) = SUM/B(K,K)
  300 CONTINUE
C
C Now back substitution to calculate X
C Solve UX = Y
C
      X(NORDER) = Y(NORDER)/B(NORDER,NORDER)
      DO 110 J = 2,NORDER
         NJ1 = NORDER-J+1
         JM1 = J-1
         SUM = Y(NJ1)
         DO 200 K = 1,JM1
            NK1 = NORDER-K+1
            SUM = SUM-X(NK1)*B(NJ1,NK1)
  200    CONTINUE
         X(NJ1) = SUM/B(NJ1,NJ1)
  110 CONTINUE
      WRITE(6,181)(X(I),I = 1,NORDER)
  181 FORMAT(' UNKNOWN POTENTIAL='/(1X,10F7.4))
C
C R is re-used for nodal potentials
C
      DO 75 I = 1,MORDER
         R(I) = 0.0
   75 CONTINUE
C
C Calculate Potentials
C
      DO 80 I = 1,NORDER
         IJ = KB(I)
         R(IJ) = X(I)
   80 CONTINUE
C
C Known nonzero potentials
C
      DO 85 I = 1,NS1
         IJ = KN(I)
         R(IJ) = 1.0
   85 CONTINUE
      WRITE(6,164)(R(I),I = 1,MORDER)
  164 FORMAT(//////' TOTAL POTENTIAL='/(1X,10F7.3))
C
C Calculate capacitance
C
      C = 0.0
      DO 150 I = 1,MORDER
         DO 151 J = 1,MORDER
            C = C+R(I)*A(I,J)*R(J)
  151    CONTINUE
  150 CONTINUE
      WRITE(6,191)C
```

```
  191 FORMAT(/' CAPACITANCE=',F12.5/)
      RETURN
      END
C-----------------------------------------------------------------
     Sample data:
     4.
     2.
     2.
     1.
     4.
     1.
     9.
     3.
     4
     6
     8
     9
     3

     and associated output:


TOTAL POTENTIAL=
  0.000  0.000  0.000  0.000  0.000  0.000  0.000  0.000  0.000  0.000
  0.000  0.028  0.057  0.089  0.125  0.163  0.198  0.221  0.233  0.236
  0.000  0.053  0.110  0.174  0.248  0.330  0.409  0.453  0.474  0.480
  0.000  0.076  0.157  0.250  0.362  0.500  0.654  0.709  0.729  0.735
  0.000  0.093  0.192  0.307  0.451  0.652  1.000  1.000  1.000  1.000
  0.000  0.092  0.191  0.304  0.437  0.598  0.773  0.842  0.869  0.876
  0.000  0.086  0.177  0.279  0.397  0.528  0.654  0.725  0.758  0.767
  0.000  0.044  0.091  0.142  0.200  0.261  0.317  0.354  0.372  0.378
  0.000  0.000  0.000  0.000  0.000  0.000  0.000  0.000  0.000  0.000

CAPACITANCE=      8.43815

STOP
```

REFERENCES

1. D. S. Jones, *Methods in Electromagnetic Wave Propagation*, Oxford University Press (Clarendon), London and New York, 1979.
2. R. F. Harrington, *Field Computation by Moment Methods*, reprint ed., Robert Krieger, Malabar, FL, 1985; orig. ed. 1968.
3. J. N. Reddy, *An Introduction to the Finite Element Method*, McGraw-Hill, New York, 1984.
4. M. Becker, *The Principles and Applications of Variational Methods*, M.I.T. Press, Cambridge, MA, 1964.
5. G. Strang and G. J. Fix, *An Analysis of the Finite Element Method*, Prentice-Hall, Englewood Cliffs, NJ, 1973.
6. A. D. Berk, "Variational principles for electromagnetic resonators and waveguides," *IRE Trans. Antennas Propag.*, vol. *AP-4*, pp. 104–111, (Apr. 1956.
7. K. Morishita and N. Kumagai, "Unified approach to the derivation of variational expression for electromagnetic fields," *IEEE Trans. Microwave Theory Tech.*, vol. *MTT-25*, pp. 34–40, Jan. 1977.

8. R. E. Collin, *Field Theory of Guided Waves*, McGraw-Hill, New York, 1960.
9. L. Cairo and T. Kahan, *Variational Techniques in Electromagnetism*, Blackie, London, 1965; *Techniques Variationelles en Radioélectricité*, orig. French ed., Dunod, Paris, 1962.
10. P. Hammond, *Energy Methods in Electromagnetism*, Oxford University Press (Clarendon), London and New York, 1981.
11. P. P. Silvester and R. L. Ferrari, *Finite Elements for Electrical Engineers*, Cambridge University Press, New York, 1983.
12. B. M. Rahman and J. B. Davies, "Finite-element analysis of optical and microwave waveguide problems," *IEEE Trans. Microwave Theory Tech.*, vol. *MTT-32*, pp. 20–28, Jan. 1984.
13. O. C. Zienkiewicz and K. Morgan, *Finite Elements and Approximation*, Wiley, New York, 1983.
14. P. Daly, "Upper and lower bounds to the characteristic impedance of transmission lines using the finite element method," *Int. J. Comput. Math. Electr. Electron. Eng.* (*COMPEL*), vol. 3, pp. 65–78, 1984.
15. J. Moore and R. Pizer (Ed.), *Moment Methods in Electromagnetics*, Research Studies Press Ltd., Letchworth, Hertfordshire, England, 1984.
16. J. B. Davies and C. A. Muilwyk, "Numerical solution of uniform hollow waveguides with boundaries of arbitrary shape," *Proc. Inst. Electr. Eng.*, vol. 113, pp. 277–284, Feb. 1966.
17. P. Silvester, "A general high-order finite-element waveguide analysis program," *IEEE Trans. Microwave Theory Tech.*, vol. MTT-17, pp. 204–210, Apr. 1969.
18. A. Konrad, "Vector variational formulation of electromagnetic fields in anisotropic media," *IEEE Trans. Microwave Theory Tech.*, vol. MTT-24, pp. 553–559, Sept. 1976.
19. B. M. Rahman and J. B. Davies, "Finite-element solution of integrated optical waveguides," *IEEE J. Lightwave Technol.*, vol. LT-2, pp. 682–687, Oct. 1984.
20. H. G. Unger, *Planar Optical Waveguides and Fibers*, Oxford University Press (Clarendon), London and New York, 1977.
21. N. Mabaya, P. E. Lagasse, and P. Vandenbulcke, "Finite element analysis of optical waveguides," *IEEE Trans. Microwave Theory Tech.*, vol. MTT-29, pp. 600–605, June 1981.
22. W. H. Press, B. P. Flannery, S. A. Teukolsky, and W. T. Vetterling, *Numerical Recipes: The Art of Scientific Computing*, Cambridge University Press, New York, 1986.
23. A. Jennings, *Matrix Computation for Engineers and Scientists*, Wiley, Chichester, 1977.
24. I. S. Duff, A. M. Erisman, and J. K. Reid, *Direct Methods for Sparse Matrices*, Oxford University Press (Clarendon), London and New York, 1986.

3

Integral Equation Technique

Juan R. Mosig
Laboratoire d'Électromagnétisme et d'Acoustique (LEMA)
École Polytechnique Fédérale de Lausanne
Lausanne, Switzerland

1. INTRODUCTION

This chapter deals with the application of integral equation techniques to microwave and millimeter-wave integrated circuits. To illustrate these techniques, we have selected the general microstrip structure, which represents one of the most commonly used technologies for microwave circuits. The word *microstrip* is used here to describe a multilayered medium sandwiched between air and a ground plane that contains several conducting objects which usually reside on the interfaces between layers.

These conducting objects are assumed to have a finite size and an irregular shape. In most cases, they will be formed by conducting sheets of negligible thickness that can exhibit ohmic losses.

The layers will be considered as being composed of homogeneous, isotropic, lossy materials, and in all the examples presented we will assume that they are also nonmagnetic. However, the theory developed here can be generalized without particular effort to anisotropic or magnetic materials. The general purpose of applying a mathematical model to a multiconductor structure is to provide an equivalent circuit that can be used to predict the structure's electrical performance.

The simplest models are based on a generalization of the transmission line concept and yield lumped *LC* circuits valid at low frequency. Improvement of these models introduces ohmic losses (a series resistance) and dielectric losses (a parallel conductance). At higher frequencies, a microstrip-like structure can no longer be represented by a classical *RLC* circuit. More elaborate models that include dispersion are then needed.

These models, most of which are described elsewhere in this book, lead directly to the scattering matrix or the complex impedance matrix of the structure and are able to predict the frequency behavior of a microstrip structure until radiation effects appear.

A model including radiation must predict impedance parameters whose real parts are not zero even in the absence of ohmic and dielectric losses. The real parts of the impedance parameters grow with frequency and become dominant when an internal resonance of the structure is reached.

Radiation is the crucial word in this chapter. We assume that the structures under study are open, and we postulate from the very beginning that the surface currents existing on the conducting parts will radiate. As we will see, radiation in a multilayered inhomogeneous structure is a complex phenomenon, and the radiated fields include both space and surface waves. Since our model does not impose a lateral wall closing the structure, fringing fields are automatically taken into account, and the accuracy of the theoretical results is thus independent of the layer's thickness.

The theory presented in this chapter is essentially three-dimensional, and it is not intended for application to propagation phenomena in uniform lines. On the other hand, this theory is especially well suited to the study of finite size conducting patches and the discontinuities that appear when they are connected through short sections of microstrip lines.

The main goal here is the development of mathematical tools that allow accurate prediction of the relevance of radiation in a given structure. As a byproduct, this model leads to equivalent circuits that include radiation losses and mutual coupling due to surface waves. For instance, coupling factors between microstrip resonators of arbitrary shape can be accurately predicted at any frequency.

From a mathematical point of view, the model presented here is characterized by the use of integral equations where the fundamental unknown is the surface electric current flowing in the conductors.

This feature is shared by other models, including techniques that employ the so-called spectral domain analysis, that are also described in this book. The distinctive points in our approach are the use of the real space domain for solving the integral equation and the formulation of this equation, whenever possible, in terms of vector and scalar potentials.

After a short digression on notation, the chapter begins with the presentation of the integral equations and associated Green's functions (Section 3). Several choices for the potentials are compared, and Sommerfeld's choice is adopted throughout the chapter.

In Section 4, the Green's functions associated with a layered medium are developed and presented as Sommerfeld integrals. A matrix chain formalism allows an easy computer implementation.

Detailed expressions of the Green's functions are given in Section 5 for single- and double-layer microstrip structures. Section 6 discusses the nature of the surface waves arising in a stratified medium and their mathematical treatment as the singular part of Sommerfeld integrals. Numerical results for the propagation constants are also given. Section 7 is a short account of the simplifications that can be made in the near- and far-field zones, or, equivalently, at low and high frequencies. In particular, it is shown that our integral equation reduces, in the low-frequency case, to the well-known static and quasi-static equations commonly used for obtaining the equivalent capacitances and inductances of microstrip-like structures. In Section 8, we develop powerful numerical algorithms for computing Green's functions, and the results are given for several practical cases in Section 9, which also provides a comparison with near- and far-field approximations. The numerical solution of the integral equation with a method of moments is discussed in Section 10. The excitation and loading of microstrip structures considered as N-port devices is studied in Section 11. Section 12 gives the results of the method applied to several of the many possible practical examples, and finally Section 13 presents some general remarks.

2. CONVENTIONS AND NOTATIONS

In this chapter, vectors are denoted by boldface roman type and dyadics by boldface italic type with a double bar above the corresponding symbol. The unit vector along the coordinate t is denoted $\mathbf{e}_t$. Dyadics are used for representing Green's functions relating an arbitrarily oriented elementary source (Hertz dipole) to the fields and vector potentials that it creates. For instance, we have for the magnetic field

$$\mathbf{dH(r)} = \bar{\bar{\boldsymbol{G}}}_H(\mathbf{r}\,|\,\mathbf{r}') \cdot I(\mathbf{r}')\,\mathbf{dl}' \tag{1}$$

and

$$\bar{\bar{\boldsymbol{G}}}_H = \sum_{s=x,y,z} \sum_{t=x,y,z} \mathbf{e}_s G_H^{st} \mathbf{e}_t \tag{2}$$

Thus, the scalar component G_H^{st} gives the s component of the magnetic field created at the point $\mathbf{r}$ by an elementary t-directed dipole of unit moment $I\,dl' = 1$ and located at the point $\mathbf{r}'$.

Let us recall that with this definition the scalar product $\bar{\bar{\boldsymbol{G}}}_H \cdot \mathbf{e}_t$ is a true vector giving the total magnetic field created by a t-directed dipole. On the other hand, the vector $\mathbf{e}_t \cdot \bar{\bar{\boldsymbol{G}}}_H$ has no physical meaning as a whole but its components give the t components of three different magnetic fields corresponding to the three possible orientations of the source dipole.

A time dependence $\exp(j\omega t)$ is assumed and suppressed throughout this chapter.

The Spectral Domain. The problem of computing the fields in a stratified medium has been tackled in many textbooks and research papers (see, for instance, Kong [1]). The most efficient technique formulates the problem in a transformed spectral domain, where the transverse Cartesian coordinates x, y are replaced by their spectral counterparts k_x, k_y according to the double Fourier transform:

$$\tilde{f}(k_x, k_y) = \frac{1}{2\pi} \iint_{-\infty}^{\infty} f(x, y) \exp(-jk_x x - jk_y y)\, dx\, dy \tag{3}$$

$$f(x, y) = \frac{1}{2\pi} \iint_{-\infty}^{\infty} \tilde{f}(k_x, k_y) \exp(jk_x x + jk_y y)\, dk_x\, dk_y \tag{4}$$

It is useful to introduce the polar vector $\boldsymbol{\rho} = x\mathbf{e}_x + y\mathbf{e}_y$ and the radial spectral variable $\mathbf{k}_\rho = \mathbf{e}_x k_x + \mathbf{e}_y k_y$. Then the "del" operator $\boldsymbol{\nabla}$ can be split into its transverse and normal parts as $\boldsymbol{\nabla} = \boldsymbol{\nabla}_t + \mathbf{e}_z\, \partial/\partial z$ or, in the spectral domain:

$$\tilde{\boldsymbol{\nabla}} = j\mathbf{k}_\rho + \mathbf{e}_z \frac{\partial}{\partial z} \tag{5}$$

Since the only spatial derivative remaining in the spectral domain is the normal one, we will use the shorter dot notation for derivatives, for example, $\partial \Psi / \partial z = \dot{\Psi}$.

We will later find many functions that depend on x, y exclusively through the radial distance ρ. For such functions we can write the element of surface in polar coordinates and integrate analytically along the polar angle ϕ.

After introducing the Bessel function J_0 we can express the transforms of eqs. (3) and (4) as

$$\tilde{f}(k_\rho) = \int_0^\infty J_0(k_\rho \rho) f(\rho) \rho\, d\rho \tag{6}$$

$$f(\rho) = \int_0^\infty J_0(k_\rho \rho) \tilde{f}(k_\rho) k_\rho\, dk_\rho \tag{7}$$

and this is the Fourier–Bessel or Hankel integral transform pair. Inverse Hankel transforms like (7) are best known among radio and microwave engineers as Sommerfeld integrals, and we will keep this tradition. Within the Sommerfeld framework, if $\tilde{f}(k_\rho)$ is the transform of $f(\rho)$, the transverse derivative of $f(\rho)$ can be expressed as

$$\frac{\partial f}{\partial x} = -\cos\phi \int_0^\infty J_1(k_\rho \rho) \tilde{f} k_\rho^2\, dk_\rho \tag{8}$$

The usefulness of the spectral domain approach is clearly illustrated by the fact that if a quantity Ψ satisfies the homogeneous Helmholtz equation

$$(\nabla^2 + k^2)\Psi = 0 \tag{9}$$

then the spectral transform is a solution of the ordinary differential equation

$$\left(\frac{d^2}{dz^2} - u^2\right)\tilde{\Psi} = 0 \tag{10}$$

where the parameter u in the traditional notation of Sommerfeld is given by

$$u^2 = -k_z^2 = k_x^2 + k_y^2 - k^2 = k_\rho^2 - k^2 \tag{11}$$

For the sake of clarity, we will drop from now on the tilde (˜) denoting a spectral quantity. This should be no source of error, the type of a given quantity being clearly determined by the nature of its mathematical expression and by the context. Finally, we will consider complex values of the spectral variable $k_\rho = \lambda + j\nu$, where $\lambda = \mathrm{Re}[k_\rho]$ should not be confused with the free space wavelength λ_0.

3. INTEGRAL EQUATIONS FOR LAYERED MEDIUM PROBLEMS

3.1. The Integral Equation Formulation

Let us consider the geometry depicted in Fig. 1, where several three-dimensional conducting objects are embedded in a stratified medium. This medium is composed of N parallel layers sandwiched between the half-space $z > 0$ (usually air) and the terminal plane $z = -d_N$. We shall assume that this plane can be modeled as an impedance wall forcing a simple relationship between the tangential electric and magnetic fields existing on it, namely

$$\mathbf{E}_t = \mathbf{e}_z \times Z_s \mathbf{H}_t \tag{12}$$

Electric and magnetic walls are included in eq. (1) as particular cases given, respectively, by $Z_s = 0$ or $Z_s = \infty$. The existence of an impedance wall implies that the problem can be solved without knowing the actual fields that can exist below the terminal plane $z = -d_N$.

Each layer is assumed to be an isotropic homogeneous and possibly lossy material with complex permittivity ε and complex permeability μ.

Similarly, we assume that the embedded conductors are characterized by a boundary condition on their surfaces:

$$\mathbf{n} \times \mathbf{E} = Z_s \mathbf{n} \times \mathbf{J}_s \tag{13}$$

where Z_s is again a surface impedance (zero for perfect electric conductors), $\mathbf{n}$ is the outwards unit normal vector (Fig. 1), and $\mathbf{J}_s$ is the electric surface current existing in the conductor.

This current is excited by an excitation field $\mathbf{E}^e$ and, in turn, creates a scattered or diffracted field $\mathbf{E}^d$. The total field in eq. (13) is the sum of the excitation and diffracted fields.

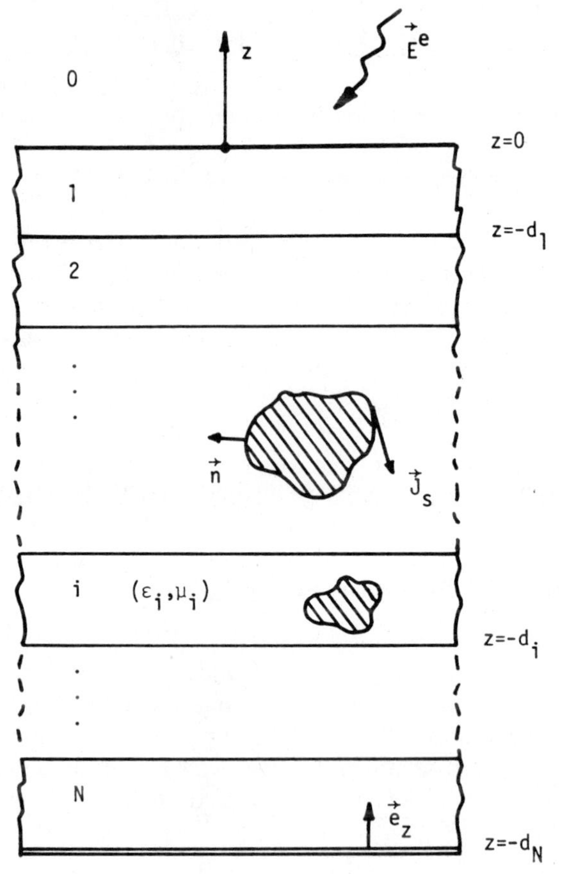

Fig. 1 Multilayered medium that includes several embedded conducting objects. The N layers are sandwiched between a half-space medium (usually the air), $0 < z < \infty$ and an impedance wall at $z = -d_N$.

The diffracted field $\mathbf{E}^d$ can be expressed with the help of a dyadic Green's function $\bar{\bar{G}}_E$ which is defined by the linear relation

$$\mathbf{dE}^d(\mathbf{r}) = \bar{\bar{G}}_E(\mathbf{r} \mid \mathbf{r}') \cdot I(\mathbf{r}')\, \mathbf{dl}' \tag{14}$$

The total diffracted field is now obtained by superposition, integrating eq. (14) over the surface S of the conductors.

In a rigorous approach, the embedded conductors should be considered as dielectrics exhibiting very high ohmic losses, and a portion of $\mathbf{E}^d$ would arise from the equivalent magnetic currents defined on the surfaces of the conductors. However, if conductivity is very high, these magnetic currents can be neglected. Finally, the boundary condition (13) becomes

$$\mathbf{n} \times \mathbf{E}^e(\mathbf{r}) = -\mathbf{n} \times \int_S \bar{\bar{G}}_E(\mathbf{r}|\mathbf{r}') \cdot \mathbf{J}_s(\mathbf{r}')\, ds' + Z_s \mathbf{n} \times \mathbf{J}_s \tag{15}$$

which is a generalized form of the electric field integral equation (EFIE) for the unknown $\mathbf{J}_s$.

An alternative integral equation for $\mathbf{J}_s$ using a boundary condition on the magnetic field can also be derived [2]. However, it has been shown [3] that the magnetic field integral equation (MFIE) is numerically unstable when the embedded conductors are thin and fails completely for zero-volume conductors. Since integrated circuits often include very thin conducting sheets, this integral equation will not be considered here.

Returning to the EFIE, there are two fundamental ways of interpreting the current $\mathbf{J}_s$ and the Green's function $\bar{\bar{G}}_E$ appearing in eq. (15).

1. In a first approach, we apply equivalence theorems for replacing the different layers by ficticious electric and magnetic currents existing in the boundaries between layers (the so-called polarization currents). In this manner, we solve an equivalent problem where the conductors are embedded in an unbounded homogeneous medium. Consequently, the Green's function in (14) and (15) takes the simple form

$$\bar{\bar{G}}_E = (k^2 \bar{\bar{I}} + \nabla\nabla)\, \frac{\exp(-jkr)}{4\pi r} \tag{16}$$

where $k^2 = \omega^2 \mu\varepsilon$ and $\bar{\bar{I}}$ is the unit dyadic. On the other hand, the unknowns in (15) include fictive as well as true currents and the integration domain must be extended to the set of parallel planes $z = -d_1$ (Fig. 1). This approach has been successfully used in static problems [4].

2. The second possibility deals directly with the real problem. The integral equation applies only to the surfaces of the conductors, and the sole unknown is the true surface current. But the boundary conditions between layers must be included in the Green's functions whose expressions are now by no means evident. In spite of this fact, this second approach has been found to be numerically efficient [5, 6], and we will deal exclusively with it throughout this chapter.

For stratified media problems, the spectral domain technique can be applied in two different ways:

1. The integral equation (15) is written and solved in the spectral domain. Consequently, we need to deal with the spectral transform $\mathbf{J}_s$ of the surface current. This approach is frequently termed spectral domain analysis (SDA) and is very useful for problems including conductors of simple shape, where the current can be expanded in a series of functions having analytic Fourier transforms. The reader is referred for this technique to Chapter 5.

2. The spectral domain is used only during the calculation of the Green's

functions, but inverse transforms are taken at this level and the integral equation is solved in the space domain. Therefore, no Fourier transforms are needed for the current. This alternative constitutes the topic of this chapter and is believed to be more flexible when dealing with odd conductor shapes while providing more physical insight into the problem.

It must be finally pointed out that both approaches differ only in the level at which the space domain is recovered, and that the results for a given problem would be identical if the algorithms used were free of numerical errors.

3.2. The Potentials

The six scalar components of the electromagnetic fields are not independent but are linked by Maxwell equations. It can be shown that in a source-free region, two scalar qualities suffice for determining the fields completely [7]. These quantities are called the potentials, and we will discuss in this section several possible choices.

3.2.1. Normal Components of the Fields

For a stratified medium it is possible to use E_z and H_z as potentials [1]. This is perhaps the conceptually simplest choice, since we do not need to introduce any new quantity. The normal components E_z, H_z satisfy homogeneous Helmholtz equations like (9) and (10) in a source-free region. The transverse components are given in the spectral domain by

$$k_\rho^2 E_x = jk_x \dot{E}_z - \omega\mu k_y H_z \qquad k_\rho^2 E_y = jk_y \dot{E}_z + \omega\mu k_x H_z \tag{17}$$

$$k_\rho^2 H_x = jk_x \dot{H}_z + \omega\varepsilon k_y E_z \qquad k_\rho^2 H_y = jk_y \dot{H}_z - \omega\varepsilon k_x E_z \tag{18}$$

Continuity of tangential components across the layers is satisfied if εE_z, $\dot{E}_z$, μH_z, and $\dot{H}_z$ are continuous. Hence, boundary conditions do not introduce coupled equations between E_z and H_z that can be calculated separately.

From the point of view of the Green's functions, we use as potentials the normal components

$$\mathbf{e}_z \cdot \bar{\bar{G}}_E = G_E^{zx}\mathbf{e}_x + G_E^{zy}\mathbf{e}_y + G_E^{zz}\mathbf{e}_z \tag{19a}$$

$$\mathbf{e}_z \cdot \bar{\bar{G}}_H = G_H^{zx}\mathbf{e}_x + G_H^{zy}\mathbf{e}_y \tag{19b}$$

Note that $G_H^{zz} = 0$ because a vertical source in a stratified medium does not produce a vertical magnetic field.

Now, the transverse components of $\bar{\bar{G}}_E$ and $\bar{\bar{G}}_H$ are obtained from the normal components by using eqs. (17) and (18) with E_s, H_s $(s = x, y)$ formally replaced by $\mathbf{e}_s \cdot \bar{\bar{G}}_E$, $\mathbf{e}_s \cdot \bar{\bar{G}}_H$.

Free-Space Values. The normal fields appearing in expressions (19a) and (19b) can be obtained by elementary techniques in the free space case. Taking the location of the source as origin of coordinates ($\mathbf{r}' = 0$), we have in the space domain.

$$G_H^{zx} = \frac{\partial \psi}{\partial y} \qquad G_H^{zy} = -\frac{\partial \psi}{\partial x} \qquad G_H^{zz} = 0 \tag{20}$$

$$j\omega\varepsilon G_E^{zx} = \frac{\partial}{\partial x}\left(\frac{\partial \psi}{\partial z}\right) \quad j\omega\varepsilon G_E^{zy} = \frac{\partial}{\partial y}\left(\frac{\partial \psi}{\partial z}\right) \quad j\omega\varepsilon G_E^{zz} = \left(k^2 + \frac{\partial^2}{\partial z^2}\right)\psi \tag{21}$$

where $\psi = \exp(-jkr)/4\pi r$.

Developing now expressions (20) and (21), it is an easy matter to show that $\mathbf{e}_z \cdot \bar{\bar{G}}_H$ and $\mathbf{e}_z \cdot \bar{\bar{G}}_E$ have singularities of type r^{-3} at the origin. We expect the same kind of singularities when free space is replaced by a layered medium.

For the sake of completeness, we also list the corresponding expressions in the spectral domain:

$$G_H^{zx} = jk_y E \qquad G_H^{zy} = -jk_x E \qquad G_H^{zz} = 0 \tag{22}$$

$$j\omega\varepsilon G_E^{zx} = jk_x \dot{E} \quad j\omega\varepsilon G_E^{zy} = jk_y \dot{E} \quad j\omega\varepsilon G_E^{zz} = k_\rho^2 E \tag{23}$$

with $E = \tilde{\psi} = \exp(-u_0|z|)/4\pi u_0$.

It must be also said that the square root defining u_0 [eq. (11)] must be chosen according to the condition $-\pi/2 < \arg(u_0) \leq \pi/2$, in order to satisfy the radiation condition and to avoid an exponential growth of the fields at infinity.

Symmetry Properties. The introduction of a layered medium does not remove the symmetry of revolution along the z axis. Therefore, the symmetry properties shown by the Green's functions in free space remain valid. In particular, we must have in the spectral domain

$$\frac{G_H^{zx}}{jk_y} = \frac{-G_H^{zy}}{jk_x} = f_H(k_\rho) \qquad \frac{G_E^{zx}}{jk_x} = \frac{G_E^{zy}}{jk_y} = f_E(k_\rho) \tag{24}$$

valid for any stratified medium with layers perpendicular to the z axis.

3.2.2. Vector Potentials

The first quantities used as potentials were introduced by H. Hertz and they are known now as the vector potential of the electric type $\mathbf{F}$ and the vector potential of the magnetic type $\mathbf{A}$.

Note. There is no universally accepted notation for the potentials. The original quantities introduced by Hertz were $\mathbf{\Pi}^e = \mathbf{A}/j\omega\mu\varepsilon$, $\mathbf{\Pi}^h = \mathbf{F}/j\omega\mu\varepsilon$ and they are still known as Hertz potentials of electric and magnetic type. Other authors (for instance, Harrington [7]) use $\mathbf{A}, \mathbf{F}$ to denote the quantities termed $\mathbf{A}/\mu$ and $\mathbf{F}/\varepsilon$ in our notation.]

These potentials satisfy the Helmholtz equations

$$(\nabla^2 + k^2)\mathbf{A} = -\mu\mathbf{J} \tag{25a}$$

$$(\nabla^2 + k^2)\mathbf{F} = -\varepsilon\mathbf{M} \tag{25b}$$

and the fields are given in the space domain by:

$$j\omega\mu\varepsilon\mathbf{E} = k^2\mathbf{A} + \nabla\cdot\nabla\mathbf{A} - j\omega\mu\nabla\times\mathbf{F} \tag{26}$$

$$j\omega\mu\varepsilon\mathbf{H} = k^2\mathbf{F} + \nabla\cdot\nabla\mathbf{F} + j\omega\varepsilon\nabla\times\mathbf{A} \tag{27}$$

The potentials $\mathbf{A}, \mathbf{F}$ must not be exclusively associated with, respectively, the currents $\mathbf{J}, \mathbf{M}$. For instance, in our problem $\mathbf{M} = 0$ and yet we can define a nonzero $\mathbf{F}$, solution of a homogeneous Helmholtz equation. Potentials are somewhat artificial quantities, and there is a great deal of arbitrariness in them. The associated dyadic Green's functions $\bar{\bar{G}}_A$ and $\bar{\bar{G}}_F$ are defined by

$$\mathbf{dA}(\mathbf{r}) = \bar{\bar{G}}_A(\mathbf{r}|\mathbf{r}')\cdot I(\mathbf{r}')\,\mathbf{dl}' \qquad \mathbf{dF}(\mathbf{r}) = \bar{\bar{G}}_F(\mathbf{r}|\mathbf{r}')\cdot I(\mathbf{r}')\,\mathbf{dl}' \tag{28}$$

For a given direction t of the source, we have to define six possible components of the dyadics related to the potentials, namely G_A^{st} and G_F^{st} ($s = x, y, z$). However, we know that two will suffice for determining the fields. We shall now review several possible choices for these components.

Hertz-Debye Potentials. The choice $\mathbf{A} = \mathbf{e}_z A_z$, $\mathbf{F} = \mathbf{e}_z F_z$ is traditionally associated with the name of Debye. With this simplification in the potentials, we can write for the fields in the spectral domain

$$j\omega\mu\varepsilon E_z = k_\rho^2 A_z \tag{29}$$

$$j\omega\mu\varepsilon H_z = k_\rho^2 F_z \tag{30}$$

It is clear now that in the spectral domain there are simple relationships between the potentials A_z, F_z and the normal fields E_z, H_z. As before, continuity of the transverse fields is guaranteed if εA_z, μF_z, $\dot{A}_z$, $\dot{F}_z$ are continuous quantities.

The fields of a horizontal electric dipole (HED) must be derived from the two potentials A_z, F_z, while A_z suffices for a vertical electric dipole (VED). Therefore, we have for the Green's functions

$$\bar{\bar{G}}_A = \mathbf{e}_z(G_A^{zx}\mathbf{e}_x + G_A^{zy}\mathbf{e}_y + G_A^{zz}\mathbf{e}_z) \tag{31}$$

$$\bar{\bar{G}}_F = \mathbf{e}_z(G_F^{zx}\mathbf{e}_x + G_F^{zy}\mathbf{e}_y + G_F^{zz}\mathbf{e}_z) \tag{32}$$

and these are related in the spectral domain to the field dyadics by the expressions

$$\mathbf{e}_z \cdot \bar{\bar{G}}_A = \frac{\mathbf{e}_z \cdot \bar{\bar{G}}_E}{k_\rho^2} \tag{33}$$

$$\mathbf{e}_z \cdot \bar{\bar{G}}_F = \frac{\mathbf{e}_z \cdot \bar{\bar{G}}_H}{k_\rho^2} \tag{34}$$

which show that there is no essential difference between using the Debye potentials or the normal fields in the spectral domain. However, no simple expressions exist for the dyadics $\bar{\bar{G}}_A$ and $\bar{\bar{G}}_F$ in the space domain, even in the free space case. For instance, according to the free space values for the normal fields [eqs. (22) and (23)] and using the spectral equivalences [eqs. (33) and (34)], we get

$$G_A^{zz} = \frac{\mu \exp(-jkr)}{4\pi r} \qquad G_F^{zz} = 0 \tag{35}$$

but the remaining components are given by inverse Hankel transforms with no analytical solution.

Sommerfeld Potentials. Since its introduction in 1909, Sommerfeld's choice for the potentials has become the most popular approach to solving stratified media problems [8].

In the absence of magnetic currents, Sommerfeld assumes $\mathbf{F} = 0$. A VED is described as before with one component A_z, while an HED needs in addition a second component parallel to the source. Hence, the dyadic $\bar{\bar{G}}_A$ is given in this approach by

$$\bar{\bar{G}}_A = (\mathbf{e}_x G_A^{xx} + \mathbf{e}_z G_A^{zx})\mathbf{e}_x + (\mathbf{e}_y G_A^{yy} + \mathbf{e}_z G_A^{zy})\mathbf{e}_y + \mathbf{e}_z G_A^{zz}\mathbf{e}_z \tag{36}$$

If the normal fields E_z, H_z are now calculated in the spectral domain according to the general formulas (26, 27) we obtain the following equivalent relationships

$$G_A^{xx} = -\frac{\mu G_H^{zx}}{jk_y} \qquad k_\rho^2 G_A^{zx} = j\omega\mu\varepsilon G_E^{zx} + \frac{k_x \mu \dot{G}_H^{zx}}{k_y} \tag{37a}$$

$$G_A^{yy} = \frac{\mu G_H^{zy}}{jk_x} \qquad k_\rho^2 G_A^{zy} = j\omega\mu\varepsilon G_E^{zy} - \frac{k_y \mu \dot{G}_H^{zy}}{k_x} \tag{37b}$$

$$k_\rho^2 G_A^{zz} = j\omega\mu\varepsilon G_E^{zz} \tag{37c}$$

Returning now to the fields as given by eqs. (26) and (27) it can be easily shown that, using Sommerfeld's choice, the continuity of the transverse fields implies that A_z and A_z/ε must be continuous for a VED. For a HED, say, in the x direction, we need continuity of A_x, $\dot{A}_x$, A_z, and $\nabla\cdot\mathbf{A}/\varepsilon$. This last condition couples the values of A_x and A_z, and therefore the corresponding Green's functions cannot be calculated independently.

It is worth mentioning that, according to the symmetry relations (24), we have here

$$G_A^{xx} = G_A^{yy} \tag{38}$$

and

$$\frac{G_A^{zx}}{jk_x} = \frac{G_A^{zy}}{jk_y} \tag{39}$$

and both expressions show an azimuthal symmetry. In particular in the free space case we have, in the space domain

$$\bar{\bar{G}}_A = \bar{\bar{I}}\mu_0 \exp(-jk_0 r)/4\pi r$$

and consequently $\bar{\bar{G}}_A$ exhibits only weak singularities of the r^{-1} type.

Transverse Potentials. A choice for the potentials first suggested by Erteza and Park [9] and recently fully developed for the two half-spaces problem [10], is $\bar{\bar{G}}_F = 0$ and

$$\bar{\bar{G}}_A = (\mathbf{e}_x G_A^{xx} + \mathbf{e}_y G_A^{yx})\mathbf{e}_x + (\mathbf{e}_x G_A^{xy} + \mathbf{e}_y G_A^{yy})\mathbf{e}_y + \mathbf{e}_z G_A^{zz}\mathbf{e}_z \tag{40}$$

The most salient feature of this choice is that transverse directed sources create only transverse potentials. On the other hand, for a VED the formulation is identical to that of Sommerfeld.

Here are the relationships with the normal fields.

$$jk_\rho^2 G_A^{xt} = j\omega\mu\varepsilon k_x \dot{G}_E^{zt}/u^2 - \mu k_y G_H^{zt}, \qquad t = x, y \tag{41a}$$

$$jk_\rho^2 G_A^{yt} = j\omega\mu\varepsilon k_y \dot{G}_E^{zt}/u^2 + \mu k_x G_H^{zt}, \qquad t = x, y \tag{41b}$$

$$k_\rho^2 G_A^{zz} = j\omega\mu\varepsilon G_E^{zz} \tag{41c}$$

In free space, $\bar{\bar{G}}_A$ reduces to a diagonal form identical to that obtained with the Sommerfeld potentials.

3.3. The Scalar Potential and the Mixed Potential Integral Equation

It is customary in electromagnetic theory to introduce a scalar potential V related to $\mathbf{A}$ by the Lorentz's gauge [11]

$$j\omega\mu\varepsilon V + \nabla \cdot \mathbf{A} = 0 \tag{42}$$

Then, in absence of an electric vector potential $\mathbf{F}$, eq. (26) for the electric field becomes

$$\mathbf{E} = -j\omega\mathbf{A} - \nabla V \tag{43}$$

Introducing in the Lorentz's gauge (42) the integral expression for $\mathbf{A}$ we have

$$j\omega\mu\varepsilon V(\mathbf{r}) = -\int_S [\nabla \cdot \bar{\bar{G}}_A(\mathbf{r}|\mathbf{r}')] \cdot \mathbf{J}_s(\mathbf{r}')\, ds' \tag{44}$$

Now, we know that in electrostatics the scalar potential depends only on the charge density q_s through a scalar Green's function G_V:

$$V(\mathbf{r}) = \int_S G_V(\mathbf{r}|\mathbf{r}')q_s(\mathbf{r}')\, ds' \tag{45}$$

In a nonstatic situation, there is also a surface charge on the surface of the conductors linked to the surface current $\mathbf{J}_s$ via the continuity equation

$$\nabla \cdot \mathbf{J}_s + j\omega q_s = 0 \tag{46}$$

The crucial point now is to determine whether definition (45) of the scalar potential can be extended to a layered medium under dynamic excitation. A thorough study of this problem in the general case of three-dimensional conductors embedded in a multilayered medium is beyond the scope of this presentation [12]. However, if the conducting surfaces are restricted to horizontal planes, a quick answer can be easily given. For then we need to consider only the transverse part of the vector $\nabla \cdot \bar{\bar{\mathbf{G}}}_A$. The expression $-\nabla \cdot \bar{\bar{G}}_A \cdot \mathbf{e}_t/j\omega\mu\varepsilon$ gives the scalar potential of a transverse t directed dipole and, physically we can think of it as created by two point charges $\pm 1/j\omega$ at both ends of the dipole. Hence, we can try to write this potential as

$$\frac{-\nabla \cdot \bar{\bar{G}}_A \cdot \mathbf{e}_t}{j\omega\mu\varepsilon} = \frac{\nabla' G_V \cdot \mathbf{e}_t}{j\omega} = \frac{-\nabla G_V \cdot \mathbf{e}_t}{j\omega} \tag{47}$$

Using the values of $\bar{\bar{G}}_A$ given either by (37) or (41) and applying the symmetry properties (20) and (21), we conclude that for transverse distributions of current and charge a scalar Green's function G_V exists and is given in the spectral domain by

$$G_V = \frac{j\omega}{k_\rho^2}\left(\frac{\dot{G}_E^{zx}}{jk_x}\right) - \left(\frac{k}{k_\rho}\right)^2\left(\frac{G_H^{zx}}{jk_y\,\varepsilon}\right) \tag{48}$$

in the Sommerfeld approach, or by

$$G_V = \frac{j\omega}{u^2}\left(\frac{\dot{G}_E^{zx}}{jk_x}\right) \tag{49}$$

if transverse potentials are used.

Note that the terms in parentheses in eqs. (48) and (49) depend only on k_ρ. Therefore, G_V will exhibit in the space domain a symmetry of revolution as could be expected from the scalar potential of a point charge. In particular, for the free-space case, (48) and (49) both reduce to the same expression, which in the space domain takes the familiar value $\exp(-jk_0 r)/4\pi\varepsilon r$.

Finally, the validity of (45) allows us to recast the original integral equation (15) as

$$\mathbf{e}_z \times \mathbf{E}^e = \mathbf{e}_z \times \left[j\omega \int_S \bar{\bar{G}}_A \cdot \mathbf{J}_s \, ds' + \nabla \int_S G_V q_s \, ds' + Z_s \mathbf{J}_s \right] \tag{50}$$

which is the mixed potential integral equation (MPIE).

Under this form, the integral equation for the currents leads to particularly efficient and accurate algorithms [13, 14]. In the coming sections, we will describe some of them as specialized to microstrip structures and related problems, but first we need a systematic procedure for deriving the Green's functions in a multilayered medium.

4. THE FIELDS OF AN ELEMENTARY SOURCE IN A LAYERED MEDIUM

In this section, we calculate in the spectral domain the normal fields created by a unit Hertz dipole embedded in a layered medium. If the dipole is directed along the t axis ($t = x, y, z$), these fields give directly the components G_E^{zt} and G_H^{zt} of the Green's functions. Once these components are known, the scalar and vector potentials and their associated Green's functions can be found by a straightforward application of eqs. (37) and (48) in the Sommerfeld approach or of eqs. (41) and (49) if transverse potentials are used.

A direct computation of the potentials is possible, but the calculation of normal fields is easier and more systematic for multilayered cases, because the boundary conditions have a simpler formulation and free-space solutions are well known. Moreover, normal fields provide the best way of formulating an integral equation when the currents include components perpendicular to the layers.

4.1. Matrix Formulation

4.1.1. Statement of the Problem

Let us consider the problem of Fig. 2, where the source is located inside the ith layer and we want the fields in the jth layer.

We shall follow here essentially the formulation of Kong [1], with several modifications aimed at simplifying the calculations in our particular problem.

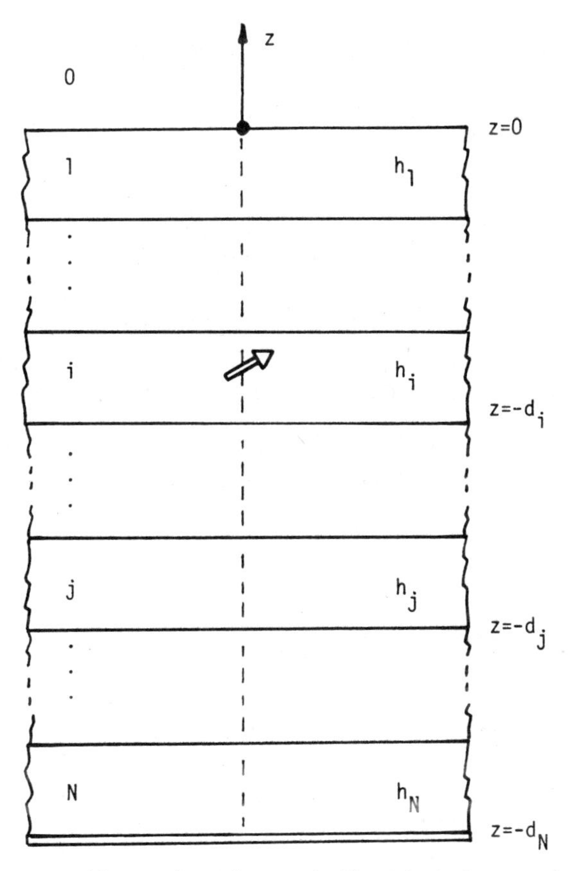

Fig. 2 The same multilayered medium as in Fig. 1 including a point source (Hertz dipole) in the ith layer.

We begin by considering the problem of calculating a function ψ satisfying the Helmholtz equation in each layer and subject to the following boundary conditions at the interface between layers i and $i+1$:

$$\alpha_i \psi_i = \alpha_{i+1} \psi_{i+1} \tag{51}$$

$$\dot{\psi}_i = \dot{\psi}_{i+1} \tag{52}$$

It is readily apparent from the previous section that ψ can represent E_z (with $\alpha_i = \varepsilon_i$) or H_z (with $\alpha_i = \mu_i$).

The thickness of the ith layer is denoted h_i, and the interface between layers i and $i+1$ is the plane $z = -d_i$, with $d_0 = 0$. Therefore, we can infer from Fig. 2 the geometrical equivalences:

$$h_i = d_i - d_{i-1} \quad \text{and} \quad d_i = \sum_{j=1}^{i} h_j \tag{53}$$

4.1.2. Two Contiguous Layers without Sources

To simplify the calculations, let us introduce local coordinates attached to each layer and defined by (Fig. 3):

$$z_i = z + d_i \tag{54}$$

Then, in absence of sources, we can write for each layer:

$$\psi_k = a_k \cosh u_k z_k + b_k \sinh u_k z_k \tag{55}$$

Notice that we prefer to formulate ψ in terms of hyperbolic functions rather than using the traditional combination of exponentials. Hyperbolic functions

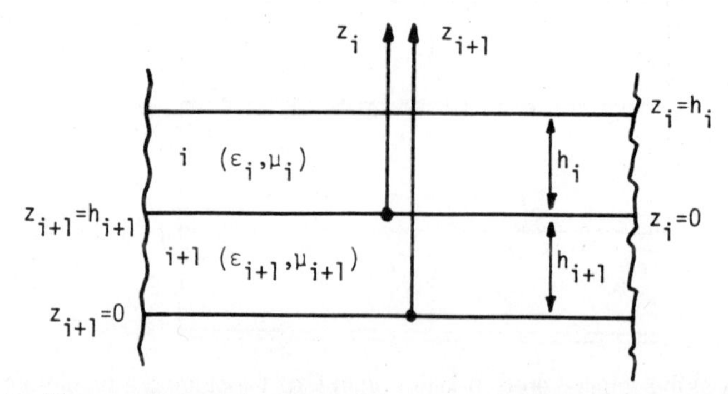

Fig. 3 Two contiguous layers showing the local coordinates employed.

lead in our problem to a more compact presentation of results, ready for computer evaluation.

Application of boundary conditions (51) and (52) leads now to the matrix equation

$$\mathbf{v}_i = \boldsymbol{T}_{i,i+1}\mathbf{v}_{i+1} \tag{56}$$

In this equation, $\mathbf{v}_i$ is a column vector containing the two amplitudes a_i and b_i of the hyperbolic terms and $\boldsymbol{T}_{i,i+1}$ is a transmission matrix between layers i and $i+1$, given by

$$\boldsymbol{T}_{i,i+1} = \begin{pmatrix} \alpha_{i+1}c_{i+1}/\alpha_i & \alpha_{i+1}s_{i+1}/\alpha_i \\ u_{i+1}s_{i+1}/u_i & u_{i+1}c_{i+1}/u_i \end{pmatrix} \tag{57}$$

where c_i and s_i stand, respectively, for $\cosh u_ih_i$ and $\sinh u_ih_i$. The matrix $\boldsymbol{T}_{i+1,i}$ relating layers in the "backwards" sense is given by the inverse of $\boldsymbol{T}_{i,i+1}$, namely:

$$\boldsymbol{T}_{i+1,i} = (\boldsymbol{T}_{i,i+1})^{-1} = \begin{pmatrix} \alpha_ic_{i+1}/\alpha_{i+1} & -u_is_{i+1}/u_{i+1} \\ -\alpha_is_{i+1}/\alpha_{i+1} & u_ic_{i+1}/u_{i+1} \end{pmatrix} \tag{58}$$

4.1.3. One Layer with the Source Inside

Let us consider now the ith layer with a point source (depicted in Fig. 4 as a HED) at the level $z_i = D$. This source splits the layer into two sublayers above and below it.

Let ψ^∞ be the solution corresponding to the same source imbedded in an unbounded homogeneous medium of the same characteristics as the ith layer. In the spectral domain, we can always put the solution into the form

$$\psi_i^\infty = \begin{cases} U_i \exp[-u_i(z_i - D)] \,, & D \lesssim z_i \lesssim h_i \\ L_i \exp[+u_i(z_i - D)] \,, & 0 \lesssim z_i \lesssim D \end{cases} \tag{59}$$

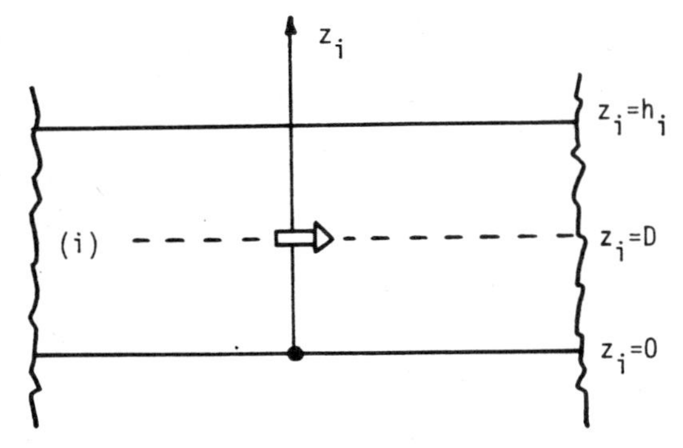

Fig. 4 The ith layer with an embedded horizontal electric dipole.

Table 1 Values of the Coefficients U_i and L_i Associated with the Upper and Lower Parts of the Layer Containing the Source

	G_H^{zx}	G_H^{zy}	G_E^{zx}	G_E^{zy}	G_E^{zz}
$U_i =$	$-jk_y/4\pi u_0$	$jk_x/4\pi u_0$	$-jk_x/4\pi j\omega\varepsilon$	$-jk_y/4\pi j\omega\varepsilon$	$k_\rho^2/4\pi j\omega\varepsilon u_0$
$L_i =$	U_i	U_i	$-U_i$	$-U_i$	U_i

where the amplitude coefficients in the upper and lower sublayers U_i and L_i depend on the physical quantity represented by ψ (see Table 1).

We can now write for the finite thickness layer with the source inside:

$$\psi_i = \psi_i^\infty + a_i \cosh u_i z_i + b_i \sinh u_i z_i \tag{60}$$

This expression can be recast in our standard form:

$$\psi_i^U = a_i^U \cosh u_i(z_i - D) + b_i^U \sinh u_i(z_i - D) \tag{61}$$

$$\psi_i^L = a_i^L \cosh u_i z_i + b_i^L \sinh u_i z_i \tag{62}$$

and we obtain the matrix relationship

$$2\mathbf{v}_i^U = \begin{pmatrix} E + 1/E & E - 1/E \\ E - 1/E & E + 1/E \end{pmatrix} \mathbf{v}_i^L + 2\mathbf{s}_i \tag{63}$$

where $E = \exp(u_i D)$, and the "source vector" $\mathbf{s}_i$ is given by

$$\mathbf{s}_i = \begin{bmatrix} -L_i + U_i \\ -L_i - U_i \end{bmatrix}$$

Of particular interest is the case where the source resides in the lower boundary of the layer ($D = 0$). In this case, eq. (63) simplifies to

$$\mathbf{v}_L^U = \mathbf{v}_i^L + \mathbf{s}_i \tag{64}$$

Keeping in mind that the amplitudes in the lower sublayer are related to those of the $i + 1$ layer by the transmission matrix $\boldsymbol{T}_{i,i+1}$, we get the sought-for relationship between two layers i and $i + 1$ of different electromagnetic properties with a point source in their common boundary:

$$\mathbf{v}_i = \boldsymbol{T}_{i,i+1} \mathbf{v}_{i+1} + \mathbf{s}_i \tag{65}$$

4.1.4. *The Upper Half-Space*

In the upper semi-infinite medium extending above the sources, we must have for the function ψ a dependence of the type

$$\psi = a_0 \exp(-u_0 z) \tag{66}$$

with the condition $-\pi/2 < \arg(u_0) \leq \pi/2$ in order to satisfy the radiation condition at infinity. Hence if we use hyperbolic functions as in eq. (55), the coefficients are linked by the simple relationship $b_0 = -a_0$.

4.1.5. *The Terminal Layer*

The Nth layer is backed by an impedance wall. Therefore we have $E_y = Z_s H_x$ and $E_x = -Z_s H_y$, or in terms of normal fields:

$$\omega\mu H_z = jZ_s \dot{H}_z \qquad \text{and} \qquad \omega\varepsilon Z_s E_z = j\dot{E}_z \tag{67}$$

This means that, for normal electric fields, the coefficients a_N, b_N satisfy the boundary condition

$$b_N = (\omega\varepsilon_N Z_s / ju_N) a_N = \eta_E a_N \tag{68}$$

while for normal magnetic fields we have

$$a_N = (Z_s u_N / j\omega\mu_N) b_N = \eta_H b_N \tag{69}$$

In particular, for an electric wall we simply obtain $\eta_E = \eta_H = 0$.

4.2. Generalization to *N* Layers

Going back to the general problem of Fig. 2, we can now use recursively the equations (57) and (65) and obtain the general matrix relation:

$$\mathbf{v}_0 = \boldsymbol{T}_{0,1}\boldsymbol{T}_{1,2} \cdots \boldsymbol{T}_{i-1,i}(\mathbf{s}_i + \boldsymbol{T}_{i,i+1}\boldsymbol{T}_{i+1,i+2} \cdots \boldsymbol{T}_{N-1,N})\mathbf{v}_N \tag{70}$$

or in a reduced notation

$$\mathbf{v}_0 = \mathbf{e}_i + \boldsymbol{T}\mathbf{v}_N \tag{71}$$

where $\boldsymbol{T}$ is the total transmission matrix,

$$\boldsymbol{T} = \prod_{i=0}^{N} \boldsymbol{T}_{i,i+1}$$

and the total excitation vector is

$$\mathbf{e}_i = \left(\prod_{k=1}^{i-1} \boldsymbol{T}_{k,k+1} \right) \mathbf{s}_i$$

Recalling now the relation between the amplitudes a_i, b_i included in the vector $\mathbf{v}_i$, for $i = 0$ (upper half-space) and $i = N$ (terminal layer), we can solve eq. (71) and obtain the amplitudes in the regions $i = 0$ and $i = N$. For instance, we obtain for the normal magnetic field, according to Table 1:

$$\begin{aligned} b_N &= -\frac{e_1 + e_2}{t_{12} + t_{22} + \eta_H(t_{11} + t_{21})} \\ a_0 &= e_1 + (t_{12} + \eta_H t_{11}) b_N \end{aligned} \tag{72}$$

and for the normal electric field:

$$\begin{aligned} a_N &= -\frac{e_1 + e_2}{t_{11} + t_{21} + \eta_E(t_{12} + t_{22})} \\ a_0 &= e_1 + (t_{11} + \eta_E t_{12}) a_N \end{aligned} \tag{73}$$

where the terms t_{ij} are the elements of the total chain matrix $\boldsymbol{T}$.

If we are now interested in the amplitudes in the jth layer, we have for $j > i$:

$$\mathbf{v}_j = \left(\prod_{k=j+1}^{N} \boldsymbol{T}_{k-1,k} \right) \mathbf{v}_N \tag{74}$$

and for $j \leq i$:

$$\mathbf{v}_j = \left(\prod_{k=j}^{i} \boldsymbol{T}_{k-1,k} \right) \mathbf{s}_i + \left(\prod_{k=j+1}^{N} \boldsymbol{T}_{k-1,k} \right) \mathbf{v}_N \tag{75}$$

Alternative expressions can be easily constructed with the backwards inverse matrices $\boldsymbol{T}_{i+1,i}$.

5. TWO PRACTICAL SUBSTRATES

In this section, we give explicit formulas for the fields and potentials for one- or two-layer substrates backed by perfect conducting planes. In both cases, we shall assume that the layers are nonmagnetic lossy dielectrics. Hence, we have for the ith layer

$$\mu_i = \mu_0 \qquad \text{and} \qquad \varepsilon_i = \varepsilon_0 \varepsilon'_{ri}(1 - j \tan \delta_i)$$

5.1. Single-Layer Microstrip

For a single layer substrate, the matrix $\boldsymbol{T}$ in eq. (71) reduces simply to $\boldsymbol{T}_{0,1}$. Normal fields can then be obtained by application of eqs. (72) and (73) and the potentials derived from eqs. (37) and (48). As a useful example, Table 2

Table 2 Spectral Domain Green's Functions for a Single-Layer Microstrip When Source and Observer Are Both at the Air–Dielectric Interface

Transverse Potentials	Sommerfeld Potentials
$\frac{2\pi G_A^{xx}}{\mu_0} = \frac{(k_x/k_\rho)^2(u_0 \tanh uh)}{uD_{TM}} + \frac{(k_y/k_\rho)^2}{D_{TE}}$	$\frac{2\pi G_A^{xx}}{\mu_0} = \frac{1}{D_{TE}}$
$\frac{2\pi G_A^{yx}}{\mu_0} = \frac{k_x k_y}{k_\rho^2}\left(\frac{u_0 \tanh uh}{uD_{TM}} - \frac{1}{D_{TE}}\right)$	$\frac{2\pi G_A^{zx}}{\mu_0} = \frac{jk_x(\varepsilon_r - 1)}{D_{TE}D_{TM}}$
$2\pi\varepsilon_0 G_V = \frac{u_0 \tanh uh}{uD_{TM}}$	$2\pi\varepsilon_0 G_V = \frac{u_0 + u \tanh uh}{D_{TE}D_{TM}}$

$$D_{TE} = u_0 + u \coth uh\,, \quad D_{TM} = \varepsilon_r u_0 + u \tanh uh$$

gives the spectral domain expressions of the relevant Green's functions for the potentials when source and observer are both in the air–dielectric interface.

Components G_A^{sy} of the dyadic Green's functions are obtained by taking into account the symmetry of revolution around the z axis.

Integral expressions in the space domain for the fields and potentials are obtained by taking the inverse transforms of the spectral values compiled in Table 2. It must be pointed out that inverse transforms as given by eqs. (4) and (7) assume a source located at the origin of coordinates. General expressions for the Green's functions, directly suitable for the integral equation formulation of the problem, are obtained by using the translational symmetry of any Green's function in the transverse directions

$$G(x, y \,|\, x', y') = G(x - x', y - y' \,|\, 0, 0)$$

Let us recall that if Sommerfeld potentials are used, we need only the component A_x and the scalar potential V for determining the transverse electric field of an x-directed HED. On the other hand, we need A_x, A_y, and V if transverse potentials are used. Moreover, expressions for the vector potential components are more complicated in the case of transverse potentials (Table 2). Consequently, we will follow from now on the Sommerfeld's traditional choice, even if transverse potentials have some advantages, particularly regarding the scalar potential V [10].

5.2. Double-Layer Microstrip

For a two-layer substrate there are two interesting locations for the source:

1. Interface between air and upper layer
2. Interface between upper and lower layer

In both cases, the expressions for the spectral Green's functions giving the normal components of the fields created by an x-directed HED are given by:

$$G_H^{zx} = \frac{C_H}{D_{TE}} \begin{Bmatrix} a_0 \exp(-u_0 z) \\ a_1 \cosh u_1(z+h_1) + b_1 \sinh u_1(z+h_1) \\ b_2 \sinh u_2(z+h_1+h_2) \end{Bmatrix} \tag{76}$$

$$G_E^{zx} = \frac{C_E}{D_{TM}} \begin{Bmatrix} a_0 \exp(-u_0 z) \\ a_1 \cosh u_1(z+h_1) + b_1 \sinh u_1(z+h_1) \\ a_2 \cosh u_2(z+h_1+h_2) \end{Bmatrix} \tag{77}$$

with the three expressions inside brackets applying, respectively, to observers in the air, the upper layer, and the lower layer. In the above formulas, the amplitude coefficients are given by $C_H = -jk_y/2\pi$ and $C_E = -k_x/2\pi\omega\varepsilon_0$ and the denominators D_{TE} and D_{TM} by the expressions:

$$\begin{aligned} D_{\mathrm{TE}} &= u_0 c_1 s_2 + u_0 u_2 s_1 c_2/u_1 + u_1 s_1 s_2 + u_2 c_1 c_2 \\ D_{\mathrm{TM}} &= \varepsilon_2 u_0 c_1 c_2/\varepsilon_0 + \varepsilon_1 u_0 u_2 s_1 s_2/\varepsilon_0 u_1 + \varepsilon_2 u_1 s_1 c_2/\varepsilon_1 + u_2 c_1 s_2 \end{aligned} \tag{78}$$

with $c_i = \cosh u_i h_i$ and $s_i = \sinh u_i h_i$. The parameters a_0, a_1, b_1, a_2, b_2 depend on the position of the source and are given in Table 3 for the two locations mentioned above.

As for the single-layer case, Green's functions for the potentials are obtained by application of eqs. (37)–(48).

It is clear from eqs. (76)–(78) and from Table 3 that the complexity of

Table 3 Parameters Appearing in the Green's Functions for the Normal Fields in a Double-Layer Substrate

	G_H^{zx}		G_E^{zx}	
	$z'=0$	$z'=-h_1$	$z'=0$	$z'=-h_1$
a_0	$c_1 s_2 + \frac{u_2 s_1 c_2}{u_1}$	s_2	$\frac{\varepsilon_2}{\varepsilon_1} u_1 s_1 c_2 + u_2 c_1 s_2$	$u_2 s_2$
a_1	s_2	$\frac{u_0 s_1 s_2}{u_1} + c_1 s_2$	$-\frac{\varepsilon_2 u_0 c_2}{\varepsilon_1}$	$u_2 s_2\left(\frac{\varepsilon_0 c_1}{\varepsilon_1} + \frac{u_0 s_1}{u_1}\right)$
b_1	$u_2 c_2/u_1$	$-\frac{u_0 c_1 s_2}{u_1} - s_1 s_2$	$-u_0 u_2 s_2/u_1$	$-u_2 s_2\left(\frac{\varepsilon_0 s_1}{\varepsilon_1} + \frac{u_0 c_1}{u_1}\right)$
a_2	0	0	$-u_0$	$-\left(u_0 c_1 + \frac{\varepsilon_0 u_1 s_1}{\varepsilon_1}\right)$
b_2	1	$c_1 + \frac{u_0 s_1}{u_1}$	0	0

the mathematical expressions grows very quickly with the number of layers. Fortunately enough, we do not need in practice to know the explicit formulas for the fields and the potentials, if we are mainly interested in evaluating them numerically. The computer can construct automatically the matrices $\boldsymbol{T}_{i,i+1}$ from the physical data and perform directly the required matrix calculations.

6. THE SURFACE WAVES

In the spectral domain, the fields and potentials are given by quotients of expressions including transcendental functions (see Tables 2 and 3). If for a given value of the spectral variables, the denominators in these expressions vanish, a pole appears in the functions to be integrated for obtaining the corresponding quantity in the space domain. These poles are physically understood as surface waves whose features are directly linked to the nature of the layers. In this section, we study the surface waves that can exist in some particular cases of stratified media.

6.1. Single-Layer Microstrip

For a microstrip structure including only one dielectric layer, the potentials of an x-directed HED, located at $\boldsymbol{\rho}' = \mathbf{0}$, are given in the space domain by

$$G_A^{xx}(\boldsymbol{\rho}\,|\,\mathbf{0}) \equiv A_x(\boldsymbol{\rho}) = \frac{\mu_0}{2\pi} \int_0^\infty dk_\rho\; J_0(k_\rho \rho)\, \frac{k_\rho}{D_{\mathrm{TE}}} \tag{79}$$

$$G_V(\boldsymbol{\rho}\,|\,\mathbf{0}) \equiv V(\boldsymbol{\rho}) = \frac{1}{2\pi\varepsilon_0} \int_0^\infty dk_\rho\; J_0(k_\rho \rho) k_\rho\, \frac{u_0 + u \tanh uh}{D_{\mathrm{TE}} D_{\mathrm{TM}}} \tag{80}$$

where the source and observer are both at the air–dielectric interface and we have assumed, for the sake of simplicity, a perfect ground plane (see Table 2). We will not consider the component G_A^{zx}, since it does not appear in the formulation of the integral equation for a transverse current distribution. It can be shown [7] that the equations $D_{\mathrm{TE}} = 0$ and $D_{\mathrm{TM}} = 0$ are the characteristic equations for the surface waves of, respectively, TE and TM type propagating in a dielectric layer backed by a perfect conductor. Hence, the zeros of D_{TE} and D_{TM} give the phase constant of the surface waves existing in a microstrip structure.

The surface waves appear as poles of the integrands in the complex plane $k_\rho = \lambda + j\nu$ (see Fig. 5). It can be shown [15] that if $k_0 h(\varepsilon_r' - 1)^{1/2} < \pi/2$ then D_{TE} has no zeros and D_{TM} has only one corresponding to the dominant zero-cutoff TM surface wave. This condition is equivalent to the restriction:

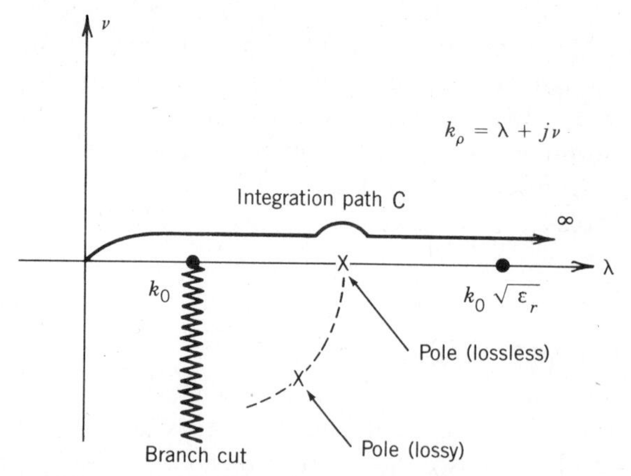

Fig. 5 Topology of the complex plane k_ρ with the original integration path from zero to infinity. The figure also shows the geometrical locus of the pole as a function of dielectric losses.

$$f\,[\mathrm{GHz}] \leq \frac{75}{h\,[\mathrm{mm}]\sqrt{\varepsilon_r' - 1}} \tag{81}$$

which gives for a practical case (alumina 1/40 in. thick and $\varepsilon_r' = 9.6$) $f < 40$ GHz.

For the sake of simplicity we shall assume from now on that this inequality holds. Higher frequencies would add new poles but the analysis made in the single pole case remains qualitatively valid.

For lossless substrates, it can be shown [7] that the pole is real ($k_\rho = \lambda_{p0}$) and lies inside the segment of the real axis $1 < \lambda_{p0}/k_0 < \sqrt{\varepsilon_r}$. If the substrate is electrically thin, the pole is expected to be near the branch point $k_\rho/k_0 = 1$. An estimation of its location can then be obtained by assuming $\lambda_{p0}/k_0 = 1 + \delta$. With this substitution the equation $D_{\mathrm{TM}} = 0$ becomes

$$\varepsilon_r\sqrt{2\delta + \delta^2} - \sum_{n=0}^{\infty} \alpha_n \delta^n = 0 \tag{82}$$

where the infinite sum is the Taylor's series for

$$\sqrt{\varepsilon_r - z^2}\,\tan(k_0 h\sqrt{\varepsilon_r - z^2}) \tag{83}$$

around the point $z = k_\rho/k_0 = 1$. Keeping only the dominant terms in eq. (82), we get the sought-for expression:

$$\lambda_{p0}/k_0 \simeq 1 + (k_0 h)^2\,\frac{(\varepsilon_r - 1)^2}{2\varepsilon_r^2} \tag{84}$$

Now, if the substrate has moderate losses the pole will migrate slightly below the lossless case location λ_{p0} and is given by the complex value $\lambda_p - j\nu_p$ with $\nu_p > 0$.

To obtain an estimation of the pole location in a lossy case, we consider D_{TM} as a complex function of two complex variables z and ε_r, and we again perform a Taylor's expansion around the point $z_0 = \lambda_{p0}/k_0$, $\varepsilon_r = \varepsilon_r'$. Keeping only the dominant terms, we obtain:

$$D_{\mathrm{TM}}(z, \varepsilon_r) = D_{\mathrm{TM}}(z_0, \varepsilon_r') + (z - z_0)A + (\varepsilon_r - \varepsilon_r')B + \cdots \tag{85}$$

Here A and B are the partial derivatives of D_{TM} with respect to, respectively, z and ε_r calculated at the point $z_0 = \lambda_{p0}/k_0$, $\varepsilon_r = \varepsilon_r'$. Replacing now the variable z by the complex value of the pole $\lambda_p - j\nu_p$, we get the final result

$$\begin{aligned} \lambda_p &\simeq \lambda_{p0} \\ \nu_p &\simeq (\varepsilon_r' - 1) \tan \delta \left(\frac{k_0 h}{\varepsilon_r'} \right)^2 \end{aligned} \tag{86}$$

which shows that, to the first order, the real part of the pole is independent of losses, while the imaginary part is proportional to the loss tangent.

The integration path for the Sommerfeld integrals (79) and (80) is, in general, the real positive axis λ. But, if we consider the theoretical case of a lossless substrate, then $\nu_p = 0$ and the pole is on the real axis. Since, by continuity, the integration path must remain above the pole, the integral from zero to infinity in the lossless case is interpreted as (Fig. 5)

$$\int_0^\infty = PV \int_0^\infty - j\pi R \tag{87}$$

where PV stands for Cauchy's principal value and R is the residue of the integrand at the pole.

Finally, it is worth mentioning that the function $u_0^2 = k_\rho^2 - k_0^2$ introduces a branch point at $k_\rho = k_0$ (Fig. 5). However, the integral remains bounded here, and no deformation of the integration path around the branch point is needed, provided that the correct Riemann sheet defined by $\mathrm{Re}(u_0) \geq 0$ is used.

Numerical computations of the first TM pole have been carried out by using a modified form of the Newton–Raphson algorithm for complex quantities. If approximation (86) is used as the initial point, this algorithm leads to very good precision (typically better than 10^{-8}) in a few iterations. In some isolated cases, mainly for thick substrates and high losses, the Newton–Raphson algorithm fails to converge. As an alternative, we then define a new variable $w = u_0/k_0$ and rewrite the equation $D_{\mathrm{TM}} = 0$ as

$$\varepsilon_r w = \sqrt{\varepsilon_r - 1 - w^2} \tan(k_0 h \sqrt{\varepsilon_r - 1 - w^2}) \tag{88}$$

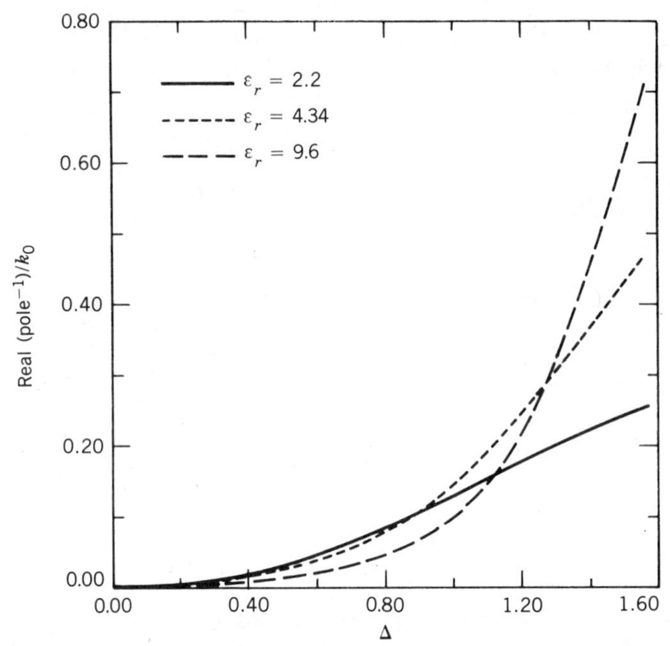

Fig. 6 Normalized real part $(\lambda_p - 1)/k_0$ of the pole associated with the first TM surface wave in a single-layer microstrip structure as a function of the normalized thickness, $\Delta = k_0 h\sqrt{\varepsilon_r - 1}$). (···) $\varepsilon_r = 2.2$; (– – –) $\varepsilon_r = 4.34$; (——) $\varepsilon_r = 9.6$.

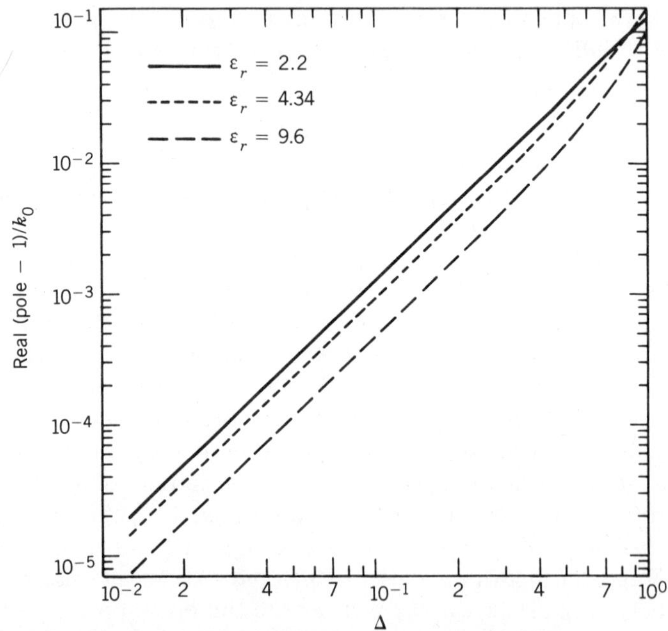

Fig. 7 Logarithmic plot of Fig. 6 stressing the proportionality of $\lambda_p - 1$ with the square of the normalized thickness for thin substrates. (···) $\varepsilon_r = 2.2$; (– – –) $\varepsilon_r = 4.34$; (——) $\varepsilon_r = 9.6$.

which is solved iteratively starting with $w = 0$ as the initial guess in the right-hand side. Figure 6 gives the normalized location of the pole $(\lambda_{p0} - 1)/k_0$ for a lossless substrate as a function of the normalized substrate thickness

$$\Delta = k_0 h \sqrt{\varepsilon_r - 1} \tag{89}$$

and for various practical dielectric constants. The maximum value considered for Δ is $\pi/2$, to ensure that only one surface wave exists [eq. 81]. If these results are plotted in a logarithmic scale (Fig. 7), the quadratic behavior of $(\lambda_{p0}/k_0) - 1$ with Δ is readily evident, particularly for thin substrates.

A logarithmic scale has been also used to show the imaginary part of the pole in a lossy case with the loss tangent as parameter (Fig. 8). The results confirm the validity of our approximation for thin substrates, which predicts a linear behavior of ν_p with Δ and with the loss tangent.

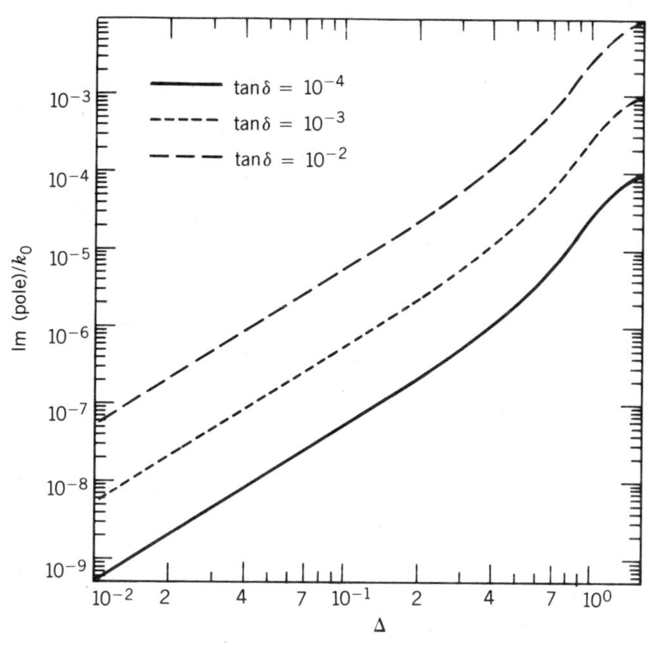

Fig. 8 Normalized imaginary part of the first TM surface wave pole for a lossy single-layer substrate $\varepsilon_r = 4.34$ as a function of the normalized thickness, $\Delta = k_0 h \sqrt{\varepsilon_r - 1}$. The logarithmic scale enhances the linear dependence of the imaginary part with Δ and with $\tan \delta$. $(\cdots)$ $\tan \delta = 10^{-4}$; $(---)$ $\tan \delta = 10^{-3}$; $(——)$ $\tan \delta = 10^{-2}$.

Finite Conductivity Ground Plane. The ohmic losses in the ground plane have the same effect on the pole location as the dielectric losses, tending to move the pole away from the real axis. However, in practical situations, dielectric losses are usually more relevant, and ohmic losses in the ground plane can be taken into account by artifically increasing the loss tangent of the substrate.

6.2. Double-Layer Microstrip

For a substrate with two different layers, surface waves are given by the zeros of the more complicated denominators D_{TE} and D_{TM} given in eq. (78) of Section 5.2. Again, we can guarantee the existence of only one surface wave of TM type and no surface waves of TE type if

$$k_0(h_1 + h_2)\sqrt{\max(\varepsilon'_{r_1}, \varepsilon'_{r2}) - 1} < \pi/2 \tag{90}$$

An approximate analytical estimation for a lossless structure is obtained following the techniques outlined in the single-layer case. For instance, if both layers are thin and lossless, we have

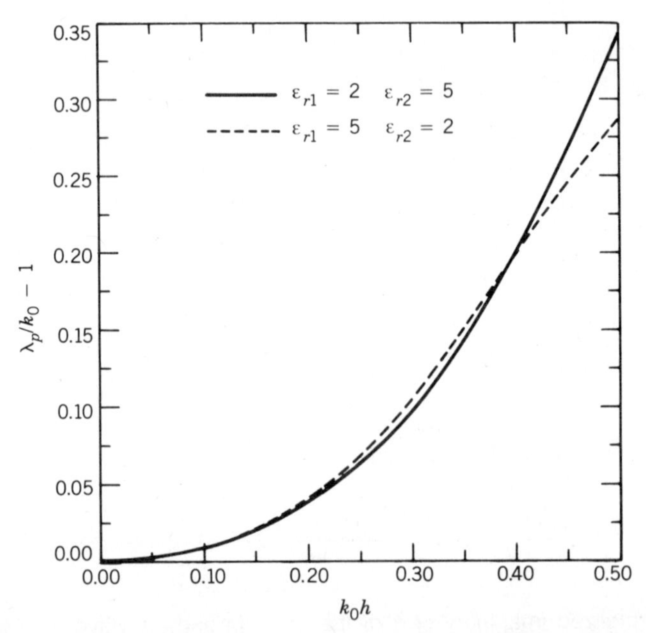

Fig. 9 First surface wave pole for a lossless two-layer medium as a function of the normalized thickness k_0h. Both layers have the same thickness. Solid line, $\varepsilon_{r1} = 2$, $\varepsilon_{r2} = 5$; dotted line, $\varepsilon_{r1} = 5$, $\varepsilon_{r2} = 2$.

$$\frac{\lambda_{p0}}{k_0} = 1 + \frac{1}{2}\left(\frac{k_0 h_1 \varepsilon_{r2}(\varepsilon_{r1} - 1) + k_0 h_2 \varepsilon_{r1}(\varepsilon_{r2} - 1)}{\varepsilon_{r1}\varepsilon_{r2}}\right)^2 \tag{91}$$

and this value can be improved by numerical techniques. Figure 9 gives the results obtained for two layers of identical thickness and dielectric constants

$$\varepsilon_{r1} = 2\,, \quad \varepsilon_{r2} = 5 \qquad \text{and} \qquad \varepsilon_{r1} = 5\,, \quad \varepsilon_{r2} = 2$$

We remark that for the same pair of dielectric constants the results are slightly asymmetrical, the pole being closer to the branch point if the lower permittivity layer is above the higher-permittivity layer, a situation commonly encountered in practice.

This asymmetry is barely visible for electrically thin substrates as evidenced by approximation (91), which is symmetrical in ε_{r1}, ε_{r2} if both layers have the same thickness. In the more general case of an N-layer substrate the surface waves of TE and TM types are associated, respectively, to the zeros of the denominators of b_N [eq. (72)] and a_N [eq. (73)].

7. ZERO, LOW, AND HIGH FREQUENCIES

The essential elements of the model proposed in this chapter are the integral equation and the Green's functions forming its kernel. In this section, we introduce several simplifications for both elements corresponding to some particular values of the frequency.

7.1. Limiting Cases of the Integral Equation

Figure 10*a* shows a microstrip structure with two access ports. The excitation fields are here produced by an ac generator connected to the input port, while the output port is loaded by an arbitrary impedance. Surface currents and charges exist in the upper conductor and from them the port impedance matrix can be determined.

From a circuit point of view, two particular cases deserve consideration. In the first one (Fig. 10*c*) the generator is a dc battery and the load is an open circuit. No current flow exists, and the sole unknown is the charge density, whose determination allows the computation of the capacitance of the microstrip structure. In the second one, we have a low-frequency current generator at the input port and a short circuit at the output port (Fig. 10*b*). A divergenceless surface current flows through the closed circuit. There is no surface charge and $I_1 = -I_2$. From the surface current, an inductance associated to the microstrip structure can be determined.

Both cases are included in the mixed potential integral equation (MPIE) and give rise, respectively, to static and quasistatic specializations of the MPIE.

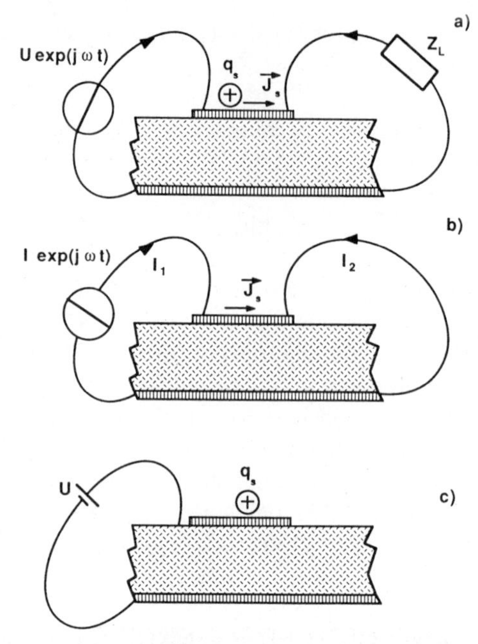

Fig. 10 Three possible excitations of a microstrip circuit: (*a*) dynamic (time harmonic); (*b*) quasi-static; (*c*) static. After [39] © 1988 IEEE.

7.1.1. *The Static Case*

In the absence of currents, eq. (50) becomes

$$\mathbf{e}_z \times \nabla \int_s ds' \, G_V(\boldsymbol{\rho}|\boldsymbol{\rho}')q_s(\boldsymbol{\rho}') = \mathbf{e}_z \times \mathbf{E}^{(e)}(\boldsymbol{\rho}) \tag{92}$$

In many practical situations, it is customary to assume that the excitation field is created by some charge distribution $q_s^{(e)}$ via the same Green's function. Then, eq. (92) is rewritten as

$$\mathbf{e}_z \times \nabla \int_s ds' \, G_V(\boldsymbol{\rho}|\boldsymbol{\rho}')[q_s^{(e)}(\boldsymbol{\rho}') + q_s(\boldsymbol{\rho}')] = 0 \tag{93}$$

which implies by integration over the tangential coordinates that

$$\int_s ds' \, G_V(\boldsymbol{\rho}|\boldsymbol{\rho}')[q_s^{(e)}(\boldsymbol{\rho}') + q_s(\boldsymbol{\rho}')] = \text{constant} = U \tag{94}$$

Instead of starting with an excitation charge, solving (93) for the "scattered charge" q_s and finally computing the voltage U with integral (94), it will be frequently easier to start by assuming the voltage U known and considering (94) as an integral equation for the total charge $q_s^{(e)} + q_s$. This last approach follows closely the circuit representation of Fig. 10*e* and corresponds to the well-known static integral equation for the evaluation of capacitances. The Green's function to be used in eq. (94) can be found by setting $k_0 = 0$ in the corresponding dynamic expression. For instance, in the case of a single layer, eq. (80) becomes with $\lambda = \text{Re}\,[k_\rho]$:

$$2\pi\varepsilon_0 G_V(\boldsymbol{\rho}|\mathbf{0}) = \int_0^\infty d\lambda\; J_0(\lambda\rho)(1 + \varepsilon_r \coth \lambda h)^{-1} \tag{95a}$$

or, expanding the sum inside the parentheses into powers of $\exp(-2\lambda h)$ and integrating the resulting infinite series term by term:

$$2\pi\varepsilon_0 G_V(\boldsymbol{\rho}|\mathbf{0}) = (1-\eta)\left[\frac{1}{R_0} - (1+\eta)\sum_{n=1}^{\infty}(-\eta)^{n-1}\frac{1}{R_n}\right] \tag{95b}$$

with η and R_n defined as

$$\eta = (\varepsilon_r - 1)/(\varepsilon_r + 1)\,; \qquad R_n^2 = \rho^2 + 4n^2h^2$$

The series (95b) is the well-known partial image representation of the static Green's function, given by Silvester and Benedek [16] while the integral representation (95a) has been first used by Patel [17]. Generalizations of (95a) to multilayered substrates can be obtained as outlined in Section 4. Some explicit values are given in Crampagne et al. [18].

7.1.2. The Quasi-Static Case

The classical technique to obtain an approximated integral equation useful at low frequencies implies neglecting losses and displacement current. Taking the divergence of (50) with $Z_s = 0$ gives

$$j\omega\mathbf{e}_z \cdot \nabla \times \int_s ds'\, \bar{\bar{G}}_A(\boldsymbol{\rho}|\boldsymbol{\rho}') \cdot \mathbf{J}_s(\boldsymbol{\rho}') = \mathbf{e}_z \cdot \nabla \times \mathbf{E}^{(e)}(\boldsymbol{\rho})$$

$$= -j\omega\mu_0 \mathbf{e}_z \cdot \mathbf{H}^{(e)}(\boldsymbol{\rho}) \tag{96}$$

where the equivalence $\nabla \cdot (\mathbf{e}_z \times \mathbf{H}) = \mathbf{e}_z \cdot (\nabla \times \mathbf{H})$ has been used. Introducing now the del operator under the integration sign leads to

$$\mathbf{e}_z \cdot \int_s ds'\, \bar{\bar{G}}_H(\boldsymbol{\rho}|\boldsymbol{\rho}') \cdot \mathbf{J}_s(\boldsymbol{\rho}') + \mathbf{e}_z \cdot \mathbf{H}^{(e)}(\boldsymbol{\rho}) = 0 \tag{97}$$

This equation simply expresses the fact that the total normal magnetic field must vanish on the surface of lossless conductors, and it is valid for any frequency. However, since (97) is a scalar equation, it does not suffice in general to determine the two components of the surface current. The second scalar equation is obtained by neglecting the displacement current in Maxwell's equations. Then $\nabla \times \mathbf{H} = \mathbf{J}$ and, consequently $\mathbf{J}_s$ is solenoidal, i.e.,

$$\nabla \cdot \mathbf{J}_s = 0 \tag{98}$$

The set of equations (97) and (98) defines the quasi-static model.

As in the static case, sometimes it will be convenient to introduce an excitation current $\mathbf{J}_s^{(e)}$. Then (97) is transformed into

$$\mathbf{e}_z \cdot \int_S ds' \, \bar{\bar{G}}_H(\boldsymbol{\rho}|\boldsymbol{\rho}') \cdot [\mathbf{J}_s(\boldsymbol{\rho}') + \mathbf{J}_s^{(e)}(\boldsymbol{\rho}')] = 0 \tag{99}$$

and the system of equations (98) and (99) is to be solved taking into account the additional condition:

$$\int_C dl \, \mathbf{e}_n \cdot \mathbf{J}_s = 0 \tag{100}$$

which relates the excitation surface current to the total current entering the structure in Fig. 10b.

Since displacement currents are neglected, the current distribution satisfies a static Poisson equation. Consequently, to ensure the internal coherence of the model, the Green's functions arising in (96), (97), and (99) must be static too. At zero frequency, the Sommerfeld integral (179) becomes

$$\begin{aligned} \frac{4\pi}{\mu_0} G_A^{xx}(\boldsymbol{\rho}|\mathbf{0}) &= 2\int_0^\infty d\lambda \, J_0(\lambda\rho)(1 + \coth \lambda h)^{-1} \\ &= \left(\frac{1}{\rho} - \frac{1}{\sqrt{\rho^2 + 4h^2}}\right) \end{aligned} \tag{101}$$

which is the steady-state solution to the problem of a point source above a ground plane. Therefore, the quasi-static model is independent of the substrate permittivity.

7.2. Approximations for the Green's Functions

For the cases where no simplification is possible in the expression of the integral equation, the Sommerfeld integrals giving the Green's functions have, in general, no analytical solution. However, it is possible to obtain the solution approximately in the low-frequency range and in the far-field region. This will be shown now for the single layer case with source and observer both in the air–dielectric interface ($G_A^{xx} \equiv A_x$, $G_V \equiv V$).

7.2.1. *Near Field*

Low-frequency approximations, valid in the near-field zone, are obtained by setting $u = u_0$ in the Sommerfeld integrals. This simplification is based on

desired at low frequency. With this assumption, the scalar potential can be expanded following the technique outlined in Section 7.1.1 and we obtain the result [19]

$$4\pi\varepsilon_0 V = (1-\eta)\left[\frac{\exp(-jk_0R_0)}{R_0} - (1+\eta)\sum_{i=1}^{\infty}(-\eta)^{i-1}\frac{\exp(-jk_0R_i)}{R_i}\right] \tag{102}$$

where, as before, $\eta = (\varepsilon_r - 1)/(\varepsilon_r + 1)$ and $R_i^2 = \rho^2 + (2ih)^2$. This approximation can be viewed as a generalization of the static potential of Silvester and Benedek [16] which we have already seen in the static integral equation [eq. (95)].

Similarly, the x component of the vector potential is approximated as

$$\frac{4\pi}{\mu}A_x = \frac{\exp(-jk_0R_0)}{R_0} - \frac{\exp(-jk_0R_1)}{R_1} \tag{103}$$

Analogous expressions can be derived for A_z and the fields. However, low-frequency approximations are of little interest for the study of microstrip antennas, and this subject will not be developed further here (see, for example, Mosig and Sarkar [19]). We shall stress only the fact that at low frequency the vector potential is independent of the permittivity and corresponds to the homogeneous ($\varepsilon_r = 1$) case.

7.2.2. *Far Field*

Far-field approximations are useful to calculate the coupling between distant conductors residing on the same microstrip substrate. Also, they provide a simple check of the validity of numerical integration techniques.

To obtain asymptotic expansions of the Sommerfeld integrals in the far field region, we must resort to rather involved analytical techniques such as saddle-point and steepest-descent techniques in the complex plane [5].

Provided that no surface wave of the TE type is excited, the asymptotic behavior of the vector potential is mainly determined by the branch point at $k_\rho = k_0$. The dominant term in the far field is

$$\frac{4\pi}{\mu_0}A_x \approx \left(\frac{\tan\Delta}{\Delta}\right)\frac{2j}{k_0}\frac{\exp(-jk_0\rho)}{\rho^2} \tag{104}$$

with the parameter Δ defined as before by $\Delta = k_0h\sqrt{\varepsilon_r - 1}$. It is of interest to check that by letting $\varepsilon_r = 1$ in eq. (104) we obtain the asymptotic expansion for the vector potential of an HED over a ground plane. As a matter of fact, if we define the complex ratio Q between the x component of the vector potential and its homogeneous counterpart,

the fact that the difference $u^2 - u_0^2 = k_0^2(\varepsilon_r - 1)$ can be made as small as

$$Q = \frac{A_x}{A_x(\varepsilon = 1)} \tag{105}$$

we obtain $Q = 1$ in the near field [see eq. (103)] and $Q = \tan^2 \Delta / \Delta^2$ (a real quantity) in the far field.

The dominant TM surface wave always propagates in a microstrip structure and determines the asymptotic behavior of the scalar potential in the air–dielectric interface. For a single-layer lossless substrate, we obtain

$$4\pi\varepsilon_0 V \sim -2\pi j P H_0^{(2)}(\lambda_p \rho) \tag{106}$$

where P is the residue of the function $f = k_\rho N / D_{\mathrm{TE}} D_{\mathrm{TM}}$ and $H_0^{(2)}$ is the Hankel function.

Consequently, in a lossless case, the scalar potential decreases very slowly (as $1/\sqrt{r}$) in the air–dielectric interface and can produce a strong coupling between the conductors of a microstrip structure. The asymptotic behavior predicted by eq. (106) has been obtained assuming that the contribution of the pole is not affected by the nearby branch point. Hence, eq. (106) is more accuratge for thick substrates and high permittivities. Otherwise, the pole is too close to the branch point and their contributions can no longer be separated. The calculation of a more precise uniform asymptotic expansion, valid in this situation is beyond the scope of this work (see Felsen and Marcuvitz [20]).

8. NUMERICAL EVALUATION OF SOMMERFELD INTEGRALS

When a microstrip structure is analyzed with an integral equation technique, it is necessary to evaluate the interaction between points separated by distances ranging from zero to several wavelengths. For most of these distances the accuracy of near field and far-field approximations does not suffice and the potentials must be numerically evaluated. For a single-layer microstrip structure, the source and the observation point are both on the interface. Hence we must set $z = z' = 0$ in the integral expressions for the fields and the potentials and the exponential function that ensures the convergence of the integrands disappears. This is numerically the most difficult case, and we will pay particular attention to it.

Even though many deformations of the original patch C of Fig. 5 have been tried (see, for instance, Mosig and Gardiol [5] and Michalski [21]), we feel that the integration along C (the real positive axis λ of the complex plane k_ρ) provides the most efficient algorithm for evaluating the Sommerfeld integrals appearing in multilayered microstrip structures.

8.1. Numerical Integration on the Real Axis

The functions to be integrated oscillate along the real axis due to the Bessel functions. The square root $u_0 = \sqrt{\lambda^2 - k_0^2}$ introduces a discontinuity of the derivative at $\lambda = k_0$, which corresponds to a branch point in the complex plane. If the integrand contains the denominator D_{TM}, there is a pole just below the real axis (or on the axis itself for a lossless substrate). This pole produces very strong variations of the integrands. Finally, many of the oscillating integrands have an envelope that converges very slowly (G_A^{xx}, G_V) or even diverges at infinity (G_H^{zx}). In the second case, the integral has no meaning in the Riemann sense since the area under the curve representing the integrand fails to converge to a finite value as the upper bound goes to infinity. However, the integral can be interpreted in the sense of distributions as the limiting value:

$$\int F \, d\lambda = \lim_{z \to 0} \int F \exp(-v_0 z) \, d\lambda \tag{107}$$

and the exponential guarantees the convergence. This means, physically, that the potentials at the interface can be considered as the limiting case of the potentials in the air when the height of the observation point above the interface goes to zero.

All these facts are clearly depicted in Fig. 11*a* which corresponds to the integrand of the scalar potential V_q for $\varepsilon_{r1} = 5$, $\tan\delta = 0.01$, $k_0 h = 0.2$, and $k_0 \rho = 3$. An enlarged view of the interval $[0.9k_0, 1.4k_0]$ is given in Fig. 11*b*.

For the sake of clarity, we have chosen for these figures a rather thick substrate where the pole is not too close to the branch point. For thinner substrates, pole and branch point are much closer, but the techniques developed in this section remain unaffected by this proximity.

The integration interval is decomposed into three subintervals, $[0, k_0]$, $[k_0, k_0\sqrt{\varepsilon_r}]$ and $[k_0\sqrt{\varepsilon_r}, \infty]$. In the region $[0, k_0]$ the infinite derivative in k_0 is eliminated with a change of variables $\lambda = k_0 \cos t$. The resulting smoother function is integrated numerically. In the interval $[k_0, k_0\sqrt{\varepsilon_r}]$, the singularity is first extracted (if the integrand includes the denominator D_{TM}). By writing the function under the integral sign in the form $F(\lambda) = J_n(\lambda\rho) f(\lambda)$, we have

$$F(\lambda) = [J_n(\lambda\rho) f(\lambda) - F_{\text{sing}}(\lambda)] + F_{\text{sing}}(\lambda) \tag{108}$$

where

$$F_{\text{sing}}(\lambda) = \frac{P}{\lambda - (\lambda_p - j\nu_p)}$$

Here $\lambda_p - j\nu_p$ is the complex pole ($\nu_p > 0$) and P the residue of F at the pole. The function F_{sing} is integrated analytically as

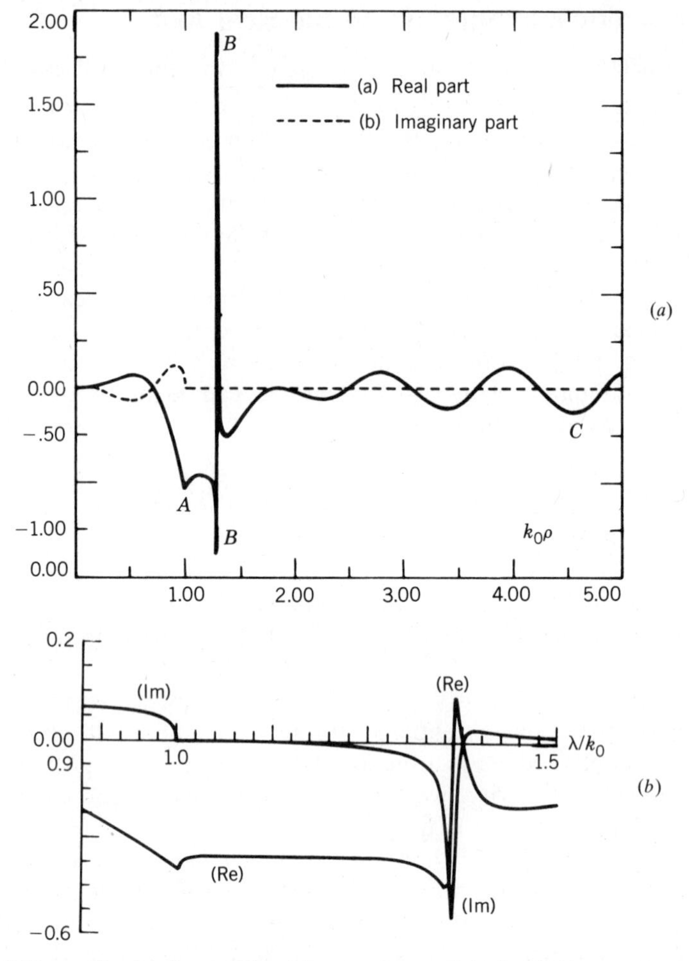

Fig. 11 (*a*) Normalized values of the integrand associated with the scalar potential V of an HED on microstrip. $\varepsilon_r = 5$; $k_0 h = 0.2$; $k_0 \rho = 3$. *A*, Discontinuities in the derivative; *B*, sharp peaks due to the pole; *C*, oscillatory and divergent behavior at infinity. Dotted line, real part; dashed line, imaginary part. (*b*) Enlarged view of Fig. 11*a* in the interval $\lambda \in [0.9k_0, 1.4k_0]$. After [19] © 1986 IEEE.

$$\int_{k_0}^{k_0\sqrt{\varepsilon_r'}} F_{\text{sing}}\, d\lambda = \frac{P}{2} \ln \frac{\nu_p^2 + (k_0\sqrt{\varepsilon_r'} - \lambda_p)^2}{\nu_p^2 + (k_0 + \lambda_p)^2} + jP \arctan \frac{k_0\sqrt{\varepsilon_r'} - \lambda_p}{\nu_p}$$
$$+ jP \arctan \frac{\lambda_p - k_0}{\nu_p} \tag{109}$$

It is worth mentioning that in the lossless case ($v_p = 0$) the above integral becomes

$$\int_{k_0}^{k_0\sqrt{\varepsilon_r}} \frac{P}{\lambda - \lambda_p}\, d\lambda - j\pi P = P \ln \frac{k_0\sqrt{\varepsilon_r} - \lambda_p}{\lambda_p - k_0} - j\pi P \qquad (110)$$

and therefore the principal-value formulation [eq. (87)] of the lossless case is included as a limiting case in this numerical technique.

Figure 12 depicts the real part of the original function $F(\lambda)$ (curve A) and the difference $F(\lambda) - F_{\text{sing}}(\lambda)$ (curve B), where the singularity has been extracted. There is still an infinite derivative in the curve B at $\lambda = k_0$. With a change of variable $\lambda = k_0 \cosh t$ one finally obtains a very smooth integrand (the discontinuous curve in Fig. 12), which is integrated by a Gaussian quadrature. The same procedure is applied to the imaginary part of $F(\lambda)$ to eliminate in a similar way the sharp peak and the infinite derivative.

Finally, in the region $[k_0\sqrt{\varepsilon_r}, \infty]$ we first extract the dominant static term defined by $F(\lambda, k_0 = 0)$. Figure 13 gives the integrand $F(\lambda, k_0)$ (curve A) and the difference $F(\lambda, k_0) - F(\lambda, 0)$ (curve B). It can be shown that the static term has the form $F(\lambda, k_0 = 0) = CJ_n(\lambda\rho)\lambda^m$, and hence it can be integrated analytically.

The remaining part is a slowly convergent oscillating function over a semi-infinite interval that is integrated with specially tailored techniques.

Multilayered substrates give rise to functions that qualitatively do not differ from those depicted in Figs. 11–13, except for an eventual exponential decrease when source and observer are at different levels.

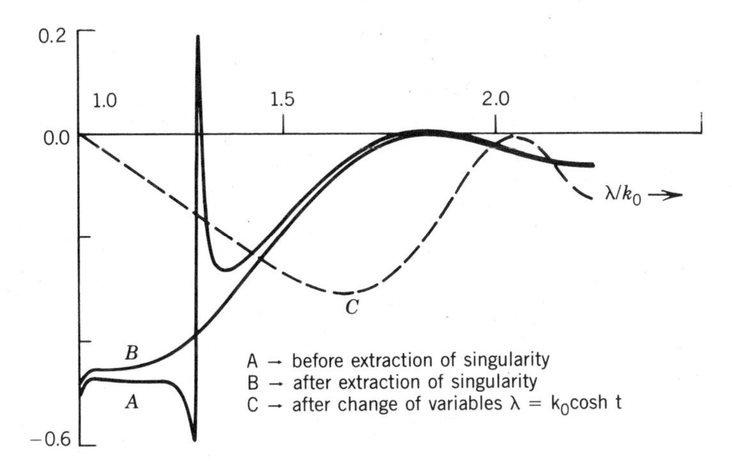

Fig. 12 The real part of the integrand of Fig. 11 for the lossy case in the interval $[k_0, k_0\sqrt{\varepsilon_r}]$. Curve A, before the extraction of the singularity; B, after the extraction of the singularity; C, after the change of variables: $\lambda = k_0 \cosh t$. After [19] © 1986 IEEE.

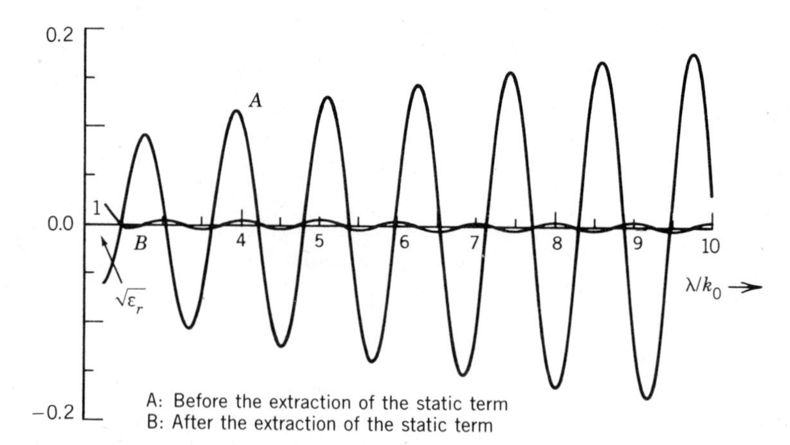

Fig. 13 The real part of the integrand of Fig. 11 for the lossy case in the interval $[k_0\sqrt{\varepsilon_r}, \infty]$. Curve *A*, before the extraction of the static term; *B*, after the extraction of the static term. After [19] © 1986 IEEE.

8.2. Integrating Oscillating Functions over Unbounded Intervals

Sommerfeld integrals, as given by eqs. (79) and (80) can be grouped in a more general class of integrals defined by:

$$I(\rho) = \int_a^\infty g(\lambda\rho) f(\lambda)\, d\lambda \tag{111}$$

where

1. $g(\lambda\rho)$ is a complex function whose real and imaginary parts oscillate with a strictly periodic behavior (sin, cos) or behave asymptotically as the product of a periodic function and monotonic function. A typical example of this class of functions, which will be termed from now on quasi-periodic, are Bessel functions of the first kind.
2. $f(\lambda)$ is a smooth, nonoscillating function that behaves asymptotically as $\lambda^\alpha \exp(-\lambda\beta)$, where β is the difference in height between source and observation points. Therefore, when both points are on the interface ($\beta = 0$) the function $f(\lambda)$ decreases very slowly or even diverges at infinity if $\alpha < 0$. This situation, commonly found in practical microstrip problems, is the most difficult from a numerical point of view. For the sake of simplicity, $f(\lambda)$ is assumed to be real. Complex functions can be handled by working successively with their real and imaginary parts.
3. The lower integration bound a has been chosen conveniently to ensure that the interval $[a, \infty]$ is far enough from the possible singularities of $f(\lambda)$. For instance, in our problem, we shall take $a = \sqrt{\varepsilon_r}$.

It is worth mentioning that the general expression (111) includes many integral transforms such as Fourier and Hankel transforms. Hence, the following techniques can be applied to many other problems in numerical analysis.

The classical problem involving Sommerfeld integrals is the problem of radiowave propagation above a lossy ground, where the comprehensive monograph of Lytle and Lager [22] is the classic reference. These authors have found an iterative Romberg integration satisfactory, since here the integrand displays an exponential convergence and its poles have been removed from the real axis. In microstrip problems, Romberg integration has also been used, but its effectiveness decreases considerably in the absence of a well-behaved integrand.

In recent years, there has been a considerable amount of work published on the numerical evaluation of Fourier transforms, which are included in eq. (111) as a particular case. The techniques involved can be classified into three groups.

1. The decomposition $[a, \infty] = [a, A] + [A, \infty]$. Here, Filon's algorithm is applied to the finite interval $[a, A]$, while an asymptotic expression of the integrand is used to estimate the integral's value over $[A, \infty]$ [23]. The most serious drawbacks of this approach are the choice of A and the analytical work required, features that are difficult to incorporate in an automatic computation routine.

2. Another approach applies if $g(\lambda\rho)$ is a strictly periodic function; then the following decomposition is used:

$$I(\rho) = \int_a^{a+p} g(\lambda\rho) \sum_{n=0}^{\infty} f(\lambda + np)\, d\lambda \tag{112}$$

where p is the period of the function g. The infinite sum under the integration sign can be evaluated by standard devices, such as Euler's transformation. Also, a more involved technique using theoretical Fourier-transform concepts has been described in connection with a problem on quantum-mechanics impact cross section [24]. These methods work very well for large values of ρ and an exponentially decreasing integrand. Unfortunately, their extension to quasi-periodic diverging integrands seems problematic.

3. The third group of techniques, introduced by Hurwitz and Zweifel [25], are defined by the decomposition

$$I(\rho) = \sum_{n=0}^{\infty} \int_{a+np/2}^{a+(n+1)p/2} g(\lambda\rho) f(\lambda)\, d\lambda \tag{113}$$

The integration over each half cycle is performed prior to the series' summation. As before, an accelerating device, such as the nonlinear transformations of Shanks (Alaylioglu et al. [26] or Sidi [27] can be used to sum the infinite series.

We feel that the decomposition (113) is particularly well suited for the Sommerfeld integrals encountered in microstrip problems, and we have used it extensively throughout this work. However, instead of the sophisticated nonlinear techniques mentioned above, we have developed a new simple technique based on the concept of a weighted average between the half-cycle integrals [5]. This accelerating device has proven to be faster and very accurate for Sommerfeld integrals.

8.3. The Weighted-Averages Algorithm

For the sake of clarity we shall illustrate the proposed algorithm for the kernel $g(\lambda\rho) = \exp(j\lambda\rho)$.

Let us introduce the partial integrals

$$I_0(n) = \int_a^{a_n} g(\lambda\rho) f(\lambda)\, d\lambda \tag{114}$$

where $a_n = a + n\pi/\rho$. If we approximate the true integral (111) by an estimation of the above type, the truncation error is given by

$$I - I_0(n) = \int_{a_n}^{\infty} g(\lambda\rho) f(\lambda)\, d\lambda \tag{115}$$

This integral can be expanded into an infinite series of inverse powers of ρ by integration by parts, with the result

$$I - I_0(n) = \frac{j}{\rho} \exp(ja_n\rho)\left[f_n + \frac{j}{\rho} f'_n + \left(\frac{j}{\rho}\right)^2 f''_n + \cdots \right] \tag{116}$$

where $f_n, f'_n, f''_n, \ldots,$ are the values of $f(\lambda)$ and its successive derivatives at $\lambda = a_n$.

Assuming that the series on the right-hand side of (116) is convergent, it is clear that the dominant error term is of order $O(\rho^{-1})$.

Let us consider now two successive estimations obtained for the integer values n and $n+1$. Since $\exp(ja_{n+1}\rho) = -\exp(ja_n\rho)$, the weighted average

$$I_1(n) = \frac{f_{n+1} I_0(n) + f_n I_0(n+1)}{f_n + f_{n+1}} \tag{117}$$

is a better estimation of I, since the truncation error is now of order $O(\rho^{-2})$:

$$I - I_1(n) = \left(\frac{j}{\rho}\right)^2 \exp(ja_n\rho)\left[\frac{f'_n f_{n+1} - f'_{n+1} f_n}{f_n + f_{n+1}}\right] + \cdots \tag{118}$$

If, as postulated, $f(\lambda)$ shows an asymptotic behavior of the type $C\lambda^{\alpha}\exp(-\beta\lambda)$, we can approximate f_n, f_{n+1} in eq. (117) and rewrite it in the simpler form

$$I_1(n) = \frac{I_0(n) + \eta_0 I_0(n+1)}{1+\eta_0} \tag{119}$$

where $\eta_0 = [n/(n+1)]^{\alpha}\exp(\beta\pi/\rho)$.

Using the same assumptions, a careful evaluation of the bracketted term in the right-hand side of (118) shows that it behaves asymptotically as $C\lambda^{\alpha-2}\exp(-\beta\lambda)$. Hence, we can define a new weighted average as

$$I_2(n) = \frac{I_1(n) + \eta_1 I_1(n+1)}{1+\eta_1} \tag{120}$$

with $\eta_1 = [(n-0.5)/(n+0.5)]^{\alpha-2}\exp(\beta\pi/\rho)$, which yields a better approximation of I. By applying recursively this algorithm to the $k+1$ values $I_0(n), \ldots, I_0(n+k)$, decreasing the exponent α by 2 each time, it is possible to extract, with little effort, a very good estimate $I_k(n)$ of the true value I.

It is worth pointing out that this algorithm can also be obtained if we write (116) for N different values of a and then eliminate the "variables" $(j/\rho)^n$, $n = 0, 1, \ldots, N-2$, in the resulting system of equations. However, this approach, while very powerful from a theoretical point of view, leads to less efficient algorithms.

The above algorithm remains valid for Hankel functions, i.e., $g(\lambda\rho) = H_{\nu}^{(2)}(\lambda\rho)$ if, according to the asymptotic expansions of these functions for large arguments, we employ $a_n = a + \pi(0.25 + 0.5\nu + n)$ and we replace $f(\lambda)$ by $f(\lambda)/\sqrt{\lambda}$. Hence, the initial value of α will be a half-unit lower.

In addition, it is clear that the algorithm applies to the Bessel functions J_n and Y_n, since they are merely the real and imaginary parts of the Hankel function.

We must finally stress the fact that the parameter α reflects an asymptotic behavior, and consequently the result that this parameter must be reduced by two units at each iteration may be too optimistic for the first iterations. In practice, reducing α by one unit at each iteration is frequently a better choice. On the other hand, if $g(\lambda\rho)$ is a real function, it is possible to reduce by one unit the initial value of α, provided that the points a_n are taken as the zeros of the derivative of $g(\lambda\rho)$.

To check the validity of the algorithm, we have selected the integral

$$I = \int_0^{\infty} J_1(\lambda)\lambda\, d\lambda = 1 \tag{121}$$

which appears in the spectral domain formulation of G_H^{zx} at zero frequency in free space [eq. (20)]. each half-period integral $I_0(n)$ was calculated with a

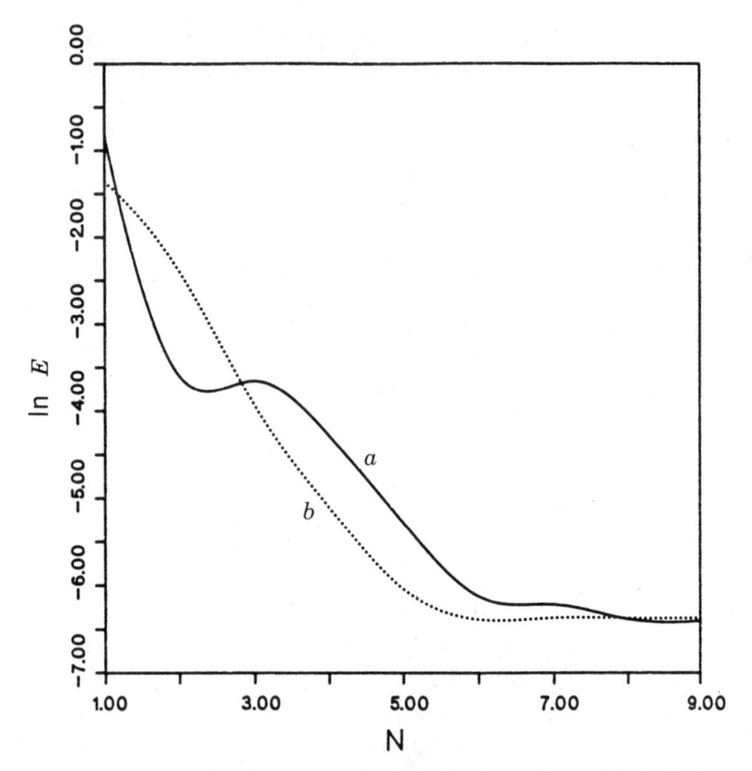

Fig. 14 Relative error E in the numerical evaluation of the integral given in eq. (121) as a function of the number of iterations for two different algorithms: (a) Weighted averages and (b) Sidi's method.

12-point Gauss–Legendre numerical rule, ensuring a precision better than 10^{-6}. Figure 14 gives the relative error when (101) is computed with two different techniques. Weighted means give a precision of 10^{-6} after only five iterations (60 functions evaluations), while, the more sophisticated and time-consuming nonlinear technique of Sidi [27], gives a precision of the same order (10^{-7}) after eight iterations.

Also, the commonly used Shanks extrapolation was found to give a precision of only 10^{-4} for the same number of points.

In a second example, we have tested the performances of our algorithm against the normalized distance $k_0\rho$. We consider a Sommerfeld integral arising in the study of wave propagation above earth. This integral has an analytical solution due to Van der Pol (Banos [28]) and can be written:

$$\int_0^\infty J_0(\lambda\rho)\,\frac{2\lambda}{u_1+u_2}\,d\lambda = \frac{2}{\rho}\,\frac{1}{k_1^2-k_2^2}\,\frac{\partial}{\partial\rho}\left(\frac{e^{-jk_2\rho}-e^{-jk_1\rho}}{\rho}\right) \tag{122}$$

Figure 15 gives the real part of the integrand $J_0(\lambda\rho)f(\lambda)$ and of the

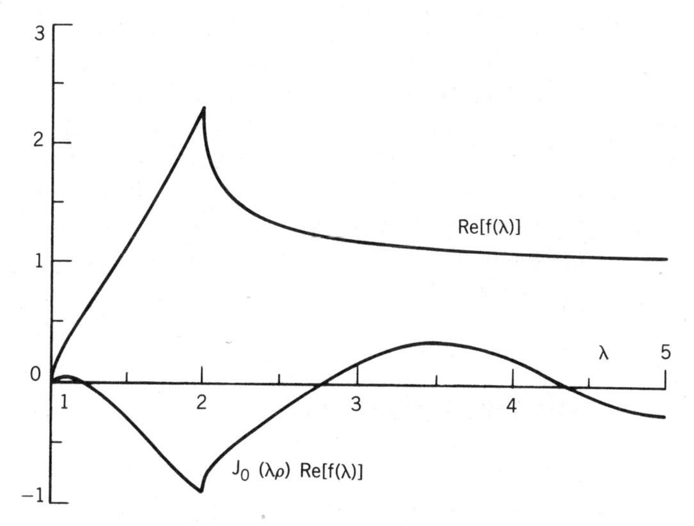

Fig. 15 Real part of Van der Pol's integrand (zero for $\lambda/k_0 < 1$), $k = k_0$, $k_0\rho = 2$.

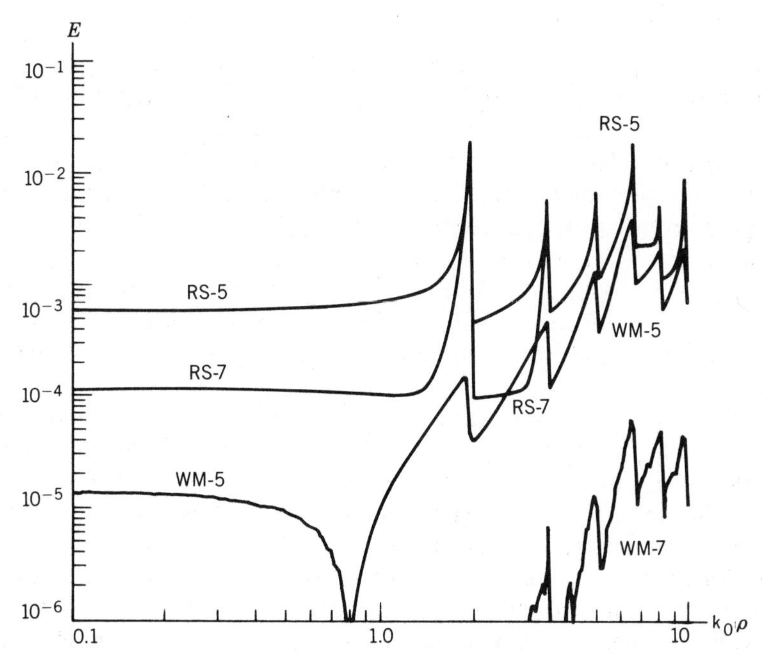

Fig. 16 Relative error E in the numerical evaluation of Van der Pol's integral as a function of the normalized distance $k_0\rho$. WM-5, WM-7, weighted averages after 5 and 7 iterations; RS-5, RS-7, Romberg–Shanks algorithm after 5 and 7 iterations.

envelope $f(\lambda)$ for $k_1 = 2k_0$ and $\rho = 2$. Infinite derivatives at $\lambda = k_0$ and $\lambda = k_1$ are readily observable. The relative errors obtained when evaluating the integral (122) by various numerical techniques are presented in Fig. 16. For the same number of function evaluations, the weighted-means technique gives far more accurate results that the commonly used Romberg–Shanks algorithm [22]. Moreover, the results improve more rapidly with the number of integration points using the weighted-means technique. Even if accuracy deteriorates as the parameter ρ increases, the precision remains better than 0.01% in the interval $0 < \rho < 10$ for only 72 evaluations of the integrand.

9. RESULTS FOR THE POTENTIALS

The above numerical algorithms have been extensively applied to the computation of the Sommerfeld integrals giving the potentials and the fields. As an illustration of the theory, we present here graphical results concerning the potentials A_x and V.

9.1. One-Layer Microstrip

9.1.1. The Vector Potential

The vector potential A_x of an x-directed HED, as given by eq. (79), is a typical example of a Sommerfeld integral without a dominant pole. Hence, we do not expect large deviations from its homogeneous ($\varepsilon_r = 1$) value. Indeed, we can assert on physical grounds that A_x and the corresponding magnetic field are only slightly affected by changes in permittivity. It is therefore more instructive to consider the complex ratio Q between the true and homogeneous values as defined by eq. (105).

The curves of Figs. 17 and 18 give, respectively, the modulus and the phase of Q as a function of the radial distance measured in free space wavelengths. The cases $\varepsilon_r = 2$, 10 and $(h/\lambda_0)\sqrt{\varepsilon_r - 1} = 0.05$, 0.125, 0.20 are shown.

The numerical results validate the near- and far-field approximations [eqs. (103) and (104)]. The modulus of Q increases with distance from unity to the asymptotic limit $(\tan^2 \Delta)/\Delta^2$. The phase of Q goes to zero in the near and far fields and reaches a maximum in the middle range. The value of this maximum depends on substrate thickness and permittivity. For thin substrates (for instance $\varepsilon_r = 2$ and $h = 0.05\lambda_0$) the maximum phase is only 0.2 degrees while the modulus of Q ranges between 1 and 1.034. It can be said, in conclusion, that for thin substrates the true vector potential can be replaced by its homogeneous value, and therefore no evaluation of Sommerfeld integrals is needed.

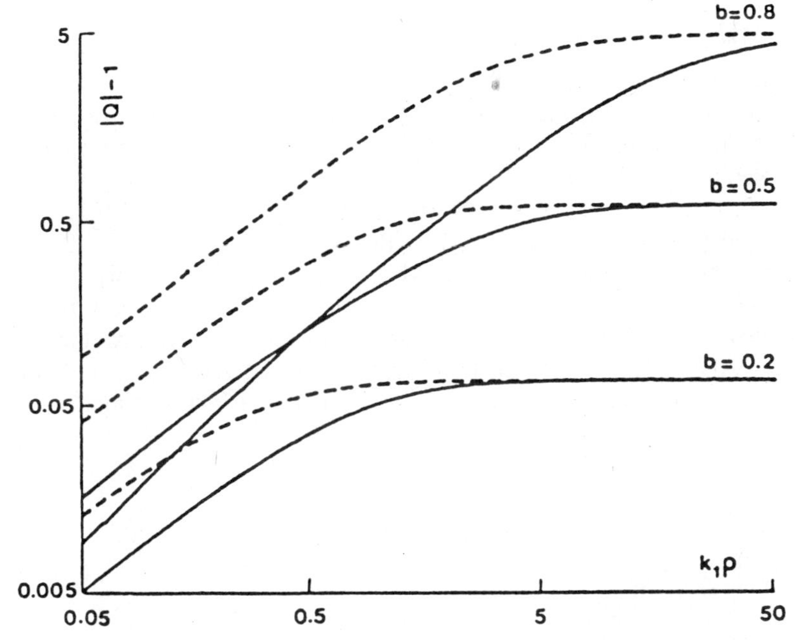

Fig. 17 Modulus of the complex ratio Q between the component A_x in a single-layer microstrip and its value in a homogeneous medium as a function of the normalized radial distance. Solid lines, $\varepsilon_r = 9.6$; dashed lines, $\varepsilon_r = 2.2$. After [15].

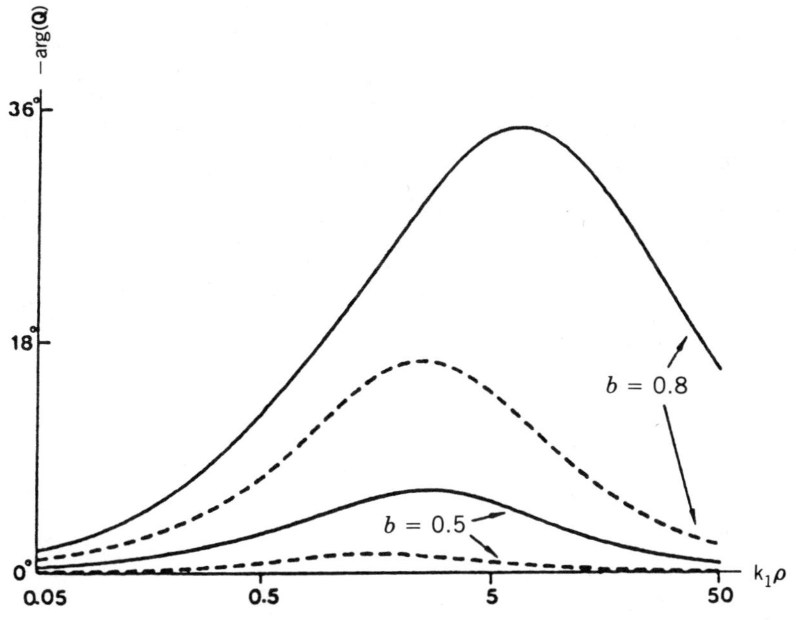

Fig. 18 Phase of the complex ratio Q between the component A_x in a single-layer microstrip and its value in a homogeneous medium as a function of the normalized radial distance. Solid lines, $\varepsilon_r = 9.6$; dashed lines, $\varepsilon_r = 2.2$. After [15].

9.1.2. The Scalar Potential

The situation is very different for the scalar potential of a point charge V, as defined by (80). In Figs. 19 and 20, the modulus and phase of the normalized scalar potential, i.e. $4\pi\varepsilon_0\lambda_0 V$, have been plotted against the normalized radial distance for $\varepsilon_r = 10$ and normalized substrate thicknesses $h/\lambda_0 = 0.05$, 0.125, and 0.2. The near- and far-field approximations [eqs. (102) and (106)], are represented by discontinuous lines in Fig. 19.

It can be seen that the near field approximation gives only reasonable results for $\rho/\lambda_0 < 0.1$ and small values of h/λ_0. On the other hand, the surface-wave behavior is reached more rapidly for electrically thicker substrates. Also, there is a transition zone between static (ρ^{-1}) and surface-wave ($\rho^{-1/2}$) dependencies that becomes more marked for small values of

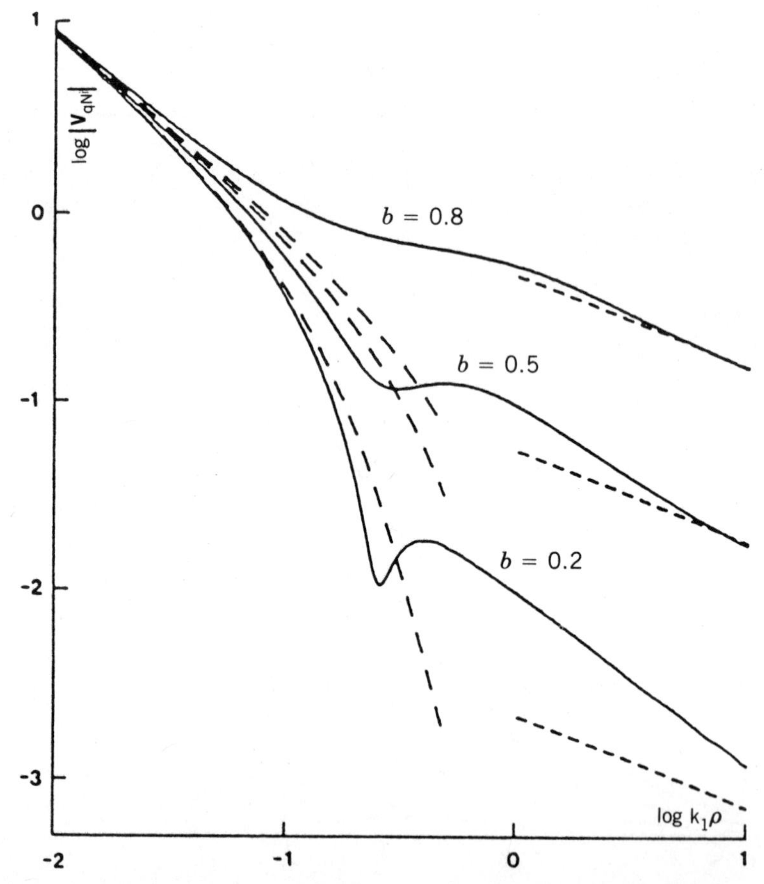

Fig. 19 Modulus of the normalized scalar potential in a single-layer ($\varepsilon_r = 10$) microstrip structure as a function of the normalized distance. (———) Numerical value; (—–—) near-field approximation; (– – –) far-field approximation. After [15].

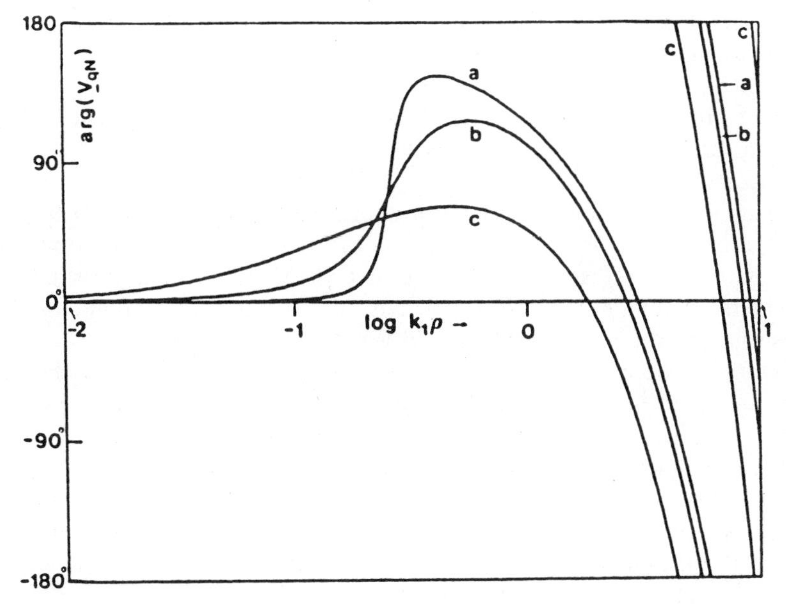

Fig. 20 Phase of the normalized scalar potential in a single layered ($\varepsilon_r = 10$) microstrip structure as a function of the normalized distance. (———) Numerical value; (——) near-field approximation; (– – –) far-field approximation. After [15].

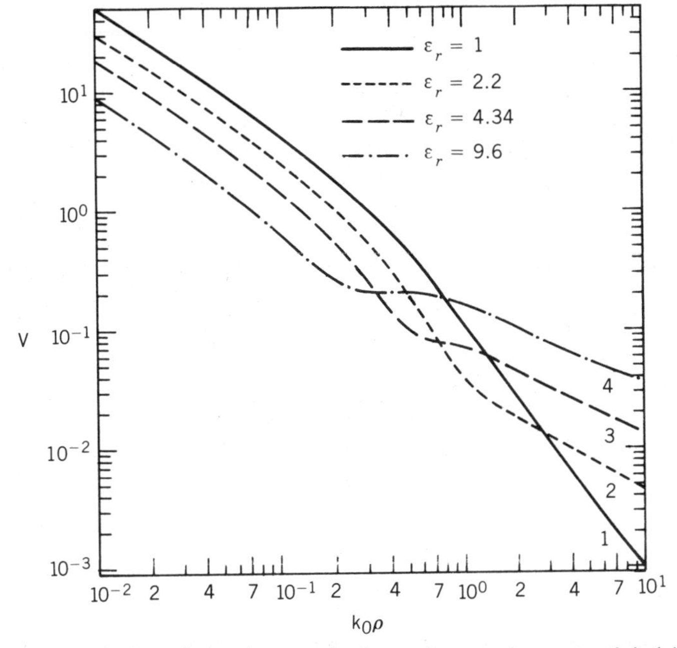

Fig. 21 Effect of the dielectric constant on the scalar potential (single-layer substrate). (1) $\varepsilon_r = 1$; (2) $\varepsilon_r = 2.2$; (3) $\varepsilon_r = 4.34$; (4) $\varepsilon_r = 9.6$.

h/λ_0. This transition zone appears as a rapid variation of the phase (Fig. 20) that rises sharply from small positive values and afterwards follows a typical surface-wave pattern.

More insight into these phenomena can be obtained by keeping h/λ_0 constant and changing ε_r. The cases $h/\lambda_0 = 0.05$ and $\varepsilon_r = 1$, 2.2, 4.34, and 9.6 are presented in Fig. 21. Except for a scaling factor $(\varepsilon_r + 1)/2$, the curves for $\varepsilon_r \neq 1$ follow closely the homogeneous behavior up to a transition zone, after which the surface wave becomes dominant. A point worth mentioning is that for $\varepsilon_r = 10$, the near-field approximation cannot be extended beyond $\rho/\lambda_0 = 0.03$, which is the distance corresponding to the transition between static and surface wave behaviors.

Figure 22 depicts simultaneously the modulus and phase of the scalar potential in a polar diagram. For the same substrate of thickness $h/\lambda_0 =$ 0.05, we have plotted the cases $\varepsilon_r = 1$ and $\varepsilon_r = 2$. The dots and stars represent values of the scalar potential ranging from $k_0\rho = 0.5$ to $k_0\rho = 10$ in steps of 0.5. Both curves start at phase zero and in the homogeneous case the phase decreases steadily yielding a clockwise spiral that converges toward the origin. On the other hand, the curve starts counterclockwise in

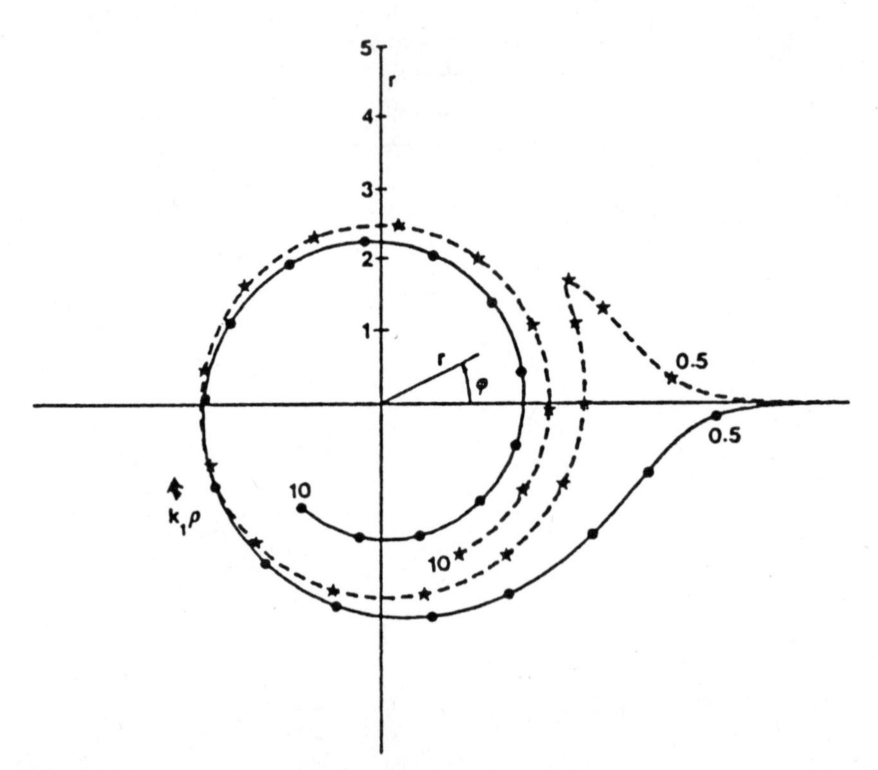

Fig. 22 Polar logarithmic plot of the normalized scalar potential for the cases (●) $\varepsilon_r = 1$ and (*) $\varepsilon_r = 2.2$ of Fig. 21. After [15].

an inhomogeneous structure. This is an anomalous situation where the phase increases with distance, due to a complex interaction between space and surface waves. After a turning point ($k_0\rho = 1.5$) the curve resumes the behavior of an outward propagating wave, spiraling clockwise, while its distance to the origin (the modulus of the scalar potential) decreases slowly as expected from a surface wave.

In conclusion, the scalar potential is dominated by the surface wave and no analytical approximations are useful in the practical range $0.01\lambda_0 < \rho < \lambda_0$. Hence, we must resort to careful numerical evaluation of the corresponding Sommerfeld integrals.

9.2. Two-Layer Microstrip

For a substrate including two layers, we obtain essentially the same behavior for the potentials. As in the single-layer case, we can again avoid the numerical evaluation of the Sommerfeld integral associated to the vector potential, provided the layers are electrically thin. Exact expressions will then be replaced by homogeneous approximations.

For the scalar potential, we can hardly use static or homogeneous approximations. Figures 23 and 24 give the modulus and phase of the

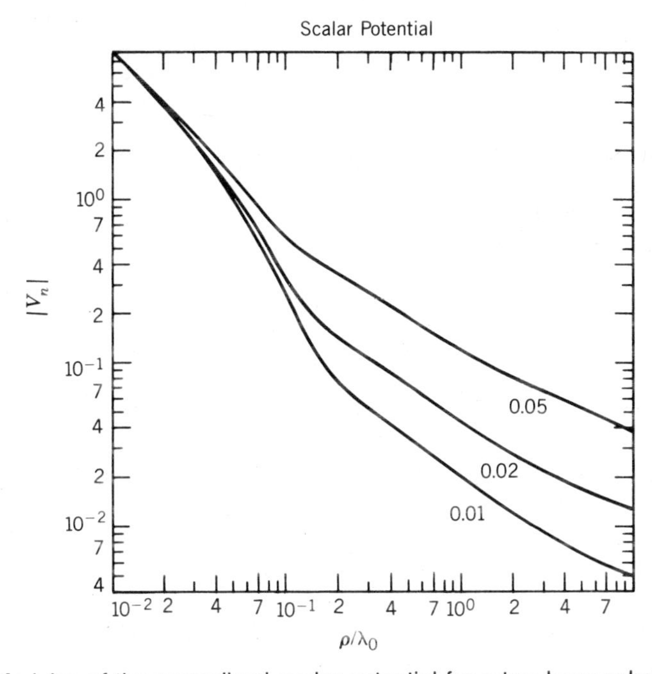

Fig. 23 Modulus of the normalized scalar potential for a two-layer substrate when the source and the observer are both on the air–upper layer interface. Upper layer: $\varepsilon_{r1} = 2$, $\tan \delta_1 = 0$, $k_0 h_1 = 0.05$. Lower layer: $\varepsilon_{r2} = 5$, $\tan \delta_1 = 0$, $k_0 h_2$ variable.

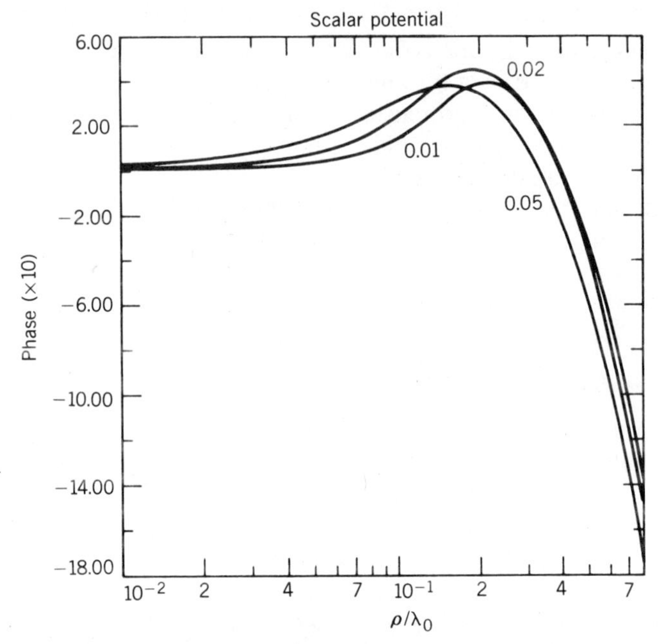

Fig. 24 Phase of the normalized scalar potential for a two-layer substrate when the source and the observer are both on the air–upper layer interface. Upper layer: $\varepsilon_{r1} = 2$, $\tan \delta_1 = 0$, $k_0 h_1 = 0.05$. Lower layer: $\varepsilon_{r2} = 5$, $\tan \delta_1 = 0$, $k_0 h_2$ variable.

normalized scalar potential $4\pi\varepsilon_0 \lambda_0 V$ as a function of the normalized radial distance. Layer's parameters are $\varepsilon_{r1} = 2$, $\varepsilon_{r2} = 5$, $k_0 h_1 = 0.05$ and $k_0 h_2 =$ 0.01, 0.025, 0.05. As could be expected, the surface-wave behavior is reached faster for thicker substrates. In these two figures, source and observer are both at the air–dielectric interface.

In fact, there are four situations of practical interest, depending on the relative position of source and observer:

1. Source and observer are both at the air–dielectric interface ($z = z' = 0$).
2. Source and observer are both at the interface between dielectrics ($z = z' = -h_1$).
3. Source at $z' = 0$ and observer at $z = -h_1$.
4. Source at $z' = -h_1$ and observer at $z = 0$.

The last two possibilities give identical results for the potentials as can be shown by reciprocity.

Figures 25 and 26 give the modulus and phase of the scalar potential, calculated in cases 1–3 for a substrate of parameters $\varepsilon_{r1} = 2$, $\varepsilon_{r2} = 5$,

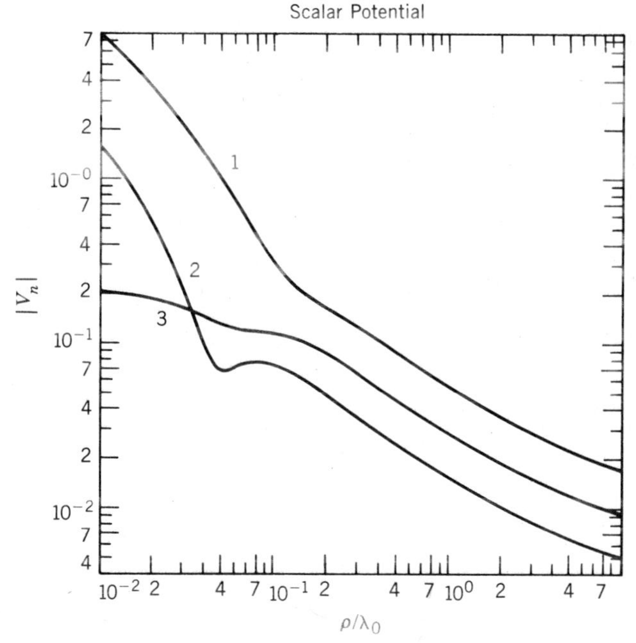

Fig. 25 Modulus of the normalized scalar potential for a two-layer substrate. Upper layer: $\varepsilon_{r1} = 2$, $\tan\delta_1 = 0$, $k_0 h_1 = 0.05$. Lower layer: $\varepsilon_{r2} = 5$, $\tan\delta_1 = 0$, $k_0 h_2 = 0.025$. (1) $z' = 0$, $z = 0$; (2) $z' = -h_1$, $z = -h_1$; (3) $z' = -h_1$, $z = 0$.

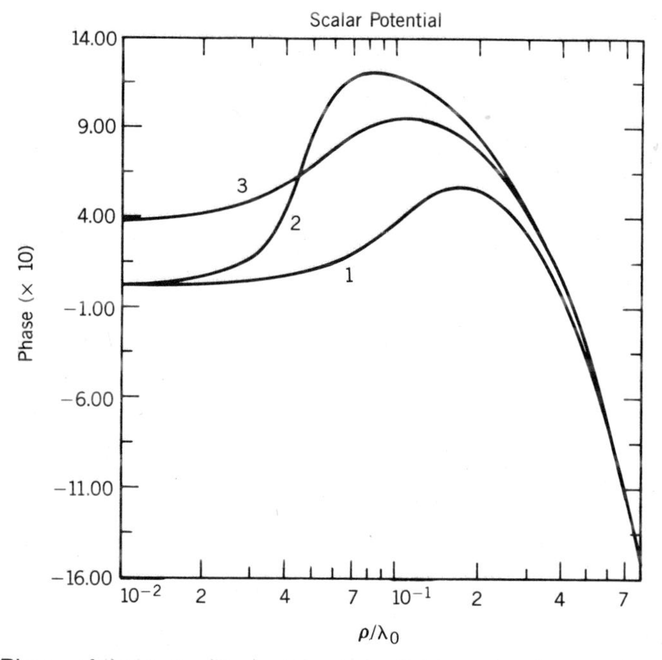

Fig. 26 Phase of the normalized scalar potential for a two-layer substrate. Upper layer: $\varepsilon_{r1} = 2$, $\tan\delta_1 = 0$, $k_0 h_1 = 0.05$. Lower layer: $\varepsilon_{r2} = 5$, $\tan\delta_1 = 0$, $k_0 h_2 = 0.025$. (1) $z' = 0$, $z = 0$; (2) $z' = -h_1$, $z = -h_1$; (3) $z' = -h_1$, $z = 0$.

$k_0h_1 = 0.05$, and $k_0h_2 = 0.025$. Curves 1 and 2 have a similar shape, but the values are higher in case 1 because the dielectric constant of the upper layer is smaller. As expected, the potential remains bounded in case 3 when the radial distance goes to zero. This fact is easily understood because source and observer are located at different levels. For the same reason, the phase starts with a nonzero value in case 3 (Fig. 26).

10. THE METHOD OF MOMENTS

In order to apply the mixed potential integral equation (MPIE) to irregular microstrip shapes, we need a very flexible numerical technique and the method of moments (MoM) has been selected. This technique [13] transforms the integral equation into a matrix algebraic equation that can be easily solved on a computer. The method of moments is among the most widely used numerical techniques in electromagnetics and a detailed account of the underlying principles will not be given here. For microstrip structures, two particular versions of the MoM deserve attention: the subsectional basis functions approach and the use of entire domain basis functions.

10.1. Subsectional Basis Functions

If no a priori assumptions about the shape of the patches are made, a successful technique must decompose the patch into small elementary cells and define simple approximations for the surface current on each cell. The most commonly used shapes for the elementary cells are the triangle [29] and the rectangle. Even though the triangular shape is more flexible, rectangular cells involve simpler calculations and suffice for many microstrip problems.

The most frequent choice is a method of moments with subsectional basis functions [13]. In this approach, the upper conductor is divided into elementary domains (cells) and the basis functions are defined over each cell. We have chosen the rectangular cell as the simplest shape that is able to provide good approximations for many practical structures. More sophisticated shapes for the cells, such as triangles [29] and quadrangles, have been used in scattering problems and could be applied to microstrips.

The size of the cells depends on the nature and geometry of the problem to be solved. In any case the linear size of the cell should not exceed one tenth of the guided wavelength.

We also have to select the basis functions. In general, each component of the surface current will depend on the two coordinates x, y, but it is possible to use basis functions that are, inside each cell, constant along the transverse coordinate. This yields expansions for J_{sx} and J_{sy} that are discontinuous along y and x, respectively, but have an associated charge that is nonsingular. Basis functions ensuring continuity of the current in any direction, such

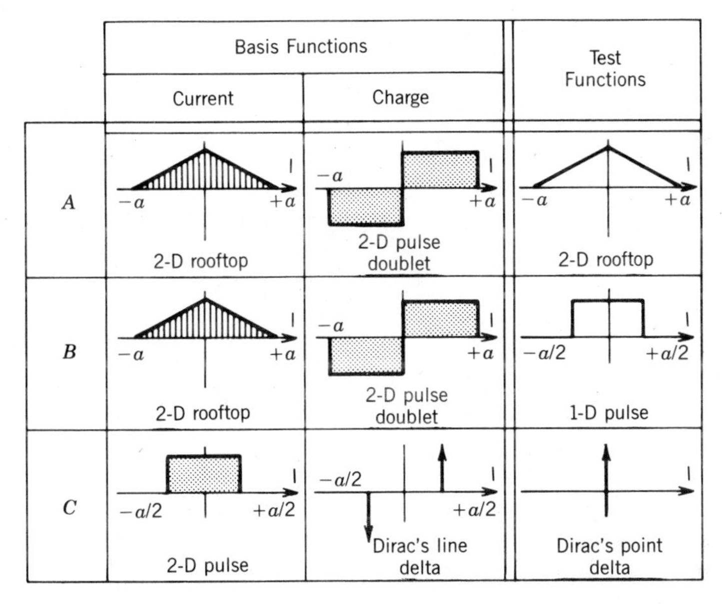

Fig. 27 Some possible choices for subsectional basis functions defined over rectangular domains. All the two-dimensional functions considered are independent of the transverse coordinate. After [39] © 1988 IEEE.

as bilinear expansions, may be used, but the improved accuracy of the results is offset by the increased difficulty of the computations.

The choice of test functions is also a crucial matter. To illustrate this, three possible combinations of basis and test functions will be described (Fig. 27).

10.1.1. Case A: Rooftop Functions and Galerkin

An interesting possibility is to use overlapping rooftop functions for the two components of the surface current [14]. Then, according to continuity equation, the basis functions for q_s are two-dimensional pulse doublets (Fig. 27). The MPIE is tested using the same rooftop functions, yielding a Galerkin procedure.

Let us define $\mathbf{T}_i$ as the vector rooftop function associated with two adjacent cells S_i^+ and S_i^- (Fig. 28). The union of these two cells will be simply denoted by S_i. In general, we need to consider N_x x-directed functions and N_y y-directed functions, the total number of basis functions is thus $N = N_x + N_y$.

$$\mathbf{T}_i = \begin{cases} \mathbf{e}_x T_{ix}\,, & i = 1, \ldots, N_x \\ \mathbf{e}_y T_{iy}\,, & i = N_x + 1, \ldots, N \end{cases} \tag{123}$$

The current and the charge are expanded as

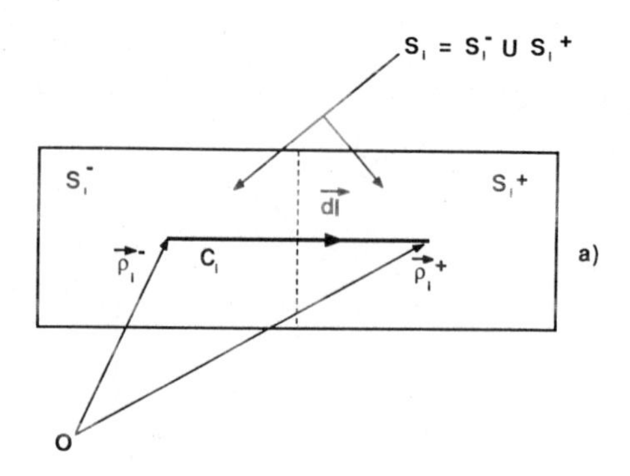

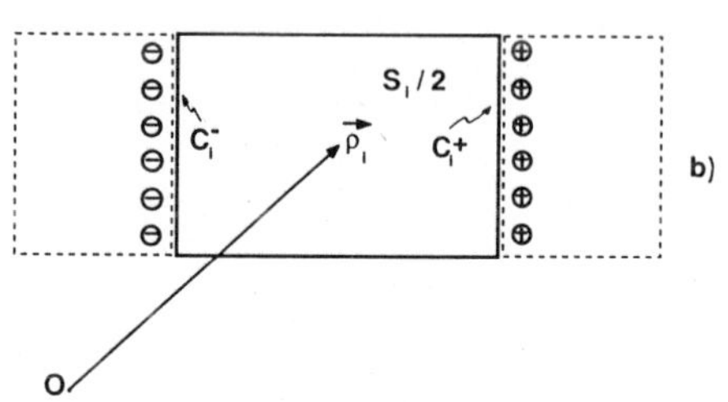

Fig. 28 Longitudinal testing segments C_i linking the centers of adjacent cells S_i^+ and S_i^-. Transverse segments C_i^+ and C_i^- containing the line charge densities in the point-matching approach. After [39] © 1988 IEEE.

$$\mathbf{J}_s = \sum_{i=1}^{N} \alpha_i \mathbf{T}_i \qquad q_s = \sum_{i=1}^{N} \alpha_i \Pi_i \tag{124}$$

where the α_i are unknown coefficients and the functions $\Pi_i = -\nabla \cdot \mathbf{T}_i / j\omega$ correspond to the pulse doublets.

Standard application of the method of moments now yields a matrix equation

$$C\boldsymbol{\alpha} = \mathbf{b} \tag{125}$$

with the elements of the matrix given by:

$$z_{ij} = a_{ij} + v_{ij} + l_{ij} \tag{126}$$

where the contributions from $\mathbf{A}$, V, and the ohmic losses are, respectively:

$$a_{ij} = j\omega \int_{S_i} \mathbf{T}_i(\boldsymbol{\rho}) \cdot \int_{S_j} \bar{\bar{G}}_A(\boldsymbol{\rho}|\boldsymbol{\rho}') \cdot \mathbf{T}_j(\boldsymbol{\rho}')\, dS'\, dS \tag{127}$$

$$v_{ij} = \frac{1}{j\omega} \int_{S_i} \Pi_i(\boldsymbol{\rho}) \int_{S_j} G_V(\boldsymbol{\rho}|\boldsymbol{\rho}') \Pi_j(\boldsymbol{\rho}')\, dS'\, dS \tag{128}$$

$$l_{ij} = Z_s \int_{S_i} \mathbf{T}_i(\boldsymbol{\rho}) \cdot \mathbf{T}_j(\boldsymbol{\rho})\, dS \tag{129}$$

The element b_i of the vector of independent terms is

$$b_i = \int_{S_i} \mathbf{T}_i(\boldsymbol{\rho}) \cdot \mathbf{E}^e(\boldsymbol{\rho})\, dS \tag{130}$$

where the second expression for v_{ij} in eq. (128) has been obtained by using the continuity equation, the edge condition for surface currents and the identity $\mathbf{y} \cdot \nabla x = \nabla \cdot (x\mathbf{y}) - x\nabla \cdot \mathbf{y}$.

Notice that a_{ij} vanishes if $\mathbf{T}_i$ is perpendicular to $\mathbf{T}_j$ and $l_{ij} = 0$ is there is no intersection between S_i and S_j. In general, the evaluation of each matrix element requires the computation of fourfold integrals.

If the Green's functions are expressed as inverse double Fourier transforms, the matrix elements include terms of the form

$$I = \int_{S_i} f_i \int_{S_j} f_j \iint_{-\infty}^{\infty} g\, dk_x\, dk_y\, dS'\, dS \tag{131}$$

The surface integrals can now be evaluated analytically and after a change to polar spectral coordinates we have

$$I = \int_0^{\pi/2} \int_0^{\infty} F_{ij}\, dk_\rho\, dk_\phi \tag{132}$$

where the functions F_{ij} are different for each matrix element. This last integral must be evaluated numerically [30]. On the other hand, if we use Hankel inverse transforms we have

$$I = \int_{S_i} f_i \int_{S_j} f_j \int_0^{\infty} G\, dk_\rho\, dS'\, dS \tag{133}$$

The double surface integral can be reduced to a single surface integral with the change of variables $x - x' = u$, $y - y' = v$, which takes advantage of the translational invariance of the Green's functions along the transverse coordinates. Finally, we have

$$I = \int_{S_{ij}} f_{ij} \int_0^{\infty} G\, dk_\rho\, du\, dv \tag{134}$$

where S_{ij} is a new domain in the uv plane, and f_{ij} is a combination of f_i and f_j. This last form of the integral seems more appropriate than (132) in spite of the fact that now we have a surface integral instead of a one-dimensional integral on k_ϕ. The reason is that in (134) the integral along k_ρ does not depend on the indices i, j. Hence, it can be precomputed, stored, and retrieved with a simple interpolation scheme.

10.1.2. Case B: Rooftop Functions and Razor Testing

This modification was suggested by [14] and successfully applied to microstrip resonators and antennas [31].

The basis functions are the same as in case A, but testing is done along the segment C_i linking the centers of cells S_i^+ and S_i^- (Fig. 28). Thus, we get, instead of (127)–(130),

$$a_{ij} = j\omega \int_{C_i} d\mathbf{l} \cdot \int_{S_j} \bar{\bar{G}}_A(\boldsymbol{\rho} | \boldsymbol{\rho}') \cdot \mathbf{T}_j(\boldsymbol{\rho}')\, dS' \tag{135}$$

$$v_{ij} = \frac{1}{j\omega} \int_{S_j} [G_v(\boldsymbol{\rho}_i^+ | \boldsymbol{\rho}') - G_v(\boldsymbol{\rho}_i^- | \boldsymbol{\rho}')]\Pi_j(\boldsymbol{\rho}')\, dS' \tag{136}$$

$$l_{ij} = Z_s \int_{C_i} \mathbf{T}_j(\boldsymbol{\rho}) \cdot d\mathbf{l} \tag{137}$$

$$b_i = \int_{C_i} \mathbf{E}^e(\boldsymbol{\rho}) \cdot d\mathbf{l} \tag{138}$$

where $P_i^{+,-}$ denote the centers of the cells $S_i^{+,-}$. These expressions, simpler than (127)–(130), can be brought to effective numerical evaluation [31].

In Section 7 we mentioned the fact that MPIE remains valid at low frequency and tends to the static integral equation. Numerically, the condition of the matrix of moments worsens when the frequency decreases, thus preventing accurate results. This drawback can be removed by testing along the segments belonging to an open tree and replacing the remaining segments by closed loops [31, 32]. According to Faraday's law, a null circulation of the electric field along closed loops is equivalent to enforcing a zero average value of the normal magnetic field inside the loop. Hence, the quasi-static integral equation of Section 7 is included in the MPIE.

10.1.3. Case C: Two-Dimensional Pulses and Point Matching

The simplest, but still meaningful, combination of basis and test functions expands the components of the current in a set of two-dimensional pulses. In order to approximately satisfy the condition for currents normal to the edges, these pulses are defined over domains that do not conicide with the original cells. Rather, each domain, symbolically denoted by $S_i/2$, is a combination of two cell's halves and can be considered a two-dimensional extension of the segment C_i (fig. 28).

The associated charges are now line charges (Dirac's delta functions) distributed along two segments C_i^+ and C_i^- (Fig. 28). The MPIE is tested here by point matching at the centers of segments C_i. Only the component of the electric field parallel to the segment is tested. A general matrix element is still given by the decomposition (126), but now we have

$$a_{ij} = j\omega \mathbf{e}_i \cdot \int_{S_j/2} \bar{\bar{G}}_A(\boldsymbol{\rho} | \boldsymbol{\rho}') \cdot \mathbf{e}_j\, dS' \tag{139}$$

$$v_{ij} = \frac{1}{j\omega} \int_{C_i^+} G_v(\boldsymbol{\rho}_i | \boldsymbol{\rho}')\, dl' - \frac{1}{j\omega} \int_{C_i^-} G_v(\boldsymbol{\rho}_i | \boldsymbol{\rho}')\, dl' \tag{140}$$

$$l_{ij} = Z_s S_{ij} \tag{141}$$

$$b_i = \mathbf{e}_i \cdot \mathbf{E}^e(\boldsymbol{\rho}_i) \tag{142}$$

where δ_{ij} is the Kronecker's symbol and $\mathbf{e}_i$ ($\mathbf{e}_j$) is a unit vector parallel to C_i (C_j).

10.1.4. The Numerical Integration Problem

The differences between the various combinations of basis and test functions disappear if careless numerical integration is used. For instance, it is meaningless to apply a Galerkin approach of the case A type and then perform the integrations in eqs. (127)–(129) by using the means theorem, because then the resulting algorithm will be more likely a point-matching technique. In this sense, the case B technique can be considered a particular version of that of case A using a rather loose integration technique. A similar result is obtained if we combine two-dimensional pulses for the current and two-dimensional pulse doublets for the charge [13, 14]. The continuity equation is no longer satisfied, but the approach can be justified on numerical grounds as being the case B technique with an approximate surface integration.

10.2. Entire Domain Basis Functions

If the microstrip patch has a simple regular shape (circular, rectangular) we can consider the equivalent electromagnetic cavity obtained when the patch is enclosed by a lateral magnetic wall. If the eigenmodes have a simple analytical expression it is reasonable to use them as a set of entire domain basis functions. For thin substrates, the surface current distribution on a microstrip patch at resonance follows closely the behavior of the corresponding eigenmode except for a slight disturbance due to the antenna's excitation. Therefore, meaningful results can frequently be obtained by using only one entire domain basis function or perhaps two or three. This fact makes possible the analysis of microstrip arrays having hundreds of elements. The size of the linear system to be solved will be equal to twice or three times the number of patches.

It is clear that if only one basis function is allowed per patch, we cannot use a poor testing procedure like point matching and match the boundary condition only at the center of the patch. The best alternative is provided by Galerkin's method where the test functions are identical to the basis functions and the inner product includes a surface integration over the patches [13].

To be clear, let us consider a single rectangular patch of dimensions L, W. The eigenmodes of the corresponding cavity with lateral magnetic walls are [33]

$$\mathbf{f}_j = \mathbf{e}_x \sin\frac{m\pi x}{L}\cos\frac{n\pi y}{W} + \mathbf{e}_y \cos\frac{m\pi x}{L}\sin\frac{n\pi y}{W} \tag{143}$$

for the surface currents, and

$$h_j = -\nabla\cdot\mathbf{f}_j/j\omega \tag{144}$$

for the surface charge.

In these expressions, the coordinate origin is in the lower left corner of the patch. The vectors $\mathbf{f}_j$ correspond to the modes TM_{mn0} in the eqivalent cavity. The choice of the modes TM_{mn0} intervening in the expansion of $\mathbf{J}_s$ is rather arbitrary and depends on the problem considered. For instance, they can be ordered according to their resonant frequencies, or we can consider only the TM_{m00} subset if variations along the coordinate y can be neglected. In any case, the relation between the integers m, n and the ordinal number j in (143) should be clearly defined.

We proceed now as discussed in Section 10.1.1 and obtain the system of linear equations specified by eqs. (125)–(130) with $\mathbf{T}_j$ replaced by $\mathbf{f}_j$, and Π_j by h_j. With entire domain basis functions, the numerical simplifications expressed by eqs. (132) and (134) are still valid. The domains S_i, S_j always coincide with a physical patch and are identical for a single-patch problem.

10.3. Computational Details

In all the versions of the MoM discussed, the integrals giving the diagonal terms c_{ii} exhibit a singularity of the Green's functions when $\mathbf{r} = \mathbf{r}'$.

Fortunately enough, this singularity is of the type $|\mathbf{r} - \mathbf{r}'|^{-1}$ for the potentials $\mathbf{A}, V$ as discussed in Section 3. For these cases, it is then recommended that the singular part G_s of a given Green's function be extracted, i.e.

$$G = G_s + (G - G_s) \tag{145}$$

This singular part corresponds to the dominant term at zero frequency (static case), and its value is given by

$$G_s = \frac{\mu}{4\pi|\mathbf{r} - \mathbf{r}'|} \tag{146}$$

for the vector potential and

$$G_s = \frac{1}{2\pi\varepsilon_0(\varepsilon_r + 1)|\mathbf{r} - \mathbf{r}'|} \tag{147}$$

for the scalar potential.

In most cases, the singular part G_s can be analytically integrated over the patch's or cell's surface.

Another possibility, easy to implement for rectangular domains, is the introduction of polar coordinates $x - x' = R\cos\alpha$, $y - y' = R\sin\alpha$. Thus, the singularity $1/R$ in the Green's functions is compensated by the element of surface $R\,dR\,d\alpha$ in polar coordinates.

Other common situations arise when the domains S_i and S_j are separated by a distance much greater than the linear dimensions of the domains.

For these cases the approximation

$$\int_{S_i} f_i \int_{S_j} f_j G(\mathbf{r}|\mathbf{r}')\,dS'\,dS \simeq G(\mathbf{r}_i|\mathbf{r}_j)\int_{S_i} f_i\,dS \int_{S_j} f_j\,dS' \tag{148}$$

where $\mathbf{r}_i$, $\mathbf{r}_j$ denote the centers of the patches, will be helpful.

More accurate approximations can be obtained by expanding the Green's functions in a Taylor series.

It must also be pointed out that for matrix elements corresponding to close or overlapping surfaces, in the subsectional basis approach, we can use rough approximations for the contribution a_{ij} of the vector potential, which is much weaker than that of the scalar potential v_{ij} provided that the cells are electrically small

10.3.1. Interpolation Among Green's Functions

The evaluation of the matrix in eq. (125) requires a large amount of computation. For a rectangular patch divided into 10×10 cells, the order of the matrix is 180; hence the number of elements in the matrix is $180^2 =$ 32,400. Even when a simple 4×4 Gaussian quadrature, combined with a point-matching procedure, is used, the number of Sommerfeld integrals that need to be evaluated would exceed half a million.

Fortunately, for a given structure these integrals depend only on the distance from source to observer. It is thus possible to tabulate the integrals for a small number (typically 50 to 200) of source–observer distances, and then interpolate between the tabulated values. The distances to be considered range from zero to the maximum linear dimension of the patch. Among the several interpolation schemes that have been tried, a good solution was obtained by separating the singular part of Green's functions according to eqs. (145)–(147), and then using linear or parabolic Lagrange interpolation for the regular part. Sampling points can be evenly distributed in the range of distances considered, or they can be concentrated in the lower end of the range, since errors in Green's functions are more critical for small distances.

An even better method uses the fact that for free space the logarithm of the modulus of Green's functions is a linear function of the logarithm of the source–observer distance, and the phase of Green's functions is a linear

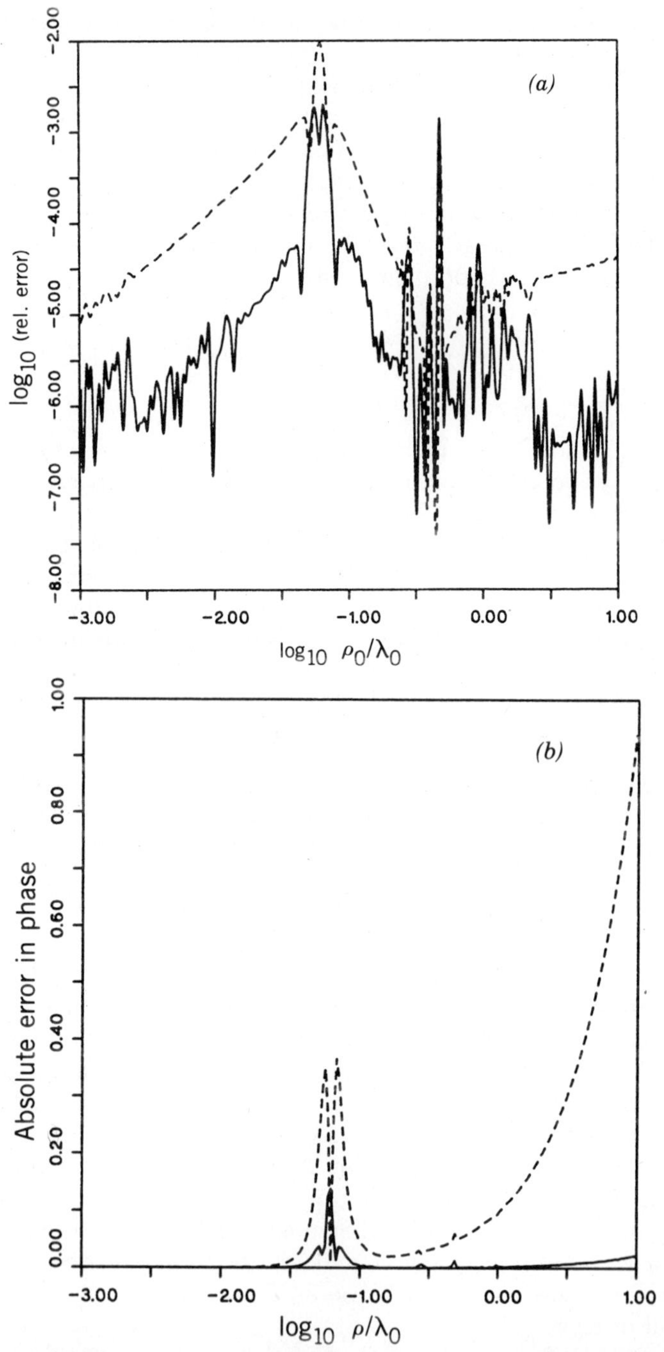

Fig. 29 Error introduced by an interpolation made on the logarithm of the scalar potential if the logarithm of the normalized distance is the independent variable linearly sampled. $\varepsilon_{r1} = 5$, $\tan \delta_1 = 0.01$, $h/\lambda_0 = 0.05$, and 200 interpolating points in the range $0.001 < \rho/\lambda_0 < 10$. (*a*) Relative error in the modulus; (*b*) absolute error (degrees) in the phase.

function of the distance. Consequently, the interpolation scheme is performed with these quantities and there is no need to extract the singular part. Another improvement results from the use of the function $\rho G(\rho)$, which remains bounded at $\rho = 0$.

Figures 29*a* and 29*b* show the error introduced in the modulus and phase of the scalar potential when the interpolation is made on its logarithm. The interpolating points are sampled for distances ranging from 0.001 to 10 free-space wavelengths and the number of points is 200 (50 per decade). If a simple linear interpolation between points is used (dashed curves), the relative error in the modulus is below 0.1% except in the transition zone (Section 9), where the potential changes abruptly and interpolation error can reach 1%. The absolute error for the phase is less than 1° throughout the range. If a more sophisticated parabolic interpolation is made, these figures are reduced by a factor of 10 (solid curves in Fig. 29).

11. EXCITATION AND MULTIPORT ANALYSIS

In practice, microstrip structures are connected to the outer world using many different techniques. A good survey of these has recently been given by Pozar [31]. We shall now describe some of the most commonly found.

11.1. Practical Excitations for Microstrip Structures

The upper conductors (patches) of a microstrip structure in a single layer configuration are usually excited by a directly connected microstrip line (Fig. 30*a*). A thorough treatment of this excitation would need the analysis of the incident and reflected quasi-TEM current waves existing in the line that is considered to extend from the patch to infinity. However, microstrip line feeds are in practice hardly straight and terminate after a usually short length into a connector or another microstrip device. Therefore, a realistic approach considers the microstrip line as having finite length and belonging to the patch and introduces a mathematical excitation (for instance, a filament of vertical current or a voltage series generator) at the end of the line (Fig. 30*b*). As we shall see soon, the vertical filament of current is also a good model for the coaxial excitation and on a physical basis is preferred to the series voltage generator. The microstrip line then must be cut at a point where uniform line conditions can be assumed. In this way the discontinuity created by the junction line patch and, possibly, discontinuities of the line itself (Fig. 30*a*) are included in the analysis. If a vertical filament of unit current is used as excitation, the input impedance is numerically equal to the voltage at the insertion point. When other mathematical excitations are used, the section of line included in the analysis must be long enough to support a standing wave pattern and estimate from it the

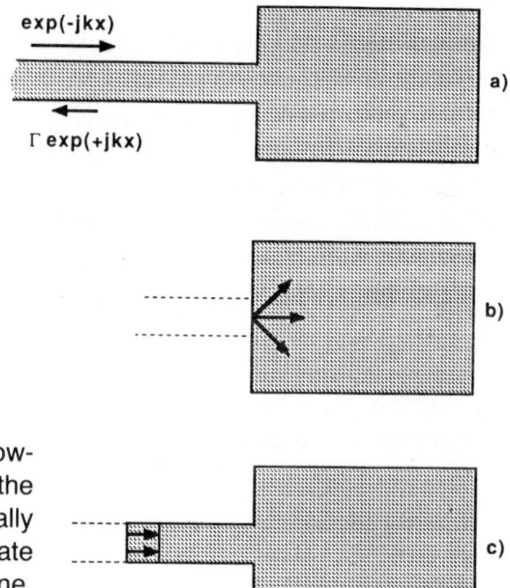

Fig. 30 (*a*) Microstrip discontinuity showing the incident and reflected waves in the feed line. (*b*) Approximate model totally neglecting the feed line. (*c*) Approximate model partially neglecting the feed line. After [39] © 1988 IEEE.

reflection coefficient and hence the input impedance at any point of the microstrip line.

Finally, it is worth mentioning that replacing the microstrip line by a coaxial probe at the corresponding point in the edge of the patch (Fig. 30*c*) is a first-order approximation that while neglecting the effect of the line width, gives surprisingly good results in many practical cases.

For two-layer substrates, a more sophisticated excitation can be used [35, 36]. The microstrip line is now under the patch and couples electromagnetically to it. The presence of two dielectric layers gives an additional degree of freedom to the design. The considerations made for the line directly connected to the patch also apply here. The microstrip line can be cut far from the patch and terminated in a coaxial probe or excited by some other mathematical device.

Another interesting way to provide excitation considers a slot in the ground plane that couples energy to the patch from, presumably, a triplate transmission line [36]. The mathematical treatment replaces the slot by a distribution of surface magnetic current. Hence, we would need the fields of horizontal magnetic dipoles, which can be related to those of an electric dipole derived in Section 4 by duality [1]. The modifications would affect only the excitation field $\mathbf{E}^{(e)}$ or, equivalently, the independent terms b_i in the matrix equation.

11.1.1. Coaxial Probes under the Patch

A coaxial line attachment under the patch (Fig. 31*a*) is also a practical way of feeding a microstrip device. From a theoretical point of view, coaxial

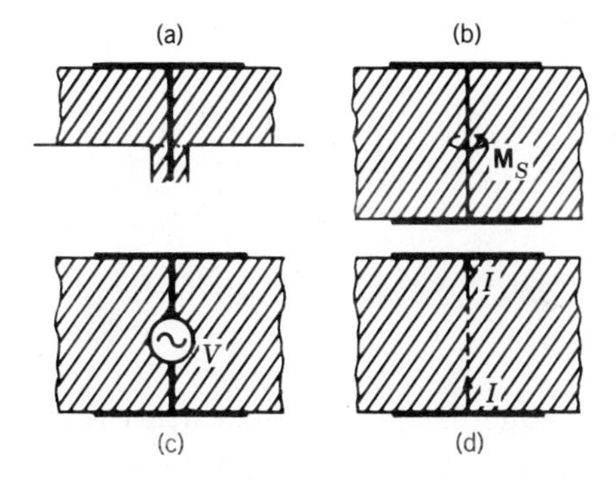

Fig. 31 Modeling a coaxial probe feed. (*a*) Actual geometry; (*b*) image theory and frill of magnetic currents; (*c*) series voltage generator; (*d*) filament of unit current.

excitation is of great interest, since it leads to the simplest mathematical models and can be used to describe other excitations as shown above.

The most accurate treatment of a coaxial probe [37, 38] assumes that the portion of the inner coaxial conductor embedded in the substrate belongs to the patch. Then the whole structure is excited by a frill of magnetic current existing between the inner and outer conductors at the patch's level (Fig. 31*b*). A possible simplification replaces the magnetic frill by a lumped series voltage generator. This approximation has been successfully used for wire antenna modeling. The excitation terms b_i are zero everywhere except at the cell occupied by the generator (Fig. 31*c*).

In the above models, the coaxial probe belongs to the structure, and the method of moments must also be applied to it. However, for thin substrates, replacing the coaxial by a vertical filament of current (Fig. 31*d*) gives results accurate enough for engineering purposes. The advantage of this simplification is that we do not need to enforce the continuity of the tangential electric field at the probe's surface. Therefore, this model is also the most adequate when the coaxial probe is just a mathematical excitation but does not correspond to the physical reality.

If the Galerkin technique is used, we can assume a filament of zero diameter ending in a point charge at the junction with the patch. The calculation of the excitation fields would require a knowledge of the fields created by vertical electric dipoles embedded in the substrate. However, the reciprocity theorem allows evaluation of the terms b_i in the matrix equation by using only formulas related to horizontal electric dipoles. We have

$$b_i = \int_{S_i} \mathbf{J}_{s_i} \cdot \mathbf{E}^e \, dS = \int_{S_e} \mathbf{J}^e \cdot \mathbf{E}_i \, dS \tag{149}$$

where $\mathbf{J}_{s_i}$ is the ith basis function for the surface current and $\mathbf{E}_i$ the field created by it.

Now the unit excitation current entering the point x_e, y_e is given by a product of Dirac's delta functions

$$\mathbf{J}^e = \mathbf{e}_z \delta(x - x_e)\delta(y - y_e)\,, \qquad -h < z < 0 \tag{150}$$

and its domain S_e reduces to a segment of length h in the case of zero-diameter filament. Consequently, we have in terms of Green's functions

$$b_i = -\int_{S_i} G_V(x_e, y_e \,|\, x', y') q_{s_i}\, dS' - j\omega \int_{-h}^{0} \mathbf{e}_z \cdot \int_{S_i} \bar{\bar{G}}_A \cdot \mathbf{J}_{s_i}\, dS'\, dz \tag{151}$$

Using now the relationships between potentials and normal fields [eqs. (37) and (48)] we can cast the above expression in an easier form as

$$b_i = -\int_{S_i} G_V^*(x_e, y_e \,|\, x', y') q_{s_i}\, dS' \tag{152}$$

where G_V^* is the scalar Green's function when transverse potentials are used [eq. (49)]. Physically, we can say that the model for our coaxial excitation is a point charge located at the point where the coaxial is attached to the patch. However, this charge belongs to a vertical dipole, and its scalar potential is not necessarily identical to that obtained when considering horizontal dipoles.

If simplified test functions, such as those described in cases B and C of Section 11.1, are used, the coaxial excitation cannot be reduced to a point charge because some integrals would diverge. Rather, this charge as well as the current entering the patch must be spread in a reasonable way around the attachment point [31].

Now, once the vector of excitation terms b_i is known, the amplitudes of the basis functions are obtained as (125) $\boldsymbol{\alpha} = \boldsymbol{C}^{-1}\mathbf{b}$. The input impedance is finally given by

$$Z_{\mathrm{IN}} = -\sum_{i=1}^{N} \alpha_i b_i = -\mathbf{b} \cdot \boldsymbol{C}^{-1} \cdot \mathbf{b} \tag{153}$$

This is the impedance at the patch level, i.e., $z = 0$. If the real feeding structure is a microstrip line of characteristic impedance Z_c, the reflexion coefficient Γ in the line can be estimated as $(Z_{\mathrm{IN}} - Z_c)/(Z_{\mathrm{IN}} + Z_c)$. Alternatively, we can infer the value of G by an inspection of the standing waveform that appears in the surface current circulating on the line.

If the patch is really fed by a coaxial probe, the input impedance at the ground plane level is obtained by adding to Z_{IN} a small term accounting for the self-reactance of the portion of the probe embedded in the substrate [31].

11.2. Multiport Analysis

In many practical situations the microstrip device is excited simultaneously at M different points of the upper conductor (Fig. 32). In this case the

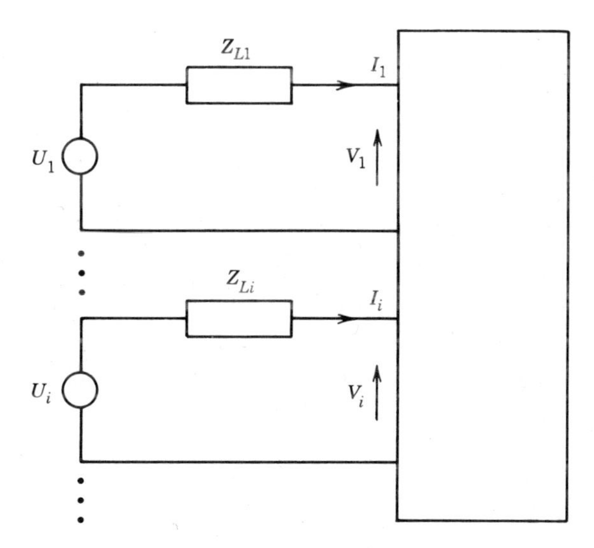

Fig. 32 The microstrip structure as a multiport device.

structure can be considered an M-port device, and standard circuit theory can be applied for a complete description.

The first step is to solve the linear system $\boldsymbol{C\alpha} = \mathbf{b}$, obtained by application of the method of moments, M times for M different excitation vectors $\mathbf{b}_j$. These vectors correspond to a physical situation in which a unit current is entering the jth port while the remaining $M - 1$ ports are open circuited. After solving the matrix equation, we get the vector $\boldsymbol{\alpha}_i = \boldsymbol{C}^{-1}\mathbf{b}_j$ containing the amplitudes of the N basis functions. Also, by computing the voltage at each port we get the quantities Z_{ij}, $i = 1, 2, \ldots, M$, which determine the jth column of the port impedance matrix.

Therefore, the complete determination of the impedance matrix needs the solution of M linear systems but, fortunately, the matrix $\boldsymbol{C}$ of the system remains unchanged. Once the impedance matrix is known, the scattering matrix is deduced by a standard transformation.

The elements Z_{jj} have been previously termed "input impedances." It must be pointed out now that these are input impedances corresponding to a single port excitation (the remaining $M - 1$ ports are open-circuited) and can be quite different from the true input impedances seen when all the ports are simultaneously loaded.

Let us consider that each port is connected to a voltage generator U_i with an internal impedance Z_{Li} (Fig. 32). This arrangement includes the case of a passive load Z_{Li} by allowing $U = 0$.

We define a vector $\mathbf{U}$ with elements U_i and a diagonal matrix $\boldsymbol{Z}_L$ with elements Z_{Li}. The equations relating currents I_i and voltages V_i at each port (Fig. 32) are

$$\mathbf{U} = \mathbf{Z}_L\mathbf{I} + \mathbf{V} = (\mathbf{Z}_L + \mathbf{Z})\mathbf{V} \qquad (154)$$

where **Z** is the matrix of impedances previously calculated. The vector of currents is then given by

$$\mathbf{I} = (\mathbf{Z}_L + \mathbf{Z})^{-1}\mathbf{U} \tag{155}$$

and the vector of voltages is $\mathbf{V} = \mathbf{Z}(Z_L + \mathbf{Z})^{-1}\mathbf{U}$.

Finally, the input impedance at each port is

$$Z_{\text{IN}} = \frac{V_i}{I_i} = \frac{\mathbf{Z}(\mathbf{Z}_L + \mathbf{Z})^{-1}\mathbf{U}]_i}{(\mathbf{Z}_L + \mathbf{Z})^{-1}\mathbf{U}]_i} \tag{156}$$

and it is clear that for a one-port device this input impedance equals the parameter Z_{11} which is directly given by eq. (142).

12. PRACTICAL APPLICATIONS

In order to check the validity of the proposed model, we have performed extensive numerical tests and compared the theoretical predictions with other published theories and with measured values. We give now two practical examples: the L-shaped patch and four coupled rectangular patches.

12.1. The L-Shaped Patch [39]

To illustrate the performance of MPIE when dealing with irregular shapes, we have selected an L-shaped patch (Fig. 33*a*). Its dimensions are $a = b = 56$ mm, and $c = d = 28$ mm. The patch has been built on a cheap substrate currently used for low-frequency integrated circuits. Estimated substrate parameters in the microwave range are $\varepsilon_r = 4.34$, $\tan\delta = 0.02$, and $h = 0.8$ mm. As depicted in Fig. 33*a*, the patch has been divided into 75 square elementary cells.

We looked first at internal resonances, considering three possible locations of the coaxial probe having the following coordinates:

A: $x = 8.4$ mm, $y = 47.6$ mm
B: $x = 47.6$ mm, $y = 8.4$ mm
C: $x = 2.8$ mm, $y = 2.8$ mm

These values always correspond to centers of square cells. This choice is not mandatory but simplifies the numerical calculations somewhat.

The three first resonances were detected as sharp minima in the determinant of the matrix of moments, at 0.998, 1.555, and 2.536 GHz. The input impedances seen from the three coaxial locations show at these frequencies a maximum of their real part, as expected in resonances.

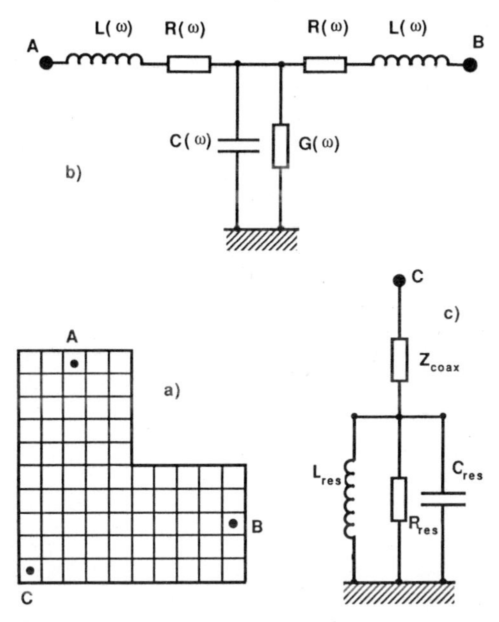

Fig. 33 The L-shaped microstrip patch. (*a*) Decomposition into cells showing the coaxial-fed ports *A*, *B*, and *C*. (*b*) Equivalent circuit at low frequency for a two-port excitation. (*c*) Equivalent circuit near resonance for a one-port excitation. After [39] © 1988 IEEE.

However, no resonant behavior of the input impedance is detected at the first resonance when the coaxial is located at the symmetrical position *C*. Figure 34 gives a vector representation for the real and imaginary part of the surface current density when the total excitation current entering the patch is normalized to $1 + j_0$ A and the frequency corresponds to the first resonance. As in any resonating situation, the imaginary part is stronger and its pattern, independent of the coaxial location, follows closely the eigenmode pattern of the equivalent cavity. On the other hand, the real part, neglected in many microstrip models, corresponds to the near-field effects created by the coaxial probe. This real part can modify greatly the input reactance values, mainly in weak resonances.

Figure 35 gives the imaginary parts (eigenmodes) associated with the second and third resonances.

The input impedance at the second resonance is given in Fig. 36 and compared with measurements. The theoretical predictions are quite good, with an error of less than 1% in the resonant frequency and about 4% in the maximum resistance. There is also a slight difference in the reactance values, which are more inductive in the measurements. The patch behaves as a parallel resonant circuit with a small series reactance due to the probe

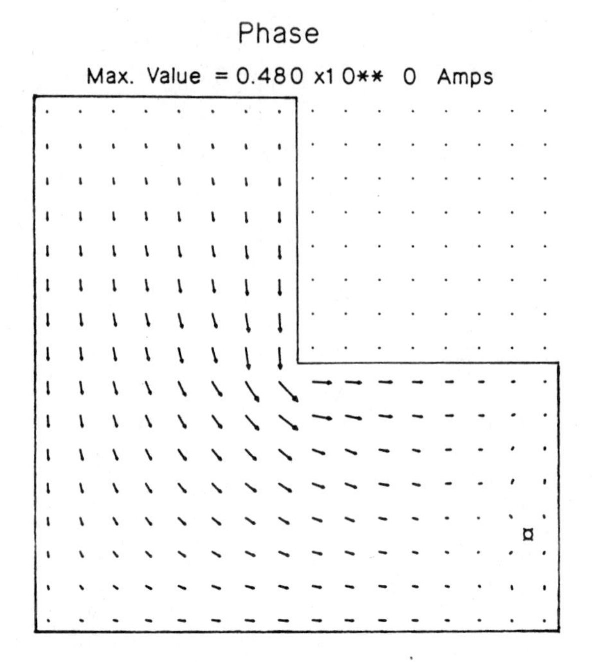

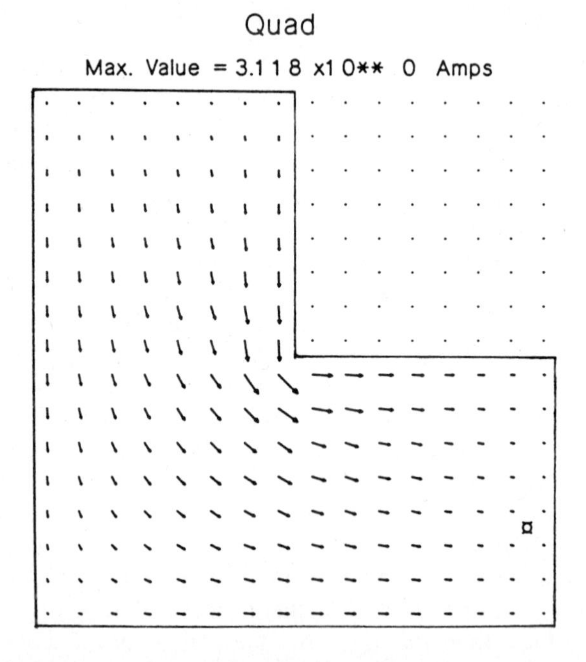

Fig. 34 L patch: Real (in-phase) and imaginary (in quadrature) parts of the surface current at the first resonance. The given maximum value corresponds to the longest arrow. Excitation current in the coaxial is $1 + j0$ A.

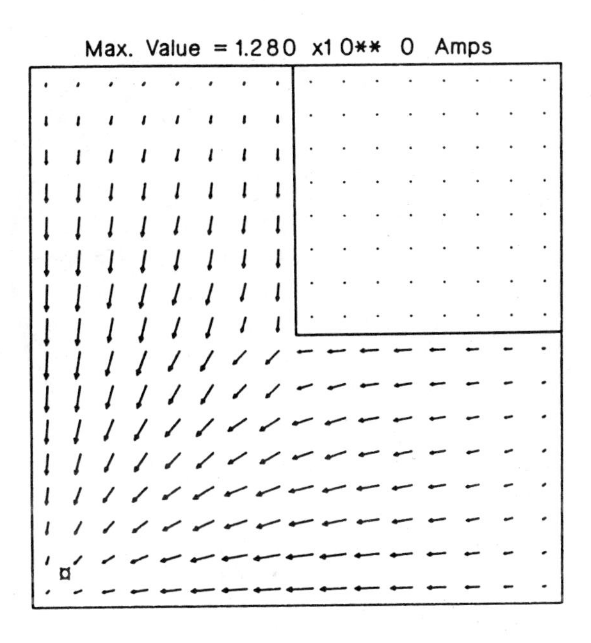

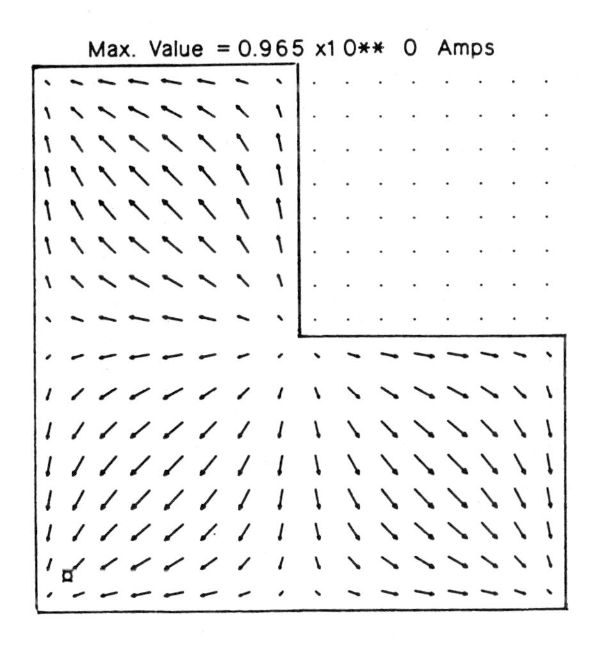

Fig. 35 L patch: Imaginary parts of the surface current for the second and third resonance.

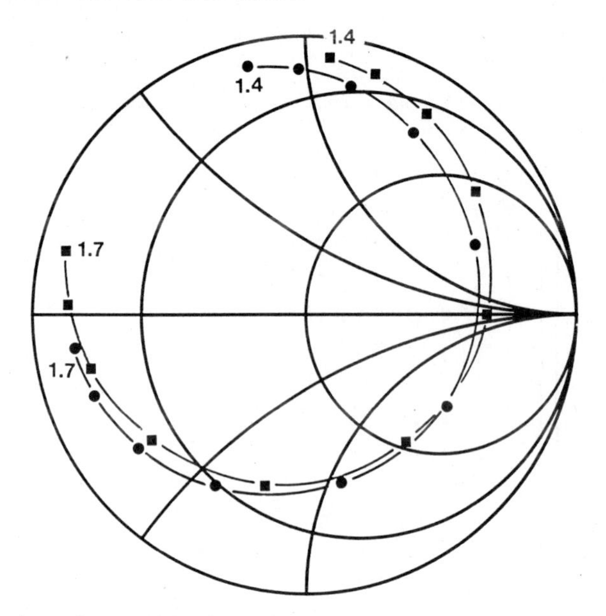

Fig. 36 Input impedance of the L patch near the second resonance (1.4–1.7 GHz). (●) Theoretical; (■) measured. After [39] © 1988 IEEE.

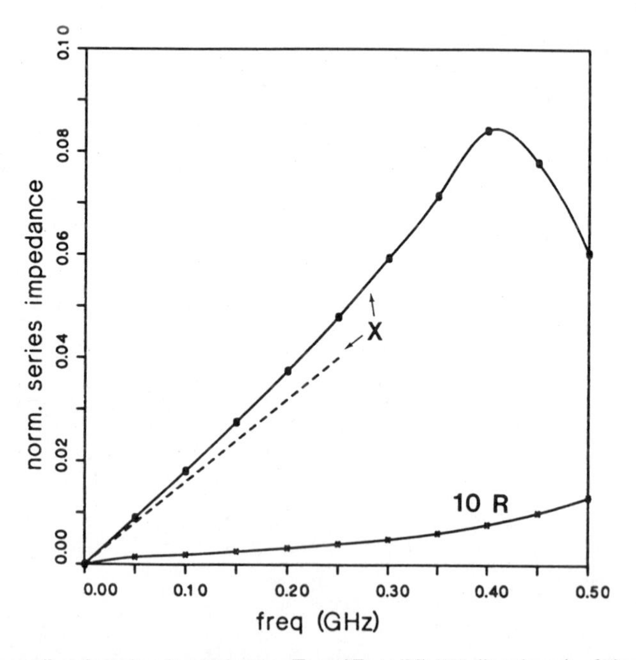

Fig. 37 Normalized series impedance $Z_s = (R + jX)/50$ (in ohms) of the L patch at low frequency (see Fig. 33*b*). The dashed line gives the quasi-static approximation $Z_s = j\omega L_0 = 0/50$ ohms. After [39] © 1988 IEEE.

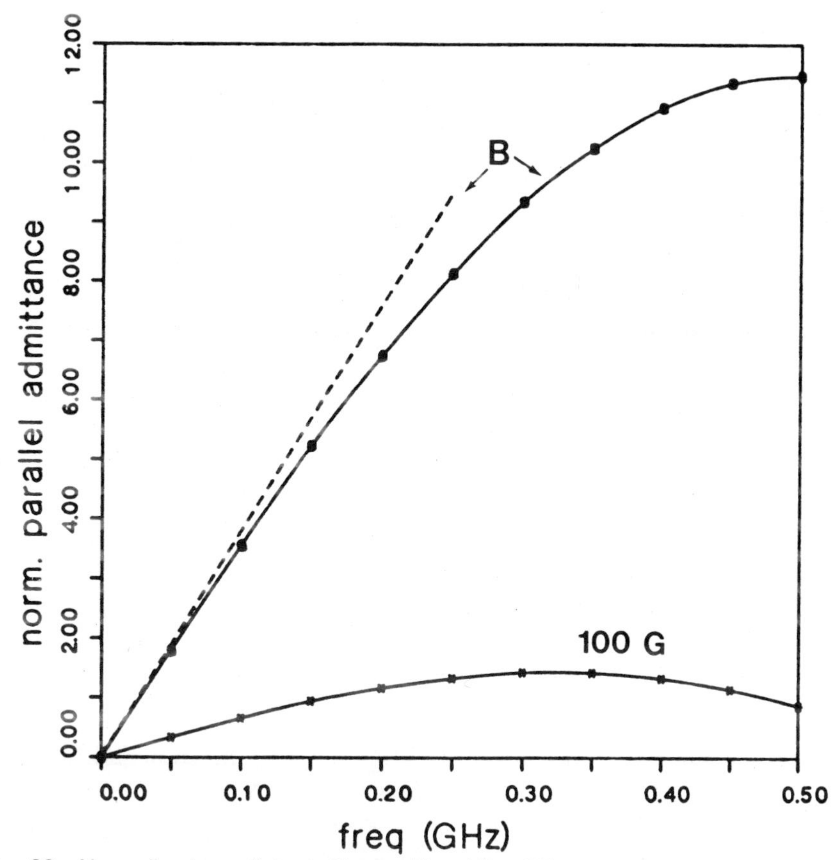

Fig. 38 Normalized parallel admittance $Y_p = (G + jB) \times 50$ ohms of the L patch at low frequency (see Fig. 33*b*). The dashed line gives the quasi-static approximation $Y_p = j\omega C_0 \times 50$ ohms. After [39] © 1988 IEEE.

(Fig. 33*c*). The elements of this equivalent circuit and the loaded quality factor are easily deduced from the impedance values.

We have also considered the L patch as a two-port network with coaxial excitations at points A and B (see Fig. 33*a*). at low frequencies, the patch behaves as a microstrip bend discontinuity, and we can assume the equivalent circuit of Fig. 33*b*. The normalized values of the series impedance $R(\omega) + jX(\omega)$ and of the parallel admittance $G(\omega) + jB(\omega)$ are given in Figs. 37 and 38. Again, the MPIE predicts correctly the behavior of the structure. In particular, as the frequency goes to zero, the reactance and susceptance values tend toward the quasi-static values ωL_0 and ωC_0, respectively, obtained with the simplified integral equations of Section 7.

12.2. Four Coupled Rectangular Patches

Four identical patches of dimension 60×40 mm have been constructed on the same substrate as above. The patches are disposed in a 2×2 rectangular lattice. The x axis is taken parallel to the longer sides and the patches resonate along this direction.

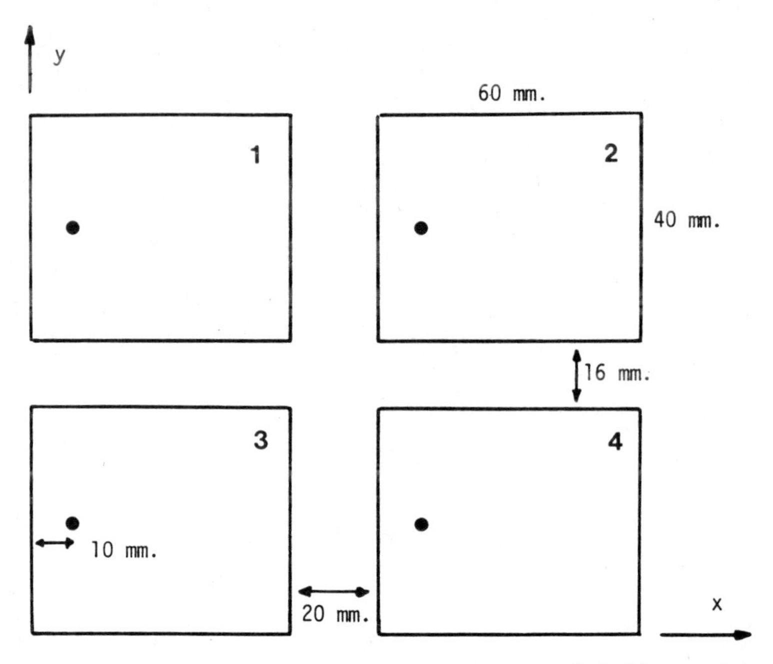

Fig. 39 Four coupled microstrip patches, numbered 1–4, fed by coaxial probes.

The separation between patches is 20 mm in the x-direction and 16 mm in the y direction. At the resonant frequency of an isolated patch (about 1.2 GHz) these gaps correspond to, respectively, 0.08 and 0.064 free-space wavelengths. The four patches are coaxially fed, with the probes contered on y and at 10 mm from the short edge (Fig. 39). The whole structure has been analyzed with two of the described techniques:

1. Rooftop functions with 40 (8×5) cells per patch. The total number of unknown basis functions is then 268.
2. Entire domain harmonics and Galerkin's method. Only five modes per patch were used (hence the total number of unknowns is 20).

Figure 40 gives the modulus in decibels of the four scattering parameters s_{1j} ($j = 1, 2, 3, 4$). Theoretical predictions made with both numerical techniques match very well, the difference being a few tenths of a decibel in modulus and less than 10 MHz in frequency. Both techniques agree closely with measurements, and they may be considered to be of identical accuracy. Therefore, entire domain basis functions are more adequate for this example, needing the resolution of a smaller linear system, even if the computation of matrix elements is more cumbersome. It must finally be said that, in this example, the stronger coupling (-19 dB) takes place between patches coupled by the resonant side. This is a typical feature of coupling between

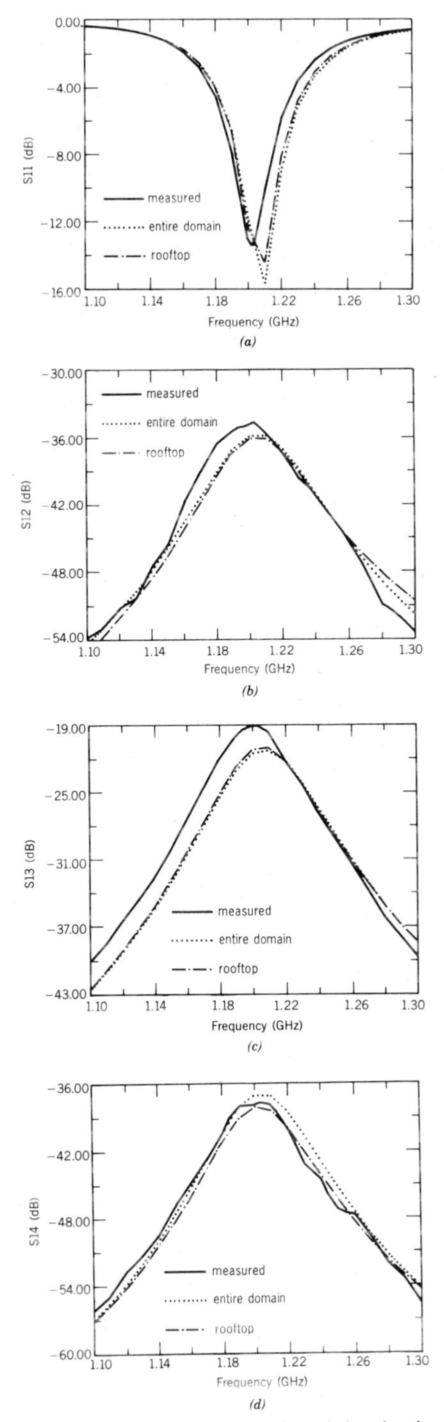

Fig. 40 (*a*)–(*d*) Some scattering parameters (modulus in decibels) for the configuration presented in Fig. 39 near resonance.

close structures, but when the distance between patches increases, the end-coupling in the x direction (parallel to the surface currents) decreases slowly, due to the surface wave, and becomes dominant.

12.3. A Slot in a Microstrip Line

As the last example we consider a microstrip line with a slot in the upper conductor (Fig. 41). The slot can be viewed as two close step discontinuities that cannot be analyzed separately. The microstrip line is printed on a single-layer substrate with $\varepsilon_r = 2.33$, $h = 0.51$ mm, $\tan\delta = 0.001$. The line width is $w = 1.53$ mm, which corresponds to a quasi-TEM characteristic impedance of 50 Ω. The slot is $s = 1.223$ mm wide and $t = 0.918$ mm deep.

In the previous examples, we obtained a scattering matrix and an equivalent circuit of the whole structure, including the coaxial probes. Therefore, the predicted theoretical values can be directly compared with measurements. However, in this example it is interesting to extract from the overall results an equivalent circuit for the geometrical region occupied by the discontinuity and limited by well-defined reference planes.

For this purpose, it is convenient to analyze the structure without a discontinuity, i.e., to remove the portion inside the planes PP' in Fig. 41 and to study the resulting uniform microstrip line section of length $2L_0$ excited by the two coaxial probes. The field analysis will then provide a total chain matrix $\boldsymbol{T}_0$ which can be decomposed in two identical parts $\boldsymbol{T}_h$, each one accounting for a half-section of length L_0:

$$\boldsymbol{T}_0 = \boldsymbol{T}_h \boldsymbol{T}_h \tag{157}$$

If we now study the structure including the discontinuity and obtain a

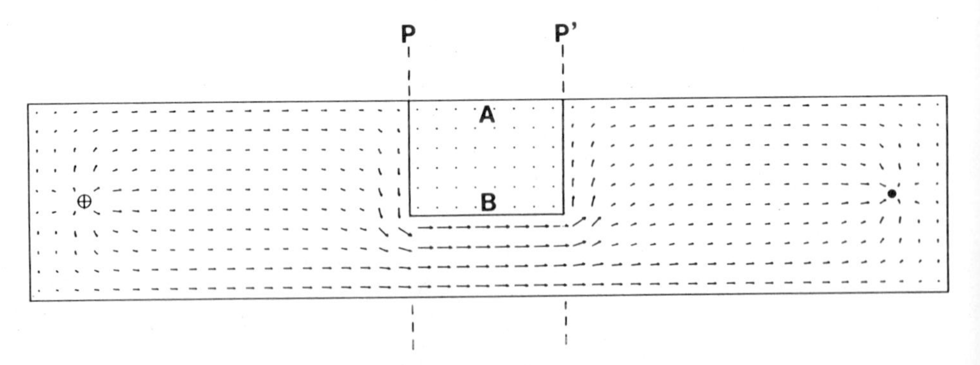

Fig. 41 A slot in the upper conductor of a 50 Ω microstrip line ($h = 0.51$ mm, $w = 1.53$ mm, $\varepsilon_r = 2.33$). The line is excited by two coaxial probes with currents I and $-I$. The figure shows the amplitude pattern of the surface current at 1 GHz. (**A**) Unperturbed contour. (**B**) Slot contour.

chain matrix $\boldsymbol{T}$, it is reasonable to assume that $\boldsymbol{T} = \boldsymbol{T}_h \boldsymbol{T}_d \boldsymbol{T}_h$ and therefore the chain matrix of the discontinuity is

$$\boldsymbol{T}_d = \boldsymbol{T}_h^{-1} \boldsymbol{T} \boldsymbol{T}_h^{-1} \tag{158}$$

In this way, the effects of the coaxial probes are eliminated and the equivalent circuit will bc, as required, independent of the excitation, provided that the line sections are long enough to see between excitation and discontinuity and typical current distribution of a uniform line.

To check the validity if the model, we first consider a slot with zero deep (contour A in Fig. 41). Then, eq. (157) should give the equivalent circuit of an unperturbed 50 Ω line section of length $2L_0$. Therefore, we can compare the impedance matrix derived from (157) with the theoretical impedance matrix provided by standard transmission line theory. In particular the complex characteristic impedance is given by

$$Z_c = \sqrt{Z_{11}^2 - Z_{12}^2} \tag{159}$$

and the complex propagation factor γ by

$$\gamma_s = \operatorname{arcosh}\left(\frac{Z_{11}}{Z_{12}}\right) \tag{160}$$

Preliminary calculation showed that with a decomposition into 5×48 cells, the real part of Z_c and the imaginary part of γ agreed with the quasi-TEM values within 1% at 1 GHz. Actually, the frequency dependence of characteristic impedance, effective dielectric constant, and attenuation factor can be obtained with this technique, but easier methods are available for propagation problems in uniform structures.

The slot was now introduced by removing the appropriate cells (contour B in Fig. 41). The current distribution resulting from the field analysis at 1 GHz is also depicted in Fig. 41. It is readily apparent how the current stems radially from the coaxial probe and then follows a uniform line pattern before flowing across the discontinuity. In particular, the transverse dependence of the current in a uniform line is well represented. If transverse currents are neglected, it would be possible to include the transverse dependence of the longitudinal current in the basis functions, but this simplification fails for complex geometries.

The chain matrices can be easily transformed into scattering matrices and we can compute the quantity

$$P = |s_{11}|^2 + |s_{21}|^2 \tag{161}$$

which is related to the total dissipated power and equals unity in a lossless two port.

Figure 42 gives the value of P corresponding to the slot and an unperturbed line section of the same width. While there is almost no dissipation in the unperturbed line ($P = 0.95$ at 10 GHz), a large amount of power is radiated and launched in the form of surface waves by the discontinuity ($P < 0.75$ at 10 GHz).

Finally it is also possible to transform the chain matrix of the discontinuity into an impedance matrix and define a $\boldsymbol{T}$ equivalent circuit with a lumped series impedance $Z_s = z_{11} - z_{12}$ and a lumped parallel admittance $Y_p = 1/z_{12}$.

The real part of these quantities account for losses due to radiation. The imaginary parts, which are depicted in Figs. 43 and 44 show a linear

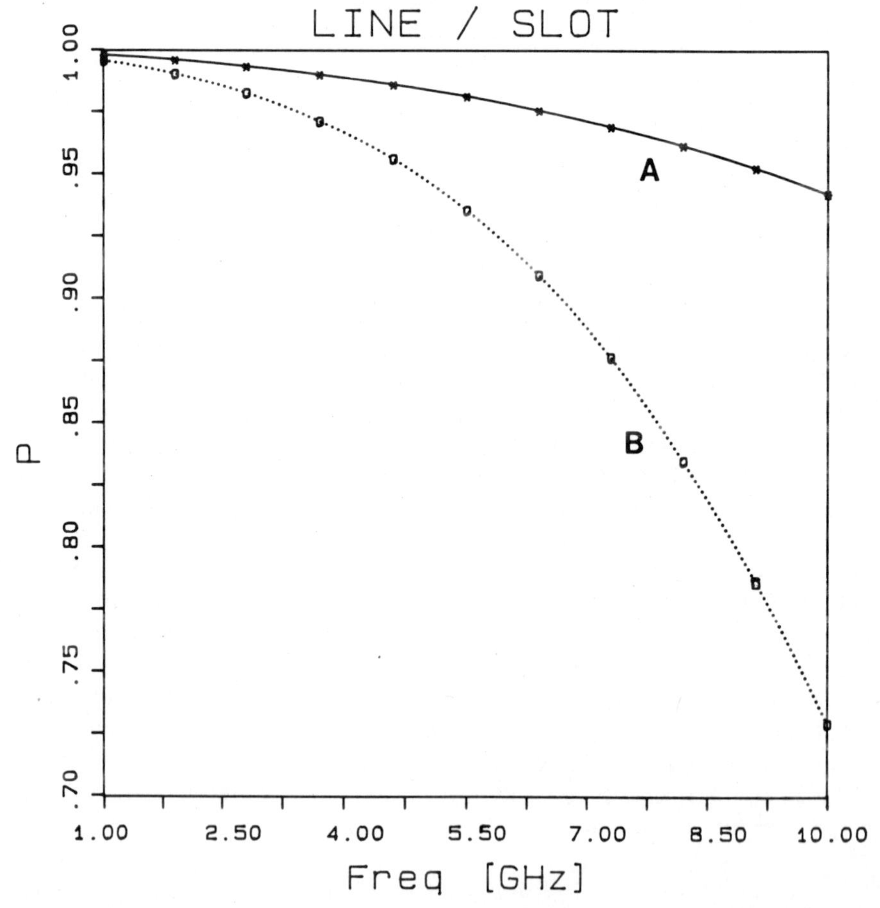

Fig. 42 Values of $P = |s_{11}|^2 + |s_{21}|^2$, as a function of frequency for (**A**), the unperturbed line section between planes PP'. (**B**) The slot between the same planes (Fig. 41).

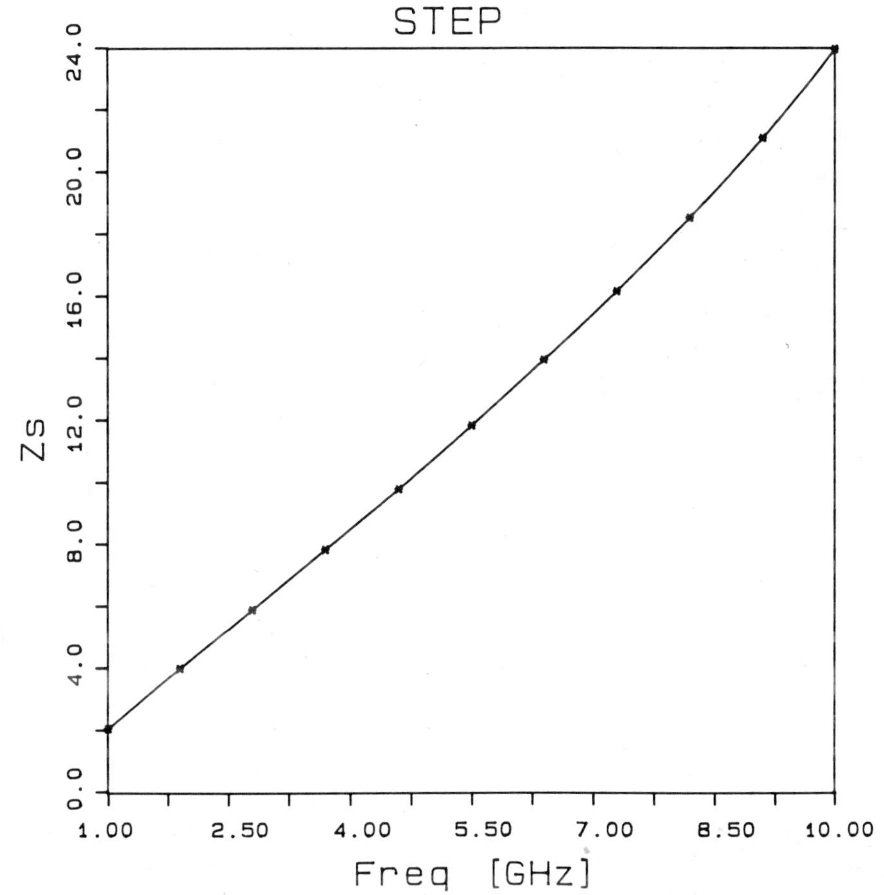

Fig. 43 Imaginary part (reactance in ohms) of the series impedance associated to the ***T***-equivalent circuit of a microstrip slot discontinuity (Fig. 41). The quasi-linear dependence with frequency shows that this reactance behaves approximately like a lumped inductance.

dependence on the frequency in the low range and therefore confirm the quasistatic predictions $Z_s \sim j\omega L_s\ \Omega$ and $Y_p \sim j\omega C_p\ \Omega^{-1}$ valid for these discontinuities at low frequencies. In addition, our analysis provides real parts for Z_s and Y_p accounting for surface waves and radiation.

13. CONCLUDING REMARKS

The mixed potential integral equation has been found to be a very powerful and flexible tool for the numerical analysis of microwave and millimeter wave integrated circuits, including planar conductors embedded in a layered

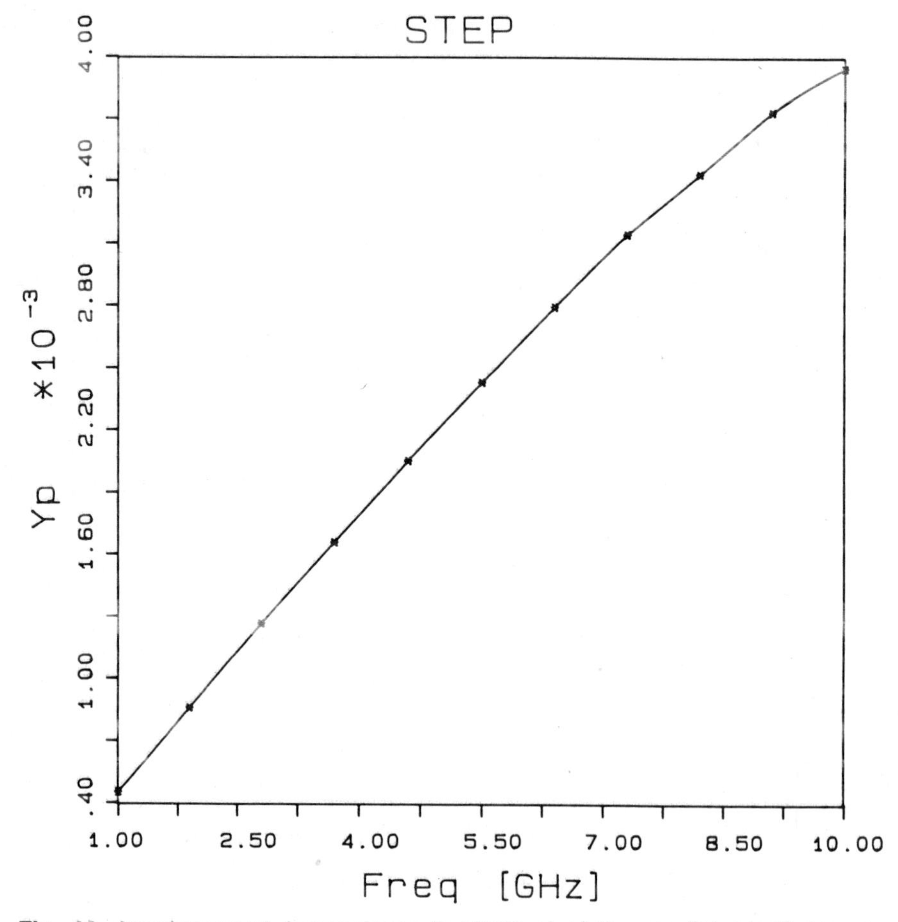

Fig. 44 Imaginary part (susceptance in siemens) of the parallel admittance associated to the ***T***-equivalent circuit of the microstrip slot discontinuity (Fig. 41). The quasi-linear dependence with frequency shows that this susceptance behaves approximately like a lumped capacitance.

medium. Combined with a method of moments using a subsectional basis, this technique can deal with conductors of quite irregular shape. Also the MPIE remains valid at any frequency and can be used for studying higher-order resonances as well as for characterizing geometrical discontinuities at relatively low frequencies. Thus, the techniques described in this chapter are particularly useful for problems where the frequency is too high for assuming a quasi-static situation but still too low for computing the fields as expansions over the unperturbed resonant modes of the structure.

The MPIE includes surface waves and radiation. Multilayered substrates can be accommodated by suitable modifications of the Green's functions. Handling conductors at different levels (stacked patches) is only a matter of increasing the number of unknowns. Even though we have concentrated on microstrip structures, the model is easily adapted to other configurations such as suspended structures, finline discontinuities, and microslot–microstrip transitions by again conveniently modifying the Green's functions.

Finally, there is no theoretical restriction on the electrical thickness of the substrate, though some of the excitation models discussed should be improved to maintain good accuracy of numerical results.

The mathematical apparatus included in this model can seem rather sophisticated and cumbersome. Nevertheless, algorithms based on this model have been successfully implemented in personal computers, and run time is surprisingly fast. On bigger computers, such as a VAX-780, computer time for analyzing the L-shaped patch at one frequency was less than two minutes.

In short, the model discussed here is perhaps not the most useful for everyday design of simple geometries at relatively low frequencies but works at its best for more complicated structures where radiation, surface waves, and fringing fields play an essential role.

ACKNOWLEDGMENTS

The author wishes to thank all his colleagues in the Laboratoire d'Electromagnétisme et d'Acoustique for their support. In particular, he is grateful to A. Skrivervik, R. Hall, B. Roudot, and L. Barlatey, for their invaluable help in preparing this chapter, to Professor Fred E. Gardiol of the École Polytechnique Fédérale de Lausanne for fruitful discussions and useful comments, and to Mrs. Mary Hall for careful typing of the manuscript.

REFERENCES

1. J. A. Kong, *Theory of Electromagnetic Waves*, Wiley, New York, 1975.
2. A. J. Poggio and E. K. Miller, "Integral equation solutions of three-dimensional scattering problems," in *Computer Techniques for Electromagnetics* (R. Mittra, ed.), Pergamon, Oxford, pp. 159–264, 1973.
3. R. Mittra, Y. Rahmat-Samii, D. V. Jamnejad, and W. A. Davis, "A new look at the thin plate scattering problem," *Radio Sci.*, vol. 8, pp. 869–875, 1973.
4. Cao Wei, R. F. Harrington, J. R. Mautz, and T. K. Sarkar, "Multiconductor transmission lines in multilayered dielectric media," *IEEE Trans. Microwave Theory Tech.*, vol. MTT-32, pp. 439–450, 1984.
5. J. R. Mosig and F. E. Gardiol, "Analytic and numerical techniques in the Green's function treatment of microstrip antennas and scatterers," *IEE Proc., Part H: Microwaves, Opt. Antennas*, vol. 130, pp. 175–182, 1983.
6. N.G. Alexopoulos and D.R. Jackson, "Fundamental superstrate (cover) effects on printed circuit antennas," *IEEE Trans. Antennas Propag.*, vol. AP-32, pp. 807–816, 1984.

7. R. F. Harrington, *Time Harmonic Electromagnetic Fields*, McGraw-Hill, New York, 1961.
8. A. Sommerfeld, *Partial Differential Equations in Physics*, Academic Press, New York, 1949 (reprint 1964).
9. A. Erteza and B. K. Park, "Nonuniqueness of resolution of Hertz vector in presence of a boundary, and the horizontal problem," *IEEE Trans. Antennas Propag.*, vol. AP-17, pp. 376–378, 1969.
10. K. A. Michalski, "On the scalar potential of a point charge associated with a time-harmonic dipole in a layered medium," *IEEE Trans. Antennas Propag.* Accepted for publication and scheduled for 1988.
11. J. A. Stratton, *Electromagnetic Theory*, McGraw-Hill, New York, 1941.
12. K. A. Michalski, "The mixed-potential electric field integral equation for objects in layered media," *Arch. Elektron, Ubertragungstech.*, vol. 39, pp. 317–322, Sept./Oct. 1985.
13. R. F. Harrington, *Field Computation by Moment Methods*, Macmillan, New York, 1968.
14. A. W. Glisson and D. R. Wilton, "Simple and efficient numerical methods for problems of electromagnetic radiation and scattering form surfaces," *IEEE Trans. Antennas Propag.*, vol. AP-29, pp. 593–603, 1980.
15. J. R. Mosig and F. E. Gardiol, "A dynamical radiation model for microstrip structures," *Adv. Electron. Electron Phys.*, vol. 59, pp. 139–237, 1982.
16. P. Silvester and P. Benedek, "Electrostatics of microstrip: Revisited," *IEEE Trans. Microwave Theory Tech.*, vol. MTT-20, pp. 756–758, 1972.
17. P. D. Patel, "Calculation of capacitance coefficients for a system of irregular finite conductors on a dielectric sheet, *IEEE Trans. Microwave Theory Tech.*, vol. MTT: 19, pp. 862–869, 1971.
18. R. Crampagne, M. Ahamdpanah, and J. L. Guiraud, "A simple method for determining the Green's function for a large class of MIC lines having multilayered dielectric structures, *IEEE Trans. Microwave Theory Tech.*, vol. MTT: 26, pp. 82–87, 1978.
19. J. R. Mosig and T. K. Sarkar, "Comparison of quasi-static and exact electromagnetic fields from a horizontal electric dipole above a lossy dielectric backed by an imperfect ground plane," *IEEE Trans. Microwave Theory Tech.*, vol. MTT-34, pp. 379–387, 1986.
20. L. B. Felsen and N. Marcuvitz, *Radiation Scattering of Waves*, Prentice-Hall, Englewood Cliffs, NJ, 1973.
21. K. A. Michalski, "On the efficient evaluation of integrals arising in the Sommerfeld halfspace problem," *IEE Proc., Part H: Microwaves, Opt. Antennas*, vol. 132, pp. 312–318, 1985.
22. R. J. Lytle and D. L. Lager, *Numerical Evaluation of Sommerfeld Integrals*, Rep. UCRL-52423, Lawrence Livermore Lab., University of California, 1974.
23. G. Pantis, "The evaluation of integrals with oscillatory integrands,"*J. Comput. Phy.*, vol. 17, pp. 229–233, 1975.
24. J. P. Boris and E. S. Oran, "Evaluation of oscillatory integrals," *J. Comput. Phys.*, vol. 17, pp. 425–433, 1975.
25. H. Hurwitz and P. F. Zweifel, "Numerical quadrature of Fourier transform integrals," *Math. Tables Aids Comput.*, vol. 10, pp. 140–149, 1956.

26. A. Alaylioglu, G. Evans, and J. Hyslop, "The evaluation of oscillatory integrals with infinite limits," *J. Comput. Phy.*, vol. 13, pp. 433–438, 1973.
27. A. Sidi, "The numerical evaluation of very oscillatory infinite integrals by extrapolation," *Math. Comput.*, vol. 38, pp. 517–529, 1982.
28. A. Banos, *Dipole Radiation in the Presence of a Conducting Half-Space*, Pergamon Press, New York, 1966.
29. S. M. Rao, D. R. Wilton, and A. W. Glisson, "Electromagnetic scattering by surfaces of arbitrary shape," *IEEE Trans. Antennas Propagat.*, vol. AP-30, pp. 409–418, 1982.
30. D. M. Pozar, "Improved computational efficiency for the moment method solution of printed dipoles and patches," *Electromagnetics*, vol. 3, pp. 299–309, 1983.
31. J. R. Mosig and F. E. Gardiol, "General integral equation formulations for microstrip antennas and scatterers," *IEE Proc., Part H: Microwaves, Opt. Antennas*, vol. 132, pp. 424–432, 1985.
32. D. R. Wilton and A. W. Glisson, "On improving the stability of electric field integral equation at low frequency," *IEEE AP-S Int. Symp.*, Los Angeles, CA, June 1981.
33. R. E. Collin, *Foundations for Microwave Engineering*, McGraw-Hill, New York, 1966.
34. D. M. Pozar, *New Architectures for Millimeter Wave Phased Array Antennas*, pp. 168–179, Journées International de Nice sur les Antennas (JINA), Nice, France, 1986.
35. P. B. Katehi and N. G. Alexopoulos, "Frequency-dependent characteristics of microstrip discontinuities in millimeter-wave integrated circuits," *IEEE Trans. Microwave Theory Tech.*, vol. MTT-33, pp. 1029–1035, 1985.
36. R. W. Jackson and D. M. Pozar, "Full-wave analysis of microstrip open-ended discontinuities," *IEEE Trans. Microwave Theory Tech.*, vol. MTT-33, pp. 1036–1042, 1985.
37. B. D. Popovic, M. B. Dragovic, and A. R. Djordjevic, *Analysis and Synthesis of Wire Antennas*, Wiley, New York, 1982.
38. R. C. Hall, J. R. Mosig, and F. E. Gardiol, "Analysis of microstrip antenna arrays with thick substrates," *17th Eur. Microwave Conf.*, pp. 951–956, Rome, Italy, Sept. 1987.
39. J. R. Mosig, "Arbitrarily shaped microstrip structures and their analysis with a mixed potential integral equation," *IEEE Trans. Microwave Theory Tech.*, vol. MTT-36, pp. 314–323, 1988.

4

Planar Circuit Analysis

K. C. Gupta
Department of Electrical and Computer Engineering
University of Colorado
Boulder, Colorado

M. D. Abouzahra
Lincoln Laboratory
Massachusetts Institute of Technology
Lexington, Massachusetts

1. INTRODUCTION

In the past, microwave engineers and researchers have generally focused on three principal categories of electric circuits. The first and most familiar category consists of lumped elements with physical dimensions much smaller than the wavelength. This class of lumped elements may be referred to as *zero-dimensional* components. Electric circuits whose dimensions are much smaller in two directions but comparable to the wavelength in the third direction constitute the second principal category. This category of distributed elements includes transmission line type circuitry and could be referred to as *one-dimensional*. The third principal category comprises the so-called *three-dimensional* elements whose three dimensions are comparable to the wavelength. Waveguide circuits fall under this category.

Complementary to these three categories is a fourth category of components that have two dimensions comparable to the wavelength and a third dimension much smaller than the wavelength. Examples of this *two-dimensional* category are the reduced-height waveguide components that have been known since the early development of microwave technology [1, 2] and the disk-shaped resonators that have been in use in microstrip configuration ever since microstrip circuits first became popular [3–5]. Electric circuits in this category are widely known as planar circuits and are the subject of this chapter. Figure 1 illustrates the four categories of electrical components described above.

The concept of two-dimensional planar circuits was first introduced in

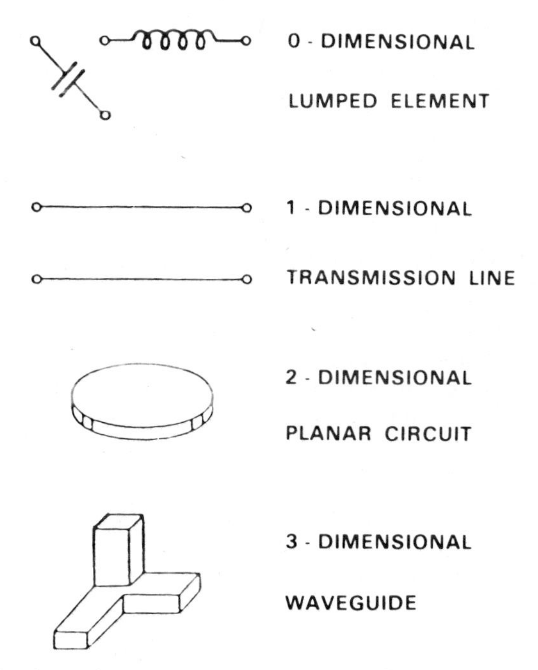

Fig. 1 Classification of electrical components. (© 1985 IEEE. Reprinted with permission from Reference 14.)

Japan in 1969 by Okoshi and co-workers [6–9] as an approach for analyzing two-dimensional microwave integrated circuitry (MIC). Almost in parallel, Ridella [10], Bianco [11], and their co-workers in Italy presented the fundamentals and the general theory for a class of distributed structures called *planar distributed N-ports*. Since then many scientists worldwide (Japan, Italy, India, the United States, Canada, Brazil, Germany, USSR, Sweden, United Kingdom, and elsewhere) have joined the research on planar circuits, and a few have written review articles and detailed books on this subject [12–16].

In this chapter we summarize the main features of the planar circuits approach. The wave equation and associated boundary conditions, the various methods of excitation, and the different methods of solving the wave equation are discussed in Section 2. In Section 3, the Green's function approach to the analysis of planar circuits is described, and in Section 4 two different methods for deriving Green's functions of simple configurations are described. In addition, Section 4 includes Green's function expressions for planar circuits with open, shorted, and mixed boundaries. The contour-integral approach that is used to analyze and/or synthesize planar circuits of arbitrary shapes with open or shorted boundaries is described in Section 5. Planar circuits having geometrical shapes that are neither simple nor completely arbitrary but are composites of simple configurations are analyzed by using segmentation/desegmentation techniques. These techniques are addressed in Section 6. Analysis of planar circuits having anisotropic spacing media such as ferrite is described in Section 7. Finally, various known applications of the planar circuit approach are described in Section 8 followed by the chapter summary.

2. PLANAR CIRCUIT ANALYSIS

The basic equations for the analysis of N-port planar circuits having isotropic or anisotropic spacing medium are described in this section. Derivation of the wave equation and its reduction into the two-dimensional form are outlined. Three boundary conditions describing the behavior of the surface current along the periphery are considered, and definitions for the terminal voltages and currents at the periphery are given. In addition, various methods for exciting the electromagnetic fields inside a planar structure are described. Finally, several techniques for solving the arising two-dimensional wave equation are described.

2.1. The Wave Equation

Figure 2 is a schematic for an N-port planar circuit of arbitrary shape. Along the periphery of this circuit several ports exist. The widths of these ports are denoted by W_i, W_j, etc. The rest of the periphery is either open, shorted, or

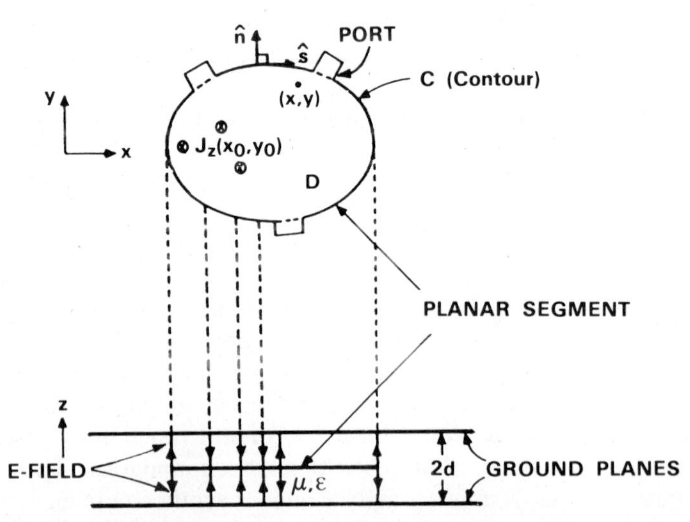

Fig. 2 A triplate-type planar circuit. Location of exciting current sources and the field point (x, y) on the circuit. (© 1981 Artech House. Reprinted with permission from Reference 15.)

terminated by an impedance wall. The excitation of the electromagnetic fields inside this structure is carried out either through some apertures at the periphery coupling the planar circuit to the external circuit (i.e., edge-fed by a microstrip) or via an internal current source such as a probe or a coaxial line center conductor.

The coordinate axes are chosen such that the planar element lies in the xy plane and is perpendicular to the z axis. Thus, while the dimensions along the x and y coordinates are comparable to the wavelength, the thickness along the z direction is much smaller than the wavelength. Therefore, the fields inside this circuit can be assumed to be uniform along the z direction, and hence we may set $\partial/\partial z = 0$ and $H_z = E_x = E_y = 0$. When the spacing material is homogeneous and isotropic, the wave equation (Helmholtz equation) governing the electromagnetic fields of this source-free structure reduces to

$$(\nabla_T^2 + k^2)E_z = 0 \tag{1}$$

where

$$\nabla_T^2 = \frac{\partial^2}{\partial x^2} + \frac{\partial^2}{\partial y^2} \tag{2}$$

$$k = \omega\sqrt{\mu\varepsilon} \tag{3}$$

and where ω, ε, and μ denote, respectively, the angular frequency, permittivity, and permeability of the spacing material, and k denotes the wavenumber in the spacing material.

2.2. Boundary Conditions

The characteristics of the planar circuit are determined by the solution of the two-dimensional Helmholtz equation (1) and the given boundary conditions. The boundary conditions for a given planar circuit are determined by the behavior of the surface current at the periphery. In general, the surface current on a conducting sheet can be obtained from

$$\mathbf{J}_s = \hat{\mathbf{n}} \times (\mathbf{H}_1 - \mathbf{H}_2) \tag{4}$$

where $\hat{\mathbf{n}}$ is a unit vector normal to the sheet and $\mathbf{H}_1$ and $\mathbf{H}_2$ are the magnetic fields on the two sides of the conducting sheet. For the central conductor of a stripline type planar circuit, $\mathbf{H}_1 = -\mathbf{H}_2$, and thus

$$\mathbf{J}_s = -2\hat{\mathbf{n}} \times \mathbf{H}_2 \text{ A/m} \tag{5}$$

Using Maxwell's equation the magnetic field can be written as

$$\mathbf{H} = -\frac{1}{j\omega\mu} \nabla \times \mathbf{E} \tag{6}$$

which upon using

$$\mathbf{E} = \hat{\mathbf{a}}_z E_z(x, y) \tag{7}$$

reduces to

$$\mathbf{H} = \frac{1}{j\omega\mu}\left(-\frac{\partial E_z}{\partial y}\hat{\mathbf{a}}_x + \frac{\partial E_z}{\partial x}\hat{\mathbf{a}}_y\right) \tag{8}$$

where $\hat{\mathbf{a}}_x$, $\hat{\mathbf{a}}_y$, and $\hat{\mathbf{a}}_z$ are unit vectors along the x, y, and z coordinates, respectively. Substituting (8) into (5) yields

$$\mathbf{J}_s = \frac{2}{j\omega\mu}\left(-\frac{\partial E_z}{\partial x}\hat{\mathbf{a}}_x + \frac{\partial E_z}{\partial y}\hat{\mathbf{a}}_y\right) \text{A/m} \tag{9}$$

The expression of $\mathbf{J}_s$ in (5) is valid at all points on the central patch of the stripline type of planar circuit (shown in Fig. 2) including the periphery. For a microstrip type of planar circuit (with periphery extended to take care of the fringing fields), the upper ground plane of Fig. 2 is not present, and hence there is no magnetic field above the central patch and the factor of 2 in (5), (9) and (10) is not needed.

For points on the periphery C, $\mathbf{J}_s$ can be rewritten in terms of components that are normal and tangential to the periphery C:

$$\mathbf{J}_s = \frac{2}{j\omega\mu}\left(\frac{\partial E_z}{\partial s}\hat{\mathbf{s}} + \frac{\partial E_z}{\partial n}\hat{\mathbf{n}}\right) \text{A/m} \tag{10}$$

where $\hat{\mathbf{s}}$ and $\hat{\mathbf{n}}$ are unit vectors tangential and normal to the periphery, respectively, as shown in Fig. 2. Apart from the coupling ports located at certain points along the periphery, the rest of the planar circuit periphery is either open (infinite impedance), shorted (zero wall impedance), or terminated by an impedance wall of finite value. On an open-boundary periphery the normal component of the surface current must be zero, or

$$\frac{\partial E_z}{\partial n} = 0 \tag{11}$$

This condition is widely referred to as the *magnetic wall boundary condition*. On the other hand, for points along the short-circuited periphery of a planar circuit, the tangential component of the electric field must be zero,

$$E_z = 0 \tag{12}$$

which is known as the *electric wall boundary condition*. Finally, a planar circuit whose periphery is terminated by an arbitrary impedance Z_w must satisfy the wall impedance boundary condition given by

$$Z_w = E_z / H_s \tag{13}$$

The electric field E_z is used to define a voltage v between the planar circuit conductor and the ground plane. This rf voltage is given by

$$v = -E_z d \tag{14}$$

where d represents the spacing between the two conductors.

2.3. Methods of Excitation

The electromagnetic fields inside a planar circuit are usually excited either by microstrip lines (or striplines) or by coaxial lines. In a coaxially fed microstrip line planar circuit, the coaxial feed is perpendicular to the circuit, with the outer conductor of the line connected to the ground plane and the inner conductor protruding through the substrate and touching the top conductor. The excitation current $i(x_0, y_0)$ in this case is given directly by the current flowing in the center conductor of the coaxial line.

In an edge-fed planar circuit, the strip conductor of the feeding line is connected to the periphery of the conducting patch. The excitation current in this case flows (at the coupling port) normal to the periphery of the planar circuit. An expression for this current is obtained from (10) by integrating the normal component of the current density over the coupling port width W to get

$$i = -\frac{p}{j\omega\mu} \int_W \frac{\partial E_z}{\partial n} \, ds \tag{15}$$

where $p = 1$ for microstrip-type circuits and $p = 2$ for stripline-type circuits, and ds is the incremental distance along the periphery. The negative sign in (15) implies that the current i flows inwards whereas $\hat{\mathbf{n}}$ in (10) points outwards. The characterization of planar components with open boundaries can now be carried out in terms of the rf voltage v on the conducting patch. Substituting eq. (14) into eqs. (1), (11), and (15) gives

$$(\nabla_T^2 + k^2)v = 0 \tag{16}$$

with

$$\frac{\partial v}{\partial n} = 0 \tag{17}$$

for points on the periphery where there are no coupling ports. The current flowing in at a coupling port is expressed as

$$i = \frac{p}{j\omega\mu d} \int_W \frac{\partial v}{\partial n} \, ds \tag{18}$$

The solution of (16), with (17) and (18) as the boundary conditions, leads to the characterization of the planar component (stripline or microstrip line). Equations governing other types of planar circuits (such as reduced-height waveguide circuits) can be obtained using similar procedures.

A solution for the two-dimensional wave equation given by (16) can be obtained by using any of the many analysis techniques available. In the next section some of these analysis techniques will be described.

2.4. Analysis Techniques

A formal solution for the wave equation and the boundary conditions governing planar circuits with isotropic spacing media can be obtained by using various analytical and numerical techniques. The selection of one approach over another is primarily driven by the desire to have compact analytical expressions and/or to have more efficient computer use.

For instance, when the planar circuit has a simple geometrical shape (rectangle, circle, etc.), the Green's function approach [12] and the eigenmode expansion approach [17] discussed in Section 2.5 are the most convenient. Using the Green's functions that are already available in the literature, the impedance matrix characterization of the circuit can be obtained for the specified port locations.

When the geometrical shape of a planar circuit is neither simple nor completely arbitrary but is a composite of simple shapes for which the Green's functions are available, the segmentation method [18, 19] can be used. Using the segmentation method, the characteristics of the overall circuit can be obtained from those of the various simple segments. In a complementary analysis technique called the desegmentation method [20], some simple shapes for which Green's functions are available are added to the configuration to be analyzed so that the resulting shape is also simple. In such a case, the characterization of the original circuit can be obtained from those of the shapes added and that of the resultant shape. Segmentation and desegmentation methods are discussed later in this chapter.

When the geometrical shape of the planar circuit becomes completely arbitrary, neither the segmentation nor the desegmentation method can be applied. Instead, numerical methods such as the finite element approach [21] and the contour integral method [12] are used to characterize the planar circuit. The former is an extension of the well-known finite elements method [22] and involves the integration of basis functions over the entire conducting patch, which is divided into numerous subsections. This approach requires more computer time.

The contour integral method (also called the boundary integral method) proposed for analyzing planar circuits involves only a line integral along the periphery. This method is based on Green's theorem in cylindrical coordinates. The wave equation is first converted to an integral equation along the circuit periphery, and the equivalent circuit parameters are then derived from this contour integral equation. The computer time required for this method is appreciably less than the computer time needed by the finite elements approach and other similar numerical techniques in which the field must be solved over the entire circuit area. The contour integral method is described in detail in Section 5.

Planar circuits with anisotropic spacing media are analyzed by using the contour integral approach [23], the Green's function approach [24], or the mode-matching approach [25]. A detailed description of these methods is given in Section 7.

3. GREEN'S FUNCTION APPROACH

This method can be employed when the shape of the planar structure is relatively simple, such as a rectangle, triangle, or circle [12]. The Green's function, which gives the voltage at any point on the planar circuit for a unit current source excitation elsewhere, is first obtained analytically. When the locations of the ports are specified, the impedance matrix characterizing the planar component can then be easily derived using the Green's function.

If the planar component is excited by a current density J_z in the z direction at any arbitrary point (x_0, y_0) inside the periphery (as shown in Fig. 2), the wave equation can be written as

$$(\nabla_T^2 + k^2)v = -j\omega\mu d J_z \tag{19}$$

where ∇_T and k are defined in (2) and (3), respectively. On the other hand, when the planar circuit is edge-fed or excited by a stripline, the term J_z denotes a fictitious rf current density injected normally into the circuit. In this case the current density $[J_n = (2/j\omega\mu d)\partial v/\partial n]$ injected into the circuit at coupling ports located along the periphery can be equivalently considered as fed normal to the circuit (along the z direction) with the magnetic wall condition $\partial v/\partial n = 0$ being imposed all along the periphery. The equivalent fictitious surface current $\boldsymbol{J}_s$ (in the z direction) obtained from the boundary condition (4) may be written as

$$\mathbf{J}_s = \frac{\cdot 1}{j\omega\mu d}\frac{\partial v}{\partial n}\,\hat{\mathbf{a}}_z \text{ A/m} \tag{20}$$

Thus, for both methods of excitation, we can consider the planar circuit as being excited by z-directed line currents located at the coupling ports. In addition, we can also impose the magnetic wall boundary condition all along the periphery.

The Green's function $G(r/r_0)$ for (19) is obtained by applying a unit line current source $\delta(r - r_0)$ flowing along the z direction in the region below the central patch and located at $r = r_0$. The Green's function $G(r/r_0)$ is a solution of

$$(\nabla_T^2 + k^2)G(r/r_0) = -j\omega\mu d\delta(r - r_0) \tag{21}$$

with the boundary condition at the periphery given by

$$\frac{\partial G}{\partial n} = 0 \tag{22}$$

The voltage at any point on the planar element can be written as

$$v(x, y) = \iint_D G(x, y \,|\, x_0, y_0) J_z(x_0, y_0)\, dx_0\, dy_0 \tag{23}$$

where $J_z(x_0, y_0)$ denotes the fictitious source current normally injected into the circuit and D denotes the two-dimensional region of the planar component enclosed by magnetic walls.

When the source current is injected only at the periphery, the voltage v at the periphery can be written as

$$v(s) = \int_C G(s \,|\, s_0) J_s(s_0)\, ds_0 \tag{24}$$

where $J_s(s_0)$ is the z-directed line current source given by (20), s and s_0 are the distances measured along the periphery, and the integral on the right-hand side is over the entire periphery C. Since the line current $J_s(s_0)$ is present only at the coupling ports, which represent a discrete number of sections on the periphery, we may write (24) as

$$v(s) = \sum_j \int_{W_j} G(s \,|\, s_0) J_s(s_0)\, ds_0 \tag{25}$$

where the summation is over all the coupling ports and W_j indicates the width of the jth coupling port. From eqs. (18) and (20), however, we have seen that the current i_j fed in at the jth port can be written in terms of the z-directed equivalent line current as

$$i_j = p \int_{W_j} J_s(s_0)\, ds_0 \tag{26}$$

If the widths of the coupling ports are assumed to be small so that the line current density J_s is distributed uniformly over the width of the port, we have from (26)

$$J_s(s_0)\big|_{\text{for } j\text{th port}} = \frac{i_j}{pW_j} \tag{27}$$

Upon substituting (27) into (25), the expression of $v(s)$, the rf voltage at any point on the periphery, becomes

$$v(s) = \sum_j \frac{i_j}{pW_j} \int_{W_j} G(s \mid s_0)\, ds_0 \tag{28}$$

In order to obtain the voltage v_i at the ith coupling port, we take the average of $v(s)$ over the width of the port as follows:

$$v_i = \frac{1}{W_i} \int_{W_i} v(s)\, ds \tag{29}$$

Substituting (28) into (29) yields

$$v_i = \sum_j \frac{i_j}{pW_iW_j} \int_{W_i} \int_{W_j} G(s \mid s_0)\, ds_0\, ds \tag{30}$$

By dividing both sides of this expression by i_j, the elements of the impedance matrix of the planar circuit can be written as

$$z_{ij} = \frac{1}{pW_iW_j} \int_{W_i} \int_{W_j} G(s \mid s_0)\, ds_0\, ds \tag{31}$$

where $p = 1$ for microstrip-type planar circuits and $p = 2$ for stripline-type planar circuits.

Note that when the shape of the planar circuit is relatively simple [e.g., rectangular, circular, ring, sector, or any shape for which the Green's function $G(s \mid s_0)$ is available in the literature], the solution approach described above is referred to as the Green's function approach. However, the entire approach is also known as the resonant mode expansion approach [14] when the Green's function of the planar circuit is not available explicitly and hence has to be derived. In the next section two methods for deriving the Green's function of a planar circuit are described. In addition, Green's functions for some planar shapes are given. Each of these Green's functions is in the form of a double series involving the resonant modes of the planar structure.

Using eq. (31), one can construct the impedance matrix $\boldsymbol{Z}$ of the planar element and then determine the scattering matrix $\boldsymbol{S}$ by using the relation

$$S = \sqrt{Y_0}(Z - Z_0)(Z + Z_0)^{-1}\sqrt{Z_0} \tag{32}$$

where $\mathbf{Z}_0$, $\sqrt{\mathbf{Z}_0}$ and $\sqrt{\mathbf{Y}_0}$ are diagonal matrices with diagonal elements given by $Z_{01}, Z_{02}, \ldots, Z_{0n}$; $\sqrt{Z_{01}}, \sqrt{Z_{02}}, \ldots, \sqrt{Z_{0n}}$; and $1/\sqrt{Z_{01}}, 1/\sqrt{Z_{02}}, \ldots, 1/\sqrt{Z_{0n}}$, respectively. The matrix elements $Z_{01}, Z_{02}, \ldots, Z_{0n}$ represent the normalizing impedances at the various ports of the planar circuit. The impedance matrix elements for several planar segments with regular shapes and open boundaries are listed in Appendix A.

In the above analysis, we have assumed that the widths W of the coupling ports are small (in comparison to the wavelength and the dimensions of the planar component) so that the injected current in each port is distributed uniformly over the width of that port. For cases where this assumption is not valid, each coupling port is divided into a number of subports over which the injected current can be assumed to be uniformly distributed. Consequently, the impedance matrix of the planar component is constructed with all the subports included. Before the scattering matrix is derived, however, the multiple subports at each coupling port are combined [19]. The method used to combine multiple subports into a single port is based on the observation that the voltages at the two connected subports are equal even though the currents may be different.

Since the voltages at the subports are equal, one can consider the subports to be connected in parallel. This implies that the total current injected in a port is divided into its various subports. The combination of subports requires inversion of the **Z** matrix to obtain the admittance matrix. In general, if ports i and j are divided into n and m subports, respectively, such that $i = \{i_1, i_2, \ldots, i_n\}$ and $j = \{j_1, j_2, \ldots, j_n\}$, then the term Y_{ij} of the overall admittance matrix is given as

$$Y_{ij} = \sum_{k=1}^{n} \sum_{l=1}^{m} Y_{kl} \tag{33}$$

where Y_{kl} are the elements of the admittance matrix with multiple subports. The overall scattering matrix can then be obtained from the impedance matrix using (32) or directly from the admittance matrix using

$$S = \sqrt{Z_0}(Y_0 - Y)(Y_0 + Y)^{-1}\sqrt{Y_0} \tag{34}$$

where $\mathbf{Y}_0$ is a diagonal matrix of normalizing admittances at various ports with diagonal elements given by $1/Z_{01}, 1/Z_{02}, \ldots, 1/Z_{0n}$.

4. IMPEDANCE GREEN'S FUNCTIONS

4.1. Evaluation of Green's Functions

The evaluation of Green's functions, for a given shape of two-dimensional component, requires solution of (21) with the boundary condition of (12) or

(22). There are two methods for obtaining the Green's functions: (1) the method of images, and (2) the expansion of Green's function in terms of eigenfunctions.

4.1.1. *Method of Images*

An analytical solution of the differential equation (21) can be obtained if the right-hand side is a periodic function [26]. For this purpose, additional current sources of the type $\delta(r - r_s)$ are placed at points r_s outside the region of the planar component. These additional sources can be obtained by taking multiple images of the line source at r_0 with respect to the various magnetic walls or electric walls of the planar component. The source term in (21) is modified, and the boundary condition is satisfied by the voltage v produced by the source and its images. It should be noted that the additional sources are all outside the region of the planar component and therefore the solution G still represents the Green's function for the geometrical shape of the planar component. For a rectangular planar component, the positions of additional line current sources are shown in Fig. 3.

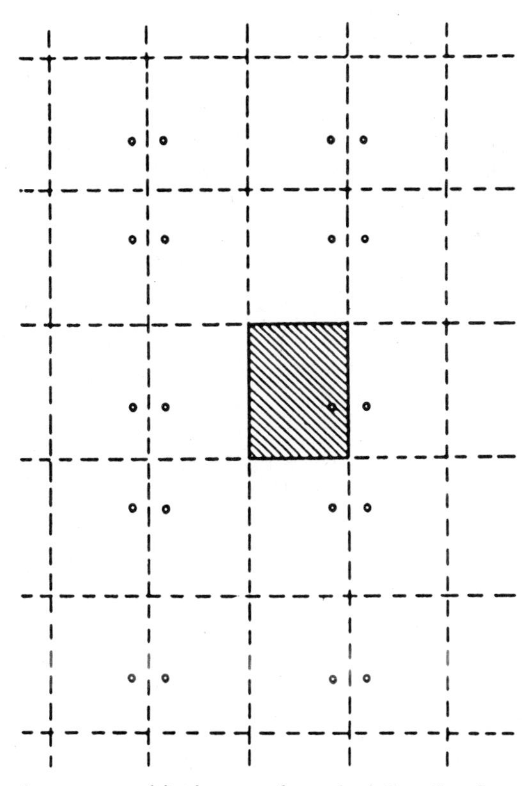

Fig. 3 Line current source and its images for calculating the Green's function of a rectangular component. (© 1981 Artech House. Reprinted with permission from Reference 15.)

The source pattern used in (21) is now periodic, and therefore its Fourier series expansion can be obtained. The Green's function can then be expressed as an infinite series summation of the functions obtained in the Fourier series expansion. These are the eigenfunctions. The coefficients in the series summation for the Green's function can then be obtained by substituting the series summation in (21).

The method of images is restricted to the shapes enclosed by boundaries that are straight lines. This is because the only mirror that gives a point image for a point source is a plane mirror. Even for polygonal shapes, the images can be uniquely specified in the two-dimensional space only if the internal angle at each vertex of the polygon is a submultiple of π. Thus the method is restricted to rectangles and some types of triangular shapes.

4.1.2. Expansion of Green's Function in Eigenfunctions

In this method [26], the Green's function is expanded in terms of known eigenfunctions of the corresponding Helmholtz's equation given by (16) and the boundary condition such as that given by (17).

Let the eigenfunctions of (16) that satisfy the boundary condition be ψ_n, and the corresponding eigenvalues be k_n^2, so that

$$\nabla_T^2 \psi_n + k_n^2 \psi_n = 0 \tag{35}$$

where n represents all the required indices defining a particular ψ_n. Also, the eigenfunctions ψ_n can be shown to form an orthonormal set, such that

$$\iint_D \psi_n^* \psi_m \, dx\, dy = \begin{cases} 1 & \text{if } n = m \\ 0 & \text{otherwise} \end{cases} \tag{36}$$

where the asterisk (*) denotes a complex conjugate and the region of integration D is bounded by the periphery of the planar component. At the periphery, ψ_n satisfies the boundary condition $\partial\psi_n/\partial n = 0$ for the open boundary or $\psi_n = 0$ for the shorted boundary. It should be noted that the Green's function $G(r|r_0)$ satisfies boundary conditions similar to those satisfied by the eigenfunctions and the total electric field. Assuming that the functions ψ_n form a complete set of orthonormal functions, it is possible to expand $G(r|r_0)$ in a series of ψ_n. Let

$$G(r|r_0) = \sum_m A_m \psi_m(r) \tag{37}$$

Substituting (37) into (21) and using (35), we get

$$\sum_m A_m (k^2 - k_m^2) \psi_m(r) = -j\omega\mu\, d\delta(r - r_0) \tag{38}$$

By multiplying both sides of eq. (38) by $\psi_n^*(r)$ and integrating it over the region D, we obtain

$$\sum_m A_m(k^2 - k_m^2) \iint_D \psi_m(r)\psi_n^*(r)\, dx\, dy = -j\omega\mu d\psi_n^*(r_0) \tag{39}$$

which, by virtue of the orthonormal property (36), reduces to

$$A_n(k^2 - k_n^2) = -j\omega\mu d\psi_n^*(r_0) \tag{40}$$

We then have

$$A_n = \frac{j\omega\mu d\psi_n^*(r_0)}{k_n^2 - k^2} \tag{41}$$

so that

$$G(r \mid r_0) = j\omega\mu d \sum_n \frac{\psi_n(r)\psi_n^*(r_0)}{k_n^2 - k^2} \tag{42}$$

is the required Green's function expansion. For a lossless circuit, the ψ_n are real and the complex conjugate is not needed in (42).

It may be noted that although the Green's functions given by (42) contain an intrinsic singularity at $r = r_0$, this does not cause any error in evaluating the impedance matrices using (31).

This method is restricted to the cases where the eigenfunctions are known. For shapes enclosed by straight edges, eigenfunctions can be obtained only if the internal angle at each vertex is a submultiple of π. Eigenfunctions can also be obtained for shapes like circles, rings, and for some sectors of circles and of circular rings.

4.2. Green's Functions for Planar Segments with Open Boundaries

In this section Green's functions for some planar shapes with open boundaries (magnetic walls), shown in Fig. 4, are given. In the expressions that follow, σ_i is given by

$$\sigma_i \stackrel{\Delta}{=} \begin{cases} 1 & \text{if } i = 0 \\ 2 & \text{otherwise} \end{cases} \tag{43}$$

(*a*) *A Rectangle*: The Green's function for the rectangle shown in Fig. 4*a* is given as [27]

$$G(x, y \mid x_0, y_0) = \frac{j\omega\mu d}{ab} \sum_{n=0}^{\infty} \sum_{m=0}^{\infty} \frac{\sigma_m \sigma_n \cos(k_x x_0)\cos(k_y y_0)\cos(k_x x)\cos(k_y y)}{k_x^2 + k_y^2 - k^2} \tag{44}$$

where $k_x = m\pi/a$ and $k_y = n\pi/b$.

(*b*) *A 30°–60° Right-Angled Triangle*: The Green's function for the triangle shown in Fig. 4*b* is given as [28]

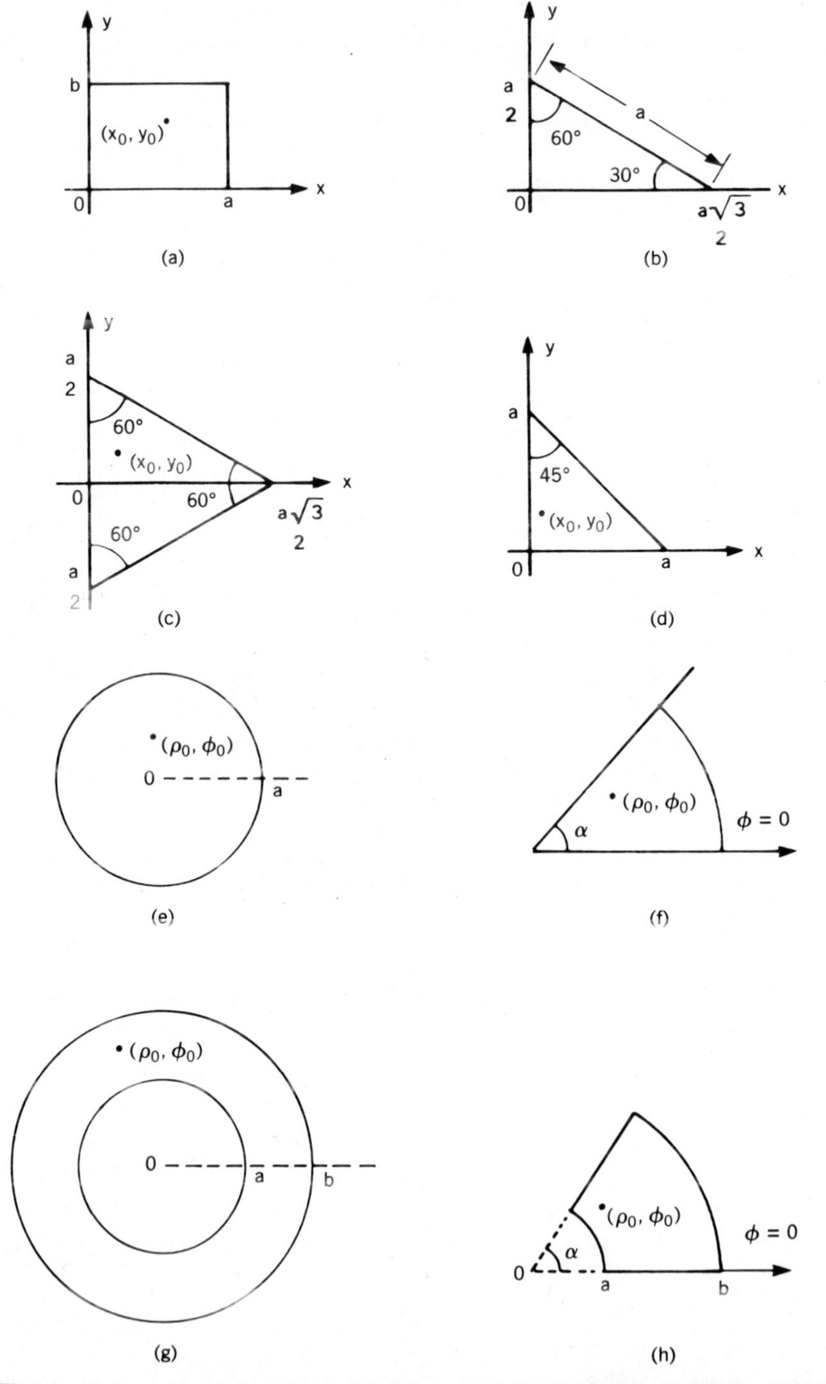

Fig. 4 Various planar circuit configurations for which Green's functions are available. (© 1981 Artech House. Reprinted with permission from Reference 15.)

$$G(x, y \mid x_0, y_0) = 8j\omega\mu d \sum_{m=-\infty}^{\infty} \sum_{n=-\infty}^{\infty} \frac{T_1(x_0, y_0) T_1(x, y)}{16\sqrt{3}\pi^2(m^2 + mn + n^2) - 9\sqrt{3}a^2k^2} \tag{45}$$

where

$$\begin{aligned} T_1(x, y) = (-1)^l \cos \frac{2\pi l x}{\sqrt{3}a} \cos \frac{2\pi(m-n)y}{3a} \\ + (-1)^m \cos \frac{2\pi m x}{\sqrt{3}a} \cos \frac{2\pi(n-l)y}{3a} \\ + (-1)^n \cos \frac{2\pi n x}{\sqrt{3}a} \cos \frac{2\pi(l-m)y}{3a} \end{aligned} \tag{46}$$

with the condition that

$$l = -(m + n) \tag{47}$$

(*c*) *An Equilateral Triangle*: The Green's function for the equilateral triangle shown in Fig. 4*c* is given as [28]

$$G(x, y \mid x_0, y_0) = 4j\omega\mu d \sum_{m=-\infty}^{\infty} \sum_{n=-\infty}^{\infty} \frac{T_1(x_0, y_0) T_1(x, y) + T_2(x_0, y_0) T_2(x, y)}{16\sqrt{3}\pi^2(m^2 + mn + n^2) - 9\sqrt{3}a^2k^2} \tag{48}$$

where $T_1(x, y)$ is given by (46) and

$$\begin{aligned} T_2(x, y) = (-1)^l \cos \frac{2\pi l x}{\sqrt{3}a} \sin \frac{2\pi(m-n)y}{3a} \\ + (-1)^m \cos \frac{2\pi m x}{\sqrt{3}a} \sin \frac{2\pi(n-l)y}{3a} \\ + (-1)^n \cos \frac{2\pi n x}{\sqrt{3}a} \sin \frac{2\pi(l-m)y}{3a} \end{aligned} \tag{49}$$

As for $T_1(x, y)$, the integer l in $T_2(x, y)$ is given by (47).

(*d*) *A Right-Angled Isosceles Triangle*: The Green's function for the right-angled isosceles triangle shown in Fig. 4*d* is given by [28]

$$G(x, y \mid x_0, y_0) = \frac{j\omega\mu d}{2} \sum_{m=0}^{\infty} \sum_{n=0}^{\infty} \frac{\sigma_m \sigma_n T(x_0, y_0) T(x, y)}{(m^2 + n^2)\pi^2 - a^2k^2} \tag{50}$$

where

$$T(x, y) = \cos\frac{m\pi x}{a}\cos\frac{n\pi y}{a} + (-1)^{m+n}\cos\frac{n\pi x}{a}\cos\frac{m\pi y}{a} \tag{51}$$

(*e*) *A Circle*: The Green's function for the circle shown in Fig. 4*e* is given by [29]

$$G(\rho, \phi \mid \rho_0, \phi_0) = \frac{d}{j\omega\varepsilon\pi a^2} + j\omega\mu d \sum_{n=0}^{\infty}\sum_{m=1}^{\infty} \frac{\sigma_n J_n(k_{mn}\rho_0) J_n(k_{mn}\rho)\cos[n(\phi - \phi_0)]}{\pi(a^2 - n^2/k_{mn}^2)(k_{mn}^2 - k^2)J_n^2(k_{mn}a)} \tag{52}$$

where $J_n(\cdot)$ represents Bessel's function of the nth order, and k_{mn} satisfy

$$\frac{\partial}{\partial\rho} J_n(k_{mn}\rho)\bigg|_{\rho=a} = 0 \tag{53}$$

The subscript m in k_{mn} denotes the mth root of (53). For a zeroth-order Bessel's function, the first root of (53) is taken to be the nonzero root.

(*f*) *A Circular Sector*: The Green's functions for circular sectors are available only when the sector angle α is a submultiple of π. For the circular sector shown in Fig. 4*f* for which $\alpha = \pi/l$, the Green's function is given as [30]

$$G(\rho, \phi \mid \rho_0, \phi_0) = \frac{2ld}{j\omega\varepsilon\pi a^2} + 2jl\omega\mu d \sum_{n=0}^{\infty}\sum_{m=1}^{\infty} \frac{\sigma_n J_{n_i}(k_{mn_i}\rho_0) J_{n_i}(k_{mn_i}\rho)\cos(n_i\phi_0)\cos(n_i\phi)}{\pi[a^2 - n_i^2/k_{mn_i}^2](k_{mn_i}^2 - k^2)J_{n_i}^2(k_{mn_i}a)} \tag{54}$$

when $n_i = nl$ and the k_{mn_i} are given by

$$\frac{\partial}{\partial\rho} J_{n_i}(k_{mn_i}\rho)\bigg|_{\rho=a} = 0 \tag{55}$$

(*g*) *An Annular Ring*: The Green's function for the annular ring shown in Fig. 4*g* is given as [30]

$$G(\rho, \phi \mid \rho_0, \phi_0) = \frac{d}{j\omega\varepsilon\pi(b^2 - a^2)} + j\omega\mu d \sum_{n=0}^{\infty}\sum_{m=1}^{\infty} \frac{\sigma_n F_{mn}(\rho_0) F_{mn}(\rho)\cos[n(\phi - \phi_0)]}{\pi[(b^2 - n^2/k_{mn}^2)F_{mn}^2(b) - (a^2 - n^2/k_{mn}^2)F_{mn}^2(a)](k_{mn}^2 - k^2)} \tag{56}$$

where

$$F_{mn}(\rho) = N'_n(k_{mn}a)J_n(k_{mn}\rho) - J'_n(k_{mn}a)N_n(k_{mn}\rho) \tag{57}$$

and k_{mn} are solutions of

$$\frac{J'_n(k_{mn}a)}{N'_n(k_{mn}a)} = \frac{J'_n(k_{mn}b)}{N'_n(k_{mn}b)} \tag{58}$$

In the above relations $N_n(\cdot)$ denotes Neumann's function of order n and $J'_n(\cdot)$ and $N'_n(\cdot)$ denote first derivatives with respect to the arguments.

(*h*) *An Annular Sector*: As in the case of circular sectors, the Green's functions for annular sectors are available only if the sector angle α is a submultiple of π. For the annular sector shown in Fig. 4*h* for which $\alpha = \pi/l$, the Green's function is given as [30]

$$G(\rho, \phi \mid \rho_0, \phi_0) = \frac{2ld}{j\omega\varepsilon\pi(b^2 - a^2)} + 2jl\omega\mu d \sum_{n=0}^{\infty} \sum_{m=1}^{\infty} \frac{\sigma_n F_{mn_i}(\rho) F_{mn_i}(\rho_0) \cos(n_i\phi_0)\cos(n_i\phi)}{\pi[(b^2 - n_i^2/k_{mn_i}^2)F_{mn_i}^2(b) - (a^2 - n_i^2/k_{mn_i}^2)F_{mn_i}^2(a)](k_{mn_i}^2 - k^2)} \tag{59}$$

where $n_i = nl$ and $F_{mn_i}(\cdot)$ is defined in (57). The values of k_{mn_i} are obtained from (58).

4.3. Green's Functions for Planar Segments with Shorted Boundaries

The two methods for evaluating Green's functions have been discussed in Section 4.1. Both of these techniques are also useful for finding the Green's function for planar components with shorted boundary.

4.3.1. *Rectangles and Triangles*

As before, the method of images can be used for rectangular and triangular geometries. Multiple images that need to be considered for a rectangular geometry are shown in Fig. 5. Note that because of reflections at the electric walls, adjacent image sources have opposite polarities. As a consequence, in the expression for Green's function given in (44), cosine terms are replaced by sine terms when electric wall boundaries exist. Thus for a rectangle with all four sides on shorted boundaries we have

$$G(x, y \mid x_0, y_0) = \frac{j\omega\mu d}{ab} \sum_{m=-\infty}^{\infty} \sum_{n=-\infty}^{\infty} \frac{\sin(k_x x_0)\sin(k_y y_0)\sin(k_x x)\sin(k_y y)}{k_x^2 + k_y^2 - k^2} \tag{60}$$

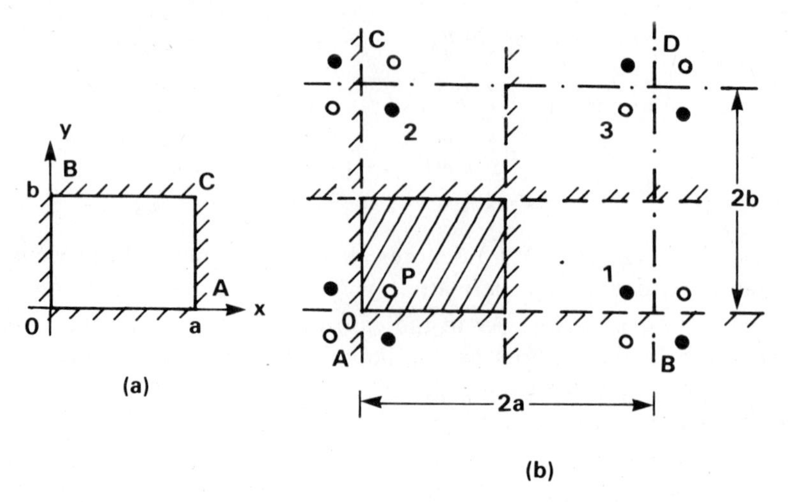

Fig. 5 (*a*) Rectangular segment with electric wall boundary. (*b*) Locations of image sources.

where $k_x = m\pi/a$ and $k_y = n\pi/b$. Green's functions for triangular geometry with shorted boundaries have also been obtained [31] by the method of images. Locations of images for a 30°–60° right-angled triangle are shown in Fig. 6. The expression for the Green's function may be written as

$$G(x, y \mid x_0, y_0) = 8j\omega\mu d \sum_{m=-\infty}^{\infty} \sum_{n=-\infty}^{\infty} \frac{T_2(x_0, y_0)T_2(x, y)}{16\sqrt{3}\pi^2(m^2 + mn - n^2) - 9\sqrt{3}a^2k^2} \tag{61}$$

where the function T_2 is given by

$$\begin{aligned} T_2(x, y) = &-(-1)^l \sin\frac{2\pi lx}{\sqrt{3}a} \sin\frac{2\pi(m-n)y}{3a} \\ &+(-1)^m \sin\frac{2\pi mx}{\sqrt{3}a} \sin\frac{2\pi(n-l)y}{3a} \\ &-(-1)^n \sin\frac{2\pi nx}{\sqrt{3}a} \sin\frac{2\pi(l-m)y}{3a} \end{aligned} \tag{62}$$

with $l = -(m+n)$.

Green's function for a right-angled isosceles triangle with a shorted boundary has also been obtained [32] by the method of images. For this case, locations of images are shown in Fig. 7, and the corresponding Green's function may be written as

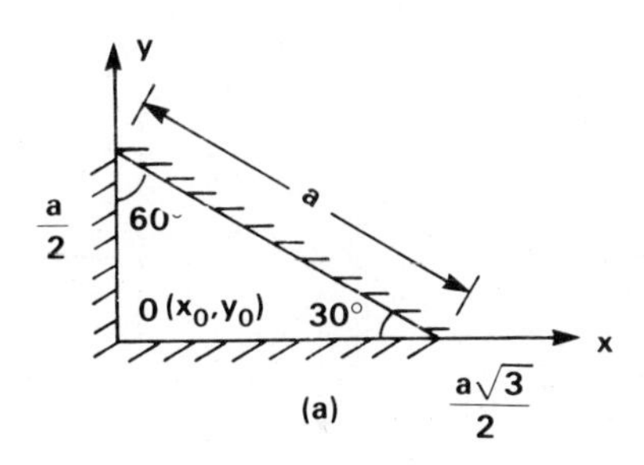

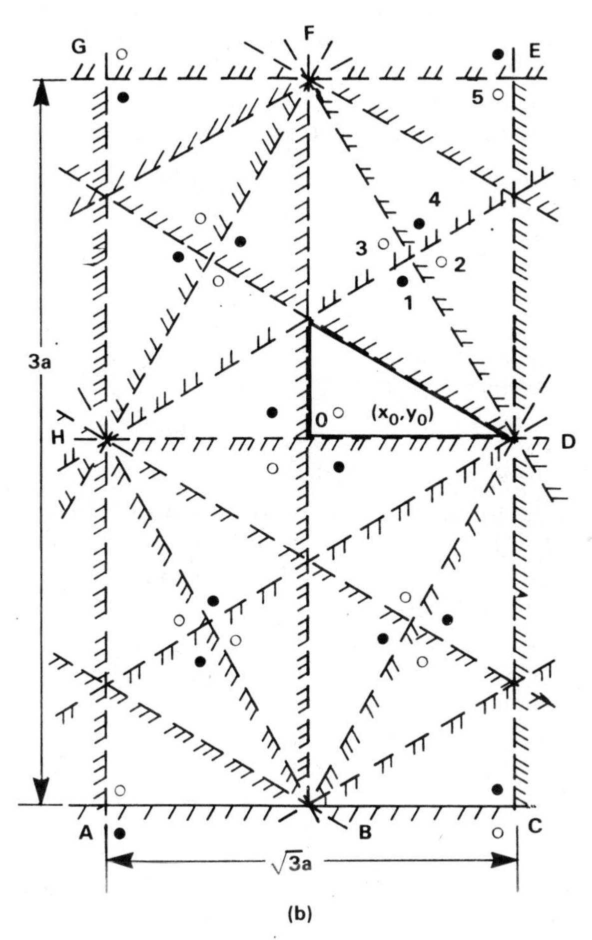

Fig. 6 (*a*) A 30°–60° right-angled triangle with all three sides as electric walls. (*b*) Location of images.

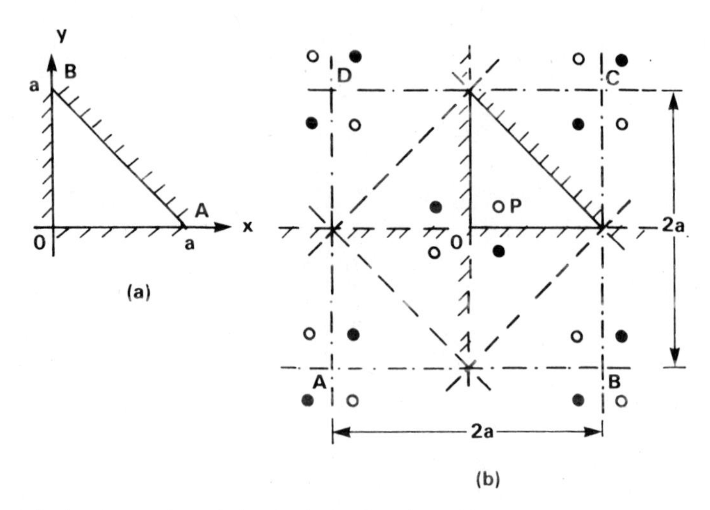

Fig. 7 (*a*) An isosceles right-angled triangle with electric wall boundary. (*b*) Location of images.

$$G(x, y \mid x_0, y_0) = j\omega\mu d \sum_{m=-\infty}^{\infty} \sum_{n=-\infty}^{\infty} \frac{S(x_0, y_0)S(x, y)}{2[(m^2 + n^2)\pi^2 - a^2k^2]} \tag{63}$$

where the function S is given by

$$S(x, y) = \sin\frac{m\pi x}{a}\sin\frac{n\pi y}{a} + (-1)^{m+n}\sin\frac{n\pi x}{a}\sin\frac{m\pi y}{a} \tag{64}$$

Green's function for an equilateral triangle with shorted boundary is obtained by considering the modes in the equilateral triangular geometry as a summation of odd and even modes corresponding to two 30°–60° right-angled triangular geometries shown in Fig. 8. The 30°–60°–90° triangle shown in Fig. 8*b* has been discussed earlier and has a Green's function given by (61). Green's function for the triangle in Fig. 8*c* (with the side along the x axis being a magnetic wall) is also obtained by the method of images and is discussed in Section 4.4. Green's function for the equilateral geometry is obtained by superposing these two Green's functions for odd- and even-mode half-sections. Following this procedure, we get [31],

$$G(x, y \mid x_0, y_0) = 8j\omega\mu d \sum_{m=-\infty}^{\infty} \sum_{n=-\infty}^{\infty} \frac{T_2(x_0, y_0)T_2(x, y) + T_3(x_0, y_0)T_3(x, y)}{16\sqrt{3}\pi(m^2 + mn + n^2) - 9\sqrt{3}a^2k^2} \tag{65}$$

where $T_2(x, y)$ is defined in (62) and the function $T_3(x, y)$ is given by

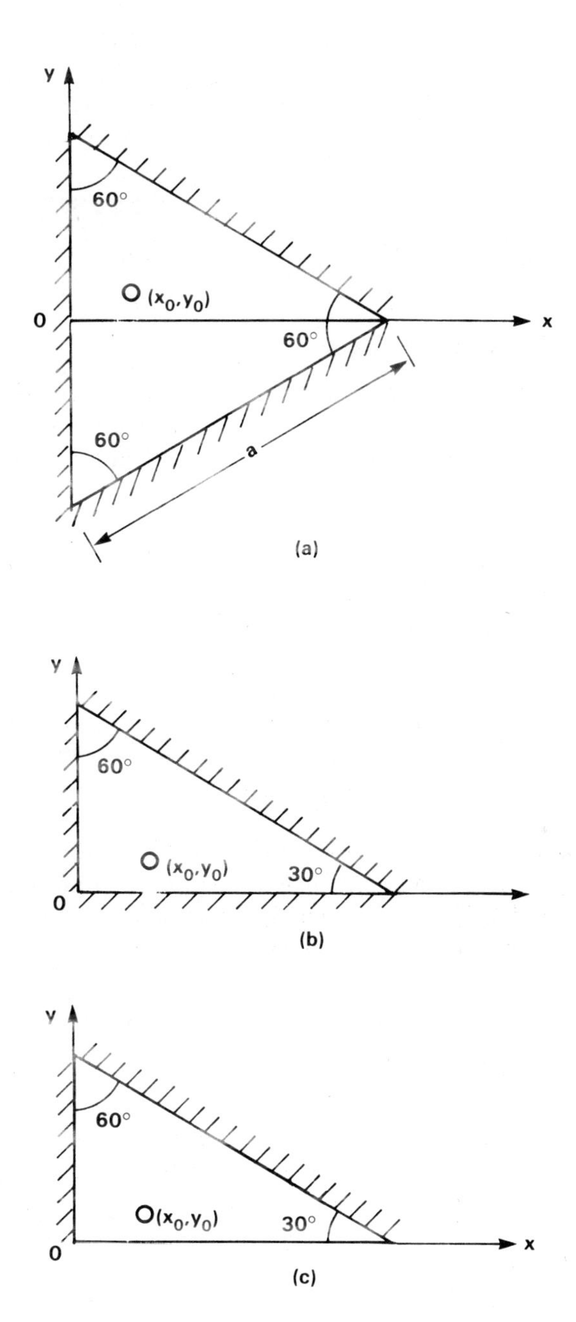

Fig. 8 (*a*) An equilateral triangular segment with shorted boundary. (*b*) Its odd-mode section. (*c*) Its even-mode section.

$$T_3(x, y) = -(-1)^l \sin\frac{2\pi lx}{\sqrt{3}a}\cos\frac{2\pi(m-n)y}{\sqrt{3}a} + (-1)^m \sin\frac{2\pi mx}{\sqrt{3}a}\cos\frac{2\pi(n-l)y}{3a} + (-1)^n \sin\frac{2\pi nx}{\sqrt{3}a}\cos\frac{2\pi(l-m)y}{3a} \tag{66}$$

with $l = -(m+n)$.

4.3.2. Circles and Circular Sectors

Green's function for a circular segment with a shorted boundary along the circumference may be obtained by expansion into eigenfunctions as discussed in Section 4.1. The result is analogous to the Green's function given in (52) and may be expressed as

$$G(\rho, \phi \mid \rho_i, \phi_i) = j\omega\mu d \sum_{n=0}^{\infty}\sum_{m=1}^{\infty} \frac{\sigma_n J_n(k_{mn}\rho_i)J_n(k_{mn}\rho)\cos\{n(\phi-\phi_i)\}}{\pi(a^2 - n^2/k_{mn}^2)(k_{mn}^2 - k^2)J_n^2(k_{mn}a)} \tag{67}$$

where k_{mn} (for the case of the shorted boundary) is given by

$$J_n(k_{mn}a) = 0 \tag{68}$$

Green's function for a circular sector with shorted boundary is also obtained by using (42). The eigenfunction ψ_n may be written as a product of angular and radial factors. The normalized angular eigenfunction (for the geometry shown in Fig. 9) that satisfies the boundary condition at $\phi = 0$ and $\phi = \phi_0$ is given by

$$\phi_m = \sqrt{\frac{2}{\phi_0}}\sin\nu\phi \tag{69}$$

where ν is an integer given by π/ϕ_0. As in the case of circular sectors with open boundaries, Green's functions are available for shorted boundary sections only when the sector angle ϕ_0 is a submultiple of π radians (180°).

The normalized radial eigenfunction that satisfy the boundary condition at $\rho = a$ are given by

$$R_n = \frac{\sqrt{2}J_\nu(k_{\nu n}\rho)}{aJ_{\nu+1}(k_{\nu n}a)}$$

where $k_{\nu n}a$ is the nth zero of a Bessel function of the first kind and order ν.

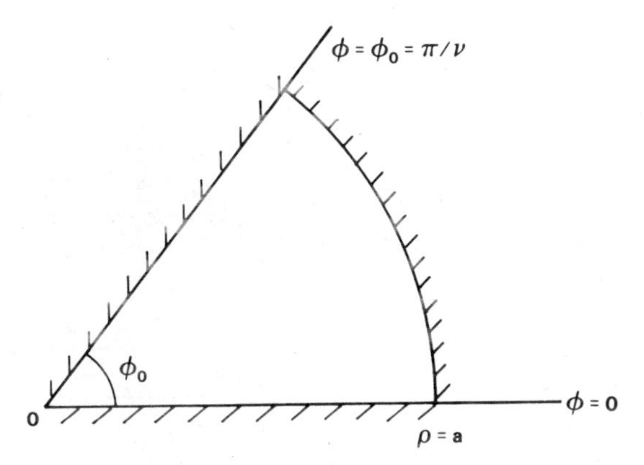

Fig. 9 A circular sector planar segment with shorted boundaries.

Then, using (42), the Green's function for the sector with all electric walls is obtained as

$$G(r|r_i) = \frac{4j\omega\mu d}{\phi_0 a^2} \sum_{m=1}^{\infty} \sum_{n=1}^{\infty} \frac{J_\nu(k_{\nu n}\rho) J_\nu(k_{\nu n}\rho_i) \sin \nu\phi \sin \nu\phi_i}{(k_{\nu n}^2 - k^2) J_{\nu+1}^2(k_{\nu n} a)} \tag{70}$$

4.4. Green's Functions for Planar Segments with Mixed Boundaries

4.4.1. Rectangles

Green's functions for rectangular planar segments are available [32] for four types of mixed boundary conditions: (1) when one side is shorted and the other three are open, (2) when two of the adjacent sides of a rectangle are shorted (with the other two being open), (3) when two parallel sides are shorted, and (4) when three sides are shorted (the fourth one being open). The method of images discussed in Section 4.1 has been used for all of these cases. The configurations and locations of doubly infinite images in the four cases are shown in Figs. 10*a–d*. Finally, the expressions for Green's functions are summarized below.

Rectangle with One Side Shorted (Fig. 10a)

$$G(x, y|x_0, y_0) = \frac{j\omega\mu d}{ab} \sum_{m=-\infty}^{\infty} \sum_{\substack{n=-\infty \\ \text{odd } n}}^{\infty} \frac{\cos k_x x_0 \sin k_y y_0 \cos k_x x \sin k_y y}{k_x^2 + k_y^2 - k^2} \tag{71}$$

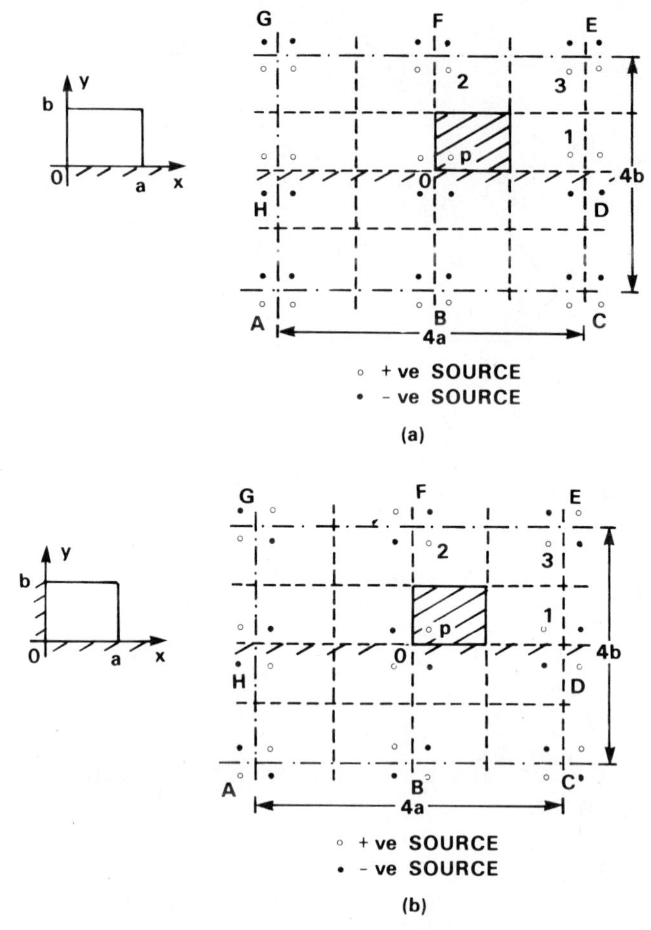

Fig. 10 (*a*) Rectangular segment with one side as electric wall and locations of image sources. (*b*) Rectangular segment with two sides as electric walls and locations of image sources.

where $k_x = m\pi/a$, $k_y = n\pi/2b$. In (71) the summation over n is for odd values only.

Rectangle with Two Adjacent Sides Shorted (Fig. 10b)

$$G(x, y | x_0, y_0) = \frac{j\omega\mu d}{ab} \sum_{\substack{-\infty \\ \text{odd}(m,n)}}^{\infty} \sum_{-\infty}^{\infty} \frac{\sin k_x x_0 \sin k_y y_0 \sin k_x x \sin k_y y}{k_x^2 + k_y^2 - k^2} \tag{72}$$

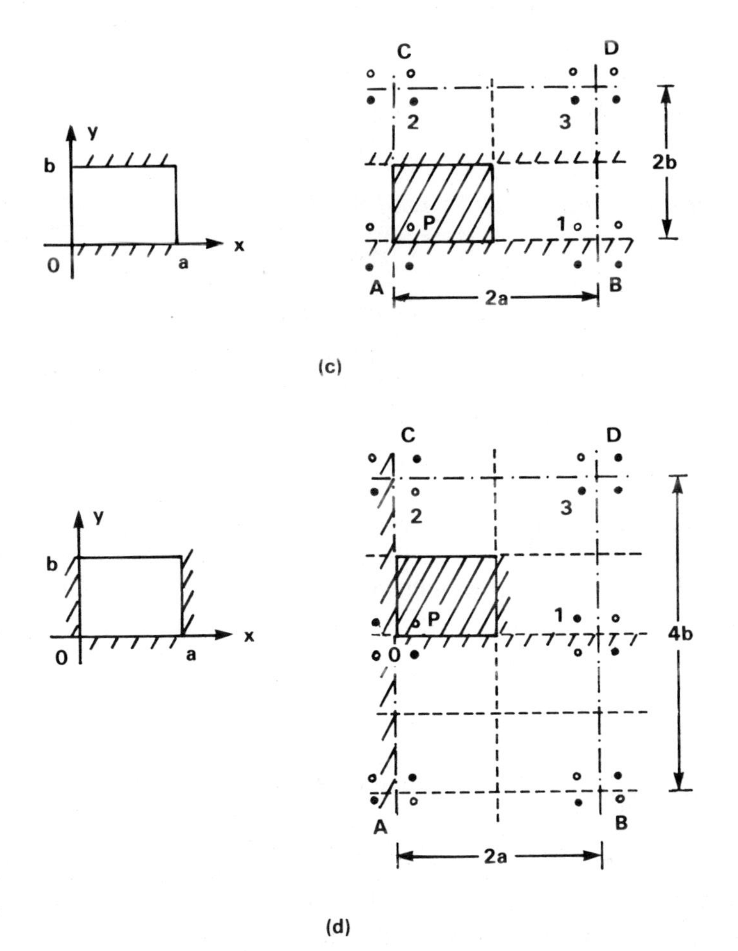

Fig. 10 continued (*c*) Rectangular segment with two parallel sides as electric walls and locations of image sources. (*d*) Rectangular segment with three sides as electric walls and locations of image sources.

where $k_x = m\pi/2a$ and $k_y = n\pi/2b$. The Green's function given by (72) is identical to that for the rectangle with all four sides shorted as given in (60), except for the fact that in the present case summations are carried over only odd values of m and n.

Rectangle with Two Parallel Sides Shorted (Fig. 10(c))

$$G(x, y \mid x_0, y_0) = \frac{j\omega\mu d}{ab} \sum_{m=-\infty}^{\infty} \sum_{n=-\infty}^{\infty} \frac{\cos k_x x_0 \sin k_y y_0 \cos k_x x \sin k_y y}{k_x^2 + k_y^2 - k^2} \tag{73}$$

where $k_x = m\pi/a$ and $k_y = n\pi/b$. It may be noted that the Green's function in (73) is similar to that of a rectangle with only one side shorted [given by (71)]. However, in the present case the summation is over all the possible values of m and n.

Rectangle with Three Sides Shorted (Fig. 10d)

$$G(x, y | x_0, y_0) = \frac{j\omega\mu d}{ab} \sum_{m=-\infty}^{\infty} \sum_{\substack{n=-\infty \\ \text{odd } n}}^{\infty} \frac{\sin k_x x_0 \sin k_y y_0 \sin k_x x \sin k_y y}{k_x^2 + k_y^2 - k^2} \tag{74}$$

where $k_x = m\pi/a$ and $k_y = n\pi/2b$.

Again, the Green's function in (74) is similar to that of a rectangle with shorted adjacent walls (72) and also similar to that of a rectangle with four shorted walls (60), except that in the present case the summation over n is for odd values only while the summation over m is for all possible integer values.

4.4.2. Triangles

Green's functions for isosceles right-angled triangular segments are available [32] for four types of mixed boundary conditions: (1) when one side (not the hypotenuse) is shorted, (2) when the hypotenuse is shorted, (3) when the hypotenuse and one of the other two sides are shorted, and (4) when the two non-hypotenuse sides are shorted. Green's functions for 30°–60° right-angled triangles are also available [31] for two types of mixed boundary conditions: (1) when the side opposite the 60° angle is shorted, and (2) when the side opposite the 60° angle is open and the other two sides are shorted.

The method of images is applicable in all of these cases, and the locations of the image sources are shown in Figs. 11–15.

Isosceles Right-Angled Triangle with One Side (Not Hypotenuse) Shorted. The Green's function for this case (Fig. 11a) is

$$G(x, y | x_0, y_0) = 2j\omega\mu d \sum_{-\infty}^{\infty} \sum_{\substack{-\infty \\ \text{odd}(m,n)}}^{\infty} \frac{U(x_0, y_0)U(x, y)}{[(m^2 + n^2)\pi^2 - 4a^2k^2]} \tag{75}$$

where

$$U(x, y) = \cos\frac{m\pi x}{2a} \sin\frac{n\pi y}{2a} - (-1)^{(m+n)/2} \cos\frac{n\pi x}{2a} \sin\frac{m\pi y}{2a} \tag{76}$$

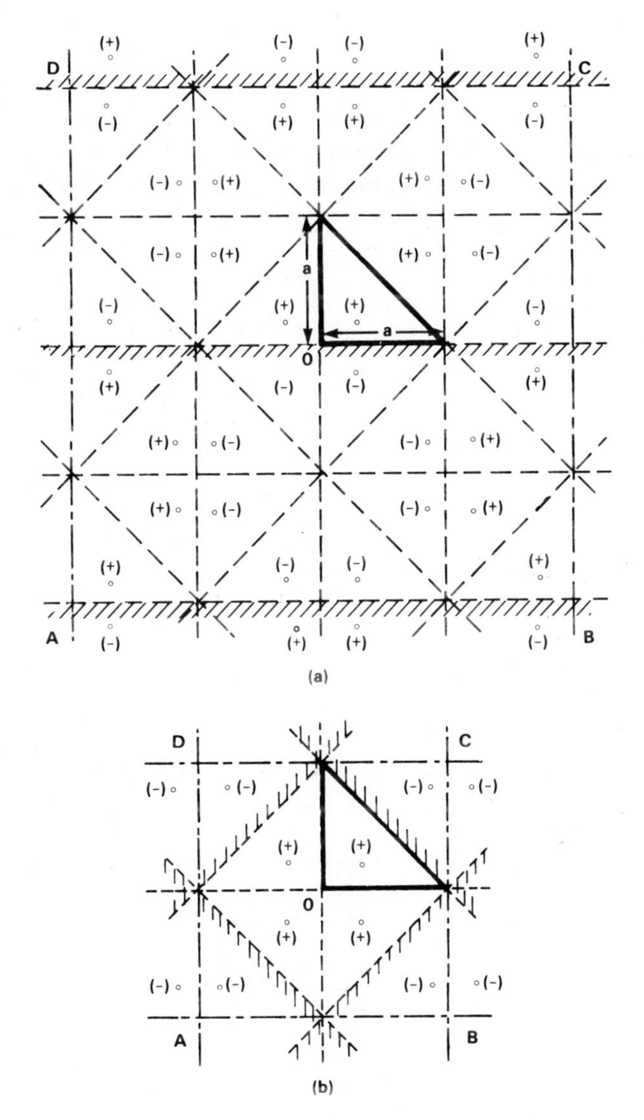

Fig. 11 Image sources of an isosceles right-angled triangle. (*a*) With one side (not hypotenuse) as an electric wall; (*b*) with hypotenuse as electric wall.

Isosceles Right-Angled Triangle with Shorted Hypotenuse. This configuration is shown in Fig. 11*b*, and its Green's function is given by

$$G(x, y \mid x_0, y_0) = \sum_{m=-\infty}^{\infty} \sum_{n=-\infty}^{\infty} \frac{j\omega\mu\, dW(x_0, y_0)W(x, y)}{2[(m^2 + n^2)\pi^2 - a^2k^2]} \tag{77}$$

where

$$W(x, y) = \cos \frac{m\pi x}{a} \cos \frac{n\pi y}{a} - (-1)^{m+n} \cos \frac{n\pi x}{a} \cos \frac{m\pi y}{a} \tag{78}$$

Isosceles Right-Angled Triangle with Hypotenuse and One Other Side Shorted. The locations of images, when the hypotenuse and one of the sides are electrical walls, are shown in Fig. 12. In this case, the Green's function is found to be

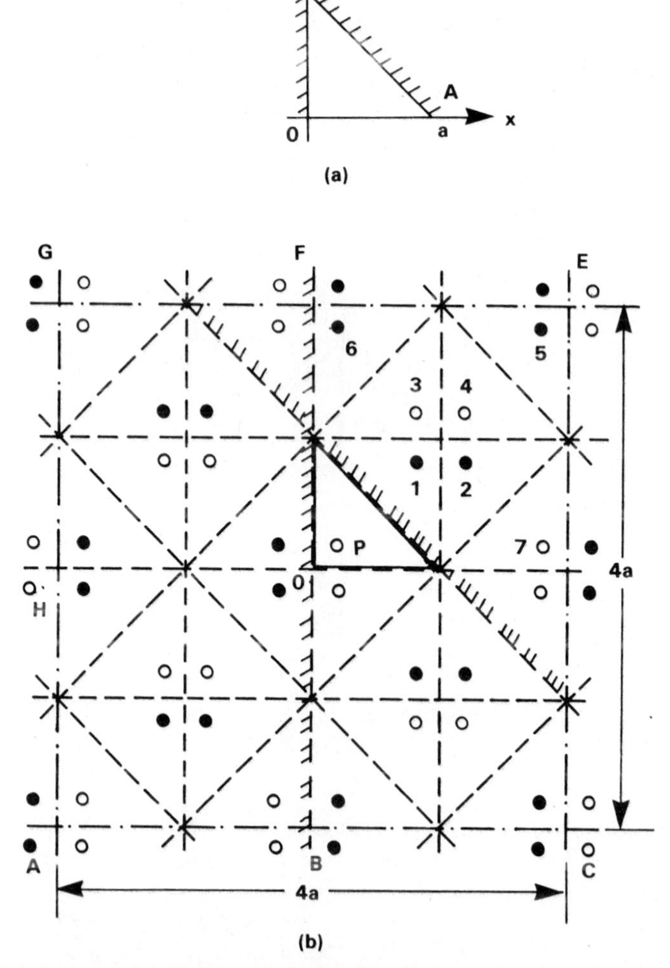

Fig. 12 (*a*) An isosceles right-angled triangle with hypotenuse and one of the other sides as electric wall. (*b*) Location of image sources.

$$G(x, y \mid x_0, y_0) = 2j\omega\mu d \sum_{m=-\infty}^{\infty} \sum_{n=-\infty}^{\infty} \frac{P(x_0, y_0)P(x, y)}{(m^2 + n^2)\pi^2 - 4a^2k^2} \tag{79}$$

where the summations are only over the odd and unequal values of m and n, and the function $P(x, y)$ is given by

$$P(x, y) = \sin \frac{m\pi x}{2a} \cos \frac{n\pi y}{2a} + (-1)^{(m+n)/2} \sin \frac{n\pi x}{2a} \cos \frac{m\pi y}{2a} \tag{80}$$

Isosceles Right-Angled Triangle with Two Nonhypotenuse Sides Shorted. In the case when the hypotenuse is a magnetic wall and the other two sides are electric walls, the locations of images are shown in Fig. 13. For this configuration, the Green's function obtained is

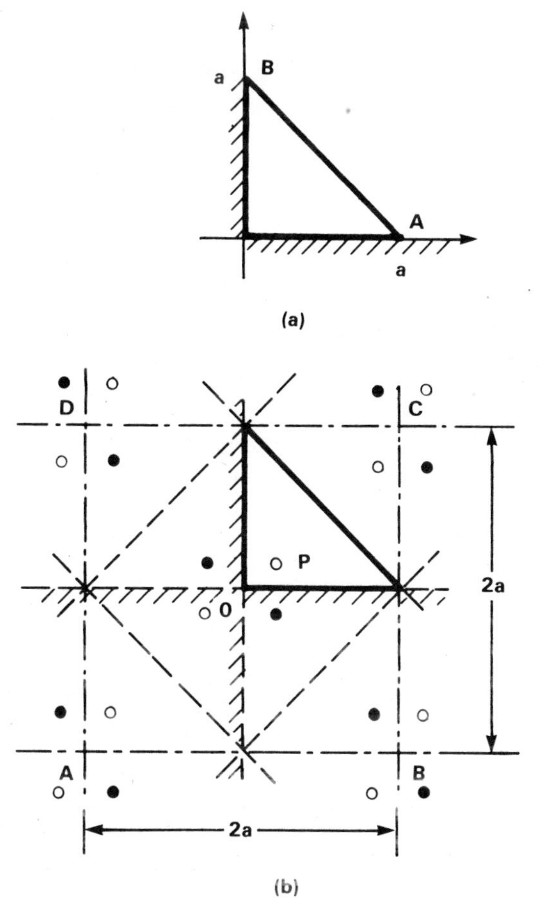

Fig. 13 (*a*) An isosceles right-angled triangle with hypotenuse as magnetic wall and the other two sides as electric walls. (*b*) Location of image sources.

$$G(x, y | x_0, y_0) = j\omega\mu d \sum_{m=-\infty}^{\infty} \sum_{n=-\infty}^{\infty} \frac{Q(x_0, y_0)Q(x, y)}{2[(m^2 + n^2)\pi^2 - a^2k^2]} \tag{81}$$

where

$$Q(x, y) = \sin \frac{m\pi x}{a} \sin \frac{n\pi y}{a} + (-1)^{m+n} \sin \frac{n\pi x}{a} \sin \frac{m\pi y}{a} \tag{82}$$

30°–60° Right-Angled Triangle with Side Opposite 60° Shorted. The Green's function for the configuration shown in Fig. 14 is given by

$$G(x, y | x_0, y_0) = 8j\omega\mu d \sum_{m=-\infty}^{\infty} \sum_{n=-\infty}^{\infty} \frac{T_1(x_0, y_0)T_1(x, y)}{16\sqrt{3}\pi^2(m^2 + n^2 + mn) - 9\sqrt{3}a^2k^2} \tag{83}$$

where the function T_1 is given by

$$\begin{aligned} T_1(x, y) = (-1)^l \cos \frac{2\pi lx}{\sqrt{3}a} \sin \frac{2\pi(m-n)y}{3a} \\ + (-1)^m \cos \frac{2\pi mx}{\sqrt{3}a} \sin \frac{2\pi(n-l)y}{3a} \\ - (-1)^n \cos \frac{2\pi nx}{\sqrt{3}a} \sin \frac{2\pi(l-m)y}{3a} \end{aligned} \tag{84}$$

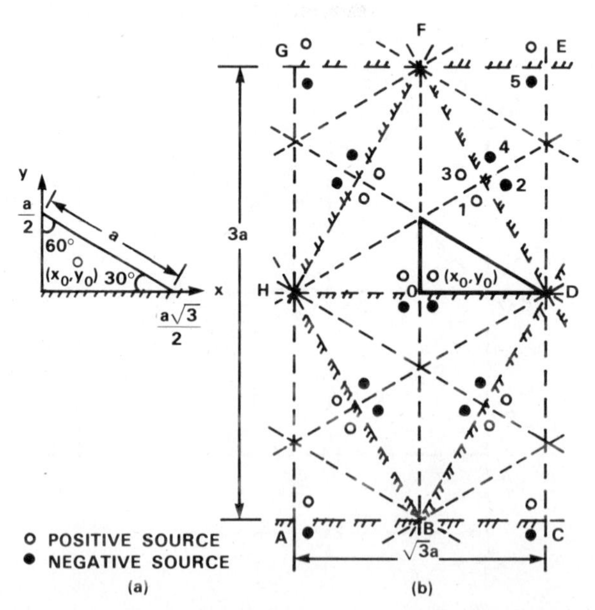

Fig. 14 (*a*) A 30°–60° right-angled triangle with the side opposite to 60° as an electric wall. (*b*) Location of image sources.

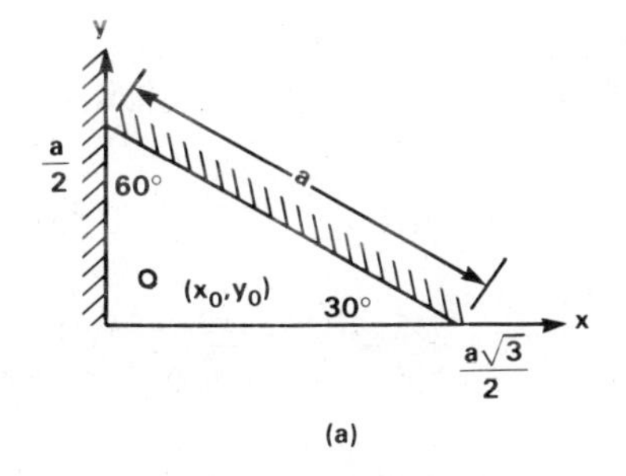

(a)

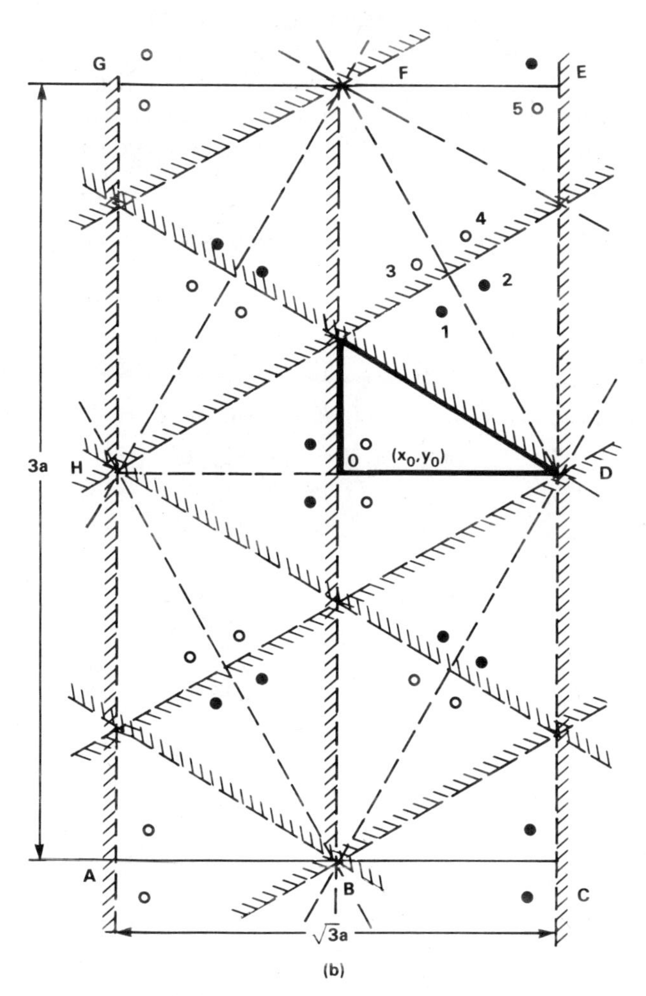

(b)

Fig. 15 (*a*) A 30°–60° right-angled triangle with the sides opposite 30° and 90° as electric walls. (*b*) Location of image source.

30°–60° Right-Angled Triangle with Side Opposite 30° and Hypotenuse Shorted. In this case (Fig. 15), the Green's function is given by

$$G(x, y \mid x_0, y_0) = 8j\omega\mu d \sum_{m=-\infty}^{\infty} \sum_{n=-\infty}^{\infty} \frac{T_3(x_0, y_0)T_3(x, y)}{16\sqrt{3}\pi^2(m^2 + mn + n^2) - 9\sqrt{3}a^2k^2} \tag{85}$$

where the function T_3 is defined in (66).

4.4.3. Circular Sectors

Green's functions for circular sectors with four different kinds of mixed boundary conditions have been derived in Reference 31. These configurations are shown in Fig. 16. In all these cases, Green's functions are obtained by expanding them in terms of appropriate eigenfunctions as discussed in Section 4.1. Expressions for the Green's functions in various cases are summarized below.

Green's Function for Sector with Straight Edges Shorted. For this case (Fig. 16*a*) the Green's function is

$$G(r \mid r_i) = \frac{4j\omega\mu d}{\phi_0} \sum_{m=1}^{\infty} \sum_{n=1}^{\infty} \frac{J_\nu(k'_{\nu n}\rho)J_\nu(k'_{\nu n}\rho_i)\sin \nu\phi \sin \nu\phi_i}{(k'^2_{\nu n} - k^2)(a^2 - \nu^2/k'^2_{\nu n})J^2_\nu(k'_{\nu n}a)} \tag{86}$$

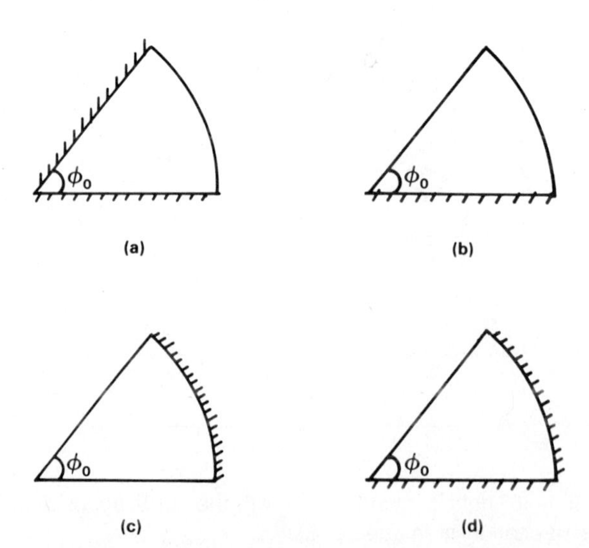

Fig. 16 Circular sectors with mixed boundaries.

As before, ν is an integer given by $m\pi/\phi_0$. The eigenvalues $k'_{\nu n}a$ are roots of

$$\frac{\partial}{\partial \rho} J_\nu(k'_{\nu n}\rho)\bigg|_{\rho=a} = 0 \tag{87}$$

Green's Function for Sector with One Straight Edge Shorted. For this case (Fig. 16*b*), the Green's function is given by [31]

$$G(r|r_i) = \frac{4j\omega\mu d}{\phi_0} \sum_{m=0}^{\infty} \sum_{n=1}^{\infty} \frac{J_\nu(k'_{\nu n}\rho)J_\nu(k'_{\nu n}\rho_i) \sin \nu\phi \sin \nu\phi_i}{(k'^2_{\nu n} - k^2)(a^2 - \nu^2/k'^2_{\nu n})J^2_\nu(k'_{\nu n}a)} \tag{88}$$

where the integer $\nu = (m + \frac{1}{2})\pi/\phi_0$. In this case also, $k'_{\nu n}a$ are given by the roots of (87).

Green's Function for Sectors with Circular Periphery Shorted. This configuration is shown in Fig. 16*c*, and the Green's function can be derived as

$$G(r|r_i) = \frac{4j\omega\mu d}{\phi_0 a^2} \sum_{m=0}^{\infty} \sum_{n=1}^{\infty} \frac{J_\nu(k_{\nu n}\rho)J_\nu(k_{\nu n}\rho_i) \cos \nu\phi \cos \nu\phi_i}{\sigma_m(k^2_{\nu n} - k^2)J^2_{\nu+1}(k_{\nu n}a)} \tag{89}$$

where $\nu = m\pi/\phi_0$, $\sigma_m = 2$ (for $m = 0$), and $\sigma_m = 1$ (for $m \neq 0$). The parameter $k_{\nu n}a$ is given by (68).

Green's Function for Sectors with One Radial Wall and Circular Wall Shorted. In this case (Fig. 16*d*), Green's function is written as

$$G(r|r_i) = \frac{4j\omega\mu d}{\phi_0 a^2} \sum_{m=0}^{\infty} \sum_{n=1}^{\infty} \frac{J_\nu(k_{\nu n}\rho)J_\nu(k_{\nu n}\rho_i) \sin \nu\phi \sin \nu\phi_i}{(k^2_{\nu n} - k^2)J^2_{\nu+1}(k_{\nu n}a)} \tag{90}$$

where the eigenvalues $k_{\nu n}a$ are determined via (68), and $\nu = (m + \frac{1}{2})\pi/\phi_0$.

All the Green's functions listed in Sections 4.2–4.4 are useful for the analysis and design of planar circuits consisting of segments of these shapes.

5. CONTOUR INTEGRAL APPROACH

5.1. Formulation of the Integral

If a planar circuit has an arbitrary shape for which a Green's function does not exist, the impedance matrix of that circuit can be obtained by using the contour integral method [12]. This method can be used to characterize planar segments with shorted, opened, or mixed boundaries and with isotropic or anisotropic spacing media. In this approach, Green's theorem expressed in cylindrical coordinates and Weber's solution for cylindrical waves are used to convert the two-dimensional wave equation given by (16) into a contour integral. The resulting contour integral relates the voltages and currents along the planar circuit periphery.

This approach proceeds by considering the two functions ν and ω, which satisfy

$$(\nabla_T^2 + k^2)\upsilon = 0 \tag{91}$$

in D, where D is a two-dimensional region inside a contour C as shown in Fig. 17, and υ stands for ν and ω. Upon setting $\upsilon = \nu$ and $\omega = H_0^{(2)}(kr)$ (with r being the distance between a point P and a point on the contour C, and $H_0^{(2)}$ denoting the zeroth-order Hankel's function of the second kind) into Green's theorem, which is given by

$$\int_C \left(\nu \frac{\partial \omega}{\partial n} - \omega \frac{\partial \nu}{\partial n} \right) ds_0 = \iint_D (\nu \nabla^2 \omega - \omega \nabla^2 \nu)\, dS = 0 \tag{92}$$

We obtain

$$\int_C \left[\upsilon(s_0) \frac{\partial H_0^{(2)}(kr)}{\partial n} - H_0^{(2)}(kr) \frac{\partial \upsilon}{\partial n} \right] ds_0 = 0 \tag{93}$$

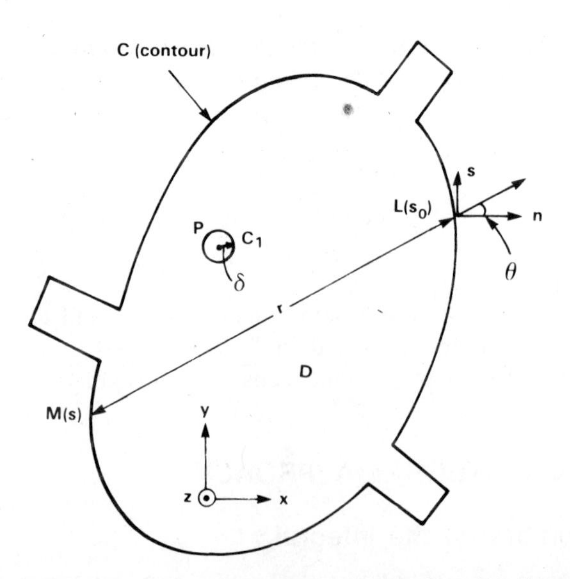

Fig. 17 Configuration of a planar component to be analyzed by the contour integral method. (© IEEE 1977. Reprinted with permission from Reference 23.)

when the point P is located outside the contour C. If point P is inside the contour C, P becomes a singular point for ω, and the area integral in (92) does not vanish. Upon applying Green's theorem to the region that excludes the singular point we get

$$\int_C \left[v(s_0) \frac{\partial H_0^{(2)}(kr)}{\partial n} - H_0^{(2)}(kr) \frac{\partial v}{\partial n} \right] ds_0$$
$$= \int_{C_1} \left[v(s_0) \frac{\partial H_0^{(2)}(kr)}{\partial r} - H_0^{(2)}(kr) \frac{\partial v}{\partial r} \right] ds_0 \tag{94}$$

where C_1 stands for the contour surrounding the excluded small circular area whose radius is chosen to be δ. Here we have also used $\partial n = -\partial r$ on C_1.

When $H_0^{(2)}(kr)$ is replaced by its small argument expression, which holds for $kr < 1$, and when v and $\partial v / \partial n$ are taken to be constant on C_1, the right-hand side of (94) can be written as

$$\int_{C_1} \left[-\frac{2j}{\pi r} v(s_0) + \left(\frac{2j}{\pi} \ln \frac{kr}{2} \right) \frac{\partial v}{\partial n} \right] ds_0 = -4jv(s) + 4j\delta \ln \frac{k\delta}{2} \frac{\partial v}{\partial n} \tag{95}$$

When the term δ tends to zero, the second term on the right-hand side of eq. (95) vanishes. Thus for a point inside the contour C we obtain

$$4jv(s) = \int_C \left[H_0^{(2)}(kr) \frac{\partial v}{\partial n} - v(s_0) \frac{\partial H_0^{(2)}(kr)}{\partial n} \right] ds_0 \tag{96}$$

This equation gives the potential at a point P in the region D in terms of v and $\partial v / \partial n$ along the contour C that surrounds D.

When the point P is on the contour C, the right-hand side of (96) is reduced by a factor of 2 and may be written as

$$v(s) = \frac{1}{2j} \int_C \left[H_0^{(2)}(kr) \frac{\partial v}{\partial n} - v(s_0) \frac{\partial H_0^{(2)}(kr)}{\partial n} \right] ds_0 \tag{97}$$

Combining the above equation with

$$\frac{\partial v}{\partial n} = \frac{j\omega\mu d}{p} J_n \tag{98}$$

which is equivalent to (20), and

$$\frac{\partial H_0^{(2)}(kr)}{\partial n} = -k \cos\theta \, H_1^{(2)}(kr) \tag{99}$$

we obtain

$$v(s) = \frac{1}{2j} \int_C \left[k \cos \theta H_1^{(2)}(kr) v(s_0) + \frac{j\omega\mu d}{p} J_n(s_0) H_0^{(2)}(kr) \right] ds_0 \tag{100}$$

The term $H_1^{(2)}$ represents the first-order Hankel function of the second kind, $H_0^{(2)}$ denotes the zeroth order Hankel function of the second kind, J_n is the current density flowing into the segment at s_0, and, as before, $p = 1$ for microstrip and reduced-height waveguide type structures, and $p = 2$ for triplate-type planar circuits. The variables s and s_0 denote distances along the contour C, and r is the straight line distance between the two points M and L (given by s and s_0) as shown in Fig. 17. Angle θ is made by the straight line joining points M and L with the normal to the periphery at L.

Equation (100) describes the relation between the rf voltage and the rf current distributions along the periphery. A planar circuit described by (100) may have an open boundary, a shorted boundary, or a mixed boundary. In either case the integral equation given by (100) can be solved by dividing the circuit periphery into N incremental sections having arbitrary widths $W_1, W_2, \ldots, W_n$ as discussed in the next two subsections.

5.2. Open-Boundary Planar Circuit [12]

As described earlier, an impedance matrix characterization is appropriate for a planar circuit with an open boundary. In order to compute the impedance matrix, the periphery is divided into N sections numbered $1, 2, \ldots, N$ having widths $W_1, W_2, \ldots, W_N$, respectively, as illustrated in Fig. 18. The periphery of the planar circuit is divided in such a manner that each coupling port contains an integral number of sections. It is assumed that the widths of the sections are so small that magnetic and electric fields can be considered uniform over each section. These multiple sections for wide coupling ports result in increased accuracy. Next, N sampling points are set, with one sampling point at the center of each section, and it has been assumed that each section is a straight segment. Under these assumptions, the line integral in (100) can be replaced by a summation over the N sections. The resulting expression is given by

$$2jv_l = \sum_{m=1}^{N} \left[kv_m G_{lm} + \frac{j\omega\mu d}{p} i_m F_{lm} \right] \tag{101}$$

where v_l is the voltage over the lth section and i_m $(= J_n W_m)$ is the total current flowing into the mth section. The matrix elements G_{lm} and F_{lm} are given as

$$G_{lm} = \begin{cases} \int_{W_m} \cos \theta H_1^{(2)}(kr) \, ds_0 & \text{if } l \neq m \\ 0 & \text{otherwise} \end{cases} \tag{102}$$

and

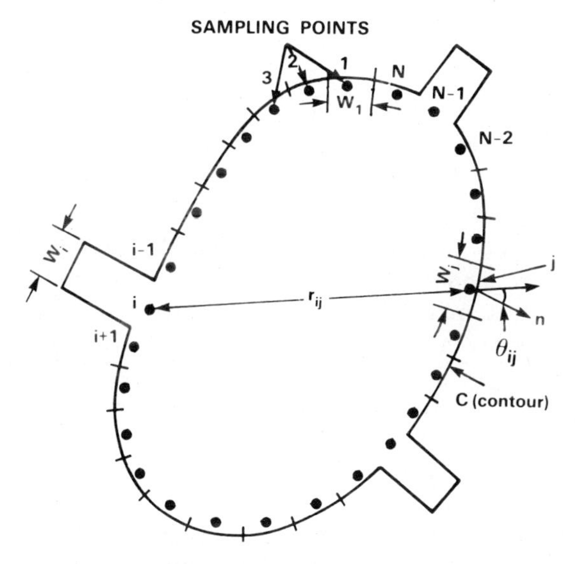

Fig. 18 Division of the periphery into N sections with a sampling point at the center of each section. (© IEEE 1977. Reprinted with permission from Reference 23.)

$$F_{lm} = \begin{cases} \dfrac{1}{W_m} \displaystyle\int_{W_m} H_0^{(2)}(kr)\, ds_0 & \text{if } l \neq m \\ 1 - \dfrac{2j}{\pi}\left(\ln \dfrac{kW_l}{4} - 1 + \gamma\right) & \text{otherwise} \end{cases} \tag{103}$$

In (103), γ (= 0.5772 . . .) denotes Euler's constant. As assumed in the above discussion, the current can be fed into the planar circuit from any of the N sections, and i_m denotes the current fed from the mth section. This yields an $N \times N$ impedance matrix for the N-port circuit. This matrix can be used to obtain the impedance matrix for any specified number and location of ports on the planar circuit being analyzed. Equation (101) is written for each section l on the periphery of the planar circuit. All these equations combined together in matrix form become

$$\boldsymbol{AV} = \boldsymbol{BI} \tag{104}$$

where $\boldsymbol{V}$ and $\boldsymbol{I}$ are column vectors consisting of the voltages and the currents at each section. $\boldsymbol{A}$ and $\boldsymbol{B}$ denote $N \times N$ matrices that are determined by the shape of the circuit. The elements of these matrices, obtained from (101), are

$$a_{lm} = -kG_{lm} \qquad \text{for } l \neq m \tag{105}$$

$$a_{ll} = 2j \tag{106}$$

and

$$b_{lm} = j \frac{\omega \mu d}{p} F_{lm} \tag{107}$$

The matrix elements a_{lm} and b_{lm} defined by (105)–(107) can be accurately calculated by using standard integral computation algorithms such as Simpson's method or Romberg's method. However, in many cases these integrals can be approximated by

$$G_{lm} = \cos\theta H_1^{(2)}(kr) W_m , \qquad l \neq m \tag{108}$$

$$F_{lm} = H_0^{(2)}(kr) , \qquad l \neq m \tag{109}$$

The selection of one calculation approach over another should be based on the required accuracy and the computer time efficiency.

From (104) we can obtain the rf voltage on each sample point as

$$\boldsymbol{V} = \boldsymbol{A}^{-1}\boldsymbol{BI} \tag{110}$$

where the matrix $\boldsymbol{A}^{-1}$ is the inverse of matrix $\boldsymbol{A}$. Assuming that all of the N sections along the periphery are coupling ports, the planar circuit can be represented by an N-port equivalent circuit whose impedance matrix is given by

$$\mathbf{Z}_N = \boldsymbol{A}^{-1}\boldsymbol{B} \tag{111}$$

In practice, however, the coupling ports are connected to only a few of the N sections while the rest of the sections are left open-circuited. Consequently, rows and columns corresponding to the open-circuited sections are deleted from $\mathbf{Z}_N$. If each coupling port covers only one section, the reduced matrix represents the required impedance matrix. If some coupling ports extend over more than one section, the sections covered by these coupling ports are treated as subports. These subports are then combined using the procedure described in Section 2, and consequently the overall impedance matrix is obtained.

The resonance frequencies of the planar components with no coupling ports connected are obtained by setting the determinant of matrix $\boldsymbol{A}$ equal to zero.

5.3. Short-Boundary Planar Circuit [33]

This category of planar circuits requires a mathematical formulation that is slightly different from that of the open-boundary case. This is merely because of the different boundary condition and the different input/output ports structure. As shown in Fig. 19, a planar circuit with a short boundary can be coupled to the external circuits through waveguide ports, coaxial ports, or a combination of the two.

The contour integral used earlier for the open-boundary case and given by (100) can also be used to compute the admittance matrix of the short-boundary circuit. The periphery of thc circuit is, as before, divided into M sections, and in addition the peripheries of the coupling ports are divided into N sections. For example, when the planar circuit has two coaxial coupling ports, p and q (Fig. 19a), the peripheries of the coupling conductors are divided into m and n sections (as shown in Fig. 20), and consequently the total number of sections becomes equal to $M + m + n$. As before, a sampling point is set at the center of each section and each section is assumed to have a straight edge. Upon replacing the line integral by a summation over the $N = m + n + M$ sections we obtain a matrix equation of the form

$$\boldsymbol{AV} = \boldsymbol{BI} \tag{112}$$

where $\boldsymbol{V}$ and $\boldsymbol{I}$ are column vectors consisting of the voltages and currents at each section. In the above equation, $\boldsymbol{A}$ and $\boldsymbol{B}$ are two $N \times N$ matrices whose elements are obtained from

$$a_{rs} = -k \int_{W_s} \cos\theta H_1^{(2)}(kr)\, ds \qquad \text{for } r \neq s \tag{113}$$

$$a_{rr} = 2j \tag{114}$$

and

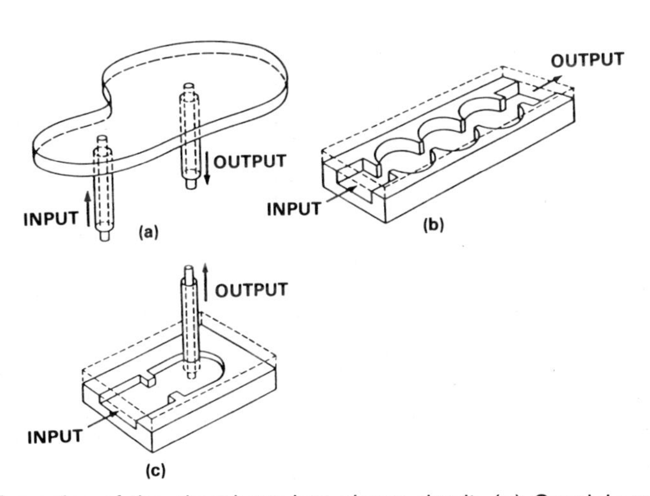

Fig. 19 Examples of the short-boundary planar circuit. (a) Coaxial-coupled type; (b) waveguide coupled type; (c) mixed type. (© IEEE 1975. Reprinted with permission from Reference 33.)

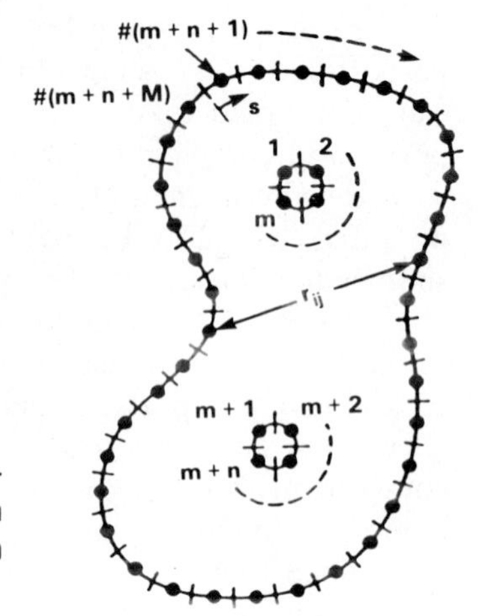

Fig. 20 Nomenclature of ports used in the analysis of a coaxial-coupled type planar circuit with shorted boundary. (© IEEE 1975. Reprinted with permission from Reference 33.)

$$b_{rs} = \begin{cases} j\omega\mu d \, \dfrac{1}{W_s} \displaystyle\int_{W_r} H_0^{(2)}(kr) \, ds_0 & r \neq s \\ j\omega\mu d \left[1 - \dfrac{2j}{\pi} \left(\ln \dfrac{kW_r}{4} - 1 + \gamma \right) \right] & \text{otherwise} \end{cases} \tag{115}$$

where $\gamma = 0.5772\ldots$ is Euler's constant, and W_r, W_s denote the widths of ports r and s, respectively. Note that in eq. (115) the factor p has been replaced by 1, its value for planar circuits with a shorted boundary.

Upon imposing the shorted boundary conditions, $V = 0$, along the periphery of the circuit (M ports) we may write (112) as

$$[A] \begin{bmatrix} V_p \\ \cdot \\ \cdot \\ \cdot \\ V_p \\ \hline V_q \\ \cdot \\ \cdot \\ \cdot \\ V_q \\ \hline 0 \\ \cdot \\ \cdot \\ \cdot \\ 0 \end{bmatrix} \begin{matrix} \left.\vphantom{\begin{matrix} 1\\1\\1\\1\\1 \end{matrix}}\right\} m \\ \left.\vphantom{\begin{matrix} 1\\1\\1\\1\\1 \end{matrix}}\right\} n \\ \left.\vphantom{\begin{matrix} 1\\1\\1\\1\\1 \end{matrix}}\right\} M \end{matrix} = [B] \begin{bmatrix} I_p/m \\ \cdot \\ \cdot \\ \cdot \\ I_p/m \\ \hline I_q/n \\ \cdot \\ \cdot \\ \cdot \\ I_q/n \\ \hline I_{m+n+1} \\ \cdot \\ \cdot \\ \cdot \\ I_{m+n+M} \end{bmatrix} \begin{matrix} \left.\vphantom{\begin{matrix} 1\\1\\1\\1\\1 \end{matrix}}\right\} m \\ \left.\vphantom{\begin{matrix} 1\\1\\1\\1\\1 \end{matrix}}\right\} n \\ \left.\vphantom{\begin{matrix} 1\\1\\1\\1\\1 \end{matrix}}\right\} M \end{matrix} \tag{116}$$

where V_p, V_q, I_p, and I_q denote, respectively, the voltages and currents at the terminal ports p and q, which are divided into m and n sections, respectively. The terms I_p/m and I_q/n appear in (116) because uniform voltage and current distribution around the conductors have been assumed.

Equation (116) is composed of $m+n+M$ equations while only $M+2$ unknown variables exist. For simplicity the two matrices $\boldsymbol{A}$ and $\boldsymbol{B}$ are first reduced into $(M+2) \times (M+2)$ matrices so that (116) can be written as

$$[A'] \begin{bmatrix} \left.\begin{matrix} V_p \\ V_q \end{matrix}\right\} 2 \\ \hline \left.\begin{matrix} 0 \\ \cdot \\ \cdot \\ \cdot \\ 0 \end{matrix}\right\} M \end{bmatrix} = [B'] \begin{bmatrix} \left.\begin{matrix} I_p \\ I_q \end{matrix}\right\} 2 \\ \hline \left.\begin{matrix} I_{m+n+1} \\ \cdot \\ \cdot \\ \cdot \\ I_{m+n+M} \end{matrix}\right\} M \end{bmatrix} \tag{117}$$

where the reduced matrix $[A']$ is defined as

$$[A'] = \left[\begin{array}{cc|ccc} \sum_{i=1}^{m}\sum_{j=1}^{m} a_{ij} & \sum_{i=1}^{m}\sum_{j=m+1}^{m+n} a_{ij} & \sum_{i=1}^{m} a_{i(m+n+1)} & & \cdots \\ \sum_{i=m+1}^{m+n}\sum_{j=1}^{m} a_{ij} & \sum_{i=m+1}^{m+n}\sum_{j=m+1}^{m+n} a_{ij} & \sum_{i=m+1}^{m+n} a_{i(m+n+1)} & & \cdots \\ \hline \sum_{j=1}^{m} a_{(m+n+1)j} & \sum_{j=m+1}^{m+n} a_{(m+n+1)j} & a_{m+n+1 \atop m+n+1} & & \\ \cdot & \cdot & & \ddots & \\ \cdot & \cdot & & & a_{m+n+M \atop m+n+M} \\ \cdot & \cdot & & & \end{array}\right] \tag{118}$$

and $[\boldsymbol{B}']$ has a similar definition with a replaced by b. The admittance matrix relating current to voltage can be derived from (117) and is given by

$$\boldsymbol{Y} = [\boldsymbol{B}']^{-1}[\boldsymbol{A}'] \tag{119}$$

The desired admittance parameters Y_{pp}, Y_{pq}, Y_{qp}, and Y_{qq} are found in the top left corner of the admittance matrix $\boldsymbol{Y}$. The generality of this method allows its application to cases in which the planar circuit has an arbitrary number of ports. If all the $M+2$ sampling points are connected to coupling ports, the planar circuit under consideration can then be characterized by an $(M+2)$-port equivalent circuit.

6. ANALYSIS OF PLANAR COMPONENTS OF COMPOSITE CONFIGURATIONS

As discussed in Sections 3 and 4, the impedance matrix for a planar circuit with a regular geometrical shape can be obtained from the Green's function when port locations are specified. On the other hand, segments with completely arbitrary shapes, for which a Green's function cannot be written, necessitate the use of the contour integral method of Section 5 to compute the impedance matrix.

Between these two extreme cases, there is a class of planar components in which the shape of the planar circuit is a composite of simple configurations. In these cases, the two-dimensional composite shapes can be decomposed into either (1) all regular shapes or (2) a combination of some regular shape(s) and some arbitrary shape(s). Examples of these two types of composite planar circuit configurations are shown in Figs. 21*a* and *b*, respectively. This process of breaking down a composite shape into simpler shapes is known as *segmentation* [27].

There is also a complementary process called desegmentation [20], which is suitable for analyzing some planar circuit configurations. An example of the latter is shown in Fig. 22. This configuration cannot be decomposed into regular shapes, but if a right-angled circular sector (a regular shape for which the Green's function is available) is added to the shape shown in Fig. 22, we end up with a regular rectangle. This process has been termed "desegmentation" [20]. It has been shown that if the impedance matrices of the segments added to the original configuration (right-angled circular sector

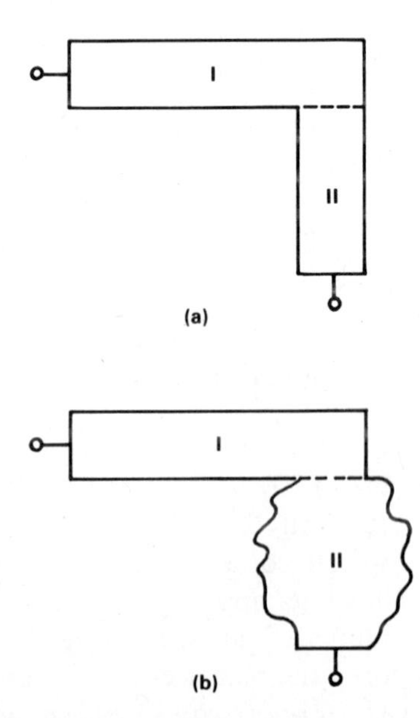

Fig. 21 Planar circuit of composite configurations. (*a*) A configuration that can be decomposed into two regularly shaped segments. (*b*) A configuration that can be decomposed into a regular and an arbitrary segment.

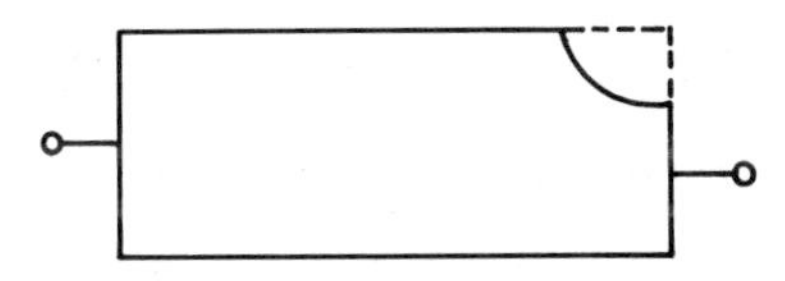

Fig. 22 Planar circuit configuration suitable for analysis by the desegmentation method.

in the present case) and the impedance matrix of the augmented configuration (rectangle in this case) are known, the impedance matrix of the original planar segment can be obtained by using "desegmentation" formulas [20, 34].

The segmentation and desegmentation procedures are described in this section.

6.1. Segmentation Method

The name segmentation has been given to this network analysis method for planar (two-dimensional) microwave circuits by Okoshi and his colleagues [12, 16, 27]. The basic idea is to divide a single complicated planar circuit configuration into simpler "segments" that have regular shapes and can therefore be characterized relatively easily. An example of such a segmentation is shown in Fig. 21*a*, where a microstrip corner geometry is broken down into two rectangular segments for which Green's functions are available. Essentially, the segmentation method gives us the overall characterization or performance of the composite network, when the characterization of each of the segments is known. Originally the segmentation method was formulated [27] in terms of ***S*** matrices of individual segments; however, it was found subsequently [19] that a ***Z*** matrix formulation is more efficient for microwave planar circuits (also for microstrip antennas). In this section, we will describe the segmentation procedure based on ***Z*** matrices.

This procedure can be illustrated by considering a multiport network consisting of only two segments A and B as shown in Fig. 23. Note that the various ports of these two segments are numbered. The external (unconnected) ports of segment A are called p_a ports, and the external unconnected ports of segment B are referred to as p_b ports. The connected ports of segment A are called q ports, and the connected ports of segment B are designated r ports. The q and r ports are numbered such that q_1 is connected to r_1, q_2 to r_2, and so on. As a result of these interconnections, we can write

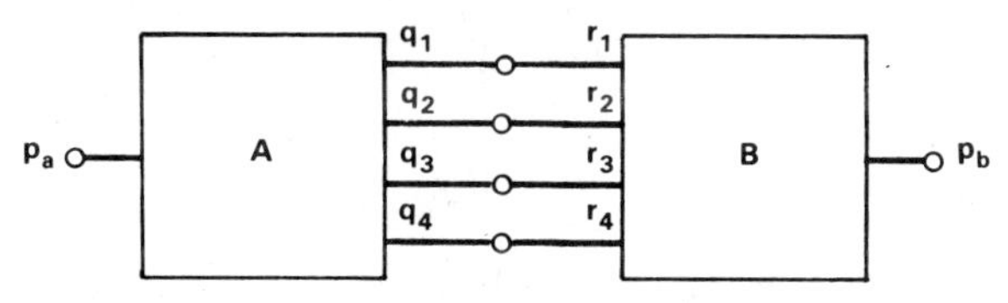

Fig. 23 Port nomenclature use in the segmentation formula.

$$V_q = V_r \qquad \text{and} \qquad i_q = -i_r \tag{120}$$

The **Z** matrices of segments A and B may be written as

$$\mathbf{Z}_A = \begin{bmatrix} \mathbf{Z}_{p_a} & \mathbf{Z}_{p_a q} \\ \mathbf{Z}_{qp_a} & \mathbf{Z}_{qq} \end{bmatrix} \qquad \mathbf{Z}_B = \begin{bmatrix} \mathbf{Z}_{p_b} & \mathbf{Z}_{p_b r} \\ \mathbf{Z}_{rp_b} & \mathbf{Z}_{rr} \end{bmatrix} \tag{121}$$

where $\mathbf{Z}_{p_a}$, $\mathbf{Z}_{p_a q}$, $\mathbf{Z}_{qp_a}$, $\mathbf{Z}_{qq}$, $\mathbf{Z}_{p_b}$, $\mathbf{Z}_{p_b r}$, $\mathbf{Z}_{rp_b}$, $\mathbf{Z}_{rr}$ are submatrices of appropriate dimensions. Since these components are reciprocal, the following relationships apply:

$$\mathbf{Z}_{p_a q} = [\mathbf{Z}_{qp_a}]^{\mathrm{t}} \qquad \text{and} \qquad \mathbf{Z}_{p_b r} = [\mathbf{Z}_{rp_b}]^{\mathrm{t}} \tag{122}$$

where the superscript t denotes the transpose of the matrix. The **Z** matrices of the segments A and B can be written together as

$$\begin{bmatrix} \mathbf{V}_p \\ \mathbf{V}_q \\ \mathbf{V}_r \end{bmatrix} = \begin{bmatrix} \mathbf{Z}_{pp} & \mathbf{Z}_{pq} & \mathbf{Z}_{pr} \\ \mathbf{Z}_{qp} & \mathbf{Z}_{qq} & \mathbf{0} \\ \mathbf{Z}_{rp} & \mathbf{0} & \mathbf{Z}_{rr} \end{bmatrix} \begin{bmatrix} \mathbf{i}_p \\ \mathbf{i}_q \\ \mathbf{i}_r \end{bmatrix} \tag{123}$$

where

$$\mathbf{V}_p = \begin{bmatrix} V_{p_a} \\ V_{p_b} \end{bmatrix} \qquad \mathbf{i}_p = \begin{bmatrix} i_{p_a} \\ i_{p_b} \end{bmatrix}$$

and

$$\mathbf{Z}_{pp} = \begin{bmatrix} \mathbf{Z}_{p_a} & \mathbf{0} \\ \mathbf{0} & \mathbf{Z}_{p_b} \end{bmatrix} \qquad \mathbf{Z}_{pq} = \begin{bmatrix} \mathbf{Z}_{p_a q} \\ \mathbf{0} \end{bmatrix} \qquad \mathbf{Z}_{pr} = \begin{bmatrix} \mathbf{0} \\ \mathbf{Z}_{p_b r} \end{bmatrix}$$

and

$$\mathbf{Z}_{pq} = [\mathbf{Z}_{qp}]^{\mathrm{t}} \qquad \mathbf{Z}_{rp} = [\mathbf{Z}_{pr}]^{\mathrm{t}}$$

where $\mathbf{0}$ denotes a null matrix of appropriate dimensions.

It may be noted that the interconnection conditions outlined in (120) have not been used for writing (123), which has been obtained by combining the individual matrices $\mathbf{Z}_A$ and $\mathbf{Z}_B$ given in (121). Relations (120) can now be substituted in the equations of (123) to eliminate $\mathbf{V}_q$, $\mathbf{V}_r$, $\mathbf{i}_q$, and $\mathbf{i}_r$. The resulting expression may be written as $\boldsymbol{V}_p = [\mathbf{Z}_{AB}]\boldsymbol{i}_p$, where

$$[\mathbf{Z}_{AB}] = \begin{bmatrix} \mathbf{Z}_{p_a} & \mathbf{0} \\ \mathbf{0} & \mathbf{Z}_{p_b} \end{bmatrix} + \begin{bmatrix} \mathbf{Z}_{p_a q} \\ -\mathbf{Z}_{p_b r} \end{bmatrix} [\mathbf{Z}_{qq} + \mathbf{Z}_{rr}]^{-1} [-\mathbf{Z}_{qp_a} \quad \mathbf{Z}_{rp_b}] \tag{124}$$

It may be noted that the size of $\mathbf{Z}_{AB}$ is $(p_a + p_b) \times (p_a + p_b)$. The second term on the right-hand side of (124) is a product of three matrices of sizes $(p_a + p_b) \times q$, $q \times q$, and $q \times (p_a + p_b)$. From the computational point of view, the most time-consuming step is the evaluation of the inverse of a matrix of size $q \times q$, where q is the number of interconnected ports.

To illustrate the above procedure for combining $\boldsymbol{Z}$ matrices of two segments together, let us consider an example of two transmission line sections of electrical lengths θ_1 and θ_2 connected in cascade (as shown in Fig. 24). The $\boldsymbol{Z}$ matrices of the individual sections A and B are given by

$$\boldsymbol{Z}_A = -\frac{jZ_0}{\sin\theta_1}\begin{bmatrix} \cos\theta_1 & 1 \\ 1 & \cos\theta_1 \end{bmatrix} \tag{125}$$

and

$$\boldsymbol{Z}_B = -\frac{jZ_0}{\sin\theta_2}\begin{bmatrix} \cos\theta_2 & 1 \\ 1 & \cos\theta_2 \end{bmatrix} \tag{126}$$

In terms of the notations of eq. (124), we have

$$\boldsymbol{Z}_{pp} = \begin{bmatrix} \boldsymbol{Z}_{p_a} & \boldsymbol{0} \\ \boldsymbol{0} & \boldsymbol{Z}_{p_b} \end{bmatrix} = \begin{bmatrix} Z_{11} & \boldsymbol{0} \\ \boldsymbol{0} & Z_{22} \end{bmatrix} = \begin{bmatrix} z_1\cos\theta_1 & 0 \\ 0 & z_2\cos\theta_2 \end{bmatrix}$$

$$\boldsymbol{Z}_{pq} = \begin{bmatrix} \boldsymbol{Z}_{p_a q} \\ \boldsymbol{0} \end{bmatrix} = z_1\begin{bmatrix} 1 \\ 0 \end{bmatrix}, \qquad z_1 = \frac{-jZ_0}{\sin\theta_1}$$

$$\boldsymbol{Z}_{pr} = \begin{bmatrix} 0 \\ \boldsymbol{Z}_{p_b r} \end{bmatrix} = z_2\begin{bmatrix} 0 \\ 1 \end{bmatrix}, \qquad z_2 = \frac{-jZ_0}{\sin\theta_2}$$

$$\boldsymbol{Z}_{qq} = Z_{33} = z_1\cos\theta_1$$

$$\boldsymbol{Z}_{rr} = Z_{44} = z_2\cos\theta_2$$

$$\boldsymbol{Z}_{qp} = [\boldsymbol{Z}_{qp_a} \quad \boldsymbol{0}] = [z_1 \quad 0]$$

$$\boldsymbol{Z}_{rp} = [\boldsymbol{0} \quad \boldsymbol{Z}_{rp_b}] = [0 \quad z_2]$$

By substituting all these submatrices into eq. (124) we get

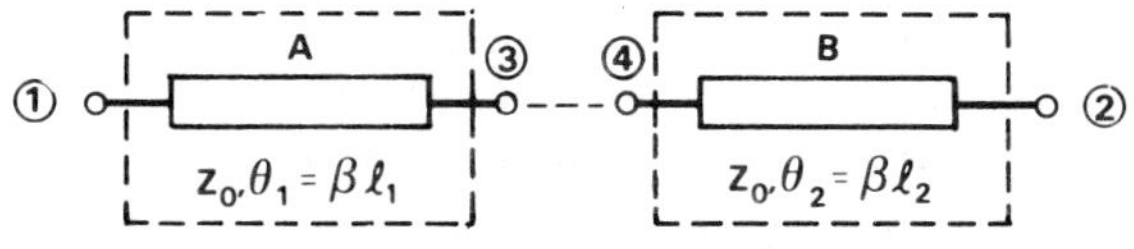

Fig. 24 Segmentation as applied to two transmission line sections.

$$\begin{aligned}\mathbf{Z}_{AB} &= \begin{bmatrix} z_1 \cos\theta_1 & 0 \\ 0 & z_2 \cos\theta_2 \end{bmatrix} + \begin{bmatrix} z_1 \\ -z_2 \end{bmatrix} [z_1 \cos\theta_1 + z_2 \cos\theta_2]^{-1} [-z_1 \quad z_2] \\ &= \begin{bmatrix} z_1 \cos\theta_1 & 0 \\ 0 & z_2 \cos\theta_2 \end{bmatrix} + \frac{1}{z_1 \cos\theta_1 + z_2 \cos\theta_2} \begin{bmatrix} z_1 \\ -z_2 \end{bmatrix} [-z_1 \quad z_2] \\ &= \begin{bmatrix} z_1 \cos\theta_1 & 0 \\ 0 & z_2 \cos\theta_2 \end{bmatrix} + \frac{1}{z_1 \cos\theta_1 + z_2 \cos\theta_2} \begin{bmatrix} -z_1^2 & z_1 z_2 \\ z_1 z_2 & -z_2^2 \end{bmatrix} \end{aligned} \tag{127}$$

By substituting for z_1 and z_2 and using trigonometric formulas for $\sin(\theta_1 + \theta_2)$ and $\cos(\theta_1 + \theta_2)$, (127) may be expressed as

$$\mathbf{Z}_{AB} = -\frac{jZ_0}{\sin(\theta_1 + \theta_2)} \begin{bmatrix} \cos(\theta_1 + \theta_2) & 1 \\ 1 & \cos(\theta_1 + \theta_2) \end{bmatrix} \tag{128}$$

which is the **Z** matrix for a uniform transmission line of length $\theta_1 + \theta_2$ and illustrates the validity of (124).

When a sensitivity analysis of the planar circuit is being carried out using the adjoint network method [15, Chapter 12], it becomes necessary to evaluate the voltages at the ports interconnecting the various segments.

Referring to Fig. 23, voltages at the connected ports (q ports) may be obtained by [15, p. 357]

$$\mathbf{V}_q = \{\mathbf{Z}_{qp} + (\mathbf{Z}_{qq} - \mathbf{Z}_{qr})(\mathbf{Z}_{qq} + \mathbf{Z}_{rr})^{-1}(\mathbf{Z}_{rp} - \mathbf{Z}_{qp})\}\mathbf{i}_p \tag{129}$$

where $\mathbf{i}_p$ is the current vector specifying the input current(s) at the external port(s) of the network.

6.2. Desegmentation Method

As mentioned earlier, there are several configurations of planar circuits that cannot be analyzed by the segmentation method discussed above. For example, the configuration shown in Fig. 22 cannot be partitioned into regular segments for which Green's functions are known. In cases like this, an alternative method called desegmentation [20, 34] is useful. The process of desegmentation can be illustrated by considering the example shown in Fig. 25. If a circular sector called segment β is added to the configuration α, the resulting configuration γ is a triangular segment. Green's functions are known for both the circular sector and the triangular shapes, and therefore **Z** matrices for characterizing both of these components may be derived. The desegmentation method allows us to derive the **Z** matrix of the original configuration α when the **Z** matrices of the triangular segment γ and the circular sector segment β are known. For deriving a relationship among the **Z** matrices of three shapes, various external and connected ports located as shown in Fig. 25 are considered. Ports $p_1, p_2, \ldots,$ are external ports of α.

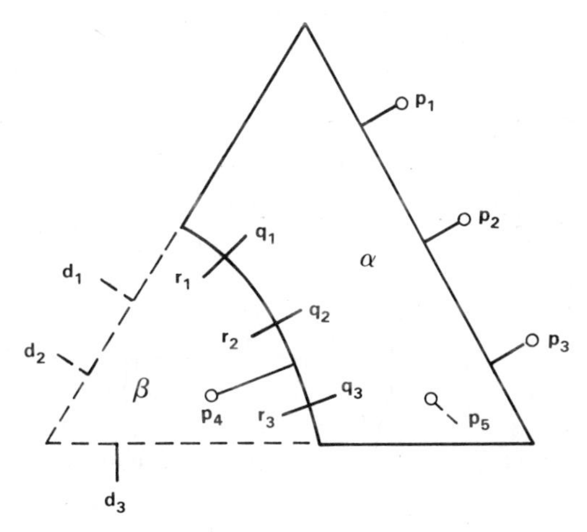

Fig. 25 Port nomenclature used in the desegmentation formula.

Characterization of α is required with respect to these ports. In general, p ports can also be located on the part of the periphery of α where the segment β is connected. An example of this is the port p_4 shown in the figure. The **Z** matrices of β and γ segments are known and may be written as

$$\mathbf{Z}_\beta = \begin{bmatrix} \mathbf{Z}_{rr} & \mathbf{Z}_{rd} \\ \mathbf{Z}_{dr} & \mathbf{Z}_{dd\beta} \end{bmatrix} \qquad \mathbf{Z}_\gamma = \begin{bmatrix} \mathbf{Z}_{pp\gamma} & \mathbf{Z}_{pd} \\ \mathbf{Z}_{dp} & \mathbf{Z}_{dd\gamma} \end{bmatrix} \tag{130}$$

As in the case of segmentation, ports q (of α) and ports r (of β) are numbered such that q_1 is connected to r_1, q_2 to r_2, etc. Ports d are the unconnected (external) port of the segment β. Evaluation of $\mathbf{Z}_\alpha$ is simplified when the number of d ports is made equal to the number of q (or r) ports. The number of q (or r) ports depends on the nature of the field variation along the α–β interface and, as in the case of segmentation, is determined by iterative computations. On the other hand, the number of d ports is arbitrary and can always be made equal to that of q (or r) ports after the number of q ports has been finalized. Under these conditions, the impedance matrix for the α segment can be expressed [20] in terms of the **Z** matrices of the β and γ segments as

$$\mathbf{Z}_\alpha = \mathbf{Z}_{pp\gamma} - \mathbf{Z}_{pd}\{\mathbf{Z}_{dd\gamma} - \mathbf{Z}_{dd\beta}\}^{-1}\mathbf{Z}_{dp} \tag{131}$$

It may be noted that the size of $\mathbf{Z}_\alpha$ is $p \times p$ since all the specified ports of α segment have been numbered as p ports.

In order to illustrate the implementation of (131), an example of two transmission line sections of electrical lengths βl_1 and βl_2 connected in cascade as shown in Fig. 24 is considered. In this case, we want to find $\mathbf{Z}_A$ when $\mathbf{Z}_B$ and $\mathbf{Z}_{AB}$ are known. The **Z** matrices of the individual sections AB

and B, which correspond to $\mathbf{Z}_\gamma$ and $\mathbf{Z}_\beta$ in (131), are respectively given by

$$\mathbf{Z}_\gamma = -\frac{jZ_0}{\sin(\theta_1+\theta_2)}\left[\begin{array}{c|c} \cos(\theta_1+\theta_2) & 1 \\ \hline 1 & \cos(\theta_1+\theta_2)\end{array}\right] = \left[\begin{array}{c|c} \mathbf{Z}_{pp\gamma} & \mathbf{Z}_{pd} \\ \hline \mathbf{Z}_{dp} & \mathbf{Z}_{dd\gamma}\end{array}\right] \tag{132}$$

and

$$\mathbf{Z}_\beta = \left[\begin{array}{c|c} \mathbf{Z}_{rr} & \mathbf{Z}_{rd} \\ \hline \mathbf{Z}_{dr} & \mathbf{Z}_{dd\beta}\end{array}\right] = \begin{bmatrix} \cos\theta_2 & 1 \\ 1 & \cos\theta_2\end{bmatrix}\left(-\frac{jZ_0}{\sin\theta_2}\right) \tag{133}$$

The impedance matrix $\mathbf{Z}_\alpha$ (1 port) of segment A can now be written using (131) as

$$\begin{aligned}
\mathbf{Z}_\alpha &= \mathbf{Z}_{pp\gamma} - \mathbf{Z}_{pd}\{\mathbf{Z}_{dd\gamma} - \mathbf{Z}_{dd\beta}\}^{-1}\mathbf{Z}_{dp} \\
&= -jZ_0\cot(\theta_1+\theta_2) - \left[\frac{-jZ_0}{\sin(\theta_1+\theta_2)}\right]\{-jZ_0\cot(\theta_1+\theta_2) \\
&\quad + jZ_0\cot\theta_2\}^{-1}\left[\frac{-jZ_0}{\sin(\theta_1+\theta_2)}\right] \\
&= -jZ_0\cot(\theta_1+\theta_2) - \left[\frac{(-jZ_0)^2}{\sin^2(\theta_1+\theta_2)}\right]\left[\frac{1}{-jZ_0[\cot(\theta_1+\theta_2)-\cot\theta_2]}\right] \\
&= -jZ_0\left\{\cot(\theta_1+\theta_2) - \frac{1}{\sin^2(\theta_1+\theta_2)[\cot(\theta_1+\theta_2)-\cot\theta_2]}\right\}
\end{aligned}$$

which in turn can be simplified to

$$\mathbf{Z}_\alpha = -jZ_0\cot\theta_1 \tag{134}$$

It can be noted that for implementing the desegmentation method, the d ports of the β segment need not be located on the periphery of the β segment. For example, consider the case of a rectangular segment with a circular hole as shown in Fig. 26. Here, no region is available on the

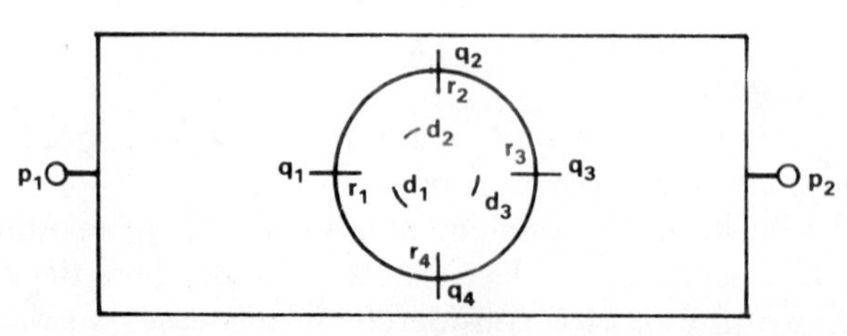

Fig. 26 Desegmentation with d ports located inside the β-segment.

periphery of the circular β segment for locating d ports. In such a case, the d ports may be located inside the circular region. In Fig. 26 three d ports d_1, d_2, d_3 are shown located inside the β segment. Although the d ports are inside the β segment, the Green's functions are still valid for any point on the segment (or on the periphery), and hence the desegmentation procedure remains unchanged.

7. PLANAR CIRCUITS WITH ANISOTROPIC SPACING MEDIA [23, 24]

Planar circuits with anisotropic spacing media (such as magnetized ferrite) are important owing to their increasing use in microwave integrated circuitry. Such planar circuits include disk-shaped wideband circulators [24, 35, 36], edge-guided mode isolators [37], edge-guided mode circulators [38], and nonreciprocal microwave dividers/combiners [39, 40]. These circuits can be analyzed by extending the techniques used in analyzing circuits with isotropic material. Before discussing these techniques, the basic equations that govern ferrite planar circuits will be described.

Figure 17 shows an arbitrarily shaped thin-center conductor of a triplate-type ferrite planar circuit of thickness $2d$. This portion of the circuit is sandwiched between two ferrite spacers and two ground conductors as shown in Fig. 2. The dc magnetic field is applied along the z direction, which is normal to the conductors. As shown in Fig. 17, several coupling ports may exist along the periphery, and the remaining periphery is open-circuited.

The wave equation that governs the rf voltage at any point in the ferrite planar circuit is given by

$$(\nabla_T^2 + \omega^2 \varepsilon \mu_{\text{eff}})v = 0 \tag{135a}$$

with

$$\nabla_T^2 = \frac{\partial^2}{\partial x^2} + \frac{\partial^2}{\partial y^2} \tag{135b}$$

$$\mu_{\text{eff}} = \frac{\mu^2 - K^2}{\mu} \tag{135c}$$

where the sign of μ_{eff} will depend upon the frequency and the static magnetic field. The quantities μ and K are the diagonal and off-diagonal elements, respectively, of the permeability tensor and are given by

$$\mu = \mu_0 + \frac{\gamma^2 H_0 M_S}{\mu^2 H_0^2 - \omega^2} \tag{136}$$

$$K = \frac{\omega \gamma M_S}{\gamma^2 H_0^2 - \omega^2} \tag{137}$$

Here μ_0 denotes the permeability of vacuum, γ the gyromagnetic ratio, H_0 the dc bias magnetic field, M_S the saturation magnetization of the ferrite, and ω the angular frequency of the signal.

The boundary condition to be met at the coupling ports is

$$j \frac{K}{\mu} \frac{\partial v}{\partial s} + \frac{\partial v}{\partial n} = -j\omega\mu_{\text{eff}} d J_n \tag{138}$$

where J_n stands for the surface current density perpendicular to the periphery, and $\partial/\partial n$ and $\partial/\partial s$ are, respectively, the normal and tangential derivatives along the periphery.

In order to derive the characteristics of a ferrite planar circuit, a solution for eqs. (135) and (138) must be obtained. This solution is obtained by using one of three different approaches. The first approach is based on the contour integral solution of the wave equation, while the second is based on the expansion of the Green's function in terms of complex eigenfunctions. The third approach is based on the mode-matching technique. In the next section each of these approaches is described briefly, and the similarities between the second and the third approaches are discussed.

7.1. Contour Integral Approach [23]

This approach starts from a contour integral equation similar to (100) but is modified to account for the gyromagnetic property of the ferrite substrate. The contour integral in the present case may take one of two forms depending on the sign of μ_{eff}.

In the first case, when $\mu_{\text{eff}} > 0$, the starting contour integral is similar to (100) with $p = 1$, differing only in the first term of the right-hand side:

$$v(s) = \frac{1}{2j} \int_C \left[k\left(\cos\theta - j\,\frac{K \sin\theta}{\mu}\right) H_1^{(2)}(kr) v(s_0) + j\omega d\mu_{\text{eff}} H_0^{(2)}(kr) J_n(s_0) \right] ds_0 \tag{139}$$

where $k = \omega\sqrt{\varepsilon\mu_{\text{eff}}}$.

When $\mu_{\text{eff}} < 0$, the starting contour integral takes a different form and is given by

$$v(s) = \frac{1}{2j} \int_C \left[k\left(\cos\theta - j\,\frac{K}{\mu}\sin\theta\right)\{K_1(hr) - j\pi I_1(hr)\} v(s_0) + j\omega d\mu_{\text{eff}}\{K_0(hr) + j\pi I_0(hr)\} J_n(s_0) \right] ds_0 \tag{140}$$

where K_1, I_1, and K_0, I_0 denote, respectively, the first-order and zeroth-order modified Bessel functions of the second kind (K_1 and K_0) and the first kind (I_1 and I_0), and

$$h = \omega\sqrt{\varepsilon \,|\, \mu_{\text{eff}}|} \tag{141}$$

The computational process in this case is essentially identical to that of the isotropic case. The periphery is divided into N incremental sections where each section is assumed to be a straight edge with a sampling point at its center. In addition, the magnetic and electric field intensities are assumed to be uniform across each section. Upon replacing the line integral in (139) or (140) by a summation over N, a matrix equation similar to (104) is obtained. From this matrix equation the resonant frequencies and the impedance matrix are obtained.

7.2. Eigenfunction Expansion Approach [16]

As in the isotropic case, this approach can be used when the geometrical shape of the planar circuit is simple. The solution to the wave equation (135) with the boundary condition (138) is slightly different from (23) and is given by

$$v(s) = -j\omega d\mu_{\text{eff}} \int_C G(s \,|\, s_0) J_n(s_0)\, ds_0 \tag{142}$$

with the Green's function $G(s \,|\, s_0)$ satisfying the following boundary condition along the periphery C:

$$j\frac{K}{\mu}\frac{\partial G}{\partial s} + \frac{\partial G}{\partial n} = 0 \tag{143}$$

Following a procedure similar to that for the isotropic case (Sections 3 and 4), the Green's function is first expanded in terms of the complex eigenfunctions ϕ_a, which satisfies (135) and (143). Upon substituting the resulting Green's function expression into (142) and then using the definitions of the voltage and current at a coupling port, the elements of the impedance matrix of the equivalent multiport network can be obtained. These impedance matrix elements are given by

$$Z_{ij} = \sum_{a=0}^{\infty} \frac{j\omega d}{2W_i W_j} \int_{W_i} \int_{W_j} \frac{\phi_a^*(s_i)\phi_a(s_j)}{\omega_a^2 - \omega^2}\, ds_i\, ds_j \tag{144}$$

where ω_a is the eigenvalue corresponding to the eigenfunction ϕ_a. The values of ω_a and ϕ_a can be obtained via the Rayleigh–Ritz variational technique using a polynomial approximation [16, 23]. It may be noted that the impedance matrix in the present case is not symmetrical, i.e., $Z_{ij} \neq Z_{ji}$, because of the anisotropy of the spacing medium. The eigenfunction ϕ_a is generally complex. However, a different relation of the form $Z_{ij} = -Z^*_{ji}$ holds. This relation is obtained from properties of lossless circuits.

7.3. Mode-Matching Approach [25]

The mode-matching approach [25] has been used for analyzing circular nonreciprocal components with ports on the circumference. In the present section this approach will be used to derive the impedance matrix of a disk-shaped planar circuit with a ferrite substrate. A comparison between the results obtained using this approach and the results obtained via the Green's function modal expansion approach will be given.

The major difference between the mode-matching approach and the modal expansion (i.e., Green's function) approach is in the way the mode terms satisfy the boundary conditions and wave equation [41]. Although both methods express the planar circuit fields in terms of an infinite set of orthogonal functions (modes), the Green's function resonant mode expansion approach requires that each mode term satisfy the boundary conditions and that the total field satisfy the inhomogeneous wave equation. On the other hand, the mode-matching approach requires that the sum of the modes satisfiy the source condition and that each mode term obey the source free wave equation.

In the case of the microstrip planar circuit configuration depicted in Fig. 27, the electric field under the metallic circuit disk has one component E_z that satisfies the homogeneous wave equation (135) in cylindrical coordinates. The general solution of the wave equation is

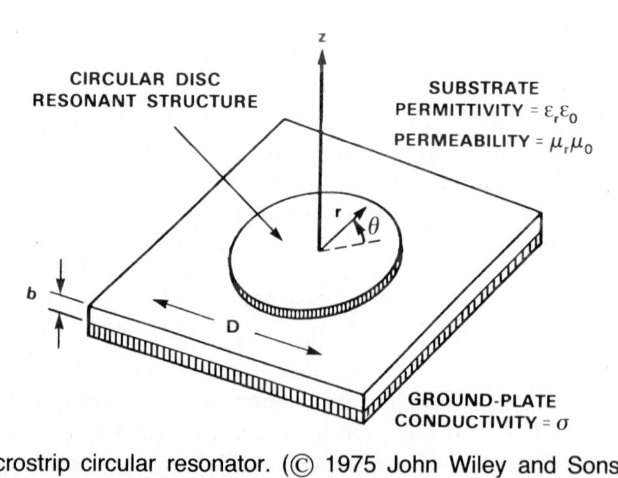

Fig. 27 Microstrip circular resonator. (© 1975 John Wiley and Sons. Reprinted with permission from Reference 25.)

$$E_z = \sum_{n=-\infty}^{\infty} A_n J_n(k_e r) e^{jn\phi} \tag{145}$$

where

$$k_e = \omega\sqrt{\mu_{\text{eff}}\varepsilon} \tag{146}$$

The magnetic field component H_ϕ can be obtained using Maxwell's equations to get

$$H_\phi = -\frac{j}{\eta_e} \sum_{n=-\infty}^{\infty} A_n \left[J_n'(k_e r) - n \frac{K}{\mu} \frac{J_n(k_e r)}{k_e r} \right] e^{jn\phi} \tag{147}$$

where

$$\eta_e = \sqrt{\frac{\mu_0 \mu_{\text{eff}}}{\varepsilon_0 \varepsilon_r}} \tag{148}$$

Enforcing the boundary condition of zero tangential magnetic field H_ϕ at $r = R$ yields

$$J_n'(k_e R) - n \frac{K}{\mu} \frac{J_n(k_e R)}{k_e R} = 0 \tag{149}$$

This equation represents the characteristic equation from which k_e, the wavenumber, is to be calculated.

In the case of N coupling ports (see Fig. 28), Bosma's approach [3] may be followed and H_ϕ is assumed to be constant over the small width of each coupling port and zero elsewhere as follows:

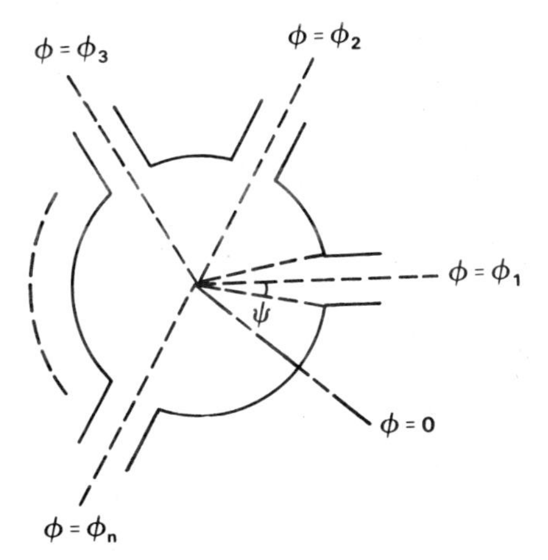

Fig. 28 Schematic of N-port circular microstrip resonator.

$$\begin{aligned} H_\phi &= H_1; \qquad (\phi_1 - \psi) < \phi < (\phi_1 + \psi) \\ &= H_2; \qquad (\phi_2 - \psi) < \phi < (\phi_2 + \psi) \\ &\;\;\vdots \\ &= H_n; \qquad (\phi_n - \psi) < \phi < (\phi_n + \psi) \end{aligned} \tag{150}$$

where ϕ_i $(i = 1, 2, \ldots, N)$ represents the angular position of the ith coupling port and

$$\sin \psi = W/2R \tag{151}$$

where W stands for the width of the coupling port.

The amplitude constant A_n in eq. (147) is now found by expanding (150) into a Fourier series with respect to ϕ:

$$H_\phi = \sum_{n=-\infty}^{\infty} b_n e^{jn\phi} \tag{152}$$

with

$$b_n = \frac{\sin n\psi}{n\pi} \sum_{l=1}^{N} H_l e^{-jn\phi_l} \tag{153}$$

and where l is the port number, ϕ_l is its angular position, and N is the total number of ports.

The amplitude constant A_n can now be obtained by matching (147) and (152) at $r = R$.

$$A_n = \frac{j\eta_e \sin n\psi}{n\pi} \frac{\sum_{l=1}^{N} H_l e^{-jn\phi_l}}{J'_n(k_e R) - n \dfrac{K}{\eta_e} \dfrac{J_n(k_e R)}{k_e R}} \tag{154}$$

Having determined A_n, we can now substitute (145) into

$$\bar{E}_i = \frac{1}{2\psi_i} \int_{\phi_i - \psi_i}^{\phi_i + \psi_i} E_z \, d\phi \tag{155}$$

to obtain $\bar{E}_i$, the average of E_z at $r = R$ and along the ith coupling port,

$$\bar{E}_i = \sum_{l=1}^{M} H_l \sum_{n=-\infty}^{\infty} T_n e^{jn(\phi_i - \phi_l)} \tag{156}$$

with

$$T_n = \frac{j\eta_e \sin^2(n\psi_i) J_n(k_e R)}{\pi n^2 \psi_i \left[J_n'(k_e R) - n \dfrac{K}{\mu} \dfrac{J_n(k_e R)}{k_e R} \right]} \tag{157}$$

Equations (156) and (157) define the wave impedance matrix of the planar circuit under consideration. In order to obtain the characteristic network impedance matrix, a relation between the voltage and current at the ports instead of the electric and magnetic fields must be developed. Having assumed E_z and H_ϕ to be uniform across the port width W_i, we may write

$$V_i = E_i d \tag{158}$$

$$I_i = H_i W_i \tag{159}$$

where d denotes the height of the magnetic substrate. Substituting (158) and (159) into (156) yields

$$V_i = d \sum_{l=1}^{N} \frac{I_l}{W_l} \sum_{n=-\infty}^{\infty} T_n e^{jn(\phi_i - \phi_l)} \tag{160}$$

This last equation defines the elements of the impedance matrix, which may be written as

$$Z_{ij} = \frac{j\eta_e d}{\pi W_j} \sum_{n=-\infty}^{\infty} \frac{\sin^2(n\psi_i) J_n(k_e R)\, e^{jn(\phi_i - \phi_j)}}{n^2 \psi_i \left[J_n'(k_e R) - n \dfrac{K}{\mu} \dfrac{J_n(k_e R)}{k_e R} \right]} \tag{161}$$

Note that the impedance matrix of the ferrite circuit is not symmetrical, i.e., $Z_{ij} \neq Z_{ji}$, but satisfies the relation $Z_{ij} = -Z_{ji}^*$. In the case of planar circuits with an isotropic dielectric substrate, the off-diagonal element K of the permeability tensor $[\mu]$ is zero, and the above expression reduces to

$$Z_{ij} = \frac{j\eta_e d}{\pi W_j} \sum_{n=0}^{\infty} \frac{\sigma_n \sin^2(n\psi_i) J_n(k_e R) \cos[n(\phi_i - \phi_j)]}{n^2 \psi_i J_n'(k_e R)} \tag{162}$$

where σ_n equals 1 when $n = 0$ and equals 2 for other values of n, $\eta_e = 120\pi\sqrt{\mu_r/\varepsilon_r}$, and the values of $k_e R$ are now determined by the condition $J_n'(k_e R) = 0$. Note that in the present case we have $Z_{ij} = Z_{ji}$, and hence the network impedance matrix is symmetrical.

Finally the characteristic impedance expressions obtained for the isotropic case from the Green's function (resonant mode expansion) approach and from the mode-matching approach can be compared for consistency. This derivation is carried out in Appendix B by comparing the Z_{ij} expressions obtained by the two methods. Using the Mittag–Leffler theorem [42] for complex variables, one can show that the two expressions are equivalent. Thus, because the Green's function result requires a double summation for the modal expansion while the mode-matching result requires a single summation, less computational effort should be required when the mode-matching procedure is used.

8. APPLICATIONS OF THE PLANAR CIRCUIT CONCEPT

The planar circuit approach discussed above has been used in designing and analyzing a variety of passive microwave and millimeter-wave components. This approach has been used to characterize circuit discontinuities [33, 43, 44]; it has also been used to design power dividers and combiners [29, 45–51], filters [17, 52–72], circulators, isolators, and other nonreciprocal components [3, 24, 25, 35–39, 73], and patch antennas [74–92]. In this section each of these applications is described.

8.1. Characterization of Discontinuities

The planar circuit approach has been used for characterizing discontinuities in rectangular waveguides [33] as well as in microstrip lines [43, 44].

8.1.1. Waveguide Discontinuities

The two-dimensional planar circuit approach is applicable to rectangular waveguides because the fields of the dominant TE_{10} mode do not vary along the narrow dimension b of the waveguide. This approach can also be used to analyze waveguide discontinuities provided that no field variation along the narrow dimension b is caused by the discontinuities. In other words, the application of the two-dimensional circuit approach to rectangular waveguide discontinuities is valid only if the higher-order evanescent modes produced by the discontinuities are of the TE_{m0} type. Three examples of such discontinuities and their equivalent circuits [33] are shown in the insets of Figs. 29–31.

A rectangular waveguide with a thick inductive window (extending over the total height b of the waveguide) is shown in Fig. 29. Each of the two uniform waveguide sections extending before and after the window is chosen to be long enough to ensure that the higher-order evanescent modes excited by the discontinuity die out (and hence only the dominant mode exists) at the input and output ports of the discontinuity configuration.

The contour integral method has been used [33] to analyze this discontinuity configuration. By following this approach a multiple number of ports are considered all around the periphery. All of the ports except those at the input and output waveguide planes are then shorted, i.e., the voltages at

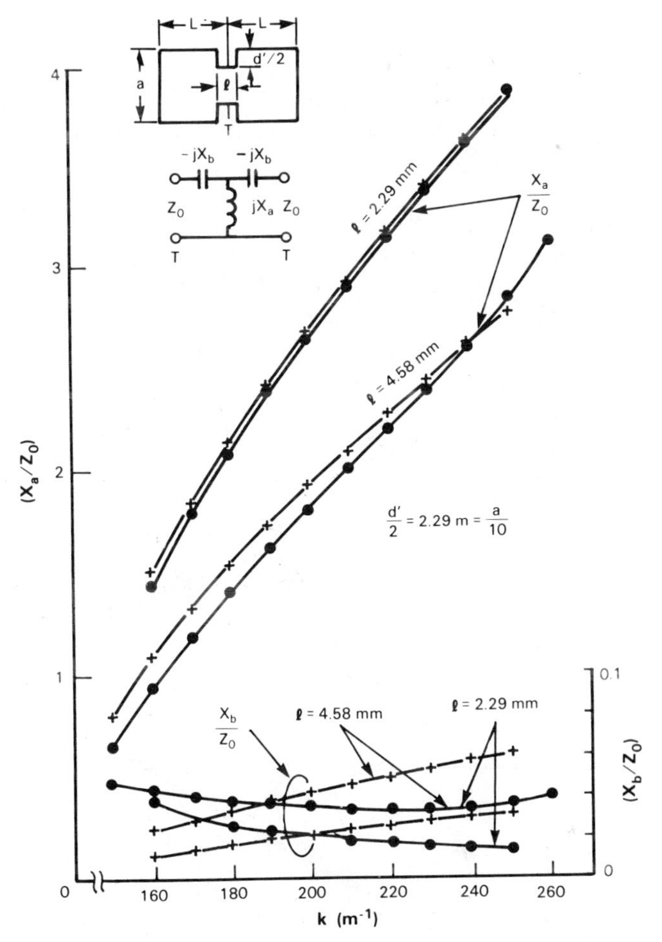

Fig. 29 The computed equivalent-circuit parameters of thick inductive window in a rectangular waveguide [33]. Crosses show the results of the approximate analysis described by N. Marcuvitz [2]. (© 1975 IEEE, reprinted with permission from Reference 33.)

these ports are set to zero. Next the voltages and currents at the input-output waveguide planes are related by the characteristic impedance of the waveguide. That is, the two waveguide sections of length L are assumed to be terminated by resistive sheets perpendicular to the waveguide axis. The surface resistance of these terminating sheets is taken to be equal to the wave impedance in the waveguide. This structure is described in terms of the equivalent circuit model shown in Fig. 29. The reactances X_a and X_b computed using this approach are plotted in Fig. 29 against the results reported by Marcuvitz [2]. A very good agreement in the case of the X_a/Z_0 curve is observed. Two other waveguide discontinuities have also been

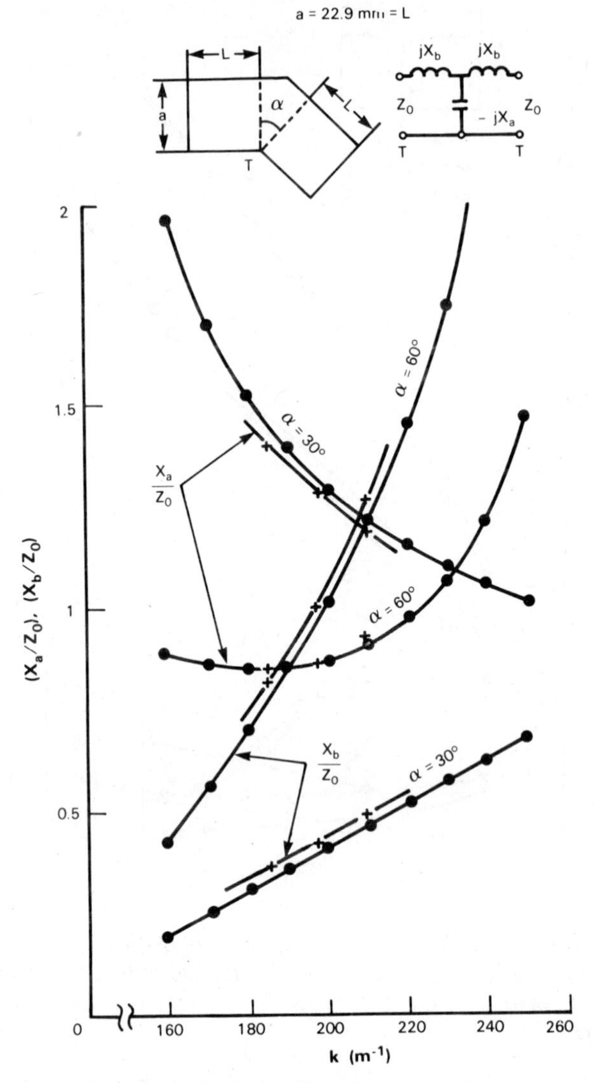

Fig. 30 The computed equivalent-circuit parameters of waveguide corners [33]. Crosses show the experimental data described by N. Marcuvitz [2]. (© 1975 IEEE. Reprinted with permission from Reference 33.)

analyzed by the contour integral approach. Their computed reactances X_a and X_b are compared with those of Marcuvitz [2] in Figs. 30 and 31.

Finally, it should be noted that the segmentation and desegmentation methods, described previously, can be utilized to analyze certain waveguide discontinuities. For instance, the discontinuity configurations shown in Figs. 29–31 (for certain values of α) can be analyzed by the segmentation/

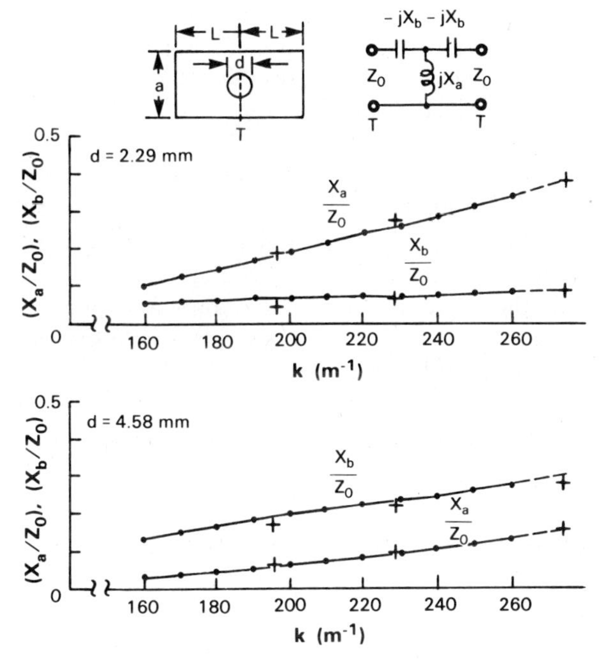

Fig. 31 The computed equivalent-circuit parameters of a post in a waveguide [33]. Crosses show the results of the approximate analysis described by N. Marcuvitz [2]. (© 1975, IEEE. Reprinted with permission from Reference 33.)

desegmentation procedures for planar circuits. However, to date no such effort has been reported.

8.1.2. Microstrip Discontinuities

The analysis of microstrip discontinuities using any planar circuits approach is based on the planar waveguide model representation [93] of microstrip lines. In the planar waveguide model, the width and the substrate permittivity ε_r of the microstrip line are replaced, respectively, by an equivalent width $W_e(f)$ and an effective dielectric constant $\varepsilon_{re}(f)$. In the planar waveguide model, two magnetic walls are located along the sides of the equivalent planar waveguide whose width is W_e. The values of W_e and ε_{re} are found by equating the characteristic impedance Z_0 and the phase velocity v_p of the microstrip line to those of the equivalent waveguide structure. Also, because Z_0 and v_p for microstrip lines are functions of frequency, W_e and ε_{re} become frequency-dependent as well, and the dispersion effects are thus incorporated in an approximate manner.

The analysis of discontinuities in microstrip lines is based on segmentation/desegmentation methods applied to equivalent planar waveguide configurations. In order to demonstrate the applicability of the segmentation/desegmentation methods to the analysis of microstrip line discontinuities, a microstrip line right-angled bend discontinuity (Fig. 32*a*) is considered. The various steps of this procedure include:

1. Write the equivalent planar waveguide model of the discontinuity configuration (Fig. 32*b*).
2. Specify the locations of the terminal planes (two in the case of the bend) where the external ports are located (these ports are located far away from the discontinuity so that only the dominant microstrip mode is present at the locations of the external ports and higher order evanescent modes have been decayed out). See Fig. 32*c*.
3. Decompose the planar configuration into regular segments suitable for analysis by the segmentation or desegmentation methods (Fig. 32*d*).
4. Compute a **Z** matrix for each multiport segment.
5. Combine the individual **Z** matrices into an overall **Z** matrix for the overall discontinuity configuration.

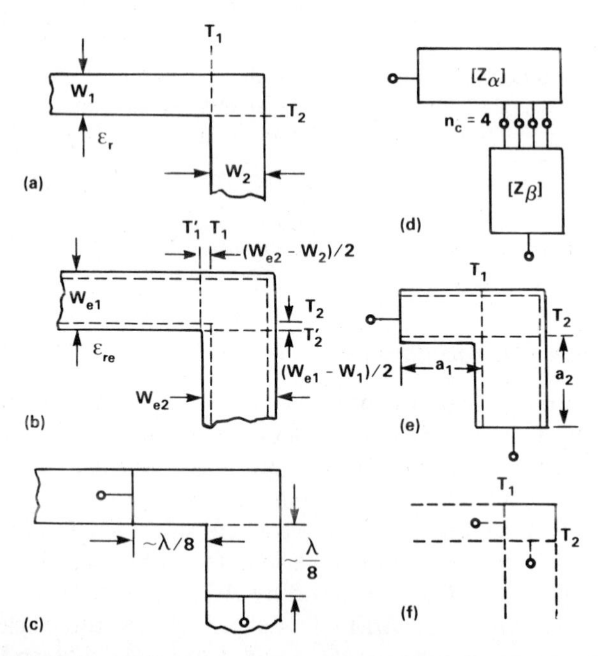

Fig. 32 An illustration of the segmentation method for analyzing microstrip discontinuities. (*a*) Physical configuration of the discontinuity; (*b*) planar model of the discontinuity configuration; (*c*) locations of reference planes for external ports; (*d*) segmentation of configuration into regular segments; (*e*) use of segmentation/desegmentation methods to find discontinuity characterization with respect to external ports; (*f*) transfer of reference planes to appropriate locations.

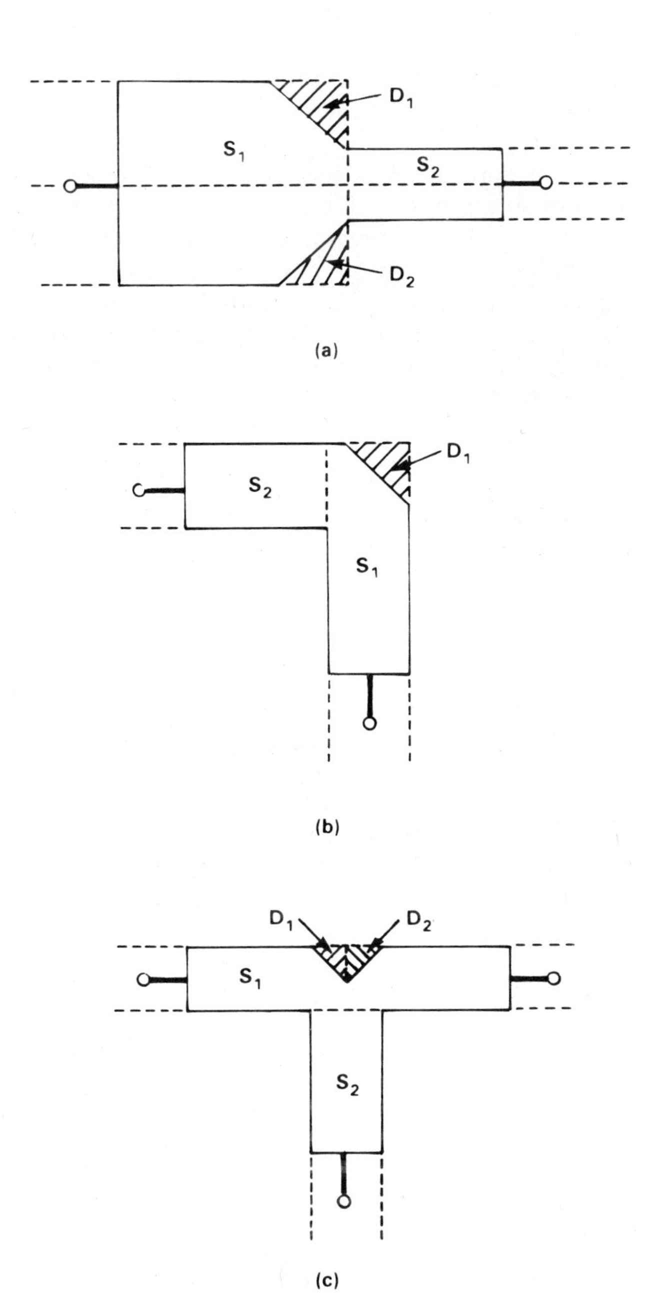

Fig. 33 An illustration of the desegmentation method for analyzing microstrip line type planar circuit discontinuities.

6. Convert the **Z** matrix into the more commonly used **S** matrix representation.
7. Transfer the reference planes to the desired locations with respect to the physical configuration of the discontinuity as shown in Fig. 32*e*.

The planar circuit approach has been used not only for characterizing microstrip discontinuities, but also for the compensation of microstrip discontinuity reactances. The compensation of discontinuity reactances [43] involves a modification of the discontinuity's geometrical configuration so as to minimize its adverse effects on the performance of microstrip circuits. This compensation has been reported [43] for the three types of discontinuities (step junctions, right-angled bends, and tee junctions) shown in Fig. 33. A combination of the segmentation and desegmentation methods has

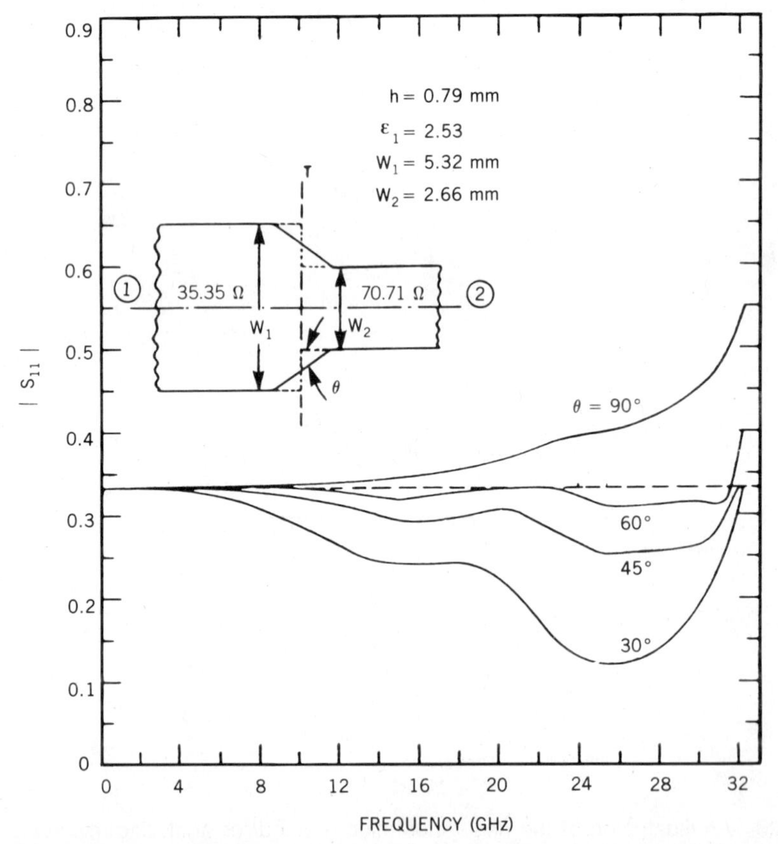

Fig. 34 Reflection coefficients for the uncompensated and compensated step discontinuities with 1:2 impedance ratio. (© 1982 IEEE. Reprinted with permission from Reference 43.)

been used in these cases. Desegmentation was used for segments identified by cross-hatching in Fig. 33. In each of the three cases, the characterization of the segment marked S1 is obtained by the desegmentation of the rectangular and triangular segments. Then segment S2 is combined with S1 by the segmentation procedure. The computed results for these three types of compensated discontinuities are shown in Figs. 34–36. For the sake of comparison, characteristics of the uncompensated discontinuities are also shown in these three cases.

Finally, it should be noted that the planar circuit approach is applicable to generalized discontinuity configurations involving circular or arbitrarily shaped boundaries. However, the circuit characterizations of such discontinuity configurations have yet to be reported.

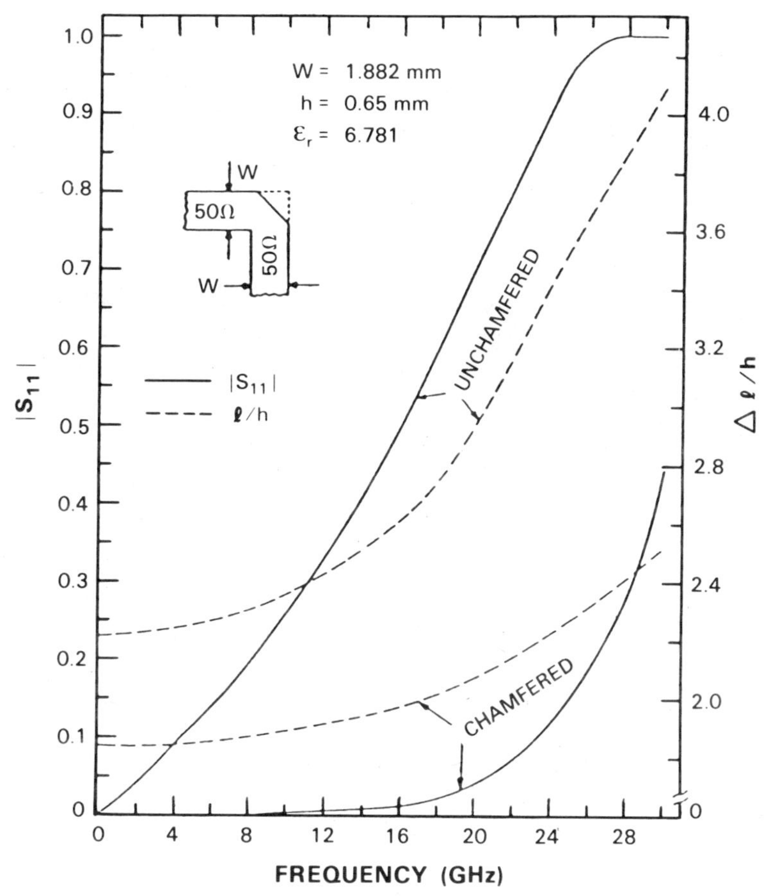

Fig. 35 Reflection coefficients and normalized electrical lengths for uncompensated and optimally compensated right-angled bends. (© 1982 IEEE. Reprinted with permission from Reference 43.)

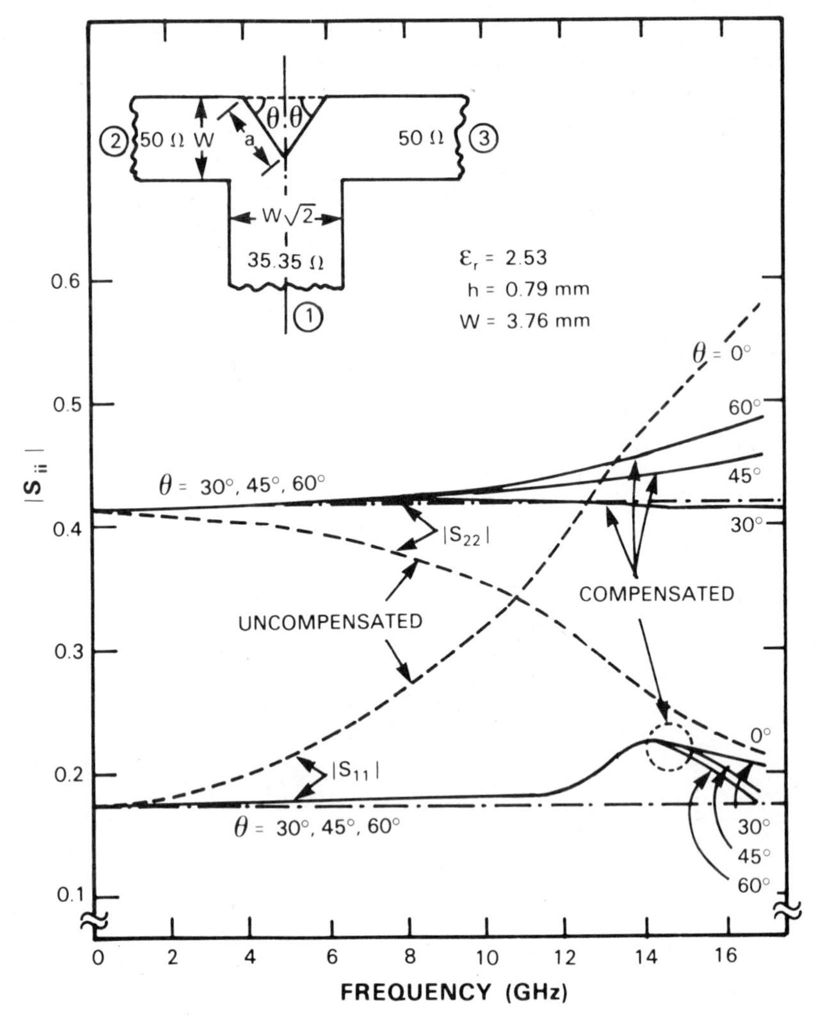

Fig. 36 Main line and branch line reflection coefficients for uncompensated and compensated T junctions (impedance ratio $1/\sqrt{2}:1:1$). (© 1982 IEEE. Reprinted with permission from Reference 43.)

8.2. Power Dividers and Combiners

Several new power divider/combiner circuit designs have emerged from the applications of the two-dimensional planar circuit approach. Circular microstrip disks have been used to design 3-dB hybrid circuits [29, 45], unequal power dividers [46], and n-way radial power dividers/combiners [47, 48]. Also, arbitrarily shaped planar circuits have been used to obtain 3-dB hybrids with optimized characteristics [49].

8.2.1. 3-dB Hybrids

The 3-dB hybrid plays an important role in microwave integrated circuits. However, as the demands for 3-dB hybrid circuits operating at higher frequencies increase, hybrid circuits based on the transmission-line concept begin to encounter manufacturing and design difficulties. Issues like parasitic reactances and unwanted line-to-line coupling start to arise. In view of these difficulties, the possibility of realizing a compact 3-dB hybrid with a simple structure has been investigated, and a four-port disk-shaped resonator circuit has been proposed [29]. Figure 37 illustrates two disk-shaped geometries with two different port arrangements. In either case, when the input signal is fed to port 1 no power is coupled to port 2, whereas the output powers at ports 3 and 4 are equal in amplitude and phase for the circuit of Fig. 37*a* and equal in amplitude but 180° out of phase for the circuit of Fig. 37*b*. On the other hand, if the input wave is fed to port 2, no power is coupled to port 1, whereas the output powers at ports 3 and 4 are equal in amplitude but of opposite phase for the circuit of Fig. 37*a* and equal in amplitude and phase for the circuit of Fig. 37*b*.

The basic operating principle of this circuit can be understood by considering the dipolar mode excitation of the circular disk resonator. When the dipolar mode is present, the voltage distribution has a null along a diameter located normal to the excitation port. Furthermore, the voltage is positive in one semicircular region and negative in the other as shown in Fig. 38. Thus, near its dipolar resonance a circular disk with the four ports located as shown in Fig. 37 acts as a 180° 3-dB hybrid. Figures 39*a* and *b* exhibit the computed characteristics [45] of the 3-dB hybrid configurations shown in Figs. 37*a* and *b*, respectively. These characteristics were computed for a circular disk of radius $R = 7.89$ mm and dielectric substrate with a height of 1.53 mm and $\varepsilon_r = 2.53$. The impedance of the coupling ports was taken to be 50 Ω. As is evident from the computed characteristics, both

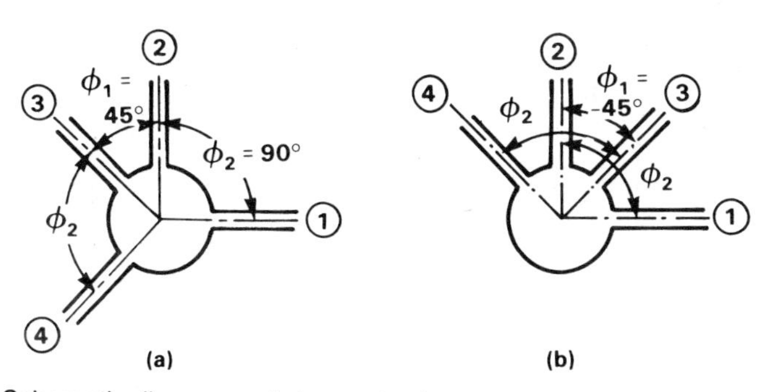

Fig. 37 Schematic diagrams of planar-circuit-type hybrid. (*a*) The original hybrid by Okoshi et al. [29]; (*b*) another type [45]. (© 1984 Scripta Technica. Reprinted with permission from Reference 45.)

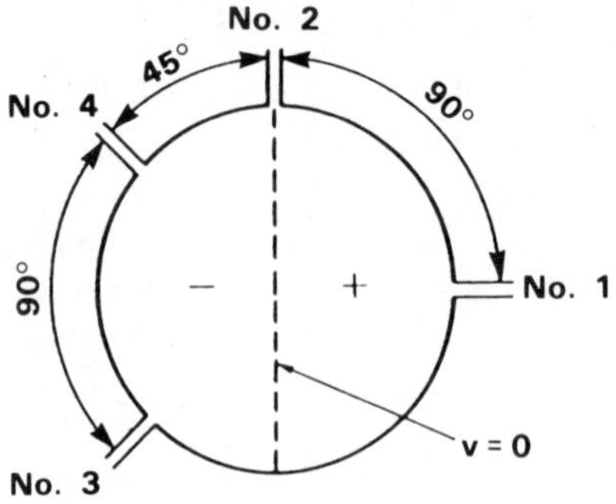

Fig. 38 Voltage distribution of the (1, 1) mode of a planar 3-dB hybrid. (© 1975 Scripta Technica. Reprinted with permission from Reference 29.)

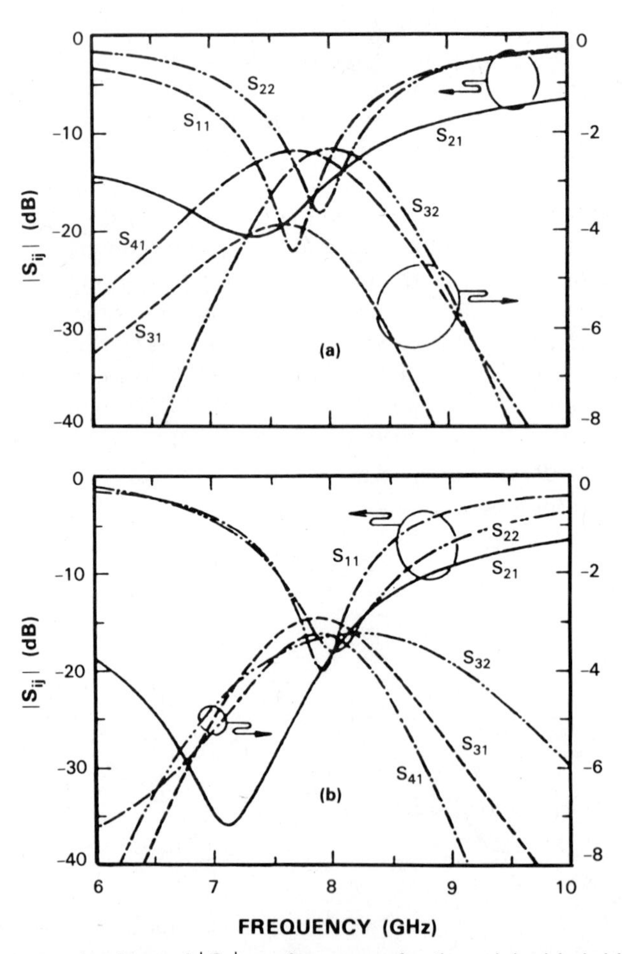

Fig. 39 Computed values of $|S_{ij}|$ vs. frequency for the original hybrids in Fig. 37. (*a*) $\phi_1 = 45°$, $\phi_2 = 90°$; (*b*) $\phi_1 = -45°$, $\phi_2 = 90°$. (© 1984 Scripta Technica. Reprinted with permission from Reference 45.)

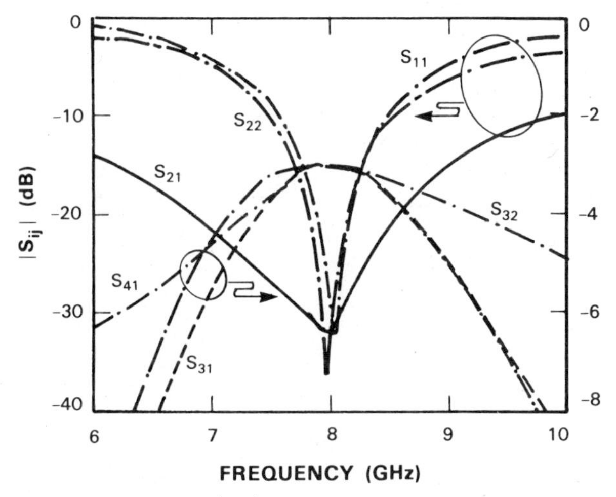

Fig. 40 Computed values of $|S_{ij}|$ vs. frequency for an improved hybrid with rearranged ports $\phi_1 = -53.33°$, $\phi_2 = 101.03°$. (© 1984 Scripta Technica. Reprinted with permission from Reference 45.)

circuits suffer from unbalanced power division, insufficient coupling, increase in return loss, and the separation of optimum frequencies for each performance parameter. The deterioration in the hybrid property of the circuit is probably caused by the presence of the higher-order cavity modes.

An improvement in the characteristics of this hybrid was obtained [45] by rearranging the locations of the four ports while preserving the symmetry of

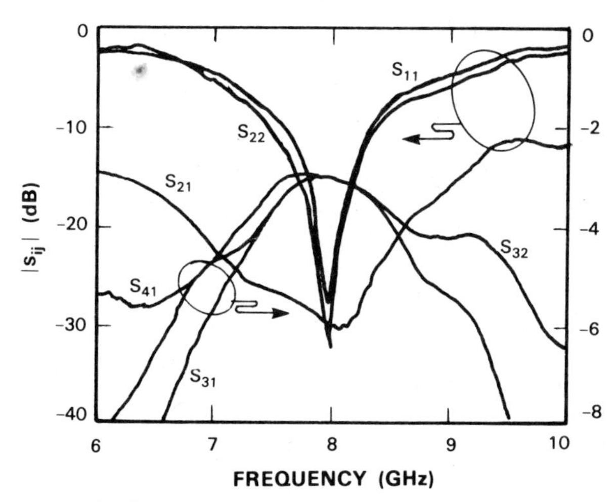

Fig. 41 Measured $|S_{ij}|$ for the hybrid shown in Fig. 37*b*. (© 1984 Scripta Technica. Reprinted with permission from Reference 45.)

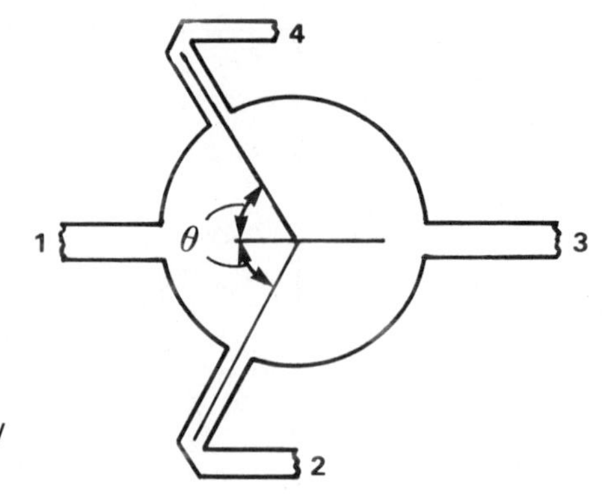

Fig. 42 Configuration of a three-way power divider.

the circuit. Figure 40 shows the computed characteristics for a 3-dB hybrid with $\phi_1 = -53.33°$, $\phi_2 = 101.03°$, and $R = 7.92$ mm. These characteristics show a great improvement over those in Fig. 39. Figure 41 demonstrates the measured characteristics of the modified circuit. These measured results are in remarkable agreement with the theoretical results shown in Fig. 40.

8.2.2. Power Dividers/Combiners

Another interesting planar circuit power divider is shown in Fig. 42. As in the case of the 3-dB hybrid circuit, this circuit is composed of a circular microstrip disk resonator connected to four coupling ports. The coupling properties of this circuit are governed by the locations of the coupling ports. By positioning port 3 diametrically opposite to port 1, and placing ports 2

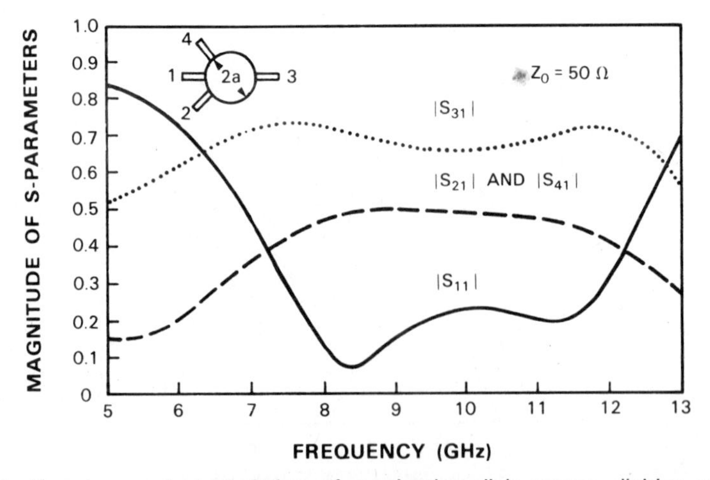

Fig. 43 Frequency characteristics of a circular disk power divider with $a = 7.60$ mm, $\varepsilon_r = 2.20$, and $d = \frac{1}{32}$ in. The characteristic impedance of the output ports is 50 Ω. (© 1985 IEEE. Reprinted with permission from Reference 46.)

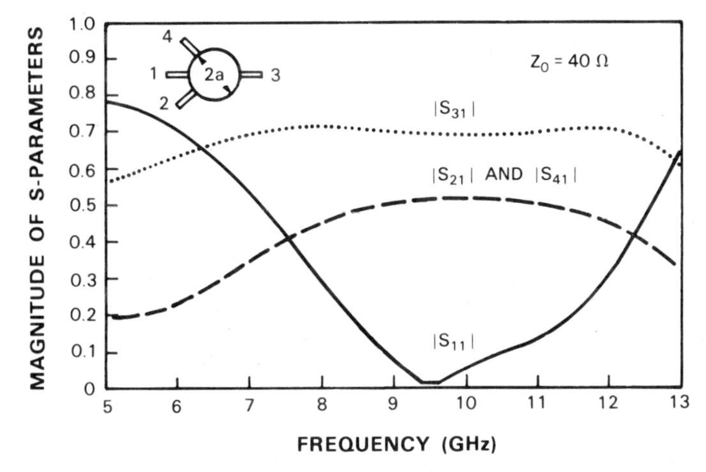

Fig. 44 Frequency characteristics of a circular disk power divider with $a =$ 7.60 mm, $\varepsilon_r = 2.20$, and $d = \frac{1}{32}$ in. The characteristic impedance of the outgoing ports is 40 Ω (© 1985 IEEE. Reprinted with permission from Reference 46.)

and 4 at θ and $-\theta$ relative to port 1, an unequal power divider can be obtained [46]. For example, when $\theta = 60°$ and the input wave is fed to port 1, half the power is coupled to port 3 whereas the other half is split equally between ports 2 and 4. Figure 43 shows the computed characteristics of this circuit for a disk radius of 7.60 mm, $\varepsilon_r = 2.20$, and substrate thickness $d = \frac{1}{32}$ in. The characteristic impedance of the port is taken to be 50 Ω. A usable bandwidth (defined for $|S_{11}| < 0.25$ or VSWR < 1.7) of 41% around 9.5 GHz is obtained.

The operating principle of this circuit can be understood by noting the voltage distribution of the (2, 2) mode and the (2, 1) mode resonances, which occur at 8.4 and 11.2 GHz, respectively. Better circuit characteristics can be obtained by changing the characteristic impedance of the various coupling ports to 40 Ω as shown in Fig. 44. The best $|S_{11}|$ value now is 0.0134 compared to the previous value of 0.078 for $Z_0 = 50\ \Omega$. Of course, this design needs the use of external impedance transformation at the output ports if the usual 50 Ω impedance level is desired. Figure 45 exhibit a comparison between the computed and the measured characteristics of the 50 Ω unequal power divider circuit. Again a good agreement between the measured and computed results confirms the validity of the planar circuit design approach.

Circuit configurations leading to other power division ratios are also possible. Table 1 lists various levels of output power at output ports 2, 3, and 4 for different values of the angle θ. Of particular interest is the $\theta = 90°$ case in which the four coupling ports are symmetrically spaced around the disk periphery. Figure 46 illustrates the computed and measured characteristics of this circuit. These characteristics show that when the input wave is fed

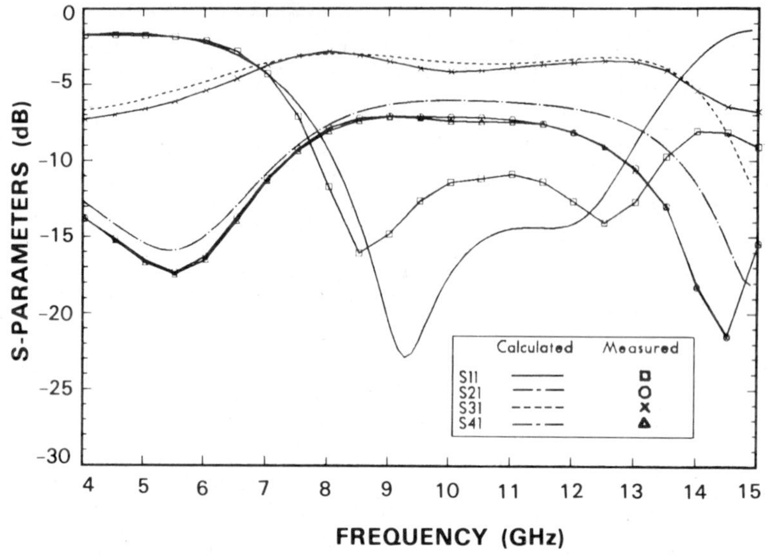

Fig. 45 Measured and calculated scattering parameters of a microstrip disk-shaped unequal three-way power divider.

to port 1, all of the input power is coupled to port 3 whereas ports 2 and 4 remain isolated. Once again the operating principle of this circuit can be understood by noting the voltage distribution of the (1, 1) mode.

Another class of power divider/combiner circuits that have been analyzed by the planar approach is the N-way radial power divider/combiner networks shown in Fig. 47. Figure 48 is a photograph of 3-, 4-, 5-, 8-, and 10-way center-fed microstrip disk power divider/combiner circuits. This class of symmetric N-way power divider/combiner circuits has received considerable attention recently [47, 48]. Due to their geometrical symmetry, these circuits do not exhibit any imbalance in either the amplitude or the phase of the output signals at any frequency. This property has made this class of circuits very attractive in many microwave and millimeter-wave

Table 1 Output Power Values vs. Port Locations for a Circular Four-Port Circuit

θ	P_3	P_2	P_4
45°	3/4	1/8	1/8
50°	2/3	1/6	1/6
60°	1/2	1/4	1/4

Source: Reprinted with permission from Reference 46. © 1985 IEEE.

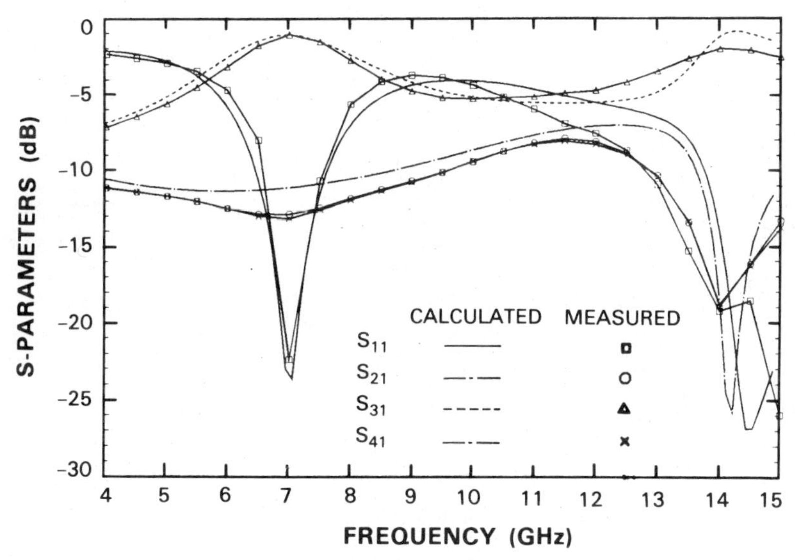

Fig. 46 Measured and calculated scattering parameters of microstrip disk with four equally spaced and symmetrically located ports.

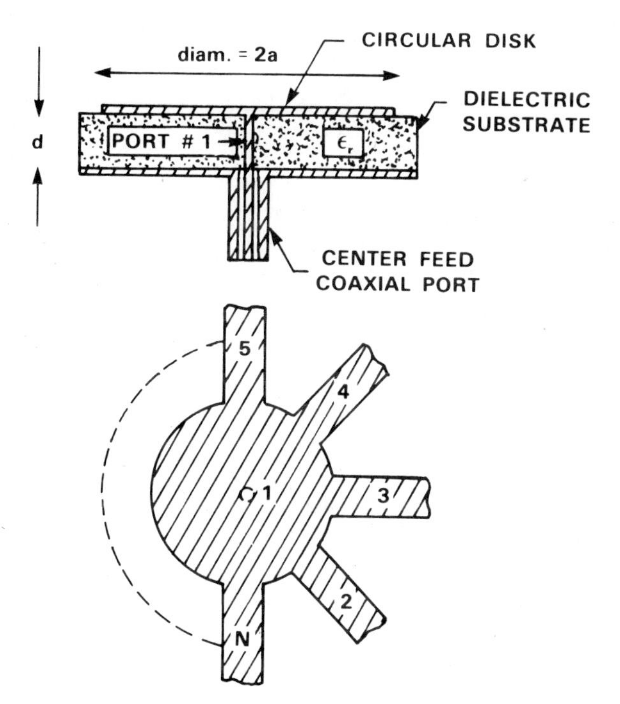

Fig. 47 Circular microstrip disk structure with a single coaxial port located at the center and $N - 1$ microstrip ports along the circumference. (© 1987 IEEE. Reprinted with permission from Reference 48.)

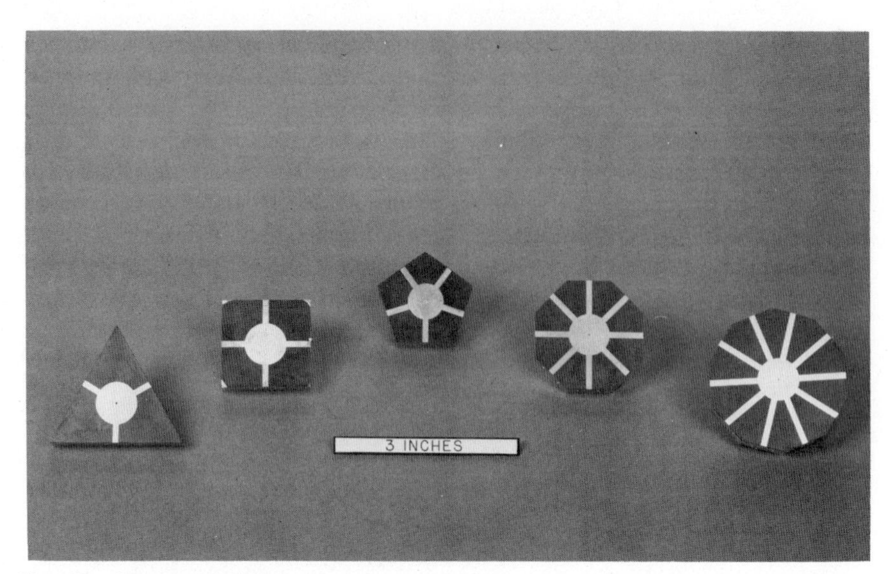

Fig. 48 Photo of various multiport power divider/combiner circuits. (© 1987 IEEE. Reprinted with permission from Reference 48.)

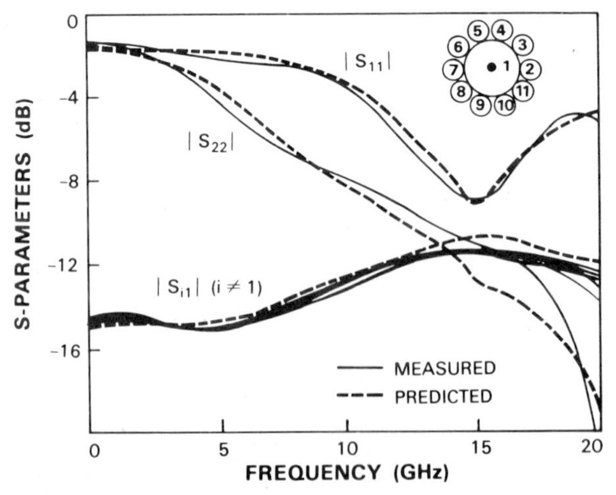

Fig. 49 Predicted and measured performance of a 10-way power divider/combiner center-fed microstrip disk circuit with disk radius 8.65 mm. (© 1987 IEEE. Reprinted with permission from Reference 48.)

applications. Indeed, such circuits have found extensive use in the design of multielement antenna feed systems [50]. They are also used extensively as multiple-port power combiners to combine multiple amplifiers (or oscillators) in a single module [48], thus yielding higher output power capabilities. The calculated and measured characteristics of the 10-way power divider/combiner circuit shown in Fig. 48 is illustrated in Fig. 49. The calculated results are obtained by following the impedance Green's function approach described earlier [48]. The measured return loss at any circumferential port and the isolation between any two circumferential ports (adjacent, straight across, or otherwise) are no worse than −14 dB (VSWR = 1.5) and −11 dB, respectively, at 17 GHz. Over a frequency range of 15–20 HGz, the corresponding values are better than −12.6 dB and −9 dB, respectively. It should be noted that this performance was achieved without using any computer or experimental optimization. Hence a better performance may be achieved if a design optimization process is followed. Finally, it should be stated that the level of isolation reported above can be realized without using external resistors. Indeed it is this property that makes this design geometry substantially more attractive than other radial combiners [51].

8.2.3. Circuit Optimization

Another advantage of the planar circuit approach is the fact that it is better suited for performance optimization than the transmission-line approach. This is because the planar circuit approach offers an increased design flexibility in the sense that circuit elements need not be restricted to uniform sections of transmission lines but could be arbitrarily shaped. In addition, discontinuity reactances and dispersion effects are inherently accounted for in this approach.

By taking advantage of the larger degree of freedom inherent to the planar circuit approach, Okoshi and his co-workers [49] reported an optimum pattern obtained by following a fully computer-oriented synthesis scheme. Figure 50 shows a schematic of this optimum circuit pattern. The external periphery of this pattern is taken to be circular rather than arbitrarily shaped. This is done to prevent having too much variation in the circuit shape and consequently make it difficult to uniquely determine the optimum circuit pattern. The external diameter, the position of the ports,

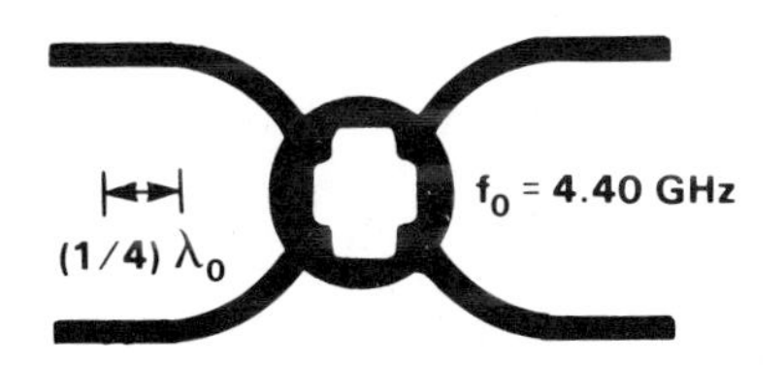

Fig. 50 Circuit pattern of a 3-dB stripline hybrid circuit. (© 1981 IEEE. Reprinted with permission from Reference 49.)

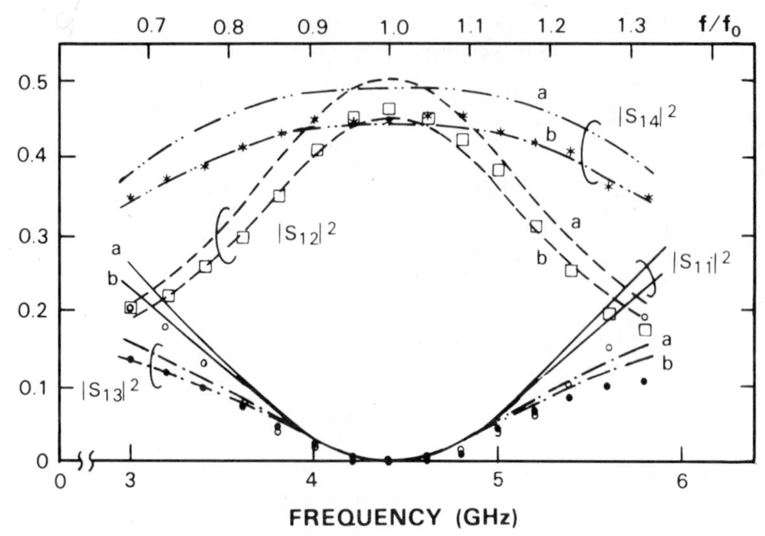

Fig. 51 Calculated and measured performance of the planar circuit shown in Fig. 50. (© 1981 IEEE. Reprinted with permission from Reference 49.)

and the shape of the internal periphery can be adjusted to obtain the best wide-band hybrid characteristics.

By combining the contour integral method of analysis with Powell's optimization method, Okoshi et al. [49] obtained a family of optimum circuit patterns. Figure 51 exhibits the measured and calculated frequency characteristics of an optimized triplate 3-dB hybrid structure. The measured values of S_{11}, S_{12}, S_{13}, and S_{14} are represented by open circles, squares, filled circles, and asterisks, respectively. The curves marked *a* represent the computed optimized characteristics, whereas the curves marked *b* are those obtained after correcting the *a* curves by a circuit loss of 0.58 dB. A further improvement in the frequency characteristics can be obtained by introducing additional broadbanding elements at all of the four ports. These broadbanding elements may take the shape of open shunted stubs, a high-(or low-)impedance section, or a combination of both. Figure 52 shows the frequency characteristics obtained with the circuit of Fig. 53 tuned at 5.5 GHz. The optimized theoretical characteristics are again shown by the *a* curves, whereas those corrected by the line loss are represented by *b* curves. Open circles, squares, filled circles, and asterisks again denote the measured

Fig. 52 Circuit pattern of an optimized 3-dB stripline hybrid component with low-impedance matching sections. (© 1981 IEEE. Reprinted with permission from Reference 49.)

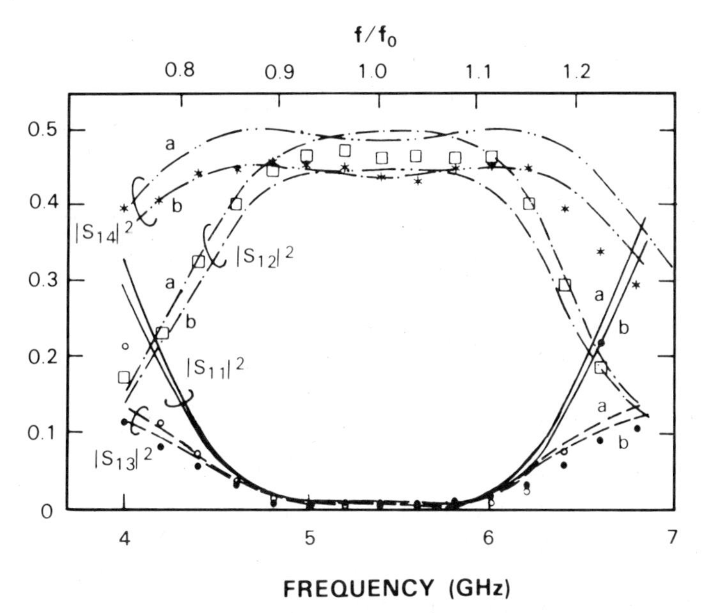

Fig. 53 Calculated and measured characteristics of the circuit shown in Fig. 52. (© 1981 IEEE. Reprinted with permission from Reference 49.)

values of S_{11}, S_{12}, S_{13}, and S_{14}, respectively. The measured and calculated differential phase angles are shown in Fig. 54. The difference between theoretical and experimental results is less than 10° over more than 50% of the bandwidth.

8.3. Filters

Several new filtering structures have emerged from the applications of the planar circuit approach. Workers in this field have emphasized the marked filtering properties of planar circuits with rectangular [17, 52–58], annular

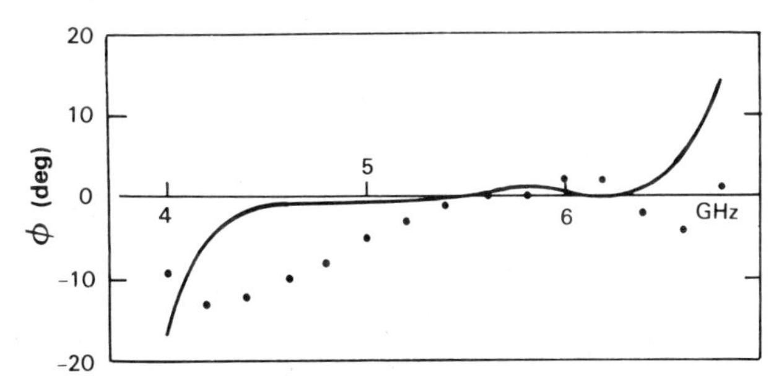

Fig. 54 Theoretical and measured phase characteristics of the circuit shown in Fig. 52. (© 1981 IEEE. Reprinted with permission from Reference 49.)

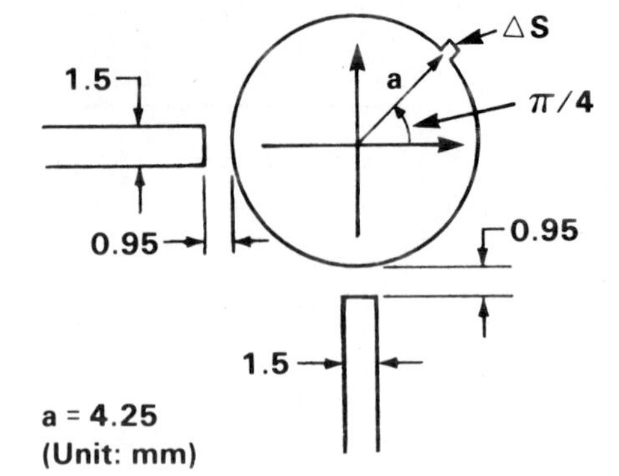

Fig. 55 Dimensions of a planar circuit experimental coupled-mode filter. (© 1972 Scripta Technica. Reprinted with permission from Reference 60.)

[59], circular [60, 61], and radial [62–70] geometries. The filtering properties of some of these structures are described in this section.

The filtering properties of circular two-port planar structures have been studied by many authors [60, 61]. Miyoshi and Okoshi [60] used the planar circuit approach to describe the properties of the mode-coupled filter shown in Fig. 55. This circuit is composed of a circular disk having three circum-

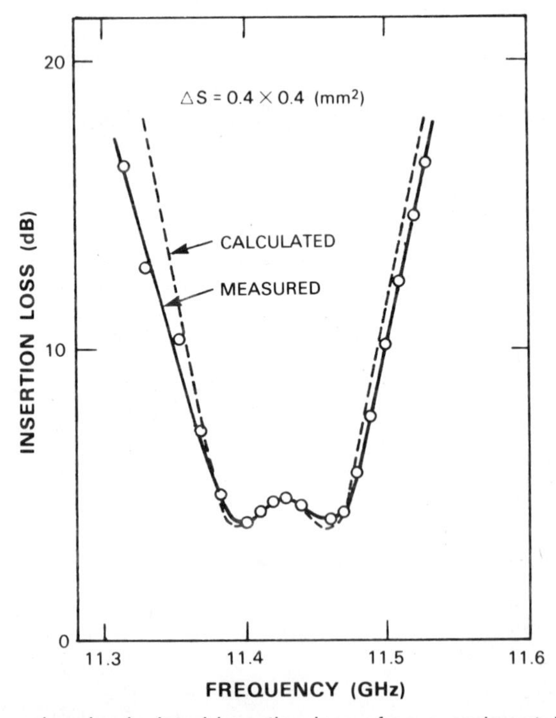

Fig. 56 Measured and calculated insertion loss of an experimental coupled-mode filter. (© 1971 Scripta Technica. Reprinted with permission from Reference 60.)

ferential ports. Ports 1 and 2 (input and output) are orthogonally located so that they couple with the two orthogonal and degenerate dipolar modes. The third port is located symmetrically with respect to ports 1 and 2. This port is connected to a short stub length that is left open-circuited. The purpose of the stub at the third port is to introduce some perturbation into the circular resonator so that a coupling takes place between the two dipolar modes. Figure 56 illustrates the measured and calculated filtering characteristics of a mode-coupled triplate filter that has been fabricated on a Rexolite substrate ($\varepsilon_r = 2.62$). The characteristic impedance of both the input and output striplines is $50\,\Omega$, and the thickness of each dielectric spacer is 1.45 mm. The size of the perturber stub is $0.4 \times 0.4\ \text{mm}^2$, and the disk radius is 4.25 mm. There is reasonably good agreement between the measured and calculated results.

The filtering properties of annular structures have also been analyzed by using the planar circuit approach [59, 60]. As has been done in the circular disk case, the filtering properties of the annular structure are related to its resonances. Figure 57 shows the measured and calculated scattering parameters of a two-port annular microstrip. The input and output ports are

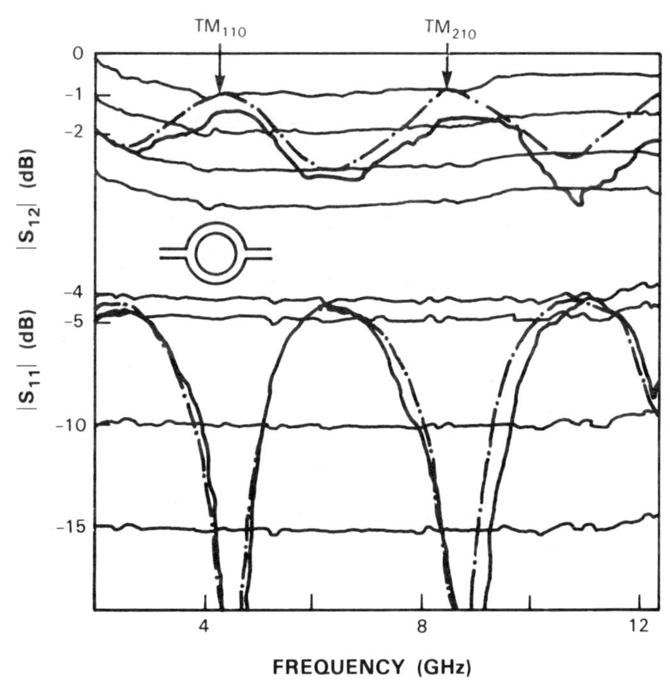

Fig. 57 Experimental and theoretical scattering parameters for an annular microstrip filter; $\varepsilon_r = 10$, $h = 0.635$ mm, $r_{in} = 4.0$ mm, $r_{out} = 4.6$ mm, $W = 0.6$ mm for both input and output lines. (© 1978 Electronics Letters. Reprinted with permission from Reference 59.)

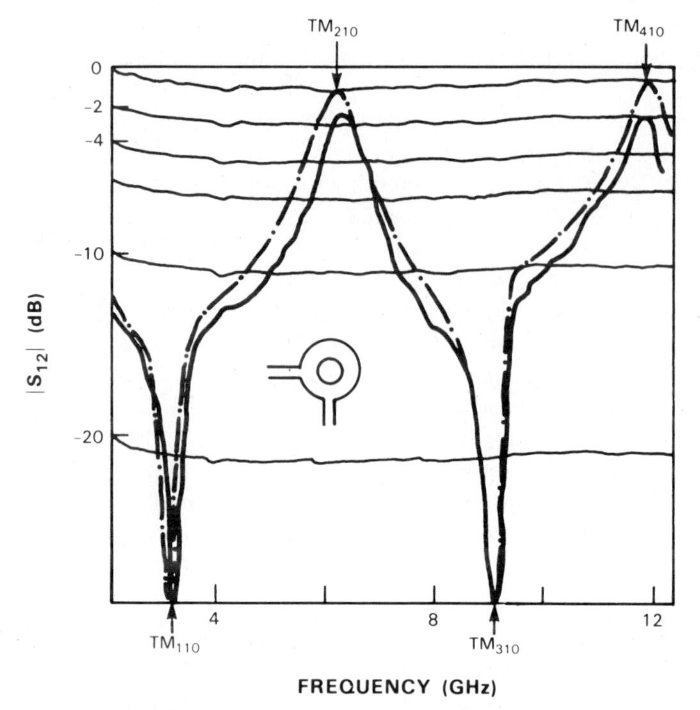

Fig. 58 Experimental and theoretical scattering parameter S_{21} for an annular microstrip filter; $\varepsilon_r = 10$, $h = 0.635$ mm, $r_{in} = 4.0$ mm, $r_{out} = 7.0$ mm, and $W = 0.6$ mm for both input and output lines. (© 1978 Electronics Letters. Reprinted with permission from Reference 59.)

diametrically opposed. Figure 58 exhibits the measured and calculated characteristics of another annular structure in which the angle between the input and output lines is 90°. Once again the degree of agreement between the theoretical and experimental results confirms the effectiveness of the planar circuit approach. The above-mentioned circuits were fabricated on dielectric substrates having a dielectric constant $\varepsilon_r = 10$ and thickness $h =$ 0.635 mm. The inner and outer radii of the structure are 4.0 mm and 7.0 mm, respectively. The widths of the input and the output lines are 0.6 mm.

The general filtering properties of rectangular structures have also been studied by many authors [17, 52–58]. Figure 59 shows the geometry of a typical two-port rectangular structure connected to two uniform input–output lines. An accurate description of the low-pass filtering behavior of this circuit can be obtained by applying the planar circuit approach. Several low-pass filters have been designed [57, 58] using this approach and have been realized in microstrip configuration on alumina substrate ($h =$ 0.0635 cm, $\varepsilon_r = 10$). Figure 60 illustrates the measured and computed scat-

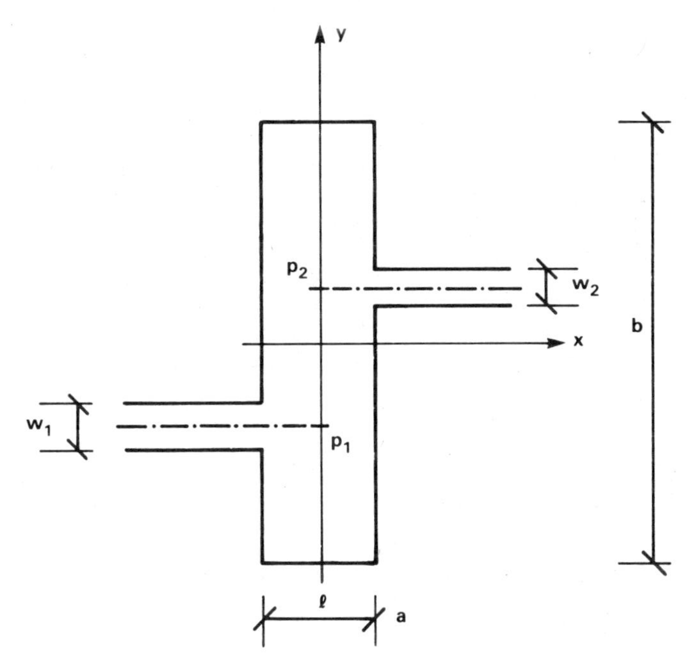

Fig. 59 Geometry of a rectangular two-port filtering structure. (© 1978 Electronics Letters. Reprinted with permission from Reference 57.)

tering parameters of a low-pass filter as a function of frequency [57]. The computed results are based on a five-mode approximation. As can be seen, the degree of agreement between the measured and computed results is quite remarkable. This filter has a 30-dB stopband over 1.55 GHz (about 17%) centered at 9 GHz. It produces a second passband at 10 GHz, and its passband attenuation is less than 0.5 dB. The performance of this filter is considered highly satisfactory up to 10 GHz.

The radial stub is another type of planar filter structure that has been extensively explored. This structure is of interest because of its low input impedance level [62–66] and its peculiar filtering properties [62, 70]. Indeed radial line stubs have been found to work better than low-impedance rectangular stubs when accurate localization of a zero-point impedance is needed. The double radial line stub (butterfly stub) has also been found to offer a distinctive advantage when used in designing high-performance low-pass filters [70]. Figure 61 illustrates the geometry and the exceptional performance of a low-pass filter design using butterfly stubs. As can be seen, the measured and theoretical results show a stopband width of two octaves with a minimum attenuation of 38 dB. The microstrip radial-line stubs utilized in this circuit have been characterized by the planar circuit approach. The sector angles of the two butterfly stubs utilized in the design of

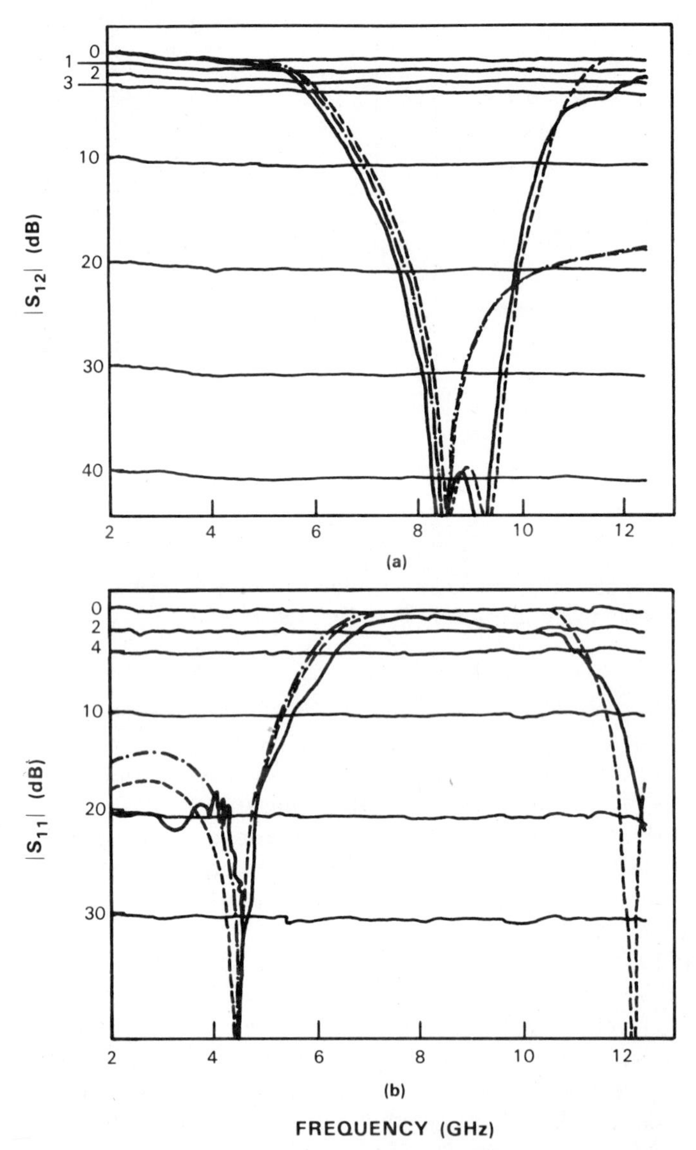

Fig. 60 Measured and calculated characteristics of the low-pass filter shown in Fig. 59. Filter dimensions $l = 0.0269$ cm, $b = 0.995$ cm, $p_2 = -p_1 = 0.176$ cm, $W_1 = W_2 = 0.06$ cm. (© 1978 Electronics Letters, reprinted with permission from Reference 57.)

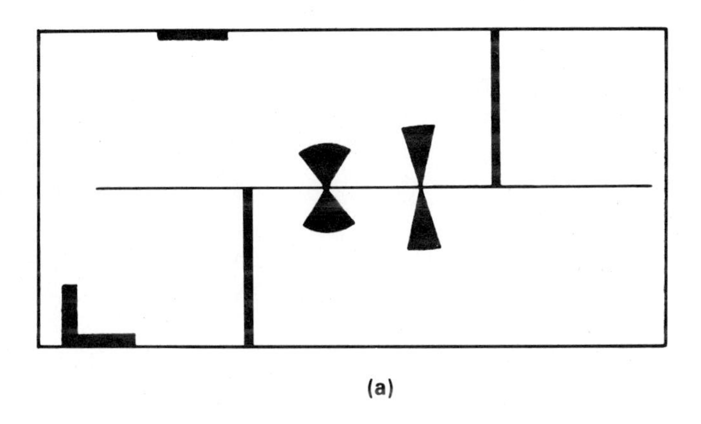

(a)

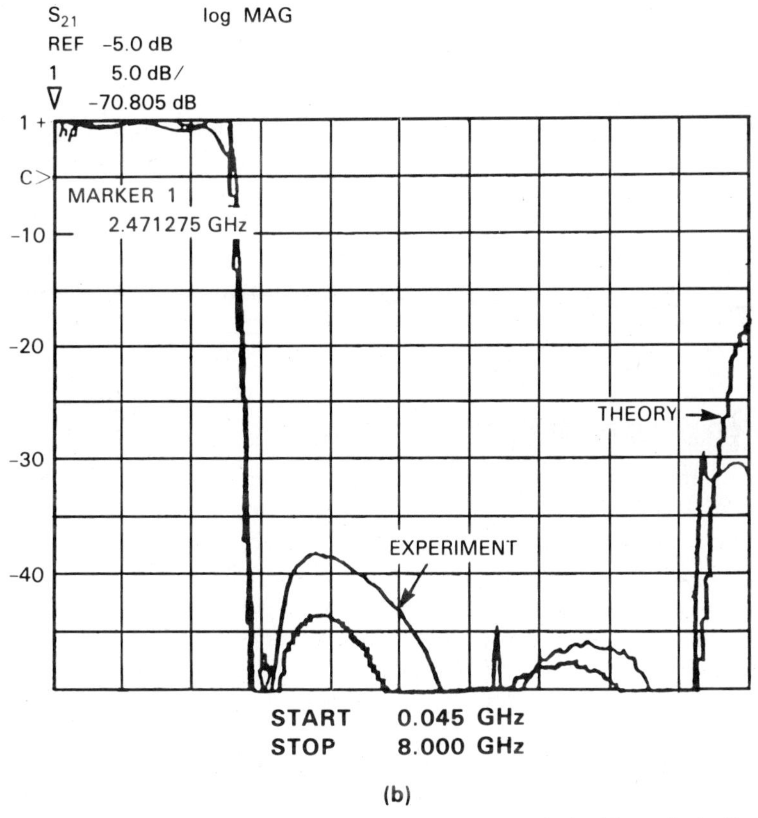

(b)

Fig. 61 Theoretical and measured characteristics of double-stub butterfly-type low-pass filter. (© 1986 Microwave Exhibitions Publishers. Reprinted with permission from Reference 70.)

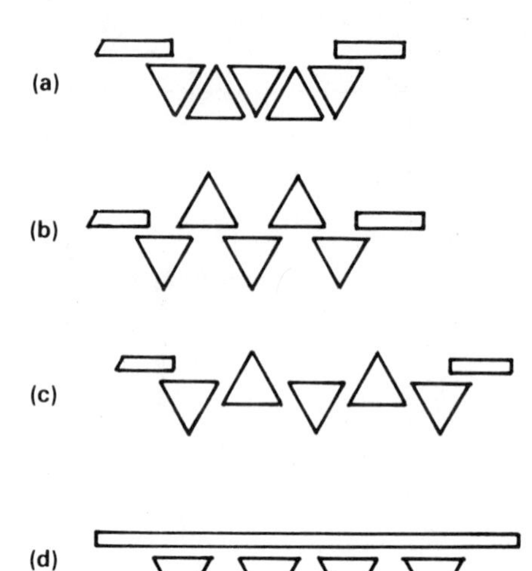

Fig. 62 Bandpass (*a*, *b*, and *c*) and bandstop (*d*) filter configurations using triangular two-dimensional components. (© 1978 IEEE. Reprinted with permission from Reference 71.)

this filter are 70.0° and 30.6°, respectively. The outer radii are 0.38 and 0.53 mm, respectively, and the overall length of the filter is 45.6 mm.

Triangular planar resonators have also been proposed [71] as prototype elements in bandpass and bandstop filters. Some of the suggested configurations are shown in Fig. 62. These circuits use gap coupling between the two-dimensional components and between the stripline and the two-dimensional components. Such circuits can be analyzed by a gap-coupled microstrip antenna technique developed by Kumar and Gupta [72, 79, 89].

8.4. Non-Reciprocal Ferrite Components

The circuit applications described in the previous sections are fabricated on isotropic dielectric substrates. In this section, planar circuits fabricated on anisotropic spacing media such as magnetized ferrite are described. Examples of ferrite planar circuits include wideband circulators [24, 35, 71], edge-guided mode isolators [37], edge-guided mode circulators [38], and nonreciprocal microwave power dividers/combiners [39, 73].

A two-dimensional theoretical treatment of planar disk resonators with ferrite substrates was first reported by Bosma [3]. Bosma used the two-dimensional planar approach to calculate the circulation parameters and to evaluate the frequency characteristics of a Y circulator. A general method for broadbanding circulators was also discussed. Later Helszajn and James [71] utilized the two-dimensional planar circuit approach to analyze triangular resonators with ferrite substrates. Helszajn's [71] experimental results indicated that circulators built with triangular resonators have lower radiation losses and a broader bandwidth than those conventional circulators built with circular disk resonators.

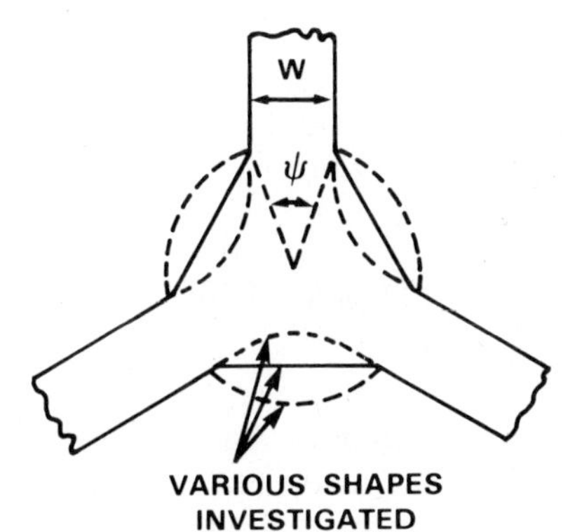

Fig. 63 Two-dimensional optimization of a circulator for maximum bandwidth.

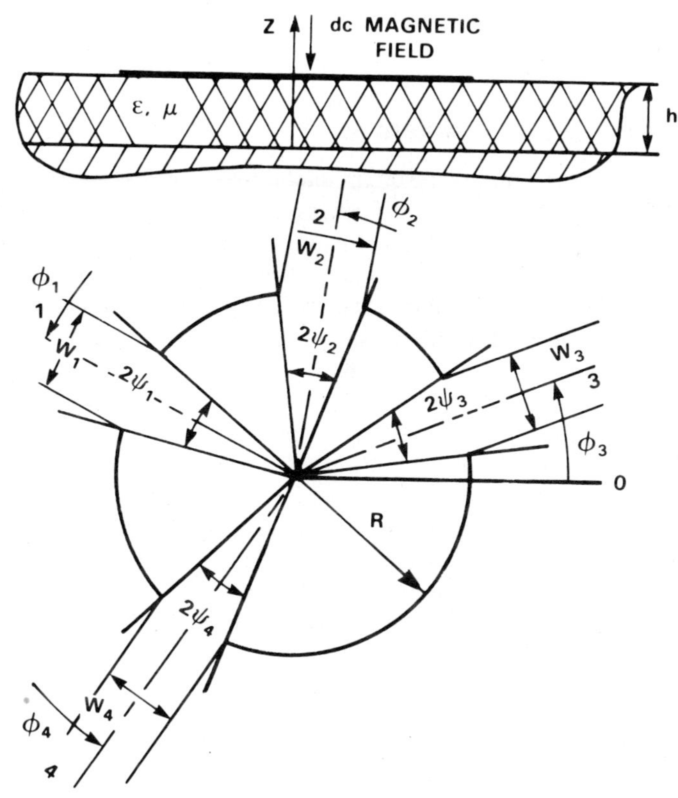

Fig. 64 Geometry of a two-channel nonreciprocal microstrip line splitter/combiner. (© 1985 Allerton Press. Reprinted with permission from Reference 73.)

Miyoshi and Miyauchi [24] applied the contour integral approach to the design of wideband circulators. By optimizing the parameters of a triangular circulator, 20 dB of isolation over 50% fractional bandwidth was obtained. The coupling port angle ψ depicted in Fig. 63 was found to be between 35° and 45°. Upon changing the straight sides of the optimized triangular circulator to various curved sides, Miyoshi and Miyauchi [24] found that triangular circulators having slightly concave sides and with $\psi = 36°$ offer the best fractional bandwidth (= 52%) for a 20-dB isolation.

Planar circuits with ferrite substrates have also been used to design two-channel nonreciprocal splitter/combiner (NSC) circuits. An NSC circuit consists of a disk ferrite resonator to which four microstrip lines of certain widths are connected (Fig. 64). For operating as a splitter, port 1 is the input and ports 2 and 4 are outputs, and a matched load is connected to port 3. When used as a combiner, ports 2 and 4 serve as inputs and port 3 is the output; here the load is connected to port 1. The angular locations of the coupling ports are determined by the splitting (or combining) ratio. A 3-dB nonreciprocal combiner, designed to operate in the centimeter-wave region, was found to have an insertion loss of 0.6–0.8 dB and more than 15 dB of isolation (between the input and output ports) over a 10–12% of bandwidth [73].

8.5. Microstrip Patch Antennas

Microstrip antennas employ resonant metallic patches over thin dielectric substrates. Since the substrate thickness of microstrip antennas is much smaller than the wavelength, the fields underneath the patch are essentially two-dimensional, and thus the planar circuit approach can be used. The analysis procedure can be illustrated by considering a rectangular patch antenna fed by a microstrip line as shown in Fig. 65*a*. A network model for this configuration is shown in Fig. 65*b*. The patch itself is modeled by a rectangular planar segment with a number of ports located around the periphery. The **Z** matrix of this multiport planar network can be obtained by the methods discussed earlier in this chapter. A section of the feeding microstrip line is also considered as a planar segment connected to the patch at (typically) five interconnecting ports. As in the case of microstrip discontinuity characterization (discussed in Section 8.1), the length of the feedline planar segment is taken to be large enough that the higher-order evanescent modes decay out at the input end. Thus only one external port is required at the input end. This modeling procedure allows the parasitic reactance at the junction between the feedline and the patch to be taken into account.

In typical microstrip antenna configurations, the field external to the patch can be considered to consist of three parts: (1) fringing fields at the edges of the patch, (2) far-zone radiation fields, and (3) fields of the surface waves excited on the dielectric substrates. For the thin substrates normally

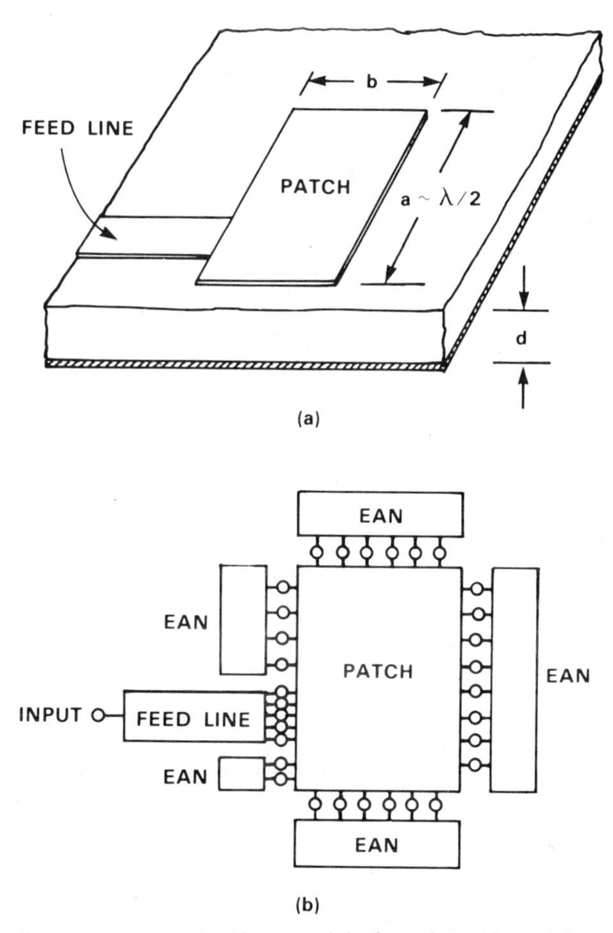

Fig. 65 Geometry and planar model of a microstrip patch antenna.

used, the surface wave fields are negligible and thus can be ignored. The other two components of the external field are incorporated in the planar model of microstrip antennas by including equivalent lumped networks, called edge admittance networks (EAN), shown in Fig. 65*b*. The computation of the edge admittance depends on the nature of the edge fields, and several approximate formulations are available [74–76].

Once a network model for the microstrip antenna is formulated, a planar circuit analysis technique (segmentation, desegmentation, etc.) can be used to compute the driving point impedance at the input port. The evaluation of the input impedance as a function of frequency yields the resonance frequency at which the microstrip antenna should be operated. Also, the input impedance bandwidth of the antenna can be found. In addition, the planar network analysis is used to determine the voltage distribution at the ports connecting the patch to the edge admittance networks. This voltage distribution is expressed in terms of an equivalent magnetic current distribution around the periphery of the patch antenna. Considering this equivalent magnetic current as an aperture source distribution, the far-zone radiation field produced by the antenna is evaluated. Then the radiation patterns and other radiation characteristics of the antenna (beam width, side-lobe level, cross-polarization level, etc.) are computed from the far-zone field.

The planar analysis approach has been used to analyze, design, and optimize a variety of microstrip radiators [77–92]. These applications are discussed in three groups: (1) circularly polarized microstrip patches; (2) broadband multiresonator microstrip antennas; and (3) multiport microstrip antennas and series-fed arrays.

8.5.1. Circularly Polarized Microstrip Patches

The single-fed circularly polarized microstrip patches analyzed by the planar analysis approach include diagonally fed, nearly square patches [78]; truncated corner square patches [78]; square patches with a diagonal slot [78];

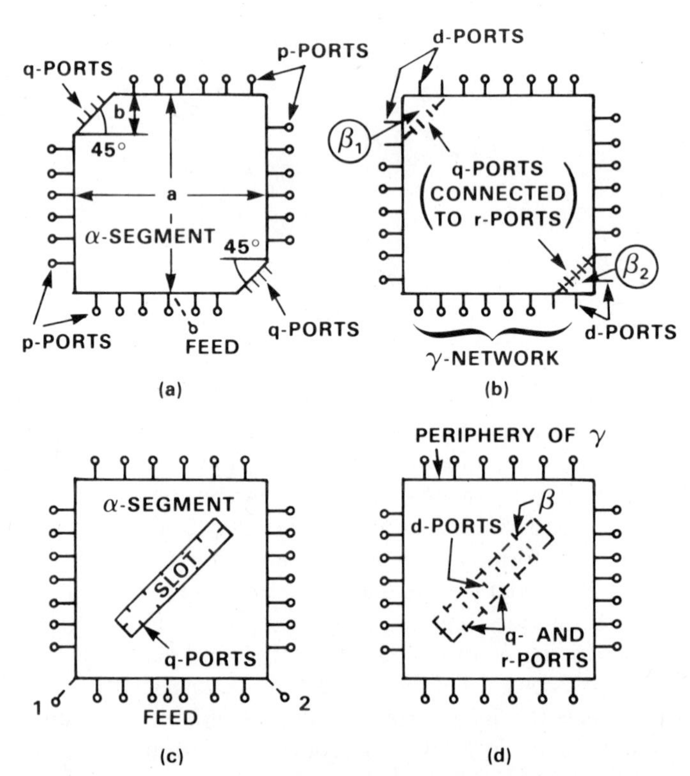

Fig. 66 Desegmentation method as applied to the truncated corner square patch and the square patch with diagonal slot microstrip antennas.

pentagonal patches [77]; square ring patches [85]; and cross-shaped patches [85].

The desegmentation method has been used for analyzing truncated corner square patches and square patches with a diagonal slot. The implementation of the desegmentation procedure in these two cases is illustrated in Fig. 66. For a truncated corner patch, the α segment (compare with Fig. 21) is shown in Fig. 66*a*. To apply the desegmentation procedure to this case, two triangular β segments, β_1 and β_2, are used. The addition of β_1 and β_2 to α segments results in the perfectly square γ segment shown in Fig. 66*b*. The configuration of the α segment for a square with a diagonal slot is shown in Fig. 66*c*. In this case the β segment is a rectangle of the size of the

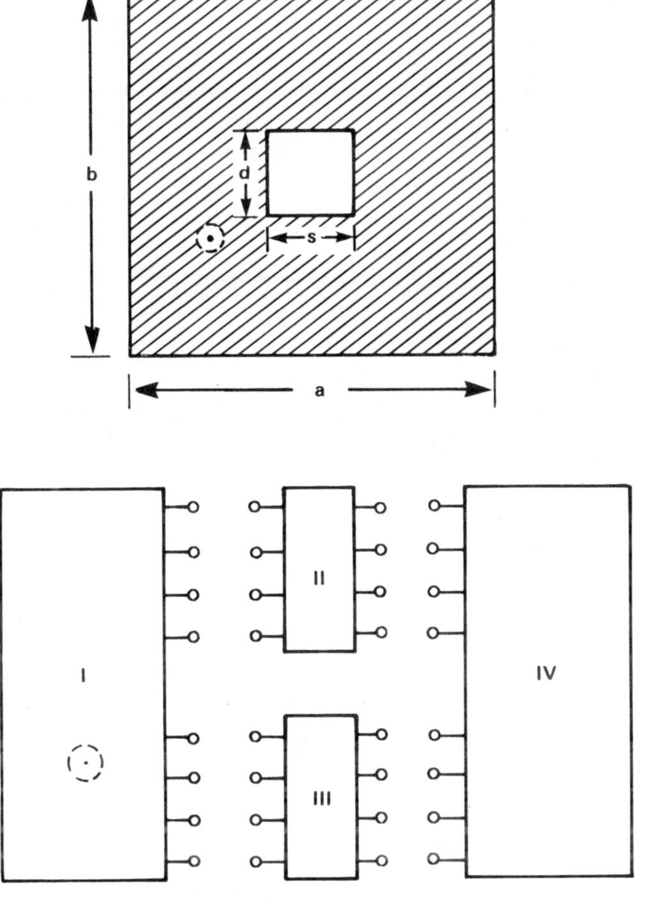

Fig. 67 Geometry and segmentation procedure of a square ring patch microstrip antenna.

slot, and the d ports are located inside the rectangle as shown in Fig. 66d (this configuration can be compared with that of Fig. 26). Detailed results for these two configurations are given in Reference 78.

Square ring patches and cross-shaped patches have been analyzed [85] by the segmentation method. The segmentation of the square ring into four rectangular segments is shown in Fig. 67, while the segmentation of a cross-shaped patch into three rectangular segments is illustrated in Fig. 68. Detailed results for these two configurations are given in Reference 85. A comparison of five different shapes (square ring, crossed strip, almost square patch, corner-chopped square patch, and a square with a diagonal slot patch) shows that the maximum axial ratio bandwidth (about 5.2% for $f = 3.0$ GHz, $\varepsilon_r = 2.5$, and $h = 0.159$ cm) is obtained by using a square ring configuration.

Another interesting antenna configuration analyzed by the multiport network approach is a pentagonal patch [77]. This antenna configuration is shown in Fig. 69a, and two different methods of analysis are depicted in Figs. 69b and c. Desegmentation with two triangular segments, β_1 and β_2, yields a 90°–60°–30° triangular segment for which the Green's function is available. The second approach illustrated in Fig. 69c employs desegmentation with one segment β_1 to yield a kite-shaped geometry. The kite shape is then segmented into two identical 90°–60°–30° triangles as shown. These two approaches yield identical results.

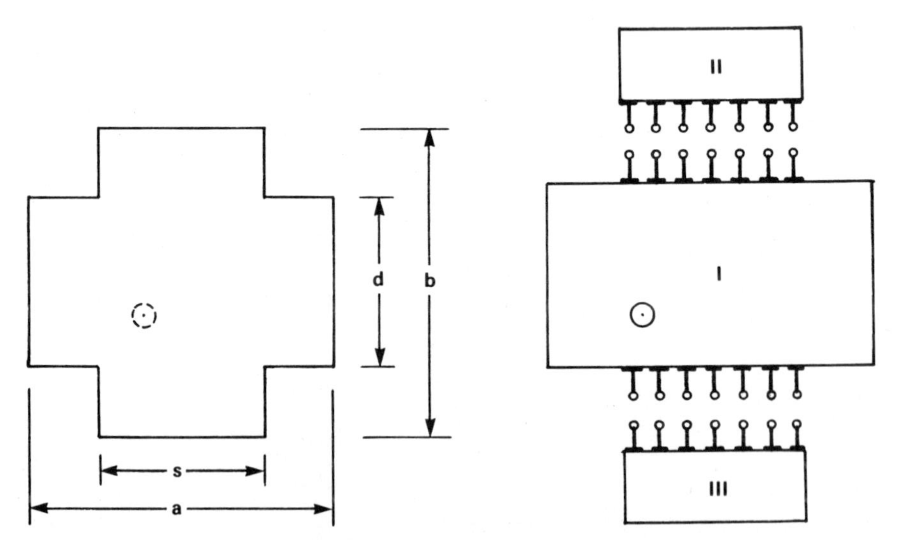

Fig. 68 Geometry and segmentation procedure of a cross-shaped patch microstrip antenna.

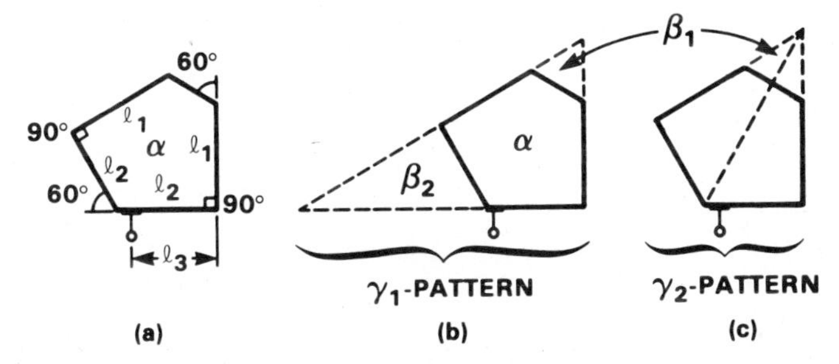

Fig. 69 Geometry and desegmentation procedure of a pentagonal patch.

8.5.2. Broadband Multiresonator Microstrip Antennas

Another group of microstrip antenna configurations that have been analyzed using the planar analysis approach are broadband microstrip antennas using coupled resonators [79, 81]. All of these configurations use multiple resonators with slightly different resonant frequencies as shown in Fig. 70 and 71. These configurations show that different resonators are coupled to each other while only one resonator (usually the central one) is connected to the feedline. Also, two different coupling mechanisms have been used. Note that the three configurations shown in Fig. 70 use capacitive coupling across the gaps between the closely spaced edges, and the three configurations of

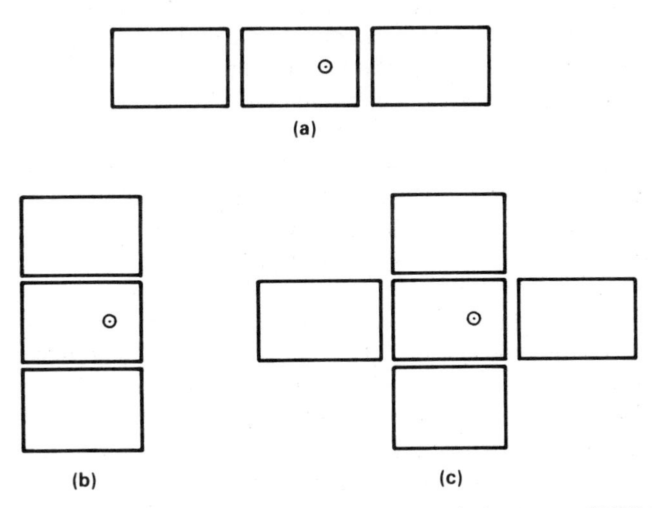

Fig. 70 Configuration of (*a*) radiating edge gap-coupled antenna (FEGCOMA), (*b*) nonradiating edge gap-coupled antenna (NEGCOMA), and (*c*) four-edge gap-coupled antenna (REGCOMA).

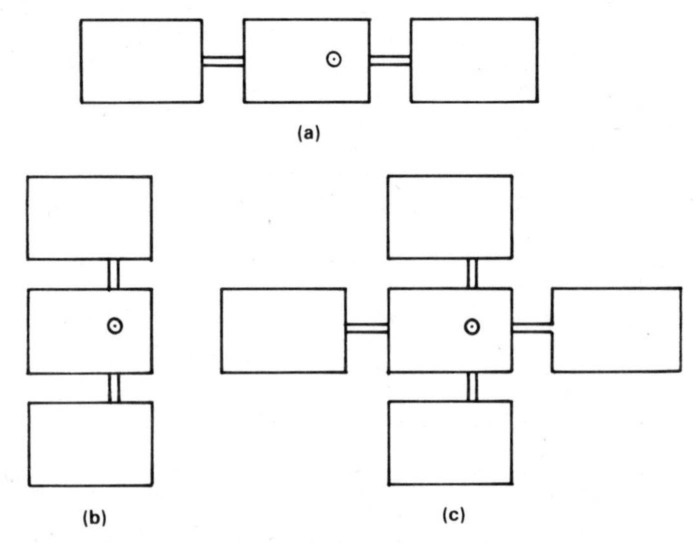

Fig. 71 Geometry of (*a*) radiating edge directly coupled antenna (REDCOMA), (*b*) nonradiating edge directly coupled antenna (NEDCOMA), and (*c*) four-edge directly coupled antenna (FEDCOMA).

Fig. 71 employ short sections of microstrip lines for providing the necessary coupling.

An analytical procedure for a gap-coupled multiresonator antenna configuration is illustrated in Fig. 72. The coupling gaps are modeled by a multiport lumped *RC* network (called a gap-coupling network or GCN) as shown in Fig. 72*b*. In modeling the GCNs, the values of C_1, C_2, and C_g are obtained from coupled microstrip transmission line analysis. The term G in Fig. 72*b* represents the radiation conductance, and its value is obtained by treating the gap fields as a line source of equivalent magnetic current. Since the feedpoint is located along the centerline *xx* (Fig. 72*a*), the symmetry of the configuration can be used to simplify the computations; then only one half of the antenna configuration, shown in Fig. 72*c*, need be analyzed. A multiport network model is shown in Fig. 72*d*.

A planar network model for a directly coupled three-resonator antenna configuration is shown in Fig. 73. In this case also, one can make use of the geometrical symmetry, and only one half of the antenna configuration (with a magnetic wall placed along the plane *xx*) need be analyzed. The multiport network model is shown in Fig. 73*c*. The interconnecting microstrip line sections are also modeled by two planar rectangular segments, and as shown there are nine edge admittance networks (EAN's). The segmentation formula is then used to find the input impedance at the feedport, and also to evaluate the voltage distribution at the edges of the radiating patches.

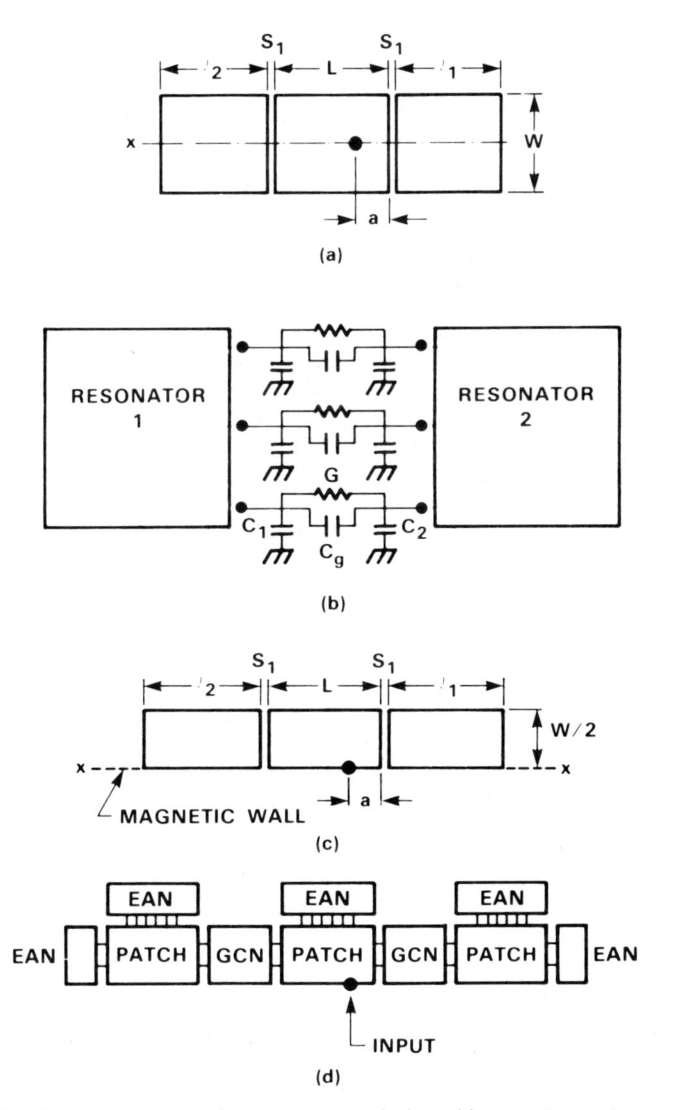

Fig. 72 Analysis procedure for a gap-coupled multiresonator antenna configuration.

A fairly wide impedance bandwidth can be achieved by using the multiple resonator microstrip antenna configurations shown in Figs. 70 and 71. Typical values for the six configurations fabricated on substrates having $\varepsilon_r = 2.55$ are summarized in Table 2 (from [89]). The various acronyms in Table 2 (REGCOMA, etc.) are defined in Figs. 70 and 71. The factor M shown gives the bandwidth (BW) as a multiple of the corresponding single

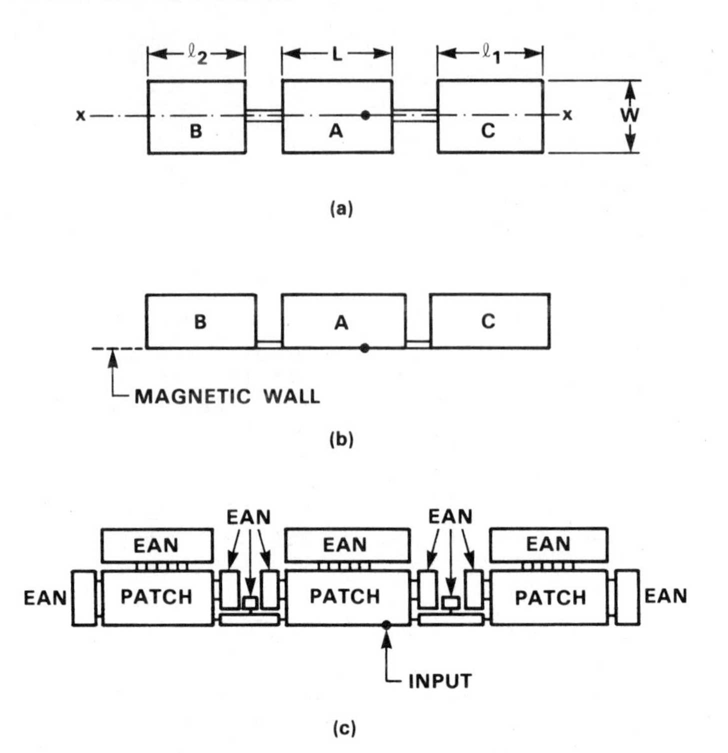

Fig. 73 Analysis procedure for a directly coupled three-resonator antenna configuration.

Table 2 Typical Impedance Bandwidth Values for Microstrip Antennas Using Multiple Coupled Resonators

Configuration	*d* (cm)	*f* (GHz)	BW (MHz)	BW (%)	*M*
REGCOMA	0.159	3.29	331	10	5.3
NEGCOMA	0.318	3.110	480	15.4	4.0
FEGCOMA	0.318	3.160	815	25.8	6.7
REDCOMA	0.318	3.200	548	17.1	5.0
NEDCOMA	0.318	3.310	605	18.3	5.5
FEDCOMA	0.318	3.380	810	24.0	7.36

Source: Reference 89.

rectangular patch antenna bandwidth, f is the center frequency, and d is the thickness of the substrate.

8.5.3. Multiport Microstrip Patches and Series-Fed Arrays

Series-fed linear arrays of microstrip patches (such as the ones shown in Fig. 74) use two-port radiators as basic building blocks. For this application, both two-port rectangular patches [82] and two-port circular patches [83] have been analyzed by using the multiport network modeling approach.

Two-Port Rectangular Patch. A network model for a rectangular patch with microstrip line ports located along the nonradiating edges is shown in Fig. 75. The segments labeled PATCH and FLN (feedline network) are rectangular planar segments. The FLN segments represent small sections (typically $\lambda/8$ long) of microstrip lines connected to the two ports. The widths of the FLNs are equal to the effective widths of the two lines, respectively. Multiple interconnections between the FLNs and the patch ensure that the parasitic reactances associated with the feeding–patch junctions are taken into account. The radiating edge admittance networks (R-EAN) and nonradiating edge admittance networks (NR-EAN) are obtained by modeling the fringing fields at the edges. The mutual coupling network (MCN) represents the external interaction between the two radiating edges.

For two-port patches with ports along the nonradiating edges as shown in Fig. 75, transmission from port 1 to port 2 can be controlled by suitable location of the ports. An example of this feature is presented in Fig. 76

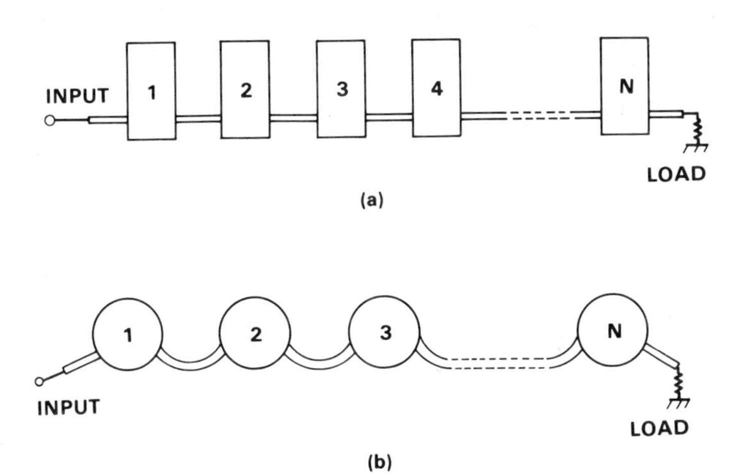

Fig. 74 Configuration of series-fed linear arrays of circular and rectangular microstrip patches.

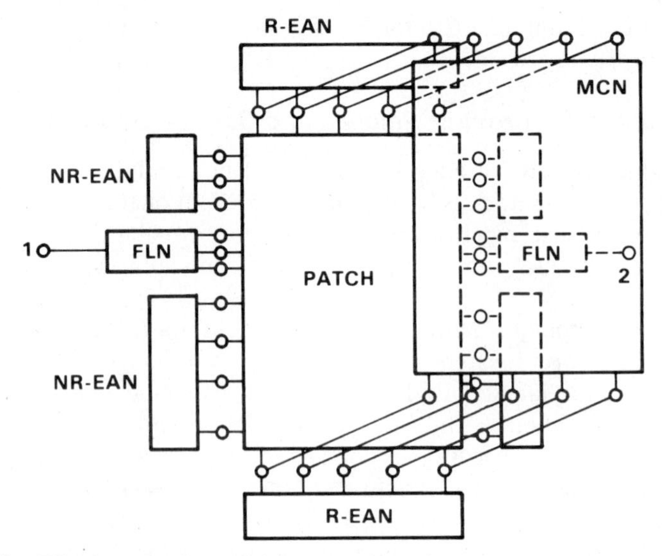

Fig. 75 A network model for a rectangular patch with two ports.

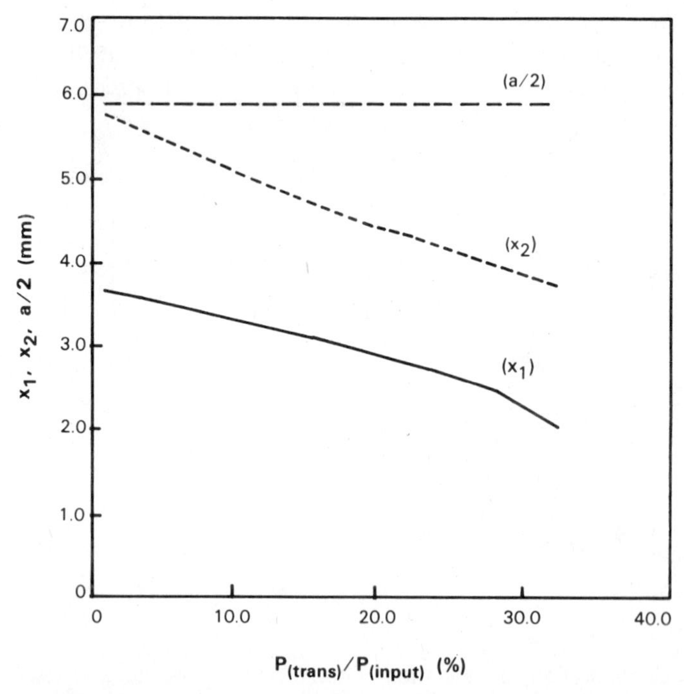

Fig. 76 Transmitted power variation with the external port locations in a rectangular microstrip patch antenna.

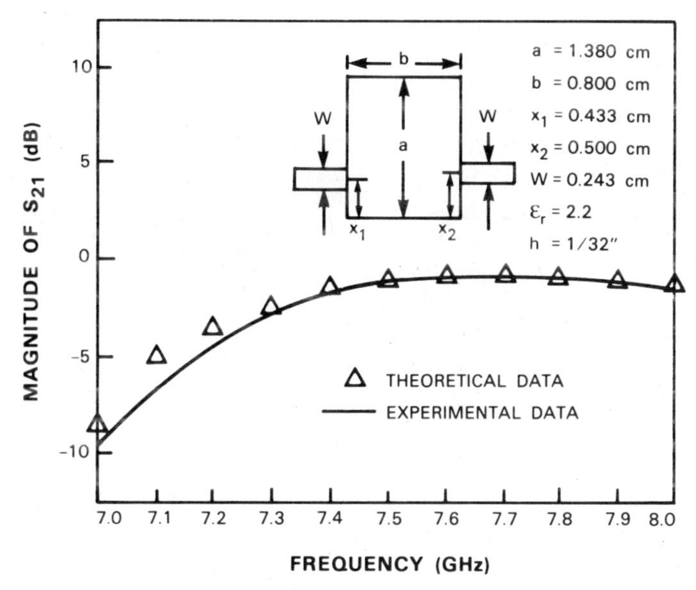

Fig. 77 Measured and calculated transmission coefficient of a two-port rectangular microstrip patch antenna.

(from Reference 90). This figure shows the variation of the transmitted power (to port 2) relative to the external port location. The values of x_1 are chosen to ensure a match at the input port ($S_{11} = 0$). When the port locations are altered, the associated change in the junction reactances causes the patch resonance frequency to shift slightly. The corresponding change in the resonant dimension a of the patch is also plotted in Fig. 76. The results shown are for a substrate with $\varepsilon_r = 2.48$, $d = \frac{1}{32}$ in., $\tan\delta = 0.002$, and a resonant frequency of 7.5 GHz. A comparison of theoretical and experimental results, for S_{21} of a two-port patch, is shown in Fig. 77 (from Reference 91). The design parameters of the two-port patch are also listed in this figure. The results shown verify the validity of the planar circuit analysis approach for S-parameter characterization of the radiating patches.

Two-Port Circular Patches. Circular microstrip patches having two ports located along the circumference have also been analyzed [83]. A planar circuit model for the circular patch and for the input/output microstrip feedlines is illustrated in Fig. 78. The physical radius a of the disk and its loss tangent δ are replaced, respectively, by the effective values a_e and δ_e. The effective radius a_e takes into account the fringing capacitance around the circumference. The effective loss tangent δ_e includes "loss" because of the power radiated from the patch.

Approximate results [83] based on using the dominant mode only and ignoring the feed junction reactances point out that for a match at the input

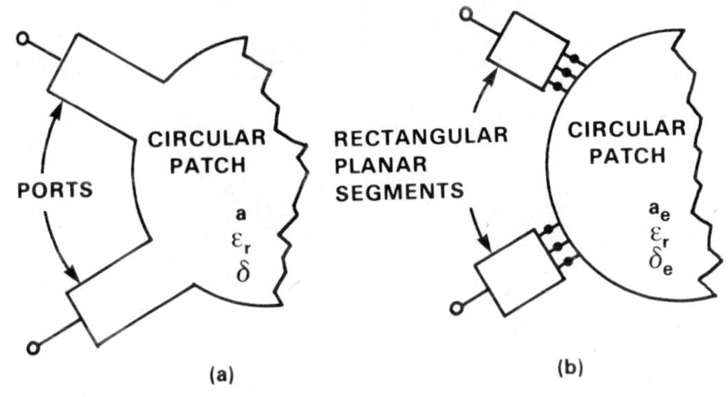

Fig. 78 Geometry of a circular microstrip patch with two ports located along the circumference.

port ($S_{11} = 0$), the impedance Z_0 of the feedline at the input port is related to the Z_{11} element of the **Z** matrix by

$$Z_0 = Z_{11} \sin \phi_{12} \tag{163}$$

where ϕ_{12} is the angular separation between the two ports and

$$Z_{11} = 1.674 \frac{d}{\lambda_0} \left(\frac{\eta_0}{\delta_e} \right) \tag{164}$$

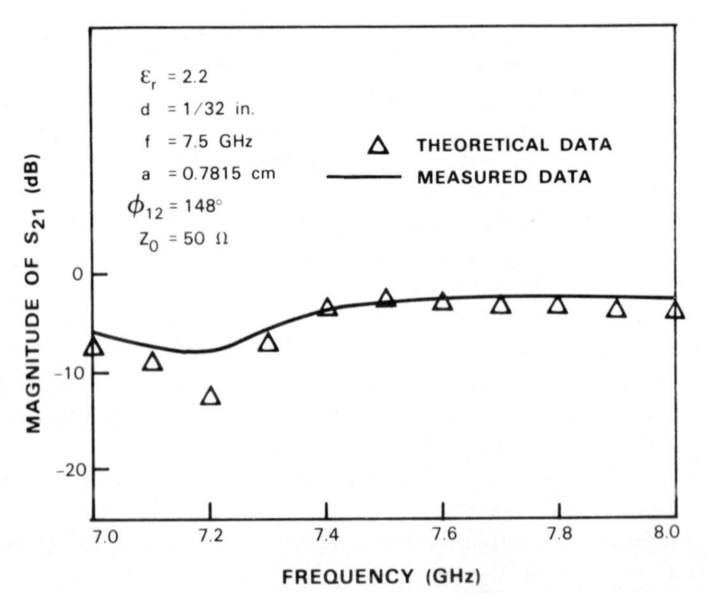

Fig. 79 Measured and calculated transmission coefficient of a two-port circular microstrip patch antenna with $\phi_{12} = 148°$. (From Reference 92.)

where d is the substrate thickness. The corresponding transmission coefficient S_{21} is given by

$$S_{21} = \frac{\cos \phi_{12}}{1 + \sin \phi_{12}} \tag{165}$$

Equation (165) suggests that for the dominant mode the transmission coefficient S_{21} varies from 1 to 0 as the angular separation ϕ_{12} changes from 0° to 90°. Also note that for high values of S_{21} close to unity, the input port characteristic impedance becomes very small and makes the design impractical.

Results based on the above method have been verified experimentally, and a sample comparison of theoretical and measured S_{21} values is presented in Fig. 79 (from Reference 92). The two-port characterization of rectangular and circular patches discussed above is used in the design of series arrays shown in Fig. 74.

9. SUMMARY

This chapter provides an overview of the concept of two-dimensional planar components, circuits, and antennas used at microwave and millimeter-wave frequencies.

For two-dimensional structures, the basic field equations and the wave equation can be reduced to a two-dimensional form that can be solved relatively easily. Analytical methods are available for planar structures with different complex geometrical configurations. For example, two-dimensional impedance Green's functions are available for simple shapes such as rectangles, three types of triangles, circles, annular rings, and sectors of circles and rings (with specific sector angles). These geometrical shapes are the simplest to analyze and are used frequently. However, since most of these Green's functions are in the form of doubly infinite series, the computer-aided evaluation of these planar structures is time consuming unless the summations can be carried out analytically. It has been shown that for rectangular shapes one of the summations can be carried out analytically [94] and only the remaining one need be obtained numerically. Similarly, analytical simplifications may be possible for other shapes. For example, by using the mode-matching approach to analyze the circular disk structure discussed in Section 7 and Appendix B, it has been shown that only one numerical summation is needed.

Equations giving Green's functions and Z matrices for various shapes of planar components are listed in Tables 3 and 4.

For planar structures of composite shapes, very powerful network analysis techniques called segmentation and desegmentation have been developed. These methods have been used for the analysis and design of the planar components that have been discussed in this chapter.

Table 3 Green's Functions and *Z* Matrices for Planar Segments with Open Boundaries

Shape	Green's Function (Equation No.)	***Z*** Matrix (Equation No.)
Rectangle	(44)	(166), (170), (178)
30°–60°–90° triangle	(45)	(216)
Equilateral triangle	(48)	(207)
45°–45°–90° triangle	(50)	(212)
Circle	(52)	(182), (185), (186)
Circular sector	(54)	(187), (189), (193), (194)
Annular ring	(56)	(195), (199), (200)
Annular sector	(59)	(201), (205), (206)

Table 4 Green's Functions for Planar Segments with Shorted and Mixed Boundaries

Shape	Shorted Boundary (Equation No.)	Mixed Boundary (Equation No.)
Rectangle	(60)	(71)–(74)
30°–60°–90° triangle	(61)	(83), (85)
Equilateral triangle	(65)	—
45°–45°–90° triangle	(63)	(75), (77), (79), (81)
Circle	(67)	—
Circular sector	(70)	(86), (88)–(90)

For planar components of arbitrary shapes, the contour integral approach (or the boundary integral approach) has been shown to be necessary. However, this type of planar component has not been used extensively, since the contour integral approach requires more computer time. Perhaps there is a need to develop a computationally efficient algorithm with a built-in contour optimization subroutine to analyze these circuits. More CAD-related activities are still needed in this area.

The range of the planar circuit concept applications (as discussed in Section 8) is indeed very wide and impressive. Undoubtedly more topics will be added to this list in the near future. Microstrip patch antenna analysis is an area where substantial effort has been put forth in recent years. Computer-aided design of printed or integrated microstrip patch antennas appears to be possible only by the use of the planar circuit concepts discussed in this chapter.

Innovative extensions of this planar (two-dimensional) electromagnetic concept are likely to evolve. The proposal for surface-wave planar circuits [95] presented by Hsu and his colleagues at the 1986 MTT-S symposium appears to be promising. This concept may have applications for two-dimensional radiating structures as well as for two-dimensional circuits. Such circuits are suitable for integration with dielectric waveguides used in the millimeter-wave frequency range.

ACKNOWLEDGMENTS

This work was supported in part by the Department of the Air Force. The views expressed are those of the authors and do not reflect the official policy or position of the U.S. Government.

M. D. Abouzahra gratefully acknowledges the support and encouragement of W. J. Ince and W. M. Brown. The considerable task of typing the manuscript was undertaken by Kimberly Telley, to whom we extend our gratitude. We would also like to thank Kathleen Abouzahra for her detailed editing of this chapter.

APPENDIX A. IMPEDANCE MATRICES FOR PLANAR SEGMENTS WITH REGULAR SHAPES

In this appendix the impedance matrix elements of several microstrip type planar segments with open boundaries are listed. These expressions may be used directly in the design of planar components.

A.1. Z Matrix for Rectangular Segments

Green's functions for various geometries, discussed in Section 4, appear as double infinite summations. In the numerical computations of the **Z** matrix elements, the corresponding Green's function is integrated over the port widths. The order of integrations and summations can be interchanged. For rectangular segments, the integrals involved can be carried out analytically [94]. When the rectangle's sides are oriented along the x and y axes, the impedance matrix element Z_{pq} can be written in the form [94]

$$Z_{pq} = \frac{j\omega\mu d}{ab} \frac{\sum_{m=0}^{\infty} \sum_{n=0}^{\infty} \sigma_m \sigma_n \phi_{mn}(x_p, y_p) \phi_{mn}(x_q, y_q)}{k_x^2 + k_y^2 - k^2} \tag{166}$$

where ϕ_{mn}, for ports oriented along the y direction, is

$$\phi_{mn}(x, y) = \cos k_x x \cos k_y y \operatorname{sinc} \frac{k_y W}{2} \tag{167}$$

and for ports oriented along the x direction is

$$\phi_{mn}(x, y) = \cos k_x x \cos k_y y \operatorname{sinc} \frac{k_x W}{2} \tag{168}$$

The function sinc z is defined as $(\sin z)/z$, and

$$\begin{aligned} k_x &= \frac{m\pi}{a} \qquad k_y = \frac{n\pi}{b} \\ \sigma_m &= \begin{cases} 1 & m = 0 \\ 2 & m \neq 0 \end{cases} \\ k^2 &= \omega^2 \mu \varepsilon_0 \varepsilon_r (1 - j\delta) \end{aligned} \tag{169}$$

where δ is the loss tangent of the dielectric, a is the rectangle's length, b is its width, and d is the substrate's height. Points (x_p, y_p) and (x_q, y_q) denote the locations of ports p and q, respectively.

It has been shown [94] that the doubly infinite series in (166) along with (167) and (168) can be reduced to a singly infinite series by summing the inner sum. The choice of summation over n or m depends on the relative locations of ports p and q and also on the aspect ratio of the rectangular segment. Next, we consider two cases.

Case I. Ports p and q Are Oriented along the Same Direction

When both ports (p and q) are oriented along the same direction (x or y), we can write Z_{pq} as

$$\begin{aligned} Z_{pq} = &-CF \sum_{l=0}^{L} \sigma_l \cos k_u u_p \cos k_u u_q \cos \gamma_l z_> \cos \gamma_l z_< \\ &\cdot \frac{\operatorname{sinc}(k_u W_p/2) \operatorname{sinc}(k_u W_q/2)}{\gamma_l \sin \gamma_l F} \\ &- jCF \sum_{l=L+1}^{\infty} \cos k_u u_q \cos k_u u_p \sin \frac{k_u W_p}{2} \\ &\cdot \operatorname{sinc}\left(\frac{k_u W_q}{2}\right) \frac{\exp[-j\gamma_l(\nu_> - \nu_<)]}{\gamma_l} \end{aligned} \tag{170}$$

where

$$(\nu_>, \nu_<) = \begin{cases} (y_>, x_<) & l = m \\ (x_>, x_<) & l = n \end{cases} \tag{171}$$

and $C = j\omega\mu d/ab$.

When the two ports are oriented along the y direction, $l = n$; and when they are oriented along the x direction, $l = m$. Also,

$$F = \begin{cases} b & l = m \\ a & l = n \end{cases} \tag{172}$$

$$(u_p, u_q) = \begin{cases} (x_p, x_q) & l = m \\ (y_p, y_q) & l = n \end{cases} \tag{173}$$

$$\gamma_l = \pm\sqrt{k^2 - k_u^2} \tag{174}$$

$$k_u = \begin{cases} m\pi/a & l = m \\ n\pi/b & l = n \end{cases} \tag{175}$$

$$(z_>, z_<) = \begin{cases} (y_> - b, y_<) & l = m \\ (x_> - a, x_<) & l = n \end{cases} \tag{176}$$

The sign of γ_l is chosen such that $\mathrm{Im}(\gamma_l)$ is negative. W_p and W_q denote the widths of ports p and q, respectively. Additionally, the notation used in (171) is defined by

$$y_> = \max(y_p, y_q) \qquad y_< = \min(y_p, y_q) \tag{177}$$

Similar notation applies for $x_>$ and $x_<$ when $l = n$. The choice of the integer L in (170) becomes a trade-off between computational speed and accuracy. A compromise is to select L such that $\gamma_l F \leq 100$.

Case II. Ports p and q Are Oriented along Different Directions

When the two ports (p and q) are oriented along different directions (x and y), various elements of the **Z** matrix may be written as

$$\begin{aligned} Z_{pq} = &-CF \sum_{l=0}^{L} \sigma_l \cos k_u u_p \cos k_u u_q \cos \gamma_l z_< \\ &\cdot \cos(\gamma_l z_>) \frac{\mathrm{sinc}(k_u W_i/2)\,\mathrm{sinc}(\gamma_l W_j/2)}{\gamma_l \sin \gamma_l F} \\ &-CF \sum_{l=L+1}^{\infty} \cos k_u u_p \cos k_u u_q \\ &\cdot \mathrm{sinc}\left(\frac{k_u W_i}{2}\right) \frac{\exp[-j\gamma_l(\nu_> - \nu_< - W_j/2)]}{\gamma_l^2 W_j} \end{aligned} \tag{178}$$

The choice of l is determined by the convergence of the last summation in eq. (178). This is realized when

$$\nu_> - \nu_< - \frac{W_j}{2} > 0 \tag{179}$$

The index of the inner summation is chosen so that this condition is satisfied. This condition can be written more explicitly as

$$l = m, \qquad \text{if } \{\max(y_p, y_q) - \min(y_p, y_q) - W_j/2\} > 0 \tag{180}$$

and

$$l = n, \qquad \text{if } \{\max(x_p, x_q) - \min(x_p, x_q) - W_j/2\} > 0 \tag{181}$$

When both of these conditions are satisfied, any choice of l will ensure convergence.

If $l = n$, W_i corresponds to the port oriented along the y direction and W_j corresponds to the port oriented along the x direction. On the other hand, if $l = m$, W_i is used for the port along the x direction and W_j for the port along the y direction.

A.2. Z Matrix for Circular Segments

For circular patches, the impedance Green's function is given by (52) and (53) (Section 4.2). When the ports are located along the disk circumference (as shown in Fig. 47), the planar component's **Z** matrix can be determined by a single expression. The diagonal (self-impedance) terms and the off-diagonal (transfer-impedance) terms of the impedance matrix are given by

$$Z_{ij} = \frac{2j\omega\mu d a^2}{\pi W_i W_j} \sum_{n=0}^{\infty} \sum_{m=1}^{\infty} \frac{\sigma_n \cos(n\phi_{ij})\{\cos[(n/2)(\Delta_i - \Delta_j)] - \cos[(n/2)(\Delta_i + \Delta_j)]\}}{n^2(a^2 - n^2/k_{mn}^2)(k_{mn}^2 - k^2)} \tag{182}$$

where the k_{mn}s are solutions of

$$\frac{\partial}{\partial \rho} J_n(k_{mn}\rho)\bigg|_{\rho=a} = 0 \tag{183}$$

and

$$\Delta_{i,j} = W_{i,j}/a \tag{184}$$

where a represents the radius of the disk and W_i is the curvilinear width of the port i measured along the circumference.

When a circular planar component's port is located at the center (a probe-type feed), the **Z** matrix element Z_{11} corresponding to this port is given by [48]

$$Z_{11} = \frac{j\omega\mu d}{\pi a^2} \sum_{m=1}^{\infty} \frac{J_0^2(k_{m0}\rho_0)}{[k_{m0}^2 - k^2] J_0^2(k_{m0}a)} \tag{185}$$

where ρ_0 denotes the radius of the feeding probe center conductor.

The off-diagonal terms in the **Z** matrix relating the center port to any of the circumferential ports are given by

$$Z_{1j} = \frac{j\omega\mu d}{\pi a^2} \sum_{m=1}^{\infty} \frac{J_0(k_{m0}\rho_0)}{(k_{m0}^2 - k^2)J_0(k_{m0}a)} \tag{186}$$

A.3. *Z* Matrix for Circular Sectorial Segments

The impedance matrix elements for a circular sectorial segment (with sector angle $\alpha = \pi/l$, l being an integer) have been derived [96] recently. When all the ports are located on the circumference, the self- and transfer impedance elements are given by

$$Z_{ij} = \frac{2j\omega\mu da^2}{\alpha W_i W_j} \sum_{n=0}^{\infty} \sum_{m=1}^{\infty} \frac{\sigma_n\{\cos(n_s\phi_{ij}) + \cos[n_s(\phi_i + \phi_j)]\}}{n_s^2\{a^2 - n_s^2/k_{mn_s}^2\}\{k_{mn_s}^2 - k^2\}} \cdot \left\{\cos\left[\frac{n_s}{2}(\Delta_i - \Delta_j)\right] - \cos\left[\frac{n_s}{2}(\Delta_i + \Delta_j)\right]\right\} \tag{187}$$

where the values of k_{mn_s} are solutions of

$$\frac{\partial}{\partial\rho} J_{n_s}(k_{mn_s}\rho)\bigg|_{\rho=a} = 0 \tag{188}$$

and $n_s = n\pi/\alpha$, $l = \pi/\alpha$, $\Delta_i = W_i/a$, with W_i being the curvilinear width of the port measured along the circumference (see Fig. 80).

When some of the ports are located along the radial edges (see Fig. 80), three other **Z** matrix elements are needed. These elements are

1. Z_{rt}, when ports r and t are located along the same radial edge of the sector
2. Z_{ir}, when the i port is located along a radial edge and the r port is located along the circumferential edge
3. Z_{rp}, when port r is located along one radial edge and port p is located along the second radial edge.

When ports r and t are located along the same radial edge at $\phi = \alpha$, the self- and transfer impedance elements are determined from

$$Z_{rt} = \frac{2j\omega\mu ld}{\pi W_r W_t} \sum_{n=0}^{\infty} \sum_{m=1}^{\infty} \frac{\sigma_n I(r)I(t)}{k_{mn_s}^2(a^2 - n_s^2/k_{mn_s}^2)(k_{mn_s}^2 - k^2)J_{n_s}^2(k_{mn_s}a)} \tag{189}$$

where

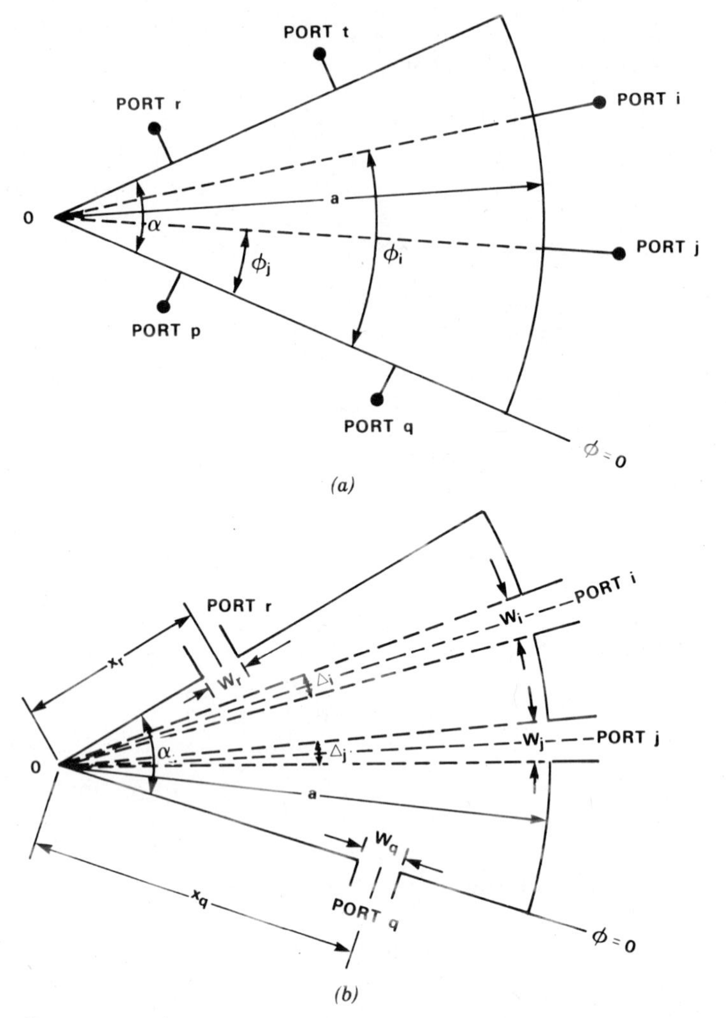

Fig. 80 Geometry and nomenclature for circular sectorial segments with radial and circumferential ports.

$$I(r) = \begin{cases} \displaystyle\int_{t_1}^{t_2} J_0(t)\, dt - 2 \sum_{k=0}^{\frac{n_s-2}{2}} \{J_{2k+1}(t_2) - J_{2k+1}(t_1)\} \\ \text{for even } n_s \\ \\ \displaystyle J_0(t_1) - J_0(t_2) + 2 \sum_{k=1}^{\frac{n_s-1}{2}} \{J_{2k}(t_1) - J_{2k}(t_2)\} \\ \text{for odd } n_s \end{cases} \tag{190}$$

with

$$t_1 = k_{mn_s}(x_r - \Delta_r/2) \tag{191}$$

and

$$t_2 = k_{mn_s}(x_r + \Delta_r/2) \tag{192}$$

The term x_r in (191) and (192) represents the radial distance between port r and the origin as illustrated in Fig. 80. In addition, Δ_r represents the linear width of port r measured along the radial edge of the sector. Finally, the integral appearing in (190) cannot be evaluated in closed form and thus will have to be numerically computed. The term $I(t)$ in (189) is obtained by replacing r with t in (191) and (192). When ports r and p are located at the two different radial edges of the sector (i.e., at $\phi = \alpha$ and $\phi = 0$, respectively), the transfer impedance element Z_{rp} is given by

$$Z_{rp} = \frac{2j\omega\mu l d}{\pi W_r W_p} \sum_{n=0}^{\infty} \sum_{m=1}^{\infty} \frac{\sigma_n \cos(n_s \alpha) I(r) I(p)}{k_{mn_s}^2 (a^2 - n_s^2/k_{mn_s}^2)(k_{mn_s}^2 - k^2) J_{n_s}^2(k_{mn_s} a)} \tag{193}$$

Similarly, when port i is located along the radial edge at $\phi = \alpha$, and port r is located on the circumferential edge at $\rho = a$, the transfer impedance element Z_{ir} is given by

$$Z_{ir} = \frac{4j\omega\mu l d a}{\pi W_i W_r} \sum_{n=0}^{\infty} \sum_{m=1}^{\infty} \frac{\sigma_n \cos n_s \alpha \cos n_s \phi_i \sin(n_s \Delta_i/2)\, I(r)}{n_s k_{mn_s} (a^2 - n_s^2/k_{mn_s}^2)(k_{mn_s}^2 - k^2) J_{n_s}(k_{mn_s} a)} \tag{194}$$

When some ports are located along the radial edge at $\phi = 0$, the corresponding self- and transfer impedance elements of the **Z** matrix are determined from (189). Similarly, the transfer impedance between a port i that is located along the circumferential edge at $\rho = a$ and a second port p that is located at the radial edge at $\phi = 0$ is determined from (194) by setting $\alpha = 0$.

A.4. *Z* Matrix for an Annular Ring Planar Segment

The configuration of the circular ring planar segment is shown in Fig. 4*g*. When all the ports are located on the outer circumference ($\rho = b$), the self- and transfer impedance elements of the **Z** matrix are determined from

$$Z_{ij} = \frac{2j\omega\mu d b^2}{\pi W_i W_j} \sum_{n=0}^{\infty} \sum_{m=1}^{\infty} \frac{\sigma_n \cos(n\phi_{ij}) \left\{ \cos\left[\frac{n}{2}(\Delta_i - \Delta_j)\right] - \cos\left[\frac{n}{2}(\Delta_i + \Delta_j)\right] \right\}}{n^2 [(b^2 - n^2/k_{mn}^2) - (a^2 - n^2/k_{mn}^2) A_{mn}^2](k_{mn}^2 - k^2)} \tag{195}$$

where

$$A_{mn} = F_{mn}(a)/F_{mn}(b) \tag{196}$$

$$F_{mn}(\rho) = N_n'(k_{mn}a)J_n(k_{mn}\rho) - J_n'(k_{mn}a)N_{mn}(k_{mn}\rho) \tag{197}$$

and the k_{mn}s are solutions of

$$J_n'(k_{mn}a)N_n'(k_{mn}b) - J_n'(k_{mn}b)N_n'(k_{mn}a) = 0 \tag{198}$$

where the prime (′) indicates the derivative of the function. In addition, the term ϕ_{ij} denotes the angular spacing between ports i and j, W_i represents the curvilinear width of port i measured along the outer circumference, and Δ_i denotes the angular width of port i.

When all the ports are located on the inner circumference (i.e., $\rho = a$), the impedance matrix is determined by

$$Z_{ij} = \frac{2j\omega\mu da^2}{\pi W_i W_j} \sum_{n=0}^{\infty} \sum_{m=1}^{\infty} \frac{\sigma_n \cos(n\phi_{ij})\left\{\cos\left[\frac{n}{2}(\Delta_i - \Delta_j)\right] - \cos\left[\frac{n}{2}(\Delta_i + \Delta_j)\right]\right\}}{n^2[(b^2 - n^2/k_{mn}^2)B_{mn}^2 - (a^2 - n^2/k_{mn}^2)](k_{mn}^2 - k^2)} \tag{199}$$

where $B_{mn} = 1/A_{mn}$. When some of the ports are located along the edge of the inner circumference ($\rho = a$) and the others are located along the edge of the outer circumference ($\rho = b$), the expression of the impedance element Z_{ij} reduces to

$$Z_{ij} = \frac{2j\omega\mu dab}{\pi W_i W_j} \sum_{n=0}^{\infty} \sum_{m=1}^{\infty} \frac{\sigma_n \cos(n\phi_{ij})\left\{\cos\left[\frac{n}{2}(\Delta_i - \Delta_j)\right] - \cos\left[\frac{n}{2}(\Delta_i + \Delta_j)\right]\right\}}{n^2[(b^2 - n^2/k_{mn}^2)B_{mn} - (a^2 - n^2/k_{mn}^2)A_{mn}](k_{mn}^2 - k^2)} \tag{200}$$

where $\Delta_i = W_i/a$, $\Delta_j = W_j/b$, and W_i and W_j are the curvilinear widths of ports i and j, respectively, measured along the ring circumference.

A.5. *Z* Matrix for Annular Sectorial Segments

The impedance matrix elements for annular sectorial segments (with sector angle $\alpha = \pi/l$, with l an integer) are derived from the Green's function given by eq. (59). As in the previous case, the ports may be located on the outer circumference, on the inner circumference, or on both. Figure 81 describes the geometry and the nomenclature used.

The self- and transfer impedance terms for ports located along the outer circumference $\rho = b$ are given by

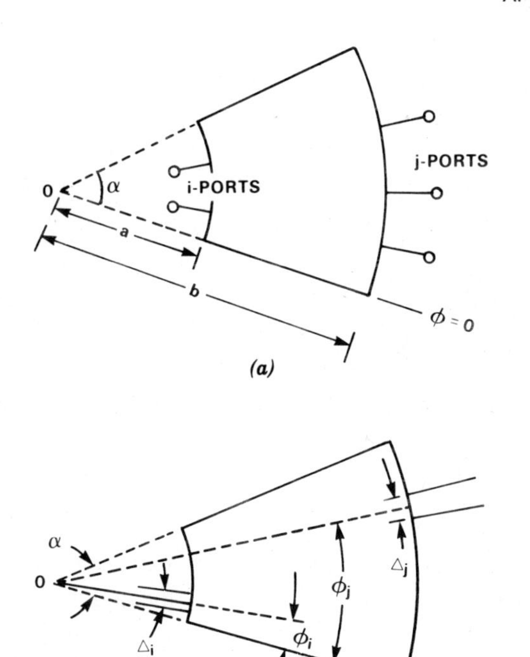

Fig. 81 Geometry and nomenclature for annular sectorial segments with circumferential ports.

$$Z_{ij} = \frac{2j\omega\mu\, db^2}{\alpha W_i W_j} \sum_{n=0}^{\infty} \sum_{m=1}^{\infty} \frac{\sigma_n \{\cos n_s \phi_{ij} + \cos[n_s(\phi_i + \phi_j)]\}}{n_s^2[(b^2 - n_s^2/k_{mn_s}^2) - (a^2 - n_s^2/k_{mn_s}^2)A_{mn_s}^2](k_{mn}^2 - k^2)} \cdot \left\{\cos\left[\frac{n_s}{2}(\Delta_i - \Delta_j)\right] - \cos\left[\frac{n_s}{2}(\Delta_i + \Delta_j)\right]\right\} \tag{201}$$

where

$$A_{mn_s} = F_{mn_s}(a)/F_{mn_s}(b) \tag{202}$$

$$F_{mn_s}(\rho) = N'_{n_s}(k_{mn_s}a)J_{n_s}(k_{mn_s}\rho) - J'_{n_s}(k_{mn_s}a)N_{n_s}(k_{mn_s}\rho) \tag{203}$$

Furthermore, the values of the wavenumber k_{mn_s} are solutions of

$$J'_{n_s}(k_{mn_s}a)N'_{n_s}(k_{mn_s}b) - J'_{n_s}(k_{mn_s}b)N'_{n_s}(k_{mn_s}a) = 0 \tag{204}$$

and $n_s = n\pi/\alpha$, $\Delta_{i,j} = W_{i,j}/b$, where $W_{i,j}$ represents the curvilinear widths of the ports measured along the outer circumference and ϕ_{ij} denotes the angular distance between ports i and j.

When some ports are located along the inner circumference $\rho = a$, the corresponding self- and transfer impedance elements of the **Z** matrix are determined from

$$Z_{ij} = \frac{2j\omega\mu da^2}{\alpha W_i W_j} \sum_{n=0}^{\infty} \sum_{m=1}^{\infty} \frac{\sigma_n\{\cos n_s\phi_{ij} + \cos[n_s(\phi_i + \phi_j)]\}}{n_s^2[(b^2 - n_s^2/k_{mn_s}^2)B_{mn_s}^2 - (a^2 - n_s^2/k_{mn_s}^2)](k_{mn_s}^2 - k^2)} \cdot \left\{\cos\left[\frac{n_s}{2}(\Delta_i - \Delta_j)\right] - \cos\left[\frac{n_s}{2}(\Delta_i + \Delta_j)\right]\right\} \tag{205}$$

where $B_{mn_s} = 1/A_{mn_s}$. When some of the ports are located at the edge of the inner circumference ($\rho = a$) and the others are located along the edge of the outer circumference ($\rho = b$), all the elements of the impedance matrix for this planar segment can be determined from (201), (205), and

$$Z_{ij} = \frac{2j\omega\mu dab}{\alpha W_i W_j} \sum_{n=0}^{\infty} \sum_{m=1}^{\infty} \frac{\sigma_n\{\cos n_s\phi_{ij} + \cos[n_s(\phi_i + \phi_j)]\}}{n_s^2[(b^2 - n_s^2/k_{mn_s}^2)B_{mn_s} - (a^2 - n_s^2/k_{mn_s}^2)A_{mn_s}](k_{mn_s}^2 - k^2)} \cdot \left\{\cos\left[\frac{n_s}{2}(\Delta_i - \Delta_j) - \cos\left[\frac{n_s}{2}(\Delta_i + \Delta_j)\right]\right]\right\} \tag{206}$$

where in (206) W_i represents ports located along the inner circumference ($\rho = a$) and W_j represents ports located along the outer circumference ($\rho = b$).

A.6. *Z* Matrices for Triangular Segments

As in other cases, **Z** matrices for triangular shapes are obtained by integrating the corresponding Green's functions given in Section 4.2. The results are summarized in this section of the Appendix.

A.6.1. *Z* Matrix for Equilateral Triangular Segments

In this case a general expression for the impedance matrix elements Z_{ij} can be written as

$$Z_{ij} = \frac{4j\omega\mu d}{W_i W_j} \sum_{m=-\infty}^{\infty} \sum_{n=-\infty}^{\infty} \frac{I_{T_1}(i)I_{T_1}(j) + I_{T_2}(i)I_{T_2}(j)}{16\sqrt{3}\pi^2(m^2 + mn + n^2) - 9\sqrt{3}a^2k^2} \tag{207}$$

where $I_{T_1}(i)$ and $I_{T_1}(j)$ are the integrals of the eigenfunction $T_1(x, y)$ [given by eq. (46)] over the widths of ports i and j, respectively. The values of these integrals depend on the location of the ports (x_i, y_i). When the coordinate system shown in Fig. 82 is used, the following expressions apply.

For ports along the side AB:

$$\begin{aligned} I_{T_1}(i)/W_i = (-1)^l \cos[\beta(m-n)] \operatorname{sinc}[\alpha(m-n)] \\ + (-1)^m \cos[\beta(n-l)] \operatorname{sinc}[\alpha(n-l)] \\ + (-1)^n \cos[\beta(l-m)] \operatorname{sinc}[\alpha(l-m)] \end{aligned} \tag{208}$$

where $\alpha = \pi W/3a$, and $\beta = 2\pi s/3$, where s is between $-\frac{1}{2}$ and $\frac{1}{2}$ and its value determines the location of the port as shown in Fig. 82.

For ports along the side AC:

$$\begin{aligned} I_{T_1}(i)/W_i = \cos[\beta(m-n)] \operatorname{sinc}[\alpha(m-n)] \\ + \cos[\beta(n-l)] \operatorname{sinc}[\alpha(n-l)] \\ + \cos[\beta(l-m)] \operatorname{sinc}[\alpha(l-m)] \end{aligned} \tag{209}$$

where α and β are the same as in (208) and $0 < s < 1$.

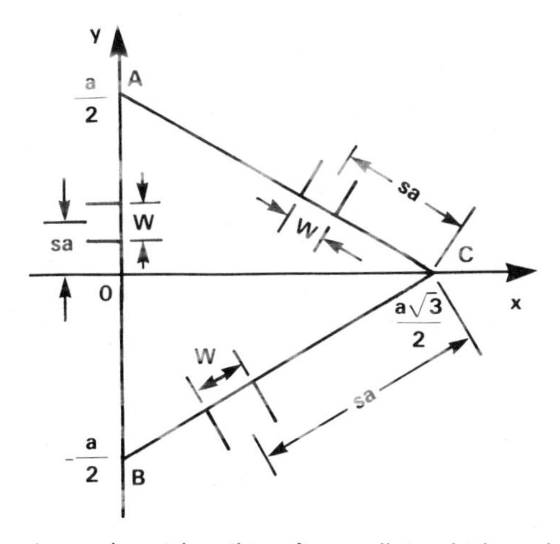

Fig. 82 Geometry and port locations for equilateral triangular segments.

For ports along the side BC, the expression for the integral is the same as that for ports along the side AC. The normalized integrals $I_{T_2}(i)$ and $I_{T_2}(j)$ for the eigenfunction $T_2(x, y)$ are given below.

For ports along the side AB:

$$\begin{aligned} I_{T_2}(i)/W_i = (-1)^l \sin[\beta(m-n)] \operatorname{sinc}[\alpha(m-n)] \\ + (-1)^m \sin[\beta(n-l)] \operatorname{sinc}[\alpha(n-l)] \\ + (-1)^n \sin[\beta(l-m)] \operatorname{sinc}[\alpha(l-m)] \end{aligned} \tag{210}$$

where α and β are the same as before and $-\frac{1}{2} < s < \frac{1}{2}$.

For ports along the side AC:

$$\begin{aligned} I_{T_2}(i)/W_i = -\sin[\beta(m-n)] \operatorname{sinc}[\alpha(m-n)] \\ - \sin[\beta(n-l)] \operatorname{sinc}[\alpha(n-l)] \\ - \sin[\beta(l-m)] \operatorname{sinc}[\alpha(l-m)] \end{aligned} \tag{211}$$

where s is between 0 and 1.

For ports along side BC, the expression for the integral is the negative of that for the ports along AC.

A.6.2. Z Matrix for Right-Angled Isosceles Triangular Segment

For a right-angled isosceles triangle, the Green's function is given by eq. (50) and the elements Z_{ij} of the impedance matrix are given by

$$Z_{ij} = \frac{j\omega\mu d}{2W_iW_j} \sum_{m=0}^{\infty} \sum_{n=0}^{\infty} \frac{\sigma_m \sigma_n I_T(i) I_T(j)}{(m^2 + n^2)\pi^2 - a^2k^2} \tag{212}$$

where $I_T(i)$ and $I_T(j)$ are integrals of the function $T(x, y)$ [given by (51)] over the widths of the respective ports. The values of these integrals depend on the locations of the ports. Referring to Fig. 83, the following expressions apply.

For ports along the side OA,

$$\frac{I_T(i)}{W_i} = \cos n\pi s \operatorname{sinc} \frac{n\pi W}{2a} + (-1)^{m+n} \cos m\pi s \operatorname{sinc} \frac{m\pi W}{2a} \tag{213}$$

For ports along the side OB,

$$\frac{I_T(i)}{W_i} = \cos m\pi s \operatorname{sinc} \frac{m\pi W}{2a} + (-1)^{m+n} \cos n\pi s \operatorname{sinc} \frac{n\pi W}{2a} \tag{214}$$

For ports along side AB,

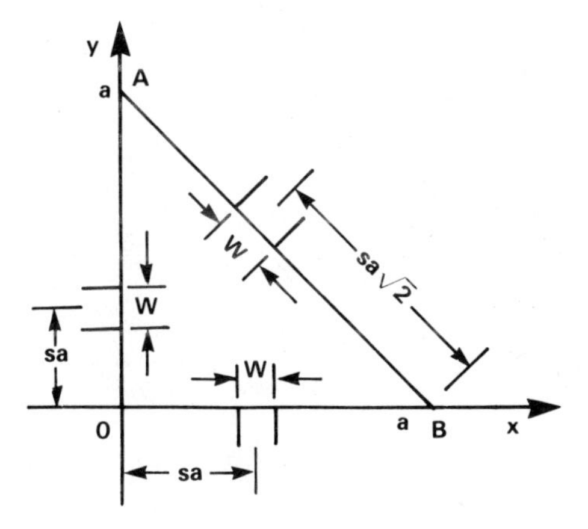

Fig. 83 Geometry and port locations for right-angled isosceles triangular segments.

$$\frac{I_T(i)}{W_i} = (-1)^m \{\cos[(m+n)\pi s] \operatorname{sinc}[(m+n)\alpha] + \cos[(m-n)\pi s] \operatorname{sinc}[(m-n)\alpha]\} \qquad (215)$$

where $\alpha = \pi W / 2\sqrt{2}a$ and $0 < s < 1$.

A.6.3. Z Matrix for 90°–60°–30° Triangular Segment

In this case, the Green's function is given by eq. (45), and the element Z_{ij} of the impedance matrix is given by

$$Z_{ij} = \frac{8j\omega\mu d}{W_i W_j} \sum_{m=-\infty}^{\infty} \sum_{n=-\infty}^{\infty} \frac{I_{T_1}(i) I_{T_1}(j)}{16\sqrt{3}\pi^2(m^2+mn+n^2) - 9\sqrt{3}a^2k^2} \qquad (216)$$

where $I_{T_1}(i)$ is the integral of the function $T_1(x, y)$ given by (46) over the width of port i. When the ports are located along the sides OA and AC (Fig. 84), the integrals of $T_1(x, y)$ are identical to those given earlier for the equilateral triangle case. For ports located along side OC, the integral $I_{T_1}(i)$ is given by

$$\frac{I_{T_1}(i)}{W_i} = (-1)^l \cos \alpha l s \operatorname{sinc} \frac{\alpha l W}{2} + (-1)^m \cos \alpha m s \operatorname{sinc} \frac{\alpha m W}{2} + (-1)^n \cos \alpha n s \operatorname{sinc} \frac{\alpha n W}{2} \qquad (217)$$

where $\alpha = 2\pi/(\sqrt{3}a)$, $0 < s < 1$, and l, m, n are related by $l + m + n = 0$.

In order to keep this appendix to a reasonable length, this list includes only expressions for the impedance matrix element Z_{ij} for planar segments

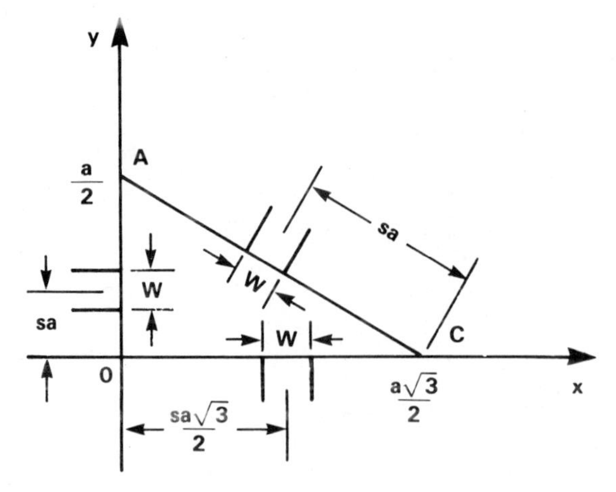

Fig. 84 Geometry and port locations for 30°–60° right-angled triangular segments.

with open-boundary conditions. Also, the results have been limited to the most frequently used port locations. Expressions for other port locations and/or other boundary conditions can be derived but are not given here.

APPENDIX B. CONSISTENCY OF THE MODE-MATCHING APPROACH WITH THE GREEN'S FUNCTION APPROACH

Although the Z matrix expressions derived from the Green's function approach (182) and from the mode-matching approach (162) are not identical in appearance, numerically they are consistent. The Z_{ij} expressions obtained by the Green's function (resonant mode) approach and by the mode-matching approach are given, respectively, by

$$Z_{ij} = \frac{2j\omega\mu da^2}{\pi W_i W_j} \sum_{n=0}^{\infty} \sum_{m=1}^{\infty} \frac{\sigma_n (1 - \cos n\Delta_i) \cos n\phi_{ij}}{n^2 (a^2 - n^2/k_{mn}^2)(k_{mn}^2 - k^2)} \tag{218}$$

and

$$Z_{ij} = \frac{j\eta d}{\pi W_j} \sum_{n=0}^{\infty} \frac{\sigma_n \sin^2(n\psi_i)\, J_n(k_e a) \cos[n(\phi_i - \phi_j)]}{n^2 \psi_i J_n'(k_e a)} \tag{219}$$

with

$$\Delta_i = 2\psi_i = 2 \sin^{-1} \frac{W_i'}{2a} = \frac{W_i}{a} \tag{220}$$

$$\phi_{ij} = \phi_i - \phi_j \tag{221}$$

$$\eta = 120\pi\sqrt{\frac{\mu_r}{\varepsilon_r}} \tag{222}$$

$$k = k_e = \omega\sqrt{\mu\varepsilon_r\varepsilon_0} \tag{223}$$

and where a stands for the radius of the microstrip disk, W_i is the effective width of the coupling port, and W_i' is the effective width of the microstrip lines connected to the coupling ports. Upon equation (218) with (219), using (220)–(223), and noting that

$$\sqrt{\frac{\mu}{\varepsilon_r}} = \frac{\omega\mu}{k_e} \tag{224}$$

$$2\sin^2 n\psi = 1 - \cos 2n\psi \tag{225}$$

we find that (218) and (219) are equal provided that

$$\sum_{m=1}^{\infty} \frac{2}{(a^2 - n^2/k_{mn}^2)(k_{mn}^2 - k^2)} = \frac{1}{k_e a}\frac{J_n(k_e a)}{J_n'(k_e a)} \tag{226}$$

Since $k_e a$ represents an arbitrary argument and $k_{mn}a$ denotes the zeros of $J_n'(k_{mn}a)$, we can rewrite (226) as

$$\frac{J_n(x)}{J_n'(x)} = \sum_{m=1}^{\infty} \frac{2x}{(1 - n^2/j_{mn}'^2)(j_{mn}'^2 - x^2)} \tag{227}$$

where we have set

$$x = k_e a \tag{228}$$

$$j_{mn}' = k_{mn}a \tag{229}$$

In order to prove that (227) is true, we can use Mittag–Leffler's theorem [42]. According to this theorem a function $f(z)$ that is analytic everywhere in the finite **Z** plane except for simple poles at $z = a_1, a_2, \ldots, a_n$ with residues $b_1, b_2, \ldots, b_n$, respectively, and is finite as $|z|$ approaches infinity (except near poles) may be written as

$$f(z) = f(0) + \sum_{n=1}^{\infty} b_n\left(\frac{1}{z - a_n} + \frac{1}{a_n}\right) \tag{230}$$

where the summation extends over all the poles of $f(z)$. Upon setting

$$f(x) = J_n(x)/J_n'(x) \tag{231}$$

and recalling that $J_n'(x)$ is the derivative of Bessel's function that has simple roots at $x = \pm j_{mn}'$, we can use (230) to write

$$\frac{J_n(x)}{J'_n(x)} = \sum_{m=1}^{\infty} \left(\frac{b_n}{x - j'_{mn}} + \frac{b_n}{x + j'_{mn}} \right) \tag{232}$$

where b_n represents the residues of $f(x)$ at $\pm j'_{mn}$ and is given by

$$b_n = -\frac{1}{1 - n^2/j'^2_{mn}} \tag{233}$$

Substituting (233) into (232) gives

$$\frac{J_n(x)}{J'_n(x)} = \sum_{m=1}^{\infty} \frac{2x}{(1 - n^2/j'^2_{mn})(j'^2_{mn} - x^2)} \tag{234}$$

which is the relation we intended to prove.

REFERENCES

1. C. G. Montgomery, R. H. Dicke, and E.M. Purcell, *Pricniples of Microwave Circuits*, McGraw-Hill, New York, 1948.
2. N. Marcuvitz, *Waveguide Handbook*, McGraw-Hill, New York, 1951.
3. H. Bosma, "On stripline Y-circulation at UHF," *IEEE Trans. Microwave Theory Tech.*, MTT-12, pp. 61–72, Jan. 1964.
4. S. Mao, S. Jones, and G. D. Vendelin, "Millimeter-wave integrated circuits," *IEEE Trans. Electron Devices*, vol. ED-15, pp. 517–523, July 1968.
5. Y. Tajima and I. Kuru, "An integrated Gunn oscillator," *Rec. Prof. Groups, Inst. Electron. Commun. Eng. Jpn.*, Pap. MW70-9, June 26, 1970.
6. T. Okoshi, "Microwave planar circuits," *Rep. Tech. Group, Inst. Electron. Commun. Eng.* Jpn., Pap. MW68-69, Feb. 17, 1969 (in Japanese).
7. T. Okoshi, "Microwave planar circuits," *Rec. Jt. Nat. Conv. Four EE Inst.*, Pap. No. 1468, Mar. 1969 (in Japanese).
8. T. Okoshi, "Planar circuits," *J. Inst. Electron. Commun. Eng.* Jpn., vol. 52, pp. 1430–1433, 1969 (in Japanese).
9. T. Okoshi, M. Migitaka, and N. Miyazaki, "Gunn oscillator using planar circuit resonator," *Rec. Nat. Conv., Inst. Electron. Commun. Eng.* Jpn., Pap. No. 708, Nov. 1969 (in Japanese).
10. S. Ridella, "Analysis of three-layer distributed structures with N terminal pairs," *Proc. Int. Symp. Network Theory*, pp. 687–707, 1968.
11. B. Bianco and P. P. Civalleri, "Basic theory of three-layer N-ports," *Alta Freq.*, vol. 30, pp. 623–631, Aug. 1969.
12. T. Okoshi and T. Miyoshi, "The planar circuit—An approach to microwave integrated circuitry," *IEEE Trans. Microwave Theory Tech.*, vol. MTT-20, pp. 245–252. Apr. 1972.
13. K. C. Gupta, "Two-dimensional analysis of microstrip circuit and antennae," *J. Inst. Electron. Telecommun. Eng.* (*New Delhi*), vol. 28, pp. 346–364, July 1982.

14. R. Sorrentino, "Planar circuits, waveguide models and segmentation method," *IEEE Trans. Microwave Theory Tech.*, vol. MTT-33, pp. 1057–1066, Oct. 1985.
15. K. C. Gupta, R. Garg, and R. Chadha, *Computer-Aided Design of Microwave Circuits*, Artech House, Dedham, MA, 1981.
16. T. Okoshi, *Planar Circuits for Microwaves and Lightwaves*, Springer-Verlag, New York, 1985.
17. P. P. Civalleri and S. Ridella, "Impedance and admittance of distributed three-layer N-ports," *IEEE Trans. Circuit Theory*, vol. CT-17, pp. 392–398, Aug. 1970.
18. T. Okoshi, Y. Uehara, and T. Takeuchi, "The segmentation method—An approach to the analysis of microwave planar circuits," *IEEE Trans. Microwave Theory Tech.*, vol. MTT-24, pp. 662–668, Oct. 1976.
19. R. Chadha and K. C. Gupta, "Segmentation method using impedance matrices for analysis of planar microwave circuits," *IEEE Trans. Microwave Theory Tech.*, vol. MTT-29, pp. 71–74, Jan. 1981.
20. P. C. Sharma and K. C. Gupta, "Desegmentation method for analysis of two-dimensional microwave circuits," *IEEE Trans. Microwave Theory Tech.*, vol. MTT-29, pp. 1094–1098, Oct. 1981.
21. P. Silvester, "Finite element analysis of planar microwave networks," *IEEE Trans. Microwave Theory Tech.*, vol. MTT-21, pp. 104–108, Feb. 1973.
22. O. C. Zienkiewicz, *The Finite Elements Method*, McGraw-Hill, London, 1977.
23. T. Miyoshi, S. Yamaguchi, and S. Goto, "Ferrite planar circuits in microwave integrated circuits," *IEEE Trans. Microwave Theory Tech.*, vol. MTT-25, pp. 593–600, July 1977.
24. T. Miyoshi and S. Miyauchi, "The design of planar circulators for wide-band operation," *IEEE Trans. Microwave Theory Tech.*, vol. MTT-28, pp. 210–214, Mar. 1980.
25. J. Helszajn, *Non-Reciprocal Microwave Junctions and Circulators*, Wiley, New York, 1975.
26. P. M. Morse and H. Feshbach, *Methods of Theoretical Physics*, Chapter 7, McGraw Hill, New York, 1953.
27. T. Okoshi and T. Takeuchi, "Analysis of planar circuits by segmentation method," *Electron. Commun., Jpn.*, vol. 58-B, pp. 71–79, Aug. 1975.
28. R. Chadha and K. C. Gupta, "Green's functions for triangular segments in microwave planar circuits," *IEEE Trans. Microwave Theory Tech.*, vol. MTT-20, pp. 1139–1143, Oct. 1980.
29. T. Okoshi et al., "Planar 3-dB hybrid circuit," *Electron. Commun., Jpn.*, vol. 58-B, pp. 80–90, Aug. 1975.
30. R. Chadha and K. C. Gupta, "Green's functions for circular sectors, annular rings and annular sectors in planar microwave circuits," *IEEE Trans. Microwave Theory Tech.*, vol. MTT-29, pp. 68–71, Jan. 1981.
31. Eswarappa, "Study of mixed boundary sectorial microstrip antennas," M. Tech. Thesis, Dept. Electr. Eng., Indian Inst. Technol., Kanpur, India, Dec. 1982.
32. H. W. Sreekantaswamy, "Study of mixed boundary rectangular microstrip antennas," M. Tech. Thesis, Dep. Electr. Eng., Indian Inst. Technol., Kanpur, India, Dec. 1982.

33. T. Okoshi and S. Kitazawa, "Computer analysis of short-boundary planar circuits," *IEEE Trans. Microwave Theory Tech.*, vol. MTT-23, pp. 299–306, Mar. 1975.
34. P. C. Sharma and K. C. Gupta, "An alternative procedure for implementing desegmentation method," *IEEE Trans. Microwave Theory Tech.*, vol. MTT-32, pp. 1–4, Jan. 1984.
35. Y. S. Wu and F. J. Rosenbaum, "Wide-band operation of microstrip circulators," *IEEE Trans. Microwave Theory Tech.*, vol. MTT-22, pp. 849–856, Oct. 1974.
36. Y. Ayasli, "Analysis of wide-band stripline circulators by integral equation technique," *IEEE Trans. Microwave Theory Tech.*, vol. MTT-28, pp. 200–209, Mar. 1980.
37. M. E. Hines, "Reciprocal and non-reciprocal modes of propagation in ferrite stripline and microstrip devices," *IEEE Trans. Microwave Theory Tech.*, vol. MTT-19, pp. 442–451, May 1971.
38. P. deSantis and F. Pucci, "The edge-guided wave circulator," *IEEE Trans. Microwave Theory Tech.*, vol. MTT-23, pp. 516–519, June 1975.
39. M. V. Vamberskii, V. P. Usachov, and S. A. Shelukhin, "Nonreciprocal two-channel microwave dividers and adders," *Izv. Vyssh. Uchebn. Zaved., Radioelektron.*, vol. 27, pp. 14–19, 1984.
40. V. P. Usachov, B. A. Gapeev, and D. I. Ogloblin, "Nonreciprocal ferrite dividers-adders," *Tr. MVTU*, No. 397, pp. 55–62, 1983.
41. Y. T. Lo, D. D. Harrison, and H. R. Richards, *An Analysis of the Disk Microstrip Antenna*, Part II, Interim Rep., RADC-TR-79-132, University of Illinois, Urbana, May 1979.
42. E. T. Whittaker and G. N. Watson, *Modern Analysis*, Cambridge University Press, London and New York, 1962.
43. R. Chadha and K. C. Gupta, "Compensation of discontinuities in planar transmission lines," *IEEE Trans. Microwave Theory Tech.*, vol. MTT-30, pp. 2151–2156, Dec. 1982.
44. K. C. Gupta, "CAD oriented characterization of discontinuties in microstrip circuits," *1984 Nat. Radio Sci. (URSI) Meet.*, June 1984.
45. I. Ohta, T. Yamashita, and I. Hagino, "Optimum ports arrangement for a planar-circuit-type 3-dB hybrid," *Trans. Inst. Electron. Commun. Eng. Jpn.*, [*Part*] *E*, vol. 67, pp. 287–288, May 1984.
46. K. C. Gupta and M. D. Abouzahra, "Analysis and design of four port and five port microstrip disc circuits," *IEEE Trans. Microwave Theory Tech.*, vol. MTT-33, pp. 1422–1428, Dec. 1985.
47. A. Fathy and D. Kalokitis, "Analysis and design of a 30-way radial combiner for Ku-band applications," *RCA Rev.*, vol. 47, pp. 487–508, Dec. 1986.
48. M. D. Abouzahra and K. C. Gupta, "Multiple-port power divider/combiner circuits using circular microstrip disk configuration," *IEEE Trans.. Microwave Theory Tech.*, vol. MTT-35, pp. 1296–1302, Dec. 1987.
49. T. Okoshi, T. Imai, and K. Ito, "Computer-oriented synthesis of optimum circuit pattern of 3-dB hybrid ring by the planar circuit approach," *IEEE Trans. Microwave Tech.*, vol. MTT-29, pp. 194–202, Mar. 1981.

50. S. Hagelin and B. Carlegrim, "Planar multiport network with rotational symmetry," in *Advanced Electronic Warfare Technology* (J. Clarke, ed.), Microwave Exhibitions and Publishers Limited, Tunbridge Wells (U.K.), pp. 198–203, 1984.
51. S. J. Foti, R. P. Flam, and W. J. Scharpf, "60-way radial combiner uses no isolator," *Microwaves & RF*, vol. 23, pp. 96–118, July 1984.
52. B. Bianco and S. Ridella, "Nonconventional transmission zeros in distributed rectangular structures," *IEEE Trans. Microwave Theory Tech.*, vol. MTT-20, pp. 297–303, May 1972.
53. B. Bianco and S. Ridella, "Analysis of a three-layer rectangular structure," in *Network Theory* (R. Boite, ed.), pp. 215–227, Gordon & Breach, New York, 1972.
54. B. Bianco, M. Granara, and S. Ridella, "Filtering properties of two-dimensional lines discontinuities," *Alta Freq.*, vol. 42, pp. 140E–148E, June 1973.
55. G. D'Inzeo, G. Giannini, C. M. Sodi, and R. Sorrentino, "Method of analysis and filtering properties of microwave planar networks," *IEEE Trans. Microwave Theory Tech.*, vol. MTT-26, pp. 462–471, July 1978.
56. G. D'Inzeo, F. Giannini, P. Matlese, and R. Sorrentino, "On the double nature of transmission zeros in microstrip structures," *Proc. IEEE*, vol. 66, pp. 800–802, July 1978.
57. G. D'Inzeo, F. Giannini, and R. Sorrentino, "Novel microwave integrated lowpass filters," *Electron. Lett.*, vol. 15, pp. 258–260, Apr. 1979.
58. G. D'Inzeo, F. Giannini, and R. Sorrentino, "Wide-band equivalent circuits of microwave planar networks," *IEEE Trans. Microwave Theory Tech.*, vol. MTT-28, pp. 1107–1113, Oct. 1980.
59. G. D'Inzeo, F. Giannini, and R. Sorrentino, "Microwave planar networks: The annular structure," *Electron. Lett.*, vol. 14, pp. 526–528, Aug. 1978.
60. T. Miyoshi and T. Okoshi, "Analysis of Microwave Planar Circuits," *Electron. Commun. Jpn.*, vol. 55-B, no. 8, pp. 24–31, 1972.
61. G. D'Inzeo, F. Giannini, and R. Sorrentino, "Design of circular planar networks for bias filter elements in microwave integrated circuits," *Alta Freq.*, Vol. 48, pp. 251E–257E, July 1979.
62. J. B. Vinding, "Radial line stubs as elements in strip line circuits," *NEREM Rec.*, pp. 108–109, 1967.
63. M. De Lina Coimbra, "A new kind of radial stub and some applications," *Proc. 14th Eur. Microwave Conf.*, pp. 516–521, 1984.
64. H. A. Atwater, "The design of the radial line stub: A useful microstrip circuit element," *Microwave J.*, vol. 28, pp. 149–153, Nov. 1985.
65. S. L. March, "Analysing lossy radial-line stubs," *IEEE Trans. Microwave Theory Tech.*, vol. MTT-33, pp. 269–271, March 1985.
66. F. Giannini, R. Sorrentino, and J. Vrba, "Planar circuit analysis of microstrip radial stub," *IEEE Trans. Microwave Theory Tech.*, vol. MTT-32, pp. 1652–1655, Dec. 1984.
67. F. Giannini and C. Paoloni, "Broadband lumped equivalent circuit for shunt-connected radial stub," *Electron. Lett.*, vol. 22, pp. 485–487, Apr. 1986.

68. F. Giannini, M. Ruggieri, and J. Vrba, "Shunt-connected microstrip radial stubs," *IEEE Trans. Microwave Theory Tech.*, vol. MTT-34, pp. 363–366, Mar. 1986.
69. F. Giannini, C. Paoloni, and J. Vrba, "Losses in microstrip radial stubs," *Proc. 16th Eur. Microwave Conf.*, pp. 523–528, 1986.
70. F. Giannini, M. Salerno, and R. Sorrentino, "Two-octave stopband microstrip low-pass filter design," *Proc. 16th Eur. Microwave Conf.*, pp. 292–297, 1986.
71. J. Helszajn and D. S. James, "Planar triangular resonators with magnetic walls," *IEEE Trans. Microwave Theory Tech.*, vol. MTT-26, pp. 95–100, Feb. 1978.
72. G. Kumar and K. C. Gupta, "Gap coupled microstrip antennas," *Proc. Int. Symp.—Microwaves Commun. Kharagpur (India)*, pp. 12–15, Dec. 1981.
73. M. V. Vamberskii, V. P. Usachov, and S. A. Shelukhin, "Technical design of two-channel non-reciprocal microstrip line splitters-combiners," *Izvest. VU, Radioelectron.* [*Radioelectron. Commun. Syst.*], vol. 27, No. 12, pp. 17–20, Allerton Press, New York, 1984.
74. A. Gogoi and K. C. Gupta, "Wiener–Hopf computation of edge admittances for microstrip patch radiators," *AEU*, vol. 36, pp. 247–251, 1982.
75. A. Van de Capelle et al., "A simple accurate formula for the radiation conductance of a rectangular microstrip antenna," *IEEE AP-S Int. Symp. Antennas Propagat., Dig.*, pp. 23–26, 1981.
76. E. F. Kuester et al., "The thin-substrate approximation for reflection from the end of a slab-loaded parallel plate waveguide with application to microstrip patch antenna," *IEEE Trans. Antennas Propagat.*, vol. AP-30, pp. 910–917, 1982.
77. K. C. Gupta and P. C. Sharma, "Segmentation and desegmentation techniques for analysis of two-dimensional microstrip antennas," *IEEE AP-S Int. Antennas Propagat. Symp. Dig.*, pp. 19–22, 1981.
78. P. C. Sharma and K. C. Gupta, "Analysis and optimized design of single feed circularly polarized microstrip antennas," *IEEE Trans. Antennas Propagat.*, vol. AP-31, pp. 949–955, 1983.
79. G. Kumar and K. C. Gupta, "Broadband microstrip antennas using additional resonators gap-coupled to radiating edges," *IEEE Trans. Antennas Propagat.*, vol. AP-32, pp. 1375–1379, 1984.
80. G. Kumar and K. C. Gupta, "Non-radiating edges and four-edges gap-coupled multiple resonator, broadband microstrip antennas," *IEEE Trans. Antennas Propagat.*, vol. AP-33, pp. 173–178, 1985.
81. G. Kumar and K. C. Gupta, "Directly coupled multiple resonator wideband microstrip antennas," *IEEE Trans. Antennas Propagat.*, vol. AP-33, pp. 588–593, 1985.
82. K. C. Gupta, "Two-port transmission characteristics of rectangular microstrip patch radiators," *IEEE AP-S Int. Antennas Propagat. Symp. Dig.*, pp. 71–74, 1985.
83. K. C. Gupta and A. Benalla, "Two-port transmission characteristics of circular microstrip patch antennas," *IEEE AP-S Int. Symp. Antennas Propagat. Dig.*, pp, 821–824, 1986.

84. V. Palanisamy and R. Garg, "Analysis of arbitrary shaped microstrip patch antennas using segmentation technique and cavity model," *IEEE Trans. Antennas Propagat.*, vol. AP-34, pp. 1208–1213, 1986.
85. V. Palanisamy and R. Garg, "Analysis of circularly polarized square ring and crossed-strip microstrip antennas," *IEEE Trans. Antennas Propagat.*, vol. AP-34, pp. 1340–1346, 1986.
86. A. Benalla and K. C. Gupta, "A method for sensitivity analysis of series-fed arrays of rectangular microstrip patches," *Nat. Radio Sci. Meet. (URSI), Dig.*, p. 65, Jan. 1987.
87. K. C. Gupta, "Multiport network modeling approach for computer-aided design of microstrip patches and arrays," *IEEE AP-S Int. Symp. Antennas Propagat. Dig.*, vol. II, pp. 786–789, June 1987.
88. K. C. Gupta and A. Benalla, "Computer-aided design of microstrip patches and arrays," (invited paper), *1987 SBMO Int. Microwave Symp. Brazil*, vol. I, pp. 591–596, July 1987.
89. G. Kumar and K. C. Gupta, "Broadband microstrip antennas using coupled resonators," *IEEE AP-S Int. Antenna Propagat. Symp. Dig.*, pp. 67–70, 1983.
90. A. Benalla and K. C. Gupta, *Two-dimensional analysis of one-port and two-port microstrip antennas*, Electromagnetics Laboratory, Sci. Rep. 85, University of Colorado (Boulder), p. 48, May 1986.
91. A. Benalla and K. C. Gupta, "Multiport network model and transmission characteristics of two-port rectangular microstrip patch antennas," *IEEE Trans. Antennas Propagat.*, vol. AP-36, pp. 1337–1342, Oct. 1988.
92. A. Benalla, unpublished experimental results, 1986.
93. G. Kompa and R. Mehran, "Planar waveguide model for calculating microstrip components," *Electron. Lett.*, vol. 11, pp. 459–460, 1975.
94. A. Benalla and K. C. Gupta, "Faster computation of *Z*-matrices for rectangular segments in planar microwave circuits", *IEEE Trans. Microwave Theory Tech.*, vol. MTT-34, pp. 733–736, 1986.
95. J.-P. Hsu et al., "Proposal of surface-wave planar circuit, formulation of its planar circuit equations and its practical applications," *IEEE MTT-S Int. Microwave Symp. Dig.*, pp. 797–800, June 1986.
96. M. D. Abouzahra and K. C. Gupta, "Use of circular sector shaped planar networks for multiport power divider-combiner circuits," *IEEE Trans. Microwave Theory Tech.*, vol. MTT-36, pp. 1747–1751, Dec. 1988.

5

Spectral Domain Approach

Tomoki Uwano
Wireless Research Laboratory
Matsushita Electric Industrial Company
Osaka, Japan

Tatsuo Itoh
Department of Electrical and Computer Engineering
The University of Texas at Austin
Austin, Texas

1. INTRODUCTION

Planar transmission line analysis in the Fourier transform domain (or spectral domain) is superior to many numerical methods in the spatial domain. The analysis in the Fourier transform domain was first introduced by Yamashita and Mittra [1] for computation of the characteristic impedance and the phase velocity of a microstrip line based on a quasi-TEM approximation. A variational method has been used in the Fourier transform domain to calculate the line capacitance from the assumed charge density. This is a low-frequency approximation neglecting longitudinal electric and magnetic fields supported by the microstrip.

As the operating frequency is increased, dispersion characteristics of the microstrip become important for precise designs. This requirement has led to the full wave analysis of microstrip lines, represented by the work of Denlinger [2], who solved the integral equations using a Fourier transform technique. The solution by his method, however, strongly depends on the assumed current distributions on the strip in the process of solution. To avoid this difficulty and permit systematic improvement of the solution for the current components to a desired degree of accuracy, a new method was presented by Itoh and Mittra [3], now commonly called the spectral domain approach (SDA). In SDA, Galerkin's method is used to yield a homogeneous system of equations to determine the propagation constant and the amplitude of current distributions from which the characteristic impedance is derived.

In each of these methods the Fourier transform is taken along the direction parallel to the substrate and perpendicular to the strip. By virtue of the Fourier transform domain analysis and Galerkin's method, SDA has several features:

- Easy formulation in the form of a pair of algebraic equations
- Variational nature in determination of the propagation constant
- Identification of the physical nature of the mode for each solution corresponding to the basis functions

SDA is applicable to the following structures:

- Most planar transmission lines such as microstrips, finlines, and CPWs (coplanar waveguides) in multilayer configurations
- Both open and enclosed structures
- Slow-wave lines with lossy dielectric materials
- Resonators of planar configurations

The main reason the SDA is numerically efficient is that it requires a significant analytical preprocessing. This feature in turn imposes a certain restriction on the applicability of the method. One of the limitations is that SDA requires infinitesimal thickness for the strip conductor. It is also difficult to treat the structure with a strip having finite conductivity. No discontinuity in the substrate in the sideward direction is allowed. In spite of these limitations, however, SDA is one of the most popular and widely used numerical techniques.

In this chapter, two formulation methods are described: the general approach and the immittance approach. Although the general approach provides a better understanding of SDA and its formulation, it will be demonstrated that the immittance approach is much simpler for derivation of the formulations.

2. GENERAL APPROACH FOR SHIELDED MICROSTRIP LINES

To illustrate the formulation process, we will use a simple shielded microstrip line with its cross-sectional view given in Fig. 1. The structure has two dielectric layers so that the formulation process is generalized and the results are applicable to conventional finlines or CPSs with a slight modification of formulation as will be described later. The formulation for the open structure can be obtained in a similar manner.

Before the detailed formulation process is presented, let us compare the types of equations obtained by the SDA and those obtained by a typical space domain formulation. In conventional space domain analysis, the structure can be analyzed by first formulating the following coupled

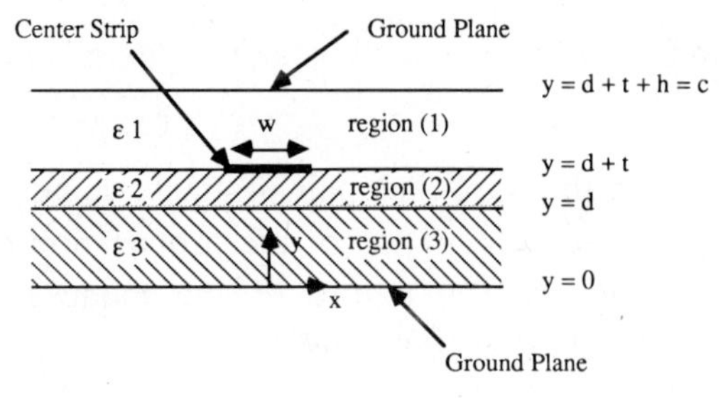

Fig. 1 Cross-sectional view of a shielded microstrip line.

homogeneous integral equations. The equations will then be solved for the unknown propagation constant β.

$$\int [Z_{zz}(x - x', y)J_z(x') + Z_{zx}(x - x', y)J_x(x')] \, dx' = E_z(x) \tag{1a}$$

$$\int [Z_{xz}(x - x', y)J_z(x') + Z_{xx}(x - x', y)J_x(x')] \, dx' = E_x(x) \tag{1b}$$

where E_z and E_x are unknown electric fields on the boundary at $y = d + t$, J_z and J_x are current components on the strip ($y = d + t$), and the Green's functions Z_{zz}, etc., are functions of β. The integration is over the strip where $E_z(x)$ and $E_x(x)$ are zero, as the strip is perfectly conducting. The left-hand sides of the equations are therefore required to be zero on the strip. These equations can be solved if Z_{zz}, etc., are given. The Green's functions Z_{zz}, etc., however, are not available in closed form for the inhomogeneous structures. As we will see shortly, the following algebraic equations, instead of the coupled integral equations, are obtained in the spectral domain formulation. These equations are Fourier transforms of the coupled integral equations.

$$\tilde{Z}_{zz}(\alpha, d + t)\tilde{J}_z(\alpha) + \tilde{Z}_{zx}(\alpha, d + t)\tilde{J}_x(\alpha) = \tilde{E}_z(\alpha, d + t) \tag{2a}$$

$$\tilde{Z}_{xz}(\alpha, d + t)\tilde{J}_z(\alpha) + \tilde{Z}_{xx}(\alpha, d + t)\tilde{J}_x(\alpha) = \tilde{E}_x(\alpha, d + t) \tag{2b}$$

where quantities with tildes ($\tilde{\ }$) are Fourier transforms of corresponding quantities. The Fourier transform is defined as

$$\tilde{\Phi}(\alpha) = \int_{-\infty}^{\infty} \Phi(x) \, e^{j\alpha x} \, dx \tag{3}$$

The right-hand side of eqs. (2) is no longer zero because the Fourier transform requires integration over all x, not only over the strip. The equations contain four unknowns J_z, J_x, E_z, E_x with unknown β. E_z and E_x, however, will be eliminated in the solution process based on the Galerkin procedure.

2.1. Field Equations

In this section, the Green's impedance functions Z_{zz}, Z_{zx}, Z_{xz}, Z_{xx} will be derived for the structure in Fig. 1. First, the hybrid fields are expressed in terms of superposition of TE-to-y and TM-to-y expressions [4] with scalar potentials $\tilde{\psi}^e$ and $\tilde{\psi}^h$ as follows:

$$
\begin{aligned}
\tilde{E}_x &= -j\frac{\alpha}{\hat{y}}\frac{\partial\tilde{\psi}^e}{\partial y} - j\beta\tilde{\psi}^h & \tilde{H}_x &= j\beta\tilde{\psi}^e - j\frac{\alpha}{\hat{z}}\frac{\partial\tilde{\psi}^h}{\partial y} \\
\tilde{E}_y &= \frac{1}{\hat{y}}\left(\frac{\partial^2}{\partial y^2} + k^2\right)\tilde{\psi}^e & \tilde{H}_y &= \frac{1}{\hat{z}}\left(\frac{\partial^2}{\partial y^2} + k^2\right)\tilde{\psi}^h \\
\tilde{E}_z &= -j\frac{\beta}{\hat{y}}\frac{\partial\tilde{\psi}^e}{\partial y} + j\alpha\tilde{\psi}^h & \tilde{H}_z &= -j\alpha\tilde{\psi}^e - j\frac{\beta}{\hat{z}}\frac{\partial\tilde{\psi}^h}{\partial y}
\end{aligned}
\tag{4}
$$

$$\hat{y} = j\omega\varepsilon \qquad \hat{z} = j\omega\mu \qquad k^2 = \omega^2\mu\varepsilon$$

where ε is permittivity, μ is permeability, the time convention $e^{j\omega t}$ is implied, and the z dependence $e^{-j\beta z}$ is assumed. Each field quantity in (4) is a Fourier transform of a corresponding quantity in the space domain. The Fourier-transformed Helmholtz equation is expressed as

$$\left(-\alpha^2 + \frac{\partial^2}{\partial y^2} - \beta^2\right)\tilde{\psi} + k^2\tilde{\psi} = 0 \tag{5}$$

The solution for this homogeneous differential equation is well known and can be described in the form of

$$\tilde{\psi} = c_1 \cosh \gamma y + c_2 \sinh \gamma y, \qquad \gamma^2 = \alpha^2 + \beta^2 - k^2 \tag{6}$$

with appropriate coefficients c_1 and c_2. When the wall boundary conditions at the top and the ground are applied, the scalar potentials in each region are given as follows:

Region 1:

$$\tilde{\psi}_1^e = A^e \cosh \gamma_1(c - y) \qquad \tilde{\psi}_1^h = A^h \sinh \gamma_1(c - y) \tag{7}$$

Region 2:

$$
\begin{aligned}
\tilde{\psi}_2^e &= B^e \sinh \gamma_2(y - d) + C^e \cosh \gamma_2(y - d) \\
\tilde{\psi}_2^h &= B^h \cosh \gamma_2(y - d) + C^h \sinh \gamma_2(y - d)
\end{aligned}
\tag{8}
$$

Region 3:

$$\tilde{\psi}_3^e = D^e \cosh \gamma_3 y \qquad \tilde{\psi}_3^h = D^h \sinh \gamma_3 y \tag{9}$$

where each subscript refers to the corresponding region and $A^e, A^h, \ldots, D^h$ are unknown coefficients. γ is the propagation constant in the y direction and may be written as $\gamma = jk_y$. In the case of an open structure, (7) may be replaced with

$$\tilde{\psi}_1^e = A^e e^{-\gamma_1 y} \qquad \tilde{\psi}_1^h = A^h e^{-\gamma_1 y} \tag{10}$$

From (5) and (6), dispersion relations are derived for each region:

$$k_i^2 = \alpha^2 + \beta^2 - \gamma_i^2, \qquad i = 1, 2, 3 \tag{11}$$

Substitution of (7)–(9) into (4) yields the field expressions in the three regions:

$$\begin{aligned}
\tilde{E}_{x1} &= j\alpha\gamma_{y1}A^e \sinh \gamma_1(c-y) - j\beta A^h \sinh \gamma_1(c-y) \\
\tilde{E}_{x2} &= -j\alpha\gamma_{y2}[B^e \cosh \gamma_2(y-d) + C^e \sinh \gamma_2(y-d)] \\
&\qquad -j\beta[B^h \cosh \gamma_2(y-d) + C^h \sinh \gamma_2(y-d)] \\
\tilde{E}_{x3} &= -j\alpha\gamma_{y3}D^e \sinh \gamma_3 y - j\beta D^h \sinh \gamma_3 y \\
\tilde{E}_{y1} &= \frac{1}{\hat{y}_1}[\gamma_1^2 + k_1^2]A^e \cosh \gamma_1(c-y) \\
\tilde{E}_{y2} &= \frac{1}{\hat{y}_2}[\gamma_2^2 + k_2^2][B^e \sinh \gamma_2(y-d) + C^e \cosh \gamma_1(y-d)] \\
\tilde{E}_{y3} &= \frac{1}{\hat{y}_3}[\gamma_3^2 + k_3^2]D^e \cosh \gamma_3 y \\
\tilde{E}_{z1} &= j\beta\gamma_{y1}A^e \sinh \gamma_1(c-y) + j\alpha A^h \sinh \gamma_1(c-y) \\
\tilde{E}_{z2} &= -j\beta\gamma_{y2}[B^e \cosh \gamma_2(y-d) + C^e \sinh \gamma_2(y-d)] \\
&\qquad + j\alpha[B^h \cosh \gamma_2(y-d) + C^h \sinh \gamma_2(y-d)] \\
\tilde{E}_{z3} &= -j\beta\gamma_{y3}D^e \sinh \gamma_3 y + j\alpha D^h \sinh \gamma_3 y \\
\tilde{H}_{x1} &= j\beta A^e \cosh \gamma_1(c-y) + j\alpha\gamma_{z1}A^h \cosh \gamma_1(c-y) \\
\tilde{H}_{x2} &= j\beta[B^e \sinh \gamma_2(y-d) + C^e \cosh \gamma_2(y-d)] \\
&\qquad - j\alpha\gamma_{z2}[B^h \sinh \gamma_2(y-d) + C^h \cosh \gamma_2(y-d)] \\
\tilde{H}_{x3} &= j\beta D^e \cosh \gamma_3 y - j\alpha\gamma_{z3}D^h \cosh \gamma_3 y \\
\tilde{H}_{y1} &= \frac{1}{\hat{z}_1}[\gamma_1^2 + k_1^2]A^h \sinh \gamma_1(c-y) \\
\tilde{H}_{y2} &= \frac{1}{\hat{z}_2}[\gamma_2^2 + k_2^2][B^h \cosh \gamma_2(y-d) + C^h \sinh \gamma_2(y-d)]
\end{aligned} \tag{12}$$

$$\tilde{H}_{y3} = \frac{1}{\hat{z}_3} [\gamma_3^2 + k_3^2] D^h \sinh \gamma_3 y$$

$$\tilde{H}_{z1} = -j\alpha A^e \cosh \gamma_1 (c - y) + j\beta \gamma_{z1} A^h \cosh \gamma_1 (c - y)$$

$$\tilde{H}_{z2} = -j\alpha [B^e \sinh \gamma_2 (y - d) + C^e \cosh \gamma_2 (y - d)]$$
$$- j\beta \gamma_{z2} [B^h \sinh \gamma_2 (y - d) + C^h \cosh \gamma_2 (y - d)]$$

$$\tilde{H}_{z3} = -j\alpha D^e \cosh \gamma_3 y - j\beta \gamma_{z3} D^h \cosh \gamma_3 y$$

$$\gamma_{yi} = \frac{\gamma_i}{y_i} \qquad \gamma_{zi} = \frac{\gamma_i}{z_i} \qquad i = 1, 2, 3$$

where each subscript refers to the corresponding region. The unknown coefficients A^e, A^h, ... D^h are eliminated by imposing the boundary conditions at each interface. The boundary conditions in the spectral domain are obtained as the Fourier transforms of those in the space domain. The latters are written as follows,

At $y = d + t$:

$$E_{x1} = E_{x2} \qquad \text{for all } x$$

$$E_{z1} = E_{z2} \qquad \text{for all } x$$

$$H_{x2} - H_{x1} = \begin{cases} J_z & |x| < w/2 \\ 0 & |x| > w/2 \end{cases}$$

$$H_{z2} - H_{z1} = \begin{cases} -J_x & |x| < w/2 \\ 0 & |x| > w/2 \end{cases}$$

At $y = d$:

$$E_{x2} = E_{x3} \qquad \text{for all } x$$

$$E_{z2} = E_{z3} \qquad \text{for all } x$$

$$H_{x2} = H_{x3} \qquad \text{for all } x$$

$$H_{z2} = H_{z3} \qquad \text{for all } x$$

where $J_x(x)$ and $J_z(x)$ are the unknown current distributions on the strip at $y = d + t$. Notice these quantities need to be introduced so that the boundary conditions are specified for the entire range of x. Otherwise, it is not possible to take Fourier transforms. In the spectral domain, the boundary conditions are now given by the following equations.

At $y = d + t$:

$$\tilde{E}_{x1} = \tilde{E}_{x2} \tag{13}$$

$$\tilde{E}_{z1} = \tilde{E}_{z2} \tag{14}$$

$$\tilde{H}_{x2} - \tilde{H}_{x1} = \tilde{J}_z \tag{15}$$

$$\tilde{H}_{z2} - \tilde{H}_{z1} = -\tilde{J}_x \tag{16}$$

At $y = d$:

$$\tilde{E}_{x2} = \tilde{E}_{x3} \tag{17}$$

$$\tilde{E}_{z2} = \tilde{E}_{z3} \tag{18}$$

$$\tilde{H}_{x3} - \tilde{H}_{x2} = 0 \tag{19}$$

$$\tilde{H}_{z3} - \tilde{H}_{z2} = 0 \tag{20}$$

where $\tilde{J}_z$ and $\tilde{J}_x$ are Fourier transforms of unknown current components $J_z(x)$ and $J_x(x)$ on the strip at $y = d + t$.

Finally, the algebraic equations are derived in matrix form as follows:

$$\begin{bmatrix} \tilde{E}_{z1} \\ \tilde{E}_{x1} \end{bmatrix} = \begin{bmatrix} \tilde{Z}_{zz} \tilde{Z}_{zx} \\ \tilde{Z}_{xz} \tilde{Z}_{xx} \end{bmatrix} \begin{bmatrix} \tilde{J}_z \\ \tilde{J}_x \end{bmatrix} \tag{21}$$

$$\tilde{Z}_{zz} = -\frac{1}{\alpha^2 + \beta^2} [\beta^2 \tilde{Z}_e + \alpha^2 \tilde{Z}_h]$$

$$\tilde{Z}_{zx} = -\frac{\alpha\beta}{\alpha^2 + \beta^2} [\tilde{Z}_e - \tilde{Z}_h]$$

$$\tilde{Z}_{xz} = \tilde{Z}_{zx}$$

$$\tilde{Z}_{xx} = -\frac{1}{\alpha^2 + \beta^2} [\alpha^2 \tilde{Z}_e + \beta^2 \tilde{Z}_h]$$

$$\tilde{Z}_e = \frac{\gamma_{y2}\mathrm{Ct}_3 + \gamma_{y3}\mathrm{Ct}_2}{\mathrm{Ct}_2\mathrm{Ct}_3 + \mathrm{Ct}_1\mathrm{Ct}_3\gamma_{y2}/\gamma_{y1} + \mathrm{Ct}_1\mathrm{Ct}_2\gamma_{y3}/\gamma_{y1} + \gamma_{y3}/\gamma_{y2}}$$

$$\tilde{Z}_h = \frac{\gamma_{z2}\mathrm{Ct}_2 + \gamma_{z3}\mathrm{Ct}_3}{\gamma_{z1}\gamma_{z2}\mathrm{Ct}_1\mathrm{Ct}_2 + \gamma_{z1}\gamma_{z3}\mathrm{Ct}_1\mathrm{Ct}_3 + \gamma_{z2}\gamma_{z3}\mathrm{Ct}_2\mathrm{Ct}_3 + \gamma_{z2}^2}$$

$$\mathrm{Ct}_1 = \coth \gamma_1 h \qquad \mathrm{Ct}_2 = \coth \gamma_2 t \qquad \mathrm{Ct}_3 = \coth \gamma_3 d$$

The derivation of these formulations is detailed in Appendix B. It should be noted that there is one more set of boundary conditions not used up to this stage. In the space domain, it is

$$E_z = E_x = 0 \qquad \text{for } |x| < w/2 \text{ at } y = d + t$$

This set of conditions is incorporated in the solution process as we will see below.

2.2. Method of Solution

In this section, a method for solving (21) is presented. Two equations in (21) contain four unknowns $\tilde{J}_z$, $\tilde{J}_x$, $\tilde{E}_z$, and $\tilde{E}_x$. The latter two unknowns $\tilde{E}_z$ and $\tilde{E}_x$, however, can be eliminated by applying Galerkin's method in the spectral domain. The first step is to expand the unknown $\tilde{J}_z$ and $\tilde{J}_x$ in terms of known basis functions $\tilde{J}_{zm}$ and $\tilde{J}_{xm}$

$$\tilde{J}_z = \sum_{m=1}^{N} c_m \tilde{J}_{zm}(\alpha) \qquad \tilde{J}_x = \sum_{m=1}^{M} d_m \tilde{J}_{xm}(\alpha) \tag{22}$$

where c_m and d_m are unknown coefficients. The basis functions must be chosen to approximate the true but unknown distributions on the strip. The current is nonzero only on the strip. Therefore, each basis function must also be chosen so that it is nonzero only on the strip. After substituting (22) into (21), one takes inner products of the resultant equations with the known basis function J_{zk}, J_{xl}, respectively, for different values of k and l. This process yields the matrix equation

$$\int_\alpha \left[\tilde{J}_{zk} \tilde{Z}_{zz} \sum_{m=1}^{N} c_m \tilde{J}_{zm} + \tilde{J}_{zk} \tilde{Z}_{zx} \sum_{m=1}^{M} d_m \tilde{J}_{xm} \right] d\alpha = 0, \qquad k = 1, 2, \ldots, N \tag{23a}$$

$$\int_\alpha \left[\tilde{J}_{xl} \tilde{Z}_{xz} \sum_{m=1}^{N} c_m \tilde{J}_{zm} + \tilde{J}_{xl} \tilde{Z}_{xx} \sum_{m=1}^{M} d_m \tilde{J}_{xm} \right] d\alpha = 0, \qquad l = 1, 2, \ldots, M \tag{23b}$$

The right-hand sides of (23) are zero by virtue of Parseval's theorem, because the currents $J_{zk}(x)$, $J_{xl}(x)$ and the field components $E_z(x, d+t)$, $E_x(x, d+t)$ are nonzero in the complementary regions of x. For instance, if the inner product of $\tilde{E}_{z1}$ on the left-hand side of (21) and $\tilde{J}_{zk}(\alpha)$ is taken, one obtains

$$\int_{-\infty}^{\infty} \tilde{J}_{zk}(\alpha) \tilde{E}_{z1}(\alpha)\, d\alpha = 2\pi \int_{-\infty}^{\infty} J_{zk}(x) E_{zl}(-x)\, dx = 0$$

In the above, $J_{zk}(x)$ is zero outside the strip and $E_{z1}(x)$ is zero on the strip. Therefore, the final boundary condition is now used. Equations (23) will be expressed in matrix form as follows,

$$\sum_{m=1}^{N} K_{km}^{(1,1)} c_m + \sum_{m=1}^{M} K_{km}^{(1,2)} d_m = 0, \qquad k = 1, 2, \ldots, N \tag{24a}$$

$$\sum_{m=1}^{N} K_{lm}^{(2,1)} c_m + \sum_{m=1}^{M} K_{lm}^{(2,2)} d_m = 0, \qquad l = 1, 2, \ldots, M \tag{24b}$$

where

$$K_{km}^{(1,1)} = \int_{-\infty}^{\infty} \tilde{J}_{zk}(\alpha)\tilde{Z}_{zz}(\alpha, \beta)\tilde{J}_{zm}(\alpha)\, d\alpha$$

$$K_{km}^{(1,2)} = \int_{-\infty}^{\infty} \tilde{J}_{zk}(\alpha)\tilde{Z}_{zx}(\alpha, \beta)\tilde{J}_{xm}(\alpha)\, d\alpha$$

$$K_{lm}^{(2,1)} = \int_{-\infty}^{\infty} \tilde{J}_{xl}(\alpha)\tilde{Z}_{xz}(\alpha, \beta)\tilde{J}_{zm}(\alpha)\, d\alpha$$

$$K_{lm}^{(2,2)} = \int_{-\infty}^{\infty} \tilde{J}_{xl}(\alpha)\tilde{Z}_{xx}(\alpha, \beta)\tilde{J}_{xm}(\alpha)\, d\alpha$$

The homogeneous system of equations is now obtained in terms of unknown coefficients c_m and d_m. In order that c_m and d_m have nontrivial solutions, the determinant of the matrix must be zero, and hence β will be determined at each frequency ω.

2.3. Propagation Constant β

Any kind of basis functions may be used as long as they are nonzero only on the strip. The efficiency and accuracy of this method, however, depend on the choice of basis functions due to the variational nature of the problem. Hence, the singular behavior of the magnetic field components normal to the stripline edge should be incorporated in the form of singular basis functions for J_z. Here, the following set of functions is employed:

$$J_{zn}(x) = \frac{\cos[2(n-1)\pi x/w]}{\sqrt{1-(2x/w)^2}}, \qquad n = 1, 2, \ldots \tag{25}$$

$$J_{xn}(x) = \frac{\sin[2n\pi x/w]}{\sqrt{1-(2x/w)^2}}, \qquad n = 1, 2, \ldots \tag{26}$$

Note that the definitions given above are only over the strip and the functions are zero elsewhere. The functions in (25) incorporate the correct edge singularity. The shapes of the first three functions are shown in Fig. 2. The Fourier transforms of (25) and (26) are

$$\tilde{J}_{zn}(\alpha) = \frac{\pi w}{4}\left[J_0\left(\left|\frac{w\alpha}{2} + (n-1)\pi\right|\right) + J_0\left(\left|\frac{w\alpha}{2} - (n-1)\pi\right|\right)\right] \tag{27}$$

$$\tilde{J}_{xn}(\alpha) = \frac{\pi w}{4j}\left[J_0\left(\left|\frac{w\alpha}{2} + n\pi\right|\right) - J_0\left(\left|\frac{w\alpha}{2} - n\pi\right|\right)\right] \tag{28}$$

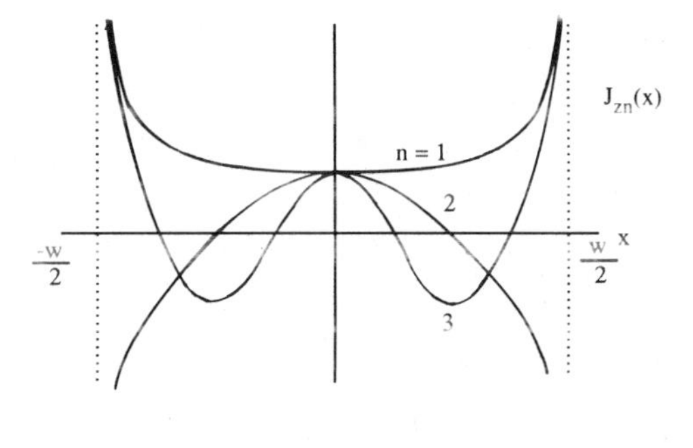

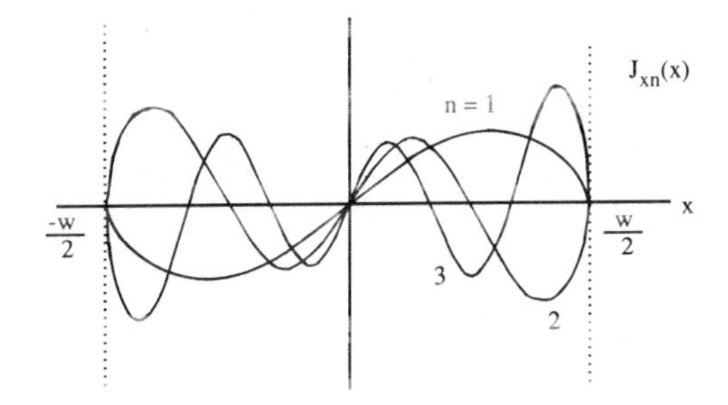

Fig. 2 Shapes of basis functions.

where J_0 denotes the zeroth-order Bessel function of the first kind. This choice is advantageous in the development of a computer program, because only one function J_0 is required as a subroutine for the current distributions.

Since the chosen basis functions approximate the current on the strip very well for conventional microstrips, only one or two basis functions are needed for β calculations. If one basis function is used for each current component, the matrix form of (24) with $N = M = 1$ becomes

$$\begin{bmatrix} \int_\alpha \tilde{J}_{z1}\tilde{Z}_{zz}\tilde{J}_{z1} & \int_\alpha \tilde{J}_{z1}\tilde{Z}_{zx}\tilde{J}_{x1} \\ \int_\alpha \tilde{J}_{x1}\tilde{Z}_{xz}\tilde{J}_{z1} & \int_\alpha \tilde{J}_{x1}\tilde{Z}_{xx}\tilde{J}_{x1} \end{bmatrix} \begin{bmatrix} c_1 \\ d_1 \end{bmatrix} = 0 \tag{29}$$

The determinant of the matrix is calculated with an assumed value of β. Then, by applying a root-seeking process, the true value of β is obtained. c_1 and d_1 are obtained as elements of an eigenvector for the true β value.

2.4. Characteristic Impedance

Because of the non-TEM nature, the definition of characteristic impedance is not unique. One possible choice of definition is

$$Z_0 = P/I_0^2$$

where P is the time-average Poynting power flow along the z axis and I_0 is the effective current flow along the z axis on the strip. P is given by

$$P = \mathrm{Re} \int_0^c \int_{-\infty}^{\infty} \mathbf{E} \times \mathbf{H}^* \cdot \hat{\mathbf{z}} \, dx \, dy \tag{30}$$

where $\hat{\mathbf{z}}$ is the unit vector in the z direction and $\mathbf{E} \times \mathbf{H}^*$ is a time-averaged quantity. By applying Parseval's theorem to the above equation, the integral part of (30) is expressed as

$$\begin{aligned}\int_s \mathbf{E} \times \mathbf{H}^* \cdot \hat{\mathbf{z}} \, dx \, dy &= \frac{1}{2\pi} \int_{-\infty}^{\infty} \int_0^c \tilde{\mathbf{E}} \times \tilde{\mathbf{H}}^* \cdot \hat{\mathbf{z}} \, dy \, d\alpha \\ &= \frac{1}{2\pi} \int_{-\infty}^{\infty} [E_{h1} + E_{h2} + E_{h3}] \, d\alpha \end{aligned} \tag{31}$$

where

$$E_{h1} = \int_{d+t}^{c} (\tilde{E}_{x1} \tilde{H}_{y1}^* - \tilde{E}_{y1} \tilde{H}_{x1}^*) \, dy$$

$$E_{h2} = \int_{d}^{d+t} (\tilde{E}_{x2} \tilde{H}_{y2}^* - \tilde{E}_{y2} \tilde{H}_{x2}^*) \, dy$$

$$E_{h3} = \int_{0}^{d} (\tilde{E}_{x3} \tilde{H}_{y3}^* - \tilde{E}_{y3} \tilde{H}_{x3}^*) \, dy$$

E and H field components in (31) are derived from (12) because A^e, $A^h, \ldots, D^h$ are already known as the functions of $\tilde{J}_z$ and $\tilde{J}_x$ after β is obtained. As the y dependence of the expressions in each equation is simple, the integration along y may be taken analytically. The results are as follows.

$$E_{h1} = \pm \frac{1}{2} A_1 A_4^* \left(\frac{S_1 C_1}{\gamma_1} - h \right) - \frac{1}{2} A_2 A_3^* \left(\frac{S_1 C_1}{\gamma_1} + h \right) \tag{32}$$

$$\begin{aligned} E_{h2} = &\pm \frac{1}{2} (C_1 C_4^* - B_2 B_3^*) \left(\frac{S_2 C_2}{\gamma_2} - t \right) + \frac{1}{2} (B_1 B_4^* - C_2 C_3^*) \left(\frac{S_2 C_2}{\gamma_2} + t \right) \\ &+ \frac{1}{2} [C_1 B_4^* - B_2 B_3^* \pm (B_1 C_4^* - C_2 B_3^*)] \frac{S_2^2}{\gamma_2} \end{aligned} \tag{33}$$

$$E_{h3} = \pm \frac{1}{2} D_1 D_4^* \left(\frac{S_3 C_3}{\gamma_3} - d \right) - \frac{1}{2} D_2 D_3^* \left(\frac{S_3 C_3}{\gamma_3} + d \right) \tag{34}$$

where

$$C_1 = \cosh \gamma_1 h \qquad C_2 = \cosh \gamma_2 t \qquad C_3 = \cosh \gamma_3 d$$

$$S_1 = \sinh \gamma_1 h \qquad C_2 = \sinh \gamma_2 t \qquad S_3 = \sinh \gamma_3 d$$

and $\pm$ denotes $+$ if γ_i is real and $-$ if γ_i is imaginary. The coefficients A_1, $A_2, \ldots, D_4$ are detailed in Appendix C. The current I_0 is obtained analytically by integrating $J_z(x)$ over the strip.

3. THE IMMITTANCE APPROACH

A formulation process called the immittance approach [5], which enables an easy solution for multilayer structures by decoupling the TE and the TM components, will be discussed in this section. In the general approach of the previous section, we derived the field formulation by solving eight unknown coefficients from eight coupled equations for a two-dielectric-layer structure. In the immittance approach, field formulation for β calculation is bypassed and a direct formulation of the eigenvalue equation is possible without knowledge of the field coefficient.

3.1. Illustration of the Formulation Process

The basic concept can be understood if we observed the inverse Fourier transform for the field:

$$\Phi(x, y)\, e^{-j\beta z} = \frac{1}{2\pi} \int_{-\infty}^{\infty} \tilde{\Phi}(\alpha, y)\, e^{-j(\alpha x + \beta z)}\, d\alpha$$

From this expression, we recognize that all the field components are a superposition of inhomogeneous (in y) plane waves that are propagating in the direction of θ from the z axis as illustrated in Fig. 3, where $\theta =$

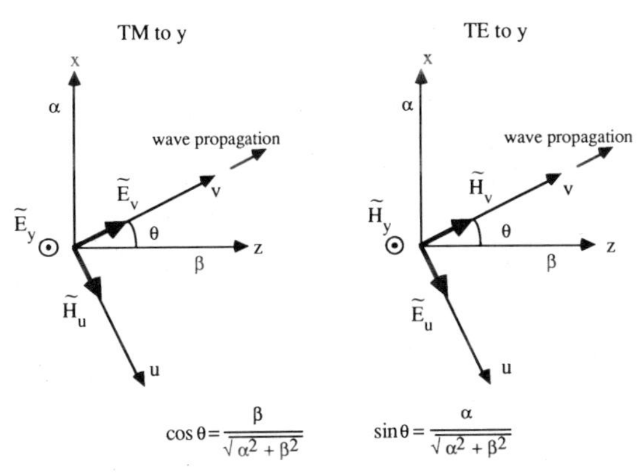

Fig. 3 Decomposition of spectral waves in the (u, v) coordinate system.

$\cos^{-1}[\beta/(\alpha^2+\beta^2)^{1/2}]$. For each θ, waves may be decomposed into TM-to-y $(\tilde{E}_y, \tilde{E}_v, \tilde{H}_u)$ and TE-to-y $(\tilde{H}_y, \tilde{E}_u, \tilde{H}_v)$ in the (u, v) coordinate system. Coordinate relations between (u, v) and (x, z) are

$$u = z\sin\theta - x\cos\theta \qquad v = z\cos\theta + x\sin\theta \tag{35}$$

We recognize that the current $\tilde{J}_v$ creates only the TM fields because it is concerned with $\tilde{H}_u$, and likewise $\tilde{J}_u$ creates the TE fields. Hence, equivalent circuits for the TM and TE fields are obtained from the transmission line analogy as shown in Fig. 4. In Fig. 4, all dielectric layers are represented as region 2 for simplicity of explanation. For the TM-to-y equivalent circuit, circuit equations are

$$\tilde{H}_{u1} - \tilde{H}_{u2} = \tilde{J}_v \qquad \frac{-\tilde{H}_{u1}}{\tilde{E}_{v1}} = Y_1^e \qquad \frac{\tilde{H}_{u2}}{\tilde{E}_{v2}} = Y_2^e$$

Thus

$$\tilde{E}_v = -\tilde{Z}_e \tilde{J}_v \qquad \tilde{E}_{v1} = \tilde{E}_{v2} = \tilde{E}_v \tag{36}$$

where

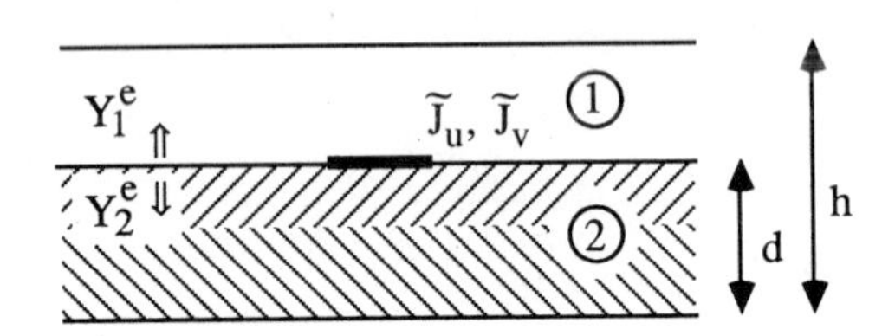

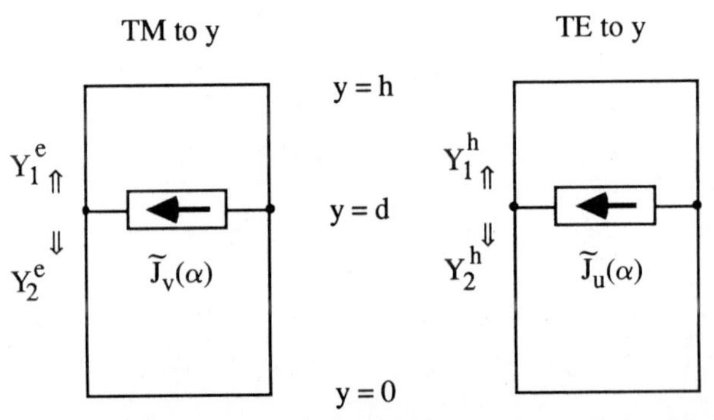

Fig. 4 Equivalent transmission lines for the TM and the TE fields.

$$\tilde{Z}_e = \frac{1}{Y_1^e + Y_2^e} \tag{37}$$

where Y_1^e and Y_2^e are the wave admittances looking from the location of the current, and these are found from the equivalent transmission line technique. Similarly, the circuit equations for the TE fields are

$$\tilde{E}_u = -\tilde{Z}_h \tilde{J}_u \qquad \tilde{E}_{u1} = \tilde{E}_{u2} = \tilde{E}_u \tag{38}$$

where

$$\tilde{Z}_h = \frac{1}{Y_1^h + Y_2^h} \tag{39}$$

where Y_1^h and Y_2^h denote input wave admittances for the TE wave. Thus, $\tilde{Z}_e$ and $\tilde{Z}_h$ may be defined as Green's functions for the TM and the TE waves, respectively.

Once $\tilde{Z}_e$ and $\tilde{Z}_h$ are known, the field formulation is obtained by mapping from the (u, v) to the (x, z) coordinate system for the spectral wave corresponding to each θ given by α and β. Because of the coordinate transform in (35), $\tilde{E}_x$ and $\tilde{E}_z$ are linear combinations of $\tilde{E}_u$ and $\tilde{E}_v$, and likewise $\tilde{J}_x$ and $\tilde{J}_z$ are those of $\tilde{J}_u$ and $\tilde{J}_v$. The results are as follows:

$$\tilde{Z}_{zz} = -\frac{1}{\alpha^2 + \beta^2}(\beta^2 \tilde{Z}_e + \alpha^2 \tilde{Z}_h) \tag{40a}$$

$$\tilde{Z}_{zx} = -\frac{\alpha\beta}{\alpha^2 + \beta^2}(\tilde{Z}_e - \tilde{Z}_h) \tag{40b}$$

$$\tilde{Z}_{zz} = -\frac{1}{\alpha^2 + \beta^2}(\alpha^2 \tilde{Z}_e + \beta^2 \tilde{Z}_h) \tag{40c}$$

Now we derive $\tilde{Z}_e$ and $\tilde{Z}_h$ for the given structures. First, by conventional transmission line theory, we find the input impedance from

$$Z_{\text{in}} = Z_{0i} \frac{Z_L \coth \gamma_i h_i + Z_0}{Z_0 \coth \gamma_i h_i + Z_L} \tag{41}$$

where Z_L is a load impedance, h_i is the thickness of the layer, and Z_{0i} is the wave characteristic impedance defined as

$$Z_{0_{\text{TM}i}} = -\frac{\tilde{E}_v}{\tilde{H}_u} = \frac{\gamma_i}{j\omega\varepsilon_i} \quad (=\gamma_{yi}) \tag{42}$$

$$Z_{0_{\text{TE}i}} = \frac{\tilde{E}_u}{\tilde{H}_v} = \frac{j\omega\mu_i}{\gamma_i} \quad \left(= \frac{1}{\gamma_{zi}}\right) \tag{43}$$

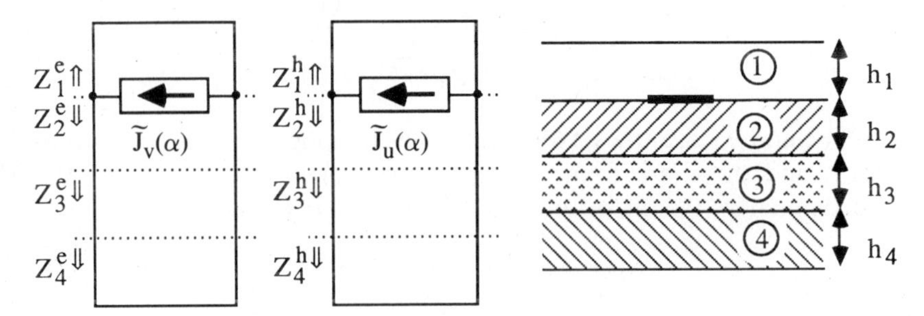

Fig. 5 Equivalent transmission lines for a three-layer microstrip line.

where i refers to the corresponding region. γ_i is obtained from (11). In the shielded structure shown in Fig. 4, the wave is all reflected with a 180° phase shift at the wall, and the wave impedance there is zero (short circuit). If an open structure is assumed, there is no reflection, and the wave impedance is indentical to the characteristic impedance. As for a multilayer structure, the impedance at each interface is obtained according to the relation in (41).

Thus, for the structure with a single-dielectric layer having the dimensions shown in Fig. 4, and similarly for two- and three-layer structures in Fig. 1 and Fig. 5, $\tilde{Z}_e$ and $\tilde{Z}_h$ are obtained as follows:

Single-layer Substrate:

$$\tilde{Z}_e = \frac{\gamma_{y1}\gamma_{y2}}{\gamma_{y1}\mathrm{Ct}_2 + \gamma_{y2}\mathrm{Ct}_1} \qquad \tilde{Z}_h = \frac{1}{\gamma_{z1}\mathrm{Ct}_1 + \gamma_{z2}\mathrm{Ct}_2} \tag{44}$$

$$\mathrm{Ct}_1 = \coth \gamma_1(h-d) \qquad \mathrm{Ct}_2 = \coth \gamma_2\, d$$

Two-layer Substrate:

$$\tilde{Z}_e = \frac{\gamma_{y2}\mathrm{Ct}_3 + \gamma_{y3}\mathrm{Ct}_2}{\mathrm{Ct}_2\mathrm{Ct}_3 + \mathrm{Ct}_1\mathrm{Ct}_3\gamma_{y2}/\gamma_{y1} + \mathrm{Ct}_1\mathrm{Ct}_2\gamma_{y3}/\gamma_{y1} + \gamma_{y3}/\gamma_{y2}} \tag{45a}$$

$$\tilde{Z}_h = \frac{\gamma_{z2}\mathrm{Ct}_2 + \gamma_{z3}\mathrm{Ct}_3}{\gamma_{z1}\gamma_{z2}\mathrm{Ct}_1\mathrm{Ct}_2 + \gamma_{z1}\gamma_{z3}\mathrm{Ct}_1\mathrm{Ct}_3 + \gamma_{z2}\gamma_{z3}\mathrm{Ct}_2\mathrm{Ct}_3 + \gamma_{z2}^2} \tag{45b}$$

$$\mathrm{Ct}_1 = \coth \gamma_1 h \qquad \mathrm{Ct}_2 = \coth \gamma_2 t \qquad \mathrm{Ct}_3 = \coth \gamma_3\, d$$

Three-layer Substrate:

$$Z_4^e = \gamma_{y4}/\mathrm{Ct}_4 \qquad Z_4^h = 1/\gamma_{z4}\mathrm{Ct}_4$$

$$Z_3^e = \gamma_{y3}\frac{\gamma_{y4}\mathrm{Ct}_3 + \gamma_{y3}\mathrm{Ct}_4}{\gamma_{y3}\mathrm{Ct}_3\mathrm{Ct}_4 + \gamma_{y4}} \qquad Z_3^h = \frac{1}{\gamma_{z3}}\frac{\gamma_{z3}\mathrm{Ct}_3 + \gamma_{z4}\mathrm{Ct}_4}{\gamma_{z4}\mathrm{Ct}_3\mathrm{Ct}_4 + \gamma_{z3}}$$

$$Z_2^e = \gamma_{y2} \frac{Z_3^e \mathrm{Ct}_2 + \gamma_{y2}}{\gamma_{y2}\mathrm{Ct}_2 + Z_3^e} \qquad Z_2^h = \frac{1}{\gamma_{z2}} \frac{\gamma_{z2} Z_3^h \mathrm{Ct}_2 + 1}{\mathrm{Ct}_2 + \gamma_{z2} Z_3^h}$$

$$Z_1^e = \gamma_{y1}/\mathrm{Ct}_1 \qquad Z_1^h = 1/\gamma_{z1}\mathrm{Ct}_1$$

$$\tilde{Z}_e = \left(\frac{1}{Z_1^e} + \frac{1}{Z_2^e}\right)^{-1} \qquad \tilde{Z}_h = \left(\frac{1}{Z_1^h} + \frac{1}{Z_2^h}\right)^{-1} \tag{46}$$

$$\mathrm{Ct}_i = \coth \gamma_i h_i, \qquad i = 1, 2, 3, 4$$

Notice that the results in (45) are identical to those in (21).

3.2. Characteristic Impedance

In the immittance approach the fields are decomposed to the TM and the TE components. Here, we derive the unknown coefficients needed for calculations of the characteristic impedance for the three-layer structure in Fig. 5. First, we modify the field expressions in (4) with respect to $\tilde{E}_y$ and $\tilde{H}_y$:

$$\begin{array}{ll} \text{TM-to-}y & \text{TE-to-}y \\ \tilde{E}_x = -jR \dfrac{\partial \tilde{E}_y}{\partial y} & \tilde{E}_x = -j\hat{z}Q\tilde{H}_y \\ \tilde{E}_z = -jQ \dfrac{\partial \tilde{E}_y}{\partial y} & \tilde{E}_x = j\hat{z}R\tilde{H}_y \\ \tilde{H}_x = j\hat{y}Q\tilde{E}_y & \tilde{H}_x = -jR \dfrac{\partial \tilde{H}_y}{\partial y} \\ \tilde{H}_z = -j\hat{y}R\tilde{E}_y & \tilde{H}_z = -jQ \dfrac{\partial \tilde{H}_y}{\partial y} \end{array} \tag{47}$$

where

$$R = \frac{\alpha}{\alpha^2 + \beta^2} \qquad Q = \frac{\beta}{\alpha^2 + \beta^2}$$

The field potentials in each region may be expressed in terms of $\tilde{E}_y$ and $\tilde{H}_y$:

TM-to-y

REGION 4: $-j\hat{y}_4\tilde{E}_{y4} = A^e \cosh \gamma_4 y$

REGION 3: $-j\hat{y}_3\tilde{E}_{y3} = B^e \cosh \gamma_3(y - h_4) + C^e \sinh \gamma_3(y - h_4)$

$$\text{Region 2:}\quad -j\hat{y}_2\tilde{E}_{y2} = D^e \cosh \gamma_2(y - h_3 - h_4) + E^e \sinh \gamma_2(y - h_3 - h_4)$$

$$\text{Region 1:}\quad -j\hat{y}_1\tilde{E}_{y1} = F^e \cosh \gamma_1(c - y) \tag{48}$$

TE-to-y

$$\text{Region 4:}\quad -j\hat{z}_4\tilde{H}_{y4} = A^h \sinh \gamma_4 y$$

$$\text{Region 3:}\quad -j\hat{z}_3\tilde{H}_{y3} = B^h \sinh \gamma_3(y - h_4) + C^h \cosh \gamma_3(y - h_4) \tag{49}$$

$$\text{Region 2:}\quad -j\hat{z}_2\tilde{H}_{y2} = D^h \sinh \gamma_2(y - h_3 - h_4) + E^h \cosh \gamma_2(y - h_3 - h_4)$$

$$\text{Region 1:}\quad -j\hat{z}_1\tilde{H}_{y1} = F^h \sinh \gamma_1(c - y), \qquad c = h_1 + h_2 + h_3 + h_4$$

Those modifications lead to less complexity in the derivation of the coefficients.

Boundary conditions are given for TM-to-y and TE-to-y, respectively, as follows,

At $y = h_4$:

$$\tilde{E}_{t3} = \tilde{E}_{t4} \tag{50}$$

$$\tilde{H}_{t3} = \tilde{H}_{t4} \tag{51}$$

At $y = h_3 + h_4$:

$$\tilde{E}_{t2} = \tilde{E}_{t3} \tag{52}$$

$$\tilde{H}_{t2} = \tilde{H}_{t3} \tag{53}$$

At $y = h_2 + h_3 + h_4$:

$$\tilde{E}_{t1} = \tilde{E}_{t2} \tag{54}$$

$$\tilde{H}_{t1} - \tilde{H}_{t2} = \tilde{J}_t \times \hat{\mathbf{y}}$$

or

$$\tilde{H}_{v1} - \tilde{H}_{v2} = \tilde{J}_v \qquad \text{for TM-to-}y \tag{55}$$

$$\tilde{H}_{u1} - \tilde{H}_{u2} = -\tilde{J}_u \qquad \text{for TE-to-}y \tag{56}$$

where the subscript t refers to the transverse components (x and z), and $\hat{\mathbf{y}}$ is a y-directed unit vector. It is seen from (47) that the z and x components satisfy the same boundary conditions. From (35) and (47), boundary conditions in (55) and (56) are described as

$$-j\frac{1}{\sqrt{\alpha^2+\beta^2}}(\hat{y}_1\tilde{E}_{y1}-\hat{y}_2\tilde{E}_{y2})=\tilde{J}_v \tag{57}$$

$$-j\frac{1}{\sqrt{\alpha^2+\beta^2}}\left(\frac{\partial \tilde{H}_{y1}}{\partial y}-\frac{\partial \tilde{H}_{y2}}{\partial y}\right)=-\tilde{J}_u \tag{58}$$

Substituting (48) and (49) into those boundary conditions, we obtain two sets of six independent equations with six unknowns, respectively. This implies that the derivation is easier than that in the general approach irrespective of the number of layers. After the coefficients are obtained, the Poynting power flow can be calculated in the same manner as in the general approach; the results are given in Appendix D.

4. FORMULATIONS FOR SLOTLINES, FINLINES, AND COPLANAR WAVEGUIDES

At millimeter-wave frequencies, finlines and CPWs are commonly used in the configuration of integrated circuits. The formulation for microstrips in SDA is easily modified to those for finlines and CPWs as well as for slotlines.

4.1. Slotlines (Finlines)

The cross-section view of the shielded slotline structure is shown in Fig. 6. A conventional slotlines is obtained by removing the ground planes to infinity. Also, the finline is realized by placing two side walls to enclose the structure. At the interface between regions 1 and 2, there is an infinitesimally thin conductor in a position complementary to that of the stripline in Fig. 1. The field expressions and the boundary conditions are the same as those of the suspended stripline; that is, the coupled algebraic equations derived in the SDA for this structure are the same as (21). It is anticipated, however, that a large number of terms of basis functions would be needed for the current expansion because of the wide conductor area. Hence, it is better to apply the Galerkin procedure to the unknown E field over the slot. The field equations are then modified to

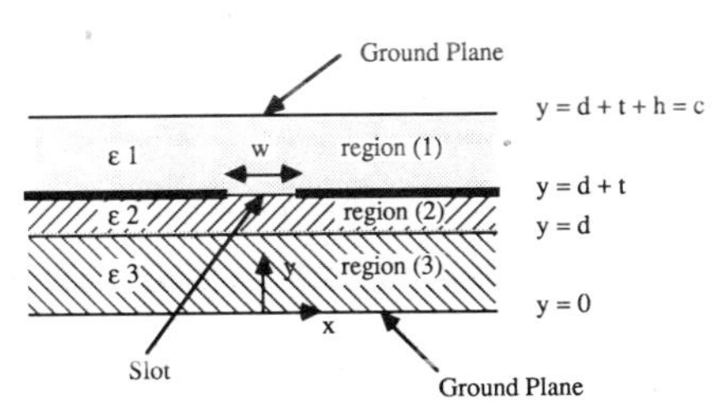

Fig. 6 Cross-sectional view of a shielded slotline.

$$\begin{bmatrix} \tilde{J}_x \\ \tilde{J}_z \end{bmatrix} = \begin{bmatrix} \tilde{Y}_{xx} & \tilde{Y}_{xz} \\ \tilde{Y}_{zx} & \tilde{Y}_{zz} \end{bmatrix} \begin{bmatrix} \tilde{E}_{x1} \\ \tilde{E}_{z1} \end{bmatrix} \tag{59}$$

where

$$\begin{bmatrix} \tilde{Y}_{xx} & \tilde{Y}_{xz} \\ \tilde{Y}_{zx} & \tilde{Y}_{zz} \end{bmatrix} = \begin{bmatrix} \tilde{Z}_{xx} & \tilde{Z}_{xz} \\ \tilde{Z}_{zx} & \tilde{Z}_{zz} \end{bmatrix}^{-1}$$

$$\tilde{Y}_{xx} = -\frac{1}{\alpha^2 + \beta^2}\left(\frac{\alpha^2}{\tilde{Z}_e} + \frac{\beta^2}{\tilde{Z}_h}\right)$$

$$\tilde{Y}_{xz} = -\frac{\alpha\beta}{\alpha^2 + \beta^2}\left(\frac{1}{\tilde{Z}_e} - \frac{1}{\tilde{Z}_h}\right)$$

$$\tilde{Y}_{zx} = \tilde{Y}_{xz}$$

$$\tilde{Y}_{zz} = -\frac{1}{\alpha^2 + \beta^2}\left(\frac{\beta^2}{\tilde{Z}_e} + \frac{\alpha^2}{\tilde{Z}_h}\right)$$

Due to the structural symmetry, an electric wall may be placed at the center of the gap for the dominant mode. *E*-field basis functions may be in the same forms as in (25) and (26) due to the duality:

$$E_{xn}(x) = \frac{\cos[2(n-1)\pi x/w]}{\sqrt{1-(2x/w)^2}}, \qquad n = 1, 2, \ldots \tag{60}$$

$$E_{zn}(x) = \frac{\sin[2n\pi x/w]}{\sqrt{1-(2x/w)^2}}, \qquad n = 1, 2, \ldots \tag{61}$$

where the functions are defined only over the slot and are zero outside. The Fourier transforms are also of the same forms as (27) and (28). Figure 7 shows the shapes of the first basis functions for each *E*-field component over the gap. When only the first basis functions for each *E*-field component are used, the Galerkin process yields the matrix equation as

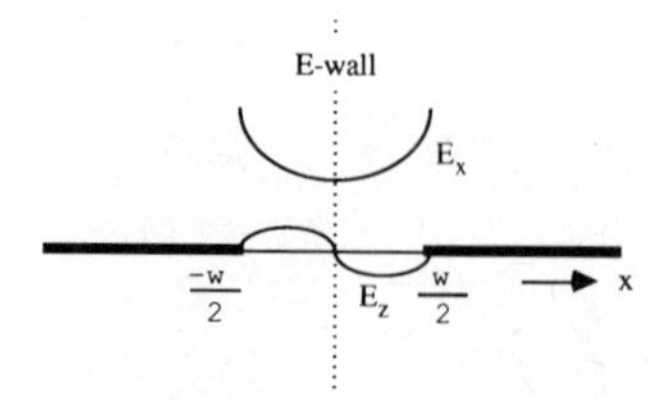

Fig. 7 Shapes of electric fields in the slot.

$$\begin{bmatrix} \int_\alpha \tilde{E}_{x1}\tilde{Y}_{xx}\tilde{E}_{x1} & \int_\alpha \tilde{E}_{x1}\tilde{Y}_{xz}\tilde{E}_{z1} \\ \int_\alpha \tilde{E}_{z1}\tilde{Y}_{zx}\tilde{E}_{x1} & \int_\alpha \tilde{E}_{z1}\tilde{Y}_{zz}\tilde{E}_{z1} \end{bmatrix} \begin{bmatrix} c_1 \\ d_1 \end{bmatrix} = 0 \tag{62}$$

β can be calculated by a root-seeking process of the characteristic equations obtained by equating the determinant to zero.

The characteristic impedance may be defined as

$$Z_0 = V_0^2/P$$

where V_0 is the effective voltage across the slot and is obtained from integration of the E_x component over the slot.

4.2. Coplanar Waveguides

The fundamental mode of a coplanar waveguide (CPW) is quasi-TEM at low frequency. The structure is symmetrical with a hypothetical magnetic wall placed at the center of the inner strip as shown in Fig. 8 for the dominant mode calculation. The structure is also regarded as the coupled slotline, and hence the E-field distribution over the slots may be expressed with the basis functions that have the same nature as that of the basis functions for the slotline. The field equations are the same as those of the slotline in (59). Since the fields over the slots are not symmetrical with respect to the center of each slot, odd basis functions for E_x field components and even basis functions for E_z field components are added. The results are as follows:

$$E_{xn}(x) = \begin{cases} \dfrac{\cos(n\pi x/w)}{\sqrt{1-(2x/w)^2}}, & n = 0, 2, \ldots \\ \dfrac{\sin(n\pi x/w)}{\sqrt{1-(2x/w)^2}}, & n = 1, 3, \ldots \end{cases} \tag{63}$$

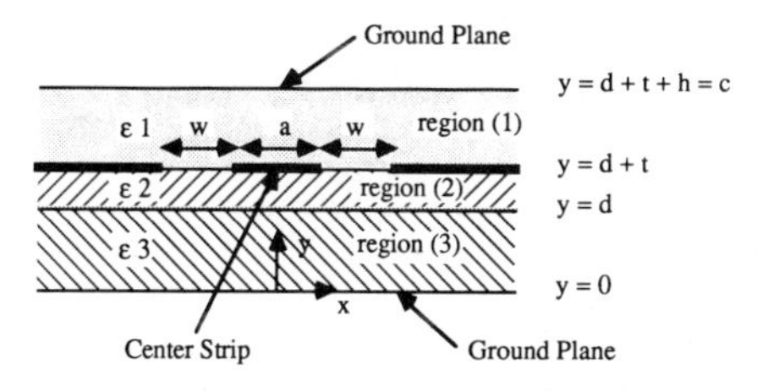

Fig. 8 Cross-sectional view of a shielded CPW.

$$E_{zn}(x) = \begin{cases} \dfrac{\cos(n\pi x/w)}{\sqrt{1-(2x/w)^2}}, & n = 1, 3, \ldots \\ \dfrac{\sin(n\pi x/w)}{\sqrt{1-(2x/w)^2}}, & n = 2, 4, \ldots \end{cases} \tag{64}$$

where the first basis functions are given with $n = 0$ for E_x and $n = 1$ for E_z. The entire expressions for the structure are

$$E_{xn}(x) = \begin{cases} \dfrac{\cos[n\pi(x+b)/w]}{\sqrt{1-[2(x+b)/w]^2}} - \dfrac{\cos[n\pi(x-b)/w]}{\sqrt{1-[2(x-b)w]^2}}, & n = 0, 2, \ldots \\ \text{for } x < 0 \qquad\qquad \text{for } x > 0 & \\ \dfrac{\sin[n\pi(x+b)/w]}{\sqrt{1-[2(x+b)/w]^2}} + \dfrac{\sin[n\pi(x-b)/w]}{\sqrt{1-[2(x-b)/w]^2}}, & n = 1, 3, \ldots \end{cases} \tag{65}$$

$$E_{xn}(x) = \begin{cases} \dfrac{\cos[n\pi(x+b)/w]}{\sqrt{1-[2(x+b)/w]^2}} + \dfrac{\cos[n\pi(x-b)/w]}{\sqrt{1-[2(x-b)/w]^2}} & n = 1, 3, \ldots \\ \text{for } x < 0 \qquad\qquad \text{for } x > 0 & \\ \dfrac{\sin[n\pi(x+b)/w]}{\sqrt{1-[2(x+b)/w]^2}} - \dfrac{\sin[n\pi(x-b)/w]}{\sqrt{1-[2(x-b)/w]^2}}, & n = 2, 4, \ldots \end{cases} \tag{66}$$

where the functions are defined only over the slots; the first terms are valid for $x < 0$ and the second terms are valid $x > 0$, and b is indicated in Fig. 9. The Fourier transforms for those functions are

$$\tilde{E}_{xn}(\alpha) = \begin{cases} \dfrac{-j\pi w}{2} \sin\alpha b\left[J_0\left(\dfrac{|w\alpha + n\pi|}{2}\right) + J_0\left(\dfrac{|w\alpha - n\pi|}{2}\right)\right], & n = 0, 2, \ldots \\ \dfrac{-j\pi w}{2} \cos\alpha b\left[J_0\left(\dfrac{|w\alpha + n\pi|}{2}\right) - J_0\left(\dfrac{|w\alpha - n\pi|}{2}\right)\right], & n = 1, 3, \ldots \end{cases} \tag{67}$$

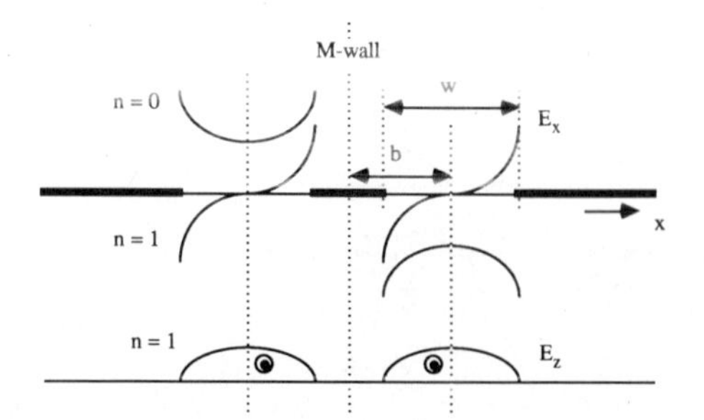

Fig. 9 Shapes of electric field basis functions for CPW.

$$\tilde{E}_{zn}(\alpha)=\begin{bmatrix} \dfrac{\pi w}{2}\cos\alpha b\left[J_0\left(\dfrac{|w\alpha+n\pi|}{2}\right)+J_0\left(\dfrac{|w\alpha-n\pi|}{2}\right)\right], & n=1,3,\ldots \\ \dfrac{-\pi w}{2}\sin\alpha b\left[J_0\left(\dfrac{|w\alpha+n\pi|}{2}\right)-J_0\left(\dfrac{|w\alpha-n\pi|}{2}\right)\right], & n=2,4,\ldots \end{bmatrix} \tag{68}$$

With the three basis functions as shown in Fig. 9, the matrix equation becomes

$$\begin{bmatrix} \int_\alpha \tilde{E}_{x1}\tilde{Y}_{xx}\tilde{E}_{x1} & \int_\alpha \tilde{E}_{x1}\tilde{Y}_{xx}\tilde{E}_{x2} & \int_\alpha \tilde{E}_{x1}\tilde{Y}_{xz}\tilde{E}_{z1} \\ \int_\alpha \tilde{E}_{x2}\tilde{Y}_{xx}\tilde{E}_{x1} & \int_\alpha \tilde{E}_{x2}\tilde{Y}_{xx}\tilde{E}_{x2} & \int_\alpha \tilde{E}_{x2}\tilde{Y}_{xz}\tilde{E}_{z1} \\ \int_\alpha \tilde{E}_{z1}\tilde{Y}_{zx}\tilde{E}_{x1} & \int_\alpha \tilde{E}_{z1}\tilde{Y}_{zx}\tilde{E}_{x2} & \int_\alpha \tilde{E}_{z1}\tilde{Y}_{zz}\tilde{E}_{z1} \end{bmatrix}\begin{bmatrix} c_1 \\ c_2 \\ d_1 \end{bmatrix}=0 \tag{69}$$

where $\tilde{E}_{x1}$ is for $n=0$, $\tilde{E}_{x2}$ is for $n=1$, and $\tilde{E}_{z1}$ is for $n=1$. β is obtained from the determinant equated to zero.

The characteristic impedance may be defined in the same way as that of the slotline.

5. NUMERICAL COMPUTATION

In actual computations, some precautions are needed. Some of them are described in this section.

5.1. Enclosed Structure

For the integration of the products in (24), it is possible that the Green's functions meet poles along the α integral path, though for the dominant microstrip mode this is not the case. The poles are encountered when the denominator of $\tilde{Z}_e$ or $\tilde{Z}_h$ is identical to zero, which gives the transverse resonant condition, namely the surface wave propagation constant of the value of $(\alpha^2+\beta^2)^{1/2}$. To avoid the poles, we may choose the integral path as illustrated in Fig. 10.

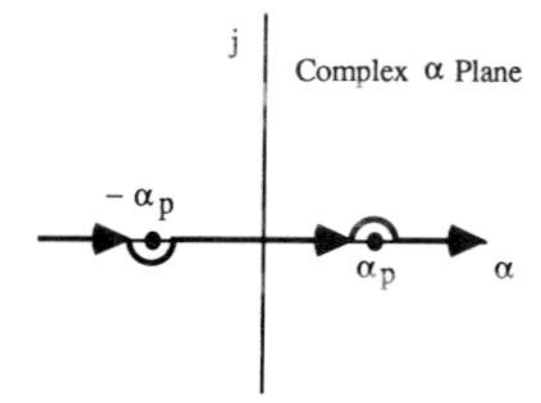

Fig. 10 Alpha integral path and poles.

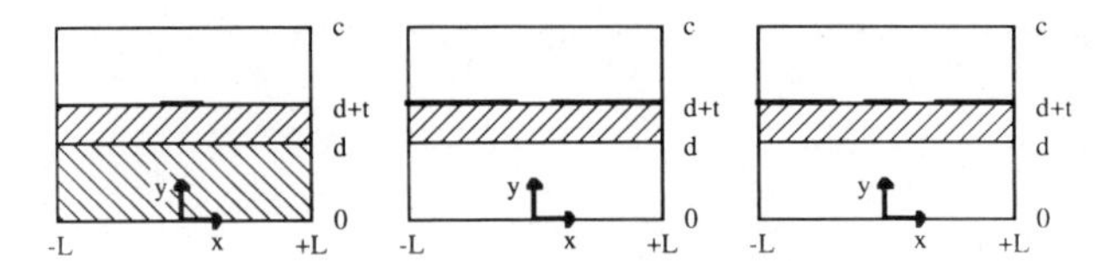

Fig. 11 Enclosed structures for each transmission line.

A convenient way to avoid poles is to use an enclosed structure with side walls as shown in Fig. 11. The advantage of enclosed structures is that chances of meeting the poles are very rare because α takes only discrete values. An open microstrip or a slotline may be simulated by choosing a large value of L and an appropriately large air region. For the structures with side walls, the Fourier integral in (3) is replaced by a discrete Fourier transform:

$$\tilde{\Phi}(\alpha) = \int_{-L}^{L} \Phi(x) \cdot e^{j\alpha x}\, dx \tag{70}$$

where

$$\alpha = \frac{n\pi}{L}, \qquad n = -\infty, \ldots, -1, 0, 1, 2, \ldots, +\infty$$

Accordingly, integrals with respect to α are all replaced by summations in terms of discrete values of $\alpha(=n\pi/L)$ from $n = -\infty$ to $+\infty$ as for eq. (23), (24), (29), (31), (62), and (69). Parseval's thorem has a different coefficient for the discrete Fourier transform, and (31) is replaced by

$$\int_s \mathbf{E} \times \mathbf{H}^* \cdot \hat{\mathbf{z}}\, dx\, dy = \frac{1}{2L} \sum_{n=-\infty}^{n=\infty} [E_{h1}(\alpha) + E_{h2}(\alpha) + E_{h3}(\alpha)]$$

From the theoretical view point, α might be chosen as $\alpha = n\pi/2L$ (n an odd number) for microstrip configurations [6], because only the fields with even E_y and odd H_y components are supported in the structure with a hypothetical magnetic wall at the center of the strip. In actual computations, however, the definition of α does not make much difference as a large number of Fourier terms are commonly used. Therefore, we may choose α as in (70) because of its availability for all structures in Fig. 11 or program writing convenience.

5.2. Some Computational Tips

5.2.1. The Nature of Functions

In Table 1, functions are itemized with respect to their even or odd symmetry. From these properties, all elements of the matrix equation are

Table 1 Function Itemization

Even	Odd
$\tilde{Z}_{zz}\tilde{Z}_{xx}$	$\tilde{Z}_{xz}\tilde{Z}_{zx}$
$\tilde{Y}_{xx}\tilde{Y}_{zz}$	$\tilde{Y}_{xz}\tilde{Y}_{zx}$
$\tilde{J}_{zn}$	$\tilde{J}_{xn}$
$\tilde{E}_{xn}$ (finline)	$\tilde{E}_{zn}$ (finline)
$\tilde{E}_{zn}$ (CPW)	$\tilde{E}_{xn}$ (CPW)

even functions of α in each case. Hence, the summation in terms of α $(=n\pi/L)$ is represented as

$$\sum_{n=-\infty}^{\infty} G(\alpha) = G(0) + 2\sum_{n=1}^{\infty} G(\alpha)$$

where $G(\alpha)$ is the product form of basis functions and Green's immittance function as shown in (24). Note that $G(0)$ is not necessarily zero.

5.2.2. Basis Functions

Unknown coefficients of the basis functions are obtained as eigenvectors of the matrix for the true β. The basis functions each may be multiplied by an arbitrary number. This does not influence the β calculation. Although c_m and d_m are then given different numbers, the characteristic impedance value remains the same, because the calculation uses the basis functions multiplied by the coefficients c_m and d_m. In the stripline case as in Section 2.3 with $N = M = 1$, the basis functions may be chosen so that the integral of J_z over the strip equals unity. For this choice, the basis function for J_{z1} is

$$J_{z1}(x) = \frac{2}{w\pi} \frac{1}{\sqrt{1-(2x/w)^2}} \tag{71}$$

The integral of (71) over the strip is 1.0. The Fourier transforms are therefore

$$\tilde{J}_{z1}(\alpha) = J_0(|w\alpha/2|) \tag{72}$$

$$\tilde{J}_{x1}(\alpha) = [J_0(|w\alpha/2 + \pi|) - J_0(|w\alpha/2 - \pi|)] \tag{73}$$

In the coplanar waveguide, the Fourier transforms for the basis functions may be modified to

$$\tilde{E}_{x1}(\alpha) = 2\sin\alpha b J_0(|w\alpha|/2), \qquad n = 0 \tag{74}$$

$$\tilde{E}_{x2}(\alpha) = \cos \alpha b \left[J_0\left(\frac{|w\alpha + \pi|}{2} \right) - J_0\left(\frac{|w\alpha - \pi|}{2} \right) \right], \qquad n = 1 \tag{75}$$

$$\tilde{E}_{z1}(\alpha) = \cos \alpha b \left[J_0\left(\frac{|w\alpha + \pi|}{2} \right) + J_0\left(\frac{|w\alpha - \pi|}{2} \right) \right], \qquad n = 1 \tag{76}$$

5.2.3. Finding c_m and d_m

Once the true β value is obtained, the matrix form remains as follows:

$$\begin{bmatrix} K_{11} & K_{12} & \cdots & K_{1N} \\ K_{21} & K_{22} & \cdots & K_{2N} \\ \vdots & \vdots & \ddots & \vdots \\ K_{N1} & K_{N2} & \cdots & K_{NN} \end{bmatrix} \begin{bmatrix} c_1 \\ c_2 \\ \vdots \\ d_1 \end{bmatrix} = 0 \tag{77}$$

where for generality, the matrix is $N \times N$ and the determinant of the matrix is zero. With $c_1 = 1.0$, (77) is modified as

$$N \overbrace{\begin{bmatrix} K_{12} & \cdots & K_{1N} \\ K_{22} & \cdots & K_{2N} \\ \vdots & \ddots & \vdots \\ K_{N2} & \cdots & K_{NN} \end{bmatrix}}^{N-1} \begin{bmatrix} c_2 \\ c_3 \\ \vdots \\ d_1 \end{bmatrix} = - \begin{bmatrix} K_{11} \\ K_{21} \\ \vdots \\ K_{N1} \end{bmatrix} \tag{78}$$

Since the matrix is not square, this may be solved by the least mean squares method.

5.2.4. Homogeneous Media

When the structure for a stripline or a coplanar waveguide is filled with homogeneous media, that is, $\varepsilon_1 = \varepsilon_2 = \varepsilon_3$, the dominant propagation mode is TEM, and $\beta = k$. In the calculation, (21) or (69) is not valid as $\gamma = 0$. This happens when $\alpha = 0$, $\beta = k$ from (11). Therefore, $\alpha = 0$ $(n = 0)$ must be omitted from the calculation. This makes sense because all Fourier-transformed entities for currents and fields should be zero for $\alpha = 0$ due to the nature of the function or the nature of the TEM wave. For the finline structure, β does not equal k except for an open slotline case. The chances that γ equals zero in the root-seeking process seem to be very rare or easily avoidable.

5.2.5. Lossy Media

If lossy dielectric media are chosen for certain layers, the slow-wave structure [7] can be analyzed. To this end, the usual convention is to let $\hat{y} = j\omega\varepsilon + \sigma$, where σ is conductivity of the media. Then the β-seeking process results in the complex β value, and accordingly the characteristic impedance shows the complex value.

A computer program for the three-layer structure with lossy media is included as Appendix E.

APPENDIX A. SOME MATHEMATICAL IDENTITIES

A Fourier transform is defined as

$$\Phi(\alpha) = \mathscr{F}\phi(x) = \int_{-\infty}^{\infty} \phi(x)\, e^{j\alpha x}\, dx$$

then

$$\mathscr{F}\{\phi(ax)\cos bx\} = \frac{1}{2a}\left[\Phi\left(\frac{\alpha+b}{a}\right) + \Phi\left(\frac{\alpha-b}{a}\right)\right]$$

$$\mathscr{F}\{\phi(ax)\sin bx\} = \frac{1}{2ja}\left[\Phi\left(\frac{\alpha+b}{a}\right) - \Phi\left(\frac{\alpha-b}{a}\right)\right]$$

$$\mathscr{F}\left\{\frac{\partial^n}{\partial x^n}\phi(x)\right\} = (-j\alpha)^n \Phi(\alpha)$$

Parseval's theorem states

$$\int_{\infty}^{\infty} \phi(x)\psi(-x)\, dx = \frac{1}{2\pi}\int_{-\infty}^{\infty} \Phi(\alpha)\Psi(\alpha)\, d\alpha\text{; Fourier integral}$$

$$\int_{-L}^{L} \phi(x)\psi(-x)\, dx = \frac{1}{2L}\sum_{n=-\infty}^{\infty} \Phi(\alpha)\Psi(\alpha)\text{; discrete Fourier transform}$$

where

$$\Phi(\alpha) = \int_{-L}^{L} \Phi(x)\cdot e^{j\alpha x}\, dx, \qquad \alpha = \frac{n\pi}{L}, \qquad n = 0, \pm 1, \pm 2, \ldots$$

Now

$$\int_{-a}^{a} \frac{1}{\sqrt{a^2 - x^2}}\, e^{j\alpha x}\, dx = \pi J_0(a|\alpha|),$$ J_0; zeroth-order Bessel function of the first kind

$$\int_{-a}^{a} \frac{1}{\pi}\frac{dx}{\sqrt{a^2 - x^2}} = 1$$

$$\int_0^h |\cosh \gamma y|^2\, dy = \begin{cases} \dfrac{1}{2}\left[\dfrac{\sinh \gamma h \cosh \gamma h}{\gamma} + h\right] & \gamma \text{ real, imaginary} \\ \mathrm{Re}(W \sinh \gamma h \cosh^* \gamma h) & \gamma \text{ complex} \end{cases}$$

where

$$W = \frac{1}{\gamma + \gamma^*} + \frac{1}{\gamma - \gamma^*}$$

$$\int_0^h |\sin \gamma y|^2 \, dy =$$

$$\begin{cases} \pm \frac{1}{2} \left[\frac{\sinh \gamma h \cosh \gamma h}{\gamma} - h \right] & +: \gamma \text{ real}; -: \gamma \text{ imaginary} \\ \text{Re} \, (W \cosh \gamma h \sinh^* \gamma h) & \gamma \text{ complex} \end{cases}$$

$$\int_0^h \sinh \gamma y \cosh^* \gamma y \, dy =$$

$$\begin{cases} \frac{1}{2\gamma} \sinh^2 \gamma h & \gamma \text{ real, immaginary} \\ \frac{1}{2} \, [W|\cosh \gamma h|^2 + W^*|\sinh \gamma h|^2 - W] & \gamma \text{ complex} \end{cases}$$

$$\int_0^h \sinh^* \gamma y \cosh \gamma y \, dy =$$

$$\begin{cases} \pm \frac{1}{2\gamma} \sinh^2 \gamma h & +: \gamma \text{ real}; -: \gamma \text{ imaginary} \\ \frac{1}{2}[W^*|\cosh \gamma h|^2 + W|\sinh \gamma h|^2 - W^*] & \gamma \text{ complex} \end{cases}$$

where the asterisk indicates complex conjugate.

APPENDIX B. DERIVATION OF EQ. (21)

Apply the boundary conditions to (12):

$\tilde{E}_{x1} = \tilde{E}_{x2}$

$$j\alpha\gamma_{y1}\text{S}_1A^e - j\beta\text{S}_1A^h = -j\alpha\gamma_{y2}\text{C}_2B^e - j\alpha\gamma_{y2}\text{S}_2C^e - j\beta\text{C}_2B^h - j\beta\text{S}_2C^h \tag{79a}$$

$\tilde{E}_{z1} = \tilde{E}_{z2}$

$$j\beta\gamma_{y1}\text{S}_1A^e + j\alpha\text{S}_1A^h = -j\beta\gamma_{y2}\text{C}_2B^e - jB\gamma_{y2}\text{S}_2C^e + j\alpha\text{C}_2B^h + j\alpha\text{S}_2C^h \tag{79b}$$

$\tilde{E}_{x2} = \tilde{E}_{x3}$

$$-j\alpha\gamma_{y2}B^e - j\beta B^h = -j\alpha y_{y3}\text{S}_3D^e - j\beta\text{S}_3D^h \tag{79c}$$

$\tilde{E}_{z2} = \tilde{E}_{z3}$

$$-j\beta\gamma_{y2}B^e + j\alpha B^h = -j\beta\gamma_3 S_3 D^e + j\alpha S_3 D^h \tag{79d}$$

$\tilde{H}_{x2} - \tilde{H}_{x1} = \tilde{J}_{z12}$

$$j\beta S_2 B^e + j\beta C_2 C^e - j\alpha\gamma_{z2} S_2 B^h - j\alpha\gamma_{z2} C_2 C^h - j\beta C_1 A^e - j\alpha\gamma_{z1} C_1 A^h = \tilde{J}_{z12} \tag{79e}$$

$\tilde{H}_{z2} - \tilde{H}_{z1} = -\tilde{J}_{x12}$

$$-j\alpha S_2 B^e - j\alpha C_2 C^e - j\beta\gamma_{z2} S_2 B^h - j\beta\gamma_{z2} C_2 C^h + j\alpha C_1 A^e - j\beta\gamma_{z1} C_1 A^h = -\tilde{J}_{x12} \tag{79f}$$

$\tilde{H}_{x3} - \tilde{H}_{x2} = \tilde{J}_{z23}$

$$j\beta C_3 D^e - j\alpha\gamma_{z3} C_3 D^h - j\beta C^e + j\alpha\gamma_{z2} C^h = \tilde{J}_{z23} \tag{79g}$$

$\tilde{H}_{z3} - \tilde{H}_{z2} = -\tilde{J}_{x23}$

$$-j\alpha C_3 D^e - j\beta\gamma_{z3} C_3 D^h + j\alpha C^e + j\beta\gamma_{z2} C^h = -\tilde{J}_{x23} \tag{79h}$$

where, for generality, the existence of current sources at the interface (2) is also assumed and the subscripts for $\tilde{J}_z$ and $\tilde{J}_x$ denote the interface regions. Multiply (79a) by α and (79b) by β and add the resulting equations; then multiply (79a) by β and (79b) by α and subtract the latter from the former. Similar procedures are also applied to (79c) and (79d). The results are

$$\gamma_{y1} S_1 A^e = -\gamma_{y2} C_2 B^e - \gamma_{y2} S_2 C^e \tag{80a}$$

$$S_1 A^h = C_2 B^h + S_2 C^h \tag{80b}$$

$$\gamma_{y3} S_3 D^e = \gamma_{y2} B^e \tag{80c}$$

$$S_3 D^h = B^h . \tag{80d}$$

From (80a)–(80d), derive B^e, B^h, C^e, C^h in terms of A^e, A^h, D^e, D^h, and substitute them into eqs. (79e)–(79h) to obtain

$$\tilde{J}_{z12} = -j\beta[Y_{12}^e A^e + P_{23} D^e] - j\alpha[Y_{12}^h A^h - Q_{23} D^h] \tag{81a}$$

$$\tilde{J}_{x12} = -j\alpha[Y_{12}^e A^e + P_{23} D^e] + j\beta[Y_{12}^h A^h - Q_{23} D^h] \tag{81b}$$

$$\tilde{J}_{z23} = j\beta[P_{12} A^e + Y_{23}^e D^e] - j\alpha[-Q_{12} A^h + Y_{23}^h D^h] \tag{81c}$$

$$\tilde{J}_{x23} = j\alpha[P_{12}A^e + Y^e_{23}D^e] + j\beta[-Q_{12}A^h + Y^h_{23}D^h] \tag{81d}$$

where

$$Y^e_{12} = \frac{S_1}{\gamma_{y2}}[\gamma_{y1}Ct_2 + \gamma_{y2}Ct_1] \qquad Y^e_{23} = \frac{S_3}{\gamma_{y2}}[\gamma_{y2}Ct_3 + \gamma_{y3}Ct_2]$$

$$Y^h_{12} = S_1[\gamma_{z1}Ct_1 + \gamma_{z2}Ct_2] \qquad Y^h_{23} = S_3[\gamma_{z2}Ct_2 + \gamma_{z3}Ct_3]$$

$$P_{12} = \gamma_{y1}S_1/\gamma_{y2}S_2 \qquad P_{23} = \gamma_{y3}S_3/\gamma_{y2}S_2$$

$$Q_{12} = \gamma_{z2}S_1/S_2 \qquad Q_{23} = \gamma_{z2}S_3/S_2$$

Multiply (81a) by β and (81b) by α and add the resulting equations; then multiply (81a) by α and (81b) by β and subtract the latter from the former. Similar processes are applied to (81c) and (81d).

$$\beta\tilde{J}_{z12} + \alpha\tilde{J}_{x12} = -j(\alpha^2 + \beta^2)[Y^e_{12}A^e + P_{23}D^e] \tag{82a}$$

$$\alpha\tilde{J}_{z12} - \beta\tilde{J}_{x12} = -j(\alpha^2 + \beta^2)[Y^h_{12}A^h - Q_{23}D^h] \tag{82b}$$

$$\beta\tilde{J}_{z23} + \alpha\tilde{J}_{x23} = j(\alpha^2 + \beta^2)[P_{12}A^e + Y^e_{23}D^e] \tag{82c}$$

$$\alpha\tilde{J}_{z23} - \beta\tilde{J}_{x23} = -j(\alpha^2 + \beta^2)[-Q_{12}A^h + Y^h_{23}D^h] \tag{82d}$$

From eqs. (82), we obtain

$$A^e = j\frac{1}{(\alpha^2 + \beta^2)\Gamma_e}[Y^e_{23}\beta\tilde{J}_{z12} + Y^e_{23}\alpha\tilde{J}_{x12} + P_{23}\beta\tilde{J}_{z23} + P_{23}\alpha\tilde{J}_{x23}] \tag{83a}$$

$$A^h = j\frac{1}{(\alpha^2 + \beta^2)\Gamma_h}[Y^h_{23}\alpha\tilde{J}_{z12} - Y^h_{23}\beta\tilde{J}_{x12} + Q_{23}\alpha\tilde{J}_{z23} - Q_{23}\beta\tilde{J}_{x23}] \tag{83b}$$

$$D^e = -j\frac{1}{(\alpha^2 + \beta^2)\Gamma_e}[P_{12}\beta\tilde{J}_{z12} + P_{12}\alpha\tilde{J}_{x12} + Y^e_{12}\beta\tilde{J}_{z23} + Y^e_{12}\alpha\tilde{J}_{x23}] \tag{83c}$$

$$D^h = j\frac{1}{(\alpha^2 + \beta^2)\Gamma_h}[Q_{12}\alpha\tilde{J}_{z12} - Q_{12}\beta\tilde{J}_{x12} + Y^h_{12}\alpha\tilde{J}_{z23} - Y^h_{12}\beta\tilde{J}_{x23}] \tag{83d}$$

where

$$\Gamma_e = Y^e_{12}Y^e_{23} - P_{12}P_{23} \qquad \Gamma_h = Y^h_{12}Y^h_{23} - Q_{12}Q_{23}$$

Substitute eqs. (83) into the next equations extracted from (12):

$$\tilde{E}_{x1} = \tilde{E}_x(\alpha, d + t) = j\alpha\gamma_{y1}S_1A^e - j\beta S_1A^h \tag{84a}$$

$$\tilde{E}_{z1} = \tilde{E}_z(\alpha, d+t) = j\beta\gamma_{y1}S_1A^e + j\alpha S_1A^h \tag{84b}$$

$$\tilde{E}_{x3} = \tilde{E}_x(\alpha, d) \qquad = -j\alpha\gamma_{y3}S_3D^e - j\beta S_3D^h \tag{84c}$$

$$\tilde{E}_{z3} = \tilde{E}_z(\alpha, d) \qquad = -j\beta\gamma_{y3}S_3D^e + j\alpha S_3D^h \tag{84d}$$

When this process is completed, we obtain the relationship between the Fourier transforms of the electric fields and those of the current distributions:

$$\begin{bmatrix} \tilde{E}_{z1} \\ \tilde{E}_{x1} \\ \tilde{E}_{z3} \\ \tilde{E}_{x3} \end{bmatrix} = [Z] \begin{bmatrix} \tilde{J}_{z12} \\ \tilde{J}_{x12} \\ \tilde{J}_{z23} \\ \tilde{J}_{x23} \end{bmatrix} \tag{85}$$

where

$$[Z] = -\frac{1}{\alpha^2+\beta^2} \begin{bmatrix} A_{11} & A_{12} & B_{11} & B_{12} \\ A_{21} & A_{22} & B_{21} & B_{22} \\ C_{11} & C_{12} & D_{11} & D_{12} \\ C_{21} & C_{22} & D_{21} & D_{22} \end{bmatrix}$$

$$A_{11} = \beta^2\gamma_{y1}\gamma_{y2}\frac{Y_2^e}{\Gamma'_e} + \alpha^2\frac{Y_2^h}{\Gamma'_h} \qquad A_{12} = \alpha\beta\left[\gamma_{y1}\gamma_{y2}\frac{Y_2^e}{\Gamma'_e} - \frac{Y_2^h}{\Gamma'_h}\right]$$

$$A_{21} = A_{12} \qquad A_{22} = \alpha^2\gamma_{y1}\gamma_{y2}\frac{Y_2^e}{\Gamma'_e} + \beta^2\frac{Y_2^h}{\Gamma'_h}$$

$$B_{11} = \beta^2\gamma_{y1}\gamma_{y2}\frac{P_2}{\Gamma'_e} + \alpha^2\frac{T}{\Gamma'_h} \qquad B_{12} = \alpha\beta\left[\gamma_{y1}\gamma_{y2}\frac{P_2}{\Gamma'_e} - \frac{T}{\Gamma'_h}\right]$$

$$B_{21} = B_{12} \qquad B_{22} = \alpha^2\gamma_{y1}\gamma_{y2}\frac{P_2}{\Gamma'_e} + \beta^2\frac{T}{\Gamma'_h}$$

$$C_{11} = \beta^2\gamma_{y3}\gamma_{y2}\frac{P_1}{\Gamma'_e} + \alpha^2\frac{T}{\Gamma'_h} \qquad C_{12} = \alpha\beta\left[\gamma_{y3}\gamma_{y2}\frac{P_1}{\Gamma'_e} - \frac{T}{\Gamma'_h}\right]$$

$$C_{21} = C_{12} \qquad C_{22} = \alpha^2\gamma_{y3}\gamma_{y2}\frac{P_1}{\Gamma'_e} + \beta^2\frac{T}{\Gamma'_h}$$

$$D_{11} = \beta^2\gamma_{y3}\gamma_{y2}\frac{Y_1^e}{\Gamma'_e} + \alpha^2\frac{Y_1^h}{\Gamma'_h} \qquad D_{12} = \alpha\beta\left[\gamma_{y3}\gamma_{y2}\frac{Y_1^e}{\Gamma'_e} - \frac{Y_1^h}{\Gamma'_h}\right]$$

$$D_{21} = D_{12} \qquad D_{22} = \alpha^2\gamma_{y3}\gamma_{y2}\frac{Y_1^e}{\Gamma'_e} + \beta^2\frac{Y_1^h}{\Gamma'_h}$$

$$\Gamma'_e = Y_1^e Y_2^e - P_1 P_2 \qquad \Gamma'_h = Y_1^h Y_2^h - T^2$$

$$Y_1^e = \gamma_{y1}\mathrm{Ct}_2 + \gamma_{y2}\mathrm{Ct}_1 \qquad Y_2^e = \gamma_{y2}\mathrm{Ct}_3 + \gamma_{y3}\mathrm{Ct}_2$$

$$Y_1^h = \gamma_{z1}\mathrm{Ct}_1 + \gamma_{z2}\mathrm{Ct}_2 \qquad Y_2^h = \gamma_{z2}\mathrm{Ct}_2 + \gamma_{z3}\mathrm{Ct}_3$$

$$P_1 = \gamma_{y1}/\mathrm{S}_2 \qquad P_2 = \gamma_{y3}/\mathrm{S}_2 \qquad T = \gamma_{z2}/\mathrm{S}_2$$

APPENDIX C. DERIVATION OF EQS. (32), (33), AND (34)

Let $\tilde{J}_{z23} = \tilde{J}_{x23} = 0$ in eqs. (83) and substitute them into (80a)–(80d). After some rearrangements, we obtain

$$A^e = j\,\frac{1}{\alpha^2+\beta^2}\,\frac{\gamma_{y2}}{\mathrm{S}_1}\,\frac{Y_2^e}{\Gamma'_e}\,[J_e] \tag{86a}$$

$$A^h = j\,\frac{1}{\alpha^2+\beta^2}\,\frac{1}{\mathrm{S}_1}\,\frac{Y_2^h}{\Gamma'_h}\,[J_h] \tag{86b}$$

$$B^e = -j\,\frac{1}{\alpha^2+\beta^2}\,\gamma_{y3}\,\frac{P_1}{\Gamma'_e}\,[J_e] \tag{86c}$$

$$B^h = j\,\frac{1}{\alpha^2+\beta^2}\,\frac{T}{\Gamma'_h}\,[J_h] \tag{86d}$$

$$C^e = -j\,\frac{1}{\alpha^2+\beta^2}\,\gamma_{y2}\mathrm{Ct}_3\,\frac{P_1}{\Gamma'_e}\,[J_e] \tag{86e}$$

$$C^h = j\,\frac{1}{\alpha^2+\beta^2}\,\frac{\gamma_3}{\gamma_2}\,\mathrm{Ct}_3\,\frac{T}{\Gamma'_h}\,[J_h] \tag{86f}$$

$$D^e = -j\,\frac{1}{\alpha^2+\beta^2}\,\frac{\gamma_{y2}}{\mathrm{S}_3}\,\frac{P_1}{\Gamma'_e}\,[J_e] \tag{86g}$$

$$D^h = j\,\frac{1}{\alpha^2+\beta^2}\,\frac{1}{\mathrm{S}_3}\,\frac{T}{\Gamma'_h}\,[J_h] \tag{86h}$$

where $[J_e] = \beta\tilde{J}_z + \alpha\tilde{J}_x$ and $[J_h] = \alpha\tilde{J}_z - \beta\tilde{J}_x$.

Substitute eqs. (86) into (12) to obtain

$$\tilde{E}_{x1} = A_1 \sinh \gamma_1(c-y) \tag{87a}$$

$$\tilde{E}_{x2} = B_1 \cosh \gamma_2(y-d) + C_1 \sinh \gamma_2(y-d) \tag{87b}$$

$$\tilde{E}_{x3} = D_1 \sinh \gamma_3 y \tag{87c}$$

$$\tilde{E}_{y1} = A_2 \cosh \gamma_1(c-y) \tag{87d}$$

$$\tilde{E}_{y2} = B_2 \sinh \gamma_2(y-d) + C_2 \cosh \gamma_2(y-d) \tag{87e}$$

$$\tilde{E}_{y3} = D_2 \cosh \gamma_3 y \tag{87f}$$

$$\tilde{H}_{x1} = A_3 \cosh \gamma_1 (c - y) \tag{87g}$$

$$\tilde{H}_{x2} = B_3 \sinh \gamma_2 (y - d) + C_3 \cosh \gamma_2 (y - d) \tag{87h}$$

$$\tilde{H}_{x3} = D_3 \cosh \gamma_3 y \tag{87i}$$

$$\tilde{H}_{y1} = A_4 \sinh \gamma_1 (c - y) \tag{87j}$$

$$\tilde{H}_{y2} = B_4 \cosh \gamma_2 (y - d) + C_4 \sinh \gamma_2 (y - d) \tag{87k}$$

$$\tilde{H}_{y3} = D_4 \sinh \gamma_3 y \tag{87l}$$

where

$$A_1 = j(\alpha \gamma_{y1} A^e - \beta A^h) \qquad B_1 = -j(\alpha \gamma_{y2} B^e + \beta B^h)$$

$$C_1 = -j(\alpha \gamma_{y2} C^e + \beta C^h) \qquad D_1 = -j(\alpha \gamma_{y3} D^e + \beta D^h)$$

$$A_2 = \frac{1}{\hat{y}_1} [\gamma_1^2 + k_1^2] A^e \qquad B_2 = \frac{1}{\hat{y}_2} [\gamma_2^2 + k_2^2] B^e$$

$$C_2 = \frac{1}{\hat{y}_2} [\gamma_2^2 + k_2^2] C^e \qquad D_2 = \frac{1}{\hat{y}_3} [\gamma_3^2 + k_3^2] D^e$$

$$A_3 = j(\beta A^e + \alpha \gamma_{z1} A^h) \qquad B_3 = j(\beta B^e - \alpha \gamma_{z2} B^h)$$

$$C_3 = j(\beta C^e - \alpha \gamma_{z2} C^h) \qquad D_3 = j(\beta D^e - \alpha \gamma_{z3} D^h)$$

$$A_4 = \frac{1}{\hat{z}_1} [\gamma_1^2 + k_1^2] A^h \qquad B_4 = \frac{1}{\hat{z}_2} [\gamma_2^2 + k_2^2] B^h$$

$$C_4 = \frac{1}{\hat{z}_2} [\gamma_2^2 + k_2^2] C^h \qquad D_4 = \frac{1}{\hat{z}_3} [\gamma_3^2 + k_3^2] D^h$$

The integrals in (31) along y are evaluated analytically by the use of the mathematical relations shown in Appendix A to yield (32)–(34):

$$\begin{aligned}
E_{h1} &= \int_{d+t}^{c} (\tilde{E}_{x1} \tilde{H}_{y1}^* - \tilde{E}_{y1} \tilde{H}_{x1}^*) \, dy \\
&= \int_{d+t}^{c} (A_1 A_4^* |\sinh \gamma_1 (c - y)|^2 - A_2 A_3^* |\cosh \gamma_1 (c - y)|^2) \, dy \\
&= \int_0^h (A_1 A_4^* |\sinh \gamma_1 y|^2 - A_2 A_3^* |\cosh \gamma_1 y|^2) \, dy \\
&= \pm \frac{1}{2} A_1 A_4^* \left(\frac{\mathrm{S}_1 \mathrm{C}_1}{\gamma_1} - h \right) - \frac{1}{2} A_2 A_3^* \left(\frac{\mathrm{S}_1 \mathrm{C}_1}{\gamma_1} + h \right)
\end{aligned}$$

$$
\begin{aligned}
E_{h2} &= \int_d^{d+t} (\tilde{E}_{x2}\tilde{H}^*_{y2} - \tilde{E}_{y2}\tilde{H}^*_{x2})\, dy \\
&= \int_d^{d+t} [\{B_1 \cosh \gamma_2(y-d) + C_1 \sinh \gamma_2(y-d)\} \\
&\quad \times \{B_4 \cosh \gamma_2(y-d) + C_4 \sinh \gamma_2(y-d)\}^* \\
&\quad - \{B_2 \sinh \gamma_2(y-d) + C_2 \cosh \gamma_2(y-d)\} \\
&\quad \times \{B_3 \sinh \gamma_2(y-d) + C_3 \cosh \gamma_2(y-d)\}^*]\, dy \\
&= \int_0^t (B_1B_4^*|\cosh \gamma_2 y|^2 + C_1C_4^*|\sinh \gamma_2 y|^2 \\
&\quad + C_1B_4^* \sinh \gamma_2 y \cosh^* \gamma_2 y + B_1C_4^* \cosh \gamma_2 y \sinh^* \gamma_2 y \\
&\quad - B_2B_3^*|\sinh \gamma_2 y|^2 - C_2C_3^*|\cosh \gamma_2 y|^2 \\
&\quad - C_2B_3^* \cosh \gamma_2 y \sinh^* \gamma_2 y - B_2C_3^* \sinh \gamma_2 y \cosh^* \gamma_2 y)\, dy \\
&= \int_0^t \{(C_1C_4^* - B_2B_3^*)|\sinh \gamma_2 y|^2 + (B_1B_4^* - C_2C_3^*)|\cosh \gamma_2 y|^2 \\
&\quad + (C_1B_4^* - B_2C_3^*) \sinh \gamma_2 y \cosh^* \gamma_2 y \\
&\quad + (B_1C_4^* - C_2B_3^*) \cosh \gamma_2 y \sinh^* \gamma_2 y\}\, dy \\
&= \pm \frac{1}{2} (C_1C_4^* - B_2B_3^*)\left(\frac{\mathrm{S}_2\mathrm{C}_2}{\gamma_2} - t\right) + \frac{1}{2} (B_1B_4^* - C_2C_3^*) \\
&\quad \times \left(\frac{\mathrm{S}_2\mathrm{C}_2}{\gamma_2} + t\right) + \frac{1}{2} [C_1B_4^* - B_2B_3^* \pm (B_1C_4^* - C_2B_3^*)] \frac{\mathrm{S}_2^2}{\gamma_2}
\end{aligned}
$$

$$
\begin{aligned}
E_{h3} &= \int_0^d (\tilde{E}_{x3}\tilde{H}^*_{y3} - \tilde{E}_{y3}\tilde{H}^*_{x3})\, dy \\
&= \int_0^d (D_1D_4^*|\sinh \gamma_3 y|^2 - D_2D_3^*|\cosh \gamma_3 y|^2)\, dy \\
&= \pm \frac{1}{2} D_1D_4^*\left(\frac{\mathrm{S}_3\mathrm{C}_3}{\gamma_3} - d\right) - \frac{1}{2} D_2D_3^*\left(\frac{\mathrm{S}_3\mathrm{C}_3}{\gamma_3} + d\right)
\end{aligned}
$$

APPENDIX D. DERIVATION OF THE POYNTING POWER FLOW FORMULATION IN THE IMMITTANCE APPROACH

By imposing boundary conditions (50)–(56), the following equations are obtained:

For TM-to-y:

$$
\frac{1}{\hat{y}_4} A^e \gamma_4 \mathrm{S}_4 = \frac{1}{\hat{y}_3} C^e \gamma_3 \qquad\qquad A^e \mathrm{C}_4 = B^e
$$

$$\frac{1}{\hat{y}_3} B^e \gamma_3 S_3 + \frac{1}{\hat{y}_3} C^e \gamma_3 C_3 = \frac{1}{\hat{y}_2} E^e \gamma_2 \qquad B^e C_3 + C^e S_3 = D^e$$

$$\frac{1}{\hat{y}_2} D^e \gamma_2 S_2 + \frac{1}{\hat{y}_2} E^e \gamma_2 C_2 = -\frac{1}{\hat{y}_1} F^e \gamma_1 S_1$$

$$\frac{1}{\sqrt{\alpha^2 + \beta^2}} F^e C_1 - \frac{1}{\sqrt{\alpha^2 + \beta^2}} (D^e C_2 + E^e S_2) = \tilde{J}_v$$

For TE-to-y:

$$A^h S_4 = C^h \qquad A^h \gamma_4 C_4 = B^h \gamma_3$$

$$B^h S_3 + C^h C_3 = E^h \qquad B^h \gamma_3 C_3 + C^h \gamma_3 S_3 = D^h \gamma_2$$

$$D^h S_2 + E^h C_2 = F^h S_1$$

$$-\frac{1}{\hat{z}\sqrt{\alpha^2 + \beta^2}} F^h \gamma_1 C_1 - \frac{1}{\hat{z}\sqrt{\alpha^2 + \beta^2}} (D^h \gamma_2 C_2 + E^h \gamma_2 S_2) = -\tilde{J}_u$$

where

$$C_i = \cosh \gamma_i h_i \qquad S_i = \sinh \gamma_i h_i \qquad i = 1, 2, 3, 4$$

and $\hat{z} = j\omega\mu_0$, where μ_0 is free space permeability.

When these equations are solved, unknown coefficients are found as follows:

For TM-to-y

$$A^e = \frac{1}{C_2 C_3 C_4} H_e \qquad B^e = \frac{1}{C_2 C_3} H_e$$

$$C^e = \frac{1}{C_2 C_3} \frac{\gamma_{y4}}{\gamma_{y3}} \frac{1}{Ct_4} H_e \qquad D^e = \frac{1}{C_2} H_{1e} H_e$$

$$E^e = \frac{1}{C_2} \frac{1}{\gamma_{y2}} H_{3e} H_e \qquad F^e = -\frac{1}{S_1} \frac{1}{\gamma_{y1}} \left(\frac{\gamma_{y2}}{Ct_2} H_{1e} + H_{3e} \right) H_e$$

where

$$H_e = -\frac{\sqrt{\alpha^2 + \beta^2} \tilde{J}_v}{H_{1e} H_{2e} + H_{3e} H_{4e}}$$

$$H_{1e} = 1 + \frac{\gamma_{y4}}{\gamma_{y3}} \frac{1}{Ct_3 Ct_4} \qquad H_{2e} = 1 + \frac{\gamma_{y2}}{\gamma_{y1}} \frac{Ct_1}{Ct_2}$$

$$H_{3e} = \frac{\gamma_{y3}}{Ct_3} + \frac{\gamma_{y4}}{Ct_4} \qquad H_{4e} = \frac{Ct_1}{\gamma_{y1}} + \frac{1}{\gamma_{y2} Ct_2}$$

For TE-to-y:

$$A^h = \frac{1}{C_2C_3S_4} H_h \qquad B^h = \frac{1}{C_2C_3} \frac{\gamma_4}{\gamma_3} Ct_4 H_h$$

$$C^h = \frac{1}{C_2C_3} H_h \qquad D^h = \frac{1}{C_2} \frac{1}{\gamma_2} H_{2h}H_h$$

$$E^h = \frac{1}{C_2} H_{4h}H_h \qquad F^h = \frac{1}{S_1}\left(\frac{1}{\gamma_2 Ct_2} H_{2h} + H_{4h}\right)H_h$$

where

$$H_h = -\frac{\hat{z}\sqrt{\alpha^2+\beta^2}\tilde{J}_u}{H_{1h}H_{2h} + H_{3h}H_{4h}}$$

$$H_{1h} = 1 + \frac{\gamma_1}{\gamma_2}\frac{Ct_1}{Ct_2} \qquad H_{2h} = \frac{\gamma_3}{Ct_3} + \gamma_4 Ct_4$$

$$H_{3h} = \gamma_1 Ct_1 + \frac{\gamma_2}{Ct_2} \qquad H_{4h} = 1 + \frac{\gamma_4}{\gamma_3}\frac{Ct_4}{Ct_3}$$

For each field component, we perform a process similar to the one in eqs. (87), coefficient expressions of A_1, B_1, etc. The Poynting power flow expressions in terms of α are given below.

Lossless Case:

$$\begin{aligned}
E_{h4} &= \pm\frac{1}{2}\,\underline{A}_1\underline{A}_4^*\left(\pm\frac{Ct_4}{|C_2C_3|^2\gamma_4} - \frac{h_4}{|C_2C_3S_4|^2}\right)\\
&\quad - \frac{1}{2}\,\underline{A}_2\underline{A}_3^*\left(\pm\frac{Ct_4}{|C_2C_3|^2\gamma_4} + \frac{h_4}{|C_2C_3S_4|^2}\right)
\end{aligned}$$

$$\begin{aligned}
E_{h3} &= \pm\frac{1}{2}\,(\underline{B}_1\underline{B}_4^* - \underline{C}_2\underline{C}_3^*)\left(\pm\frac{1}{|C_2|^2\gamma_3 Ct_3^*} - \frac{h_3}{|C_2C_3|^2}\right)\\
&\quad + \frac{1}{2}\,(\underline{C}_1\underline{C}_4^* - \underline{B}_2\underline{B}_3^*)\left(\pm\frac{1}{|C_2|^2\gamma_3 Ct_3^*} + \frac{h_3}{|C_2C_3|^2}\right)\\
&\quad + \frac{1}{2}\,[\underline{C}_1\underline{B}_4^* - \underline{B}_2\underline{C}_3^*) \pm (\underline{B}_1\underline{C}_4^* - \underline{C}_2\underline{B}_3^*)]\,\frac{1}{|C_2Ct_3|^2\gamma_3}
\end{aligned}$$

$$\begin{aligned}
E_{h2} &= \pm\frac{1}{2}\,(\underline{D}_1\underline{D}_4^* - \underline{E}_2\underline{E}_3^*)\left(\pm\frac{1}{\gamma_2 Ct_2^*} - \frac{h_2}{|C_2|^2}\right)\\
&\quad + \frac{1}{2}\,(\underline{E}_1\underline{E}_4^* - \underline{D}_2\underline{D}_3^*)\left(\pm\frac{1}{\gamma_2 Ct_2^*} + \frac{h_2}{|C_2|^2}\right)\\
&\quad + \frac{1}{2}\,[(\underline{E}_1\underline{D}_4^* - \underline{D}_2\underline{E}_3^*) \pm (\underline{D}_1\underline{E}_4^* - \underline{E}_2\underline{D}_3^*)]\,\frac{1}{|Ct_2|^2\gamma_2}
\end{aligned}$$

$$E_{h1} = \pm\frac{1}{2}\,\underline{F}_1\underline{F}_4^*\left(\pm\frac{\mathrm{Ct}_1}{\gamma_1} - \frac{h_1}{|S_1|^2}\right) - \frac{1}{2}\,\underline{F}_2\underline{F}_3^*\left(\pm\frac{\mathrm{Ct}_1}{\gamma_1} + \frac{h_1}{|S_1|^2}\right)$$

where $\pm$ signifies $+$ for γ real and $-$ for γ imaginary.

Lossy Dielectric Layer Case:

$$E_{h4} = \frac{1}{(\mathrm{C}_2^2\mathrm{C}_3^2)}\,[\underline{A}_1\underline{A}_4^*\,\mathrm{Re}(W_4\mathrm{Ct}_4) - \underline{A}_2\underline{A}_3^*\,\mathrm{Re}(W_4\mathrm{Ct}_4^*)$$

$$E_{h3} = \frac{1}{|\mathrm{C}_2|^2}\left[(\underline{B}_1\underline{B}_4^* - \underline{C}_2\underline{C}_3^*)\,\mathrm{Re}\left(\frac{W_3}{\mathrm{Ct}_3^*}\right) + (\underline{C}_1\underline{C}_4^* - \underline{B}_2\underline{B}_3^*)\,\mathrm{Re}\left(\frac{W_3}{\mathrm{Ct}_3}\right)\right]$$
$$+\frac{1}{2}\,\frac{1}{|\mathrm{C}_2|^2}\,(\underline{C}_1\underline{B}_4^* - \underline{B}_2\underline{C}_3^*)$$
$$\times\left[\left(1 - \frac{1}{\cos 2b_3}\right)W_3^* + \frac{1}{|\mathrm{Ct}_3|^2}\left(W_3 + \frac{W_3^*}{\cos 2b_3}\right)\right]$$
$$+\frac{1}{2}\,\frac{1}{|\mathrm{C}_2|^2}\,(\underline{B}_1\underline{C}_4^* - \underline{C}_2\underline{B}_3^*)$$
$$\times\left[\left(1 - \frac{1}{\cos 2b_3}\right)W_3 + \frac{1}{|\mathrm{Ct}_3|^2}\left(W_3^* + \frac{W_3}{\cos 2b_3}\right)\right]$$

$$E_{h2} = \left[(\underline{D}_1\underline{D}_4^* - \underline{E}_2\underline{E}_3^*)\,\mathrm{Re}\left(\frac{W_2}{\mathrm{Ct}_2^*}\right) + (\underline{E}_1\underline{E}_4^* - \underline{D}_2\underline{D}_3^*)\,\mathrm{Re}\left(\frac{W_2}{\mathrm{Ct}_2}\right)\right]$$
$$+\frac{1}{2}\,(\underline{E}_1\underline{D}_4^* - \underline{D}_2\underline{E}_3^*)\left[\left(1 - \frac{1}{\cos 2b_2}\right)W_2^* + \frac{1}{|\mathrm{Ct}_2|^2}\left(W_2 + \frac{W_2^*}{\cos 2b_2}\right)\right]$$
$$+\frac{1}{2}\,(\underline{D}_1\underline{E}_4^* - \underline{E}_2\underline{D}_3^*)\left[\left(1 - \frac{1}{\cos 2b_2}\right)W_2 + \frac{1}{|\mathrm{Ct}_2|^2}\left(W_2^* + \frac{W_2}{\cos 2b_2}\right)\right]$$

$$E_{h1} = \underline{F}_1\underline{F}_4^*\,\mathrm{Re}(W_1\mathrm{Ct}_1) - \underline{F}_2\underline{F}_3^*\,\mathrm{Re}(W_1\mathrm{Ct}_1^*)$$

where

$$\underline{A}_i = \mathrm{C}_2\mathrm{C}_3\mathrm{S}_4A_i \qquad \underline{B}_i = \mathrm{C}_2\mathrm{C}_3B_i \qquad \underline{C}_i = \mathrm{C}_2\mathrm{C}_3C_i \qquad i = 1, 2, 3, 4$$
$$\underline{D}_i = \mathrm{C}_2D_i, \qquad \underline{E}_i = \mathrm{C}_2E_i, \qquad \underline{F}_i = \mathrm{S}_1F_i$$

$$A_1 = \gamma_{y4}RA^e + QA^h \qquad B_1 = \gamma_{y3}RB^e + QB^h$$
$$C_1 = \gamma_{y3}RC^e + QC^h \qquad D_1 = \gamma_{y2}RD^e + QD^h$$
$$E_1 = \gamma_{y2}RE^e + QE^h \qquad F_1 = \gamma_{y1}RF^e - QF^h$$
$$A_2 = \frac{1}{-j\hat{y}_4}\,A^e \qquad B_2 = \frac{1}{-j\hat{y}_3}\,B^e \qquad C_2 = \frac{1}{-j\hat{y}_3}\,C^e$$

$$D_2 = \frac{1}{-j\hat{y}_2} D^e \qquad E_2 = \frac{1}{-j\hat{y}_2} E^e \qquad F_2 = \frac{1}{-j\hat{y}_1} F^e$$

$$A_3 = -QA^e + \gamma_{z4} RA^h \qquad B_3 = -QB^e + \gamma_{z3} RB^h$$

$$C_3 = -QC^e + \gamma_{z3} RC^h \qquad D_3 = -QD^e + \gamma_{z2} RD^h$$

$$E_3 = -QE^e + \gamma_{z2} RE^h \qquad F_3 = -QF^e - \gamma_{z1} RF^h$$

$$A_4 = \frac{1}{-j\hat{z}} A^h \qquad B_4 = \frac{1}{-j\hat{z}} B^h \qquad C_4 = \frac{1}{-j\hat{z}} C^h$$

$$D_4 = \frac{1}{-j\hat{z}} D^h \qquad E_4 = \frac{1}{-j\hat{z}} E^h \qquad F_4 = \frac{1}{j\hat{z}} F^h$$

$$W_i = \frac{1}{\gamma_i + \gamma_i} + \frac{1}{\gamma_i - \gamma_i}, \qquad i = 1, 2, 3, 4 \qquad b_i = \mathrm{Im}(\gamma_i h_i), \qquad i = 2, 3$$

APPENDIX E. COMPUTER PROGRAM EXAMPLE

```
*
      program ZIMP31
*
*                                                       1986-0717
*                                                       By Tom Uwano
*     ================================================
*       IMSL subroutine ZANLYT is required in linking.
*     ================================================
*
      Logical      Noloss
      Integer*4    N
      Complex      beta,ZBETA1,ZZ01,cn1,dn1,Z0
*
      Common   /PAR/N
      Common   /DIM/ w,h1,h2,h3,h4,rL
      Common   /Cond/ Freq,Er1,Er2,Er3,Er4,sgma1,sgma2,sgma3,sgma4
      Common   /Mat/ cn1,dn1,cn2
*
      EXTERNAL ZBETA1
*
      OPEN(FILE='ZIMP31.DAT',UNIT=7,STATUS='OLD')
*
      WRITE(*,901)
  901 FORMAT('     Slow wave stripline calculation    '
     -      //'             Jzn: n=1                   '
     -       /'             Jxn: n=1                   '
     -      //'          _________________             '
     -       /'          I        w       I h1 Er1,sgma1 '
     -       /'          I_______***______I             '
     -       /'          I________________I h2 Er2,sgma2 '
     -       /'          I////////////////I h3 Er3,sgma3 '
     -       /'          I----------------I             '
     -       /'          I________________I h4 Er4,sgma4 '
     -       /'              --- L ---                 ')
```

```
      write(*,*)'Input h1,h2,h3,h4 in [mm] : '
      read(*,*) h1,h2,h3,h4
      write(*,*)'Input w,L in [mm] : '
      read(*,*) w,rrL
      write(*,*) 'Input Er1,Er2,Er3,Er4 : '
      read(*,*) Er1,Er2,Er3,Er4
      write(*,*)'Input Conductivity of layer #1,2,3,4 in [Mho/m] : '
      read(*,*)sgma1,sgma2,sgma3,sgma4
      write(*,*) 'Input Freq. in [GHz] : '
      read(*,*) Freq
*
      w=w/1000.
      h1=h1/1000.
         if(h1.LT.1.E-10) h1=1.E-10
      h2=h2/1000.
         if(h2.LT.1.E-10) h2=1.E-10
      h3=h3/1000.
         if(h3.LT.1.E-10) h3=1.E-10
      h4=h4/1000.
         if(h4.LT.1.E-10) h4=1.E-10
      rrL=rrL/1000.
      rL=rrL/2.
      Freq=Freq*1e9
*
      BETA0=209.59411E-10*Freq
      Ermax=AMAX1(Er1,Er2,Er3,Er4)
      BETA=0.5*(SQRT(Ermax)+1.)*BETA0
      N1=50.*RL/(W*3.14)
      RK=BETA0/BETA
*
      WRITE(*,903)N1,RK
  903 FORMAT(' INPUT N (OVER',I5,'), and K (about',F6.3,') : ')
      READ(*,*)N,RK
      BETA=BETA0/RK
*
      write(7,902) freq,Er1,Er2,Er3,Er4,sgma1,sgma2,sgma3,sgma4,
     -             h1,h2,h3,h4,w,rrL
  902 FORMAT(' Freq. = ',1PE13.6/' Er1,Er2,Er3,Er4 = ',0P4F10.3
     -       /' sgma1,2,3,4   = ',1P4E13.6/' h1,h2,h3,h4 = ',4E14.3
     -       /' w      = ',E13.6/' L      = ',E13.6)
      write(7,*) ' N      = ',N
*
*
      it=30
      Call ZANLYT(ZBETA1,1.e-20,6,0,1,1,beta,it,ifn,ie)
*
*
      Noloss=     sgma1.EQ.sgma2.AND.sgma2.EQ.sgma3.AND.sgma3.EQ.sgma4
     -        .AND.sgma1.EQ.0.
      if((Noloss)) Beta=REAL(Beta)
      Rbeta=Beta0/REAL(Beta)
      att=-8.658896*AIMAG(Beta)
      write(*,904) Beta,Rbeta,1./Rbeta,att,IE
      write(7,904) Beta,Rbeta,1./Rbeta,att,IE
  904 format(' Final Beta =',2G15.8
     -       /' K (Bo/B)    =',G15.8,',      B/Bo =',G15.8
     -       /' Attenuation=',G15.8,' dB/m'
     -       /' Beta searching Error =',I3)
```

```
      if(IE.NE.0) stop
      Z0=ZZ01(Beta)
      write(*,*)'Z0 =',Z0,'Ohm'
      write(7,*)'Z0 =',Z0,'Ohm'
      write(7,*)'dn1=',dn1
      STOP
      END
*
*----------------BETA FUNCTION (1)
*                                  1986-0717
*                                  By Tom Uwano
      COMPLEX FUNCTION ZBETA1(Beta)
*
*       BASIS FUNCTION
*              Jzn:
*              Jxn:  n=1  only
*
*                             w
*                  -------------------
*       Er1,sgma1 l________***________l (1)   h1
*       Er2,sgma2 l-------------------l (2)   h2
*       Er3,sgma3 l-------------------l (3)   h3
*       Er4,sgma4 l___________________l (4)   h4
*                -L      - 2xL -     +L
*
       Logical    Homo
       Integer*4 NN,N
       Real       Mu0,L
       Complex    X,A2B2,cn1,dn1
       Complex    Beta,Kk1,Kk2,Kk3,Kk4
       Complex    J,yhat,yhat1,yhat2,yhat3,yhat4,zhat
       Complex    Gmr1,Gmr2,Gmr3,Gmr4
       Complex    Gy1,Gy2,Gy3,Gy4,Gz1,Gz2,Gz3,Gz4
       Complex    Ze,Zh,Z1e,Z2e,Z3e,Z4e,Z1h,Z2h,Z3h,Z4h
       Complex    Zzz,Zzx,Zxz,Zxx,Jz,Jx,K11,K21,K12,K22,Det
       Complex    CCOSH,CSINH
       Complex    CO1,CO2,CO3,CO4,SI1,SI2,SI3,SI4,CT1,CT2,CT3,CT4
*
       Common   /PAR/N
       Common   /DIM/ w,h1,h2,h3,h4,L
       Common   /Cond/ Freq,Er1,Er2,Er3,Er4,sgma1,sgma2,sgma3,sgma4
       Common   /Mat/ cn1,dn1,cn2
*
       Parameter (J=(0.,1.),Pai=3.1415926)
       Parameter (Epsi0=8.855e-12,Mu0=Pai*4.e-7)
*
***------ FUNCTION DEF. ---------------------------
*
       ar1(X)=SIGN(1.,REAL(X))*AMIN1(60.,ABS(REAL(X)))
       ar2(X)=AIMAG(X)
       CCOSH(X)=COSH(ar1(X))*COS(ar2(X))+J*SINH(ar1(X))*SIN(ar2(X))
       CSINH(X)=SINH(ar1(X))*COS(ar2(X))+J*COSH(ar1(X))*SIN(ar2(X))
*
*-------------------------------------------------
*
```

```
      Homo=      Er1.EQ.Er2.AND.Er2.EQ.Er3.AND.Er3.EQ.Er4
     -      .AND.sgma1.EQ.sgma2.AND.sgma2.EQ.sgma3.AND.sgma3.EQ.sgma4
*
      Omega=2.*pai*Freq
      yhat = J*Omega*Epsi0
      yhat1= Er1*yhat+sgma1
      yhat2= Er2*yhat+sgma2
      yhat3= Er3*yhat+sgma3
      yhat4= Er4*yhat+sgma4
      zhat = J*Omega*Mu0
      Kk1=-zhat*yhat1
      Kk2=-zhat*yhat2
      Kk3=-zhat*yhat3
      Kk4=-zhat*yhat4
*
      if((Homo)) then
        cn1=1.
        dn1=0.
        ZBETA1=Beta-CSQRT(Kk1)
        write(*,*) Beta,ZBETA1
        write(7,*) Beta,ZBETA1
        write(*,*)'                dn1=',dn1
        return
        endif
*
****--------- INTEGRAL OR SUMMATION IN TERMS OF Alfa
*
      K11=(0.,0.)
      K21=(0.,0.)
      K12=(0.,0.)
      K22=(0.,0.)
*
      DO 201 NN=0,N
*                     NN must be from 0,not from N to 0
*
      Alfa=FLOAT(NN)*Pai/(L)
      A2B2=Alfa*Alfa+Beta*Beta
      Gmr1=CSQRT(A2B2-Kk1)
      Gmr2=CSQRT(A2B2-Kk2)
      Gmr3=CSQRT(A2B2-Kk3)
      Gmr4=CSQRT(A2B2-Kk4)
      Gy1=Gmr1/yhat1
      Gy2=Gmr2/yhat2
      Gy3=Gmr3/yhat3
      Gy4=Gmr4/yhat4
      Gz1=Gmr1/zhat
      Gz2=Gmr2/zhat
      Gz3=Gmr3/zhat
      Gz4=Gmr4/zhat
*
      CO1=CCOSH(Gmr1*h1)
      CO2=CCOSH(Gmr2*h2)
      CO3=CCOSH(Gmr3*h3)
      CO4=CCOSH(Gmr4*h4)
      SI1=CSINH(Gmr1*h1)
```

```
      SI2=CSINH(Gmr2*h2)
      SI3=CSINH(Gmr3*h3)
      SI4=CSINH(Gmr4*h4)
      CT1=CO1/SI1
      CT2=CO2/SI2
      CT3=CO3/SI3
      CT4=CO4/SI4
*
*-----------------TM to y ;  Ze
*
      Z3e=Gy3*(Gy4*CT3+Gy3*CT4)/(Gy3*CT3*CT4+Gy4)
      Z2e=Gy2*(Z3e*CT2+Gy2)/(Gy2*CT2+Z3e)
      Z1e=Gy1/CT1
      Ze=1./(1./Z1e+1./Z2e)
*
*-----------------TE to y ;  Zh
*
      Z3h=(1./Gz3)*(Gz3*CT3+Gz4*CT4)/(Gz4*CT3*CT4+Gz3)
      Z2h=(1./Gz2)*(Z3h*Gz2*CT2+1.)/(CT2+Z3h*Gz2)
      Z1h=1./(Gz1*CT1)
      Zh=1./(1./Z1h+1./Z2h)
*
      Zzz= -(1./A2B2)*(Beta*Beta*Ze+Alfa*Alfa*Zh)
      Zzx= -(1./A2B2)*Alfa*Beta*(Ze-Zh)
      Zxz=  Zzx
      Zxx= -(1./A2B2)*(Alfa*Alfa*Ze+Beta*Beta*Zh)
*
      arg1=ABS(w*Alfa/2.)
      arg2=ABS(w*Alfa/2.+Pai)
      arg3=ABS(w*Alfa/2.-Pai)
      Jz= BSJ0(arg1)
      Jx= BSJ0(arg2)-BSJ0(arg3)
*
      K11=K11+Jz*Zzz*Jz
         if(NN.EQ.0) K11=K11/2.
      K12=K12+Jz*Zzx*Jx
      K22=K22+Jx*Zxx*Jx
*
  201 continue
*
      K21=K12
*
****------------------------------ END OF INTEGRAL
*
      det=K11*K22-K12*K21
      cn1=1.
      dn1=-cn1*0.5*(K11/K12+K21/K22)
*
      write(6,*) Beta,Det
      WRITE(7,*) Beta,det
      WRITE(*,*)'                 dn1=',dn1
*
  202 CONTINUE
      ZBETA1=det
      return
      end
```

```
*
*
*
      COMPLEX FUNCTION ZZ01(Beta)
*
*----------------
*                                         1986-0717
*
      Logical   Noloss,Homo
      Integer*4 NN,N
      Real      Mu0,L
      Complex   X,A2B2,R,Q,cn1,dn1
      Complex   Beta,Kk1,Kk2,Kk3,Kk4
      Complex   J,yhat,yhat1,yhat2,yhat3,yhat4,zhat
      Complex   Gmr1,Gmr2,Gmr3,Gmr4
      Complex   Gy1,Gy2,Gy3,Gy4,Gz1,Gz2,Gz3,Gz4
      Complex   Jz,Jx,Juab,Jvab
      Complex   CCOSH,CSINH,CJ
      Complex   CO1,CO2,CO3,CO4,SI1,SI2,SI3,SI4,CT1,CT2,CT3,CT4
      Complex   Ae,Ah,Be,Bh,Ce,Ch,De,Dh,Ee,Eh,Fe,Fh,He,Hh
      Complex   H1e,H2e,H3e,H4e,H1h,H2h,H3h,H4h
      Complex   A1,A2,A3,A4,B1,B2,B3,B4,C1,C2,C3,C4,D1,D2,D3,D4
      Complex   E1,E2,E3,E4,F1,F2,F3,F4
      Complex   Uh1,Uh2,Uh3,Uh4,Eh1,Eh2,Eh3,Eh4
      Complex   Wa1,Wa2,Wa3,Wa4,Wb1,Wb2,Wb3,Wb4
      Complex   x1,x2,x3,x4,y1,y2,y3,y4
*
      Common   /PAR/N
      Common   /DIM/ w,h1,h2,h3,h4,L
      Common   /Cond/ Freq,Er1,Er2,Er3,Er4,sgma1,sgma2,sgma3,sgma4
      Common   /Mat/ cn1,dn1,cn2
*
      Parameter (J=(0.,1.),Pai=3.1415926)
      Parameter (Epsi0=8.855e-12,Mu0=Pai*4.e-7)
*
***------ FUNCTION DEF. --------------------------
*
      ar1(X)=SIGN(1.,REAL(X))*AMIN1(40.,ABS(REAL(X)))
      ar2(X)=AIMAG(X)
      CCOSH(X)=COSH(ar1(X))*COS(ar2(X))+J*SINH(ar1(X))*SIN(ar2(X))
      CSINH(X)=SINH(ar1(X))*COS(ar2(X))+J*COSH(ar1(X))*SIN(ar2(X))
      CJ(X)=CONJG(X)
*
*-----------------------------------------------
*
      Noloss=     sgma1.EQ.sgma2.AND.sgma2.EQ.sgma3.AND.sgma3.EQ.sgma4
     -        .AND.sgma1.EQ.0.
      Homo=      Er1.EQ.Er2.AND.Er2.EQ.Er3.AND.Er3.EQ.Er4
     -      .AND.sgma1.EQ.sgma2.AND.sgma2.EQ.sgma3.AND.sgma3.EQ.sgma4
*
      Omega=2.*pai*Freq
      yhat = J*Omega*Epsi0
      yhat1= Er1*yhat+sgma1
      yhat2= Er2*yhat+sgma2
      yhat3= Er3*yhat+sgma3
      yhat4= Er4*yhat+sgma4
      zhat = J*Omega*Mu0
```

```
      Kk1=-zhat*yhat1
      Kk2=-zhat*yhat2
      Kk3=-zhat*yhat3
      Kk4=-zhat*yhat4
*
****--------- INTEGRAL OR SUMMATION IN TERMS OF Alfa
*
      Eh1=(0.,0.)
      Eh2=(0.,0.)
      Eh3=(0.,0.)
      Eh4=(0.,0.)
*
      DO 201 NN=0,N
*
      if(Homo.AND.NN.EQ.0) go to 201
      Alfa=FLOAT(NN)*Pai/(L)
*
      A2B2=Alfa*Alfa+Beta*Beta
      Gmr1=CSQRT(A2B2-Kk1)
      Gmr2=CSQRT(A2B2-Kk2)
      Gmr3=CSQRT(A2B2-Kk3)
      Gmr4=CSQRT(A2B2-Kk4)
      Gy1=Gmr1/yhat1
      Gy2=Gmr2/yhat2
      Gy3=Gmr3/yhat3
      Gy4=Gmr4/yhat4
      Gz1=Gmr1/zhat
      Gz2=Gmr2/zhat
      Gz3=Gmr3/zhat
      Gz4=Gmr4/zhat
*
      CO1=CCOSH(Gmr1*h1)
      CO2=CCOSH(Gmr2*h2)
      CO3=CCOSH(Gmr3*h3)
      CO4=CCOSH(Gmr4*h4)
      SI1=CSINH(Gmr1*h1)
      SI2=CSINH(Gmr2*h2)
      SI3=CSINH(Gmr3*h3)
      SI4=CSINH(Gmr4*h4)
      CT1=CO1/SI1
      CT2=CO2/SI2
      CT3=CO3/SI3
      CT4=CO4/SI4
*
*-------------------------------------
      arg1=ABS(w*Alfa/2.)
      arg2=ABS(w*Alfa/2.+Pai)
      arg3=ABS(w*Alfa/2.-Pai)
      Jz=cn1* BSJ0(arg1)
*              cn1 is set as value 1.0
      Jx=dn1* (BSJ0(arg2)-BSJ0(arg3))
*
      Jvab=Beta*Jz+Alfa*Jx
      Juab=Alfa*Jz-Beta*Jx
*-------------------------------------
```

```
*
      R= Alfa/A2B2
      Q= Beta/A2B2
*
      H1e=1.+(Gy4/Gy3)*(1./(CT3*CT4))
      H2e=1.+(Gy2/Gy1)*(CT1/CT2)
      H3e=Gy3/CT3+Gy4/CT4
      H4e=CT1/Gy1+1./(Gy2*CT2)
      H1h=1.+Gmr1*CT1/(Gmr2*CT2)
      H2h=Gmr3/CT3+Gmr4*CT4
      H3h=Gmr1*CT1+Gmr2/CT2
      H4h=1.+Gmr4*CT4/(Gmr3*CT3)
      He= -1./(H1e*H2e+H3e*H4e)*Jvab
      Hh=  1./(H1h*H2h+H3h*H4h)*zhat*Juab
*
      Ae= He
      Be= He
      Ce= Gy4/(Gy3*CT4)  *He
      De= H1e *He
      Ee= (1./Gy2)*H3e *He
      Fe=-(1./Gy1)*(Gy2/CT2*H1e+H3e)  *He
      Ah= Hh
      Bh= Gmr4*CT4/Gmr3 *Hh
      Ch= Hh
      Dh= (1./Gmr2)*H2h *Hh
      Eh= H4h *Hh
      Fh= (1./(Gmr2*CT2)*H2h+H4h)  *Hh
*
      A1= Gy4/CT4*R*Ae+Q*Ah
      B1= Gy3*R*Be+Q*Bh
      C1= Gy3*R*Ce+Q*Ch
      D1= Gy2*R*De+Q*Dh
      E1= Gy2*R*Ee+Q*Eh
      F1= Gy1*R*Fe-Q*Fh
      A2= 1./(-J*yhat4*CT4)*Ae
      B2= 1./(-J*yhat3)*Be
      C2= 1./(-J*yhat3)*Ce
      D2= 1./(-J*yhat2)*De
      E2= 1./(-J*yhat2)*Ee
      F2= 1./(-J*yhat1)*Fe
      A3=-1./CT4*Q*Ae+Gz4*R*Ah
      B3=-Q*Be+Gz3*R*Bh
      C3=-Q*Ce+Gz3*R*Ch
      D3=-Q*De+Gz2*R*Dh
      E3=-Q*Ee+Gz2*R*Eh
      F3=-Q*Fe-Gz1*R*Fh
      A4= 1./(-J*zhat)*Ah
      B4= 1./(-J*zhat)*Bh
      C4= 1./(-J*zhat)*Ch
      D4= 1./(-J*zhat)*Dh
      E4= 1./(-J*zhat)*Eh
      F4=-1./(-J*zhat)*Fh
*
*-------Avoiding overflow---
      if(CABS(CO2*CO3).LE.1.e10)  c2c3=CABS(CO2*CO3)
```

```
      if(CABS(CO2*CO3).GT.1.e10) c2c3=1.e10
      if(CABS(c2c3*SI4).LE.1.e10) c2c3s4=CABS(c2c3*SI4)
      if(CABS(c2c3*SI4).GT.1.e10) c2c3s4=1.e10
      Ps1=CABS(SI1*SI1)
      Pc2=CABS(CO2*CO2)
      if(Ps1.GT.1.e20) Ps1=1.e20
      if(Pc2.GT.1.e20) Pc2=1.e20
      Pcc=c2c3*c2c3
      Pcs=c2c3s4*c2c3s4
*--------------------------
*
      if((Noloss.OR.Homo)) then
*
*---------------------------Lossless case
      Gr1=REAL(A2B2-Kk1)
      Gr2=REAL(A2B2-Kk2)
      Gr3=REAL(A2B2-Kk3)
      Gr4=REAL(A2B2-Kk4)
       sgn1=SIGN(1.,Gr1)
       sgn2=SIGN(1.,Gr2)
       sgn3=SIGN(1.,Gr3)
       sgn4=SIGN(1.,Gr4)
*
      Uh4= 0.5*sgn4*A1*CJ(A4)*(sgn4*CT4/(Gmr4*Pcc)-h4/Pcs)
     -        -0.5*A2*CJ(A3)*(sgn4*CT4/(Gmr4*Pcc)+h4/Pcs)
      Uh3=
     -  0.5*sgn3*(B1*CJ(B4)-C2*CJ(C3))*(sgn3/(Gmr3*CJ(CT3)*Pc2)-h3/Pcc)
     -      +0.5*(C1*CJ(C4)-B2*CJ(B3))*(sgn3/(Gmr3*CJ(CT3)*Pc2)+h3/Pcc)
     -      +0.5*((C1*CJ(B4)-B2*CJ(C3))+sgn3*(B1*CJ(C4)-C2*CJ(B3)))
     -           /(Gmr3*CABS(CT3*CT3)*Pc2)
      Uh2=
     -  0.5*sgn2*(D1*CJ(D4)-E2*CJ(E3))*(sgn2/(Gmr2*CJ(CT2))-h2/Pc2)
     -      +0.5*(E1*CJ(E4)-D2*CJ(D3))*(sgn2/(Gmr2*CJ(CT2))+h2/Pc2)
     -      +0.5*((E1*CJ(D4)-D2*CJ(E3))+sgn2*(D1*CJ(E4)-E2*CJ(D3)))
     -           /(Gmr2*CABS(CT2*CT2))
      Uh1= 0.5*sgn1*F1*CJ(F4)*(sgn1*CT1/Gmr1-h1/Ps1)
     -        -0.5*F2*CJ(F3)*(sgn1*CT1/Gmr1+h1/Ps1)
      else
*
*---------------------------Slow wave
      Wa1= 1./(Gmr1+CJ(Gmr1)) + 1./(Gmr1-CJ(Gmr1))
      Wa2= 1./(Gmr2+CJ(Gmr2)) + 1./(Gmr2-CJ(Gmr2))
      Wa3= 1./(Gmr3+CJ(Gmr3)) + 1./(Gmr3-CJ(Gmr3))
      Wa4= 1./(Gmr4+CJ(Gmr4)) + 1./(Gmr4-CJ(Gmr4))
      Wb1= CJ(Wa1)
      Wb2= CJ(Wa2)
      Wb3= CJ(Wa3)
      Wb4= CJ(Wa4)
      Cos2b2=COS(2.*AIMAG(Gmr2*h2))
      Cos2b3=COS(2.*AIMAG(Gmr3*h3))
      Sinb2 =SIN(AIMAG(Gmr2*h2))
      Sinb3 =SIN(AIMAG(Gmr3*h3))
*
      X3=(1.-1./Cos2b3+(1.+1./Cos2b3)/CABS(CT3*CT3))*REAL(Wa3)
     -     +J*2.*(AIMAG(Wa3)*sinb3*sinb3/CABS(CO3*CO3)+1.e-15-1.e-15)
      X2=(1.-1./Cos2b2+(1.+1./Cos2b2)/CABS(CT2*CT2))*REAL(Wa2)
     -     +J*2.*(AIMAG(Wa2)*sinb2*sinb2/Pc2+1.e-15-1.e-15)
```

```
      y3=CJ(x3)
      y2=CJ(x2)
*
      Uh4= ( A1*CJ(A4)*REAL(Wa4*CT4)
     -          -A2*CJ(A3)*REAL(Wa4*CJ(CT4)) )/Pcc
      Uh3= ( (B1*CJ(B4)-C2*CJ(C3))*REAL(Wa3/CJ(CT3))
     -      +(C1*CJ(C4)-B2*CJ(B3))*REAL(Wa3/CT3)    )/Pc2
     -     +0.5*(C1*CJ(B4)-B2*CJ(C3))*x3/Pc2
     -     +0.5*(B1*CJ(C4)-C2*CJ(B3))*y3/Pc2
      Uh2= ( (D1*CJ(D4)-E2*CJ(E3))*REAL(Wa2/CJ(CT2))
     -      +(E1*CJ(E4)-D2*CJ(D3))*REAL(Wa2/CT2) )
     -     +0.5*(E1*CJ(D4)-D2*CJ(E3))*x2+0.5*(D1*CJ(E4)-E2*CJ(D3))*y2
      Uh1=  F1*CJ(F4)*REAL(Wa1*CT1)-F2*CJ(F3)*REAL(Wa1*CJ(CT1))
      endif
*
      Eh1=Eh1 +Uh1
      Eh2=Eh2 +Uh2
      Eh3=Eh3 +Uh3
      Eh4=Eh4 +Uh4
*
      if(NN.EQ.0) then
        Eh1=Eh1/2.
        Eh2=Eh2/2.
        Eh3=Eh3/2.
        Eh4=Eh4/2.
        endif
*
  201 continue
*
      Eh1=2.*Eh1
      Eh2=2.*Eh2
      Eh3=2.*Eh3
      Eh4=2.*Eh4
*---------------------------
*
      ZZ01=(Eh1+Eh2+Eh3+Eh4)/(2.*L)
      RETURN
      END
*
*
      FUNCTION BSJ0(x)
*                           Bessel function        1986-0226
      ax=ABS(x)
      if(ax.GT.3.) go to 1
      x2=x*x/9.
      x4=x2*x2
      x6=x2*x4
      x8=x4*x4
      x10=x2*x8
      x12=x6*x6
      BSJ0=1.-2.2499997d0*x2+1.2656208d0*x4-0.3163866d0*x6
     -       +0.0444479d0*x8-0.0039444d0*x10+0.00021d0*x12
      return
    1 continue
      x1=3./ax
```

```
      x2=x1*x1
      x3=x1*x2
      x4=x1*x3
      x5=x1*x4
      x6=x1*x5
      f= 0.79788456d0-0.00000077d0*x1-0.0055274d0*x2-0.00009512d0*x3
     -  +0.00137237d0*x4-0.00072805d0*x5+0.00014476d0*x6
      t=ax-0.78539816d0-0.04166397d0*x1-0.00003954d0*x2+0.00262573d0*x3
     -     -0.00054125d0*x4-0.00029333d0*x5+0.00013558d0*x6
      BSJ0=f*cos(t)/SQRT(ax)
      return
      end
*
*
```

REFERENCES

1. E. Yamashita and R. Mittra, "Variational method for the analysis of microstrip line," *IEEE Trans. Microwave Theory Tech.*, vol. MTT-16, pp. 251–256, Ap. 1968.
2. E. J. Denlinger, "A frequency dependent solution for microstrip transmission lines," *IEEE Trans. Microwave Theory Tech.*, vol. MTT-19, pp. 30–39, Jan. 1971.
3. T. Itoh and R. Mittra, "Spectral-domain approach for calculating the dispersion characteristics of microstrip lines," *IEEE Trans. Microwave Theory Tech.*, vol. MTT-21, pp. 496–499, July 1973.
4. R. F. Harrington, *Time-Harmonic Electromagnetic Fields*, McGraw-Hill, New York, 1961.
5. T. Itoh, "Spectral domain immittance approach for dispersion characteristics of generalized printed transmission lines," *IEEE Trans. Microwave Theory Tech.*, vol. MTT-28, pp. 733–736, July 1980.
6. T. Itoh and R. Mittra, "A technique for computing dispersion characteristics of shielded microstrip lines," *IEEE Trans. Microwave Theory Tech.*, vol. MTT-22, pp. 896–898, Oct. 1974.
7. H. Hasegawa, M. Furukawa, and H. Yanai, "Properties of micro-striplines on Si–SiO_2system," *IEEE Trans. Microwave Theory Tech.*, vol. MTT-19, pp. 869–881, Nov. 1971.

6

The Method of Lines

Reinhold Pregla and Wilfrid Pascher
Allgemeine und Theoretische Elektrotechnik
Fern Universität Hagen
Federal Republic of Germany

PREFACE

This treatise was written as a contribution for the development of peaceful use of microwave technology. For everyone who utilizes the method of lines for the dimensioning of circuits for peaceful purposes, it shall serve as a good instrument. Only for a usage for these purposes of the method presented here can the authors give their consent. The authors wish that every user acts in responsibility to God, the creator, according to the direction He gave us: "Subdue the earth." The authors believe that God blesses those activities which aim at the weal and not at the destruction of mankind. SDG

1. INTRODUCTION

The method of lines was developed by mathematicians in order to solve partial differential equations [1–3]. The method of lines has certain similarities with the mode-matching technique and with the finite difference method. From the latter it differs in the fact that for a given system of partial differential equations all but one of the independent variables are discretized to obtain a system of ordinary differential equations. This semianalytical procedure saves a lot of computing time. Regarding its applicability certain limitations have to be accepted. The method of lines has been applied to various problems in theoretical physics [1].

This chapter describes—based on the papers of Schulz [4], Worm [5], and Diestel [6], on earlier works of our own [7–9] and of Schmückle [36]—how this method can be used for the analysis of planar and quasiplanar waveguide structures.

The papers mentioned have shown that this class of waveguides can be analyzed accurately and nevertheless in an easy way. The relative converg-

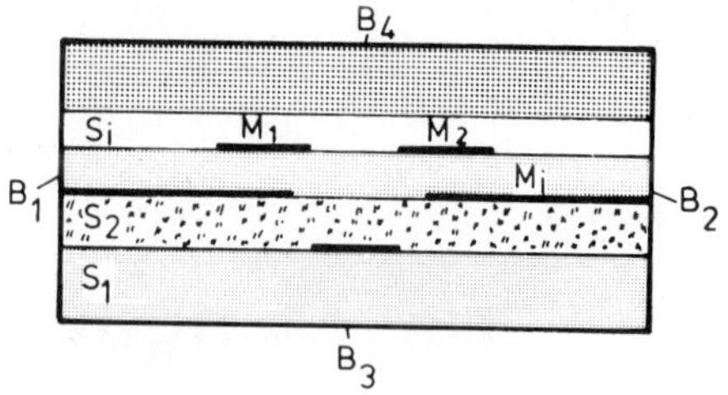

Fig. 1 Cross section of a quasiplanar waveguide. B_i = boundaries; M_i = metallizations; S_i = substrate layers.

ence phenomenon, which may turn up in the mode-matching technique as a consequence of the Fourier series truncations, does not occur in the method of lines. Optimum convergence is always assured if the simple condition is satisfied that the strip edges are located in the right way between the discretization lines [10]. It should be noted, however, that the convergence of the propagation constant, the characteristic impedance, or the resonance frequency does not critically depend on the so-called edge parameters, so that the problem of convergence on the whole is not critical.

To begin with, quasiplanar waveguides with constant cross sections as in Fig. 1 can be analyzed. The number of substrate layers S_i is arbitrary. The metallizations M_i can be included in various planes. The boundaries B_i consist of electric or magnetic walls. The lower and/or upper boundaries B_3 and B_4, respectively, may be infinite. The waveguide cross section in Fig. 1 includes the common microstrips, microslots, and finlines.

In monolithic integrated circuits the metallization thickness cannot be neglected compared with the conductor width or slot width. Thus the consideration of the finite metallization thickness is presented in another section.

In complex planar circuits not only waveguides with constant cross sections are used. For phase shifters and for couplers, periodic structures have a certain importance. As longitudinal inhomogeneities, step discontinuities occur in many circuits. Therefore it will be shown in another section how longitudinally inhomogeneous waveguides, e.g., periodic structures, discontinuities, or resonators (for examples, see Fig. 2), can be treated.

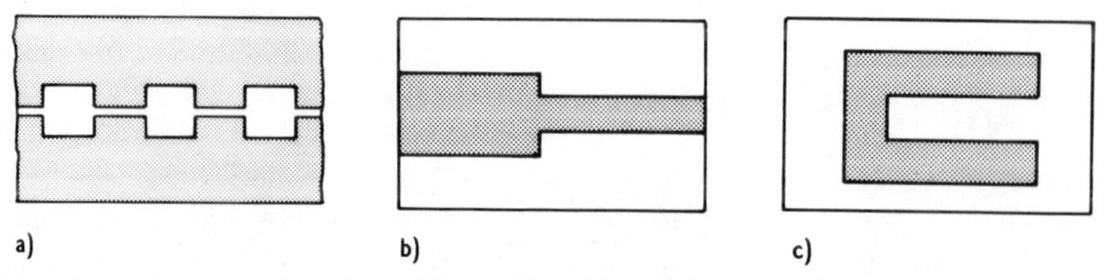

Fig. 2 Outlines of longitudinally inhomogeneous waveguides. Examples for (*a*) periodic structures, (*b*) discontinuities, (*c*) resonators.

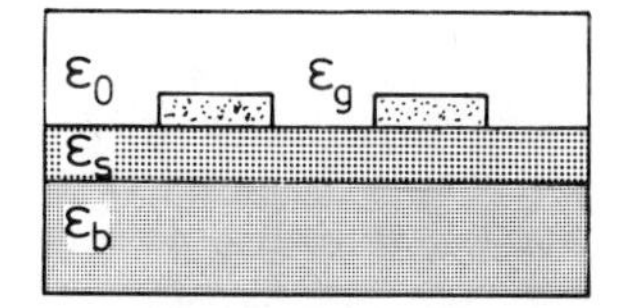

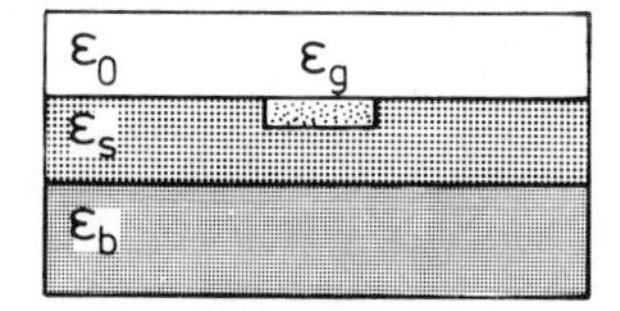

Fig. 3 Cross sections of dielectric waveguides (models for waveguides in the integrated optics).

Circuits in integrated optics are basically realized in the same way as in integrated microwave techniques. Instead of the metallization, dielectric strips with higher dielectric constant are used (Fig. 3). For the analysis of such structures and also of hybrid ones, it is described how an abrupt transition in the dielectric constant within one layer can be included in the method of lines. Finally, it is shown for an example of ferrite substrate how anisotropic layers can be treated.

The method of lines has a certain relation to the discrete Fourier transformation (DFT). Their relationship is derived, and it is pointed out at which point the approximation is made. The difference between the method of lines and the mode-matching technique is discussed.

2. THE FULL WAVE ANALYSIS OF PLANAR WAVEGUIDE STRUCTURES BY THE METHOD OF LINES

2.1. Basic Equations

The electromagnetic fields in planar waveguide structures of integrated microwave techniques and optics can be calculated from two independent potentials or two field components. Here the field components will be used directly. In certain cases this is the more sensible approach, e.g., for anisotropic substrates (especially for magnetized ferrite, (see section 3.3)). For inhomogeneous substrates, however, it is advisable to define suitable potential functions. For the analysis of the waveguide structures of Fig. 1 we start with the independent field components e_z and h_z. From Maxwell's equations we find that e_z and h_z must fulfill the Helmholtz equation

$$\frac{\partial^2\psi}{\partial x^2} + \frac{\partial^2\psi}{\partial y^2} + \frac{\partial^2\psi}{\partial z^2} + k^2\psi = L\psi = 0 \tag{1}$$

$$k^2 = \varepsilon_r k_0^2 \qquad k_0 = \omega\sqrt{\mu_0\varepsilon_0}$$

in each separate layer (for ψ we have to substitute either e_z or h_z). Moreover, e_z and h_z must fulfill the following boundary conditions:

$$\begin{aligned} &\text{ELECTRIC WALL} \quad e_z = 0 \quad (\text{D}); \quad \frac{\partial h_z}{\partial n} = 0 \quad (\text{N}) \\ &\text{MAGNETIC WALL} \quad h_z = 0 \quad (\text{D}); \quad \frac{\partial e_z}{\partial n} = 0 \quad (\text{N}) \end{aligned} \tag{2}$$

where n denotes the direction of the normal on the corresponding wall, and D stands for Dirichlet condition and N for Neumann condition. Here the independence of the two components especially expresses itself in the fact that different, namely dual boundary, conditions, are valid for each of them.

The other field components are calculated from e_z and h_z according to

$$\left(\frac{\partial^2}{\partial z^2} + k^2\right)\begin{bmatrix} e_x \\ e_y \end{bmatrix} = \begin{bmatrix} \dfrac{\partial^2}{\partial x\,\partial z} & -jk\eta\,\dfrac{\partial}{\partial y} \\ \dfrac{\partial^2}{\partial y\,\partial z} & jk\eta\,\dfrac{\partial}{\partial x} \end{bmatrix}\begin{bmatrix} e_z \\ h_z \end{bmatrix} \tag{3}$$

$$\left(\frac{\partial^2}{\partial z^2} + k^2\right)\begin{bmatrix} h_x \\ h_y \end{bmatrix} = \begin{bmatrix} j\,\dfrac{k}{\eta}\,\dfrac{\partial}{\partial y} & \dfrac{\partial^2}{\partial x\,\partial z} \\ -j\,\dfrac{k}{\eta}\,\dfrac{\partial}{\partial x} & \dfrac{\partial^2}{\partial y\,\partial z} \end{bmatrix}\begin{bmatrix} e_z \\ h_z \end{bmatrix} \tag{4}$$

with $\eta = \sqrt{\mu_0/\varepsilon} = \eta_0/\sqrt{\varepsilon_r}$; $\eta_0 = \sqrt{\mu_0/\varepsilon_0}$.

2.2. Qualitative Description of the Method of Lines

One approach for solving the partial differential equation (1) is to approximate the functions ψ in a suitable way. This is done in the mode-matching technique or in the method of moments in all its variations. Another possibility is to approximate the differential operator L. In the finite difference method it is completely substituted by the difference quotient. In the method of lines the differential operator is partially substituted by differences, but only as far as it is absolutely necessary, namely to convert the partial differential equation into an ordinary one. For the analysis of waveguides with constant cross section, this has to be done only with respect to one coordinate direction [11].

Considering the structure in Fig. 4, it becomes clear that the discretization should be done in the direction parallel to the interfaces of the layers (x direction). The individual layers are homogeneous in the y direction, but not in the x direction. The discretization of the operator L means that the field is considered on straight lines that are perpendicular to the interfaces of the layers and may be equidistant over the cross section. This is shown in Fig. 4 for the simple case of a single microstrip. Because of symmetry, only half a cross section has to be considered. In this case a magnetic wall has to be inserted in the middle of the structure. The figure shows that two separate line systems are used for e_z and h_z. There are several reasons for this. First

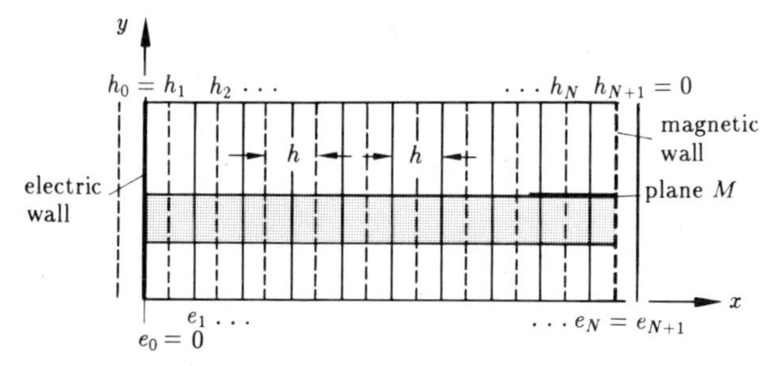

Fig. 4 Planar waveguide cross section with discretization lines. (———) Lines for e_z; (– – – –) lines for h_z.

of all, the lateral boundary conditions are immediately fulfilled if the lines are in the right position with respect to the boundaries. In order to fulfill the Dirichlet condition it is best to put a line on the lateral boundary and set the corresponding field component to zero. In the subsequent calculation it is not necessary to carry along this component. The Neumann condition is easily satisfied by including the boundary between two lines and equating the potentials.

The shifting of the two line systems has still more advantages:

–It allows an optimal edge positioning.
–It reduces the discretization error.
–It results in an easy quantitative description.

These advantages will be individually illustrated at convenient points.

2.3. Quantitative Description of the Discretization

Let the number of e_x and h_x lines in the cross section of Fig. 4 be equal to N. The field components e_z and h_z on these are combined to a vector $\mathbf{E}_z$ and $\mathbf{H}_z$, respectively. The ith component of $\mathbf{E}_z$ is the field component e_z on the ith line. Of course, $\mathbf{E}_{zi}$ is a function of y. For the discretization of eqs. (1), (3), and (4), the discretized derivatives with respect to x of the field component are needed. From eqs. (3) and (4) it can be seen that the derivative of e_z with respect to x is needed on an h_z line and the derivative of h_z is needed on an e_z line, respectively, as the portions in the individual equations can, of course, only be added or subtracted at the same place. This means, however, that the first derivatives have to be formed as difference quotients from adjacent components because of the shifting of the line systems. Seen from the center between the lines, the derivatives are thus formed using half the discretization distance, resulting in a reduction of the discretization error.

We write

$$h\frac{\partial e_z}{\partial x} \to \boldsymbol{D}\mathbf{E}_z \tag{5}$$

$$h\frac{\partial h_z}{\partial x} \to -\boldsymbol{D}^{\mathrm{t}}\mathbf{H}_z \tag{6}$$

with

$$\boldsymbol{D} = \begin{bmatrix} 1 & \ddots & & \\ -1 & & & \\ & \ddots & \ddots & \\ & & -1 & 1 \end{bmatrix} \tag{7}$$

Equations (5) and (6) make clear that the difference operator $\boldsymbol{D}$, defined for the representation of e_x in eq. (7) can be also used for the representation of the derivative of h_z. This is another consequence of the shifting of the line systems. In the difference operator $\boldsymbol{D}$ the lateral boundary conditions are also included. Moreover, the second derivatives can also be represented by means of the operator $\boldsymbol{D}$.

$$h^2\frac{\partial^2 e_z}{\partial x^2} = h^2\frac{\partial}{\partial x}\left(\frac{\partial e_z}{\partial x}\right) \to -\boldsymbol{D}^{\mathrm{t}}\boldsymbol{D}\mathbf{E}_z = -\boldsymbol{P}_{\mathrm{DN}}\mathbf{E}_z \tag{8}$$

$$h^2\frac{\partial^2 h_z}{\partial x^2} = h^2\frac{\partial}{\partial x}\left(\frac{\partial h_z}{\partial x}\right) \to -\boldsymbol{D}\boldsymbol{D}^{\mathrm{t}}\mathbf{H}_z = -\boldsymbol{P}_{\mathrm{ND}}\mathbf{H}_z \tag{9}$$

thus the difference operator for the second derivative results from the product of the difference operators for the first derivatives. This can be seen from the chain form of the second derivatives, keeping in mind that the boundary conditions for the outer derivative are dual to those for the inner derivative. The difference operator $\boldsymbol{P}$ can be represented in the following way:

$$\boldsymbol{P} = \begin{bmatrix} p_l & -1 & \ddots & & \\ -1 & 2 & & & \\ & \ddots & \ddots & \ddots & \\ & & \ddots & 2 & -1 \\ & & & -1 & p_r \end{bmatrix} \tag{10}$$

where $p_{l,r} = 2$ is valid if we take the Dirichlet condition for the left (right) wall and $p_{l,r} = 1$ for the Neumann condition.

Substituting eq. (8) or (9) in eq. (1), we obtain the ordinary differential equation

$$\frac{d^2}{dy^2}\boldsymbol{\psi} + [(k^2 - k_z^2)\boldsymbol{I} - h^{-2}\boldsymbol{P}]\boldsymbol{\psi} = \mathbf{0} \tag{11}$$

assuming wave propagation in z direction according to $e^{-jk_z z}$, where $\boldsymbol{\psi}$ is either $\mathbf{E}_z$ or $\mathbf{H}_z$ and $\boldsymbol{P}$ is the corresponding difference matrix. $\boldsymbol{I}$ is the unit matrix in this equation and in the following. In this equation there are always three components that are coupled with each other because of the tridiagonal structure of $\boldsymbol{P}$. Hence a direct solution is not possible.

Therefore we make a transformation to principal axes. With

$$\boldsymbol{\psi} = \boldsymbol{T}\bar{\boldsymbol{\psi}} \tag{12}$$

we require that

$$\boldsymbol{T}^{\mathrm{t}}\boldsymbol{P}\boldsymbol{T} = \boldsymbol{\lambda}^2 \tag{13}$$

is valid and that $\boldsymbol{\lambda}^2$ is a diagonal matrix. $\boldsymbol{\lambda}^2$ is the eigenvalue matrix and $\boldsymbol{T}$ the eigenvector matrix belonging to $\boldsymbol{P}$. The calculation of these matrices is given in Appendix A for various boundary conditions. As $\boldsymbol{P}$ is a symmetric matrix. $\boldsymbol{T}$ is orthogonal for a suitable normalization of the eigenvectors:

$$\boldsymbol{T}^{-1} = \boldsymbol{T}^{\mathrm{t}} \tag{14}$$

From eq. (11) we get with eqs. (12) and (13)

$$\left[\left(\frac{d^2}{dy^2} + k^2 - k_z^2\right)\boldsymbol{I} - h^{-2}\boldsymbol{\lambda}^2\right]\bar{\boldsymbol{\psi}} = \mathbf{0} \tag{15}$$

Let

$$k_{yi}^2 = k_0^2(\bar{\lambda}_i^2 - \varepsilon_r + \varepsilon_{re}) \tag{16}$$

with

$$\varepsilon_{re} = \frac{k_z^2}{k_0^2} \qquad \bar{\lambda}_i^2 = \frac{\lambda_i^2}{(k_0 h)^2} \tag{17}$$

Then we get the general solution for the ith component of $\bar{\boldsymbol{\psi}}$

$$\bar{\psi}_i = A_i \cosh k_{yi} y + B_i \sinh k_{yi} y \tag{18}$$

Since in most cases the components and their derivatives are only needed on the layer interfaces, we give the solution for an arbitrary layer with thickness d (see Fig. 5) also in the following form:

$$\begin{bmatrix} \bar{\boldsymbol{\psi}}'(y_1) \\ \bar{\boldsymbol{\psi}}'(y_2) \end{bmatrix} = k_y^2 \begin{bmatrix} \boldsymbol{\gamma} & \boldsymbol{\alpha} \\ \boldsymbol{\alpha} & \boldsymbol{\gamma} \end{bmatrix} \begin{bmatrix} -\bar{\boldsymbol{\psi}}(y_1) \\ \bar{\boldsymbol{\psi}}(y_2) \end{bmatrix} \tag{19}$$

with

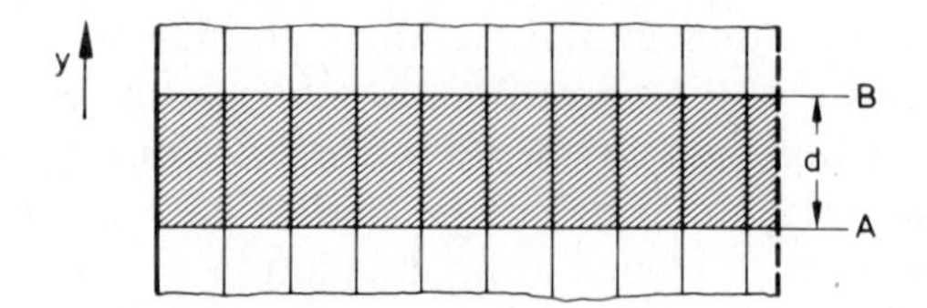

Fig. 5 Notation of the interfaces of a layer for the calculation of the field components and their derivatives.

$$\bar{\boldsymbol{\psi}}' = \frac{1}{k_0}\frac{d}{dy}\bar{\boldsymbol{\psi}}$$

$$\boldsymbol{\alpha} = \text{diag}\left(\frac{k_{yi}}{k_0}\sinh k_{yi}\, d\right)^{-1}$$

$$\boldsymbol{\gamma} = \text{diag}\left(\frac{k_{yi}}{k_0}\tanh k_{yi}\, d\right)^{-1} \tag{20}$$

$$\boldsymbol{k}_y = \text{diag}\left(\frac{k_{yi}}{k_0}\right)$$

2.4. Solutions for the Other Field Components

With the solutions for the transform components $\bar{\mathbf{E}}_z$ and $\bar{\mathbf{H}}_z$, the other components can also be determined. For this purpose eqs. (3) and (4) are discretized and transformed. To determine the propagation constant, only the tangential components are necessary to begin with. Hence we take from eqs. (3) and (4) only those for e_x and h_x, yielding

$$k_0\varepsilon_d\begin{bmatrix}\mathbf{E}_x\\ \eta_0\mathbf{H}_x\end{bmatrix} = j\begin{bmatrix}-\sqrt{\varepsilon_{re}}\,h^{-1}\boldsymbol{D} & -\boldsymbol{I}\dfrac{\partial}{\partial y}\\ \varepsilon_r\boldsymbol{I}\dfrac{\partial}{\partial y} & \sqrt{\varepsilon_{re}}\,h^{-1}\boldsymbol{D}^{\text{t}}\end{bmatrix}\begin{bmatrix}\mathbf{E}_z\\ \eta_0\mathbf{H}_z\end{bmatrix} \tag{21}$$

with

$$\varepsilon_d = \varepsilon_r - \varepsilon_{re}$$

It follows that the other two components are:

$$k_0\varepsilon_d\begin{bmatrix}\mathbf{E}_y\\ \eta_0\mathbf{H}_y\end{bmatrix} = -j\begin{bmatrix}\sqrt{\varepsilon_{re}}\,\boldsymbol{I}\dfrac{\partial}{\partial y} & h^{-1}\boldsymbol{D}^{\text{t}}\\ \varepsilon_r h^{-1}\boldsymbol{D} & \sqrt{\varepsilon_{re}}\,\boldsymbol{I}\dfrac{\partial}{\partial y}\end{bmatrix}\begin{bmatrix}\mathbf{E}_z\\ \eta_0\mathbf{H}_z\end{bmatrix} \tag{22}$$

The transformation yields

$$\varepsilon_d\begin{bmatrix} \bar{\mathbf{E}}_x \\ \eta_0\bar{\mathbf{H}}_x \end{bmatrix} = j\begin{bmatrix} -\sqrt{\varepsilon_{re}}\bar{\boldsymbol{\delta}} & -\boldsymbol{I}\dfrac{1}{k_0}\dfrac{\partial}{\partial y} \\ \boldsymbol{I}\dfrac{\varepsilon_r}{k_0}\dfrac{\partial}{\partial y} & \sqrt{\varepsilon_{re}}\bar{\boldsymbol{\delta}}^{\mathrm{t}} \end{bmatrix}\begin{bmatrix} \bar{\mathbf{E}}_z \\ \eta_0\bar{\mathbf{H}}_z \end{bmatrix} \tag{23}$$

$$\varepsilon_d\begin{bmatrix} \bar{\mathbf{E}}_y \\ \eta_0\bar{\mathbf{H}}_y \end{bmatrix} = -j\begin{bmatrix} \boldsymbol{I}\dfrac{\sqrt{\varepsilon_{re}}}{k_0}\dfrac{\partial}{\partial y} & \bar{\boldsymbol{\delta}}^{\mathrm{t}} \\ \varepsilon_r\bar{\boldsymbol{\delta}} & \boldsymbol{I}\dfrac{\sqrt{\varepsilon_{re}}}{k_0}\dfrac{\partial}{\partial y} \end{bmatrix}\begin{bmatrix} \bar{\mathbf{E}}_z \\ \eta_0\bar{\mathbf{H}}_z \end{bmatrix} \tag{24}$$

with

$$\bar{\boldsymbol{\delta}} = (k_0 h)^{-1}\boldsymbol{\delta}$$

and

$$\boldsymbol{\delta} = \boldsymbol{T}_h^{\mathrm{t}}\boldsymbol{D}\boldsymbol{T}_e \tag{25}$$

The matrix $\boldsymbol{\delta}$ is a diagonal or a quasi-diagonal matrix. Particularities are given in Appendix B. Here $\boldsymbol{T}_e$ stands as transformation matrix for $\mathbf{E}_z$ and $\boldsymbol{T}_h$ for $\mathbf{H}_z$. For the structure in Fig. 4 there is $\boldsymbol{T}_e = \boldsymbol{T}_{\mathrm{DN}}$ and $\boldsymbol{T}_h = \boldsymbol{T}_{\mathrm{ND}}$.

If the system of equations (23) is established for the interface planes A and B (see Fig. 5), we obtain, after some algebraic manipulations using eq. (19), the following system of equations for the tangential field components:

$$\eta_0\begin{bmatrix} -j\bar{\mathbf{H}}_{zA} \\ \bar{\mathbf{H}}_{xA} \\ -j\bar{\mathbf{H}}_{zB} \\ \bar{\mathbf{H}}_{xB} \end{bmatrix} = \begin{bmatrix} -\varepsilon_d\boldsymbol{\gamma}_h & \boldsymbol{\gamma}_h\tilde{\bar{\boldsymbol{\delta}}} & -\varepsilon_d\boldsymbol{\alpha}_h & \boldsymbol{\alpha}_h\tilde{\bar{\boldsymbol{\delta}}} \\ \tilde{\bar{\boldsymbol{\delta}}}^{\mathrm{t}}\boldsymbol{\gamma}_h & \boldsymbol{\gamma}_E & \tilde{\bar{\boldsymbol{\delta}}}^{\mathrm{t}}\boldsymbol{\alpha}_h & \boldsymbol{\alpha}_E \\ -\varepsilon_d\boldsymbol{\alpha}_h & \boldsymbol{\alpha}_h\tilde{\bar{\boldsymbol{\delta}}} & -\varepsilon_d\boldsymbol{\gamma}_h & \boldsymbol{\gamma}_h\tilde{\bar{\boldsymbol{\delta}}} \\ \tilde{\bar{\boldsymbol{\delta}}}^{\mathrm{t}}\boldsymbol{\alpha}_h & \boldsymbol{\alpha}_E & \tilde{\bar{\boldsymbol{\delta}}}^{\mathrm{t}}\boldsymbol{\gamma}_h & \boldsymbol{\gamma}_E \end{bmatrix}\begin{bmatrix} \bar{\mathbf{E}}_{xA} \\ -j\bar{\mathbf{E}}_{zA} \\ -\bar{\mathbf{E}}_{xB} \\ j\bar{\mathbf{E}}_{xB} \end{bmatrix} \tag{26}$$

with

$$\tilde{\bar{\boldsymbol{\delta}}} = \sqrt{\varepsilon_{re}}\bar{\boldsymbol{\delta}} \qquad \begin{Bmatrix} \boldsymbol{\alpha}_E \\ \boldsymbol{\gamma}_E \end{Bmatrix} = (\bar{\boldsymbol{\lambda}}_e^2 - \varepsilon_r\boldsymbol{I})\begin{Bmatrix} \boldsymbol{\alpha}_e \\ \boldsymbol{\gamma}_e \end{Bmatrix}$$

e or h, respectively, was chosen as subscript for $\boldsymbol{\alpha}$, $\boldsymbol{\gamma}$, $\boldsymbol{\lambda}$, depending on whether ψ stands for e_z or for h_z in eq. (19).

With the abbreviations

$$\bar{\mathbf{H}}_{A,B} = \eta_0\begin{bmatrix} -j\bar{\mathbf{H}}_{zA,B} \\ \bar{\mathbf{H}}_{xA,B} \end{bmatrix} \qquad \bar{\mathbf{E}}_{A,B} = \begin{bmatrix} \bar{\mathbf{E}}_{xA,B} \\ -j\bar{\mathbf{E}}_{zA,B} \end{bmatrix} \tag{27}$$

and

$$\bar{y}_1 = \begin{bmatrix} -\varepsilon_d \gamma_h & \gamma_h \tilde{\delta} \\ \tilde{\delta}^t \gamma_h & \gamma_E \end{bmatrix} \qquad \bar{y}_2 = \begin{bmatrix} -\varepsilon_d \alpha_h & \alpha_h \tilde{\delta} \\ \tilde{\delta}^t \alpha_h & \alpha_E \end{bmatrix} \tag{28}$$

eq. (26) can be written in a shorter way:

$$\begin{bmatrix} \bar{\mathbf{H}}_A \\ \bar{\mathbf{H}}_B \end{bmatrix} = \begin{bmatrix} \bar{y}_1 & \bar{y}_2 \\ \bar{y}_2 & \bar{y}_1 \end{bmatrix} \begin{bmatrix} \bar{\mathbf{E}}_A \\ -\bar{\mathbf{E}}_B \end{bmatrix} \tag{29}$$

In order to transform the tangential field components from one place to the other in the case of layered substrates, we convert eq. (29) to

$$\begin{bmatrix} \bar{\mathbf{E}}_B \\ \bar{\mathbf{H}}_B \end{bmatrix} = \begin{bmatrix} \bar{V} & \bar{Z} \\ \bar{Y} & \bar{V} \end{bmatrix} \begin{bmatrix} \bar{\mathbf{E}}_A \\ \bar{\mathbf{H}}_A \end{bmatrix} \tag{30}$$

The matrices $\bar{V}, \bar{Y}, \bar{Z}$ are calculated according to

$$\bar{V} = \bar{y}_2^{-1} \bar{y}_1 = \bar{y}_1 \bar{y}_2^{-1} = \text{blockdiag}(\boldsymbol{\tau}_h, \boldsymbol{\tau}_e) \tag{31}$$

$$\bar{Y} = \bar{y}_2 - \bar{y}_1 \bar{y}_2^{-1} \bar{y}_1 = -\begin{bmatrix} \mathbf{k}_{yh}^{-2} \boldsymbol{\alpha}_h^{-1} & \boldsymbol{0} \\ \boldsymbol{0} & \mathbf{k}_{ye}^{-2} \boldsymbol{\alpha}_e^{-1} \end{bmatrix} \begin{bmatrix} -\varepsilon_d \boldsymbol{I} & \tilde{\boldsymbol{\delta}} \\ \tilde{\boldsymbol{\delta}}^t & (\bar{\lambda}_e^2 - \varepsilon_r I) \end{bmatrix} \tag{32}$$

$$\bar{Z} = -\bar{y}_2^{-1} = -\frac{1}{\varepsilon_r} \begin{bmatrix} \mathbf{k}_{yh}^{-2} \boldsymbol{\alpha}_h^{-1} & \boldsymbol{0} \\ \boldsymbol{0} & \mathbf{k}_{ye}^{-2} \boldsymbol{\alpha}_e^{-1} \end{bmatrix} \begin{bmatrix} -(\bar{\lambda}_h^2 - \varepsilon_r I) & \tilde{\boldsymbol{\delta}} \\ \tilde{\boldsymbol{\delta}}^t & \varepsilon_d I \end{bmatrix} \tag{33}$$

with

$$\boldsymbol{\tau}_h = \text{diag}(\cosh k_{yhi} d) \qquad \boldsymbol{\tau}_e = \text{diag}(\cosh k_{yei} d) \tag{34}$$

The inversion of eq. (30) is

$$\begin{bmatrix} \bar{\mathbf{E}}_A \\ \bar{\mathbf{H}}_A \end{bmatrix} = \begin{bmatrix} \bar{V} & -\bar{Z} \\ -\bar{Y} & \bar{V} \end{bmatrix} \begin{bmatrix} \bar{\mathbf{E}}_B \\ \bar{\mathbf{H}}_B \end{bmatrix} \tag{35}$$

With a metallic shielding in interface 0 we obtain $\bar{\mathbf{E}}_0 = \mathbf{0}$. In the first step for interface 1, from eq. (30) it follows that

$$\bar{\mathbf{H}}_1 = \bar{Y}_t^{(1)} \bar{\mathbf{E}}_1 \tag{36}$$

with

$$\bar{Y}_t^{(1)} = \bar{V}_1 \bar{Z}_1^{-1} = -\bar{y}_1 \tag{37}$$

$\bar{Y}_1$, $\bar{Z}_1$, and $\bar{Y}_t^{(1)}$ are determined by the properties of substrate layer 1.

Substituting this result in eq. (30), which is now applied for layer 2, we obtain the tangential field component in interface 2. From

$$\begin{bmatrix} \bar{\mathbf{E}}_2 \\ \bar{\mathbf{H}}_2 \end{bmatrix} = \begin{bmatrix} \bar{V}_2 & \bar{Z}_2 \\ \bar{Y}_2 & \bar{V}_2 \end{bmatrix} \begin{bmatrix} \bar{\mathbf{E}}_1 \\ \bar{Y}_t^{(1)} \bar{\mathbf{E}}_1 \end{bmatrix} \tag{38}$$

follows

$$\bar{\mathbf{H}}_2 = \bar{Y}_t^{(2)} \bar{\mathbf{E}}_2 \tag{39}$$

with

$$\bar{Y}_t^{(2)} = (\bar{Y}_2 + \bar{V}_2 \bar{Y}_t^{(1)})(\bar{V}_2 + \bar{Z}_2 \bar{Y}_t^{(1)})^{-1} \tag{40}$$

In interface k, by generalizing this relation we obtain

$$\bar{\mathbf{H}}_k = \bar{Y}_t^{(k)} \bar{\mathbf{E}}_k \tag{41}$$

with the recurrence relation for $\bar{Y}_t^{(k)}$

$$\bar{Y}_t^{(k)} = (\bar{Y}_k + \bar{V}_k \bar{Y}_t^{(k-1)})(\bar{V}_k + \bar{Z}_k \bar{Y}_t^{(k-1)})^{-1} \tag{42}$$

$\bar{Y}_t^{(k)}$ can be represented in the following way:

$$\bar{Y}_t^{(k)} = \begin{bmatrix} \bar{y}_{11} & \bar{y}_{12} \\ \bar{y}_{12}^{t} & \bar{y}_{22} \end{bmatrix}_k \tag{43}$$

With a magnetic wall as lower boundary there is $\bar{\mathbf{H}}_0 = \mathbf{0}$. From eq. (30) follows for eq. (36), the first step of the recurrence according to eq. (42)

$$\bar{Y}_t^{(1)} = \bar{Y}_1 \bar{V}_1^{-1} \tag{44}$$

It was assumed that no metallization is to be found in any of the interfaces. If there is a metallization in the interface m, then

$$\bar{\mathbf{H}}_{m^+} - \bar{\mathbf{H}}_{m^-} = -\bar{\mathbf{J}}_m \tag{45}$$

has to be fulfilled, with $\bar{\mathbf{J}}_m = \eta_0 (j\bar{\mathbf{J}}_x^{t}, \bar{\mathbf{J}}_z^{t})^{t}$. The upper side of the metallization is marked by m^+ and the lower side by m^-. $\bar{\mathbf{J}}_m$ is the transformed current in the interface m.

For a subsequent transformation over the substrate layer $m+1$, $\bar{\mathbf{H}}_A$ on the right-hand side of eq. (30) must be substituted by

$$\bar{\mathbf{H}}_{m^+} = \bar{\mathbf{H}}_{m^-} - \bar{\mathbf{J}}_m = \bar{Y}_t^{(m)} \bar{\mathbf{E}}_m - \bar{\mathbf{J}}_m \tag{46}$$

For the components in interface $m+1$, we obtain

$$\bar{\mathbf{H}}_{m+1} = \bar{\mathbf{Y}}_t^{(m+1)}\bar{\mathbf{E}}_{m+1} + (\bar{\mathbf{Y}}^{(m+1)}\mathbf{Z}_{m+1} - \bar{\mathbf{V}}_{m+1})\bar{\mathbf{J}}_m \tag{47}$$

In a corresponding way we can start from the upper cover by means of eq. (35) and transform downwards. If we start counting the interfaces with zero at the top, we obtain for the interface l

$$\bar{\mathbf{H}}_l = -\bar{\mathbf{Y}}_t^{(l)}\bar{\mathbf{E}}_l \tag{48}$$

with

$$\bar{\mathbf{Y}}_t^{(l)} = (\bar{\mathbf{Y}}_l + \bar{\mathbf{V}}_l\bar{\mathbf{Y}}_t^{(l-1)})(\bar{\mathbf{V}}_l + \bar{\mathbf{Z}}_l\bar{\mathbf{Y}}_t^{(l-1)})^{-1} \tag{49}$$

and

$$\bar{\mathbf{Y}}_t^{(1)} = \bar{\mathbf{V}}_1\bar{\mathbf{Z}}_1^{-1} \quad \text{or} \quad \bar{\mathbf{Y}}_t^{(1)} = \bar{\mathbf{Y}}_1\bar{\mathbf{V}}_1^{-1}$$

with electric or magnetic cover, respectively. Again we have

$$\bar{\mathbf{Y}}_t^{(l)} = \begin{bmatrix} \bar{\mathbf{y}}_{11} & \bar{\mathbf{y}}_{12} \\ \bar{\mathbf{y}}_{12}^{t} & \bar{\mathbf{y}}_{22} \end{bmatrix}_l \tag{50}$$

It has been assumed that a metallization exists in none of the interfaces zero to l.

2.5. System Equation in the Transform Domain

In the next step we have to take care of fulfilling the continuity conditions at all the interfaces. At most of the interfaces provision was made for this by the special approach. In other words, the fields resulting from the transformation from above and below must be matched to each other at the remaining interfaces. The result will be an algebraic relation that connects the tangential electric fields with the currents at the interfaces with metallization. (The subscript M in the following basic equations marks this association.)

$$\bar{\mathbf{J}}_M = f(\bar{\mathbf{E}}_M) \qquad \text{or} \qquad \bar{\mathbf{E}}_M = f(\bar{\mathbf{J}}_M) \tag{51}$$

Let interface k (counted from the bottom), for example, be identical to interface l (counted from the top) and let a metallization be only at this interface (now marked M, see, e.g., Fig. 4 with $k = 2$, $l = 1$). At this interface the matching equations hold

$$\bar{\mathbf{E}}_k = \bar{\mathbf{E}}_l = \bar{\mathbf{E}}_M \qquad \bar{\mathbf{H}}_k - \bar{\mathbf{H}}_l = \bar{\mathbf{J}}_M \tag{52}$$

With eqs. (41) and (48) we get the system equations in the transform domain:

$$(\bar{Y}_t^{(k)} + \bar{Y}_t^{(l)})\bar{\mathbf{E}}_M = \bar{\mathbf{J}}_M \tag{53}$$

or

$$(\bar{Y}_t^{(k)} + \bar{Y}_t^{(l)})^{-1}\bar{\mathbf{J}}_M = \bar{\mathbf{E}}_M \tag{54}$$

As another example we look at the waveguide in Fig. 6. At interfaces A and B relations according to eqs. (41) and (48) hold, namely,

$$\bar{\mathbf{H}}_A^{\mathrm{I}} = \bar{Y}_t^{\mathrm{I}}\bar{\mathbf{E}}_A \qquad \bar{\mathbf{H}}_B^{\mathrm{III}} = -\bar{Y}_t^{\mathrm{III}}\bar{\mathbf{E}}_B \tag{55}$$

Moreover, an equation according to (29) holds for the tangential field components at interfaces A and B of layer II:

$$\begin{bmatrix} \bar{\mathbf{H}}_A^{\mathrm{II}} \\ \bar{\mathbf{H}}_B^{\mathrm{II}} \end{bmatrix} = \begin{bmatrix} \bar{y}_1^{\mathrm{II}} & \bar{y}_2^{\mathrm{II}} \\ \bar{y}_2^{\mathrm{II}} & \bar{y}_1^{\mathrm{II}} \end{bmatrix} \begin{bmatrix} \bar{\mathbf{E}}_A \\ -\bar{\mathbf{E}}_B \end{bmatrix} \tag{56}$$

In the formulation of eqs. (55) and (56), namely by using the same $\bar{\mathbf{E}}_A$ and $\bar{\mathbf{E}}_B$, respectively, in both systems of equations, the continuity condition for the tangential electric field is already fulfilled. For the magnetic field we have

$$\bar{\mathbf{H}}_A^{\mathrm{I}} - \bar{\mathbf{H}}_A^{\mathrm{II}} = \bar{\mathbf{J}}_A \qquad \bar{\mathbf{H}}_B^{\mathrm{II}} - \bar{\mathbf{H}}_B^{\mathrm{III}} = \bar{\mathbf{J}}_B \tag{57}$$

The combination of eqs. (55)–(57) leads to the following system equation in the transform domain:

$$\begin{bmatrix} \bar{y}_1^{\mathrm{II}} - \bar{Y}_t^{\mathrm{I}} & \bar{y}_2^{\mathrm{II}} \\ \bar{y}_2^{\mathrm{II}} & \bar{y}_1^{\mathrm{II}} - \bar{Y}_t^{\mathrm{III}} \end{bmatrix} \begin{bmatrix} \bar{\mathbf{E}}_A \\ -\bar{\mathbf{E}}_B \end{bmatrix} = \begin{bmatrix} -\bar{\mathbf{J}}_A \\ \bar{\mathbf{J}}_B \end{bmatrix} \tag{58}$$

This equation should be converted depending on the width of the metallizations compared with the width of the slots. If, for example, at interface B merely the metal strip in the middle exists, i.e., the width of the two outer metallizations is zero, then a conversion into

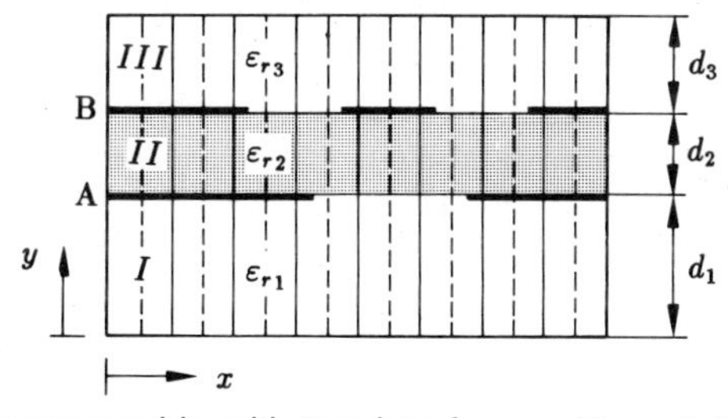

Fig. 6 Quasiplanar waveguide with two interfaces with metallization (model for a finline).

$$\begin{bmatrix} \bar{Y}_A & \bar{V}_{AB} \\ \bar{V}_{BA} & \bar{Z}_B \end{bmatrix} \begin{bmatrix} \bar{\mathbf{E}}_A \\ \bar{\mathbf{J}}_B \end{bmatrix} = \begin{bmatrix} -\bar{\mathbf{J}}_A \\ \bar{\mathbf{E}}_B \end{bmatrix} \tag{59}$$

should take place. The reason for this form of representation will be made clear by the ideas put forward in the next section.

2.6. System Equation in the Spatial Domain

Now the interface conditions in the interfaces with metallization still have to be fulfilled. This can be done only in the spatial domain. The tangential electric field components on the metallic strips and the electric current densities in the slots must be zero.

$$E_{xi}, E_{zi} = 0 \qquad J_{xk}, J_{zk} = 0 \tag{60}$$

where i takes the values of the line numbers inside the metallization and k the values outside the metallization. For interface M in Fig. 4, the quantities **E** and **J** can be calculated in the following way:

$$\mathbf{E}_{xM} = \begin{bmatrix} \mathbf{E}_{xs} \\ \mathbf{0} \end{bmatrix} \quad \mathbf{E}_{zM} = \begin{bmatrix} \mathbf{E}_{zs} \\ \mathbf{0} \end{bmatrix} \quad \mathbf{J}_{xM} = \begin{bmatrix} \mathbf{0} \\ \mathbf{J}_{xm} \end{bmatrix} \quad \mathbf{J}_{zM} = \begin{bmatrix} \mathbf{0} \\ \mathbf{J}_{zm} \end{bmatrix} \tag{61}$$

The subscript s in $\mathbf{E}_x$ and $\mathbf{E}_z$ stands for slot, and the subscript m in $\mathbf{J}_x$ and $\mathbf{J}_z$ for metallization. Now these conditions have to be substituted in eqs. (53) or (54), which have to be transformed back to the spatial domain. We assume that the width of the metallizaiton is less than the width of the slot, and therefore we use eq. (54). As the matrices have a structure according to eqs. (43) and (49), respectively, we write for eq. (54)

$$\begin{bmatrix} \bar{Z}_{11} & \bar{Z}_{12} \\ \bar{Z}^{t}_{12} & \bar{Z}_{22} \end{bmatrix} \begin{bmatrix} j\bar{\mathbf{J}}_{xM} \\ \bar{\mathbf{J}}_{zM} \end{bmatrix} = \begin{bmatrix} \bar{\mathbf{E}}_{xM} \\ -j\bar{\mathbf{E}}_{zM} \end{bmatrix} \tag{62}$$

The quantities $\bar{\mathbf{J}}_x$ and $\bar{\mathbf{E}}_x$ are now to be transformed back by $\mathbf{T}_h$, and the quantities $\bar{\mathbf{J}}_z$ and $\bar{\mathbf{E}}_z$ with $\mathbf{T}_e$. Thus the inverse transform of eq. (62) in the spatial domain is

$$\begin{bmatrix} \boldsymbol{T}_h & \boldsymbol{0} \\ \boldsymbol{0} & \boldsymbol{T}_e \end{bmatrix} \begin{bmatrix} \bar{Z}_{11} & \bar{Z}_{12} \\ \bar{Z}^{t}_{12} & \bar{Z}_{22} \end{bmatrix} \begin{bmatrix} \boldsymbol{T}^{t}_h & \boldsymbol{0} \\ \boldsymbol{0} & \boldsymbol{T}^{t}_e \end{bmatrix} \begin{bmatrix} j\mathbf{J}_{xM} \\ \mathbf{J}_{zM} \end{bmatrix} = \begin{bmatrix} \mathbf{E}_{xM} \\ -j\mathbf{E}_{zM} \end{bmatrix} \tag{63}$$

or

$$\begin{bmatrix} \mathbf{Z}_{11} & \mathbf{Z}_{12} \\ \mathbf{Z}^{t}_{12} & \mathbf{Z}_{22} \end{bmatrix} \begin{bmatrix} j\mathbf{J}_{xM} \\ \mathbf{J}_{zM} \end{bmatrix} = \begin{bmatrix} \mathbf{E}_{xM} \\ -j\mathbf{E}_{zM} \end{bmatrix} \tag{64}$$

where the $\boldsymbol{Z}_{ik}$ result from the $\bar{\boldsymbol{Z}}_{ik}$ by multiplication with the corresponding $\boldsymbol{T}$ or $\boldsymbol{T}^{\mathrm{t}}$, respectively, from the left and right. In eq. (64) we introduce the conditions according to eq. (61) and obtain

$$\begin{bmatrix} \boldsymbol{Z}_{11} & \boldsymbol{Z}_{12} \\ \boldsymbol{Z}_{12}^{\mathrm{t}} & \boldsymbol{Z}_{22} \end{bmatrix} \begin{bmatrix} \mathbf{0} \\ j\mathbf{J}_{xm} \\ \mathbf{0} \\ \mathbf{J}_{zm} \end{bmatrix} = \begin{bmatrix} \mathbf{E}_{xs} \\ \mathbf{0} \\ -j\mathbf{E}_{zs} \\ \mathbf{0} \end{bmatrix} \tag{65}$$

It is observed that in this system of equations those columns of the block matrix of $\boldsymbol{Z}_{ik}$, which in the course of multiplication with the vector $\mathbf{J}$ have to be multiplied with the subvector $\mathbf{0}$, give no contributions. Hence it is not necessary to calculate them at all. For this reason we rewrite eq. (65) in a convenient way. In the first step we omit the columns mentioned in each submatrix $\boldsymbol{Z}_{ik}$, yielding the so-called reduced matrix. In the second step we partition each of the rectangular submatrices in an upper (superscript u) and a lower (superscript l) submatrix according to the partition of the vector on the right-hand side, and we obtain from eq. (65)

$$\begin{bmatrix} \boldsymbol{Z}_{11}^{ru} & \boldsymbol{Z}_{12}^{ru} \\ \boldsymbol{Z}_{11}^{rl} & \boldsymbol{Z}_{12}^{rl} \\ \boldsymbol{Z}_{21}^{ru} & \boldsymbol{Z}_{22}^{ru} \\ \boldsymbol{Z}_{21}^{rl} & \boldsymbol{Z}_{22}^{rl} \end{bmatrix} \begin{bmatrix} j\mathbf{J}_{xM} \\ \mathbf{J}_{zM} \end{bmatrix} = \begin{bmatrix} \mathbf{E}_{xs} \\ \mathbf{0} \\ -j\mathbf{E}_{zs} \\ \mathbf{0} \end{bmatrix} \tag{66}$$

The superscript r means reduced in connection with the first step. The matrix $\boldsymbol{Z}_{12}^{\mathrm{t}}$ was termed $\boldsymbol{Z}_{21}$ here, as its reduced parts are only partly related with the reduced parts of $\boldsymbol{Z}_{12}$. Concrete: $\boldsymbol{Z}_{21}^{rl} = (\boldsymbol{Z}_{12}^{rl})^{\mathrm{t}}$ only for the lower submatrix. The system of equations (66) is divided into two systems

$$\begin{bmatrix} \boldsymbol{Z}_{11}^{rl} & \boldsymbol{Z}_{12}^{rl} \\ \boldsymbol{Z}_{12}^{rlt} & \boldsymbol{Z}_{22}^{rl} \end{bmatrix} \begin{bmatrix} j\mathbf{J}_{xm} \\ \mathbf{J}_{zm} \end{bmatrix} = \begin{bmatrix} \mathbf{0} \\ \mathbf{0} \end{bmatrix} \tag{67}$$

and

$$\begin{bmatrix} \boldsymbol{Z}_{11}^{ru} & \boldsymbol{Z}_{12}^{ru} \\ \boldsymbol{Z}_{21}^{ru} & \boldsymbol{Z}_{22}^{ru} \end{bmatrix} \begin{bmatrix} j\mathbf{J}_{xm} \\ \mathbf{J}_{zm} \end{bmatrix} = \begin{bmatrix} \mathbf{E}_{xs} \\ -j\mathbf{E}_{zs} \end{bmatrix} \tag{68}$$

System (67) is an indirect eigenvalue system. The elements $(\boldsymbol{Z}_{ij}^{rl})_{lk}$ of the system matrix contain the normalized propagation constant as $\varepsilon_{re} = (k_z/k_0)^2$. The eigenvalue ε_{re} must be varied until the determinant of this system matrix vanishes.

The current vector $[j\mathbf{J}_{xm}^{\mathrm{t}}, \mathbf{J}_{zm}^{\mathrm{t}}]^{\mathrm{t}}$ is determined as an eigenvector afterwards. If ε_{re} and the current vector are evaluated, the field vector $[\mathbf{E}_{xs}^{\mathrm{t}}, -j\mathbf{E}_{zs}^{\mathrm{t}}]^{\mathrm{t}}$ can be calculated with the system (68) and all the other field components in the various interfaces by means of the equations given in

Section 2.3. These must be transformed back to the spatial domain first. The field quantities can also be determined inside the layers, as the transformed field quantities vary in the y direction according to the line equations [compare eq. (18)].

The system matrix in eq. (67) is smaller than the matrix in eq. (65) or even considerably smaller in the case of narrow strip widths compared with the slot widths. Often a few lines on the strip are sufficient for very accurate results. If the width of the strip is greater than the slot width, eq. (53) should be transformed back into the spatial domain instead of eq. (54) and should be reduced by an approach analogous to the one above. Instead of eq. (67), an indirect eigenvalue system for the vector $[\mathbf{E}^{t}_{xs}, -j\mathbf{E}^{t}_{zs}]^{t}$ results, and a second system of equations is used to determine the vector $[j\mathbf{J}^{t}_{xm}, \mathbf{J}^{t}_{zm}]^{t}$. This corresponds to the fact that the analysis of a time-invariant network can be either done by the node or by mesh analysis or by a mixed analysis. A mixed analysis should be used here in the case of waveguides as shown in Fig. 6 with a strip in interface B and the slot in interface A under the assumption of the depicted proportions. Whereas in interface A the smaller line number is present in the slot region, the line number in interface B is smaller in the region of the strip than in rest of the interface. Thus eq. (59) should be transformed back and then reduced corresponding to those components that are different from zero in the subvectors $\mathbf{E}_A$ and $\mathbf{J}_B$. Finally the system is partitioned into two systems again, according to the parts on the right-hand side, which are either zero or not. In Section 3.1 it is shown how the reduction can be done by a reduction of the transformation matrices.

2.7. Nonequidistant Discretization

A disadvantage of the method of lines lies in the fact that for the case of extreme differences in the width of the conductors and the intermediate gaps, the number of lines, and therefore the computing time as well, increases dramatically. A possible remedy is nonequidistant discretization. The fundamental research work for this was done by Diestel [6]. A presentation with various examples is given in Reference 12. The specific differences are only sketched here.

Figure 7 visualizes the nonequidistant discretization and makes clear that

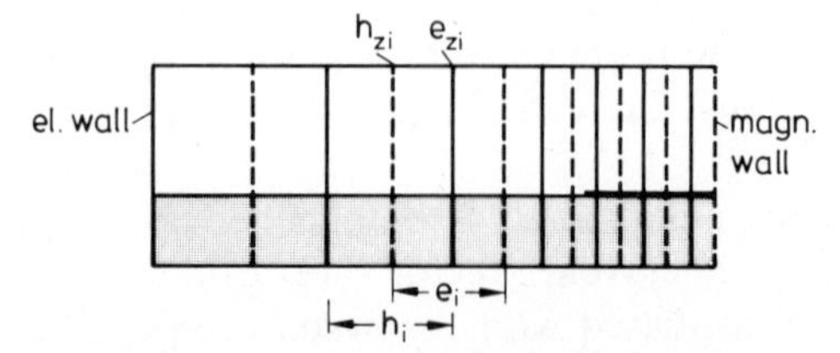

Fig. 7 Example for the positioning of the discretization lines for the field components e_z and h_z with nonequidistant discretization.

the distance of the lines is increased where the field concentration decreases. The distances of the lines for e_z and h_z are marked with e_i and h_i, respectively. In order to get symmetric difference operators for the second derivatives, normalization field components $\mathbf{E}_n$ and $\mathbf{H}_n$ are introduced according to

$$\mathbf{E}_z = \boldsymbol{r}_e \mathbf{E}_{nz} \tag{69}$$

$$\mathbf{H}_z = \boldsymbol{r}_h \mathbf{H}_{nz} \tag{70}$$

with

$$\boldsymbol{r}_e = \operatorname{diag}\left(\sqrt{\frac{h}{e_i}}\right) \qquad \boldsymbol{r}_h = \operatorname{diag}\left(\sqrt{\frac{h}{h_i}}\right) \tag{71}$$

For equidistant discretization, both $\boldsymbol{r}_e$ and $\boldsymbol{r}_h$ turn into unit matrices.

The difference quotient for the first derivative of e_z with respect to x is determined on the discretization lines for h_z. On the ith line for h_z the first derivative for e_z is approximated in the following way:

$$\frac{\partial e_z}{\partial x} \to \frac{E_{z(i+1)} - E_{zi}}{h_i} \tag{72}$$

In matrix representation and after normalization we obtain

$$h\boldsymbol{r}_h^{-1} \operatorname{diag}\left(\frac{\partial e_z}{\partial x}\bigg|_i\right) \to \boldsymbol{r}_h \boldsymbol{D} \mathbf{E}_z = \boldsymbol{r}_h \boldsymbol{D} \boldsymbol{r}_e \mathbf{E}_{nz} = \bar{\boldsymbol{D}} \mathbf{E}_{nz} \tag{73}$$

For the boundary conditions in Fig. 7, $\boldsymbol{D}$ is given according to eq. (7). For the first derivative of h_z, it holds analogously that

$$h\boldsymbol{r}_e^{-1} \operatorname{diag}\left(\frac{\partial h_z}{\partial x}\bigg|_i\right) \to -\boldsymbol{r}_e \boldsymbol{D}^{\mathrm{t}} \mathbf{H}_z = -\boldsymbol{r}_e \boldsymbol{D}^{\mathrm{t}} \boldsymbol{r}_h \mathbf{H}_{nz} = -\bar{\boldsymbol{D}}^{\mathrm{t}} \mathbf{H}_{nz} \tag{74}$$

For the second derivatives we obtain

$$h^2 \boldsymbol{r}_e^{-1} \operatorname{diag}\left(\frac{\partial^2 e_z}{\partial x^2}\bigg|_i\right) \to -\bar{\boldsymbol{D}}^{\mathrm{t}} \bar{\boldsymbol{D}} \mathbf{E}_{nz} = -\bar{\boldsymbol{P}}_{\mathrm{DN}} \mathbf{E}_{nz} \tag{75}$$

$$h^2 \boldsymbol{r}_h^{-1} \operatorname{diag}\left(\frac{\partial^2 h_z}{\partial x^2}\bigg|_i\right) \to -\bar{\boldsymbol{D}} \bar{\boldsymbol{D}}^{\mathrm{t}} \mathbf{H}_{nz} = -\bar{\boldsymbol{P}}_{\mathrm{ND}} \mathbf{H}_{nz} \tag{76}$$

The difference operators $\bar{\boldsymbol{P}}_{\mathrm{DN}}$ and $\bar{\boldsymbol{P}}_{\mathrm{ND}}$ are real symmetric tridiagonal matrices. Thus they can be diagonalized by an orthogonal transformation.

$$\boldsymbol{T}_{e,h}^{\mathrm{t}} \bar{\boldsymbol{P}} \boldsymbol{T}_{e,h} = \boldsymbol{\lambda}_{e,h}^2 \tag{77}$$

Here the eigenvalues, which form the diagonal matrices, and the eigenvectors, which form the transformation matrices, must be calculated numerically, however. Let

$$\boldsymbol{\psi}_n = \boldsymbol{T}\bar{\boldsymbol{\psi}}_n$$

where $\boldsymbol{\psi}_n$ stands either for $\mathbf{E}_{nz}$ or for $\mathbf{H}_{nz}$ and $\bar{\boldsymbol{\psi}}_n$ either for $\bar{\mathbf{E}}_{nz}$ or for $\bar{\mathbf{H}}_{nz}$. Thus we obtain with eq. (75) or eq. (76) the ordinary differential equation corresponding to eq. (15)

$$\left[\left(\frac{d^2}{dy^2} + k^2 - k_z^2\right)\boldsymbol{I} - h^{-2}\boldsymbol{\lambda}^2\right]\bar{\boldsymbol{\psi}}_n = \mathbf{0} \tag{78}$$

The solution and the subsequent flow of the analysis runs as given in Section 2. Especially

$$\boldsymbol{T}_h^{\mathrm{t}}\bar{\boldsymbol{D}}\boldsymbol{T}_e = \boldsymbol{\delta} = \boldsymbol{\lambda}_e \tag{79}$$

is valid again—as it can be shown numerically—where $\boldsymbol{\delta}$ is a diagonal or a quasidiagonal matrix and

$$\boldsymbol{\lambda}_e^2 = \boldsymbol{\delta}^{\mathrm{t}}\boldsymbol{\delta} \tag{80}$$

$$\boldsymbol{\lambda}_h^2 = \boldsymbol{\delta}\boldsymbol{\delta}^{\mathrm{t}} \tag{81}$$

is valid. For a convenient approach to carry out the nonequidistant discretization we refer to Reference 13.

2.8. About Edge Positioning

A criterion for the estimation of a numeric method is the convergence behavior of the solutions. Especially if singularities occur in some field quantities, as, e.g., at the metallic edges of the planar structures considered here, it may happen that the solutions converge toward the wrong values. This behavior is known as "relative convergence" and occurs if the edge condition [14] is violated. In the mode-matching technique this is the case if the numbers of expansion functions in the individual regions are not proportional to the geometric sizes of these regions.

In the method of lines these relations are always fulfilled. Hence the problem of relative convergence is not observed. In the vicinity of the singularities, however, larger discretization errors can arise. How these can be minimized has been demonstrated in Reference 10. To this end the optimal position of a metallic edge relative to the lines—in Fig. 8, given by the edge parameter p or q, respectively—must be examined. The field components parallel to the strip edge exhibit a regular behavior. Under the condition $\rho \to 0$ it is valid for e_z and h_z, respectively [15]:

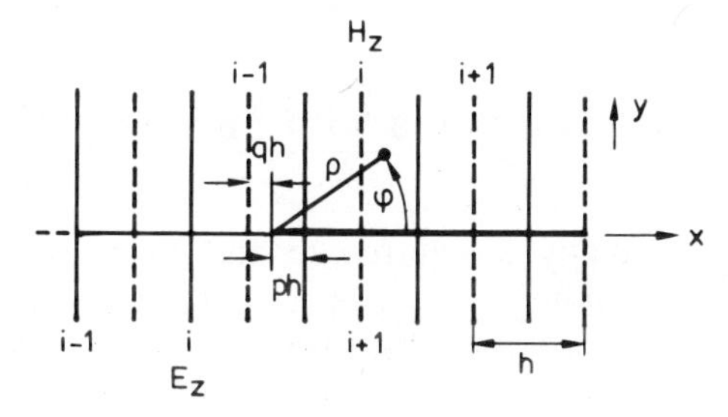

Fig. 8 Definition of the edge parameters p and q.

$$e_z \approx \rho^{1/2} \sin \frac{\varphi}{2} \tag{82}$$

$$h_z \approx \rho^{1/2} \cos \frac{\varphi}{2} \tag{83}$$

Substituting these expressions into the Helmholtz equation $F = L\psi$ with $\psi = e_x$ or $\psi = h_z$, respectively,

$$F = (k^2 - k_z^2)\psi$$

which is a finite expression, remains. The same behavior should be observed for the discretized field. We demand this on the lines i in the plane $y = 0$. This leads to the equation

$$\left.\frac{d^2\psi_i}{dy^2}\right|_{y=0} + \frac{\psi_{i-1} - 2\psi_i + \psi_{i+1}}{h^2} = 0 \tag{84}$$

Noting that $E_{x,i+1} = 0$ and $H_{z,i-1} = 0$ because of eqs. (82) and (83), respectively, this turns into an equation for $p = \zeta$ and $q = \zeta$, respectively.

$$\frac{1}{2}(1 - \zeta)^{-3/2} + 2(2 - \zeta)^{1/2} - 4(1 - \zeta)^{1/2} = 0 \tag{85}$$

with the solution $\zeta = 0.265$. It is not possible to achieve both values at the same time, as there has to be $p + q = 0.5$. For practical application $p = q = 0.25$ is chosen. The value for the edge parameter found here is valid only for infinitely long edges. In the case of finite edge length, for example, for resonator structures, the edge parameter must be determined again for each structure. We want to emphasize that the solutions of the method of lines always converge, even if the edge parameter is not chosen optimally. This convergence behavior is presented in Section 4, Fig. 16.

3. EXTENSIONS

3.1. Extension for Structures with Finite Metallization Thickness

For the analysis of many structures in the microwave technique it is sufficient to assume infinitely thin strip or metallization thickness, as has been done so far in our approach. In MMICs, however, the finite thickness must be taken into account, as it is in the same order of magnitude as the width of the strip or of the slots between the strips.

To present the analysis [36] we assume a structure according to Fig. 9. For an easier representation the metallization was considered in only one interface. Regions I and III consist of several layers of dielectrics with different ε_r. Region II contains the metallizations for the arrangements of the strips; the gaps between the strips can be also filled with different dielectrics. The edges of the metallizations do always coincide with a full line. Thus the intermediate region IIi is enclosed by metallic walls. It is interspersed with N^i_{II} e_z lines and $(N^i_{\mathrm{II}}+1)$ h_z lines. As a connection between the transformed tangential electric and magnetic field components at interfaces A and B for this intermediate region, we get analogously to eq. (26)

$$\eta_0 \begin{bmatrix} -j\bar{\mathbf{H}}_{zA} \\ \bar{\mathbf{H}}_{xA} \\ -j\bar{\mathbf{H}}_{zB} \\ \bar{\mathbf{H}}_{xB} \end{bmatrix}_{\mathrm{II}i} = \begin{bmatrix} -\varepsilon_d \boldsymbol{\gamma}_h & \boldsymbol{\gamma}_h \tilde{\boldsymbol{\delta}} & -\varepsilon_d \boldsymbol{\alpha}_h & \boldsymbol{\alpha}_h \tilde{\boldsymbol{\delta}} \\ \tilde{\boldsymbol{\delta}}^{\mathrm{t}} \boldsymbol{\gamma}_h & \boldsymbol{\gamma}_E & \tilde{\boldsymbol{\delta}}^{\mathrm{t}} \boldsymbol{\alpha}_h & \boldsymbol{\alpha}_E \\ -\varepsilon_d \boldsymbol{\alpha}_h & \boldsymbol{\alpha}_h \tilde{\boldsymbol{\delta}} & -\varepsilon_d \boldsymbol{\gamma}_h & \boldsymbol{\gamma}_h \tilde{\boldsymbol{\delta}} \\ \tilde{\boldsymbol{\delta}}^{\mathrm{t}} \boldsymbol{\alpha}_h & \boldsymbol{\alpha}_E & \tilde{\boldsymbol{\delta}}^{\mathrm{t}} \boldsymbol{\gamma}_h & \boldsymbol{\gamma}_E \end{bmatrix}_{\mathrm{II}i} \begin{bmatrix} \bar{\mathbf{E}}_{xA} \\ -j\bar{\mathbf{E}}_{zA} \\ -\bar{\mathbf{E}}_{xB} \\ j\bar{\mathbf{E}}_{zB} \end{bmatrix}_{\mathrm{II}i} \tag{86}$$

Such a system of equations has to be established for each of the intermediate regions. In general the quantities in the matrix are different from one intermediate region to the other. In particular, there are different matrices $\boldsymbol{\lambda}$ and $\boldsymbol{T}$ in these subregions, and these are again different from those of regions I and III at any rate. Collating the subvectors of the field

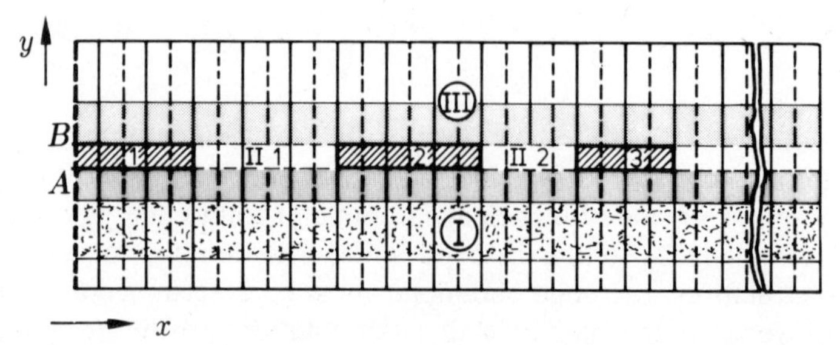

Fig. 9 Cross section of a waveguide with several strips of finite thickness.

components of the same kind and the same interface according to

$$\bar{\mathbf{E}}_{A,B,\mathrm{II}i} = \begin{bmatrix} \bar{\mathbf{E}}_{xA,B} \\ -j\bar{\mathbf{E}}_{zA,B} \end{bmatrix}_{\mathrm{II}i} \qquad \bar{\mathbf{H}}_{A,B\,\mathrm{II}i} = \eta_0 \begin{bmatrix} -j\bar{\mathbf{H}}_{zA,B} \\ \bar{\mathbf{H}}_{xA,B} \end{bmatrix}_{\mathrm{II}i} \tag{87}$$

eq. (86) can be written in a shorter way:

$$\begin{bmatrix} \bar{\mathbf{H}}_A \\ \bar{\mathbf{H}}_B \end{bmatrix}_{\mathrm{II}i} = \begin{bmatrix} \bar{\mathbf{y}}_1^{\mathrm{II}i} & \bar{\mathbf{y}}_2^{\mathrm{II}i} \\ \bar{\mathbf{y}}_2^{\mathrm{II}i} & \bar{\mathbf{y}}_1^{\mathrm{II}i} \end{bmatrix} \begin{bmatrix} \bar{\mathbf{E}}_A \\ -\bar{\mathbf{E}}_B \end{bmatrix}_{\mathrm{II}i} \tag{88}$$

with

$$\bar{\mathbf{y}}_1^{\mathrm{II}i} = \begin{bmatrix} -\varepsilon_d \boldsymbol{\gamma}_h & \boldsymbol{\gamma}_h \tilde{\boldsymbol{\delta}} \\ \tilde{\boldsymbol{\delta}}^{\mathrm{t}} \boldsymbol{\gamma}_h & \boldsymbol{\gamma}_E \end{bmatrix}_{\mathrm{II}i} \qquad \bar{\mathbf{y}}_2^{\mathrm{II}i} = \begin{bmatrix} -\varepsilon_d \boldsymbol{\alpha}_h & \boldsymbol{\alpha}_h \tilde{\boldsymbol{\delta}} \\ \tilde{\boldsymbol{\delta}}^{\mathrm{t}} \boldsymbol{\alpha}_h & \boldsymbol{\alpha}_E \end{bmatrix}_{\mathrm{II}i} \tag{89}$$

The connection between the transformed tangential field components at the upper interface A of region I can be written as

$$\begin{bmatrix} -j\bar{\mathbf{H}}_{zA} \\ \bar{\mathbf{H}}_{xA} \end{bmatrix} = \begin{bmatrix} \bar{\mathbf{y}}_{11}^{\mathrm{I}} & \bar{\mathbf{y}}_{12}^{\mathrm{I}} \\ \bar{\mathbf{y}}_{12}^{\mathrm{It}} & \bar{\mathbf{y}}_{22}^{\mathrm{I}} \end{bmatrix} \begin{bmatrix} \bar{\mathbf{E}}_{xA} \\ -j\bar{\mathbf{E}}_{zA} \end{bmatrix} \tag{90}$$

The diagonal matrices $\bar{\mathbf{y}}_{ik}^{\mathrm{I}}$ are determined in the same way as described in Section 2.4. Hence they are given by eq. (42). Analogously, the following relation holds at the lower interface B of region III:

$$\begin{bmatrix} j\bar{\mathbf{H}}_{zB} \\ -\bar{\mathbf{H}}_{xB} \end{bmatrix} = \begin{bmatrix} \bar{\mathbf{y}}_{11}^{\mathrm{III}} & \bar{\mathbf{y}}_{12}^{\mathrm{III}} \\ \bar{\mathbf{y}}_{12}^{\mathrm{IIIt}} & \bar{\mathbf{y}}_{22}^{\mathrm{III}} \end{bmatrix} \begin{bmatrix} \bar{\mathbf{E}}_{xB} \\ -j\bar{\mathbf{E}}_{zB} \end{bmatrix} \tag{91}$$

The diagonal matrices $\bar{\mathbf{y}}_{ik}^{\mathrm{III}}$ are given by eq. (49).

Field matching at interfaces A and B has to be done in the spatial domain. At interface A the following hold:

$$\mathbf{E}_{xA}^{\mathrm{I}} = \begin{bmatrix} \mathbf{0} \\ \mathbf{E}_{xA}^{\mathrm{II}\,1} \\ \mathbf{0} \\ \mathbf{E}_{xA}^{\mathrm{II}\,2} \\ \vdots \end{bmatrix} \quad \mathbf{E}_{zA}^{\mathrm{I}} = \begin{bmatrix} \mathbf{0} \\ \mathbf{E}_{zA}^{\mathrm{II}\,1} \\ \mathbf{0} \\ \mathbf{E}_{zA}^{\mathrm{II}\,2} \\ \vdots \end{bmatrix} \quad \mathbf{H}_{xA}^{\mathrm{I}} = \begin{bmatrix} \mathbf{J}_{zA}^{1} \\ \mathbf{H}_{xA}^{\mathrm{II}\,1} \\ \mathbf{J}_{zA}^{2} \\ \mathbf{H}_{xA}^{\mathrm{II}\,2} \\ \vdots \end{bmatrix} \quad \mathbf{H}_{zA}^{\mathrm{I}} = \begin{bmatrix} -\mathbf{J}_{xA}^{1} \\ \mathbf{H}_{zA}^{\mathrm{II}\,1} \\ -\mathbf{J}_{xA}^{2} \\ \mathbf{H}_{xA}^{\mathrm{II}\,2} \\ \vdots \end{bmatrix} \tag{92}$$

and analogously at interface B,

$$\mathbf{E}_{xB}^{\mathrm{III}} = \begin{bmatrix} \mathbf{0} \\ \mathbf{E}_{xB}^{\mathrm{II}\,1} \\ \mathbf{0} \\ \mathbf{E}_{xB}^{\mathrm{II}\,2} \\ \vdots \end{bmatrix} \quad \mathbf{E}_{zB}^{\mathrm{III}} = \begin{bmatrix} \mathbf{0} \\ \mathbf{E}_{zB}^{\mathrm{II}\,1} \\ \mathbf{0} \\ \mathbf{E}_{zB}^{\mathrm{II}\,2} \\ \vdots \end{bmatrix} \quad \mathbf{H}_{xB}^{\mathrm{III}} = \begin{bmatrix} -\mathbf{J}_{xB}^{1} \\ \mathbf{H}_{xB}^{\mathrm{II}\,1} \\ -\mathbf{J}_{xB}^{2} \\ \mathbf{H}_{xB}^{\mathrm{II}\,2} \\ \vdots \end{bmatrix} \quad \mathbf{H}_{zB}^{\mathrm{III}} = \begin{bmatrix} \mathbf{J}_{xB}^{1} \\ \mathbf{H}_{xB}^{\mathrm{II}\,1} \\ \mathbf{J}_{xB}^{2} \\ \mathbf{H}_{zB}^{\mathrm{II}\,2} \\ \vdots \end{bmatrix} \tag{93}$$

In these equations the superscripts 1, 2, ... in the currents mark the 1st, 2nd, ... conductor, and in the field components 1, 2, ... mark the 1st, 2nd, ... intermediate region. Transforming eq. (90) back to spatial domain and substituting eq. (92), we get

$$\begin{bmatrix} j\mathbf{J}_{xA}^{1} \\ -j\mathbf{H}_{zA}^{\mathrm{II}\,1} \\ j\mathbf{J}_{xA}^{2} \\ -j\mathbf{H}_{zA}^{\mathrm{II}\,2} \\ \vdots \\ \mathbf{J}_{zA}^{1} \\ \mathbf{H}_{xA}^{\mathrm{II}\,1} \\ \mathbf{J}_{zA}^{2} \\ \mathbf{H}_{xA}^{\mathrm{II}\,2} \\ \vdots \end{bmatrix} = \begin{bmatrix} \boldsymbol{T}_{\mathrm{DN}} & \\ & \boldsymbol{T}_{\mathrm{ND}} \end{bmatrix} \begin{bmatrix} \bar{\boldsymbol{y}}_{11}^{\mathrm{I}} & \bar{\boldsymbol{y}}_{12}^{\mathrm{I}} \\ \bar{\boldsymbol{y}}_{12}^{\mathrm{It}} & \bar{\boldsymbol{y}}_{22}^{\mathrm{I}} \end{bmatrix} \begin{bmatrix} \boldsymbol{T}_{\mathrm{DN}}^{\mathrm{t}} & \\ & \boldsymbol{T}_{\mathrm{ND}}^{\mathrm{t}} \end{bmatrix} \begin{bmatrix} \mathbf{0} \\ \mathbf{E}_{xA}^{\mathrm{II}\,1} \\ \mathbf{0} \\ \mathbf{E}_{xA}^{\mathrm{II}\,2} \\ \vdots \\ \mathbf{0} \\ -j\mathbf{E}_{zA}^{\mathrm{II}\,1} \\ \mathbf{0} \\ -j\mathbf{E}_{zA}^{\mathrm{II}\,2} \\ \vdots \end{bmatrix} \tag{94}$$

This system of equations decomposes into two independent systems, one that describes the connection between the field components in the slot regions and one that allows the calculation of the currents. To achieve this division, we introduce the following abbreviations:

$$\mathbf{H}_{A}^{\mathrm{II}} = \begin{bmatrix} -j\mathbf{H}_{zA}^{\mathrm{II}\,1} \\ -j\mathbf{H}_{zA}^{\mathrm{II}\,2} \\ \vdots \\ \mathbf{H}_{xA}^{\mathrm{II}\,1} \\ \mathbf{H}_{xA}^{\mathrm{II}\,2} \\ \vdots \end{bmatrix} \quad \mathbf{E}_{A}^{\mathrm{II}} = \begin{bmatrix} \mathbf{E}_{xA}^{\mathrm{II}\,1} \\ \mathbf{E}_{xA}^{\mathrm{II}\,2} \\ \vdots \\ -j\mathbf{E}_{zA}^{\mathrm{II}\,1} \\ -j\mathbf{E}_{zA}^{\mathrm{II}\,2} \\ \vdots \end{bmatrix} \quad \mathbf{J}_{A} = \begin{bmatrix} j\mathbf{J}_{xA}^{1} \\ j\mathbf{J}_{xA}^{2} \\ \vdots \\ \mathbf{J}_{zA}^{1} \\ \mathbf{J}_{zA}^{2} \\ \vdots \end{bmatrix} \tag{95}$$

and the reduced (superscript r) transformation matrix and its complement (superscript rc)

$$\boldsymbol{T}_{\mathrm{I}}^{r} = \begin{bmatrix} \boldsymbol{T}_{\mathrm{DN}}^{r} & \\ & \boldsymbol{T}_{\mathrm{ND}}^{r} \end{bmatrix} \quad \boldsymbol{T}_{\mathrm{I}}^{rc} = \begin{bmatrix} \boldsymbol{T}_{\mathrm{DN}}^{rc} & \\ & \boldsymbol{T}_{\mathrm{ND}}^{rc} \end{bmatrix} \tag{96}$$

The reduced transformation matrices just consist of those rows that belong to the lines in the slot. For example, from the square matrix $\boldsymbol{T}_{\mathrm{DN}}$, the non-square reduced matrix $\boldsymbol{T}^{r}_{\mathrm{DN}}$ and the non-square complementary matrix $\boldsymbol{T}^{rc}_{\mathrm{DN}}$ arise by partitioning of the rows (see Fig. 10). From eq. (94) the equations

$$\mathbf{H}_A^{\mathrm{II}} = \boldsymbol{T}_{\mathrm{I}}^{r}\bar{y}^{\mathrm{I}}\boldsymbol{T}_{\mathrm{I}}^{rt}\mathbf{E}_A^{\mathrm{II}} \tag{97}$$

$$\mathbf{J}_A = \boldsymbol{T}_{\mathrm{I}}^{rc}\bar{y}^{\mathrm{I}}\boldsymbol{T}_{\mathrm{I}}^{rt}\mathbf{E}_A^{\mathrm{II}} \tag{98}$$

result with $\bar{y}^{\mathrm{I}}$ as the abbreviation for the block matrix of the $\bar{y}_{ik}^{\mathrm{I}}$.

Completely analogously with the abbreviations

$$\mathbf{H}_B^{\mathrm{II}} = \begin{bmatrix} -j\mathbf{H}_{zB}^{\mathrm{II}\,1} \\ -j\mathbf{H}_{zB}^{\mathrm{II}\,2} \\ \vdots \\ \mathbf{H}_{xB}^{\mathrm{II}\,1} \\ \mathbf{H}_{xB}^{\mathrm{II}\,2} \\ \vdots \end{bmatrix} \qquad \mathbf{E}_B^{\mathrm{II}} = \begin{bmatrix} \mathbf{E}_{xB}^{\mathrm{II}\,1} \\ \mathbf{E}_{xB}^{\mathrm{II}\,2} \\ \vdots \\ -j\mathbf{E}_{zB}^{\mathrm{II}\,1} \\ -j\mathbf{E}_{zB}^{\mathrm{II}\,2} \\ \vdots \end{bmatrix} \qquad \mathbf{J}_B = \begin{bmatrix} j\mathbf{J}_{xB}^{1} \\ j\mathbf{J}_{xB}^{2} \\ \vdots \\ \mathbf{J}_{zB}^{1} \\ \mathbf{J}_{zB}^{2} \\ \vdots \end{bmatrix} \tag{99}$$

we obtain from eq. (91) the equations

$$-\mathbf{H}_B^{\mathrm{II}} = \boldsymbol{T}_{\mathrm{I}}^{r}\bar{y}^{\mathrm{III}}\boldsymbol{T}_{\mathrm{I}}^{rt}\mathbf{E}_B^{\mathrm{II}} \tag{100}$$

$$\mathbf{J}_B = \boldsymbol{T}_{\mathrm{I}}^{rc}\bar{y}^{\mathrm{III}}\boldsymbol{T}_{\mathrm{I}}^{rt}\mathbf{E}_B^{\mathrm{II}} \tag{101}$$

with $\bar{y}^{\mathrm{III}}$ as abbreviation for the block matrix of the $\bar{y}_{ik}^{\mathrm{III}}$. The inverse transformation of eq. (88) yields

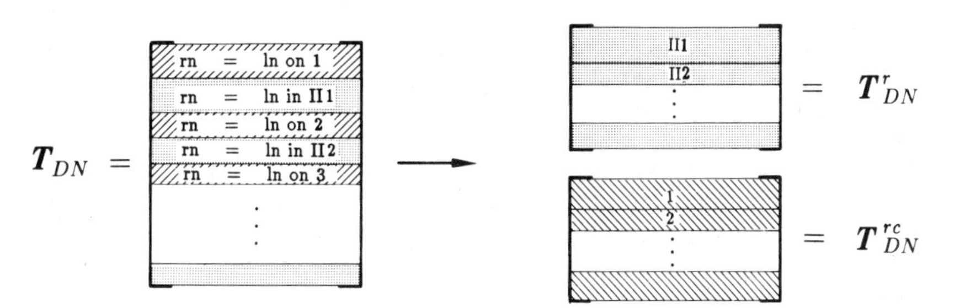

Fig. 10 Illustration of the reduction of the transformation matrices. rn = ln: row numbers are equal to line numbers.

$$\begin{bmatrix} \mathbf{H}_A^{\mathrm{II}i} \\ \mathbf{H}_B^{\mathrm{II}i} \end{bmatrix} = \begin{bmatrix} \boldsymbol{T}^i_{\mathrm{II}} & \\ & \boldsymbol{T}^i_{\mathrm{II}} \end{bmatrix} \begin{bmatrix} \bar{y}_1^{\mathrm{II}i} & \bar{y}_2^{\mathrm{II}i} \\ \bar{y}_2^{\mathrm{II}i} & \bar{y}_1^{\mathrm{II}i} \end{bmatrix} \begin{bmatrix} \boldsymbol{T}^i_{\mathrm{II}} & \\ & \boldsymbol{T}^i_{\mathrm{II}} \end{bmatrix}^{\mathrm{t}} \begin{bmatrix} +\mathbf{E}_A^{\mathrm{II}i} \\ -\mathbf{E}_B^{\mathrm{II}i} \end{bmatrix} \tag{102}$$

with

$$\boldsymbol{T}^i_{\mathrm{II}} = \begin{bmatrix} \boldsymbol{T}^i_{\mathrm{NN}} & \\ & \boldsymbol{T}^i_{\mathrm{DD}} \end{bmatrix} \tag{103}$$

where the $\boldsymbol{T}^i_{\mathrm{NN}}$ and $\boldsymbol{T}^i_{\mathrm{DD}}$ are the specific transformation matrices for each slot. In order to achieve a combination of the equations of all slot regions IIi to one total equation the vectors $\mathbf{H}^{\mathrm{II}i}$ and $\mathbf{E}^{\mathrm{II}i}$ must be divided into parts with x and z components and arranged in the same order as the components in the vectors of eqs. (95) and (99). To this end block matrices are constructed from the submatrices of $\bar{y}_1^{\mathrm{II}i}$ and $\bar{y}_2^{\mathrm{II}i}$ according to

$$\begin{aligned} \bar{y}^{\mathrm{II}}_{k,11} &= \mathrm{diag}(\bar{y}^{\mathrm{II}i}_{k,11}) & \bar{y}^{\mathrm{II}}_{k,22} &= \mathrm{diag}(\bar{y}^{\mathrm{II}i}_{k,22}) \\ \bar{y}^{\mathrm{II}}_{k,12} &= \mathrm{quasidiag}(\bar{y}^{\mathrm{II}i}_{k,12}) & \bar{y}^{\mathrm{II}}_{k,21} &= \mathrm{quasidiag}(\bar{y}^{\mathrm{II}i}_{k,21}) \end{aligned} \tag{104}$$

with $k = 1$ and $k = 2$. These four matrices are now combined to one block matrix

$$\bar{y}^{\mathrm{II}}_k = \begin{bmatrix} \bar{y}^{\mathrm{II}}_{k,11} & \bar{y}^{\mathrm{II}}_{k,12} \\ \bar{y}^{\mathrm{II}}_{k,21} & \bar{y}^{\mathrm{II}}_{k,22} \end{bmatrix} \tag{105}$$

($k = 1$ or 2). Now we form block matrices from the individual transformation matrices according to

$$\bar{\boldsymbol{T}}_{\mathrm{NN}} = \mathrm{diag}\{\boldsymbol{T}^i_{\mathrm{NN}}\} \tag{106}$$

$$\bar{\boldsymbol{T}}_{\mathrm{DD}} = \mathrm{diag}\{\boldsymbol{T}^i_{\mathrm{DD}}\} \tag{107}$$

and collate these to one (super) block matrix

$$\bar{\bar{\boldsymbol{T}}}_{\mathrm{II}} = \mathrm{diag}\{\bar{\boldsymbol{T}}_{\mathrm{NN}}, \bar{\boldsymbol{T}}_{\mathrm{DD}}\} \tag{108}$$

Now the following total equation can be written for the field components in the slot regions of the interfaces A and B

$$\begin{bmatrix} \mathbf{H}_A^{\mathrm{II}} \\ \mathbf{H}_B^{\mathrm{II}} \end{bmatrix} = \begin{bmatrix} \bar{\bar{\boldsymbol{T}}}_{\mathrm{II}} & \\ & \bar{\bar{\boldsymbol{T}}}_{\mathrm{II}} \end{bmatrix} \begin{bmatrix} \bar{y}_1^{\mathrm{II}} & \bar{y}_2^{\mathrm{II}} \\ \bar{y}_2^{\mathrm{II}} & \bar{y}_1^{\mathrm{II}} \end{bmatrix} \begin{bmatrix} \bar{\bar{\boldsymbol{T}}}_{\mathrm{II}} & \\ & \bar{\bar{\boldsymbol{T}}}_{\mathrm{II}} \end{bmatrix}^{\mathrm{t}} \begin{bmatrix} \mathbf{E}_A \\ -\mathbf{E}_B \end{bmatrix} \tag{109}$$

On the left-hand side of this equation the relations according to (97) and (99) are substituted.

Thus the indirect eigenvalue system

$$\left(\begin{bmatrix} \bar{\bar{T}}_{\text{II}} & \\ & \bar{\bar{T}}_{\text{II}} \end{bmatrix} \begin{bmatrix} \bar{y}_1^{\text{II}} & \bar{y}_2^{\text{II}} \\ \bar{y}_2^{\text{II}} & \bar{y}_1^{\text{II}} \end{bmatrix} \begin{bmatrix} \bar{\bar{T}}_{\text{II}} & \\ & \bar{\bar{T}}_{\text{II}} \end{bmatrix}^{\text{t}} - \begin{bmatrix} T_{\text{I}}^{r} & \\ & T_{\text{I}}^{r} \end{bmatrix} \begin{bmatrix} \bar{y}^{\text{I}} & \\ & \bar{y}^{\text{III}} \end{bmatrix} \begin{bmatrix} T_{\text{I}}^{r} & \\ & T_{\text{I}}^{r} \end{bmatrix}^{\text{t}} \right) \times \begin{bmatrix} \mathbf{E}_A \\ -\mathbf{E}_B \end{bmatrix} = \mathbf{0} \tag{110}$$

results, from which the propagation constant can be determined. With the two parts $\mathbf{E}_A$ and $\mathbf{E}_B$ of the eigenvector the currents $\mathbf{J}_A$ and $\mathbf{J}_B$ can be calculated from eqs. (98) and (101). Of course, it is possible to develop a homogeneous system of equations (indirect eigenvalue system) instead of eq. (110) for the currents on the metallizations.

3.2. The Method of Lines for an Inhomogeneous Dielectric Layer

3.2.1. Introduction

In this section the analysis of dielectric waveguides—for instance, optical waveguides, or planar waveguides coupled with dielectric waveguides—is described. An extensive analysis of optical waveguide structures was done by Diestel [6, 16]. The analysis presented here differs in that the Sturm–Liouville differential equation (113b) is used in the self-adjoint form [8] in contrast to the latter. In Fig. (11*a* and *b*) the dielectric waveguide is coupled with a microstrip or a microslot, respectively. In Fig. 11*c* the dielectric layer can be useful to compensate the dispersion of the microstrip wave in a certain frequency range. The structures described are only very simple examples for demonstrating the method. More complex structures, e.g., those consisting of several microstrip lines and several dielectric waveguides, can also be analyzed.

3.2.2. Discretization and Transformation for an Abrupt Transition

In order to solve the problems, the problem of a layer with a space-dependent dielectric constant must be solved. For this purpose we consider a layer as demonstrated in Fig. 12. In this layer the dielectric constant is a function of x.

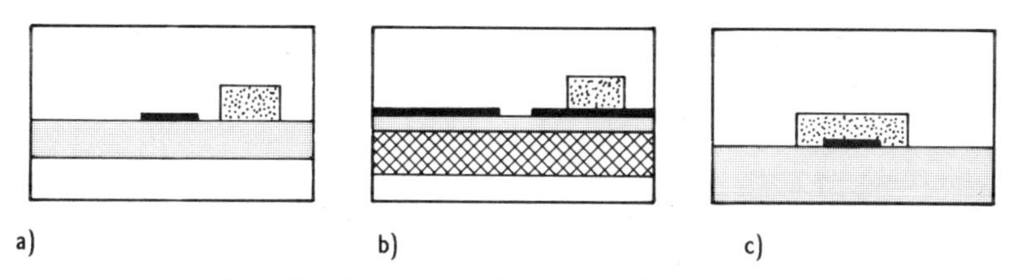

Fig. 11 Some hybrid waveguide structures.

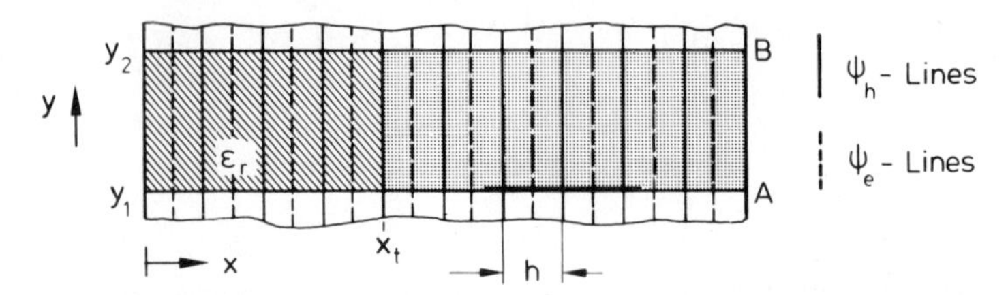

Fig. 12 An inhomogeneous layer with an abrupt transition in the dielectric constant.

After Collin [17] the wave field in this case can be determined from two vector potentials $\mathbf{\Pi}_e$ and $\mathbf{\Pi}_h$, which each have only one component in the x direction. The fields result from these potentials according to

$$\begin{aligned} \mathbf{E} &= \varepsilon_r(x)^{-1}\nabla \times \nabla \times \mathbf{\Pi}_e - jk_0 \nabla \times \mathbf{\Pi}_h \\ \eta_0 \mathbf{H} &= jk_0 \nabla \times \mathbf{\Pi}_e + \nabla \times \nabla \times \mathbf{\Pi}_h \end{aligned} \tag{111}$$

with $k_0 = \omega\sqrt{\mu_0 \varepsilon_0}$ and $\eta_0 = \sqrt{\mu_0/\varepsilon_0}$. For the potentials we set

$$\begin{aligned} \mathbf{\Pi}_e &= \psi_e e^{-jk_z z} k_0^{-2} \mathbf{a}_x \qquad \text{(LSM modes)} \\ \mathbf{\Pi}_h &= \psi_h e^{-jk_z z} k_0^{-2} \mathbf{a}_x \qquad \text{(LSE modes)} \end{aligned} \tag{112}$$

where $\mathbf{a}_x$ is the unit vector in the x direction. With this formulation we have taken into account that the wave is propagating in the z direction. The scalar potentials ψ_e and ψ_h must fulfill the Helmholtz equation and the Sturm–Liouville differential equation, respectively:

$$\begin{aligned} &\frac{\partial^2 \psi_h}{\partial x^2} + \frac{\partial^2 \psi_h}{\partial y^2} + (\varepsilon_r(x) k_0^2 - k_z^2)\psi_h = 0 \qquad (a) \\ \varepsilon_r(x)\, &\frac{\partial}{\partial x}\left(\frac{1}{\varepsilon_r(x)} \frac{\partial \psi_e}{\partial x}\right) + \frac{\partial^2 \psi_e}{\partial y^2} + (\varepsilon_r(x) k_0^2 - k_z^2)\psi_e = 0 \qquad (b) \end{aligned} \tag{113}$$

The following boundary conditions

$$\begin{aligned} &\textsc{Metallic walls:} \qquad \psi_h = 0 \qquad \frac{\partial \psi_e}{\partial x} = 0 \\ &\textsc{Magnetic walls:} \qquad \psi_e = 0 \qquad \frac{\partial \psi_h}{\partial x} = 0 \end{aligned} \tag{114}$$

are valid. Note that the boundary conditions for ψ_h are the same as for e_z in Section 2, and those for ψ_e are the same as for h_z. Hence the quantities $\boldsymbol{\lambda}_h^2$, $\boldsymbol{\lambda}_e^2$, $\boldsymbol{k}_{ye}^2$, $\boldsymbol{k}_{yh}^2$ and the derived ones each have the dual properties as compared with those equally named in Section 2.

Besides the potentials the dielectric constants must be discretized here. Because in the method of lines the field components in discretized form are determined on two systems of lines that are shifted toward each other by half the discretization distance (in the case of equidistant discretization), we must make a distinction between the two line systems:

$$\varepsilon_r(x) \rightarrow \mathrm{diag}(\varepsilon_r(x_e)) = \boldsymbol{\varepsilon}_e \qquad \varepsilon_r(x) \rightarrow \mathrm{diag}(\varepsilon_r(x_h)) = \boldsymbol{\varepsilon}_h \tag{115}$$

For ε_h at interfaces where the dielectric constant changes abruptly, we must take into account special considerations $\varepsilon_r(x_t) = [\varepsilon_r(x_t + 0) + \varepsilon_r(x_t - 0)]/2$ (see Appendix C). The discretized differential quotients can be written in the following way:

$$\begin{gathered}
h \frac{\partial \psi_h}{\partial x} \rightarrow \boldsymbol{D}\boldsymbol{\psi}_h \qquad h \frac{\partial \psi_e}{\partial x} \rightarrow -\boldsymbol{D}^{\mathrm{t}}\boldsymbol{\psi}_e \\
h^2 \frac{\partial^2 \psi_h}{\partial x^2} \rightarrow -\boldsymbol{D}^{\mathrm{t}}\boldsymbol{D}\boldsymbol{\psi}_h = -\boldsymbol{P}_h^{\varepsilon}\boldsymbol{\psi}_h \\
h^2 \varepsilon_r(x) \frac{\partial}{\partial x}\left(\frac{1}{\varepsilon_r(x)} \frac{\partial \psi_e}{\partial x}\right) \rightarrow -\boldsymbol{\varepsilon}_e \boldsymbol{D} \boldsymbol{\varepsilon}_h^{-1} \boldsymbol{D}^{\mathrm{t}} \boldsymbol{\psi}_e = -\boldsymbol{P}_e^{\varepsilon} \boldsymbol{\psi}_e
\end{gathered} \tag{116}$$

Note that in the discretized second derivative of ψ_e, i.e., in $\boldsymbol{P}_e^{\varepsilon}$, the $\boldsymbol{\varepsilon}_h$ appears because the first derivative is taken on a ψ_h line. On the other hand, $\boldsymbol{P}_h^{\varepsilon} = \boldsymbol{P}_h$. The partial differential equations (113) can now be written as follows:

$$\left\{-h^{-2}\boldsymbol{P}^{\varepsilon} + k_0^2 \boldsymbol{\varepsilon} + \left(\frac{d^2}{dy^2} - k_z^2\right)\boldsymbol{I}\right\} \boldsymbol{\psi} = \boldsymbol{0} \tag{117}$$

where either a subscript e or h has to be added to the matrix quantities.

To be able to solve the differential equations for y, a transformation according to

$$\boldsymbol{\psi} = \boldsymbol{T}^{\varepsilon}\tilde{\boldsymbol{\psi}} = \boldsymbol{T}\boldsymbol{S}\tilde{\boldsymbol{\psi}} \tag{118}$$

is performed, where $\boldsymbol{T}_h$ and $\boldsymbol{T}_e$ are the transformation matrices for a homogeneous dielectric layer and $\boldsymbol{S}_h$ and $\boldsymbol{S}_e$ are additional transformation matrices to account for the inhomogeneous dielectric. With the abbreviations

$$\boldsymbol{Q}_e = \bar{\boldsymbol{\varepsilon}}_e \bar{\boldsymbol{\delta}} \bar{\boldsymbol{\varepsilon}}_h^{-1} \bar{\boldsymbol{\delta}}^{\mathrm{t}} - \bar{\boldsymbol{\varepsilon}}_e \qquad \boldsymbol{Q}_h = \bar{\boldsymbol{\lambda}}_h^2 - \bar{\boldsymbol{\varepsilon}}_h \tag{119}$$

where $\bar{\boldsymbol{\delta}} = (k_0 h)^{-1}\boldsymbol{\delta}$; $\boldsymbol{\delta} = \boldsymbol{T}_e^{\mathrm{t}}\boldsymbol{D}\boldsymbol{T}_h$; $\bar{\boldsymbol{\lambda}}_h^2 = \bar{\boldsymbol{\delta}}^{\mathrm{t}}\bar{\boldsymbol{\delta}}$; $\bar{\boldsymbol{\varepsilon}}_e = \boldsymbol{T}_e^{\mathrm{t}}\boldsymbol{\varepsilon}_e\boldsymbol{T}_e$; $\bar{\boldsymbol{\varepsilon}}_h = \boldsymbol{T}_h^{\mathrm{t}}\boldsymbol{\varepsilon}_h\boldsymbol{T}_h$, the following eigenvalue problems are to be solved:

$$\boldsymbol{Q}_e\boldsymbol{S}_e = \boldsymbol{S}_e\tilde{\boldsymbol{\lambda}}_e^2 \qquad \boldsymbol{Q}_h\boldsymbol{S}_h = \boldsymbol{S}_h\tilde{\boldsymbol{\lambda}}_h^2 \tag{120}$$

to obtain the uncoupled ordinary differential equations for $\tilde{\boldsymbol{\psi}}_e$ and $\tilde{\boldsymbol{\psi}}_h$,

$$\left(\boldsymbol{I}\frac{1}{k_0^2}\frac{d^2}{dy^2} - \boldsymbol{k}_{ye,h}^2\right)\tilde{\boldsymbol{\psi}}_{e,h} = \boldsymbol{0} \tag{121}$$

where $\boldsymbol{k}_y = \text{diag}(k_{yi}/k_0)$ and k_{yi}^2 was substituted for $(\tilde{\lambda}_i^2 + \varepsilon_{re})k_0^2$.
The solution to eq. (121) runs:

$$\begin{bmatrix} \tilde{\boldsymbol{\psi}}'_A \\ \tilde{\boldsymbol{\psi}}'_B \end{bmatrix} = \boldsymbol{k}_y^2 \begin{bmatrix} \boldsymbol{\gamma} & \boldsymbol{\alpha} \\ \boldsymbol{\alpha} & \boldsymbol{\gamma} \end{bmatrix} \begin{bmatrix} -\tilde{\boldsymbol{\psi}}_A \\ \tilde{\boldsymbol{\psi}}_B \end{bmatrix} \tag{122}$$

with

$$\tilde{\boldsymbol{\psi}}' = \frac{1}{k_0}\frac{d}{dy}\tilde{\boldsymbol{\psi}}$$

$$\boldsymbol{\alpha} = \text{diag}\left(\frac{k_{yi}}{k_0}\sinh k_{yi}d\right)^{-1} \qquad \boldsymbol{\gamma} = \text{diag}\left(\frac{k_{yi}}{k_0}\tanh k_{yi}d\right)^{-1} \tag{123}$$

In eq. (122), which is similar to eq. (19), either the subscript e or h has to be added to $\tilde{\boldsymbol{\psi}}$ and $\boldsymbol{k}_y$.

3.2.3. Calculations of the Field Components

The field components are calculated from the potentials ψ according to eq. (111). After discretization and transformation with $\boldsymbol{T}_e^{\mathrm{t}}$ and $\boldsymbol{T}_h^{\mathrm{t}}$, respectively, we get the transformed field components necessary for the matching on the interfaces:

$$\bar{\mathbf{E}}_x = -\bar{\boldsymbol{\varepsilon}}_e^{-1}\boldsymbol{Q}_e\boldsymbol{S}_e\tilde{\boldsymbol{\psi}}_e \tag{124}$$

$$\eta_0\bar{\mathbf{H}}_x = -\boldsymbol{Q}_h\boldsymbol{S}_h\tilde{\boldsymbol{\psi}}_h \tag{125}$$

$$j\bar{\mathbf{E}}_z = -\boldsymbol{S}_h\tilde{\boldsymbol{\psi}}'_h - \bar{\boldsymbol{\varepsilon}}_h^{-1}\tilde{\boldsymbol{\delta}}^{\mathrm{t}}\boldsymbol{S}_e\tilde{\boldsymbol{\psi}}_e \tag{126}$$

$$j\eta_0\bar{\mathbf{H}}_z = \boldsymbol{S}_e\tilde{\boldsymbol{\psi}}'_e + \tilde{\boldsymbol{\delta}}\boldsymbol{S}_h\tilde{\boldsymbol{\psi}}_h \tag{127}$$

$$\tilde{\boldsymbol{\delta}} = \sqrt{\varepsilon_{re}}\,\bar{\boldsymbol{\delta}} \tag{128}$$

These equations must be used for both interfaces A and B. The $\tilde{\boldsymbol{\psi}}'$ can be substituted by means of eq. (122).

$$\begin{bmatrix} \bar{\mathbf{E}}_{xA} \\ \bar{\mathbf{E}}_{xB} \end{bmatrix} = -\bar{\boldsymbol{\varepsilon}}_e^{-1}\boldsymbol{Q}_e\boldsymbol{S}_e\begin{bmatrix} \tilde{\boldsymbol{\psi}}_{eA} \\ \tilde{\boldsymbol{\psi}}_{eB} \end{bmatrix} = -(\bar{\boldsymbol{\delta}}\bar{\boldsymbol{\varepsilon}}_h^{-1}\bar{\boldsymbol{\delta}}^{\mathrm{t}} - \boldsymbol{I}_e)\boldsymbol{S}_e\begin{bmatrix} \tilde{\boldsymbol{\psi}}_{eA} \\ \tilde{\boldsymbol{\psi}}_{eB} \end{bmatrix} \tag{129}$$

$$\eta_0\begin{bmatrix}\bar{\mathbf{H}}_{xA}\\ \bar{\mathbf{H}}_{xB}\end{bmatrix} = -\boldsymbol{Q}_h\boldsymbol{S}_h\begin{bmatrix}\tilde{\boldsymbol{\psi}}_{hA}\\ \tilde{\boldsymbol{\psi}}_{hB}\end{bmatrix} = -(\bar{\boldsymbol{\lambda}}_h^2 - \bar{\boldsymbol{\varepsilon}}_h)\boldsymbol{S}_h\begin{bmatrix}\tilde{\boldsymbol{\psi}}_{hA}\\ \tilde{\boldsymbol{\psi}}_{hB}\end{bmatrix} \tag{130}$$

$$\begin{bmatrix}j\bar{\mathbf{E}}_{zA}\\ j\bar{\mathbf{E}}_{zB}\end{bmatrix} = -\boldsymbol{S}_h\boldsymbol{k}_{yh}^2\begin{bmatrix}-\boldsymbol{\gamma}_h & \boldsymbol{\alpha}_h\\ -\boldsymbol{\alpha}_h & \boldsymbol{\gamma}_h\end{bmatrix}\begin{bmatrix}\tilde{\boldsymbol{\psi}}_{hA}\\ \tilde{\boldsymbol{\psi}}_{hB}\end{bmatrix} - \bar{\boldsymbol{\varepsilon}}_h^{-1}\tilde{\boldsymbol{\delta}}^{t}\boldsymbol{S}_e\begin{bmatrix}\tilde{\boldsymbol{\psi}}_{eA}\\ \tilde{\boldsymbol{\psi}}_{eB}\end{bmatrix} \tag{131}$$

$$\eta_0\begin{bmatrix}j\bar{\mathbf{H}}_{zA}\\ j\bar{\mathbf{H}}_{zB}\end{bmatrix} = \boldsymbol{S}_e\boldsymbol{k}_{ye}^2\begin{bmatrix}-\boldsymbol{\gamma}_e & \boldsymbol{\alpha}_e\\ -\boldsymbol{\alpha}_e & \boldsymbol{\gamma}_e\end{bmatrix}\begin{bmatrix}\tilde{\boldsymbol{\psi}}_{eA}\\ \tilde{\boldsymbol{\psi}}_{eB}\end{bmatrix} + \tilde{\boldsymbol{\delta}}\boldsymbol{S}_h\begin{bmatrix}\tilde{\boldsymbol{\psi}}_{hA}\\ \tilde{\boldsymbol{\psi}}_{hB}\end{bmatrix} \tag{132}$$

After discarding the potentials $\tilde{\boldsymbol{\psi}}_e$ and $\tilde{\boldsymbol{\psi}}_h$, we finally get a relation between the tangential components in interface A and interface B.

$$\begin{bmatrix}j\bar{\mathbf{E}}_{zA}\\ j\bar{\mathbf{E}}_{zB}\end{bmatrix} = \eta_0\boldsymbol{S}_h\boldsymbol{k}_{yh}^2\begin{bmatrix}-\boldsymbol{\gamma}_h & \boldsymbol{\alpha}_h\\ -\boldsymbol{\alpha}_h & \boldsymbol{\gamma}_h\end{bmatrix}\boldsymbol{S}_h^{-1}\boldsymbol{Q}_h^{-1}\begin{bmatrix}\bar{\mathbf{H}}_{xA}\\ \bar{\mathbf{H}}_{xB}\end{bmatrix} + \bar{\boldsymbol{\varepsilon}}_h^{-1}\tilde{\boldsymbol{\delta}}^{t}\boldsymbol{Q}_e^{-1}\bar{\boldsymbol{\varepsilon}}_e\begin{bmatrix}\bar{\mathbf{E}}_{xA}\\ \bar{\mathbf{E}}_{xB}\end{bmatrix} \tag{133}$$

$$\eta_0\begin{bmatrix}j\bar{\mathbf{H}}_{zA}\\ j\bar{\mathbf{H}}_{zB}\end{bmatrix} = -\boldsymbol{S}_e\boldsymbol{k}_{ye}^2\begin{bmatrix}-\boldsymbol{\gamma}_e & \boldsymbol{\alpha}_e\\ -\boldsymbol{\alpha}_e & \boldsymbol{\gamma}_e\end{bmatrix}\boldsymbol{S}_e^{-1}\boldsymbol{Q}_e^{-1}\bar{\boldsymbol{\varepsilon}}_e\begin{bmatrix}\bar{\mathbf{E}}_{xA}\\ \bar{\mathbf{E}}_{xB}\end{bmatrix} - \eta_0\tilde{\boldsymbol{\delta}}\boldsymbol{Q}_h^{-1}\begin{bmatrix}\bar{\mathbf{H}}_{xA}\\ \bar{\mathbf{H}}_{xB}\end{bmatrix} \tag{134}$$

The relations for homogeneous layers can be developed directly from these equations. We merely have to set

$$\bar{\boldsymbol{\varepsilon}}_e = \varepsilon_r\boldsymbol{I}_e \qquad \bar{\boldsymbol{\varepsilon}}_h = \varepsilon_r\boldsymbol{I}_h \qquad \bar{\boldsymbol{\delta}}\bar{\boldsymbol{\delta}}^{t} = \bar{\boldsymbol{\lambda}}_e^2 \qquad \boldsymbol{S}_h = \boldsymbol{I}_h \qquad \boldsymbol{S}_e = \boldsymbol{I}_e$$

$$\tilde{\boldsymbol{\lambda}}_e^2 = \bar{\boldsymbol{\lambda}}_e^2 - \varepsilon_r\boldsymbol{I}_e \qquad \tilde{\boldsymbol{\lambda}}_h^2 = \bar{\boldsymbol{\lambda}}_h^2 - \varepsilon_r\boldsymbol{I}_h$$

Using eq. (120) the two systems of equations (133) and (134) are converted to

$$\eta_0\begin{bmatrix}\bar{\mathbf{H}}_{xA}\\ \bar{\mathbf{H}}_{xB}\end{bmatrix} = \boldsymbol{S}_h\tilde{\boldsymbol{\lambda}}_h^2\begin{bmatrix}\boldsymbol{\gamma}_h & \boldsymbol{\alpha}_h\\ \boldsymbol{\alpha}_h & \boldsymbol{\gamma}_h\end{bmatrix}\boldsymbol{S}_h^{-1}\begin{bmatrix}-j\bar{\mathbf{E}}_{zA}\\ j\bar{\mathbf{E}}_{zB}\end{bmatrix}$$
$$+\,\boldsymbol{S}_h\begin{bmatrix}\boldsymbol{\gamma}_h & \boldsymbol{\alpha}_h\\ \boldsymbol{\alpha}_h & \boldsymbol{\gamma}_h\end{bmatrix}\tilde{\boldsymbol{\lambda}}_h^2\boldsymbol{S}_h^{-1}\bar{\boldsymbol{\varepsilon}}_h^{-1}\tilde{\boldsymbol{\delta}}^{t}\boldsymbol{Q}_e^{-1}\bar{\boldsymbol{\varepsilon}}_e\begin{bmatrix}\bar{\mathbf{E}}_{xA}\\ -\bar{\mathbf{E}}_{xB}\end{bmatrix} \tag{135}$$

$$\eta_0\begin{bmatrix}-j\bar{\mathbf{H}}_{zA}\\ -j\bar{\mathbf{H}}_{zB}\end{bmatrix} = \tilde{\boldsymbol{\delta}}\boldsymbol{S}_h\begin{bmatrix}\boldsymbol{\gamma}_h & \boldsymbol{\alpha}_h\\ \boldsymbol{\alpha}_h & \boldsymbol{\gamma}_h\end{bmatrix}\boldsymbol{S}_h^{-1}\begin{bmatrix}-j\bar{\mathbf{E}}_{zA}\\ j\bar{\mathbf{E}}_{zB}\end{bmatrix}$$
$$-\left(\boldsymbol{S}_e\boldsymbol{k}_{ye}^2\begin{bmatrix}\boldsymbol{\gamma}_e & \boldsymbol{\alpha}_e\\ \boldsymbol{\alpha}_e & \boldsymbol{\gamma}_e\end{bmatrix}\tilde{\boldsymbol{\lambda}}_e^{-2}\boldsymbol{S}_e^{-1}\bar{\boldsymbol{\varepsilon}}_e - \tilde{\boldsymbol{\delta}}\boldsymbol{S}_h\begin{bmatrix}\boldsymbol{\gamma}_h & \boldsymbol{\alpha}_h\\ \boldsymbol{\alpha}_h & \boldsymbol{\gamma}_h\end{bmatrix}\boldsymbol{S}_h^{-1}\bar{\boldsymbol{\varepsilon}}_h^{-1}\tilde{\boldsymbol{\delta}}^{t}\boldsymbol{Q}_e^{-1}\bar{\boldsymbol{\varepsilon}}_e\right)\begin{bmatrix}\bar{\mathbf{E}}_{xA}\\ -\bar{\mathbf{E}}_{xB}\end{bmatrix} \tag{136}$$

These equations can be collated using the definitions according to eq. (27) and

$$\bar{y}_1 = \begin{bmatrix} \tilde{\gamma}_H \boldsymbol{\rho}_e & \tilde{\boldsymbol{\delta}} \tilde{\gamma}_h \\ \tilde{\gamma}_h \boldsymbol{\rho} & \tilde{\gamma}_E \end{bmatrix} \quad \text{and} \quad \bar{y}_2 = \begin{bmatrix} \tilde{\boldsymbol{\alpha}}_H \boldsymbol{\rho}_e & \tilde{\boldsymbol{\delta}} \tilde{\boldsymbol{\alpha}}_h \\ \tilde{\boldsymbol{\alpha}}_h \boldsymbol{\rho} & \tilde{\boldsymbol{\alpha}}_E \end{bmatrix} \tag{137}$$

in the form

$$\begin{bmatrix} \bar{\mathbf{H}}_A \\ \bar{\mathbf{H}}_B \end{bmatrix} = \begin{bmatrix} \bar{y}_1 & \bar{y}_2 \\ \bar{y}_2 & \bar{y}_1 \end{bmatrix} \begin{bmatrix} \bar{\mathbf{E}}_A \\ -\bar{\mathbf{E}}_B \end{bmatrix} \tag{138}$$

The following abbreviations hold:

$$\tilde{\gamma}_h = \boldsymbol{S}_h \boldsymbol{\gamma}_h \boldsymbol{S}_h^{-1} \qquad \tilde{\boldsymbol{\alpha}}_h = \boldsymbol{S}_h \boldsymbol{\alpha}_h \boldsymbol{S}_h^{-1}$$

$$\tilde{\gamma}_E = \boldsymbol{S}_h \tilde{\boldsymbol{\lambda}}_h^2 \boldsymbol{\gamma}_h \boldsymbol{S}_h^{-1} \qquad \tilde{\boldsymbol{\alpha}}_E = \boldsymbol{S}_h \tilde{\boldsymbol{\lambda}}_h^2 \boldsymbol{\alpha}_h \boldsymbol{S}_h^{-1}$$

$$\tilde{\gamma}_e = \boldsymbol{S}_e \boldsymbol{k}_{ye}^2 \boldsymbol{\gamma}_e \boldsymbol{S}_e^{-1} \qquad \tilde{\boldsymbol{\alpha}}_e = \boldsymbol{S}_e \boldsymbol{k}_{ye}^2 \boldsymbol{\alpha}_e \boldsymbol{S}_e^{-1}$$

$$\tilde{\gamma}_H = -\tilde{\gamma}_e + \tilde{\boldsymbol{\delta}} \tilde{\gamma}_h \bar{\boldsymbol{\varepsilon}}_h^{-1} \tilde{\boldsymbol{\delta}}^{\mathrm{t}} \qquad \tilde{\boldsymbol{\alpha}}_H = -\tilde{\boldsymbol{\alpha}}_e + \tilde{\boldsymbol{\delta}} \tilde{\boldsymbol{\alpha}}_h \bar{\boldsymbol{\varepsilon}}_h^{-1} \tilde{\boldsymbol{\delta}}^{\mathrm{t}}$$

$$\boldsymbol{\rho}_e = \boldsymbol{Q}_e^{-1} \bar{\boldsymbol{\varepsilon}}_e = (\bar{\boldsymbol{\delta}} \bar{\boldsymbol{\varepsilon}}_h^{-1} \bar{\boldsymbol{\delta}}^{\mathrm{t}} - \boldsymbol{I}_e)^{-1} \qquad \boldsymbol{\rho}_h = \boldsymbol{Q}_h \bar{\boldsymbol{\varepsilon}}_h^{-1} = \bar{\boldsymbol{\lambda}}_h^2 \bar{\boldsymbol{\varepsilon}}_h^{-1} - \boldsymbol{I}_h$$

$$\boldsymbol{\rho} = \boldsymbol{\rho}_h \tilde{\boldsymbol{\delta}}^{\mathrm{t}} \boldsymbol{\rho}_e$$

With eq. (138) the present case is formally arranged in order of the representation given in Section 2.4. The matrices $\bar{y}_1$ and $\bar{y}_2$, however, are not diagonal matrices.

3.3. Planar Waveguides with Magnetized Ferrite Substrates

3.3.1. Basic Equations

The waveguides to be investigated should have cross sections as in Fig. 4 or 6, but the number of substrate layers can be more than is illustrated, and at least one of the layers should be of ferromagnetic substrate. The propagation of the waves does not depend on the structure alone, but also on the direction of the magnetization. The approach presented here can also be transferred to layers with anisotropic dielectric.

Here the case of magnetization perpendicular to the substrate is presented [7]. In the diagram of the waveguide structures this is always the y direction. This is an interesting case for practical purposes [18]. On the case of the magnetization in the x direction has also been reported [19]. The case that is going to be covered here is more complicated than the others. The permeability tensor in this case is given by

$$\bar{\bar{\mu}} = \mu_0 \begin{bmatrix} 1+\chi & 0 & j\kappa \\ 0 & 1 & 0 \\ -j\kappa & 0 & 1+\chi \end{bmatrix} \tag{139}$$

with

$$\chi = \frac{\omega_0 \omega_s}{\omega_0^2 - \omega^2} \qquad \kappa = \frac{\omega \omega_s}{\omega_0^2 - \omega^2} \qquad \omega_0 = \gamma H_0 \qquad \omega_s = \gamma M_s \tag{140}$$

where γ is the gyromagnetic ratio and M_s is the saturation magnetization. The fields must be represented in terms of components that are situated parallel to the static field [20], namely, the components e_y and h_y. Compared with the isotropic substrates there exist the following differences:

1. The wave equations for e_y and h_y are coupled with each other. It follows under the assumption that the fields are propagating in the z direction according to $\exp(-jk_z z)$ that

$$\frac{\partial^2 e_y}{\partial x^2} + \frac{\partial^2 e_y}{\partial y^2} + (\mu_f k^2 - k_z^2) e_y = \frac{\kappa k \eta}{1 + \chi} \frac{\partial h_y}{\partial y} \tag{141}$$

$$\frac{\partial^2 h_y}{\partial x^2} + \frac{1}{1 + \chi} \frac{\partial^2 h_y}{\partial y^2} + (k^2 - k_z^2) h_y = -\frac{\kappa k}{(1 + \chi)\eta} \frac{\partial e_y}{\partial y} \tag{142}$$

with

$$k = \omega \sqrt{\mu_0 \varepsilon} = \sqrt{\varepsilon_r} k_0 \qquad \eta = \sqrt{\frac{\mu_0}{\varepsilon}} \qquad \mu_f = 1 + \chi - \frac{\kappa^2}{1 + \chi} \tag{143}$$

For e_y and h_y, respectively, follow partial differential equations of the fourth order.

2. The boundary conditions on the walls are only partially homogeneous. The following expressions hold.

$$\text{Electric wall:} \quad e_y = 0 \qquad \frac{\partial h_y}{\partial x} = -j \frac{\kappa}{1 + \chi} \frac{\partial h_z}{\partial y} \tag{144}$$

$$\text{Magnetic wall:} \quad h_y = 0 \qquad \frac{\partial e_y}{\partial x} = -\kappa \mu_0 \omega h_x \tag{145}$$

It is merely the Dirichlet condition that is homogeneous and not the Neumann condition.

3. A further problem arises in eqs. (141) and (142), because e_y and h_y are coupled with each other. Nevertheless both will be discretized over each others shifted line system (see next section). It is therefore necessary to interpolate in such a manner that the components on the other line system are also known.

The remaining components will be determined from e_y and h_y with the help of the following equations:

$$\begin{bmatrix} (1+\chi)k^2 + \dfrac{\partial^2}{\partial y^2} & j\kappa k^2 \\ -j\kappa k^2 & (1+\chi)k^2 + \dfrac{\partial^2}{\partial y^2} \end{bmatrix} \begin{bmatrix} e_x \\ e_z \end{bmatrix}$$

$$= jk\eta \begin{bmatrix} 1+\chi & j\kappa \\ -j\kappa & 1+\chi \end{bmatrix} \begin{bmatrix} \dfrac{\partial h_y}{\partial z} \\ -\dfrac{\partial h_y}{\partial x} \end{bmatrix} + \frac{\partial}{\partial y} \begin{bmatrix} \dfrac{\partial e_y}{\partial x} \\ \dfrac{\partial e_y}{\partial z} \end{bmatrix} \tag{146}$$

$$\begin{bmatrix} (1+\chi)k^2 + \dfrac{\partial^2}{\partial y^2} & j\kappa k^2 \\ -j\kappa k^2 & (1+\chi)k^2 + \dfrac{\partial^2}{\partial y^2} \end{bmatrix} \begin{bmatrix} h_x \\ h_z \end{bmatrix} = j\frac{k}{\eta} \begin{bmatrix} -\dfrac{\partial e_y}{\partial z} \\ \dfrac{\partial e_y}{\partial x} \end{bmatrix} + \frac{\partial}{\partial y} \begin{bmatrix} \dfrac{\partial h_y}{\partial x} \\ \dfrac{\partial h_y}{\partial z} \end{bmatrix} \tag{147}$$

3.3.2. Periodic Boundary Conditions

Equations (144) and (145) make clear that in this case of the magnetized ferrite substrate, problems can occur. One way of overriding these kinds of problems is the assumption of periodic boundary conditions. This means it will be assumed that the strip or slot will repeat itself periodically in the x direction. The periodic length should be big enough that the field propagates on each strip independently of the others.

If N is the number of the lines within a period (see Fig. 13), then we write, according to Floquet's theorem,

$$E_{yi+N} = s^{-2N} E_{yi} \qquad H_{yi+N} = s^{-2N} H_{yi} \tag{148}$$

with

$$s = e^{jk_x h/2} \qquad s^{-2N} = e^{-jk_x Nh}$$

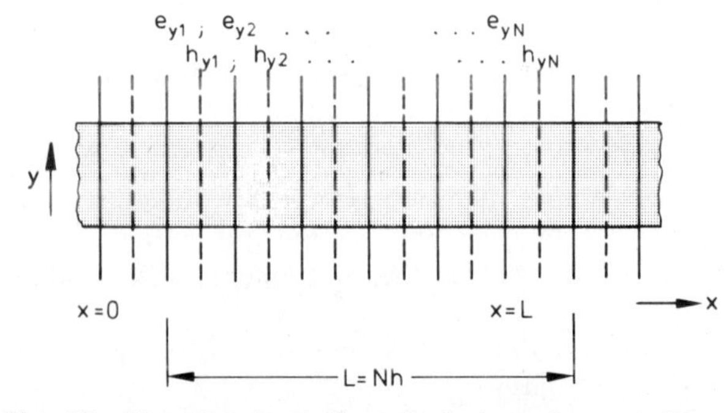

Fig. 13 Discretization with periodic boundary conditions.

because no propagation in the x direction should take place, the value of k_x is put equal to zero. The difference operator $\boldsymbol{D}$ now acquires the form

$$\boldsymbol{D} = \begin{bmatrix} -1 & 1 & & \\ & \ddots & \ddots & \\ & & \ddots & 1 \\ s^{-2N} & & & -1 \end{bmatrix} \tag{149}$$

The change with respect to eq. (7) is very small and causes not problems. Thus eq. (149) also shows the simplicity of handling periodic structures with the method of lines [21].

In the analysis of periodic structures it is convenient to normalize the field with respect to the phase in the following way [5]:

$$e^{ik_x x} e_y \qquad e^{jk_x x} h_y \tag{150}$$

These products are periodic functions in x. In discretized form we write

$$e_y e^{jk_x x} \rightarrow \mathbf{S}_e \mathbf{E}_y \qquad h_y e^{jk_x x} \rightarrow \mathbf{S}_h \mathbf{H}_y \tag{151}$$

with

$$\boldsymbol{S}_e = \text{diag}(e^{jk_x ih}) \tag{152}$$

$$\boldsymbol{S}_h = \text{diag}(e^{jk_x (i+1/2)h}) \tag{153}$$

For the phase normalized derivatives we obtain

$$\begin{aligned} he^{jk_x x} \frac{\partial e_y}{\partial x} &\rightarrow \boldsymbol{S}_h \boldsymbol{D} \boldsymbol{S}_e^* \boldsymbol{S}_e \mathbf{E}_y = \boldsymbol{D}_n \boldsymbol{S}_e \mathbf{E}_y \\ he^{jk_x x} \frac{\partial h_y}{\partial x} &\rightarrow -\boldsymbol{S}_e \boldsymbol{D}^{\mathrm{t}} \boldsymbol{S}_h^* \boldsymbol{S}_h \mathbf{H}_y = -\boldsymbol{D}_n^{*\mathrm{t}} \boldsymbol{S}_h \mathbf{H}_y \end{aligned} \tag{154}$$

The phase normalized difference operators $\boldsymbol{D}_n$ and $\boldsymbol{P}_n$ are given by

$$\boldsymbol{D}_n = \boldsymbol{S}_h \boldsymbol{D} \boldsymbol{S}_e^* = \begin{bmatrix} -s & s^* & & \\ \vdots & & \ddots & \\ & & & s^* \\ s^* & & & -s \end{bmatrix} \tag{155}$$

and

$$\boldsymbol{P}_n = \boldsymbol{D}_n \boldsymbol{D}_n^{*\mathrm{t}} = \boldsymbol{D}_n^{*\mathrm{t}} \boldsymbol{D}_n \tag{156}$$

Hence $\boldsymbol{P}_n$ is the same for both normalized field quantities.

As usual in the method of lines, the field components of the line systems, which have been collated in vector form, are transformed. Individually they are as follows:

$$\begin{aligned} \mathbf{E}_y &= \boldsymbol{T}_e \bar{\mathbf{E}}_y & \mathbf{H}_y &= \boldsymbol{T}_h \bar{\mathbf{H}}_y \\ \mathbf{E}_z &= \boldsymbol{T}_e \bar{\mathbf{E}}_z & \mathbf{H}_z &= \boldsymbol{T}_h \bar{\mathbf{H}}_z \\ \mathbf{H}_x &= \boldsymbol{T}_e \bar{\mathbf{E}}_x & \mathbf{E}_x &= \boldsymbol{T}_h \bar{\mathbf{E}}_x \end{aligned} \tag{157}$$

where the transformation matrices $\boldsymbol{T}_e$ and $\boldsymbol{T}_h$ are given in Appendix D. Because of the coupling of e_y and h_y according to eqs. (141) and (142), we need the discretized fields not only on the associated line system but also on the lines belonging to the other field component. From the representation of the ith component of, e.g., $\mathbf{E}_y$, i.e. the field e_y on line i of the e_y lines

$$e^{jk_x ih} E_{yi} = \frac{1}{\sqrt{N}} \sum_{k=1}^{N} e^{j(2\pi/N)ki} \bar{E}_{yk} \tag{158}$$

it can be noticed how the field $\mathbf{E}_y^H$ in the middle between the ith and the $(i+1)$th line, i.e. on the ith h_y line can be calculated [22]:

$$e^{jk_x(i+1/2)h} E_{yi}^H = e^{jk_x(i+1/2)h} E_{yi+1/2} = \frac{1}{\sqrt{N}} \sum_{k=1}^{N} e^{j(2\pi/N)k(i+1/2)} \bar{E}_{yk} \tag{159}$$

Therefore

$$\mathbf{S}_h \mathbf{E}_y^H = \boldsymbol{T}_h \bar{\mathbf{E}}_y \tag{160}$$

By analogy it is valid that

$$\boldsymbol{S}_e \mathbf{H}_y^E = \boldsymbol{T}_e \bar{\mathbf{H}}_y \tag{161}$$

With these equations we obtain the transformed differential equations from eqs. (141) and (142):

$$\left[\left(\frac{1}{1+\chi}\frac{1}{k_0^2}\frac{\partial^2}{\partial y^2} + \varepsilon_r - \varepsilon_{re}\right)\boldsymbol{I} - \bar{\boldsymbol{\lambda}}^2\right]\eta_0 \bar{\mathbf{H}}_y = -\frac{\kappa}{1+\chi}\frac{\varepsilon_r}{k_0}\frac{\partial}{\partial y}\bar{\mathbf{E}}_y \tag{162}$$

$$\left[\left(\frac{1}{k_0^2}\frac{\partial^2}{\partial y^2} + \mu_f \varepsilon_r - \varepsilon_{re}\right)\boldsymbol{I} - \bar{\boldsymbol{\lambda}}^2\right]\bar{\mathbf{E}}_y = \frac{\kappa}{1+\chi}\frac{1}{k_0}\frac{\partial}{\partial y}(\eta_0 \bar{\mathbf{H}}_y) \tag{163}$$

Combination yields

$$\left[\left(\mathbf{I}\frac{1}{1+\chi}\frac{1}{k_0^2}\frac{\partial^2}{\partial y^2}+\boldsymbol{\xi}^h\right)\left(\mathbf{I}\frac{1}{k_0^2}\frac{\partial^2}{\partial y^2}+\boldsymbol{\xi}^e\right)+\mathbf{I}\left(\frac{\kappa}{1+\chi}\right)^2\varepsilon_r\frac{1}{k_0^2}\frac{\partial^2}{\partial y^2}\right]\bar{\boldsymbol{\psi}}=0 \tag{164}$$

with

$$\bar{\boldsymbol{\psi}}=\bar{\mathbf{E}}_y \qquad \text{or} \qquad \bar{\boldsymbol{\psi}}=\eta_0\bar{\mathbf{H}}_y$$

and

$$\boldsymbol{\xi}^h=\operatorname{diag}(\varepsilon_r-\varepsilon_{re}-\bar{\lambda}_i^2) \qquad \boldsymbol{\xi}^e=\operatorname{diag}(\mu_f\varepsilon_r-\varepsilon_{re}-\bar{\lambda}_i^2) \tag{165}$$

The solutions for the ith components are

$$\bar{E}_{yi}=A_i\cosh k_{y1}^i y+B_i\sinh k_{y1}^i y+\rho_e^i(C_i\cosh k_{y2}^i y+D_i\sinh k_{y2}^i y) \tag{166}$$

$$\eta_0\bar{H}_{yi}=\rho_h^i(A_i\sinh k_{y1}^i y+B_i\cosh k_{y1}^i y)+C_i\sinh k_{y2}^i y+D_i\cosh k_{y2}^i y \tag{167}$$

where $k_{y1,2}^i$ are the solutions of the biquadratic characteristic equation

$$\frac{1}{1+\chi}\left(\frac{k_y}{k_0}\right)^4+\left(\xi_i^h+\frac{1}{1+\chi}\xi_i^e+\frac{\kappa^2}{(1+\chi)^2}\varepsilon_r\right)\left(\frac{k_y}{k_0}\right)^2+\xi_i^e\xi_i^h=0 \tag{168}$$

Because of the coupling of the differential equations the coefficients are not independent of each other. ρ_h^i is obtained from the differential equation (162), for $\bar{\mathbf{H}}_y$. ρ_e^i is obtained from the differential equation (163) for $\bar{\mathbf{E}}_y$.

$$\rho_e^i=\frac{\kappa}{1+\chi}\left(\frac{k_{y2}^i}{k_0}\right)\left[\left(\frac{k_{y2}^i}{k_0}\right)^2+\xi_i^e\right]^{-1} \tag{169}$$

$$\rho_h^i=-\frac{\kappa\varepsilon_r}{1+\chi}\left(\frac{k_{y1}^i}{k_0}\right)\left[\frac{1}{1+\chi}\left(\frac{k_{y1}^i}{k_0}\right)^2+\xi_i^h\right]^{-1} \tag{170}$$

For $\kappa\to 0$, the values of ρ_h and ρ_e tend toward zero and k_{y1} tends toward the solution of the characteristic equation of the uncoupled equation (163). k_{y2} tends toward the solution of the characteristic equation of the uncoupled equation (162). Equations (166) and (167) will again be converted into a form corresponding to eq. (19). For this purpose we introduce the following abbreviations:

$$\boldsymbol{\rho}_e = \operatorname{diag}(\rho_e^i) \qquad \boldsymbol{\rho}_h = \operatorname{diag}(\rho_h^i)$$

$$\bar{\mathbf{E}} = \begin{bmatrix} \bar{\mathbf{E}}_y(y_1) \\ \bar{\mathbf{E}}_y(y_2) \end{bmatrix} \qquad \bar{\mathbf{E}}' = \frac{1}{k_0}\frac{\partial}{\partial y}\begin{bmatrix} \bar{\mathbf{E}}_y(y)|_{y_1} \\ \bar{\mathbf{E}}_y(y)|_{y_2} \end{bmatrix}$$

$$\bar{\mathbf{H}} = \eta_0\begin{bmatrix} \bar{\mathbf{H}}_y(y_1) \\ \bar{\mathbf{H}}_y(y_2) \end{bmatrix} \qquad \bar{\mathbf{H}}' = \frac{\eta_0}{k_0}\frac{\partial}{\partial y}\begin{bmatrix} \bar{\mathbf{H}}_y(y)|_{y_1} \\ \bar{\mathbf{H}}_y(y)|_{y_2} \end{bmatrix}$$

$$\boldsymbol{k}_{y1,2} = \operatorname{diag}\left(\frac{k_{y1,2}^i}{k_0}\right)$$

$$\boldsymbol{\Gamma}_{1,2} = \begin{bmatrix} -\tilde{\boldsymbol{\gamma}}_{1,2} & \tilde{\boldsymbol{\alpha}}_{1,2} \\ -\tilde{\boldsymbol{\alpha}}_{1,2} & \tilde{\boldsymbol{\gamma}}_{1,2} \end{bmatrix} \tag{171}$$

$$\tilde{\boldsymbol{\gamma}}_{1,2} = \operatorname{diag}[(\tanh k_{y1,2}^i d)^{-1}] \qquad \tilde{\boldsymbol{\alpha}}_{1,2} = \operatorname{diag}[(\sinh k_{y1,2}^i d)^{-1}]$$

As result, after suitable algebraic manipulations, we obtain

$$\bar{\mathbf{E}}' = (\boldsymbol{\Gamma}_2 - \boldsymbol{\rho}_e\boldsymbol{\rho}_h\boldsymbol{\Gamma}_1)^{-1}[(\boldsymbol{k}_{y1} - \boldsymbol{\rho}_e\boldsymbol{\rho}_h\boldsymbol{k}_{y2})\bar{\mathbf{E}} + \boldsymbol{\rho}_e(\boldsymbol{k}_{y2}\boldsymbol{\Gamma}_1 - \boldsymbol{k}_{y1}\boldsymbol{\Gamma}_2)\bar{\mathbf{H}}] \tag{172}$$

$$\bar{\mathbf{H}}' = (\boldsymbol{\Gamma}_1 - \boldsymbol{\rho}_e\boldsymbol{\rho}_h\boldsymbol{\Gamma}_2)^{-1}[(\boldsymbol{k}_{y2} - \boldsymbol{\rho}_e\boldsymbol{\rho}_h\boldsymbol{k}_{y1})\bar{\mathbf{H}} + \boldsymbol{\rho}_h(\boldsymbol{k}_{y1}\boldsymbol{\Gamma}_2 - \boldsymbol{k}_{y2}\boldsymbol{\Gamma}_1)\bar{\mathbf{E}}] \tag{173}$$

For the conversions the following relation was used in particular:

$$\boldsymbol{\Gamma}_{1,2}^2 = \boldsymbol{I} \tag{174}$$

With $\boldsymbol{\rho}_e = \boldsymbol{\rho}_h = \mathbf{0}$ the result is equal to the isotropic case, namely, to eq. (19). Equations (172) and (173) are now reduced to eqs. (146) and (147), respectively determining the tangential field components necessary for the field matching. The subsequent analysis follows the approach in Section 2. The system matrices, however, are not real in this case.

3.4. The Method of Lines for Longitudinally Inhomogeneous Planar Structures. Two-Dimensional Discretization

The extensions in the preceding sections covered various types of waveguides with constant cross sections. Now we make another type of extension, namely to structures that do not have a metallization of constant width. These are resonators, longitudinally periodic structures, and discontinuities. As the metallization contours are two-dimensional, all three types of structures require discretization in two directions, which was introduced by Worm and applied to various problems [23]. Nonequidistant discretization (see Section 2.7) was also used to increase the flexibility of the method, resulting in greater efficiency [12].

In this section a different formulation of the two-dimensional discretization is presented [9], which was introduced in order to optimize the computation of the eigenvalues. For this purpose we consider the difference-differential equations and their solutions for a general "two-dimensional structure," as we call a structure with two-dimensional metallization, and present a clear mathematical formulation based on the Kronecker product. The discretization of resonators, periodic structures, and discontinuities exhibits a number of common features. However, the structures differ in the boundary conditions and hence in the difference operators $\boldsymbol{D}$ and $\boldsymbol{P}$ as well as the transformation matrices $\boldsymbol{T}$. While these particularities are substituted into the general formulas, e.g., from Section 3.3 and Appendix D for periodic structures, the formal properties of the general approach remain the same. For the equidistant periodic case it is shown that the characteristic matrix consists of block Toeplitz submatrices. A fast inversion algorithm is introduced for this matrix structure.

For discontinuities a special treatment according to Worm [5] is outlined, namely, the consideration of the sources as inhomogeneous boundary conditions, which leads to an additional term in the derivatives in the z direction and to special treatment in the transform domain.

This general approach to two-dimensional structures mainly runs along the lines of the one-dimensional approach presented in Section 2. Hence we only give the differences in the following.

3.4.1. Description of the Discretization

As shown in Fig. 14, the considered region is discretized in the x and z directions with the distances h_x and h_z, respectively (not to be confused with the continuous magnetic field components h_x and h_z!). The discretization lines for h_z and e_z are shifted by $\mathrm{h}_x/2$ and $\mathrm{h}_z/2$, which has the same advantages as in Section 2.2. Note that the edge parameters p and q are not

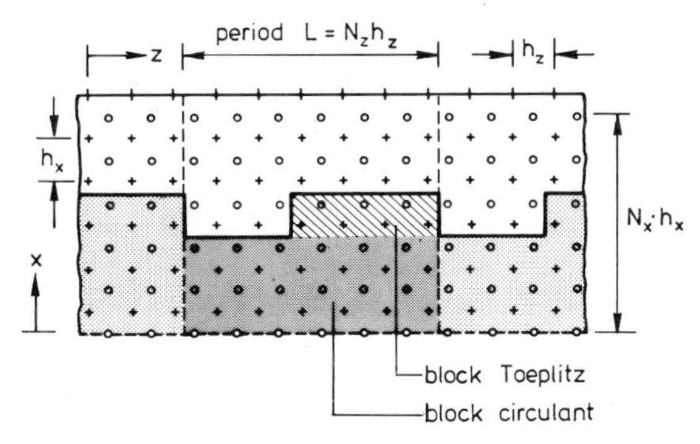

Fig. 14 Position of the discretization lines for a typical periodic structure ((∘) e_z lines, (+) h_z lines).

necessarily equal to 0·25 as in the case of infinitely long edges in Section 2.8. The e_z lines are linearly numbered by

$$n = i + (k-1)N_x \qquad 1 \le i \le N_x, \qquad 1 \le k \le N_z \tag{175}$$

Thus $\mathbf{E}_z$ is a vector of dimension $N_x N_z$. The dimension of $\mathbf{H}_z$ may differ according to the dual boundary conditions.

To use the approach of Section 2 for two-dimensional structures, we replace $\boldsymbol{D}$, $\boldsymbol{P}$, and $\boldsymbol{T}$ by the two-dimensional operators $\hat{\boldsymbol{D}}$, $\hat{\boldsymbol{P}}$, and $\hat{\boldsymbol{T}}$ in eqs. (5)–(14), by means of the Kronecker product [24]. If $\boldsymbol{A}$ and $\boldsymbol{B}$ are $m \times n$ and $p \times q$ matrices, respectively, the Kronecker product is an $mp \times nq$ matrix defined by

$$\boldsymbol{A} \otimes \boldsymbol{B} = \begin{bmatrix} a_{11}\boldsymbol{B} & \cdots & a_{1n}\boldsymbol{B} \\ \vdots & & \vdots \\ a_{m1}\boldsymbol{B} & \cdots & a_{mn}\boldsymbol{B} \end{bmatrix} \tag{176}$$

Two important properties of the Kronecker product are

$$(\boldsymbol{A} \otimes \boldsymbol{B})(\boldsymbol{C} \otimes \boldsymbol{D}) = (\boldsymbol{AC}) \otimes \boldsymbol{BD} \qquad (\boldsymbol{A} \otimes \boldsymbol{B})^{\mathrm{t}} = \boldsymbol{A}^{\mathrm{t}} \otimes \boldsymbol{B}^{\mathrm{t}}$$

Thus the two-dimensional difference operators are

$$\boldsymbol{D}_x \rightarrow \hat{\boldsymbol{D}}_x = \boldsymbol{I}_z \otimes \boldsymbol{D}_x \tag{177}$$

$$\boldsymbol{D}_z \rightarrow \hat{\boldsymbol{D}}_z = \boldsymbol{D}_z \otimes \boldsymbol{I}_x \tag{178}$$

where $\boldsymbol{D}_z$ is obtained in the same way as $\boldsymbol{D}_x$ but using the derivative with respect to z, and $\boldsymbol{I}_x$ and $\boldsymbol{I}_z$ are identity matrices of a dimension of N_x and N_z, respectively. $\hat{\boldsymbol{P}}_x$ and $\hat{\boldsymbol{P}}_z$ and all the following matrices with hats (^) and subscripts x or z will be constructed in the same way as $\hat{\boldsymbol{D}}_x$ and $\hat{\boldsymbol{D}}_z$, respectively.

Substituting the two-dimensional forms of eq. (8) or (9) or their generalizations of sections 2.7, 3.2, or 3.3 in eq. (1), we obtain the following ordinary differential equation instead of eq. (11):

$$\frac{d^2}{dy^2}\boldsymbol{\psi} + (k^2\hat{\boldsymbol{I}} - \mathrm{h}_x^{-2}\hat{\boldsymbol{P}}_x - \mathrm{h}_z^{-2}\hat{\boldsymbol{P}}_z)\boldsymbol{\psi} = \boldsymbol{0} \tag{179}$$

where $\hat{\boldsymbol{I}}$ is the unit matrix $\boldsymbol{I}_z \otimes \boldsymbol{I}_x$.

Equation (12) is generalized with the two-dimensional tranformation matrix

$$\hat{\boldsymbol{T}} = \boldsymbol{T}_z \otimes \boldsymbol{T}_x \tag{180}$$

where $\boldsymbol{T}_z$ in the general case fulfills

$$\boldsymbol{T}_z^{*t}\boldsymbol{P}_z\boldsymbol{T}_z = \boldsymbol{\lambda}_z^2 \qquad \boldsymbol{T}_z^{-1} = \boldsymbol{T}_z^{*t} \tag{181}$$

where the $*$ denotes complex conjugate. From eq. (179) with eqs. (12), (13), and (181) we obtain the discretized Helmholtz equation:

$$\left[\left(\frac{d^2}{dy^2} + k^2\right)\hat{\boldsymbol{I}} - \mathrm{h}_x^{-2}\hat{\boldsymbol{\lambda}}_x^2 - \mathrm{h}_z^{-2}\hat{\boldsymbol{\lambda}}_z^2\right]\bar{\boldsymbol{\psi}} = \boldsymbol{0} \tag{182}$$

Equations (18)–(20) are still valid if we use k_{yn}^2 instead of k_{yi}^2 from eq. (16):

$$k_{yn}^2 = k_0^2(\bar{\lambda}_{xi}^2 + \bar{\lambda}_{zk}^2 - \varepsilon_r) \tag{183}$$

where

$$\bar{\lambda}_{xi}^2 = \frac{\lambda_i^2|_{N=N_x}}{(k_0\mathrm{h}_x)^2} \qquad \bar{\lambda}_{zk}^2 = \frac{\lambda_k^2|_{N=N_z}}{(k_0\mathrm{h}_z)^2} \tag{184}$$

The resulting diagonal matrices $\boldsymbol{\alpha}$, $\boldsymbol{\gamma}$, and $\boldsymbol{k}_y$ have a dimension of N_xN_z.

3.4.2. Solutions for the Other Field Components

The analysis here is done in exactly the same way as in Section 2.4. The matrices $\bar{\boldsymbol{Y}}$, $\bar{\boldsymbol{Z}}$ for the two-dimensional structures exhibit a greater amount of symmetry than in eqs. (32) and (33). This originates from replacing $-jk_0\sqrt{\varepsilon_{re}}\boldsymbol{I}$ by $\mathrm{h}_z^{-1}\hat{\boldsymbol{D}}_z$ or $-\mathrm{h}_z^{-1}\hat{\boldsymbol{D}}_z^{*t}$. For the discretized field components we obtain from eqs. (3) and (4):

$$\begin{bmatrix} -\mathrm{h}_z^{-2}\hat{\boldsymbol{P}}_{zh} + k^2\hat{\boldsymbol{I}} & \\ & -\mathrm{h}_z^{-2}\hat{\boldsymbol{P}}_{ze} + k^2\hat{\boldsymbol{I}} \end{bmatrix}\begin{bmatrix} \mathbf{E}_x \\ \eta_0\mathbf{H}_x \end{bmatrix} = \begin{bmatrix} \mathrm{h}_x^{-1}\hat{\boldsymbol{D}}_x\mathrm{h}_z^{-1}\hat{\boldsymbol{D}}_z & -jk_0\hat{\boldsymbol{I}}\dfrac{\partial}{\partial y} \\ jk_0\varepsilon_r\hat{\boldsymbol{I}}\dfrac{\partial}{\partial y} & \mathrm{h}_x^{-1}\hat{\boldsymbol{D}}_x^{*t}\mathrm{h}_z^{-1}\hat{\boldsymbol{D}}_z^{*t} \end{bmatrix}\begin{bmatrix} \mathbf{E}_z \\ \eta_0\mathbf{H}_z \end{bmatrix} \tag{185}$$

$$\begin{bmatrix} -\mathrm{h}_z^{-2}\hat{\boldsymbol{P}}_{zh} + k^2\hat{\boldsymbol{I}} & \\ & -\mathrm{h}_z^{-2}\hat{\boldsymbol{P}}_{ze} + k^2\hat{\boldsymbol{I}} \end{bmatrix}\begin{bmatrix} \mathbf{E}_y \\ \eta_0\mathbf{H}_y \end{bmatrix} = \begin{bmatrix} \dfrac{\partial}{\partial y}\mathrm{h}_z^{-1}\hat{\boldsymbol{D}}_z & -jk_0\mathrm{h}_x^{-1}\hat{\boldsymbol{D}}_x^{*t} \\ -jk_0\varepsilon_r\mathrm{h}_x^{-1}\hat{\boldsymbol{D}}_x & -\dfrac{\partial}{\partial y}\mathrm{h}_z^{-1}\hat{\boldsymbol{D}}_z^{*t} \end{bmatrix}\begin{bmatrix} \mathbf{E}_z \\ \eta_0\mathbf{H}_z \end{bmatrix} \tag{186}$$

The transformation yields

$$\begin{bmatrix} -(\hat{\boldsymbol{\lambda}}_{zh}^2 - \varepsilon_r \hat{\boldsymbol{I}}) & \\ & -(\hat{\boldsymbol{\lambda}}_{ze}^2 - \varepsilon_r \hat{\boldsymbol{I}}) \end{bmatrix} \begin{bmatrix} \bar{\mathbf{E}}_x \\ \eta_0 \bar{\mathbf{H}}_x \end{bmatrix} = \begin{bmatrix} \hat{\boldsymbol{\delta}}_x \hat{\boldsymbol{\delta}}_z & -j\hat{\boldsymbol{I}} \dfrac{1}{k_0} \dfrac{\partial}{\partial y} \\ j\hat{\boldsymbol{I}} \dfrac{\varepsilon_r}{k_0} \dfrac{\partial}{\partial y} & \hat{\boldsymbol{\delta}}_x^{*t} \hat{\boldsymbol{\delta}}_z^{*t} \end{bmatrix} \begin{bmatrix} \bar{\mathbf{E}}_z \\ \eta_0 \bar{\mathbf{H}}_z \end{bmatrix} \tag{187}$$

$$\begin{bmatrix} -(\hat{\boldsymbol{\lambda}}_{zh}^2 - \varepsilon_r \hat{\boldsymbol{I}}) & \\ & -(\hat{\boldsymbol{\lambda}}_{ze}^2 - \varepsilon_r \hat{\boldsymbol{I}}) \end{bmatrix} \begin{bmatrix} \bar{\mathbf{E}}_y \\ \eta_0 \bar{\mathbf{H}}_y \end{bmatrix}$$

$$= \begin{bmatrix} \dfrac{1}{k_0} \dfrac{\partial}{\partial y} \hat{\boldsymbol{\delta}}_z & -j\hat{\boldsymbol{\delta}}_x^{*t} \\ -j\varepsilon_r \hat{\boldsymbol{\delta}}_x & -\dfrac{1}{k_0} \dfrac{\partial}{\partial y} \hat{\boldsymbol{\delta}}_z^{*t} \end{bmatrix} \begin{bmatrix} \bar{\mathbf{E}}_z \\ \eta_0 \bar{\mathbf{H}}_z \end{bmatrix} \tag{188}$$

with

$$\begin{aligned} \bar{\boldsymbol{\delta}}_x &= (k_0 \mathrm{h}_x)^{-1} \boldsymbol{\delta}_x & \bar{\boldsymbol{\delta}}_z &= (k_0 \mathrm{h}_z)^{-1} \boldsymbol{\delta}_z \\ \boldsymbol{\delta}_x &= \boldsymbol{T}_{xh}^{*t} \boldsymbol{D}_x \boldsymbol{T}_{xe} & \boldsymbol{\delta}_z &= \boldsymbol{T}_{zh}^{*t} \boldsymbol{D}_z \boldsymbol{T}_{ze} \end{aligned} \tag{189}$$

If the system of equations (187) is established for the interface planes A and B (see Fig. 5), we obtain the systems of equations (29) and (30) for the tangential field components after some algebraic manipulations similar to the one-dimensional case, but we have to redefine the $\bar{y}_i$ and $\bar{Y}$, $\bar{Z}$:

$$\bar{y}_1 = \begin{bmatrix} \gamma_H & \gamma_h \tilde{\boldsymbol{\delta}} \\ \tilde{\boldsymbol{\delta}}^{*t} \gamma_h & \gamma_E \end{bmatrix} \qquad \bar{y}_2 = \begin{bmatrix} \boldsymbol{\alpha}_H & \boldsymbol{\alpha}_h \tilde{\boldsymbol{\delta}} \\ \tilde{\boldsymbol{\delta}}^{*t} \boldsymbol{\alpha}_h & \boldsymbol{\alpha}_E \end{bmatrix} \tag{190}$$

$$\bar{\boldsymbol{Y}} = -\begin{bmatrix} \boldsymbol{k}_{yh}^{-2} \boldsymbol{\alpha}_h^{-1} & \boldsymbol{0} \\ \boldsymbol{0} & \boldsymbol{k}_{ye}^{-2} \boldsymbol{\alpha}_e^{-1} \end{bmatrix} \begin{bmatrix} (\hat{\boldsymbol{\lambda}}_{zh}^2 - \varepsilon_r \hat{\boldsymbol{I}}) & \hat{\boldsymbol{\delta}} \\ \hat{\boldsymbol{\delta}}^{*t} & (\hat{\boldsymbol{\lambda}}_{xe}^2 - \varepsilon_r \hat{\boldsymbol{I}}) \end{bmatrix} \tag{191}$$

$$\bar{\boldsymbol{Z}} = -\frac{1}{\varepsilon_r} \begin{bmatrix} \boldsymbol{k}_{yh}^{-2} \boldsymbol{\alpha}_h^{-1} & \boldsymbol{0} \\ \boldsymbol{0} & \boldsymbol{k}_{ye}^{-2} \boldsymbol{\alpha}_e^{-1} \end{bmatrix} \begin{bmatrix} -(\hat{\boldsymbol{\lambda}}_{xh}^2 - \varepsilon_r \hat{\boldsymbol{I}}) & \tilde{\boldsymbol{\delta}} \\ \tilde{\boldsymbol{\delta}}^{*t} & -(\hat{\boldsymbol{\lambda}}_{ze}^2 - \varepsilon_r \hat{\boldsymbol{I}}) \end{bmatrix} \tag{192}$$

where the following two-dimensional quantities are used:

$$\tilde{\boldsymbol{\delta}} = j\hat{\boldsymbol{\delta}}_x \hat{\boldsymbol{\delta}}_z \qquad \begin{Bmatrix} \boldsymbol{\alpha}_E \\ \gamma_E \end{Bmatrix} = (\hat{\boldsymbol{\lambda}}_{xe}^2 - \varepsilon_r \hat{\boldsymbol{I}}) \begin{Bmatrix} \boldsymbol{\alpha}_e \\ \gamma_e \end{Bmatrix} \qquad \begin{Bmatrix} \boldsymbol{\alpha}_H \\ \gamma_H \end{Bmatrix} = (\hat{\boldsymbol{\lambda}}_{zh}^2 - \varepsilon_r \hat{\boldsymbol{I}}) \begin{Bmatrix} \boldsymbol{\alpha}_h \\ \gamma_h \end{Bmatrix}$$

Now we are able to use eqs. (30) and (35) and the rest of Section 2.4 as in the one-dimensional case. Hence we formally get the same equations in the transform domain (see Section 2.5).

3.4.3. The System Equation in the Spatial Domain. Fast Inversion

Now the system matrices in the transform domain $\bar{Z}_{ik}$ according to eq. (62) are block diagonal, e.g.,

$$\bar{Z}_{11} = \text{diag}(\bar{A}_1, \ldots, \bar{A}_{N_z}) \tag{193}$$

with diagonal $\bar{A}_i$. Using the properties of block diagonal matrices and of the Kronecker product, we get Z_{11} as the result of two successive inverse transformations:

$$Z_{11} = \hat{T}_h \bar{Z}_{11} \hat{T}_h^{*t} = \hat{T}_{zh} \quad \text{diag}(T_{xh}\bar{A}_1 T_{xh}^{*t}, \ldots, T_{xh}\bar{A}_{N_z} T_{xh}^{*t}) \quad \hat{T}_{zh}^{*t} \tag{194}$$

The other submatrices Z_{ik} are calculated similarly according to the two-dimensional version of eq. (63), using $\hat{T}$ and $\hat{T}^{*t}$ instead of T and T^t, respectively. The reduction of the system of equations and its solution are done as in Section 2.6. We end up with an indirect eigenvalue system in the case of resonators and periodic structures.

For periodic structures we obtain special system matrices Z_{ik} that can be inverted fast. D_z is circulant, which means that the elements of each row of D_z are identical to those in the previous row but are moved one position to the right and wrapped around [24]. Hence T_z is a Fourier matrix and Z_{11} is block circulant. The same relations apply for the other submatrices Z_{ik}.

The idea of the fast algorithm is demonstrated for Z_{11} in the case of a step discontinuity (see Fig. 14). Discard of rows and columns gives $Z_{11,\text{red}}$:

$$Z_{11,\text{red}} = \begin{bmatrix} C & R^* \\ R & T \end{bmatrix} \tag{195}$$

where C is a block circulant, R is rectangular block circulant, and T is a block Toeplitz matrix. In a Toeplitz matrix every diagonal parallel to the principal diagonal consists of identical elements. The circulant corresponds to the long rectangle in Fig. 14, the Toeplitz matrix to the short one, and the rectangular block circulant to the interaction between both of them. We compute

$$\det(Z_{11,\text{red}}) = \det(C) \cdot \det(T') \tag{196}$$

where

$$T' = T - R \cdot C^{-1} \cdot R^* \tag{197}$$

T' has the same structure as T, because products and inverses of block circulants are block circulants again [24]. The time consumption of the whole algorithm is $O(N_z^2)$ multiplications [25]. whereas the well-known determinant of a Hermitian matrix $\boldsymbol{H}$ needs $O(N_z^3)$, which means that the fast Toeplitz inversion saves a factor of about 10 in the computing time of realistic structures.

3.4.4. Discontinuities. Inhomogeneous Boundary Conditions [5]

In this section we present the source method of analyzing discontinuities using the same approach as Worm [5], but in the formulation based on the Kronecker product (see above). A discontinuity such as the structure in Fig. 15 is discretized as other two-dimensional structures (Fig. 14). Thus the derivatives with respect to x are the same as before. Considering the sources by inhomogeneous boundary conditions [5], however, the derivatives with respect to z contain an additional term, and we obtain an inhomogeneous differential equation in the transform domain.

We assume that only the fundamental mode propagates in the transmission line part at $z = 0$ and that the distance from the discontinuity is big enough that the higher-order modes have decayed. In the transmission line there is a field distribution according to

$$\begin{bmatrix} \mathbf{e}_z \\ \mathbf{h}_z \end{bmatrix} = \frac{1}{1-r} \begin{bmatrix} \mathbf{e}_{z0} \\ \mathbf{h}_{z0} \end{bmatrix} e^{-j\beta z} + \frac{r}{1-r} \begin{bmatrix} -\mathbf{e}_{z0} \\ \mathbf{h}_{z0} \end{bmatrix} e^{j\beta z} \tag{198}$$

with the voltage reflection coefficient r. We use the inhomogeneous Dirichlet condition for the electric field and the inhomogeneous Neumann condition for the magnetic field:

$$\mathbf{e}_z \big|_{z=0} = \mathbf{e}_{z0} \tag{199}$$

$$\mathrm{h}_z \frac{\partial \mathbf{h}_z}{\partial z} \bigg|_{z=0} = -j\beta \mathrm{h}_z \mathbf{h}_{z0} = \mathbf{h}'_{z0} \tag{200}$$

Here and in the following $\mathbf{e}_z$, etc., are the one-dimensional vectors of the transmission line problem and $\mathbf{E}_z$, etc., the two-dimensional ones, especially

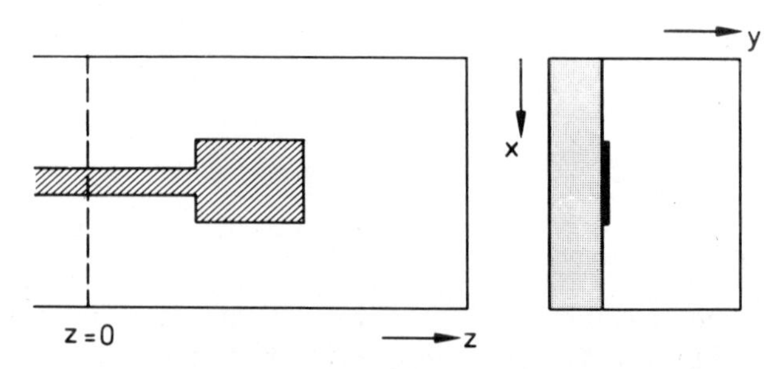

Fig. 15 Discontinuity in microstrip.

$$\mathbf{E}_{z0} = \begin{bmatrix} 1 \\ \mathbf{0} \end{bmatrix} \otimes \mathbf{e}_{z0} \qquad \mathbf{H}'_{z0} = \begin{bmatrix} 1 \\ \mathbf{0} \end{bmatrix} \otimes \mathbf{h}'_{z0} \tag{201}$$

Thus we have to discretize the derivatives with respect to z in the following way [cf. (5)–(9)]:

$$\mathrm{h}_z \frac{\partial e_z}{\partial z} \rightarrow \hat{\boldsymbol{D}}_z \mathbf{E}_z - \mathbf{E}_{z0} \tag{202}$$

$$\mathrm{h}_z \frac{\partial h_z}{\partial z} \rightarrow -\hat{\boldsymbol{D}}_z^{\mathrm{t}} \mathbf{H}_z \tag{203}$$

$$\mathrm{h}_z^2 \frac{\partial^2 e_z}{\partial z^2} = \mathrm{h}_z^2 \frac{\partial}{\partial z}\left(\frac{\partial e_z}{\partial z}\right) \rightarrow -\hat{\boldsymbol{D}}_z^{\mathrm{t}} \hat{\boldsymbol{D}}_z \mathbf{E}_z + \hat{\boldsymbol{D}}_z^{\mathrm{t}} \mathbf{E}_{z0} \tag{204}$$

$$\mathrm{h}_z^2 \frac{\partial^2 h_z}{\partial z^2} = \mathrm{h}_z^2 \frac{\partial}{\partial z}\left(\frac{\partial h_z}{\partial z}\right) \rightarrow -\hat{\boldsymbol{D}}_z \hat{\boldsymbol{D}}_z^{\mathrm{t}} \mathbf{H}_z - \mathbf{H}'_{z0} \tag{205}$$

Now the discretized forms of the Helmholtz equation (1) are inhomogeneous differential equations [cf. (182)]:

$$\left[\left(\frac{1}{k_0^2} \frac{d^2}{dy^2} + \varepsilon_r\right)\hat{\boldsymbol{I}} - \hat{\boldsymbol{\lambda}}_{xe}^2 - \hat{\boldsymbol{\lambda}}_{ze}^2\right]\bar{\mathbf{E}}_z = -(k_0 \mathrm{h}_z)^{-2} \hat{\boldsymbol{\delta}}_z^{\mathrm{t}} \bar{\mathbf{E}}_{z0} \tag{206}$$

$$\left[\left(\frac{1}{k_0^2} \frac{d^2}{dy^2} + \varepsilon_r\right)\hat{\boldsymbol{I}} - \hat{\boldsymbol{\lambda}}_{xh}^2 - \hat{\boldsymbol{\lambda}}_{2h}^2\right]\bar{\mathbf{H}}_z = (k_0 \mathrm{h}_z)^{-2} \bar{\mathbf{H}}'_{z0} \tag{207}$$

These equations are solved in the transform domain [5]. Transforming back we obtain the following deterministic equation for the current distribution:

$$\begin{bmatrix} j\mathbf{J}_x^r \\ \mathbf{J}_z \end{bmatrix} = Z^{r\ -1} Z_q^r \begin{bmatrix} j\mathbf{j}_{x0}^r \\ \mathbf{j}_{z0}^r \end{bmatrix} \tag{208}$$

where Z_q^r is the coupling matrix between the electric field and the current distribution assumed at $z = 0$. The input impedance and the scattering parameters can be calculated from the current distribution.

4. NUMERICAL RESULTS

In this section some numerical results are presented as examples. First the convergence behavior is treated. Figure 16 shows typical diagrams for infinitely thin metallization and one-dimensional discretization [4]. The convergence behavior of the normalized wavelength and of the characteristic impedance are presented for a single microstrip line. For the calculation of the characteristic impedance see Appendix E. M is the numer of e_z lines on half the strip, h the discretization distance, and p is the edge parameter as

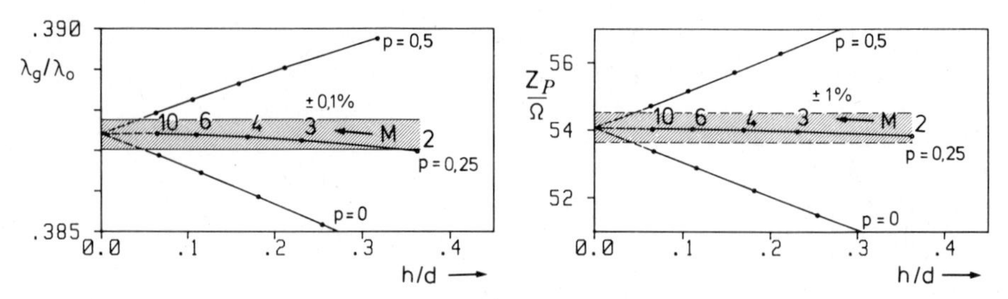

Fig. 16 Convergence behavior of λ_g and Z_P for a single microstrip line for various edge parameters p [4].

defined in Section 2.8. All these convergence curves are smooth and not oscillating, even with a parameter p not so well chosen. If $p = 0.25$ is chosen, then very accurate results are obtained, even for $M = 2$. The run of the curves makes clear that it is possible to determine the "exact" value by means of extrapolation.

Figure 17 shows the convergence behavior of the effective dielectric constant ε_{re} for a microstrip line with finite metallization thickness t [36]. $\varepsilon_{re\infty}$ is the extrapolated value for $h = 0$. N is the total number of discretization lines. Here the convergence curves are also smooth, such that an extrapolation to the final value is possible. Recently, it could be shown that the

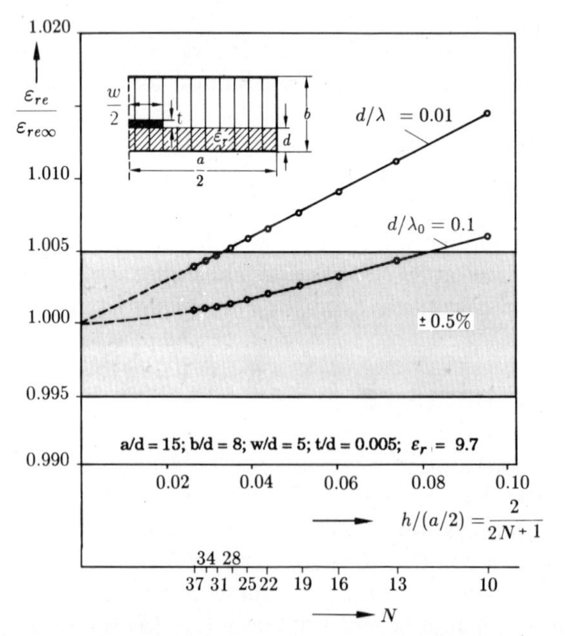

Fig. 17 Convergence behavior of ε_{re} for a microstrip with finite thickness [36].

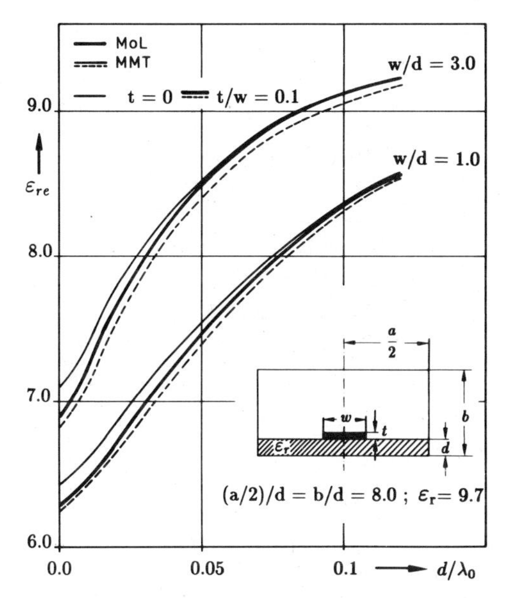

Fig. 18 Dispersion diagram for a microstrip with finite thickness. ε_{re} versus substrate thickness [36].

convergence can be still improved by a suitable choice of the edge parameter.

An example for the dispersion curves of a planar waveguide is given in Fig. 18. The comparison curves were determined with the mode-matching technique (MMT) [26]. The full curves were calculated by Schmückle with the method of lines (MoL) [36]. At this place we mention also that ridged waveguides were treated by Schulz [27] by means of the method of lines.

Figure 19 shows dispersion diagrams and characteristic impedances of a waveguide with metallization in two interfaces: a microstrip line with tuning septums [11]. The comparison curve originates from Reference [28].

The current distribution of a single microstrip line is found in Fig. 28 (see Section 5). For the current and field distributions of other geometries we refer to the literature.

Examples for dielectric waveguides can be found in Figs. 20 and 21. In Fig. 20 the dispersion diagrams of a insulated image guide are presented. The comparison curves were produced using the finite difference analysis (FD) [29], for which the numerical effort is considerably higher than it is here.

In Fig. 21 the coupling between a microstrip line and a dielectric waveguide is described [8]. The results for the unperturbed microstrip (and the dielectric waveguide without microstrip) are drawn in.

As our first example for two-dimensional structures [23], a triangular microstrip resonator is considered, demonstrating that the analysis is not

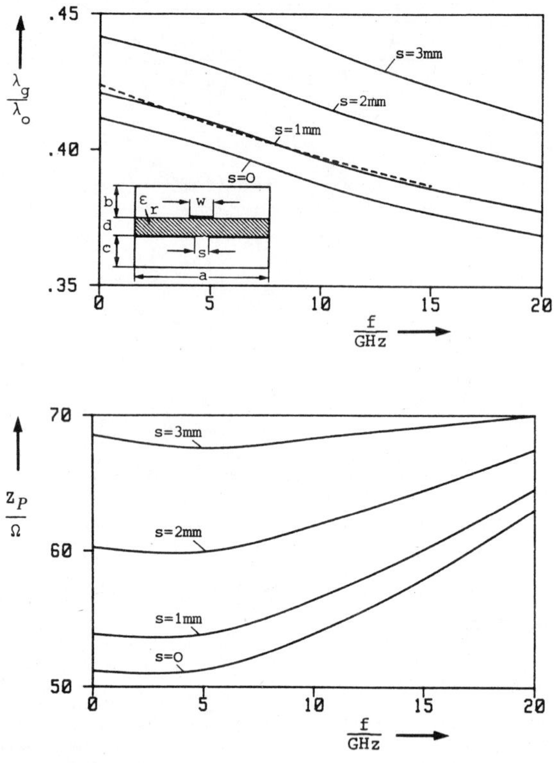

Fig. 19 Dispersion diagram of λ_g and Z_P for a microstrip line with tuning septums [11]. Copyright © AGU 1981.

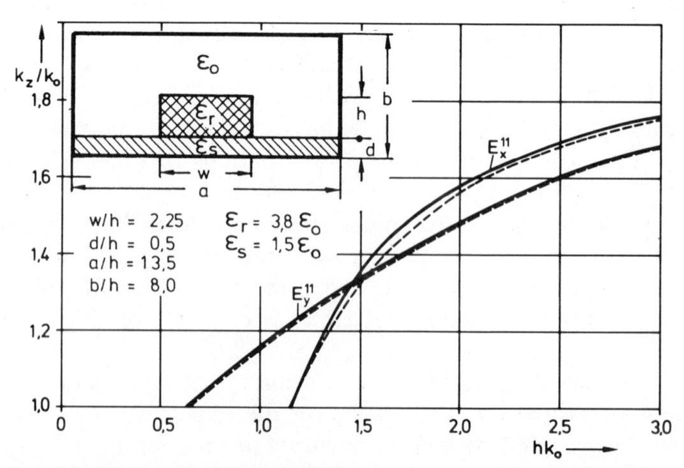

Fig. 20 Insulated image guide, normalized propagation constant versus normalized frequency (Full line, MoL; dashed line, FD [29]).

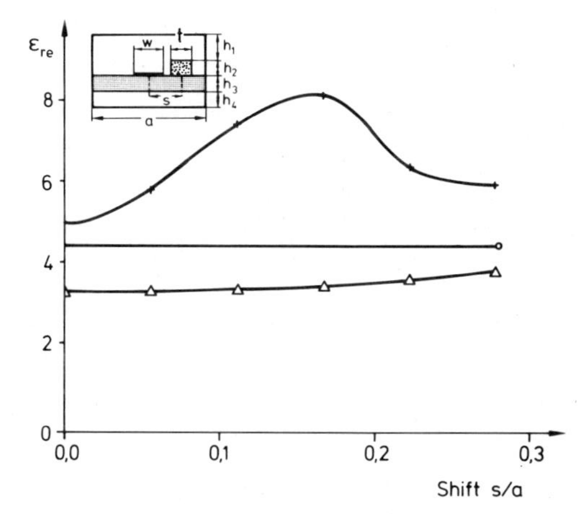

Fig. 21 Coupling of microstrip and dielectric waveguide. Effective dielectric constant versus shift between microstrip and dielectric waveguide [8]. ($f = 30$ GHz, $\varepsilon_{r2} = 16$, $\varepsilon_{r3} = 9.6$, $a = 7.112$ mm, $w = 1.6$ mm, $t = 1.422$ mm, $h_1 = \cdots = h_4 = 0.729$ mm; $M = 4$ lines on the strip). (+) Coupled guides, (∘) microstrip only, (△) dielectric waveguide only.

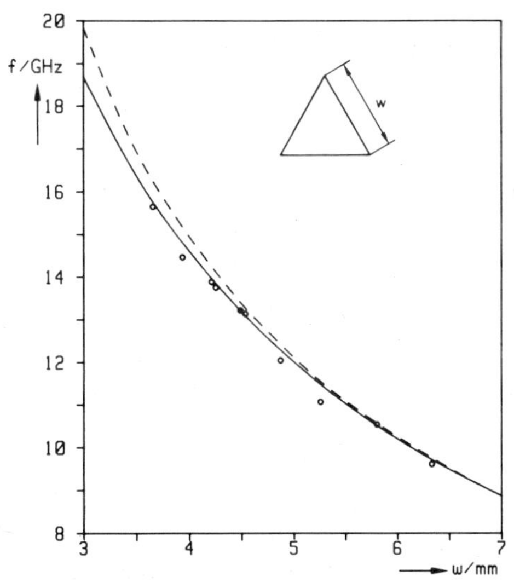

Fig. 22 Resonance frequency for a triangular microstrip resonator. $\varepsilon_r = 9.7$, $d = 0.635$, $b = 10d$. (——) Mol [23], (– – –) TRM [30], (∘) experiment [30].

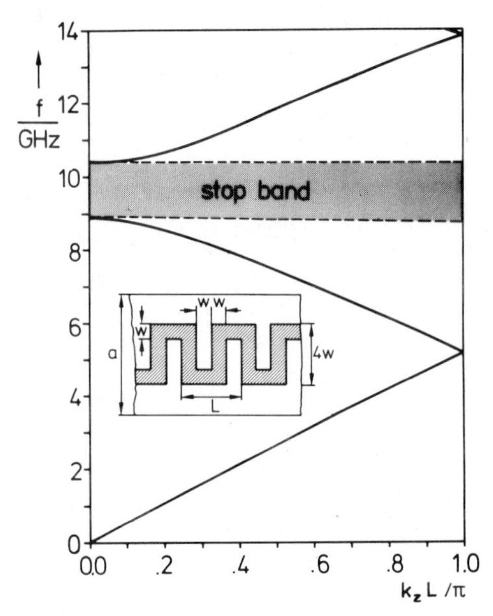

Fig. 23 Dispersion diagram of a microstrip meander line. $\varepsilon_r = 2.3$, $w = 2.37$ mm, $d = 0.79$ mm, $b = 10d$, $a = 12w$, $L = 4w$. After [23] copyright © 1984 IEEE.

restricted to rectangular structures. Figure 22 shows the resonant frequencies obtained by the method of lines in comparison with results from the transverse resonance method (TRM) and measurements, both from [30]. The dispersion diagram of a microstrip meander line is considered in Fig. 23. The next example demonstrates the use of periodic structures in coupled microstrips. Figure 24 shows that the slotting has more influence on the odd mode than on the even mode, so it is possible to achieve phase equalization.

At last we present two examples for the analysis of discontinuities by the method of lines which have been given by Worm [5]. The input reactance of a structure is calculated with either a short or an open circuit at the output. From three such calculations the scattering matrix of a microstrip step discontinuity (Fig. 25) can be determined. The agreement of the curves is very good until the cut-off frequency of the first higher-order mode at 6.2 GHz, which clearly comes out in the mode-matching technique [31] but does not show up in the magnetic-wall model [33]. The scattering parameters of a parallel coupled-line filter are determined similarly. Figure 26 shows good agreement between the method of lines, the spectral domain analysis (SDA) [34], and the experiment [5], especially in the fact that there is no frequency shift in the stop band between the experiment and the method of lines.

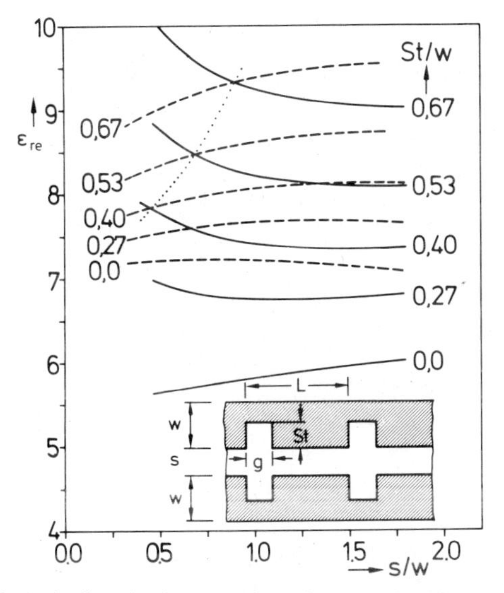

Fig. 24 Effect of periodic slotting on the phase velocities of coupled microstrip lines. $\varepsilon_r = 9.6$, $f = 6$ GHz, $w = d = 0.635$ mm, $b = 10d$, $L = 4g = 0.15$ mm. (——) Odd mode, (– – –) even mode. After [23] copyright © 1984 IEEE.

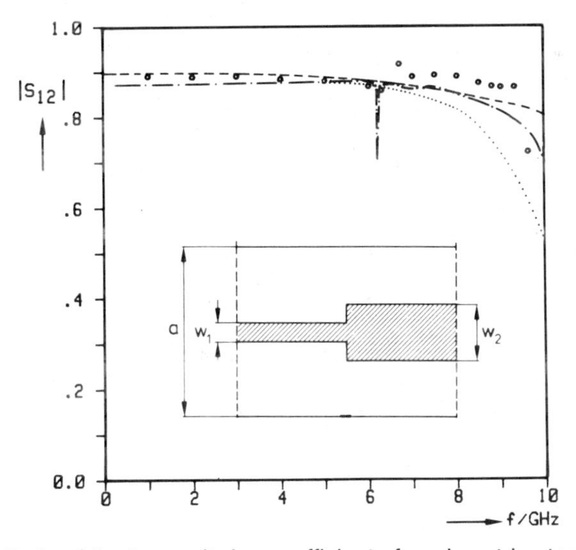

Fig. 25 Magnitude of the transmission coefficient of a microstrip step discontinuity. $\varepsilon_r = 2.32$, $w_1 = 4.5$ mm, $w_2 = 15.95$ mm, $d = 1.58$ mm, $d_l = 10$ mm, $a = 23.7$ mm. (◦) MoL [5], (—·—) MMT [31], (···) SDA [32], (– – –) magnetic-wall model [33].

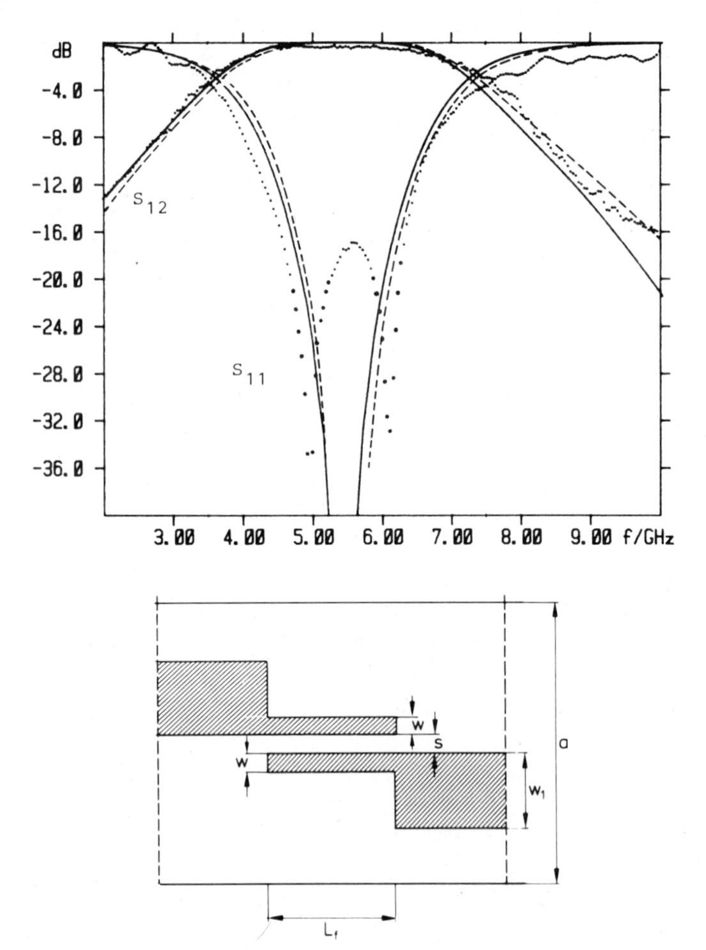

Fig. 26 Magnitude of the scattering parameters of a parallel coupled-line filter. $\varepsilon_r = 2.37$, $w = w_1 = s = 0.24$ mm, $L_f = 10$ mm, $d = 0.79$ mm, $d_l = 10d$, $a = 4.9$ mm. (———) MoL [5], (· · ·) experiment [5], (– – –) SDA [34].

5. ABOUT THE NATURE OF THE METHOD OF LINES

5.1. Introduction

In this section the question of the nature of the method of lines is investigated [35]. The relation to the discrete Fourier transformation is pointed out, and the difference between the method of lines and the mode-matching technique is clarified. First, however, the connection between the shielded and the equivalent periodic structures must be established.

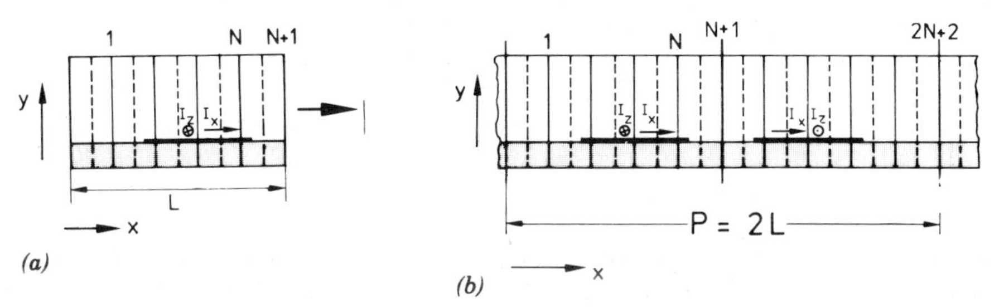

Fig. 27 Relations between a shielded microstrip line and a periodic one. Copyright © 1987 Hirzel Verlag, Stuttgart.

5.2. Relation between Shielded Structures and Periodic Structures

In the analysis of periodic structures with the method of lines it is found that the matrices representing the eigenvalue problem of, e.g., the propagation constant and current distribution have the Toeplitz structure. Toeplitz matrices are determined by their first row and column only and are therefore inverted fast and easily [9]. Thus it may be convenient to analyze shielded planar waveguides by converting them into an exactly equivalent periodic structure to take advantage of this property. The foundation of the conversion is image theory. Instead of the laterally shielded microstrip line in Fig. 27*a*, for example, the periodic structure of Fig. 27*b* can be analyzed. Naturally, the propagation constant in the x direction must be set to zero. Additional attention must be paid to determine the odd mode of the periodic structure. (The even mode corresponds to the mode in Fig. 27*a* with magnetic side walls.) The advantage of Toeplitz matrices in Fig. 27*b* is paid for by the disadvantage of a matrix twice as large. For large matrices, however, the advantage dominates over the disadvantage. The geometric conversion corresponds mathematically to the fact that matrices of the structure in Fig. 27*a* are sums of Toeplitz and Hankel matrices. The matrix sum may be converted in a block Toeplitz matrix of twice the original size.

5.3. Method of Lines and Discrete Fourier Transformation

Now we are able to tell something about the nature of the method of lines. In the field theory of planar structures, partial differential equations of the Helmholtz type

$$\frac{\partial^2 F}{\partial x^2} + \frac{\partial^2 F}{\partial y^2} + (k^2 - k_z^2)F = 0 \tag{208a}$$

must be solved. Metallic side walls must be taken into account by the corresponding boundary conditions. Because F and $F'' = \partial^2 F/\partial x^2$ are periodic functions, we use the following Fourier expansions:

$$F(x, y) = \sum_{n=-\infty}^{\infty} \tilde{F}_n e^{j2\pi nx/P} \qquad F''(x, y) = -\sum_{n=-\infty}^{\infty} k_n^2 \tilde{F}_n e^{j2\pi nx/P} \tag{209}$$

with $k_n = 2\pi n/P$.

In the mode-matching technique these series are terminated at sufficiently high $n = N'$. In the method of lines we discretize, i.e., we take F only for N points $x = mP/N = x_m$ $(m = 0, 1, \ldots, N-1)$ althogether. With $F(x_m) = F_m$, we have

$$F_m = \sum_{n=-\infty}^{\infty} \tilde{F}_n e^{j2\pi mn/N} \tag{210}$$

or with

$$\bar{F}_i = \sum_{r=-\infty}^{\infty} \tilde{F}_{i+rN} = \bar{F}_{i+N} \tag{211}$$

we get

$$F_m = \sum_{i=0}^{N-1} \bar{F}_i e^{j2\pi mi/N} \tag{212}$$

Equation (212) is an expression representing a discrete Fourier transformation (DFT). Its inverse transformation is as follows (as is immediately seen from the orthogonal properties of the exponential function):

$$\bar{F}_i = \frac{1}{N} \sum_{m=0}^{N-1} F_m e^{-j2\pi im/N} \tag{213}$$

So far no approximations have been made. Thus, if the $\bar{F}_i$ are known exactly, the values of the function F at the points x_m are also known exactly. The $\bar{F}_i$ must be determined from the differential equation according to eq. (208). Therefore, we must now look at F''_m. With

$$\bar{F}_i \lambda_i^2 = \sum_{r=-\infty}^{\infty} \tilde{F}_{i+rN} k_{i+rN}^2 \tag{214}$$

corresponding to eq. (211), or

$$\lambda_i^2 = \frac{\sum_{r=-\infty}^{\infty} \tilde{F}_{i+rN} k_{i+rN}^2}{\sum_{r=-\infty}^{\infty} \tilde{F}_{i+rN}} \tag{215}$$

we now get a DFT pair

$$F''_m = -\sum_{i=0}^{N-1} \lambda_i^2 \bar{F}_i e^{j2\pi mi/N} \tag{216}$$

$$-\lambda_i^2 \bar{F}_i = \frac{1}{N} \sum_{m=0}^{N-1} F''_m e^{-j2\pi im/N} \tag{217}$$

Substituting eqs. (212) and (216) into eq. (208), we obtain the ordinary diffential equations

$$\frac{d^2\bar{F}_i}{dy^2} + (k^2 - k_z^2 - \lambda_i^2)\bar{F}_i = 0 \tag{218}$$

for $\bar{F}_i$. The $\bar{F}_i$, which are functions of y, of course, can only be determined exactly if the λ_i^2 are known exactly. This is the only point where an approximation has to be used in the discretized representation.

5.4. Discussion

At this point we make a comparison with the mode-matching technique. Generally speaking the difference is that the mode-matching technique uses a global approximation of the field quantities F in a particular region, whereas the method of lines approximates the differential operator $\partial^2 F/\partial x^2$. If we calculate a zeroth-order approximition from eq. (215) by only using the series terms with $r = 0$, we obtain

$$\lambda_i^2 = k_i^2 \tag{219}$$

But these are exactly the values used in the mode-matching technique, where the series in eq. (209) is terminated at $n = N'$. For this approximation of λ_i^2, the $\bar{F}_i$ change into $\tilde{F}_i$, i.e., no terms exceeding N' are considered for the field representation. This means that the second derivative $\partial^2\tilde{F}_i/\partial x^2$ of the approximated function $\tilde{F}_i$ is used. In order to get accurate results, λ_i^2 must be determined accurately, which means that higher terms in $\bar{F}_i$ must also contribute.

In the method of lines, F'' and so λ_i^2 are determined from the corresponding difference quotient. According to the mean value theorem of Weierstrass, there is a place in every discretization interval where the difference quotient is exactly equal to the differential quotient. Therefore it is to be expected that these values are more accurate than the values according to eq. (219).

Thus, monotonous curves, i.e., nonscillating ones, are to be expected for the field quantities, such as the current distribution [35] presented for a microstrip line in Fig. 28. This curve is drawn using the results for 11 lines for J_z on the strip. When the current distribution for 9, 7 and 5 lines is calculated, it is noticed that the corresponding points lie very close to the

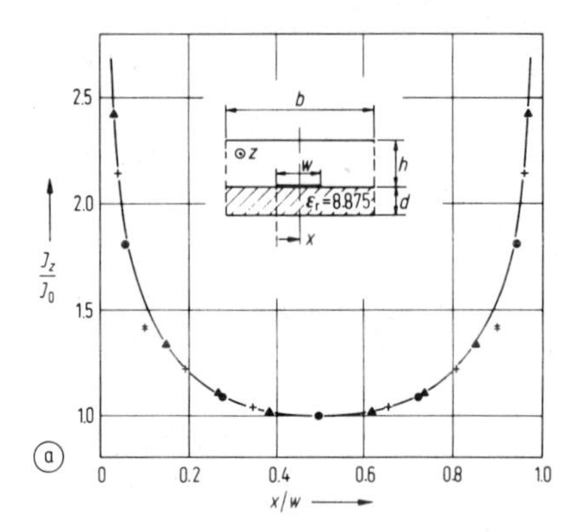

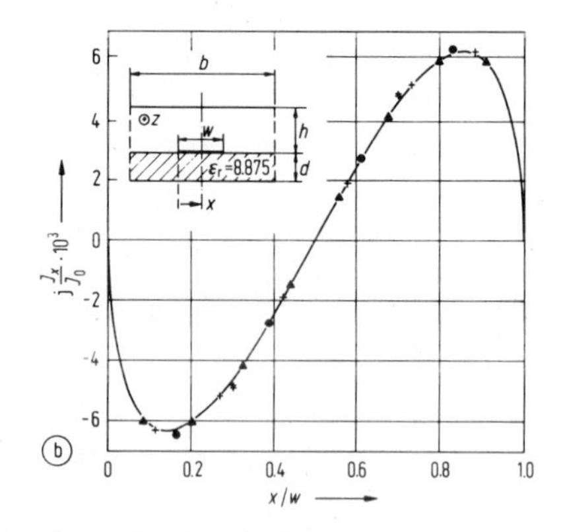

Fig. 28 Current distribution of a shielded microstrip line [35]. (*a*) J_z; (*b*) J_x. ($w = 1$ mm, $d = 1$ mm, $b = 12$ mm, $h = 9$ mm). (———) $m = 11$, (▲) $m = 9$, (+) $m = 7$, (●) $m = 5$, (∗) $m = 3$. Copyright © 1987 Hirzel Verlag, Stuttgart.

full line. Merely for three lines the two outer points show a certain deviation. The current J_x is phase-shifted by $\pi/2$ with respect to J_z. Thus jJ_x/J_0 is a real quantity. This diagram and the whole section are a proof that in the method of lines the fields on the lines are determined very accurately and hence the λ_i^2 are very well approximated.

ACKNOWLEDGMENTS

The work on which this chapter is based has been supported by several grants of the Deutsche Forschungsgemeinschaft and one of the Stiftung Volkswagenwerk.

APPENDIX A. DETERMINATION OF THE EIGENVALUES AND EIGENVECTORS OF *P*

The elements λ_k^2 of the matrix $\boldsymbol{\lambda}^2$ are the eigenvalues and the column vectors $\mathbf{t}_k$ of the transformation matrix $\boldsymbol{T}$ are the eigenvectors belonging to the matrix $\boldsymbol{P}$

$$(\boldsymbol{P} - \lambda_k^2 \boldsymbol{I})\mathbf{t}_k = 0 \tag{220}$$

As $\boldsymbol{P}$ is a tridiagonal matrix, we obtain a second-order difference equation

$$-t_{i-1}^{(k)} + (2 - \lambda_k^2)t_i^{(k)} - t_{i+1}^{(k)} = 0 \tag{221}$$

Only the first and the last equation in (220) are different from this form. With the substitution

$$t_i^{(k)} = A_k e^{ji\varphi_k} + B_k e^{-ji\varphi_k} \tag{222}$$

we obtain the characteristic equation from (221):

$$\lambda_k^2 = 2(1 - \cos\varphi_k)$$

or

$$\lambda_k^2 = 4\sin^2(\varphi_k/2) \tag{223}$$

This equation is valid for all boundary combinations. Only the φ_k depend on the boundary combinations. To determine the φ_k and the A_k and B_k in (222), the first and last equations are used. From the explanations in Section 2.2, it is known that the boundary conditions for the two components are dual to each other. Thus it is convenient to look at the solutions for the two components (here generally named ϕ_1 and ϕ_2) together. For a given problem the pair of boundary conditions is either

I. ϕ_1: Dirichlet–Neumann ϕ_2: Neumann–Dirichlet or
II. ϕ_1: Dirichlet–Dirichlet ϕ_2: Neumann–Neumann

For the easy solution of the first and last equations (220), these equations are extended by fictitious quantities X_0, X_{N+1} or even X_{N+2}, which also obey the law (222). Thus eq. (221) is fulfilled as well.

A.1. Dirichlet–Neumann (DN) Condition and Neumann–Dirichlet (ND) Condition

The number of lines is equal to N for both components. The first and the last equation, respectively, run as follows:

DN	ND
$t_0^{(k)} = 0$ $-t_N^{(k)} + t_{N+1}^{(k)} = 0$ or with (222) $\begin{bmatrix} 1 & 1 \\ e^{jN\varphi_k}(e^{j\varphi_k} - 1) & e^{-jN\varphi_k}(e^{-j\varphi_k} - 1) \end{bmatrix} \times \begin{bmatrix} A_k \\ B_k \end{bmatrix} = \mathbf{0}$	$t_0^{(k)} - t_1^{(k)} = 0$ $t_{N+1}^{(k)} = 0$ $\begin{bmatrix} 1 - e^{j\varphi_k} & 1 - e^{-j\varphi_k} \\ e^{j(N+1)\varphi_k} & e^{-j(N+1)\varphi_k} \end{bmatrix}\begin{bmatrix} A_k \\ B_k \end{bmatrix} = \mathbf{0}$

from the condition for nontrivial solutions it follows that

$$\varphi_k = \frac{k-1/2}{N+1/2}\,\pi \qquad k=1,2\ldots,N \tag{224}$$

and from (222) with

DN	ND
$A_k = -B_k$	$(1-e^{j\varphi_k})A_k = -(1-e^{-j\varphi_k})B_k$
	$e^{j\varphi_k/2}A_k = e^{-j\varphi_k/2}B_k$
$t_i^{(k)} = A_k \sin i\varphi_k$	$t_i^{(k)} = A_k \cos(i-\frac{1}{2})\varphi_k$

Thus the following transformation matrices result for $i, k = 1, 2, \ldots, N$:

For **DN** $\quad T_{\mathrm{DN}ik} = \sqrt{\dfrac{2}{N+1/2}}\,\sin\dfrac{i(k-1/2)\pi}{N+1/2}$

For **ND** $\quad T_{\mathrm{ND}ik} = \sqrt{\dfrac{2}{N+1/2}}\,\cos\dfrac{(i-1/2)(k-1/2)\pi}{N+1/2}$ (225)

The vectors are orthonormal, which means that

$$T^{\mathrm{t}}T = I$$

A.2. Dirichlet–Dirichlet (DD) Condition and Neumann–Neumann (NN) Condition

The number of lines for the boundary condition DD is N, the number of lines for NN is $N+1$. The first and the last equation, respectively, run as follows:

DD	NN
$t_0^{(k)} = 0$	$t_0^{(k)} - t_1^{(k)} = 0$
$t_{N+1}^{(k)} = 0$	$t_{N+2}^{(k)} - t_{N+1}^{(k)} = 0$
or with (222)	

DD	NN
$\begin{bmatrix} 1 & 1 \\ e^{j(N+1)\varphi_k} & e^{-j(N+1)\varphi_k} \end{bmatrix} \begin{bmatrix} A_k \\ B_k \end{bmatrix} = \mathbf{0}$	$\begin{bmatrix} 1 & 1 \\ e^{j(N+1)\varphi_k} & \overline{e^{-j(N+1)\varphi_k}} \end{bmatrix} \times \begin{bmatrix} (1-e^{j\varphi_k})A_k \\ (1-e^{-j\varphi_k})B_k \end{bmatrix} = \mathbf{0}$

From the condition for nontrivial solutions follows

$$\varphi_k = \frac{k\pi}{N+1}$$

For **DD** $\quad k = 1, 2, \ldots, N \quad$ For **NN** $\quad k = 0, 1, 2, \ldots, N\,. \quad (226)$

and from eq. (222) with

DD	NN
$A_k = -B_k$	$(1-e^{j\varphi_k})A_k = -(1-e^{-j\varphi_k})B_k$
	$e^{j\varphi_k/2}A_k = +e^{-j\varphi_k/2}B_k$
$t_i^{(k)} = A_k \sin i\varphi_k$	$t_i^{(k)} = A_k \cos(i-\frac{1}{2})\varphi_k$

we finally obtain the transformation matrices

For **DD** $\quad \boldsymbol{T}_{\mathrm{DD}i,k} = \sqrt{\frac{2}{N+1}} \sin \frac{ik\pi}{N+1} \qquad i, k = 1, 2, \ldots, N$

For **NN** $\quad \boldsymbol{T}_{\mathrm{NN}i1} = \sqrt{\frac{1}{N+1}}$

$$\boldsymbol{T}_{\mathrm{NN}i,k+1} = \sqrt{\frac{2}{N+1}} \cos \frac{(i-\frac{1}{2})k\pi}{N+1} \qquad i = 1, 2, \ldots, N+1$$

$$k = 1, 2, \ldots, N \qquad (227)$$

In this case the eigenvectors are also orthonormal, which means again

$$\boldsymbol{T}^{\mathrm{t}}\boldsymbol{T} = \boldsymbol{I}$$

The eigenvalues are equal in case I. In case II the first eigenvalue for the boundary conditions NN is equal to zero, whereas the other eigenvalues are equal to those for the boundary conditions DD. We can give the following relations;

$$\boldsymbol{\lambda}^2_{\mathrm{DN}} = \boldsymbol{\lambda}^2_{\mathrm{ND}} \tag{228}$$

$$\boldsymbol{\lambda}^2_{\mathrm{NN}} = \mathrm{blockdiag}(0, \boldsymbol{\lambda}^2_{\mathrm{DD}}) \tag{229}$$

where we have generally set

$$\boldsymbol{\lambda}^2 = \mathrm{diag}(\lambda_k^2) \tag{230}$$

APPENDIX B. CALCULATION OF THE MATRICES $\boldsymbol{\delta}$

The matrix $\boldsymbol{\delta}$ is defined as the product of the difference matrix $\boldsymbol{D}$ with the corresponding transformation matrix $\boldsymbol{T}$ from the right and with the transposed matrix $\boldsymbol{T}_d^{\mathrm{t}}$ of the dual boundary value problem from the left.

$$\boldsymbol{\delta} = \boldsymbol{T}_d^{\mathrm{t}} \boldsymbol{D} \boldsymbol{T} \tag{231}$$

Here the subscript d stands for dual. Before $\boldsymbol{\delta}$ is determined, one result can be anticipated. The eigenvalue matrix $\boldsymbol{\lambda}^2$ results from the product

$$\boldsymbol{\delta}^{\mathrm{t}}\boldsymbol{\delta} = \boldsymbol{T}^{\mathrm{t}}\boldsymbol{D}^{\mathrm{t}}\boldsymbol{T}_d\boldsymbol{T}_d^{\mathrm{t}}\boldsymbol{D}\boldsymbol{T} = \boldsymbol{T}^{\mathrm{t}}\boldsymbol{D}^{\mathrm{t}}\boldsymbol{D}\boldsymbol{T} = \boldsymbol{\lambda}^2 \tag{232}$$

Thus a relation between $\boldsymbol{\delta}$ and $\boldsymbol{\lambda}^2$ is presumed.

B.1. Boundary Conditions DD and NN

For the boundary conditions DD the difference matrix $\boldsymbol{D}$ is given by

$$\boldsymbol{D} = \begin{bmatrix} 1 & & \\ -1 & \ddots & \\ & \ddots & 1 \\ & & -1 \end{bmatrix} \tag{233}$$

the corresponding $\boldsymbol{\delta}$ is

$$\boldsymbol{\delta} = \boldsymbol{T}^{\mathrm{t}}_{\mathrm{NN}} \boldsymbol{D} \boldsymbol{T}_{\mathrm{DD}} \tag{234}$$

If we first multiply $\boldsymbol{T}^{\mathrm{t}}_{\mathrm{NN}}$ by $\boldsymbol{D}$, we find that the first row of the result vanishes, since all elements of the first column vector in $\boldsymbol{T}_{\mathrm{NN}}$ are identical. The remaining part of the resulting matrix can be written as the product of $\boldsymbol{\lambda}_{\mathrm{DD}}$ and $\boldsymbol{T}_{\mathrm{DD}}$, where $\boldsymbol{\lambda}_{\mathrm{DD}}$ is the diagonal matrix formed by the positive square roots of $\boldsymbol{\lambda}^2_{\mathrm{DD}}$. Hence there is

$$\boldsymbol{T}^{\mathrm{t}}_{\mathrm{NN}} \boldsymbol{D} = \begin{bmatrix} \mathbf{0}^{\mathrm{t}} \\ \boldsymbol{\lambda}_{\mathrm{DD}} \boldsymbol{T}_{\mathrm{DD}} \end{bmatrix} = \begin{bmatrix} \mathbf{0}^{\mathrm{t}} \\ \boldsymbol{\lambda}_{\mathrm{DD}} \end{bmatrix} \boldsymbol{T}_{\mathrm{DD}} \tag{235}$$

Since $\boldsymbol{T}_{\mathrm{DD}}$ is a symmetric matrix, $\boldsymbol{T}_{\mathrm{DD}} = \boldsymbol{T}^{\mathrm{t}}_{\mathrm{DD}}$. Consequently $\boldsymbol{\delta}$ is given by

$$\boldsymbol{\delta} = \begin{bmatrix} \boldsymbol{0}^t \\ \boldsymbol{\lambda}_{\mathrm{DD}} \end{bmatrix} \tag{236}$$

Thus $\boldsymbol{\delta}$ is a quasi-diagonal matrix.

For the boundary conditions NN, the result is $-\boldsymbol{\delta}^{\mathrm{t}}$ because

$$-\boldsymbol{\delta}^{\mathrm{t}} = \boldsymbol{T}^{\mathrm{t}}_{\mathrm{DD}}(-\boldsymbol{D}^{\mathrm{t}})\boldsymbol{T}_{\mathrm{NN}} \tag{237}$$

and $-\boldsymbol{D}^{\mathrm{t}}$ is the difference operator for the boundary conditions NN. With $\boldsymbol{\delta}$ according to eq. (236) there is

$$\boldsymbol{\delta}^{\mathrm{t}}\boldsymbol{\delta} = \boldsymbol{\lambda}^2_{\mathrm{DD}} \qquad \text{and} \qquad \boldsymbol{\delta}\boldsymbol{\delta}^{\mathrm{t}} = \boldsymbol{\lambda}^2_{\mathrm{NN}} \tag{238}$$

For the conversion of equations containing $\boldsymbol{\delta}$, we hint at a useful relation. Because of the speical structure of $\boldsymbol{\delta}$ according to eq. (236) not only are

$$\boldsymbol{\delta}\boldsymbol{\lambda}^2_{\mathrm{DD}} = \boldsymbol{\lambda}^2_{\mathrm{NN}}\boldsymbol{\delta} \qquad \text{and} \qquad \boldsymbol{\delta}^{\mathrm{t}}\boldsymbol{\lambda}^2_{\mathrm{NN}} = \boldsymbol{\lambda}^2_{\mathrm{DD}}\boldsymbol{\delta}^{\mathrm{t}} \tag{239}$$

valid, which follows from eq. (238), but, e.g., also

$$\boldsymbol{\delta}\boldsymbol{k}^2_{y\mathrm{DD}} = \boldsymbol{k}^2_{y\mathrm{NN}}\boldsymbol{\delta} \qquad \text{and} \qquad \boldsymbol{\delta}^{\mathrm{t}}\boldsymbol{k}^2_{y\mathrm{NN}} = \boldsymbol{k}^2_{y\mathrm{DD}}\boldsymbol{\delta}^{\mathrm{t}} \tag{240}$$

Instead of $\boldsymbol{k}^2_y$ other diagonal matrices determined from $\boldsymbol{\lambda}^2$ and $\boldsymbol{k}_y$ can be used.

B.2. Boundary Conditions DN and ND

For the boundary conditions DN the difference matrix $\boldsymbol{D}$ is given by

$$\boldsymbol{D} = \begin{bmatrix} 1 & & & \\ -1 & \ddots & & \\ & \ddots & \ddots & \\ & & -1 & 1 \end{bmatrix} \tag{241}$$

The corresponding $\boldsymbol{\delta}$ is

$$\boldsymbol{\delta} = \boldsymbol{T}^{\mathrm{t}}_{\mathrm{ND}}\boldsymbol{D}\boldsymbol{T}_{\mathrm{DN}} \tag{242}$$

We find that

$$\boldsymbol{\delta} = \mathrm{diag}(\lambda_k) = \boldsymbol{\lambda}_{\mathrm{DN}} = \boldsymbol{\lambda}_{\mathrm{ND}} \tag{243}$$

The λ_k are the positive square roots of λ^2_k. For the boundary conditions ND in analogy to above, we obtain $-\boldsymbol{\delta}^{\mathrm{t}}$.

APPENDIX C. THE COMPONENT OF ε_h AT AN ABRUPT TRANSITION

As numerical investigations have shown, it is best to put the interface, where the dielectric constant has an abrupt transition, on a ψ_h line (or to choose the discretization in such a way that a ψ_h line coincides with the interface line t) (see Fig. 29). Which value of ε_r is to be inserted in $\boldsymbol{\varepsilon}_h$ on this line? To answer this question we look at the behavior of ψ_e on the interface line t. ψ_e is taken because $\boldsymbol{\varepsilon}_h$ is found in $\boldsymbol{P}_e^\varepsilon$ and not in $\boldsymbol{P}_h^\varepsilon$. On the interface itself, ψ_e is continuous, but not $\partial\psi_e/\partial x$. Because of the continuity of the tangential E field and the normal D field components there is, rather,

$$\psi_t^1 = \psi_t^2 \tag{244}$$

$$\frac{1}{\varepsilon_{r1}} \frac{\partial \psi_t^1}{\partial x} = \frac{1}{\varepsilon_{r2}} \frac{\partial \psi_t^2}{\partial x} \tag{245}$$

(The subscript e is omitted here, as only ψ_e is regarded.) Consequently, in the figure the ψ curve is represented with a break on the interface, and the dashed curves are drawn without any break for the definition of the auxiliary quantities ψ_{k+1}^1 and ψ_k^2. These auxiliary quantities are necessary to compute the second derivative on the lines k and $k+1$. With the incremental formula (Taylor series up to the linear term, applied on the interface line t) we find

$$\begin{aligned}
\psi_{k+1}^1 &\approx \psi_t + \frac{\partial \psi^1}{\partial x}\left(\frac{h}{2}\right) \qquad & \psi_k &\approx \psi_t - \frac{\partial \psi^1}{\partial x}\left(\frac{h}{2}\right) \\
\psi_{k+1} &\approx \psi_t + \frac{\partial \psi^2}{\partial x}\left(\frac{h}{2}\right) \qquad & \psi_k^2 &\approx \psi_t - \frac{\partial \psi^2}{\partial x}\left(\frac{h}{2}\right)
\end{aligned} \tag{246}$$

Condition (244) was immediately taken into account in these equations. With (245) we find for the two derivatives on the interface line:

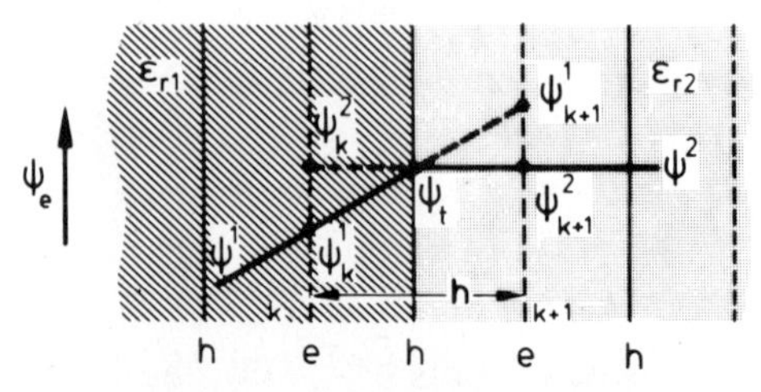

Fig. 29 Behavior of ψ_e at an abrupt transition from ε_{r1} to ε_{r2}.

$$
\begin{aligned}
h \frac{\partial \psi^1}{\partial x} &\approx \frac{2\varepsilon_{r1}}{\varepsilon_{r1} + \varepsilon_{r2}} (\psi_{k+1} - \psi_k) \\
h \frac{\partial \psi^2}{\partial x} &\approx \frac{2\varepsilon_{r2}}{\varepsilon_{r1} + \varepsilon_{r2}} (\psi_{k+1} - \psi_k)
\end{aligned}
\tag{247}
$$

With these two derivatives we are able to give the special second derivatives on the lines k and $k + 1$:

LINE k:

$$
h^2 \frac{\partial}{\partial x}\left(\frac{1}{\varepsilon_r} \frac{\partial \psi}{\partial x}\right) \approx \frac{2}{\varepsilon_{r1} + \varepsilon_{r2}} (\psi_{k+1} - \psi_k) - \frac{1}{\varepsilon_{r1}} (\psi_k - \psi_{k-1}) \tag{248}
$$

LINE $k + 1$:

$$
h^2 \frac{\partial}{\partial x}\left(\frac{1}{\varepsilon_r} \frac{\partial \psi}{\partial x}\right) \approx \frac{1}{\varepsilon_{r2}} (\psi_{k+2} - \psi_{k+1}) - \frac{2}{\varepsilon_{r1} + \varepsilon_{r2}} (\psi_{k+1} - \psi_k) \tag{249}
$$

We get the same results from eq. (116) if we fill the place for line t in $\boldsymbol{\varepsilon}_h$ with $\varepsilon_t = (\varepsilon_{r1} + \varepsilon_{r2})/2$, that is, the arithmetic mean of the two ε_r values.

APPENDIX D. EIGENVALUES AND EIGENVECTORS FOR PERIODIC BOUNDARY CONDITIONS

The eigenvalues and eigenvectors of the matrix $\boldsymbol{P}_n$ for periodic structures are determined from the eigenvalues and eigenvectors of the difference matrix $\boldsymbol{D}_n$, which is given by eq. (155) for phase-normalized field quantities or potentials. $\boldsymbol{D}_n$ is a circulant matrix. Analogous to Appendix A we find for its eigenvalues

$$
\delta_{nk} = 2je^{j\varphi_k/2} \sin \frac{\varphi_k - \beta_x h}{2} \tag{250}
$$

with

$$
\varphi_k = \frac{2\pi}{N} k \qquad k = 1, 2, \ldots, N \tag{251}
$$

As eigenvector matrix either $\boldsymbol{T}_e$ or $\boldsymbol{T}_h$, which are Fourier matrices, is chosen:

$$
\boldsymbol{T}_{e\,ik} = \frac{1}{\sqrt{N}}\, e^{ji\varphi_k} \qquad \boldsymbol{T}_{h\,ik} = \frac{1}{\sqrt{N}}\, e^{j(i+1/2)\varphi_k} \tag{253}
$$

The general phase in the eigenvector matrix is arbitrary. Because of the shifting of the discretization lines by $h/2$, a phase shifting by $\varphi_k/2$ with respect to $\boldsymbol{T}_e$ is introduced in $\boldsymbol{T}_h$:

$$T_h = T_e S_\varphi \tag{253}$$

with

$$S_\varphi = \mathrm{diag}(e^{j\varphi_k/2}) \tag{254}$$

As the matrix P_n is given as a product in eq. (156) its eigenvalues λ^2 are products as well:

$$\lambda^2 = \delta_n \delta_n^* \tag{255}$$

where

$$\lambda = \mathrm{diag}\left(2 \sin \frac{\varphi_k - \beta_x h}{2}\right) \tag{256}$$

$$\delta_n = j\lambda S_\varphi = jS_\varphi \lambda \tag{257}$$

In addition there are

$$T_{e,h}^{*t} D_n T_{e,h} = \delta_n \tag{258}$$

$$T_h^{*t} D_n T_e = j\lambda \tag{259}$$

$$T_e^{*t} D_n^{*t} T_h = -j\lambda \tag{260}$$

APPENDIX E. CALCULATION OF CHARACTERISTIC IMPEDANCE

In waveguides where hybrid waves can propagate, characteristic impedances are defined by means of the power transfer P. For example, in a microstrip line it is defined as

$$Z_P = P/I^2 \tag{261}$$

and in a microslot line,

$$Z_P = U^2/P \tag{262}$$

where I is the total current flow on the strip in the direction of wave propagation and U is the integral of the electric field from one conductor edge in the slot to the next one. The power transfer P is determined as the integral of the Poynting vector over the waveguide cross section F. For propagation in the z direction it is

$$P = \int\int_{(F)} (e_x h_y - e_y h_x)\, dx\, dy \tag{263}$$

The power transfer must be calculated for each layer of the planar structure separately and summed over all layers. For the layer between y_1 and y_2 in Fig. 5 we obtain, using the discretized fields (approximating the integral with respect to x by means of the rectangular or trapezoidal rule),

$$P_d = h \int_{y_1}^{y_2} (\mathbf{E}_x^t \mathbf{H}_y - \mathbf{E}_y^t \mathbf{H}_x)\, dy = h \int_{y_1}^{y_2} (\bar{\mathbf{E}}_x^t \bar{\mathbf{H}}_y - \bar{\mathbf{E}}_y^t \bar{\mathbf{H}}_x)\, dy \tag{264}$$

Instead of the field components in the spatial domain, the field components in the transformed domain can be used, even with advantage, because there is, e.g.,

$$\mathbf{E}_x^t \mathbf{H}_y = \bar{\mathbf{E}}_x^t \boldsymbol{T}^t \boldsymbol{T} \bar{\mathbf{H}}_y = \bar{\mathbf{E}}_x^t \bar{\mathbf{H}}_y \tag{265}$$

For the calculation of the integral over y in eq. (264) we set for each field component vector

$$\bar{\mathbf{F}} = \boldsymbol{\alpha}(\boldsymbol{S}_1 \bar{\mathbf{F}}_A + \boldsymbol{S}_2 \bar{\mathbf{F}}_B) \tag{266}$$

with $\boldsymbol{\alpha}$ according to eq. (20) and

$$\boldsymbol{S}_1 = \mathrm{diag}\left(\frac{k_{yi}}{k_0} \sinh k_{yi}(y_2 - y)\right) \qquad \boldsymbol{S}_2 = \mathrm{diag}\left(\frac{k_{yi}}{k_0} \sinh k_{yi}(y - y_1)\right) \tag{267}$$

Then we can write for the first term in eq. (264)

$$\int_{y_1}^{y_2} \bar{\mathbf{E}}_x^t \bar{\mathbf{H}}_y\, dy = \int_{y_1}^{y_2} \begin{bmatrix} \bar{\mathbf{E}}_{xA} \\ \bar{\mathbf{E}}_{xB} \end{bmatrix}^t \left(\boldsymbol{\alpha}_h^2 \begin{bmatrix} \boldsymbol{S}_1 \\ \boldsymbol{S}_2 \end{bmatrix} \begin{bmatrix} \boldsymbol{S}_1 \\ \boldsymbol{S}_2 \end{bmatrix}^t \begin{bmatrix} \bar{\mathbf{H}}_{yA} \\ \bar{\mathbf{H}}_{yB} \end{bmatrix} \right) dy \tag{268}$$

with the solution

$$\int_{y_1}^{y_2} \bar{\mathbf{E}}_x^t \bar{\mathbf{H}}_y\, dy = \frac{1}{2} \begin{bmatrix} \bar{\mathbf{E}}_{xA} \\ \bar{\mathbf{E}}_{xB} \end{bmatrix}^t \begin{bmatrix} \boldsymbol{g}_1 & \boldsymbol{g}_2 \\ \boldsymbol{g}_2 & \boldsymbol{g}_1 \end{bmatrix} \begin{bmatrix} \bar{\mathbf{H}}_{yA} \\ \bar{\mathbf{H}}_{yB} \end{bmatrix} \tag{269}$$

where

$$\begin{aligned} \boldsymbol{g}_1 &= k_0^{-1}(\boldsymbol{\gamma}_h - k_0 d\, \boldsymbol{k}_{yh}^2 \boldsymbol{\alpha}_h^2) \\ \boldsymbol{g}_2 &= k_0^{-1}(k_0 d\, \boldsymbol{k}_h^2 \boldsymbol{\alpha}_h \boldsymbol{\gamma}_h - \boldsymbol{\alpha}_h) \end{aligned} \tag{270}$$

and $\boldsymbol{\alpha}$, $\boldsymbol{\gamma}$ and $\boldsymbol{k}_y$ are according to eq. (20). A corresponding result is also obtained for the second term. Note that the quantities $\boldsymbol{k}_y^2$, $\boldsymbol{\alpha}$, $\boldsymbol{\gamma}$ then have to be marked with the subscript *e*.

REFERENCES

1. O. A. Liskovets, "The method of lines," Review, *Differ. Uravneniya*, vol. 1, pp. 1662–1678, 1965.
2. B. P. Demidowitsch et al., *Numerical Methods of Analysis* (in German), Chapter 5, VEB Wissenschaften, Berlin, 1968.
3. S. G. Michlin and C. Smolizki, *Näherungsmethoden zur Lösung von Differential- und Integralgleichungen*, pp. 238–243, Teubner, Leipzig, 1969.
4. U. Schulz, "The method of lines—A new technique for the analysis of planar microwave structures" (in German), Ph.D. Thesis, FernUniv., Hagen, Federal Republic of Germany, 1980.
5. S. B. Worm, "Analysis of planar microwave structures with arbitrary contour" (in German), Ph.D. Thesis, FernUniv., Hagen, Federal Republic of Germany, 1983.
6. H. Diestel, "A method for calculating inhomogeneous planar dielectric waveguides" (in German), Ph.D. Thesis, FernUniv., Hagen, Federal Republic of Germany, 1984.
7. R. Pregla, "Analysis of planar microwave structures on magnetized ferrite substrate," *Arch. Elektron. Uebertragungstech*, vol. 40, pp. 270–274, 1986.
8. R. Pregla, M. Koch and W. Pascher, "Analysis of hybrid waveguide structures consisting of microstrips and dielectric waveguides." *Proc. 17th Eur. Microwave Conf.*, pp. 927–932, 1987.
9. W. Pascher and R. Pregla, "Full wave analysis of complex planar microwave structures," *Radio Sci.* vol. 22, pp. 999–1002, 1987.
10. U. Schulz, "On the edge condition with the method of lines in planar waveguides," *Arch. Elektron. Uebertragungstech.*, vol. 34, pp. 176–178, 1980.
11. U. Schulz and R. Pregla, "A new technique for the analysis of the dispersion characteristics of planar waveguides and its application to microstrips with tuning septums," *Radio Sci.*, vol. 16, pp. 1173–1178, 1981.
12. H. Diestel and S. B. Worm, "Analysis of hybrid field problems by the method of lines with nonequidistant discretization, "*IEEE Trans. Microwave Theory Tech.*, vol. MTT-32, pp. 633–638, 1984.
13. H. Diestel, "Analysis of planar multiconductor transmission-line systems with the method of lines," *Arch. Elektron. Uebertragungstech.*, vol. 41, pp. 169–175, 1987.
14. R. Mittra, T. Itoh and T. Li, "Analytical and numerical studies of the relative convergence phenomenon arising in the solution of an integral equation by the moment method" *IEEE Trans. Microwave Theory Tech.*, vol. MTT-20, pp. 96–104, 1972.
15. J. Meixner, "The behaviour of electromagnetic fields at edges," *IEEE Antennas Propag.*, vol. AP–20, pp. 442–446, 1972.

16. H. Diestel, "A method for calculating the guided modes of strip-loaded optical waveguides with arbitrary index profile," *IEEE J. Quantum Electron.*, vol. QE-20, pp. 1288–1293, 1984.
17. R. E. Collin, *Field Theory of Guided Waves*, pp. 224–234, McGraw-Hill, New York. 1960.
18. M. Lemke and W. Schilz, "Broad band isolators up to 18 GHz" (in German), *Valvo Ber.* vol. 18, pp. 243–250, 1974.
19. R. Pregla and S. B. Worm, "A new technique for the analysis of planar waveguides with magnetized ferrite substrate," *Proc. 12th Eur. Microwave Conf.*, pp. 747–752, 1982.
20. I. Wolff, "A contribution to the theory of transverse magnetized ferrite," (in German), *Frequenz*, vol. 25, pp. 235–241, 1971.
21. R. Pregla and S. B. Worm, "Analysis of periodic structures in planar waveguides," *Proc. 7th Microcoll*, pp. 423–426, 1982.
22. S. B. Worm, private communication.
23. S. B. Worm and R. Pregla, "Hybrid mode analysis of arbitrarily shaped planar microwave structures by the method of lines," *IEEE Trans. Microwave Theory Tech.*, vol. MTT-32, pp. 191–196, 1984.
24. P. J. Davies, *Circulant Matrices*, Wiley, New York, 1979.
25. H. Akaike, "Block Toeplitz matrix inversion," *SIAM J. Appl. Math.*, vol. 24, pp. 234–241, 1973.
26. G. Kowalski and R. Pregla, "Dispersion characteristics of shielded microstrips with finite thickness," *Arch. Elektron. Uebertragungstech.*, vol. 25, pp. 193–196, 1971.
27. U. Schulz, "Impedance characteristics of ridge waveguides in phase controlled slotted ridge waveguide antennas" (in German), *ITG-Fachber.*, vol. 99, pp. 245–250, 1987.
28. T. Itoh, "Spectral domain immittance approach for dispersion characteristics of shielded microstrips with tuning septums," *Proc. 9th Eur. Microwave Conf.*, pp. 435–439, 1979.
29. K. Bierwirth, N. Schulz, and F. Arndt, "Finite-difference analysis of rectangular dielectric waveguide structures," *IEEE Trans. Microwave Theory Tech.*, vol. MTT-34, pp. 1104–1114, 1986.
30. W. T. Nisbet and J. Helszajn, "Mode charts for microstrip resonators on dielectric and magnetic substrates using a transverse-resonance method," *IEE J. Microwaves Opt. Acoust.*, vol. 3, pp. 69–77, 1979.
31. L. P. Schmidt, "Rigorous computation of the frequency dependent properties of filters and coupled resonators composed from transverse microstrip discontinuities", *Proc. 10th Eur. Microwave Conf.* pp. 436–440, 1980.
32. N. H. Koster and R. H. Jansen, "The microstrip step discontinuity: A revised description", *IEEE Trans. Microwave Theory Tech.*, vol. MTT-34, pp. 213–223, 1986.
33. G. Kompa, "The frequency dependent transmission properties of stripline step discontinuities, filters, and stubs in microstrip technique" (in German), Ph.D. Thesis, RWTH Aachen, Federal Republic of Germany, 1975.

34. R. H. Jansen and W. Wertgen, "A 3D field theoretical simulation tool for the CAD of millimeter wave MMICs," *Alta Freq.* Italy, vol. LVII, pp. 203–216, 1988.

35. R. Pregla, "About the nature of the method of lines," *Arch. Electron. Uebertragungstech.*, vol. 41, pp. 368–370, 1987.

36. R. Pregla and F. J. Schmückle, "The method of lines for the analysis of planar waveguide structures with finite metallization thickness" (in German), *Kleinheubacher Ber.*, vol. 31, pp. 431–438, 1988.

7

The Waveguide Model for the Analysis of Microstrip Discontinuities

Ingo Wolff
Department of Electrical Engineering and Sonderforschungsbereich 254
Universität Duisburg
Duisburg, Federal Republic of Germany

1. INTRODUCTION

Modern microwave and millimeter-wave integrated circuits are hybrid or monolithic integrated circuits based on the planar transmission line techniques. The transmission line that is used in 90% of the circuits is the microstrip line (or asymmetrical stripline). It consists of a dielectric substrate material of height h and relative dielectric constant ε_r that is metallized on its back side. On top of the substrate a metallic strip of width w and metallization thickness t is the waveguiding structure. Other planar transmission lines, as they are shown together with the microstrip line in Fig. 1, are of less importance and are used only in special applications.

Microwave integrated circuits consist of active and passive components as shown in Fig. 2. The active components are, e.g., transistors or diode structures for generating or amplifying the microwave signals. The planar, distributed, and passive components used in microwave integrated circuits can be considered to be composed of planar transmission lines and microstrip discontinuities. Figure 2 shows how a microwave integrated circuit can be divided into components that consist of subcomponents such as discontinuities and microstrip lines. It is therefore an essential task to accurately model the frequency-dependent transmission properties of these subcomponents (discontinuities) if accurate computer-aided design techniques for the analysis and synthesis of microwave integrated circuits are to be established. The microstrip line and its frequency-dependent properties in the meantime can be described by simple formulas that accurately model the results of efficient numerical methods for the calculation of the phase velocities and the characteristic impedances of the fundamental mode and higher-order modes [1, 2].

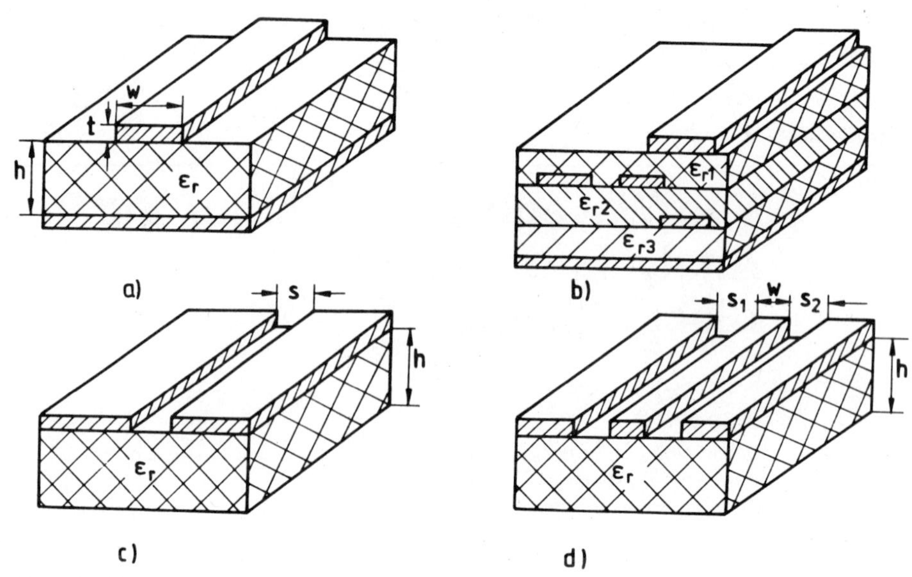

Fig. 1 Planar microwave transmission lines for application in microwave and millimeter-wave integrated circuits. (*a*) The microstrip line; (*b*) multiply coupled microstrip lines; (*c*) the slot line; (*d*) the coplanar line.

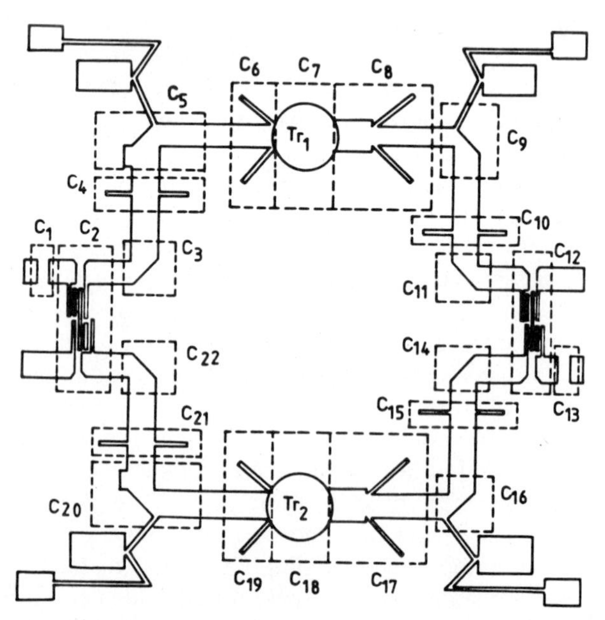

Fig. 2 A planar, integrated microwave circuit and its division into subcomponents for computer-aided analysis.

For the analysis of microstrip discontinuities, three primary model levels are available:

1. The discontinuities are described by equivalent circuits consisting of lumped elements, and the circuit elements (capacitances, inductances, and resistances) are calculated using static or stationary methods (see, e.g., References 3–9). The advantage of this method is that it leads to simple formulas describing the equivalent circuit elements; the disadvantage is that these formulas have only a restricted validity range (frequency, substrate materials, line widths) for their application.

2. If in particular the frequency-dependent properties of the discontinuities are to be described more accurately, the wave properties of the electromagnetic fields on the microstrip structures must be taken into account. A first, approximate way in which this can be done is to apply a waveguide model of the microstrip line for modeling the electromagnetic fields near microstrip discontinuities. This method, which is the subject of this chapter, is a compromise between the requirement for more accurate and more broadband models for microstrip discontinuities and the requirement for small numerical effort so that the models can be used directly in desktop computer programs.

3. An exact analysis of microstrip discontinuities can be found if the complete field distribution on a microstrip line is taken into account and the electromagnetic fields near a discontinuity are modeled using these exact field solutions. Methods have been described for simple discontinuities such as impedance steps and gaps between microstrip lines using mode-matching techniques [10] or a spectral domain approach [11]. Numerical field analysis methods that can be used to analyze arbitrary planar discontinuity structures while also considering coupling effects are available on the basis of methods that have been successfully used, for example, in microstrip antenna design [12, 13]. The first results have been presented [14], but their numerical effort is still very high.

2. THE WAVEGUIDE MODEL OF THE MICROSTRIP LINE

It is the aim of this chapter to describe a special method for calculating the transmission properties of various kinds of microstrip discontinuities. Figure 3 shows the microstrip discontinuities that can be considered using the method described here. Only two examples out of the variety of discontinuities shown in Fig. 3 will be discussed: the microstrip impedance step and the asymmetrical microstrip crossing.

As is well known, the microstrip line is an open waveguide structure, and its electromagnetic field theoretically is defined in an infinite space. For the numerical analysis of this field in most cases a metallic shielding or at least a

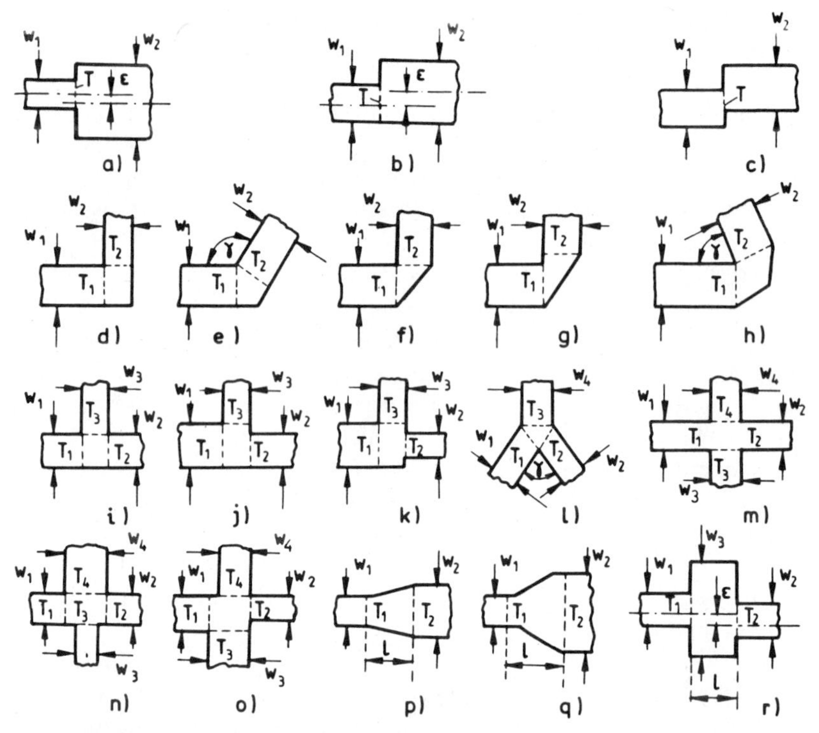

Fig. 3 Microstrip discontinuities that can be analyzed using the waveguide model technique.

metallic cover plate of infinite dimensions above the line is used to define well-determined boundary conditions for the field. Because of the inhomogeneous dielectric field region (substrate material/air space) the electromagnetic fields on the microstrip line are hybrid modes; that is, the electric field strength and magnetic field strength always have three field components. These hybrid modes can be classified as EH or HE modes. Figure 4 shows the field distributions of the electric and magnetic field strengths of the fundamental EH mode (quasi-TEM mode) on a covered microstrip line, as it has been computed by Ermert [15, 16] on the basis of rigorous field analysis.

As a consequence of their hybrid field nature, the parameters describing wave propagation on the line even in the case of the fundamental mode are frequency-dependent. Several approaches have been published to describe these line parameters by static methods [17, 18], by models of a different kind to simulate the frequency dependence of the parameters (e.g., [19–28]), or by rigorous numerical field analysis methods (e.g., References 2, 15, 16, 29–31). A rigorous solution using numerical field analysis always requires a relatively high degree of numerical effort. As a result the analysis of

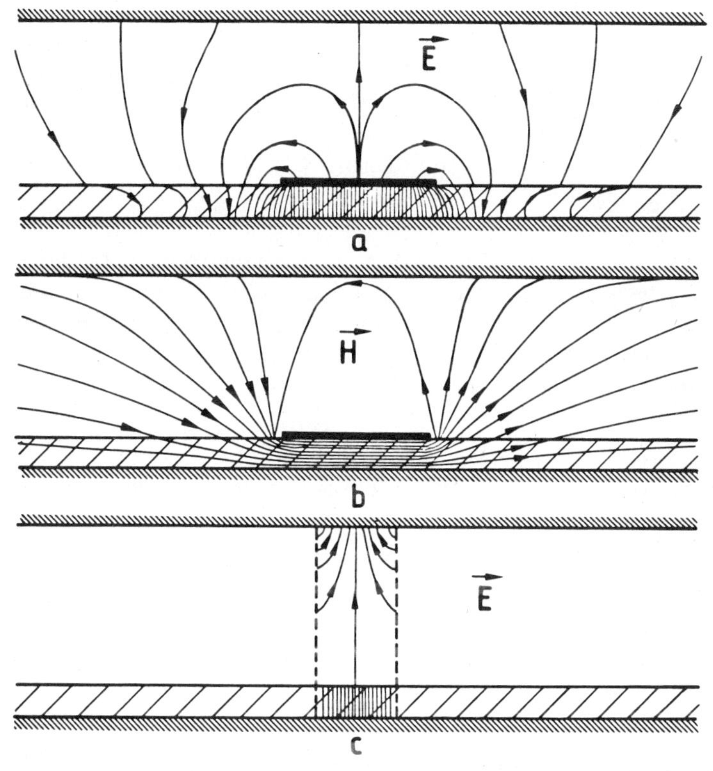

Fig. 4 Distribution of the electromagnetic field of the fundamental mode on a microstrip line calculated by Ermert [15] using a rigorous field analysis technique. (*a*) Electric field strength in the cross section; (*b*) magnetic field strength in the cross section; (*c*) electric field strength in the longitudinal direction.

microstrip discontinuities using these numerical solutions of the electromagnetic fields always leads to long computation times and normally cannot be used for direct application in computer-aided circuit design techniques. This situation may change with the further development of computer techniques. A rigorous solution, of course, can in any case make sense for developing simple approximating formulas or models for the discontinuities.

In this chapter, however, a method is presented that requires medium numerical effort and therefore can be directly applied to the circuit design even at the desktop computer level but nevertheless describes the transmission properties of microstrip discontinuities with good accuracy. Additionally the method presented here is very flexible and can be applied to a variety of microstrip discontinuities (see Fig. 3). The method is based on a waveguide model for the microstrip line introduced by Wolff and co-workers [32, 33] that can be used to describe approximately the electromagnetic fields near microstrip discontinuities. The waveguide model in principle already has been used by Oliner and Altschuler [34–37] for the analysis of discontinuities in symmetrical stripline technique, but because of the dispersive properties of the microstrip line it is more complicated than in the case of the symmetrical stripline.

The waveguide model for the microstrip line was developed to fulfill the following requirements:

1. The waveguide model must describe the electromagnetic fields and the characteristic line parameters (characteristic impedance and phase velocity) of the fundamental quasi-TEM mode on the microstrip line with high accuracy over the frequency range up to the cutoff frequency of the first higher-order mode.
2. The cutoff frequencies and the electromagnetic fields of the higher-order modes must be modeled so that the application of these higher-order modes to the analysis of microstrip discontinuities leads to acceptable accuracy.
3. The model must be simple enough that well-known numerical methods for analyzing waveguide discontinuities can be applied and, e.g., the S parameters of the microstrip discontinuities can be computed considering the energy stored near the discontinuities, thereby taking into account the frequency-dependent transmission properties of these structures.

To find a model that fulfills these requirements, the results of Wheeler's classic work [17, 18] for describing the influence of electrical stray fields of the microstrip line on the line parameters can be used. Figure 5 shows in principle how Wheeler analyzed the microstrip line considering the dielectric substrate material: If it is assumed that the fundamental EH mode on the microstrip line is a quasi-TEM mode (which is correct at least at low

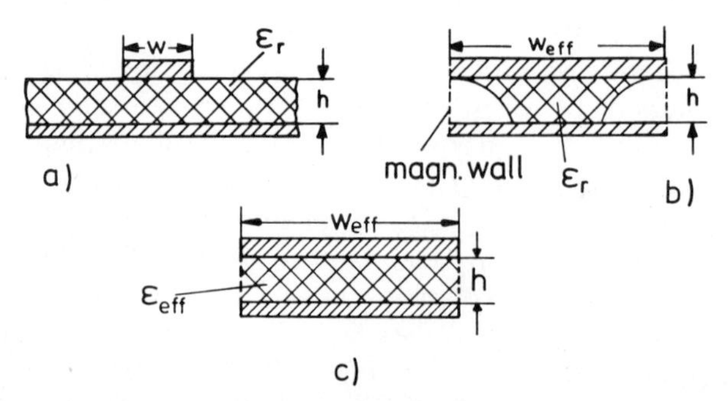

Fig. 5 The microstrip line (*a*) and its waveguide model (*c*), which has been developed using a conformal mapping technique. Figure (*b*) shows in principle how the dielectric–air interface is transformed by the conformal mapping technique.

frequencies), a conformal mapping technique can be used to transform the electric field in the cross section of the microstrip line into the field of an ideal parallel-plate waveguide as shown in Fig. 5. This ideal waveguide has no stray fields; i.e., it is closed by magnetic side walls and, of course, has electric walls on its top and bottom. The height h of the waveguide is assumed to be identical to the height of the substrate material. The width w_{eff} of this waveguide can be found from the conformal mapping technique and is given in the publications of Wheeler for the static case (i.e., for very low frequencies).

Figure 56 shows in principle how the air–dielectric interface is formed if the microstrip line is transformed into the waveguide model. The cross section of the waveguide is only partly filled by dielectric material of relative dielectric constant ε_r. To simplify the waveguide model, it is filled with a homogeneous dielectric medium (Fig. 5*c*) instead of the inhomogeneous medium (Fig. 5*b*) so that the phase velocity of the fundamental wave in the final waveguide model (Fig. 5*c*) is identical to the phase velocity of the fundamental wave on the microstrip line (Fig. 5*a*).

This waveguide model describes correctly the transmission properties of the fundamental quasi-TEM mode for very low frequencies, because at low frequencies the longitudinal components of the electromagnetic field of the microstrip line are small and the field is nearly a TEM mode and the conformal mapping technique can be applied. The effective dielectric constant $\varepsilon_{\mathrm{eff}}$ and the effective width w_{eff} can be calculated using Wheeler's [17, 18] formulas:

$$\varepsilon_{\mathrm{eff}} = \left\{ \sqrt{\varepsilon_r} + \frac{(\varepsilon_r - 1)[\ln(\pi/4)^2 + 1 - \varepsilon_r \ln(\pi e/2)(w/2h + 0.94)]}{2\sqrt{\varepsilon_r}\,\varepsilon_r\{w\pi/2h + \ln[2\pi e(w/2h + 0.94)]\}} \right\}^2 \tag{1}$$

$$w_{\mathrm{eff}} = \left\{ \frac{w}{h} + \frac{2}{\pi} \ln\left[2\pi e\left(\frac{w}{2h} + 0.92\right)\right]\right\} h \tag{2}$$

or equivalent formulas of other authors (e.g., [38, 39]). The phase velocity v_{ph} of the fundamental quasi-TEM mode and its characteristic impedance for low frequencies using (1) and (2) can easily be calculated as the parameters of the fundamental mode of the waveguide model:

$$v_{\mathrm{ph}} = \frac{c_0}{\sqrt{\varepsilon_{\mathrm{eff}}}}, \qquad c_0 = 2.997 \times 10^8\ \mathrm{m/s} \tag{3}$$

and

$$Z_0 = \sqrt{\frac{\mu_0}{\varepsilon_0}}\,\frac{h}{\sqrt{\varepsilon_{\mathrm{eff}}}\,w_{\mathrm{eff}}} \tag{4}$$

The parameters calculated in this way are in agreement with the normal definitions for the microstrip line.

As mentioned above, the hybrid nature of the microstrip fields leads to dispersive behavior of the line parameters; the phase velocity and the characteristic impedance of the fundamental mode (and all higher-order modes) are frequency-dependent. Physically this means that with increasing frequency the electromagnetic field is more and more concentrated under the strip of the microstrip line until at very high frequencies the stray field in the air region vanishes and the field is only in the dielectric substrate material. As a result the effective dielectric constant ε_{eff} converges into the relative dielectric constant ε_r and the effective width w_{eff} into the line width w for increasing frequencies.

As a consequence of this physical phenomenon, the phase velocity and therefore the effective dielectric constant become frequency-dependent. The influence of the phenomenon on the characteristic impedance and thereby on the effective width is not so easily discussed. At very low frequencies the fundamental mode on the microstrip line is a quasi-TEM mode with electromagnetic fields in the air and in the dielectric substrate material. With increasing frequency the influence of the air–dielectric interface changes the field of the fundamental mode into an EH mode with six field components until at very high frequencies the field is concentrated in the dielectric material and the field distribution now is nearly a TEM mode again. At medium and high frequencies the definition of a characteristic impedance in principle is not possible, because it is defined only for a TEM mode and not for a hybrid mode. Nevertheless, the concept of a characteristic impedance has also been used for microstrip lines at higher frequencies, but because of the mentioned difficulties there is some arbitrariness in the definition of the characteristic impedance; this has been discussed extensively in the literature (e.g., [40]).

Figure 6 shows the frequency-dependent effective dielectric constant for two different substrate materials calculated by three approximating formulas. The formulas of Kirschning and Jansen [1] are known to be in good agreement with rigorous numerical field analysis methods and therefore can be considered fairly accurate over a broad frequency range. As the figure shows, for low dielectric constant substrate materials the formulas derived by Yamashita [41] are also very accurate, and they are also much simpler than the formulas of Kirschning and Jansen and therefore lead to shorter computation times.

In Fig. 7 the frequency of the characteristic impedances of microstrip lines with different line widths is shown for an alumina substrate material of height 0.65 mm. The figure shows two solutions from rigorous field analysis methods [2] using the definitions $Z_0 = U/I$ and $Z_0 = P/I^2$, where U is the voltage between the middle of the metallic strip and the ground plane, I is the total current through the metallic strip, and P is the power transported through the cross section of the microstrip line. Additionally an approximate solution published by Bianco et al. [25] is given in Fig. 7 for comparison. Kirschning and Jansen have published approximating formulas that agree

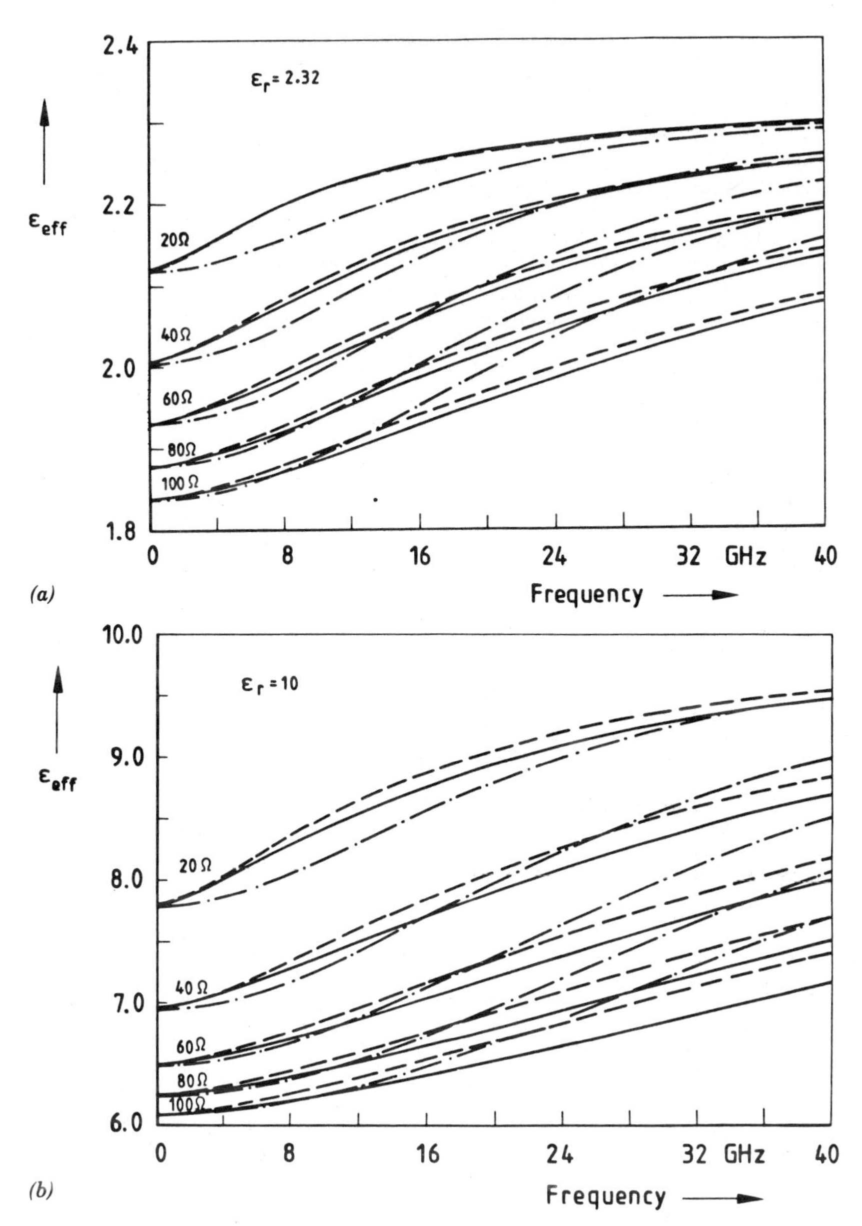

Fig. 6 The frequency dependence of the effective dielectric constant of microstrip lines on substrates with relative dielectric constants (*a*) $\varepsilon_r = 2.32$ and (*b*) $\varepsilon_r = 10.1$. (———) After Kirschning and Jansen [1]; (– – –) after Yamashita [41]; (—·—·—) after Hammerstad [39].

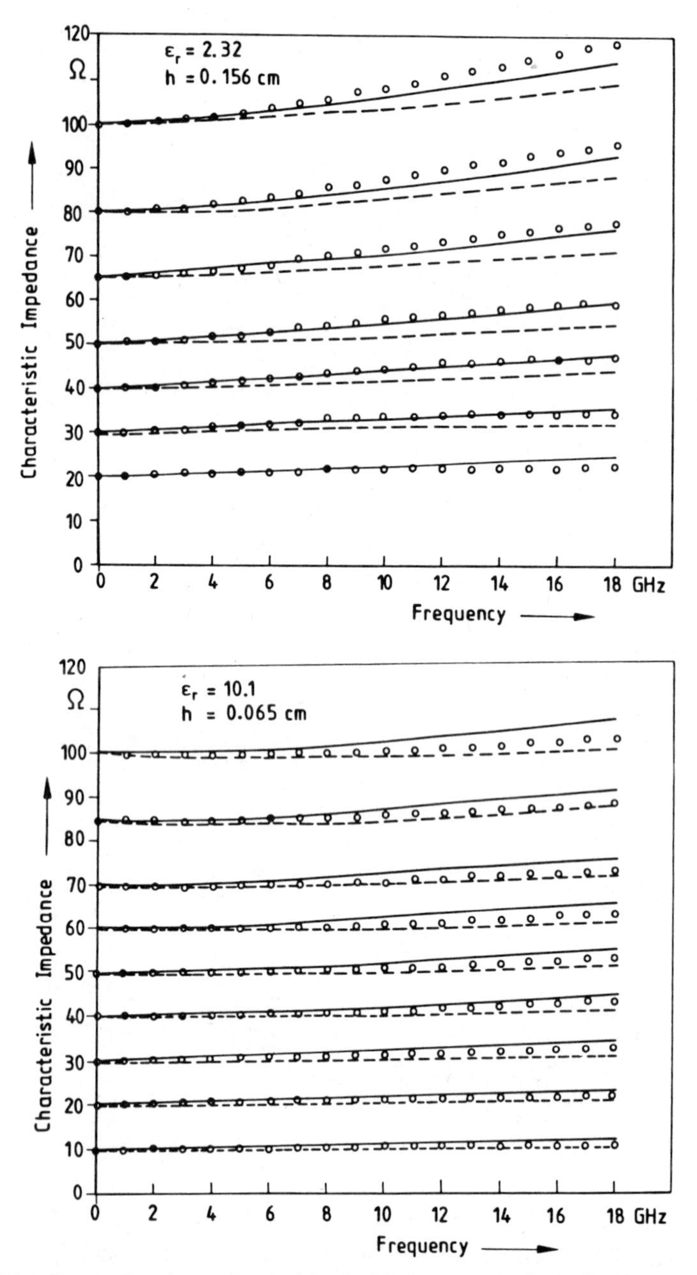

Fig. 7 The frequency dependence of microstrip characteristic impedances calculated from different definitions and numerical methods for two substrate materials with (*a*) $\varepsilon_r = 2.32$ and (*b*) $\varepsilon_r = 10.1$. (– – –) $Z_0 = U/I$ after Jansen [2]; (———) $Z_0 = P/I^2$ after Jansen [2]; (°°°) Z_0 after Bianco et al. [25].

well with the results of the $Z_0 = P/I^2$ curves in Fig. 7, but they again have the disadvantage that they use several exponential functions, which leads to a high computation time compared with, e.g., Bianco's formulas.

The effective dielectric constant and its frequency dependence are a measure for the dispersion of the phase velocity of waves on the microstrip lines. As shown above, the effective dielectric constant of the waveguide model has been chosen in such a way that the phase velocity of a wave on this line is identical with that of the waves on the original microstrip line. This means that the effective dielectric constants of the original microstrip line and its waveguide model must be identical, and therefore at higher frequencies the effective dielectric constant of the waveguide model must be replaced by the frequency-dependent effective dielectric constant as described above. If the waveguide model is to simulate the wave transmission on the microstrip line correctly, it must also be made sure that the characteristic impedances of the two structures are identical. The characteristic impedance of the waveguide model is given by

$$Z_0(f) = \sqrt{\frac{\mu_0}{\varepsilon_0}} \frac{h}{\sqrt{\varepsilon_{\text{eff}}(f)} w_{\text{eff}}(f)} \tag{5}$$

If (5) is solved for the effective line width w_{eff},

$$w_{\text{eff}}(f) = \sqrt{\frac{\mu_0}{\varepsilon_0}} \frac{h}{Z_0(f)\sqrt{\varepsilon_{\text{eff}}(f)}} \tag{6}$$

and if the effective dielectric constant and the characteristic impedance are replaced by the equivalent frequency-dependent values, a frequency-dependent effective line width $w_{\text{eff}}(f)$ is derived that, if applied to the waveguide model, makes sure that the characteristic impedance of the microstrip line is also modeled correctly at each frequency.

It has already been said that the definition of the frequency-dependent characteristic impedance of the microstrip line is somewhat arbitrary, even if some good arguments [40] give an advantage to the $Z_0 = P/I^2$ definition. The uncertainty of the characteristic impedance leads to an equivalent uncertainty of the defined frequency-dependent effective width. Mehran and Kompa [26, 42], for example, used an empirical expression for the frequency-dependent effective width:

$$w_{\text{eff}}(f) = w + \frac{w_{\text{eff}}(f=0) - w}{1 + f/f_g} \tag{7}$$

with

$$f_g = \frac{c_0}{2w\sqrt{\varepsilon_r}}, \qquad c_0 = 2.997 \times 10^8 \text{ m/s} \tag{8}$$

which they derived from the requirements that (1) the effective width for low frequencies must converge into the effective width calculated by, e.g., Wheeler's [17, 18] static theory; (2) for very high frequencies it must converge into the geometrical line width w; and (3) for all other frequencies, the effective line width must deliver good agreement between the calculated and measured system properties of microstrip discontinuities. They required in particular good agreement between the calculated and measured cutoff frequencies of the higher-order modes on the microstrip line and the waveguide model. Further investigations in the meantime have shown that the influence of the choice of the effective width model on the S parameters of the microstrip discontinuities is not great. Therefore in the following eq. (6) together with a good approximating formula for $Z_0(f)$ (e.g., [1]), is used for the definition of the frequency-dependent effective width $w_{eff}(f)$.

In Figs. 8 and 9, the frequency-dependent effective width of the waveguide model is shown for two substrate materials. The values are calculated from (6) [43] using the characteristic impedances of Jansen's method [2], formula (7), and a similar solution proposed by Owens [27]. For small line widths, all three solutions are in good agreement; for larger line widths, solution (7) seems to approximate Jansen's results better than Owens' solution [27] does.

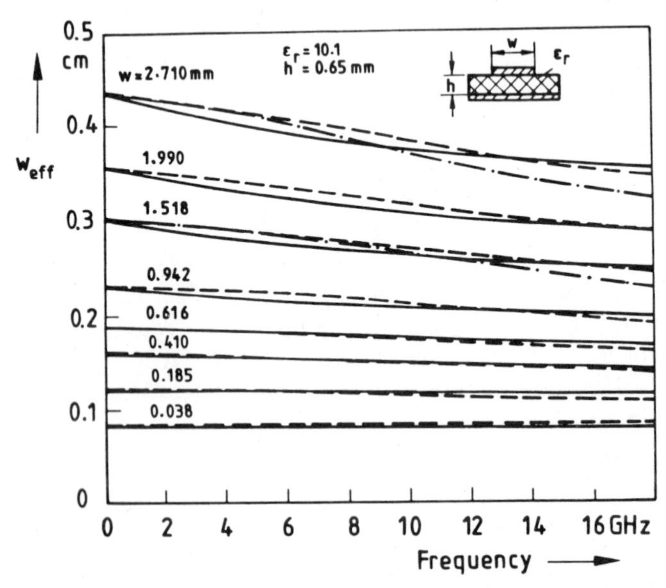

Fig. 8 The frequency-dependent effective width of the waveguide model for an alumina substrate material, calculated by three different methods: (———) using eq. (6) and $Z_0(f)$ from Jansen's work [2]; (– – –) using (7); and (· — ·) using Owens' [27] results. After Mehran [43].

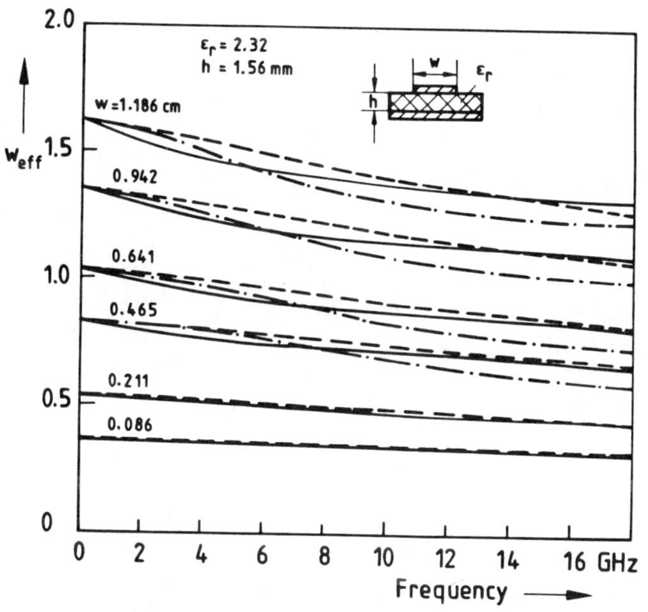

Fig. 9 The frequency-dependent effective width of the waveguide model for a low dielectric constant substrate material, calculated by three different methods: (– – –) using (6) and $Z_0(f)$ from Jansen's work [2]; (– – –) using (7); and (· —·) using Owens' [27] results. After Mehran [43].

With the defined properties, the waveguide model can be used to model the transmission properties of the fundamental mode on the microstrip accurately, if only the design formulas used for $Z_0(f)$ and $\varepsilon_{\text{eff}}(f)$ are accurate enough. The main question that still has to be discussed has to do with how far this model can also approximate the field distributions on the microstrip line. The electromagnetic field of the fundamental mode (quasi-TEM mode) and the two first higher-order modes play a particularly significant role in the modeling of microstrip discontinuities. Figure 10 shows a sketch of the electromagnetic fields of the lowest-order modes in the waveguide model, where only the modes independent of the y coordinate are considered. Compared to Fig. 4, which shows the exact field distribution of the fundamental mode, and Fig. 11, which shows the exact field distribution of the electric field of the second-higher-order mode on a microstrip line, it must be clear that the field distributions are not identical. But the fundamental structure of the fields under the metallic strip (excluding the stray fields) are quite similar. This means that it cannot be expected that models for microstrip discontinuties derived from the waveguide model of the microstrip line are perfect, but as will be shown later, using some additional information from measurements, the models can be made accurate enough for design applications. Therefore the waveguide model and its

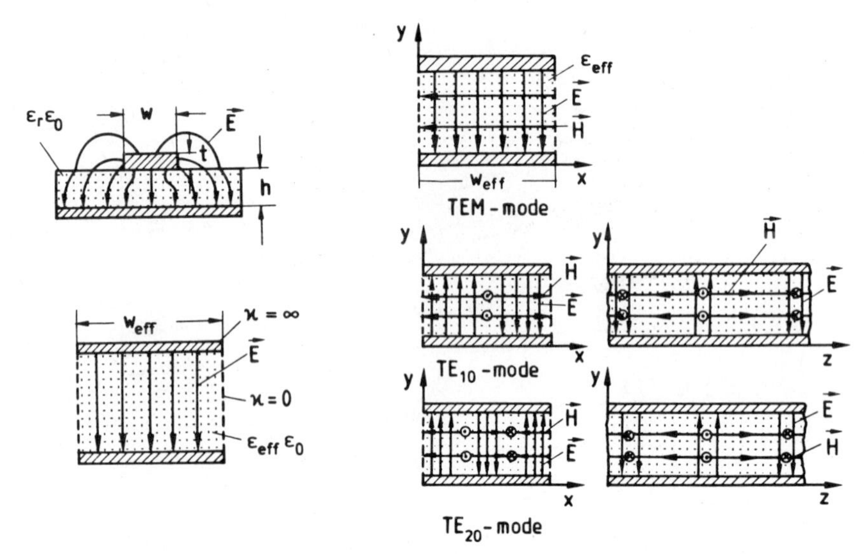

Fig. 10 The electromagnetic fields of the lowest order (y-independent) modes in the waveguide model.

electromagnetic fields are accepted as a basis for a modeling technique that can describe the frequency-dependent transmission properties of the microstrip discontinuities shown in Fig. 3.

To prove the accuracy of the waveguide model, Kompa [44] measured the guide wavelength for the fundamental quasi-TEM mode and the first two higher-order modes on microstrip lines with two different substrate materials. For this purpose he used two different measurement techniques: (1) a sliding load technique and (2) a resonator technique. If, as in the waveguide model, it is assumed that the effective dielectric constant is identical for the fundamental mode and the higher-order modes, the effective widths of the modes can be calculated from the measured guide

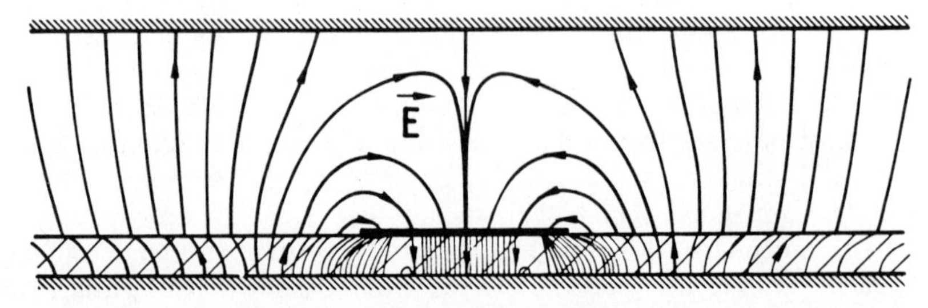

Fig. 11 The field distribution of the electric field of the second-higher-order (y-independent) mode on a microstrip line. After Ermert [15].

wavelengths from the following two equations, which are valid for the waveguide model:

$$\lambda_g = \frac{\lambda_0}{\sqrt{\varepsilon_{\mathrm{eff}}(f)}\sqrt{1-(f_g/f)^2}} \tag{9}$$

$$f_g = \frac{mc_0}{2w_{\mathrm{eff}}(f)\sqrt{\varepsilon_{\mathrm{eff}}(f)}}\,, \qquad m = 1, 2, \ldots \tag{10}$$

In Fig. 12 the measured effective line widths are compared with the theoretical results from (7) in dependence on the frequency. It must be noted that (1) the strip widths w used are quite large so as to excite the higher-order modes at low frequencies and (2) the frequency dependence of the measured effective width is not in good agreement with the theoretical results in all cases. But it should be considered that the measurement of the

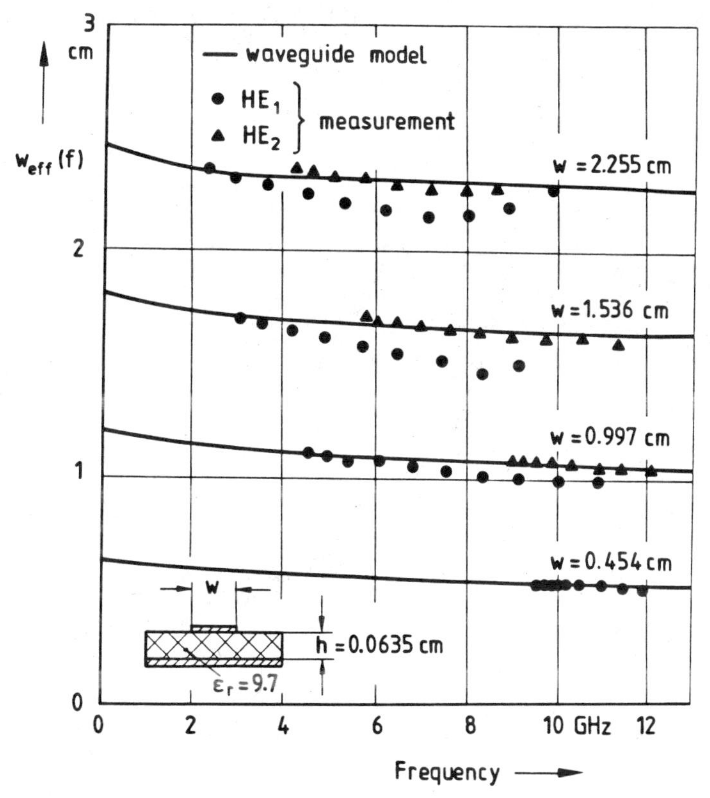

Fig. 12 Comparison between the measured and the calculated effective strip widths of the first two higher-order modes (TE_{10} and TE_{20} modes) of the waveguide model. After Kompa [44].

higher-order wavelengths in dependence on the frequency is a very difficult task and therefore the results can be declared to be acceptable. Furthermore if the measured cutoff frequencies f_c are compared to the theoretical values (Fig. 13), good agreement can be found for a large range of strip widths. Exact knowledge of the cutoff frequency is needed to describe the frequency-dependent properties of the microstrip discontinuities with good accuracy.

The solution for the electromagnetic fields of the waveguide model can be

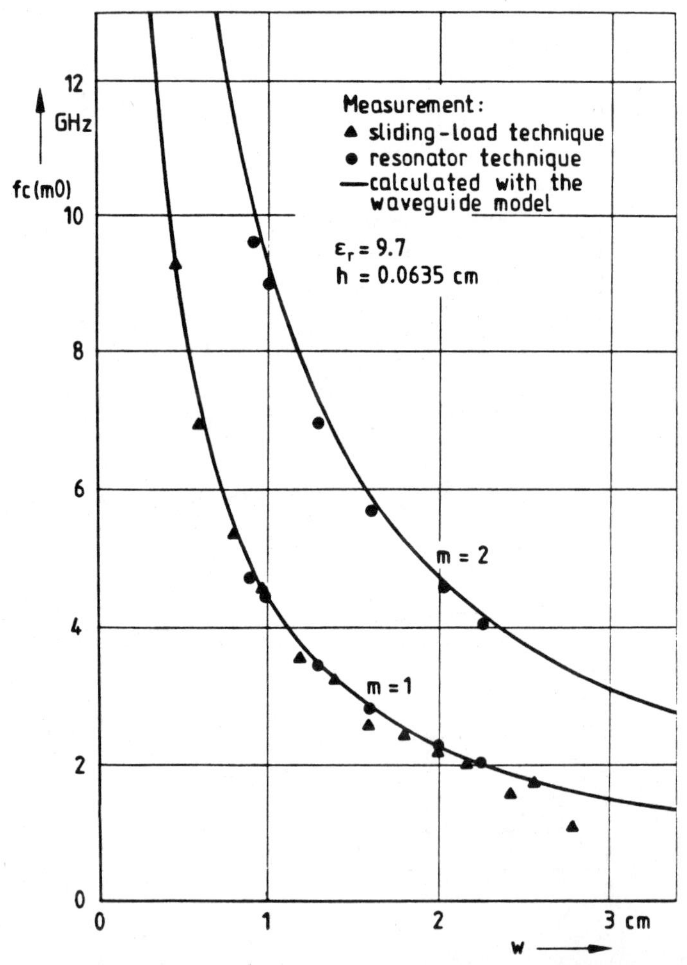

Fig. 13 Comparison of the measured and the calculated cutoff frequencies of the first two higher-order modes (TE_{10} and TE_{20} modes) of the waveguide model in dependence on the strip width of the microstrip line. After Kompa [44].

found in the same way as for metallic waveguides. The boundary conditions on the electric walls and the magnetic walls are

$$\mathbf{E} \times \hat{\mathbf{n}}_1 = 0 , \qquad \mathbf{H} \times \hat{\mathbf{n}}_2 = 0 \tag{11}$$

where $\mathbf{n}_1$ and $\hat{\mathbf{n}}_2$ are unit surface vectors on the electric walls and on the magnetic walls, respectively. The field solutions are of TEM, TE(H), and TM(E) modes; their electromagnetic fields can generally be described by forward and backward traveling waves. In the case of a forward traveling wave, the fields are given by

$$\mathbf{E}_t = A\mathbf{g}_t(x, y)\, e^{-\gamma z} \tag{12a}$$

$$\mathbf{H}_t = \frac{A}{Z_F} [\hat{\mathbf{u}}_z \times \mathbf{g}_t(x, y)]\, e^{-\gamma z} \tag{12b}$$

$$E_z = \frac{A}{\gamma} \operatorname{div}_t \mathbf{g}_t(x, y)\, e^{-\gamma z} \tag{12c}$$

$$H_z = \frac{A}{\gamma Z_F} \operatorname{div}_t [\hat{\mathbf{u}}_z \times \mathbf{g}_t(x, y)]\, e^{-\gamma z} \tag{12d}$$

where the index t describes the transverse field components, A is the field amplitude coefficient, Z_F the characteristic field impedance, γ the propagation constant, $\mathbf{g}_t$ the transverse vectorial structure function, and $\hat{\mathbf{u}}_z$ the unit vector in the z direction. The structure functions are orthogonal to each other, and in addition they will be orthonormalized:

$$\iint_A \mathbf{g}_{t\nu} \cdot \mathbf{g}_{t\mu}\, dA = \delta_{\nu\mu} = \begin{cases} 1 & \text{for } \nu = \mu \\ 0 & \text{for } \nu \neq \mu \end{cases} \tag{13}$$

The integrals have to be evaluated over the cross section of the waveguide model.

The totality of the eigensolutions and their structure functions form a complete system. An arbitrary, piecewise continuous, and transversal vector field $\mathbf{A}_t$ can be described by an infinite sum of the structure functions:

$$\mathbf{A}_t = \sum_{\nu=1}^{\infty} A_\nu \mathbf{g}_{t\nu} \qquad \text{with} \qquad A_\nu = \iint_A \mathbf{A}_t \cdot \mathbf{g}_{t\nu}\, dA \tag{14}$$

This means that the conditions for applying the method of orthogonal series expansions are fulfilled for the waveguide model and its fields.

The forward traveling electromagnetic modes of the waveguide model can be described by an electric potential ϕ_{mn} and a magnetic potential ψ_{mn}, where the indices m and n describe the field dependences in the x and y coordinate directions, respectively:

$$\mathbf{E}_t = \sum_{m=0}^{\infty} \sum_{n=0}^{\infty} A_{mn}^H \, e^{-\gamma_{mn}^H z} (\mathbf{u}_z \times \text{grad}_t \psi_{mn}) + \sum_{m=0}^{\infty} \sum_{n=1}^{\infty} A_{mn}^E \, e^{-\gamma_{mn}^E z} \, \text{grad}_t \phi_{mn} \tag{15a}$$

$$\mathbf{H}_t = \sum_{m=0}^{\infty} \sum_{n=0}^{\infty} \frac{A_{mn}^H}{Z_{mn}^H} \, e^{-\gamma_{mn}^H z} (-\text{grad}_t \psi_{mn}) + \sum_{m=0}^{\infty} \sum_{n=1}^{\infty} \frac{A_{mn}^E}{Z_{mn}^E} \, e^{-\gamma_{mn}^E z} (\hat{\mathbf{u}}_z \times \text{grad}_t \phi_{mn}) \tag{15b}$$

$$E_z = \sum_{m=0}^{\infty} \sum_{n=1}^{\infty} \frac{A_{mn}^E}{\gamma_{mn}^E} \, e^{-\gamma_{mn}^E z} \, \Delta_t \phi_{mn} \tag{15c}$$

$$H_z = \sum_{m=0}^{\infty} \sum_{n=0}^{\infty} \frac{A_{mn}^H}{Z_{mn}^H \gamma_{mn}^H} \, e^{-\gamma_{mn}^H z} (-\Delta_t \psi_{mn}) \tag{15d}$$

with

$$\gamma_{mn}^E = \gamma_{mn}^H = \alpha_{mn} + j\beta_{mn} = \left[\left(\frac{m\pi}{w_{\text{eff}}} \right)^2 + \left(\frac{n\pi}{h} \right)^2 - \omega^2 \varepsilon_0 \varepsilon_{\text{eff}} \mu_0 \right]^{1/2} \tag{16a}$$

$$Z_{mn}^E = -\frac{j\gamma_{mn}^E}{\omega \varepsilon_0 \varepsilon_{\text{eff}}} \qquad Z_{mn}^H = \frac{j\omega\mu_0}{\gamma_{mn}^H} \tag{16b}$$

and the potential functions

$$\phi_{mn} = \left(\frac{\nu_m^E}{w_{\text{eff}} h} \right)^{1/2} \frac{1}{[(m\pi/w_{\text{eff}})^2 + (n\pi/h)^2]^{1/2}} \cos\left(\frac{m\pi}{w_{\text{eff}}} x \right) \sin\left(\frac{n\pi}{h} y \right) \tag{16c}$$

$$\psi_{mn} = \begin{cases} \dfrac{1}{\sqrt{w_{\text{eff}} h}} x & \text{for } m = n = 0 \\[2ex] \left(\dfrac{\nu_m^H}{w_{\text{eff}} h} \right)^{1/2} \dfrac{1}{[(m\pi/w_{\text{eff}})^2 + (n\pi/h)^2]^{1/2}} \sin\left(\dfrac{m\pi}{w_{\text{eff}}} x \right) \cos\left(\dfrac{n\pi}{h} y \right) & \text{otherwise} \end{cases} \tag{16d}$$

The Neumann coefficients ν_m^E and ν_m^H of the TM (E) modes and the TE (H) modes are given by

$$\nu_m^E = \begin{cases} 2 & \text{for } m = 0, n \neq 0 \\ 4 & \text{for } m \neq 0, n \neq 0 \end{cases} \tag{17a}$$

$$\nu_m^H = \begin{cases} 2 & \text{for } m \neq 0, n = 0 \\ 4 & \text{for } m \neq 0, n \neq 0 \end{cases} \tag{17b}$$

The two different solutions in (16d) for the TEM ($m = 0, n = 0$) and TE modes can be written in one equation if for $m \to 0$ and $n = 0$ the value of ν_m^H is chosen equal to 1. An equivalent description of the backward traveling waves can be given.

Microstrip lines are used on the condition that the height h of the substrate material is so small that the electromagnetic fields of the waveguide model have no dependency on the y coordinate (Fig. 5). In the following no discontinuities with a change of the substrate height will be considered. All structures that will be analyzed are fed only with a quasi-TEM mode. As has been shown by Kompa [44], the TEM mode in discontinuity structures that are independent of the y coordinate couple only to higher-order modes that have a field distribution similar to that of the TEM mode in the coupling area. In particular, no coupling occurs between the TM (E) modes and the TEM mode and between the $TE_{mn}(H_{mn})$ modes which have y-dependent electromagnetic fields ($n \neq 0$). Under these conditions only electromagnetic fields that are independent of the y coordinate can exist on the microstrip line, and as a consequence only modes of the waveguide model that also are independent of the y coordinate must be considered when modeling the microstrip discontinuities. Therefore all TM (E) modes and all $TE_{mn}(H_{mn})$ modes with $n \neq 0$ are no longer considered in the following calculations.

If in addition the field amplitudes A_{mn}^H and A_{mn}^H are replaced by wave amplitudes a_m and b_m of forward and backward traveling waves, the electromagnetic fields are described by

$$\mathbf{E}_t = \sum_{m=0}^{\infty} \sqrt{Z_m}(a_m\, e^{-\gamma_m z} + b_m\, e^{+\gamma_m z})(\mathbf{u}_z \times \operatorname{grad}_t \psi_{m0}) \tag{18a}$$

$$\mathbf{H}_t = -\sum_{m=0}^{\infty} \sqrt{Y_m}(a_m\, e^{-\gamma_m z} - b_m\, e^{+\gamma_m z}) \operatorname{grad}_t \psi_{m0} \tag{18b}$$

$$H_z = -\sum_{m=1}^{\infty} \frac{\sqrt{Y_m}}{\gamma_m} (a_m\, e^{-\gamma_m z} + b_m\, e^{+\gamma_m z})\, \Delta_t \psi_{m0} \tag{18c}$$

Using the potential function ψ_{m0} as defined in (16d), the field components finally can be written as

$$E_x = 0 \tag{19a}$$

$$E_y = \sum_{m=0}^{\infty} \sqrt{Z_m}(a_m\, e^{-\gamma_m z} + b_m\, e^{+\gamma_m z})\left(\frac{\nu_m^H}{w_{\text{eff}} h}\right)^{1/2} \cos\left(\frac{m\pi}{w_{\text{eff}}}\, x\right) \tag{19b}$$

$$E_z = 0 \tag{19c}$$

$$H_x = -\sum_{m=0}^{\infty} \sqrt{Y_m}(a_m\, e^{-\gamma_m z} - b_m\, e^{+\gamma_m z})\left(\frac{\nu_m^H}{w_{\text{eff}} h}\right)^{1/2} \cos\left(\frac{m\pi}{w_{\text{eff}}}\, x\right) \tag{19d}$$

$$H_y = 0 \tag{19e}$$

$$H_z = \sum_{m=1}^{\infty} \frac{\sqrt{Y_m}}{\gamma_m} (a_m e^{-\gamma_m z} + b_m e^{+\gamma_m z}) \left(\frac{\nu_m^H}{w_{\text{eff}} h} \right)^{1/2} \frac{m\pi}{w_{\text{eff}}} \sin\left(\frac{m\pi}{w_{\text{eff}}} x \right) \tag{19f}$$

3. MATHEMATICAL ANALYSIS OF MICROSTRIP DISCONTINUITIES

Using the waveguide model of the microstrip line described in Section 2, the most important microstrip discontinuities can be analyzed (see Fig. 3) by using known mathematical methods of analyzing waveguide discontinuities or by new methods that have been developed by Wolff and his research group. Five analysis methods have been derived [43]; in each method the microstrip discontinuity structure is divided into subregions, and in each subregion the electromagnetic fields are defined using complete series or integral expansions that a priori fulfill the boundary conditions on the magnetic and electric walls. At the common interfaces between the subregions the boundary conditions are fulfilled in an integral sense, thereby defining the coupling of the field modes in the different regions. If, in addition, the structure under consideration is excited, for example, only by the fundamental TEM mode, the scattering parameters of the microstrip structure can be computed. The different methods that have been used for the analysis of microstrip discontinuities will be explained in the following using the example of a microstrip impedance step and the asymmetrical crossing.

3.1. The Microstrip Impedance Step

The first method to be described here has been used by Kompa [44, 48] for the analysis of asymmetrical microstrip impedance steps. For the analysis the microstrip impedance step is replaced by an equivalent step of a waveguide model structure.

Figure 14 shows the original microstrip structure and the equivalent waveguide model impedance step which is divided into two field regions A and B. The widths a and b are the frequency-dependent effective widths of the waveguide model, and the model is filled with a material described by the frequency-dependent effective dielectric constant as discussed in Section 2. If the reference plane (RP') of the original microstrip impedance step is at $z' = 0$ (Fig. 14*a*), this reference plane, because of the stray fields at the open ends of the structure, is shifted to position RP at $z = 0$ or $z' = -1$, Fig. 14*b*) in the waveguide model. To get the correct phase information for the microstrip impedance step, this reference plane displacement must be calculated back into the original position ($z' = 0$) after the analysis with the waveguide model (see Section 5). Two additional reference planes RP_A and

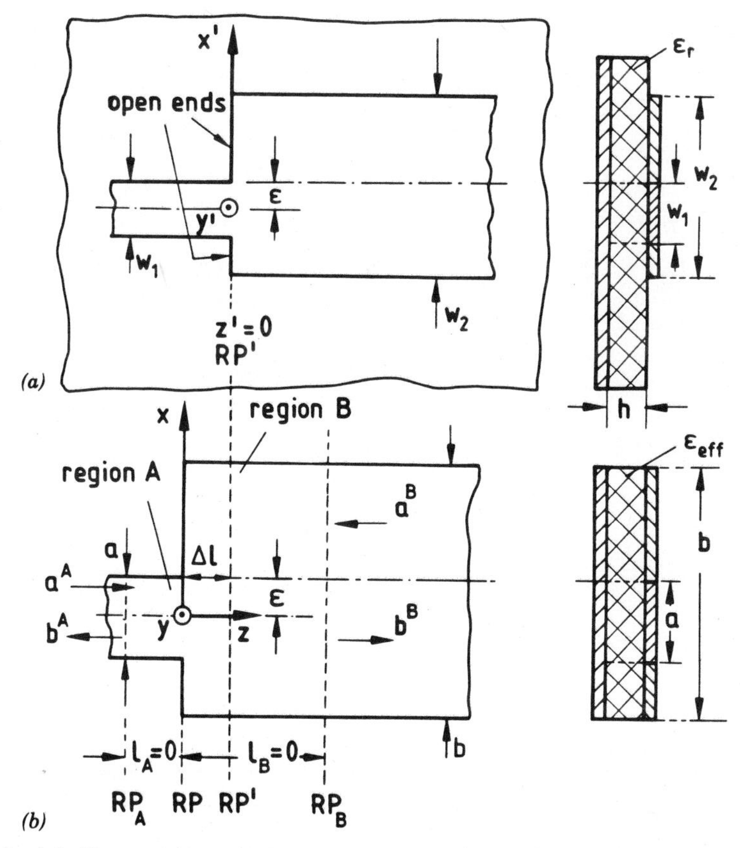

Fig. 14 (*a*) The asymmetrical microstrip impedance step; (*b*) the equivalent waveguide model structure.

RP_B are defined left and right from the step reference plane RP in the waveguide model structure. It is required that higher-order modes that are excited at RP be decreased to zero at RP_A and RP_B so that a scattering matrix of the fundamental mode can be defined without problems at these planes. The distances l_A and l_B of these reference planes from the plane RP are assumed to be zero in the final calculations.

Under these assumptions the electromagnetic fields in regions A and B are described by complete, infinite sums (discrete spectra) of eigenfunctions of forward and backward traveling waves [compare eqs. (15)–(17)].

Region A:

$$\mathbf{E}_t^A(z) = \sum_{m=0}^{\infty} U_m^A(z)(\hat{\mathbf{u}}_z \times \text{grad}_t\, \psi_m^A) \tag{20a}$$

$$\mathbf{H}_t^A(z) = \sum_{m=0}^{\infty} I_m^A(z)(-\text{grad}_t\, \psi_m^A) \tag{20b}$$

with

$$\sqrt{Y_m^A} U_m^A(z) = a_m^A \exp(-j\beta_m^A z) + b_m^A \exp(+j\beta_m^A z) \tag{20c}$$

$$\sqrt{Z_m^A} I_m^A(z) = a_m^A \exp(-j\beta_m^A z) - b_m^A \exp(+j\beta_m^A z) \tag{20d}$$

Region B:
Considering the coordinate transformation from region A to region B, the following equations are valid for the electromagnetic fields:

$$\mathbf{E}_t^B(z) = \sum_{p=0}^{\infty} U_p^B(z)(\mathbf{u}_z \times \text{grad}_t\, \psi_p^B) \tag{21a}$$

$$\mathbf{H}_t^B(z) = \sum_{p=0}^{\infty} I_p^B(z)(\text{grad}_t\, \psi_p^B) \tag{21b}$$

with

$$\sqrt{Y_p^B} U_p^B(z) = a_p^B \exp(+j\beta_p^B z) + b_p^B \exp(-j\beta_p^B z) \tag{21c}$$

$$\sqrt{Z_p^B} I_p^B(z) = a_p^B \exp(+j\beta_p^B z) - b_p^B \exp(-j\beta_p^B z) \tag{21d}$$

For the potential functions of region A, eqs. (15)–(17) with $w_{\text{eff}} = a$ and $n = 0$ and considering the changed coordinate system (compare Figs. 5 and 15) are applicable:

$$\psi_m^A = \begin{cases} \dfrac{1}{\sqrt{ah}} \left(x + \dfrac{a}{2}\right) & \text{for } m = 0 \\ \sqrt{\dfrac{2}{ah}} \left(\dfrac{a}{m\pi}\right) \sin\left(\dfrac{m\pi}{a} x + \dfrac{a}{2}\right) & \text{for } m \neq 0 \end{cases} \tag{22a}$$

For region B, because of the coordinate transformation from region A to region B (eccentricity ε), the equations are

$$\psi_p^B = \begin{cases} \dfrac{1}{\sqrt{bh}} \left(x + \dfrac{b^*}{2}\right) & \text{for } p = 0 \\ \sqrt{\dfrac{2}{bh}} \left(\dfrac{b}{p\pi}\right) \sin\left[\dfrac{p\pi}{b} \left(x + \dfrac{b^*}{2}\right)\right] & \text{for } p \neq 0 \end{cases} \tag{22b}$$

with $b^*/2 = b/2 + \varepsilon$. a_m^A, b_m^A and a_p^B, b_p^B are the normalized field (wave) amplitudes of the forward and backward traveling waves in region A and region B, respectively. In the boundary between region A and region B the electromagnetic fields must fulfill the boundary conditions

$$H_x^B = 0 \ \text{ in } A_B - A_A \qquad H_x^A = H_x^B \ \text{ in } \mathrm{A}_A \qquad E_y^A = E_y^B \ \text{ in } A_A \tag{23}$$

where A_A is the common boundary of the two waveguides and $A_B - A_A$ is the magnetic wall that closes the broad waveguide at the discontinuity. Because $A_B - A_A$ is a magnetic wall, the tangential magnetic field component must vanish in it. The boundary conditions given in (23) can be fulfilled only if the transverse magnetic field of region A is developed into the eigenfunctions of region B. With this technique the amplitude coefficients I_m^B in (21b) can be chosen so that the magnetic field component H_x^B vanishes on $A_B - A_A$ and additionally H_x^A is equal to H_x^B in A_A. Conversely, the electric field component need not fulfill any special boundary condition in $A_B - A_A$, so the electric field components in the common boundary area A_A can be developed into the eigenfunctions of region A.

The amplitude coefficients U and I can be calculated using (14) and the orthogonality relationship (13). If (21) is multiplied by $(\text{grad}_t\, \psi_m)$ and integrated over the area A_B, the result is

$$I_P^B = \iint_{A_B} \mathbf{H}_t^B \cdot \text{grad}_t\, \psi_P^B \, dA \tag{24}$$

Analogously, after multiplication of eqs. (20) with $(\hat{\mathbf{u}}_z \times \text{grad}_t\, \psi_m)$ and integration over A_A, it follows that

$$U_M^A = \iint_{A_A} \mathbf{E}_t^A \cdot (\mathbf{u}_z \times \text{grad}_t\, \psi_M^A)\, dA \tag{25}$$

If the boundary conditions are introduced into (24) and (25), the remaining amplitude coefficients can be calculated:

$$I_P^B = \iint_{A_A} \mathbf{H}_t^A \cdot \text{grad}_t\, \psi_P^B \, dA \tag{26}$$

$$U_M^A = \iint_{A_A} \mathbf{E}_t^B \cdot (\hat{\mathbf{u}}_z \times \text{grad}_t\, \psi_M^A)\, dA \tag{27}$$

The discontinuity problem of an asymmetrical microstrip impedance step is thus reduced to the solution of a multiply infinite system of coupled linear equations for the amplitudes of the electric and magnetic field strengths. The coefficients of the equation system are integrals over the products of eigenfunctions of different field regions. Their evaluation gives information on the coupling between different modes in the waveguide model. In the case that the microstrip line and therefore the waveguide model are excited by a TEM mode (or an H_{m0} mode), the result is

$$U_M^A = \sum_{p=0}^{\infty} U_p^B K_{Mp} \qquad I_P^B = -\sum_{m=0}^{\infty} I_m^A K_{mP} \tag{28}$$

The coupling integrals are given by

$$K_{00} = \sqrt{\frac{a}{b}} \qquad K_{m0} = 0 \qquad K_{0p} = \sqrt{\frac{8}{ab}}\,\frac{b}{\pi}\cos\left(\frac{p\pi}{2}\,\frac{b^*}{b}\right)\sin\left(\frac{p\pi}{2}\,\frac{a}{b}\right) \tag{29}$$

and

$$K_{mp} = \begin{cases} -\dfrac{2}{\sqrt{ab}}\,\dfrac{p\pi/b}{(m\pi/a)^2}\left\{(-1)^m \sin\left[\dfrac{p\pi}{2}\left(\dfrac{a+b^*}{b}\right)\right] + \sin\left[\dfrac{p\pi}{2}\left(\dfrac{a-b^*}{b}\right)\right]\right\} \\ \text{for } \dfrac{m}{a} \neq \dfrac{p}{b} \\ \sqrt{\dfrac{a}{b}}\cos\left[\dfrac{\pi}{2}\left(m - \dfrac{b^*}{b}\,p\right)\right] \\ \text{for } \dfrac{m}{a} = \dfrac{p}{b} \end{cases} \tag{30}$$

Introducing the wave amplitudes as given in (20c) and (21c) into (28), the equation system can be written as

$$\sqrt{Z_M^A}\{a_M^A + b_M^A\} = \sum_{p=0}^{\infty} \sqrt{Z_p^B} K_{Mp}\{a_p^B + b_p^B\} \tag{31a}$$

$$\sqrt{Y_P^B}\{a_P^B - b_P^B\} = -\sum_{m=0}^{\infty} \sqrt{Y_m^A} K_{mP}\{a_m^A - b_m^A\} \tag{31b}$$

with $m, M, p, P = 0, 1, 2, \ldots$. If this equation system is rearranged so that the amplitudes b_m^A and b_p^B of the reflected waves are connected to the amplitudes a_m^A and a_p^B of the incident waves of regions A and B, respectively, the connecting matrix is the scattering matrix:

$$\begin{bmatrix} b_0^A \\ b_1^A \\ \cdot \\ \cdot \\ \cdot \\ \hline b_0^B \\ b_1^B \\ \cdot \\ \cdot \\ \cdot \end{bmatrix} = \left[\begin{array}{ccc|ccc} S_{00}^{AA} & S_{01}^{AA} & \cdots & S_{00}^{AB} & S_{01}^{AB} & \cdots \\ S_{10}^{AA} & S_{11}^{AA} & \cdots & S_{10}^{AB} & S_{11}^{AB} & \cdots \\ \cdot & \cdot & & \cdot & \cdot & \\ \cdot & \cdot & & \cdot & \cdot & \\ \cdot & \cdot & & \cdot & \cdot & \\ \hline S_{00}^{BA} & S_{01}^{BA} & \cdots & S_{00}^{BB} & S_{01}^{BB} & \cdots \\ S_{10}^{BA} & S_{11}^{BA} & \cdots & S_{10}^{BB} & S_{11}^{BB} & \cdots \\ \cdot & \cdot & & \cdot & \cdot & \\ \cdot & \cdot & & \cdot & \cdot & \\ \cdot & \cdot & & \cdot & \cdot & \end{array}\right] \begin{bmatrix} a_0^A \\ a_1^A \\ \cdot \\ \cdot \\ \cdot \\ \hline a_0^B \\ a_1^B \\ \cdot \\ \cdot \\ \cdot \end{bmatrix} \tag{32}$$

For the numerical computation of the scattering matrix, eqs. (31) are used. The amplitude coefficients a_m^A, a_p^B of all incident waves in regions A and B are assumed to be zero with the exception of the amplitude a_E^i ($i = A$ or $i = B$) of one exciting (index E) mode in region A or B, respectively. Using (31), the amplitudes b_m^A and b_p^B of the waves can be calculated. If, in addition, it is assumed that $a_E^i = 1$, the amplitudes b_m^A and b_p^B are equal to the elements of the scattering matrix connecting the b amplitudes with the a amplitudes as shown in (32). Using the coupling integrals in (30), the following equation system for the S parameters can be found:

$$\sum_{p=0}^{\infty} M(Pp) S_{pE}^{BA} = \delta_{PE} GA(P) \tag{33a}$$

$$\sum_{p=0}^{\infty} M(Pp) S_{pE}^{BB} = \delta_{PE} GB(P) \tag{33b}$$

$$S_{ME}^{AA} = \sum_{p=0}^{\infty} Q(Mp) S_{pE}^{BA} + \delta_{ME} SA(M) \tag{33c}$$

$$S_{ME}^{AB} = \sum_{p=0}^{\infty} Q(Mp) S_{pE}^{BB} + \delta_{ME} SB(M) \tag{33d}$$

with the abbreviations

$$\delta_{ik} = \begin{cases} 1 & \text{for } i = k \\ 0 & \text{for } i \neq k \end{cases} \tag{33e}$$

$$SA(M) = -1 \tag{33f}$$

$$SB(M) = \sqrt{Z_E^B / Z_M^A}\, K_{ME} \tag{33g}$$

$$M(Pp) = \begin{cases} \sqrt{Z_0^B / Z_0^A}\, K_{00} + [\sqrt{Z_0^B / Z_0^A}\, K_{00}]^{-1} & \text{for } P = p = 0 \\ -\sqrt{Z_P^B / Z_0^B}\, K_{0P} / K_{00} & \text{for } p = 0,\ P \neq 0 \\ \sqrt{Z_p^B / Z_0^A}\, K_{0p} & \text{for } P = 0,\ p \neq 0 \\ \sum_{m=1}^{\infty} \sqrt{Z_p^B Z_P^B / Z_m^A}\, K_{mp} K_{mP} + \delta_{Pp} & \text{for } P,\ p \neq 0 \end{cases} \tag{33h}$$

$$GA(P) = \begin{cases} 2 & \text{for } P = 0 \\ 2\sqrt{Z_P^B / Z_E^A}\, K_{EP} & \text{for } P \neq 0 \end{cases} \tag{33i}$$

$$Q(Mp) = \sqrt{Z_p^B / Z_M^A} K_{Mp} \tag{33k}$$

and

$$\varepsilon(E) = \begin{cases} 2 & \text{for } E = 0 \\ 0 & \text{for } E \neq 0 \end{cases} \tag{33l}$$

3.2. The Asymmetrical Microstrip Crossing

In this section the mathematical background for calculating the transmission properties of an asymmetrical microstrip crossing is discussed. Figure 15 shows the considered microstrip structure and the equivalent structure of the waveguide model. The original reference planes of the microstrip crossing are called RP_1', RP_2', RP_3', and RP_4', whereas the reference planes of the waveguide model structure are RP_1, RP_2, RP_3, and RP_4. As in the case of the microstrip impedance step, the reference planes of the waveguide model have to be calculated back into the reference planes of the original microstrip structure later so as to correctly describe the S-parameter phases (see Section 4).

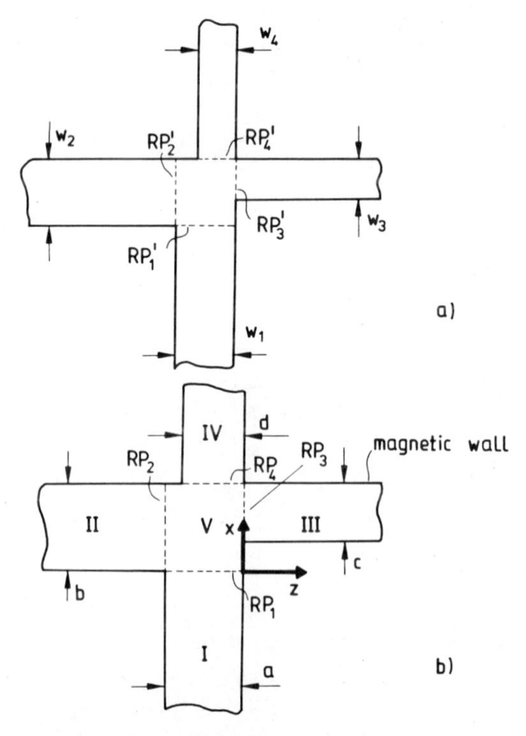

Fig. 15 (a) The asymmetrical microstrip crossing; (b) its equivalent waveguide model structure.

The method for analyzing the crossing is similar to that for the impedance step insofar as it also uses a series expansion of the electromagnetic fields in the waveguide model. As Fig. 15 shows, the crossing is divided into five field regions, and a field description as given in eqs. (15)–(17) is used to describe the fields in the four field regions I, II, III, and IV, which are the connecting waveguides, and the central region V. a, b, c, and d are the frequency-dependent effective widths of the four microstrip lines, and the equivalent regions of the waveguide structure are filled with a material of relative dielectric constant $\varepsilon_{\mathrm{eff},i}$ $(i = 1, 2, 3, 4)$ respectively. Line I will be broader than line IV, and line II broader than line III, so that the geometrical dimensions of the coupling region V are determined by the effective widths w_{eff1} and w_{eff2} of microstrip lines 1 and 2, respectively. Microstrip lines 3 and 4 can be connected to field region V at an arbitrary position. To hold the calculations simple, in this description the configuration shown in Fig. 15 is used. In region V an equivalent dielectric constant as defined in References 66 and 67 for a microstrip disk capacitor is introduced, taking into account the electric stray fields only at those sides of the region where no microstrip line is connected.

The boundary conditions for the electromagnetic field of region V are given by

$$\left.\begin{aligned} \mathbf{E}_{\tan}^{\mathrm{V}} &= \mathbf{E}_{\tan}^{\mathrm{I}} \\ \mathbf{H}_{\tan}^{\mathrm{V}} &= \mathbf{H}_{\tan}^{\mathrm{I}} \end{aligned}\right\} \text{ for } x = 0,\ -a \leqq z \leqq 0 \tag{34a}$$

$$\left.\begin{aligned} \mathbf{E}_{\tan}^{\mathrm{V}} &= \mathbf{E}_{\tan}^{\mathrm{II}} \\ \mathbf{H}_{\tan}^{\mathrm{V}} &= \mathbf{H}_{\tan}^{\mathrm{II}} \end{aligned}\right\} \text{ for } z = -a,\ 0 \leqq x \leqq b \tag{34b}$$

$$\left.\begin{aligned} \mathbf{E}_{\tan}^{\mathrm{V}} &= \mathbf{E}_{\tan}^{\mathrm{III}} \\ \mathbf{H}_{\tan}^{\mathrm{V}} &= \mathbf{H}_{\tan}^{\mathrm{III}} \end{aligned}\right\} \text{ for } z = 0,\ b - c \leqq x \leqq b \tag{34c}$$

$$\mathbf{H}_{\tan}^{\mathrm{V}} = 0 \qquad \text{for } z = 0,\ 0 \leqq x \leqq b - c \tag{34d}$$

$$\left.\begin{aligned} \mathbf{E}_{\tan}^{\mathrm{V}} &= \mathbf{E}_{\tan}^{\mathrm{IV}} \\ \mathbf{H}_{\tan}^{\mathrm{V}} &= \mathbf{H}_{\tan}^{\mathrm{IV}} \end{aligned}\right\} \text{ for } x = b,\ -d \leqq z \leqq 0 \tag{34e}$$

$$\mathbf{H}_{\tan}^{\mathrm{V}} = 0 \qquad \text{for } x = b,\ -a \leqq z \leqq -d \tag{34f}$$

Considering the position of the four waveguides with respect to the coordinate system introduced in Fig. 15, the transversal field components of field regions I–IV can be described by

$$E_y^{\mathrm{I}} = \sum_{p=0}^{\infty} \sqrt{Z_p^{\mathrm{I}}}\left(a_p^{\mathrm{I}}\, e^{-j\beta_p^{\mathrm{I}} x} + b_p^{\mathrm{I}}\, e^{+j\beta_p^{\mathrm{I}} x}\right)\sqrt{\frac{\nu_p}{ah}}\cos\left(\frac{p\pi}{a} z\right) \tag{35a}$$

$$H_z^{\mathrm{I}} = \sum_{p=0}^{\infty} \sqrt{Y_p^{\mathrm{I}}}\left(a_p^{\mathrm{I}}\, e^{-j\beta_p^{\mathrm{I}} x} - b_p^{\mathrm{I}}\, e^{+j\beta_p^{\mathrm{I}} x}\right)\sqrt{\frac{\nu_p}{ah}}\cos\left(\frac{p\pi}{a} z\right) \tag{35b}$$

$$E_y^{\mathrm{II}} = \sum_{k=0}^{\infty} \sqrt{Z_k^{\mathrm{II}}}\left(a_k^{\mathrm{II}} e^{-j\beta_k^{\mathrm{II}}(z+a)} + b_k^{\mathrm{II}} e^{+j\beta_k^{\mathrm{II}}(z+a)}\right)$$

$$\times \sqrt{\frac{\nu_k}{bh}} \cos\left(\frac{k\pi}{b} x\right) \tag{36a}$$

$$H_x^{\mathrm{II}} = -\sum_{k=0}^{\infty} \sqrt{Y_k^{\mathrm{II}}}\left(a_k^{\mathrm{II}} e^{-j\beta_k^{\mathrm{II}}(z+a)} - b_k^{\mathrm{II}} e^{+j\beta_k^{\mathrm{II}}(z+a)}\right)$$

$$\times \sqrt{\frac{\nu_k}{bh}} \cos\left(\frac{k\pi}{b} x\right) \tag{36b}$$

$$E_y^{\mathrm{III}} = \sum_{m=0}^{\infty} \sqrt{Z_m^{\mathrm{III}}}\left(a_m^{\mathrm{III}} e^{+j\beta_m^{\mathrm{III}} z} + b_m^{\mathrm{III}} e^{-\beta_m^{\mathrm{III}} z}\right)$$

$$\times \sqrt{\frac{\nu_m}{ch}} \cos\left[\frac{m\pi}{c}(x-b+c)\right] \tag{37a}$$

$$H_x^{\mathrm{III}} = \sum_{m=0}^{\infty} \sqrt{Y_m^{\mathrm{III}}}\left(a_m^{\mathrm{III}} e^{+j\beta_m^{\mathrm{III}} z} - b_m^{\mathrm{III}} e^{-j\beta_m^{\mathrm{III}} z}\right)$$

$$\times \sqrt{\frac{\nu_m}{ch}} \cos\left[\frac{m\pi}{c}(x-b+c)\right] \tag{37b}$$

$$E_y^{\mathrm{IV}} = \sum_{q=0}^{\infty} \sqrt{Z_q^{\mathrm{IV}}}\left(a_q^{\mathrm{IV}} e^{+j\beta_q^{\mathrm{IV}}(x-b)} + b_q^{\mathrm{IV}} e^{-j\beta_q^{\mathrm{IV}}(x-b)}\right)$$

$$\times \sqrt{\frac{\nu_q}{dh}} \cos\left(\frac{q\pi}{d} z\right) \tag{38a}$$

$$H_z^{\mathrm{IV}} = -\sum_{q=0}^{\infty} \sqrt{Y_q^{\mathrm{IV}}}\left(a_q^{\mathrm{IV}} e^{+j\beta_q^{\mathrm{IV}}(x-b)} - b_q^{\mathrm{IV}} e^{-j\beta_q^{\mathrm{IV}}(x-b)}\right)$$

$$\times \sqrt{\frac{\nu_q}{dh}} \cos\left(\frac{q\pi}{d} z\right) \tag{38b}$$

with the characteristic wave impedances and phase constants as defined in (16a) and (16b) using the adjoint effective widths of the microstrip lines and $\nu_p = 1$ for $p = 0$ and $\nu_p = 2$ for $p \neq 0$.

For the analysis of the asymmetrical crossing, which already is a very complicated microstrip structure, a field analysis method first developed by Kühn [55] is of great advantage. According to Kühn's method, the electromagnetic field in the connecting field region V is described by the superimposition of four standing wave solutions. Figure 16 shows the physical background of the method: The structure is divided into four substructures consisting of the connecting area and one waveguide structure in each case. The connections to the remaining waveguides are closed by magnetic walls so that in each of the so-defined substructures of region V the electromagnetic fields are standing waves. Therefore the fields of the four subregions $\mathrm{V^a}$, $\mathrm{V^b}$, $\mathrm{V^c}$, and $\mathrm{V^d}$ (Fig. 16) are given by the following equations.

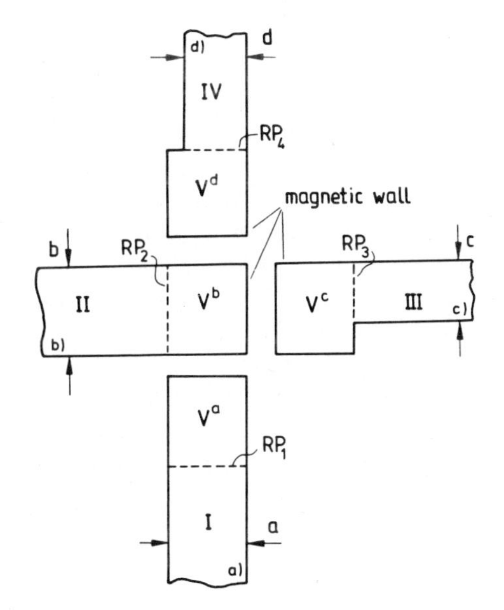

Fig. 16 The choice of subregions for superimposing the fields of the asymmetrical crossing.

Subregion V^a:

$$E_y^{\mathrm{Va}} = \sum_{p=0}^{\infty} \sqrt{\bar{Z}_p^{\mathrm{I}}} c_p^{\mathrm{Va}} \cos[\bar{\beta}_p^{\mathrm{I}}(x-b)] \sqrt{\frac{\nu_p}{ah}} \cos\left(\frac{p\pi}{a} z\right) \tag{39a}$$

$$H_z^{\mathrm{Va}} = -j \sum_{p=0}^{\infty} \sqrt{\bar{Y}_p^{\mathrm{I}}} c_p^{\mathrm{Va}} \sin[\bar{\beta}_p^{\mathrm{I}}(x-b)] \sqrt{\frac{\nu_p}{ah}} \cos\left(\frac{p\pi}{a} z\right) \tag{39b}$$

Subregion V^b:

$$E_y^{\mathrm{Vb}} = \sum_{k=0}^{\infty} \sqrt{\bar{Z}_k^{\mathrm{II}}} c_k^{\mathrm{Vb}} \cos(\bar{\beta}_k^{\mathrm{II}} z) \sqrt{\frac{\nu_k}{bh}} \cos\left(\frac{k\pi}{b} x\right) \tag{40a}$$

$$H_x^{\mathrm{Vb}} = j \sum_{k=0}^{\infty} \sqrt{\bar{Y}_k^{\mathrm{II}}} c_k^{\mathrm{Vb}} \sin(\bar{\beta}_k^{\mathrm{II}} z) \sqrt{\frac{\nu_k}{bh}} \cos\left(\frac{k\pi}{b} x\right) \tag{40b}$$

Subregion V^c:

$$E_y^{\mathrm{Vc}} = \sum_{m=0}^{\infty} \sqrt{\bar{Z}_m^{\mathrm{II}}} c_m^{\mathrm{Vc}} \cos[\bar{\beta}_m^{\mathrm{II}}(z+a)] \sqrt{\frac{\nu_m}{bh}} \cos\left(\frac{m\pi}{b} x\right) \tag{41a}$$

$$H_x^{\mathrm{Vc}} = j \sum_{m=0}^{\infty} \sqrt{\bar{Y}_m^{\mathrm{II}}} c_m^{\mathrm{Vc}} \sin[\bar{\beta}_m^{\mathrm{II}}(z+a)] \sqrt{\frac{\nu_m}{bh}} \cos\left(\frac{m\pi}{b} x\right) \tag{41b}$$

Subregion V^d:

$$E_y^{\mathrm{Vd}} = \sum_{q=0}^{\infty} \sqrt{\bar{Z}_q^{\mathrm{I}}} c_q^{\mathrm{Vd}} \cos(\bar{\beta}_q^{\mathrm{I}} x) \sqrt{\frac{\nu_p}{ah}} \cos\left(\frac{q\pi}{a} z\right) \tag{42a}$$

$$H_z^{\mathrm{Vd}} = -j \sum_{q=0}^{\infty} \sqrt{\bar{Y}_q^{\mathrm{I}}} c_q^{\mathrm{Vd}} \sin(\bar{\beta}_q^{\mathrm{I}} x) \sqrt{\frac{\nu_p}{ah}} \cos\left(\frac{q\pi}{a} z\right) \tag{42b}$$

The phase constants $\bar{\beta}^{\nu}$ ($\nu = \mathrm{I}, \mathrm{II}$) used in (39)–(42) are the phase constants of region I or II, respectively, calculated with the equivalent dielectric constant of region V instead of $\varepsilon_{\mathrm{eff1}}$ or $\varepsilon_{\mathrm{eff2}}$ of region I or II, respectively. In the same way, $\bar{Z}^{\nu}$ ($\nu = \mathrm{I}, \mathrm{II}$) is calculated from Z^{ν}.

It is the great advantage of Kühn's method that the magnetic fields of regions I–IV can be matched to the magnetic field of region V separately because the structures that are superimposed (Fig. 16) at each reference plane have only one open boundary. The remaining reference planes of the substructures are closed by magnetic walls so that the magnetic field strengths vanish here. The relations that match the magnetic fields of regions I and V^{a}, and those of II and V^{b}, can be found simply by comparison of the amplitude coefficients:

$$\sqrt{Y_p^{\mathrm{I}}}(a_p^{\mathrm{I}} - b_p^{\mathrm{I}}) = j\sqrt{\bar{Y}_p^{\mathrm{I}}} c_p^{\mathrm{Va}} \sin(\bar{\beta}_p^{\mathrm{I}} b) \tag{43}$$

and

$$\sqrt{Y_k^{\mathrm{II}}}(a_k^{\mathrm{II}} - b_k^{\mathrm{II}}) = j\sqrt{\bar{Y}_k^{\mathrm{II}}} c_k^{\mathrm{Vb}} \sin(\bar{\beta}_k^{\mathrm{II}} a) \tag{44}$$

The relationships between the fields of regions III and IV and those of region V^{c} or V^{d}, respectively, have to be determined by a normal mode-matching process. At the III–V^{c} interface this leads to

$$j\sqrt{\bar{Y}_M^{\mathrm{II}}} c_M^{\mathrm{Vc}} \sin(\bar{\beta}_M^{\mathrm{II}} a) = \sum_{m=0}^{\infty} \sqrt{Y_m^{\mathrm{III}}}(a_m^{\mathrm{III}} - b_m^{\mathrm{III}}) K_1^{m,M} \tag{45}$$

with the coupling integral

$$K_1^{m,M} = \frac{1}{h} \sqrt{\frac{\nu_m \nu_M}{bc}} \iint_{z=0} \cos\left[\frac{m\pi}{c}(x - b + c)\right] \cos\left(\frac{M\pi}{b} x\right) dx\, dy \tag{46}$$

and at the IV–V^{d} interface the result is

$$j\sqrt{\bar{Y}_Q^{\mathrm{I}}} c_Q^{\mathrm{Vd}} \sin(\bar{\beta}_a^{\mathrm{I}} b) = \sum_{q=0}^{\infty} \sqrt{Y_q^{\mathrm{IV}}}(a_q^{\mathrm{IV}} - b_q^{\mathrm{IV}}) K_2^{q,Q} \tag{47}$$

with the coupling integral

$$K_2^{q,Q} = \frac{1}{h} \sqrt{\frac{\nu_q \nu_Q}{ad}} \iint_{x=b} \cos\left(\frac{q\pi}{d} z\right) \cos\left(\frac{Q\pi}{a} z\right) dz\, dy \tag{48}$$

By the method described above, the magnetic fields of the subregions are matched separately. If now the total electromagnetic field is calculated, the four field solutions (39)–(42) taking into account the relationships (43)–(48) are superimposed, and the electric field strengths of waveguides I–IV and the total electric field in the connecting field region V are additionally matched at the reference planes. As an example, the boundary condition in reference plane RP_1 at $x = 0$ that must be fulfilled by the fields is

$$\mathbf{E}_{\tan}^{\mathrm{V}}\big|_{x=0} = \mathbf{E}_{\tan}^{\mathrm{Va}}\big|_{x=0} + \mathbf{E}_{\tan}^{\mathrm{Vb}}\big|_{x=0} + \mathbf{E}_{\tan}^{\mathrm{Vc}}\big|_{x=0} + \mathbf{E}_{\tan}^{\mathrm{Vd}}\big|_{x=0} \tag{49}$$

The result of this procedure is

$$\sqrt{Z_p^{\mathrm{I}}}(a_P^{\mathrm{I}} + b_P^{\mathrm{I}}) = \sqrt{\bar{Z}_P^{\mathrm{I}}} c_P^{\mathrm{Va}} \cos(\beta_P^{\mathrm{I}} b) + \sum_{k=0}^{\infty} \sqrt{\bar{Z}_k^{\mathrm{II}}} c_k^{\mathrm{Vb}} c_k^{\mathrm{Vb}} K_3^{k,P}$$
$$+ \sum_{m=0}^{\infty} \sqrt{\bar{Z}_m^{\mathrm{II}}} c_m^{\mathrm{Vc}} K_4^{m,P} + \sqrt{\bar{Z}_P^{\mathrm{I}}} c_P^{\mathrm{Vd}} \tag{50}$$

with

$$K_3^{k,P} = \frac{1}{h} \sqrt{\frac{\nu_k \nu_P}{ab}} \iint_{x=0} \cos(\bar{\beta}_k^{\mathrm{II}} z) \cos\left(\frac{P\pi}{a} z\right) dz\, dy \tag{51}$$

$$K_4^{m,P} = \frac{1}{h} \sqrt{\frac{\nu_m \nu_P}{ab}} \iint_{x=0} \cos[\bar{\beta}_m^{\mathrm{II}}(z + a)] \cos\left(\frac{P\pi}{a} z\right) dz\, dy$$
$$= (-1)^P K_3^{m,P} \tag{52}$$

Similar operations at the boundary between regions II and V lead to

$$\sqrt{Z_K^{\mathrm{II}}}(a_K^{\mathrm{II}} + b_K^{\mathrm{II}}) = \sum_{p=0}^{\infty} \sqrt{\bar{Z}_p^{\mathrm{I}}} c_p^{\mathrm{Va}} K_5^{p,K} + \sqrt{\bar{Z}_K^{\mathrm{II}}} c_K^{\mathrm{Vb}} \cos(\beta_K^{\mathrm{II}} a)$$
$$+ \sqrt{\bar{Z}_K^{\mathrm{II}}} c_K^{\mathrm{Vc}} + \sum_{q=0}^{\infty} \sqrt{\bar{Z}_q^{\mathrm{I}}} c_q^{\mathrm{Vd}} K_6^{q,K} \tag{53}$$

with

$$K_5^{p,K} = (-1)^p \frac{1}{h} \sqrt{\frac{\nu_p \nu_K}{ab}} \iint_{z=-a} \cos[\bar{\beta}_p^{\mathrm{I}}(x - b)] \cos\left(\frac{K\pi}{b} x\right) dx\, dy \tag{54}$$

and

$$K_6^{q,K} = (-1)^q \frac{1}{h} \sqrt{\frac{\nu_q \nu_K}{ab}} \iint_{z=-a} \cos(\bar{\beta}_q^{\mathrm{I}} x) \cos\left(\frac{K\pi}{b} x\right) dx\, dy$$
$$= (-1)^K K_5^{q,K} \tag{55}$$

The result of the field matching at the boundary between regions III and V is

$$\sqrt{Z_M^{\mathrm{III}}}(a_M^{\mathrm{III}} + b_M^{\mathrm{III}}) = \sum_{p=0}^{\infty} \sqrt{\bar{Z}_p^{\mathrm{I}}} c_p^{\mathrm{Va}} K_7^{p,M} + \sum_{k=0}^{\infty} \sqrt{\bar{Z}_k^{\mathrm{II}}} c_k^{\mathrm{Vb}} K_1^{M,k}$$

$$+ \sum_{m=0}^{\infty} \sqrt{\bar{Z}_m^{\mathrm{II}}} c_m^{\mathrm{Vc}} \cos(\bar{\beta}_m^{\mathrm{I}} a)\, K_1^{M,m} + \sum_{q=0}^{\infty} \sqrt{\bar{Z}_q^{\mathrm{I}}} c_q^{\mathrm{Vd}} K_8^{q,M} \qquad (56)$$

with

$$K_7^{p,M} = \frac{1}{h} \sqrt{\frac{\nu_p \nu_M}{ac}} \iint_{z=0} \cos[\bar{\beta}_p^{\mathrm{I}}(x-b)] \cos\left[\frac{M\pi}{c}(x-b+c)\right] dx\, dy \qquad (57)$$

and

$$K_8^{q,M} = \frac{1}{h} \sqrt{\frac{\nu_q \nu_M}{ac}} \iint_{z=0} \cos(\bar{\beta}_q^{\mathrm{I}} x) \cos\left[\frac{M\pi}{c}(x-b+c)\right] dx\, dy \qquad (58)$$

and finally, from the field matching at the boundary between regions IV and V,

$$\sqrt{Z_Q^{\mathrm{IV}}}(a_Q^{\mathrm{IV}} + b_Q^{\mathrm{IV}}) = \sum_{p=0}^{\infty} \sqrt{\bar{Z}_p^{\mathrm{I}}} c_p^{\mathrm{Va}} K_2^{Q,p} + \sum_{k=0}^{\infty} \sqrt{\bar{Z}_k^{\mathrm{II}}} c_k^{\mathrm{Vb}} K_9^{k,Q}$$

$$+ \sum_{m=0}^{\infty} \sqrt{\bar{Z}_m^{\mathrm{II}}} c_m^{\mathrm{Vc}} K_{10}^{m,Q} + \sum_{q=0}^{\infty} \sqrt{\bar{Z}_q^{\mathrm{I}}} c_q^{\mathrm{Vd}} \cos(\bar{\beta}_q^{\mathrm{I}} b)\, K_2^{Q,q} \qquad (59)$$

with

$$K_9^{k,Q} = (-1)^k \frac{1}{h} \sqrt{\frac{\nu_K \nu_Q}{bd}} \iint_{x=b} \cos(\bar{\beta}_k^{\mathrm{II}} z) \cos\left(\frac{Q\pi}{d} z\right) dz\, dy \qquad (60)$$

and

$$K_{10}^{m,Q} = (-1)^m \frac{1}{h} \sqrt{\frac{\nu_m \nu_Q}{bd}} \iint_{x=b} \cos[\bar{\beta}_m^{\mathrm{II}}(z+a)] \cos\left(\frac{Q\pi}{d} z\right) dz\, dy \qquad (61)$$

The wave amplitudes c_ν^μ of the field solutions in region V in (50)–(59) can be replaced by the results given in (43)–(47). Additionally, as has been mentioned already above, only TEM modes on the four microstrip lines are assumed as exciting waves. Under these conditions the equations given above can be written as four different infinite equation systems for the unknown amplitude coefficients b_ν^μ of the reflected waves:

$$
\begin{aligned}
&-\sqrt{Z_P^{\mathrm{I}}}\left[1 - j\,\frac{\bar{Z}_P^{\mathrm{I}}}{Z_P^{\mathrm{I}}}\cot(\bar{\beta}_P^{\mathrm{I}} b)\right] b_P^{\mathrm{I}} + \sum_{k=0}^{\infty} j\,\frac{\bar{Z}_k^{\mathrm{II}}}{\sqrt{Z_k^{\mathrm{II}}}}\,\frac{K_3^{k,P}}{\sin(\bar{\beta}_k^{\mathrm{II}} a)}\, b_k^{\mathrm{II}} \\
&+ \sum_{i=0}^{\infty}\sum_{m=0}^{\infty} j\,\frac{\bar{Z}_m^{\mathrm{II}}}{\sqrt{Z_i^{\mathrm{III}}}}\,\frac{K_1^{i,m}K_4^{m,P}}{\sin(\bar{\beta}_m^{\mathrm{II}} a)}\, b_i^{\mathrm{III}} + \sum_{q=0}^{\infty} j\,\frac{\bar{Z}_P^{\mathrm{I}}}{\sqrt{Z_q^{\mathrm{IV}}}}\,\frac{K_2^{q,P}}{\sin(\bar{\beta}_p^{\mathrm{I}} b)}\, b_q^{\mathrm{IV}} \\
&\qquad = \sqrt{Z_0^{\mathrm{I}}}\left[1 + j\,\frac{\bar{Z}_0^{\mathrm{I}}}{Z_0^{\mathrm{I}}}\cot(\bar{\beta}_0^{\mathrm{I}} b)\right]\delta_{P0}\, a_0^{\mathrm{I}} + j\,\frac{\bar{Z}_0^{\mathrm{II}}}{\sqrt{Z_0^{\mathrm{II}}}}\,\frac{K_3^{0,P}}{\sin(\bar{\beta}_0^{\mathrm{II}} a)}\, a_0^{\mathrm{II}} \\
&\qquad + j\sum_{m=0}^{\infty}\frac{\bar{Z}_m^{\mathrm{II}}}{\sqrt{Z_0^{\mathrm{III}}}}\,\frac{K_1^{0,m}K_4^{m,P}}{\sin(\bar{\beta}_m^{\mathrm{II}} a)}\, a_0^{\mathrm{III}} + j\,\frac{\bar{Z}_P^{\mathrm{I}}}{\sqrt{Z_0^{\mathrm{IV}}}}\,\frac{K_2^{0,P}}{\sin(\bar{\beta}_P^{\mathrm{I}} b)}\, a_0^{\mathrm{IV}}
\end{aligned}
\tag{62}
$$

$$
\begin{aligned}
&\sum_{p=0}^{\infty} j\,\frac{\bar{Z}_p^{\mathrm{I}}}{\sqrt{Z_p^{\mathrm{I}}}}\,\frac{K_5^{p,K}}{\sin(\bar{\beta}_p^{\mathrm{I}} b)}\, b_p^{\mathrm{I}} - \sqrt{Z_K^{\mathrm{II}}}\left[1 - j\,\frac{\bar{Z}_K^{\mathrm{II}}}{Z_K^{\mathrm{II}}}\cot(\bar{\beta}_K^{\mathrm{II}} a)\right] b_K^{\mathrm{II}} \\
&+ \sum_{m=0}^{\infty} j\,\frac{\bar{Z}_K^{\mathrm{II}}}{\sqrt{Z_m^{\mathrm{III}}}}\,\frac{K_1^{m,K}}{\sin(\bar{\beta}_K^{\mathrm{II}} a)}\, b_m^{\mathrm{III}} + \sum_{i=0}^{\infty}\sum_{q=0}^{\infty} j\,\frac{\bar{Z}_q^{\mathrm{I}}}{\sqrt{Z_i^{\mathrm{IV}}}}\,\frac{K_2^{i,q}K_6^{q,K}}{\sin(\bar{\beta}_q^{\mathrm{I}} b)}\, b_i^{\mathrm{IV}} \\
&\qquad = j\,\frac{\bar{Z}_0^{\mathrm{I}}}{\sqrt{Z_0^{\mathrm{I}}}}\,\frac{K_5^{0,K}}{\sin(\bar{\beta}_0^{\mathrm{I}} b)}\, a_0^{\mathrm{I}} + \sqrt{Z_0^{\mathrm{II}}}\left[1 + j\,\frac{\bar{Z}_0^{\mathrm{II}}}{Z_0^{\mathrm{II}}}\cot(\bar{\beta}_0^{\mathrm{II}} a)\right] a_0^{\mathrm{II}}\,\delta_{K0} \\
&\qquad + j\,\frac{\bar{Z}_K^{\mathrm{II}}}{\sqrt{Z_0^{\mathrm{III}}}}\,\frac{K_1^{0,K}}{\sin(\bar{\beta}_K^{\mathrm{II}} a)}\, a_{\mathrm{a}}^{\mathrm{III}} + j\sum_{q=0}^{\infty}\frac{\bar{Z}_q^{\mathrm{I}}}{\sqrt{Z_0^{\mathrm{IV}}}}\,\frac{K_2^{0,q}K_6^{q,K}}{\sin(\bar{\beta}_q^{\mathrm{I}} b)}\, a_0^{\mathrm{IV}}
\end{aligned}
\tag{63}
$$

$$
\begin{aligned}
&\sum_{p=0}^{\infty} j\,\frac{\bar{Z}_p^{\mathrm{I}}}{\sqrt{Z_p^{\mathrm{I}}}}\,\frac{K_7^{p,M}}{\sin(\bar{\beta}_p^{\mathrm{I}} b)}\, b_p^{\mathrm{I}} + \sum_{k=0}^{\infty} j\,\frac{\bar{Z}_k^{\mathrm{II}}}{\sqrt{Z_k^{\mathrm{II}}}}\,\frac{K_1^{M,k}}{\sin(\bar{\beta}_k^{\mathrm{II}} a)}\, b_k^{\mathrm{II}} \\
&- \left[\sqrt{Z_M^{\mathrm{III}}}\, b_M^{\mathrm{III}} - j\sum_{i=1}^{\infty}\sum_{m=0}^{\infty}\frac{\bar{Z}_m^{\mathrm{II}}}{\sqrt{Z_i^{\mathrm{III}}}}\cot(\bar{\beta}_m^{\mathrm{II}} a)\, K_1^{M,m}K_1^{i,m}\, b_i^{\mathrm{III}}\right] \\
&+ \sum_{i=0}^{\infty}\sum_{q=0}^{\infty} j\,\frac{\bar{Z}_q^{\mathrm{I}}}{\sqrt{Z_i^{\mathrm{IV}}}}\,\frac{K_2^{i,q}K_8^{q,M}}{\sin(\bar{\beta}_q^{\mathrm{I}} b)}\, b_i^{\mathrm{IV}} \\
&\qquad = j\,\frac{\bar{Z}_0^{\mathrm{I}}}{\sqrt{Z_0^{\mathrm{I}}}}\,\frac{K_7^{0,M}}{\sin(\bar{\beta}_0^{\mathrm{I}} b)}\, a_0^{\mathrm{I}} + j\,\frac{\bar{Z}_0^{\mathrm{II}}}{\sqrt{Z_0^{\mathrm{II}}}}\,\frac{K_1^{M,0}}{\sin(\bar{\beta}_0^{\mathrm{II}} a)}\, a_0^{\mathrm{II}} \\
&\qquad + \sqrt{Z_0^{\mathrm{III}}}\, a_0^{\mathrm{III}}\,\delta_{M0} + j\sum_{m=0}^{\infty}\frac{\bar{Z}_m^{\mathrm{II}}}{\sqrt{Z_0^{\mathrm{III}}}}\cot(\bar{\beta}_m^{\mathrm{II}} a)\, K_1^{M,m}K_1^{0,m}\, a_0^{\mathrm{III}} \\
&\qquad + j\sum_{q=0}^{\infty}\frac{\bar{Z}_q^{\mathrm{I}}}{\sqrt{Z_0^{\mathrm{IV}}}}\,\frac{K_2^{0,q}K_8^{q,M}}{\sin(\bar{\beta}_0^{\mathrm{I}} b)}\, a_0^{\mathrm{IV}}
\end{aligned}
\tag{64}
$$

$$\sum_{p=0}^{\infty} j \frac{\bar{Z}_p^{\mathrm{I}}}{\sqrt{Z_p^{\mathrm{I}}}} \frac{K_2^{Q,p}}{\sin(\bar{\beta}_p^{\mathrm{I}} b)} b_p^{\mathrm{I}} + \sum_{k=0}^{\infty} j \frac{\bar{Z}_k^{\mathrm{II}}}{\sqrt{Z_k^{\mathrm{II}}}} \frac{K_9^{k,Q}}{\sin(\bar{\beta}_k^{\mathrm{II}} a)} b_k^{\mathrm{II}}$$

$$+ \sum_{i=1}^{\infty} \sum_{m=0}^{\infty} j \frac{\bar{Z}_m^{\mathrm{II}}}{\sqrt{Z_i^{\mathrm{III}}}} \frac{K_1^{m,i} K_{10}^{m,Q}}{\sin(\bar{\beta}_m^{\mathrm{II}} a)} b_i^{\mathrm{III}} - \sqrt{Z_Q^{\mathrm{IV}}} b_Q^{\mathrm{IV}}$$

$$+ \sum_{i=0}^{\infty} \sum_{q=0}^{\infty} \frac{\bar{Z}_q^{\mathrm{I}}}{\sqrt{Z_i^{\mathrm{IV}}}} \cot(\bar{\beta}_q^{\mathrm{I}} b)\, K_2^{Q,q} K_2^{i,q} b_i^{\mathrm{IV}}$$

$$= j \frac{\bar{Z}_0^{\mathrm{I}}}{\sqrt{Z_0^{\mathrm{I}}}} \frac{K_2^{Q,0}}{\sin(\bar{\beta}_0^{\mathrm{I}} b)} a_0^{\mathrm{I}} + j \frac{\bar{Z}_0^{\mathrm{II}}}{\sqrt{Z_0^{\mathrm{II}}}} \frac{K_9^{0,Q}}{\sin(\bar{\beta}_0^{\mathrm{II}} a)} a_0^{\mathrm{II}}$$

$$+ j \sum_{m=0}^{\infty} \frac{\bar{Z}_m^{\mathrm{II}}}{\sqrt{Z_0^{\mathrm{III}}}} \frac{K_1^{m,0} K_{10}^{m,Q}}{\sin(\bar{\beta}_m^{\mathrm{II}} a)} a_0^{\mathrm{III}} + \sqrt{Z_0^{\mathrm{IV}}} a_0^{\mathrm{IV}} \delta_{Q0}$$

$$+ j \sum_{q=0}^{\infty} \frac{\bar{Z}_q^{\mathrm{I}}}{\sqrt{Z_0^{\mathrm{IV}}}} \cot(\bar{\beta}_q^{\mathrm{I}} b)\, K_2^{Q,q} K_2^{0,q} a_0^{\mathrm{IV}} \tag{65}$$

The infinite sums of these four equation systems are cut off to consider only a finite number of elements. If, in addition, an exciting wave is considered on only one microstrip line or the equivalent waveguide, the S parameters of the crossing,

$$S_{\nu\mu} = b_0^{\mu} / a_0^{\nu} \tag{66}$$

for the fundamental (quasi-)TEM mode can be calculated by solving eqs. (62)–(65).

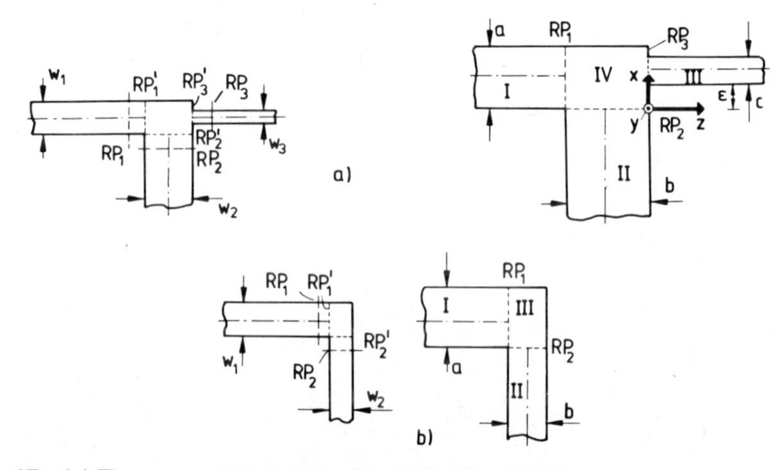

Fig. 17 (*a*) The asymmetrical microstrip T junction, (*b*) the asymmetrical microstrip bend, and their equivalent waveguide structures.

If (65) is eliminated out of the equation system and if in addition the coefficients a_0^{IV} and b_ν^{IV} are set to zero in (62)–(64), the resulting equation system can be used to analyze the transmission properties of the asymmetrical T junction (Fig. 17*a*). If eqs. (64) and (65) are eliminated out of the equation system and if in addition the amplitude coefficients a_0^{III}, a_0^{IV}, b_ν^{III}, and b_ν^{IV} are set to zero in the remaining equations, the remaining equation system describes the transmission properties of a 90° bend in microstrip technique (Fig. 17*b*). This means that the demonstrated method can be used for the analysis of a variety of complicated microstrip circuit elements.

4. CONVERGENCE AND NUMERICAL RESULTS

The final equation systems for the calculation of the scattering parameters using the waveguide model technique are infinite linear equation systems. These equation systems can be solved numerically only if the infinite sums are truncated and only a finite number of series elements are considered. The influence of the truncation on the calculated scattering parameters has to be proved to gain an insight into the convergence behavior of the methods. The good convergence of the applied method will be demonstrated with two examples: the microstrip impedance step and the asymmetrical microstrip T junction.

In both cases the ratio of the considered modes in the different field regions is kept constant while the number of modes considered is varied. Figure 18 shows the dependence of the transmission parameters of a microstrip impedance step on the number of modes used in the description of the electromagnetic fields in regions A and B (compare Fig. 14). Different curves have been evaluated for constant ratios of the mode numbers in the two regions. As can be seen from Fig. 18, the phenomenon of relative convergence [68, 69] does not occur for the example of the impedance step, but it can be seen that the convergence is fastest when the ratio of the considered mode numbers is chosen equal to the ratio of the line widths of the impedance step. If P is the mode number considered in the field area of the microstrip line of larger width and M the mode number in the field region of smaller width, the results for the absolute values of the transmission parameters for small P values are always larger than the converged results for $P/M < b/a$ (with b and a the effective widths of the microstrip lines), and they are always smaller than the converged results for $P/M > b/a$. A similar result is found for the phases of the transmission parameters (Fig. 18*b*).

A similar result is found for the microstrip T junction. Figure 19 shows the convergence of the reflection coefficients of an asymmetrical T junction with the characteristic impedances $Z_1 = Z_3 = 50\,\Omega$ and $Z_2 = 20\,\Omega$ on RT/Duroid substrate material ($\varepsilon_r = 2.32$) of height $h = 1.58$ mm. The numbers of considered field modes in field regions I (N_1) and III (N_3) have been

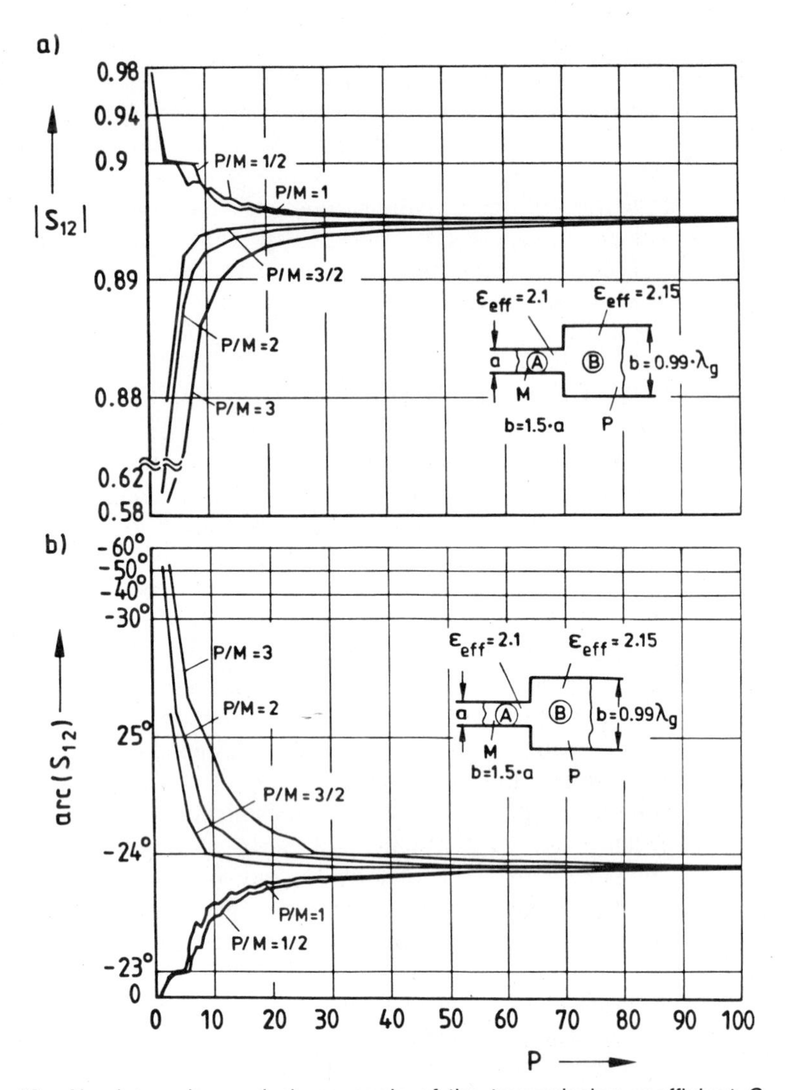

Fig. 18 Absolute value and phase angle of the transmission coefficient S_{12} of a symmetrical microstrip impedance step in dependence on the P/M ratio of modes used for the field expansion in the two field regions A and B, [44].

assumed to be equal. The ratio of the number of field modes in region II (N_2) to the number in region I is kept constant at different values, and the convergence is investigated in dependence on the mode numbers N_1 and N_2.

Again a good convergence of the scattering parameters in dependence on the mode numbers can be found, and again no relative convergence phenomena occur. In both examples (impedance step and T junction) it is found that at low frequencies (i.e., far from the cutoff frequency of the first

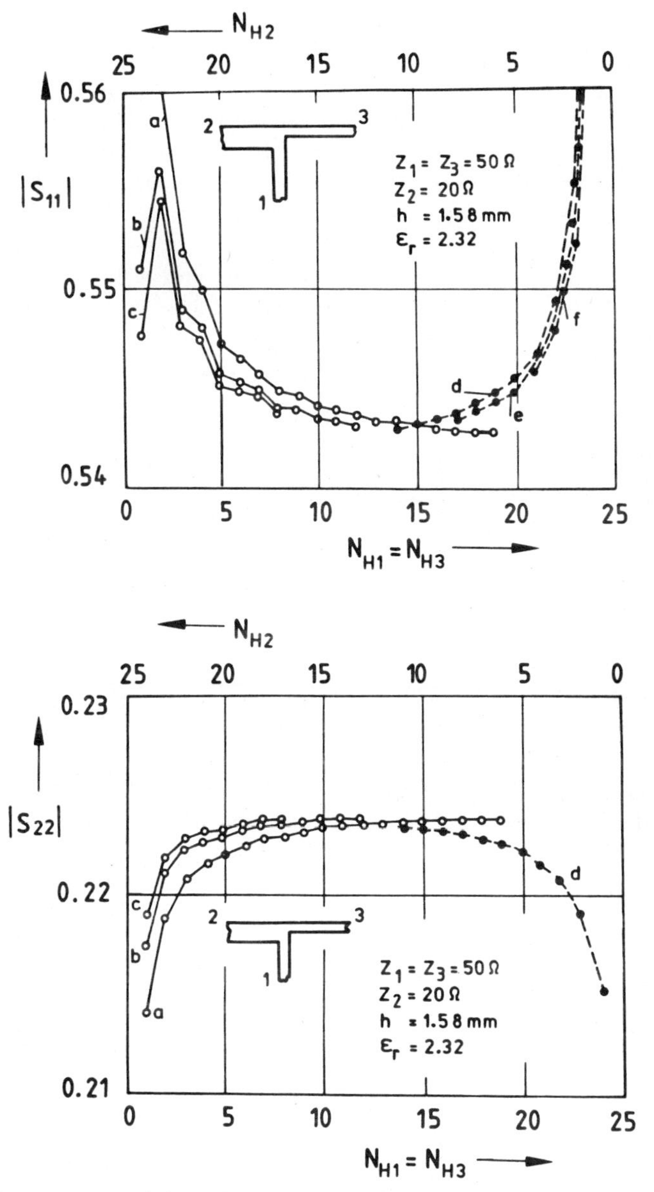

Fig. 19 Convergence of the reflection coefficients S_{11} and S_{22} of an asymmetrical microstrip T junction. (*a*) $N_{H1}/N_{H2} = 1$; (*b*) $N_{H1}/N_{H2} = \frac{1}{2}$; (*c*) $N_{H1}/N_{H2} = \frac{1}{3}$; (*d*) $N_{H1}/N_{H2} = 2$; (*e*) $N_{H1}/N_{H2} = 3$; (*f*) $N_{H1}/N_{H2} = 4$, [49].

higher-order mode on the microstrip lines) only five field modes must be considered in each field region to reach a numerical accuracy better than 1%; at higher frequencies eight to ten modes might be needed.

It has already been mentioned in Section 2 that the waveguide model is an approximate model for describing the electromagnetic fields on the microstrip line and near the microstrip discontinuities. It therefore makes no sense to require a numerical accuracy of the calculations higher than 1%, and a good compromise between the requirements for accuracy and for low computation times is a value of six to eight field modes in the different field regions.

Before presenting numerical results to demonstrate the frequency-dependent transmission properties of the microstrip discontinuities, the question of calculating the reference planes of the waveguide model back into the reference planes of the microstrip discontinuity will be discussed. As mentioned in Section 3, in the case of the microstrip impedance step the reference plane RP′ (Fig. 14*a*) has to be shifted back into the position of reference plane RP (Fig. 14*b*) because of the stray fields at the partly open microstrip line in the step plane. This is done using well-known formulas for the shift of the reference plane at an open-ended microstrip line [71]. The width of the open microstrip line that is used in the calculations is $w = w_1 - w_2$ $(w_1 > w_2)$. If this is done, the calculated phases are in quite good agreement with results from rigorous numerical methods [11] and measured results.

In the case of the microstrip crossing, the T junction, and the 90° bend, the reference planes RP'_ν $(\nu = 1, 2, 3, 4)$ of the waveguide model can theoretically be transformed back into position RP_ν $(\nu = 1, 2, 3, 4)$ of the original microstrip structure (compare Figs. 15 and 17), using the difference between the effective widths of the waveguide model and the geometrical widths of the microstrip lines. If this is done, comparisons with measurements show that this back-transformation in many cases overestimates the influence of the reference plane shift. Comparisons with measured phases of the scattering parameters show that it is a good compromise if in the process of back-transformation the phase shift is calculated using only $0.6(w_{\text{eff}} - w_\nu)/2$ $(\nu = 1, 2, 3, 4)$ instead of the total differences between the effective line widths and the geometrical line widths. This result clearly shows the model character of the magnetic wall waveguide. On the other hand the waveguide models many effects which occur inside the microstrip discontinuities on a physical basis so that even when the final results are not accurate enough for direct application, e.g., in computer-aided circuit design, the model can easily be corrected using measurement results.

Figures 20 to 23 show the typical absolute values and phases of the scattering parameters of microstrip impedance steps. The assumed substrate material is RT/Duroid with a dielectric constant of $\varepsilon_r = 2.32$ and a height of $h = 1.57$ mm. In Fig. 20 the reflection coefficients $|S_{11}|$ of microstrip impedance steps from a 50-Ω line to lines of lower characteristic impedances are

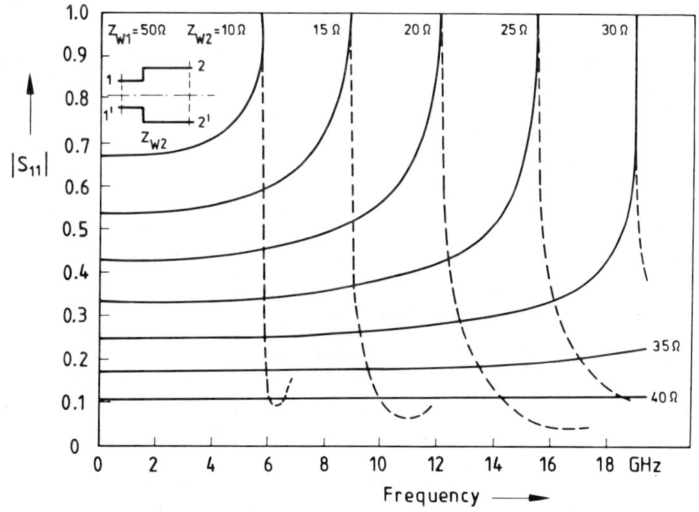

Fig. 20 Absolute values of the reflection coefficients of symmetrical microstrip impedance steps from 50 Ω to lower impedances in dependence on the frequency. Substrate material: RT/Duroid, $\varepsilon_r = 2.32$, h = 1.57 mm.

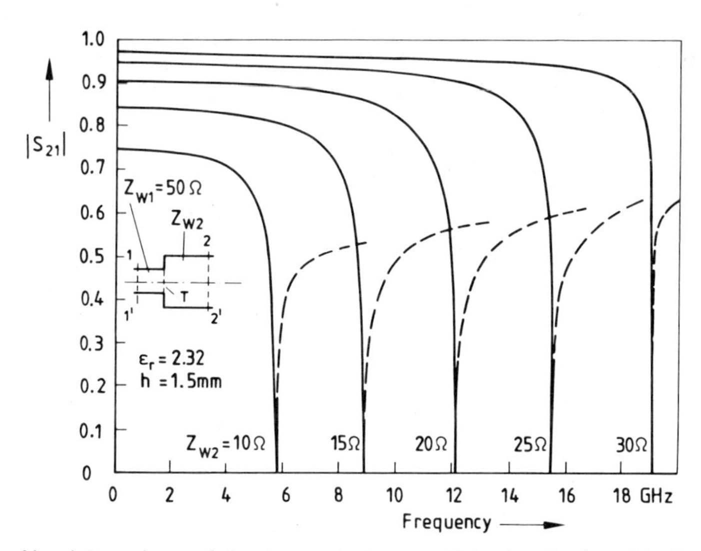

Fig. 21 Absolute values of the transmission coefficients of microstrip impedance steps from 50 Ω to lower impedances in dependence on the frequency. Substrate material: RT/Duroid, $\varepsilon_r = 2.32$, $h = 1.57$ mm.

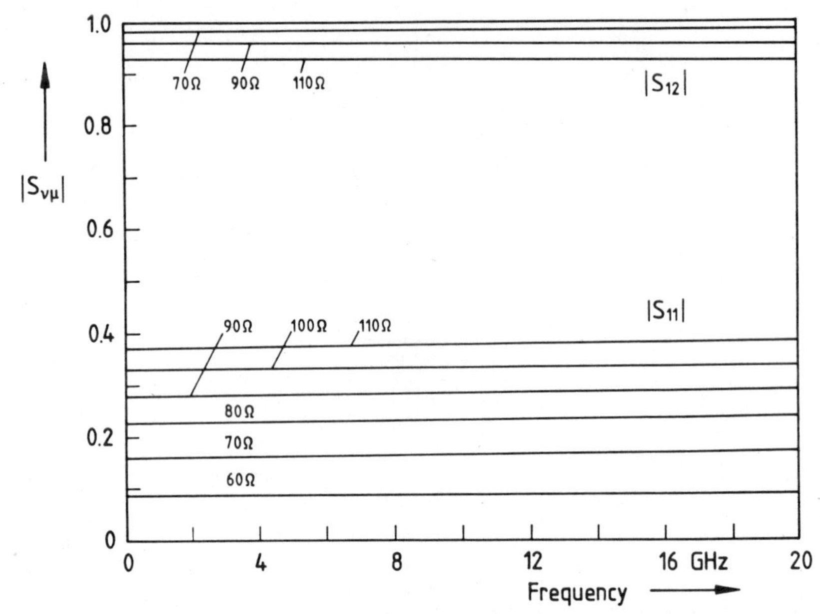

Fig. 22 Absolute values of the scattering parameters of microstrip impedance steps from 50 Ω to higher impedances in dependence on the frequency. Substrate material: RT/Duroid, $\varepsilon_r = 2.32$, $h = 1.57$ mm.

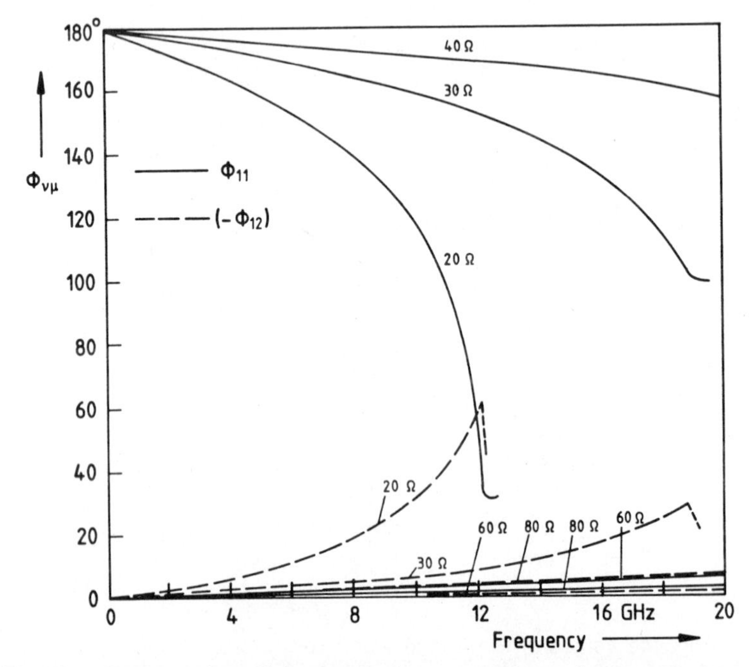

Fig. 23 The phase angles of the scattering parameters of microstrip impedance steps in dependence on the frequency. Substrate material: RT/Duroid, $\varepsilon_r = 2.32$, $h = 1.57$ mm.

shown. At very low frequencies the reflection coefficients can be calculated directly from the characteristic impedances: $r = (Z_2 - Z_1)/(Z_2 + Z_1)$. With increasing frequency the reflection coefficient increases until at the cutoff frequency of the first higher-order mode it becomes 1. The cutoff frequencies of the lines with large line widths w (i.e., the lines with low characteristic impedances) are the lowest, and as a consequence the frequency dependence of the reflection coefficients of these lines is high. In Fig. 21 the adjoint transmission coefficients $|S_{21}|$ of the impedance steps are shown. At the cutoff frequencies the transmission coefficients are zero in agreement with the results shown in Fig. 20.

In the case of impedance steps with one line impedance equal to 50 Ω and one line impedance higher than 50 Ω, the frequency dependence of the scattering parameters $|S_{11}|$ and $|S_{21}|$ is very small, as shown in Fig. 22. Over a large frequency range the absolute values of the scattering parameters are well described by the low frequency values, which again are determined by the characteristic impedances.

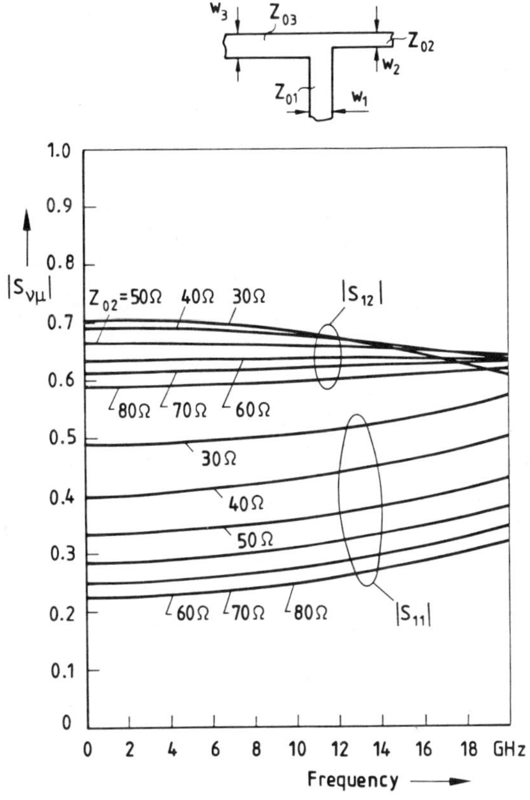

Fig. 24 The absolute values of the scattering parameters S_{11} and S_{12} of an asymmetrical microstrip T junction in dependence on the frequency. Substrate material: alumina, $\varepsilon_r = 9.9$, $h = 0.635$ mm.

In Fig. 23 the typical frequency dependence of the phases of the scattering parameters S_{11} and S_{21} is shown. At zero frequency the phase angle of S_{11} is 180° and that of S_{21} is zero. The phase angle of the reflection coefficient decreases with increasing frequency, whereas that of the transmission coefficient increases. The large frequency dependencies of the phases as they are shown in Fig. 23 are valid only for impedance steps with one low (<50 Ω) characteristic impedance. Impedance steps from 50 Ω to higher characteristic impedances again have a very low frequency dependence on the phase angles of the scattering parameters.

In Figs. 24–27, some numerical results for an asymmetrical microstrip T junction are shown. The substrate material this time is alumina with a dielectric constant of $\varepsilon_r = 9.9$ and a height of $h = 0.635$ mm. Lines 1 and 3 have a line width of 0.6 mm, which means that the characteristic impedances of these lines are nearly 50 Ω at a frequency of 10 GHz (Fig. 24). The line widths of line 2 are determined so that at a frequency of 10 GHz the characteristic impedances of this line are given by the values in the figures. As can be seen from the figures again the reflection and transmission coefficients at low frequencies can be calculated from the static values of the characteristic impedances. The frequency dependence of the scattering

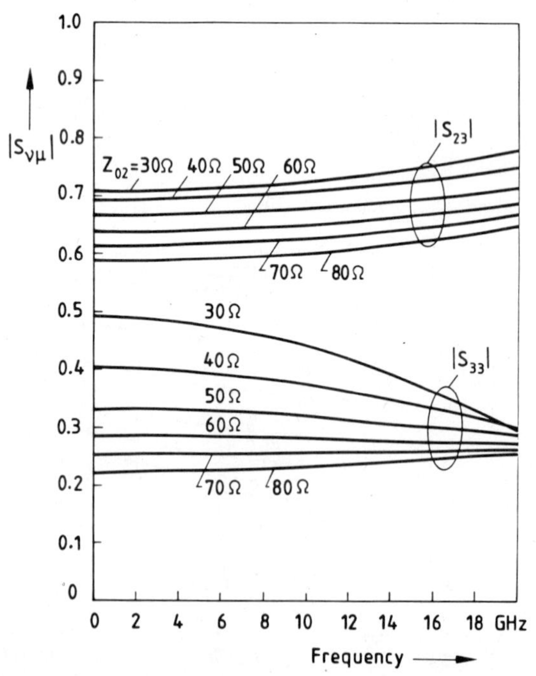

Fig. 25 The absolute values of the scattering parameters S_{23} and S_{33} of an asymmetrical microstrip T junction in dependence on the frequency. Substrate material: alumina, $\varepsilon_r = 9.9$, $h = 0.635$ mm.

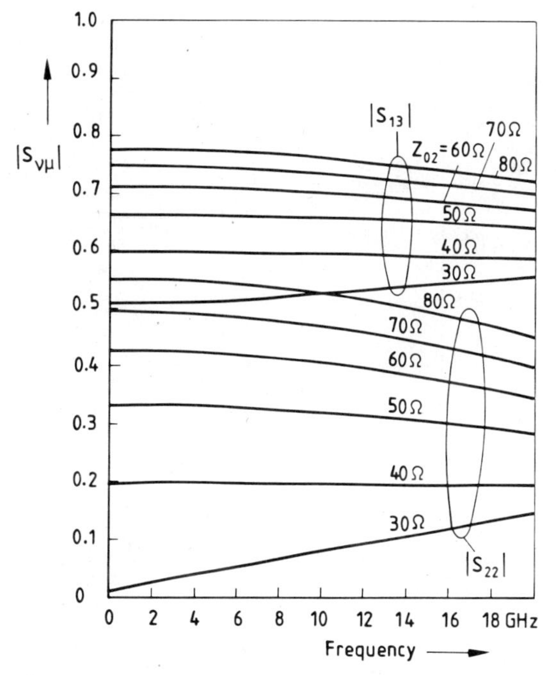

Fig. 26 The absolute values of the scattering parameters S_{13} and S_{22} of an asymmetrical microstrip T junction in dependence on the frequency. Substrate material: alumina, $\varepsilon_r = 9.9$, $h = 0.635$ mm.

parameters for this structure on a high dielectric constant substrate material is not so high as in the case of the impedance steps on the low dielectric constant substrate material.

Finally, in Fig. 28 the influence of the eccentricity e on the scattering parameters is shown. Again an asymmetrical T junction on a high dielectric constant substrate is considered. As can be seen from the figure, the T junction can be optimized in the sense that its scattering parameters underlie a minimum frequency dependence. It is not suprising that the lowest frequency dependence is found in the case where the T junction's geometrical structure does not have an additional step in the connection between line 2 and line 3.

5. SUMMARY

In this chapter a numerical technique has been described to model microstrip discontinuities of various kinds. The method is based on a magnetic wall waveguide model that is assumed to be valid for describing the electromagnetic fields in the microstrip line and in the vicinity of microstrip

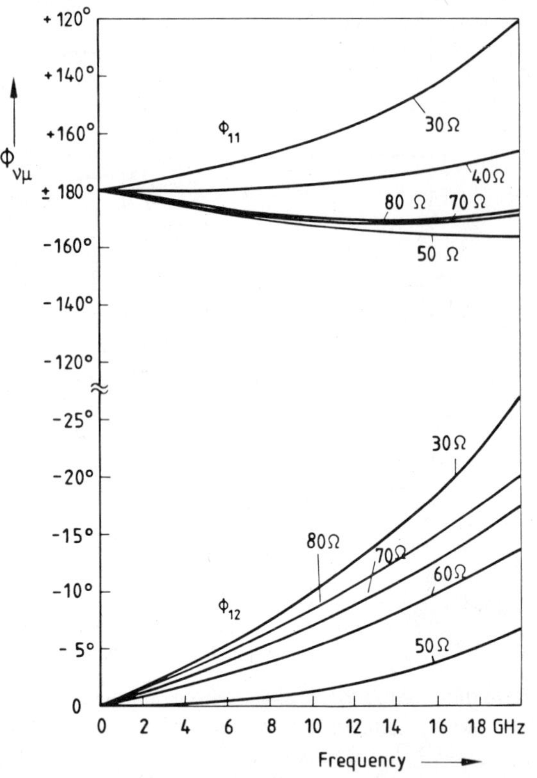

Fig. 27 The phase angles of the scattering parameters S_{11} and S_{12} of an asymmetrical microstrip T junction in dependence on the frequency. Substrate material: alumina, $\varepsilon_r = 9.9$, $h = 0.635$ mm.

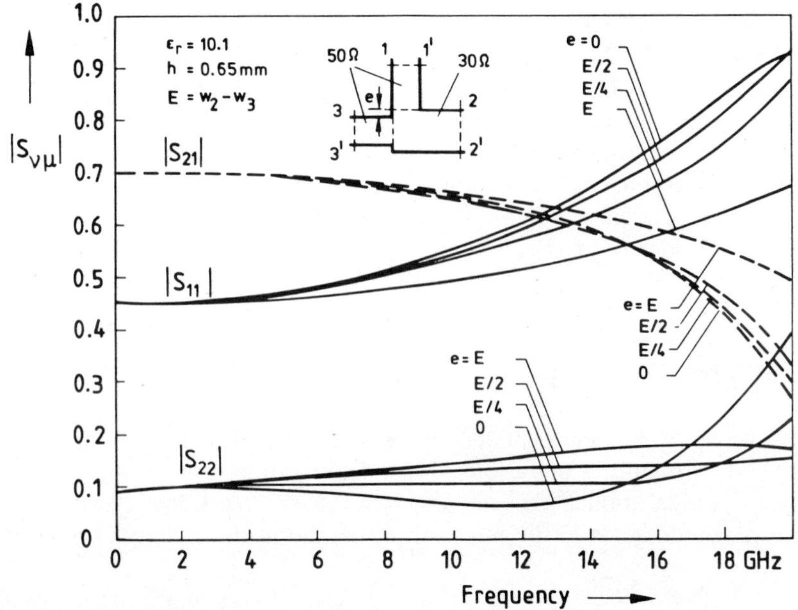

Fig. 28 The dependence of the scattering parameters of asymmetrical T junctions on the eccentricity e of the structure.

discontinuities. The validity of this model has been discussed intensively, and it has been found that the model can be used to describe the microstrip discontinuities with acceptable accuracy if additional empirical information is used to improve the accuracy of the phase calculations. This can be easily done using measured results or results of rigorous numerical methods. Because rigorous methods are not available for most of the microstrip discontinuities at this time, the empirical information has been taken from measurements. The results are numerical models for microstrip discontinuities that are of medium numerical expense and of good accuracy and can be used in computer-aided circuit analysis even with a desktop computer.

REFERENCES

1. M. Kirschning and R. H. Jansen, "Accurate model for effective dielectric constant of microstrip with validity up to millimetre-wave frequencies," *Electron. Lett.*, vol. 18, pp. 272–273, 1982.
2. R. H. Jansen, "Unified, user-oriented computation of shielded, covered and open planar microwave and millimeter-wave transmission-line characteristics," *J. Microwaves, Opt., Acoust.*, vol. 3, pp. 14–22, 1979.
3. I. M. Stephenson and B. Easter, "Resonant techniques for establishing the equivalent circuits of small discontinuities in microstrip," *Electron. Lett.*, vol. 7, p. 19,1971.
4. A. Farrar and A. T. Adams, "Computation of lumped microstrip capacities by matrix methods: Rectangular sections and end effect," *IEEE Trans. Microwave Theory Tech.*, vol. MTT-19, pp. 495–496, 1971; correction: vol. MTT-20, p. 294, 1972.
5. B. Easter, "The equivalent circuit of some microstrip discontinuities," *IEEE Trans. Microwave Theory Tech.*, vol. MTT-23, pp. 655–660, 1975.
6. P. Bendek and P. Silvester, "Microstrip discontinuity capacitance for right-angle bends, T junctions, and crossings,"*IEEE Trans. Microwave Theory Tech.*, vol. MTT-21, pp. 341–346, 1973.
7. W. H. Leighton and A. G. Milnes, "Junction reactance and dimensional tolerance effects on X-band 3-dB directional couplers," *IEEE Trans. Microwave Theory Tech.*, vol. MTT-19, pp. 818–824, 1971.
8. R. Garg and I. J. Bahl, "Microstrip discontinuities," *Int. J. Electron.*, vol. 45, pp. 81–87, 1978.
9. A. Gopinath and C. Gupta, "Capacitance parameters of discontinuities in microstrip line," *IEEE Trans. Microwave Theory Tech.*, vol. MTT-26, pp. 831–836, 1978.
10. L.-P. Schmidt, "Zur feldtheoretischen Berechnung von transversalen Diskontinuitäten in Mikrostrip-Leitungen," Doctoral Thesis, Technical University Aachen, West Germany, 1979.
11. N. H. L. Koster and R. H. Jansen, "The microstrip step discontinuity, a revised description," *IEEE Trans. Microwave Theory Tech.*, vol. MTT-34, pp. 213–223, 1986.

12. D. M. Pozar, "Finite phased array of rectangular microstrip patches," *IEEE Trans. Antennas Propag.*, vol. AP-34, pp. 658–665, 1986.
13. G. Gronau, "Theoretische und experimentelle Untersuchung der Verkopplung in Streifenleitungsantennen," Doctoral Thesis, Duisburg University, Duisburg, Federal Republic of Germany, 1987.
14. R. H. Jansen and W. Wertgen, "Modular source-type 3D analysis of scattering parameters for general discontinuities, components, and coupling effects in (M)MIC's," *Proc. 17th Eur. Microwave Conf.*, pp. 427–432, 1987.
15. H. Ermert, "Ein Verfahren zur Berechnung der Dispersion und der Feldverteilung von Wellentypen auf einer Mikrowellenstreifenleitung," Habilitationsschrift, Friedrich-Alexander-Universität, Erlangen-Nürnberg, West Germany, 1975.
16. H. Ermert, "Guiding and radiation characteristics of planar waveguides," *IEE J. Microwaves, Op. Acoust.*, vol. 3, pp. 59–62, Mar. 1979.
17. H. A. Wheeler, "Transmission line properties of parallel wide strips by conformal mapping approximation," *IEEE Trans. Microwave Theory Tech.*, vol. MTT-12, pp. 280–289, 1964.
18. H. A. Wheeler, "Transmission line properties of parallel strips separated by a dielectric sheet," *IEEE Trans. Microwave Theory Tech.*, vol. MTT-13, pp. 172–185, 1965.
19. C. P. Hartwig, D. Masse, and R. A. Pucel, "Frequency dependent behaviour of microstrip," *1968 G-MTT Int. Symp. Dig.*, pp. 110–119, 1968.
20. S. Arnold, "Dispersive effects in microstrip on alumina substrates," *Electron. Lett.*, vol. 5, pp. 673–674, 1969.
21. W. J. Chudobiak, O. P. Jain, and V. Makios, "Dispersion in microstrip," *IEEE Trans. Microwave Theory Tech.*, vol. MTT-19, pp. 783–784, 1971.
22. M. V. Schneider, "Microstrip dispersion," *Proc. IEEE*, vol. 60, pp. 144–146, 1972.
23. W. J. Getsinger, "Microstrip dispersion model," *IEEE Trans. Microwave Theory Tech.*, vol. MTT-21, pp. 34–39, 1973.
24. H. J. Carlin, "A simplified circuit model for microstrip," *IEEE Trans. Microwave Theory Tech.*, vol. MTT-21, pp. 589–591, 1973.
25. B. Bianco, A. Chiabrera, M. Granara, and S. Ridella, "Frequency dependence of microstrip parameters," *Alta Freq.*, vol. 43, pp. 413–416, 1974.
26. G. Kompa and R. Mehran, "Planar waveguide model for calculating microstrip components," *Electron. Lett.*, vol. 11, pp. 459–460, 1975.
27. R. P. Owens, "Predicted frequency dependence of microstrip characteristic impedance using the planar waveguide model," *Electron. Lett.*, vol. 12, pp. 269–270, 1976.
28. T. C. Edwards and R. P. Owens, "2–18 GHz dispersion measurements on 10–100 ohm microstrip line on sapphire," *IEEE Trans. Microwave Theory Tech.*, vol. MTT-24, pp. 506–513, 1976.
29. H. J. Schmitt and K. H. Sarges, "Wave propagation in microstrip," *Nachrichtentech. Z.*, vol. 24, pp. 260–264, 1971.
30. T. Itoh and R. Mittra, "Spectral-domain approach for calculating the dispersion characteristics of microstrip lines," *IEEE Trans. Microwave Theory Tech.*, vol. MTT-21, pp. 496–499, 1973.

31. T. Itoh and R. Mittra, "A technique for computing dispersion characteristics of shielded microstrip lines," *IEEE Trans. Microwave Theory Tech.*, vol. MTT-22, pp. 896–898, 1974.

32. I. Wolff, G. Kompa, and R. Mehran, "Calculation method for microstrip discontinuities and T-junctions," *Electron. Lett.*, vol. 8, pp. 177–179, 1972.

33. I. Wolff, G. Kompa, and R. Mehran, "Streifenleitungsdiskontinuitäten und Verzweigungen," *Nachrichtentechn. Z.*, vol. 25, pp. 217–224, 1972.

34. A. A. Oliner, "Equivalent circuits for discontinuities in balanced strip transmission line", *IRE Trans. Microwave Theory Tech.*, vol. MTT-3, pp. 134–143, 1955.

35. H. M. Altschuler and A. A. Oliner, "Discontinuities in the center conductor of symmetric strip transmission line," *IRE Trans. Microwave Theory Tech.*, vol. MTT-8, pp. 328–339, 1960.

36. H. M. Altschuler and A. A. Oliner, "Addendum to: Discontinuities in the center conductor of symmetric strip transmission line," *IRE Trans. Microwave Theory Tech.*, vol. MTT-10, p. 143, 1962.

37. A. G. Franco and A. A. Oliner, "Symmetric strip transmission line tee junction," *IRE Trans. Microwave Theory Tech.*, vol. MTT-10, pp. 118–124, 1962.

38. E. O. Hammerstad, "Equations for microstrip circuit design," *Proc. 5th Eur. Microwave Conf.*, pp. 268–272, 1975.

39. E. Hammerstad and O. Jensen, "Accurate models for microstrip computer-aided design," *1980 IEEE MTT-S Int. Microwave Symp. Dig.*, pp. 407–409, 1980.

40. R. H. Jansen and M. Kirschning, "Arguments and accurate mathematical model for the power-current formulation of microstrip characteristic impedance," *Arch. Elektron, Übertragungstech.*, vol. 37, pp. 18–112, 1983.

41. E. Yamashita, K. Atsuki, and T. Hirahata, "Microstrip dispersion in a wide frequency range," *IEEE Trans. Microwave Theory Tech.*, vol. MTT-29, pp. 610–611, 1981.

42. R. Mehran, "Die frequenzabhängigen Übertragungseigenschaften von Streifenleitungs-T-Verzweigungen und 90°-Winkeln," Doctoral Thesis, Technical University of Aachen, Aachen, West Germany, 1974.

43. R. Mehran, *Grundelemente des rechnergestützten Entwurfs von Mikrostreifenleitungs-Schaltungen*, Verlag H. Wolff, Aachen, West Germany, 1984.

44. G. Kompa, "Die frequenzabhängigen Übertragungseigenschaften von Streifenleitungs-Wellenwiderstadssprüngen, Filtern und Stichleitungen in Mikrostrip-Technik," Doctoral Thesis, Technical University of Aachen, Aachen, West Germany, 1974.

45. I. Wolff, "Computer aided design of microstrip power dividers," *Proc. 3rd Eur. Microwave Conf.*, Paper A.12.5, 1973.

46. I. Wolff, *Einführung in die Mikrostrip-Leitungstechnik*, 2nd ed., Verlag H. Wolff, Aachen, 1978.

47. E. Hammerstad, "Computer-aided design of microstrip couplers with accurate discontinuity models," *IEEE MTT-S Int. Microwave Symp. Dig.*, pp. 54–56, 1980.

48. G. Kompa, "S-matrix computation of microstrip discontinuities with a planar

waveguide model," *Arch. Elektron, Übertragungstechn.*, vol. 30, pp. 58–64, 1976.

49. W. Menzel, "Die frequenzabhängigen Übertragungseigenschaften von unsymmetrischen Kreuz- und T-Verzweigungen, Y-Verzweigungen sowie 90°- und 120°-Winkeln in Mikrostreifenleitungstechnik," Doctoral Thesis, University of Duisburg, Duisburg, Federal Republic of Germany, 1978.
50. R. Mehran, "The frequency-dependent scattering matrix of microstrip right angle bends, T-junctions and crossings," *Arch. Elektron. Übertragungstech.*, vol. 29, pp. 454–460, 1975.
51. G. Kompa, "Frequency-dependent behaviour of microstrip offset junctions," *Electron. Lett.*, vol. 11, pp. 537–538, 1975.
52. R. Mehran, "Frequency dependent equivalent circuits for microstrip right-angle bends, T-junctions and crossings," *Arch. Elektron. Übertragungstech.*, vol. 30, pp. 80–82, 1976.
53. G. Kompa, "Reduced coupling aperture of microstrip stubs provides new aspects in the stub filter design," *Proc. 6th Eur. Microwave Conf.*, pp. 39–43, 1976.
54. W. Menzel and I. Wolff, "A method for calculating the frequency dependent properties of microstrip discontinuities," *IEEE Trans. Microwave Theory Tech.*, vol. MTT-25, pp. 107–112, 1977.
55. E. Kühn, "A mode-matching method for solving field problems in waveguide and resonator circuits," *Arch. Elektron Übertragungstech.*, vol. 27, pp. 511–513, 1973.
56. R. Mehran, "The frequency dependent scattering matrix of twofold truncated microstrip bends," *Arch. Elektron. Übertragungstechn.*, vol. 31, pp. 411–415, 1977.
57. W. Menzel, "Calculation of inhomogeneous microstrip lines," *Electron. Lett.*, vol. 13, pp. 183–184, 1977.
58. R. Mehran, "Calculation of microstrip bends and Y-junctions with arbitrary angle," *IEEE Trans. Microwave Theory Tech.*, vol. MTT-26, pp. 400–405, 1978.
59. G. Kompa, "Design of stepped microstrip components," *Radio Electron. Eng.*, vol. 48, pp. 53–63, 1978.
60. R. Mehran, "Method of analysis of some microwave planar networks," *1979 IEEE MTT-S Int. Microwave Symp. Dig.*, pp. 575–577, 1979.
61. W. Menzel, "The frequency dependent transmission properties of microstrip Y-junctions and 120°-bends," *IEE J. Microwaves, Opt. Acoust.*, vol. 2, pp. 55–59, 1978.
62. R. Mehran, "Computer-aided design of microstrip filters considering dispersion, loss and discontinuity effects," *IEEE Trans. Microwave Theory Tech.*, vol. MTT-26, pp. 239–245, 1979.
63. R. Mehran, "Some annular structures in microwave planar networks and their filtering properties," *Arch. Elektron, Übertragungstech.*, vol. 33, pp. 265–268, 1979.
64. G. Kompa and R. Mehran, "Microstrip filter analysis using a waveguide model," *Radio Electron. Eng.*, vol. 50, pp. 54–58, 1980.
65. T. S. Chu and T. Itoh, "Generalized scattering matrix method for analysis of

cascaded and offest microstrip step discontinuities," *Trans. Microwave Theory Tech.*, vol. MTT-34, pp. 280–284, 1986.

66. I. Wolff, "Statische Kapazitäten von rechteckigen und kreisförmigen Mikrostrip-Schiebenkondensatoren," *Arch. Elektron. Übertragungstech.*, vol. 27, pp. 44–47, 1973.
67. I. Wolff and N. Knoppik, "Rectangular and circular microstrip disc capacitors and resonator," *IEEE Trans. Microwave Theory Tech.*, vol. MTT-22, pp. 857–864, 1974.
68. R. Mittra and S. W. Lee, *Analytical Techniques in the Theory of Guided Waves*, Chapter I, Macmillan, New York, 1971.
69. S. W. Lee, W. R. Jones, and J. J. Campbell, "Convergence of numerical solutions of iris-type discontinuity problems," *IEEE Trans. Microwave Theory Tech.*, vol. MTT-19, pp. 528–536, 1971.
70. E. O. Hammerstad and F. Bekkadal, *Microstrip Handbook*, ELAB Rep. STF 44 A 74169, University of Trondheim, 1975.
71. M. Kirschning, R. H. Jansen, and N. H. L. Koster, "An accurate model for the open end effect of microstrip line," *Electron. Lett.*, vol. 17, pp. 123–125, 1981.

8

The Transmission Line Matrix (TLM) Method

Wolfgang J. R. Hoefer
Laboratory for Electromagnetics and Microwaves
Department of Electrical Engineering
University of Ottawa
Ottawa, Ontario, Canada

1. INTRODUCTION

This chapter is devoted to a numerical technique suitable for solving electromagnetic problems of the most general kind. Such problems may involve nonlinear, inhomogeneous, anisotropic, time-dependent material properties and arbitrary geometries. It is known as the transmission line matrix or TLM method. Developed by P. B. Johns and his co-workers in the early 1970s, this method represents a computer simulation of electromagnetic fields in three-dimensional space and time. Its theoretical foundations and its most important variations are described, and the basic algorithms for simulating the propagation of fields in unbounded and bounded space are derived in detail. Sources and types of errors are discussed, and possible pitfalls are pointed out. A description of typical applications to microwave and millimeter-wave structures is also included. The conclusion of this chapter summarizes the advantages and disadvantages of the TLM method and indicates under what circumstances it is appropriate to select the TLM method rather than other numerical techniques for solving a particular problem. A two-dimensional TLM computer program for implementation on a personal computer is presented in the appendix, together with a typical application example.

With the exception of a few waveguides with simple cross sections, most modern microwave and millimeter-wave structures are not amenable to an analysis in closed form. Planar and quasi-planar circuits especially involve singularities in the field or potential functions that cannot be handled with reasonable analytical effort. It therefore becomes necessary to employ some form of numerical technique. But which technique or techniques will be the most appropriate for a given problem?

To answer this question, the circuit designer must understand the capabilities and limitations of the various available methods. The objective of this book is to provide the basis for this understanding, and this chapter illustrates the potential of the TLM method in this context.

All field solution techniques, whether analytical, graphical, or numerical, serve the same purpose, namely to find solutions of Maxwell's equations that satisfy the boundary conditions and material equations imposed by the structure under study. In their most general formulation, Maxwell's equations are four-dimensional; to solve them, one must integrate with respect to one time and three space variables. To underline the four-dimensional nature of electromagnetic fields, one can express Maxwell's equations in one single tensor equation in four-dimensional space. The general electromagnetic problem thus amounts to solving such a four-dimensional equation under specific boundary conditions.

A number of very effective numerical methods for analyzing fields in microwave and millimeter-wave structures are presented in this book. They all require some kind of discretization, either directly in space and time or in some other domain (most frequently the Fourier transform domain). In most practical problems the time dependency can be removed by Fourier transform into the frequency domain to obtain time-harmonic solutions. Furthermore, if the structure under study features only regularly shaped homogeneous subregions, where all interfaces and boundaries are parallel to the coordinate surfaces (MIC structures with multilayer stratified dielectric and thin conductor pattern), a Fourier transformation with respect to space coordinates can further reduce the number of space variables (see the spectral domain technique, Chapter 5).

There is no doubt that for a large majority of passive structures, time-harmonic and spectral domain solution methods lead to fast and efficient computer programs. Considerable analytical preprocessing reduces the numerical effort and simplifies the data input and output procedures. However, these methods have the following shortcomings:

1. Since they are based on the principle of superposition, they cannot be applied very easily to nonlinear problems, i.e., structures with electromagnetic characteristics that depend on the amplitude of the applied fields.
2. Since they operate in the frequency domain, they cannot directly account for time-dependent properties of structures, i.e., variations in material properties due to temperature changes.
3. If they employ a spatial Fourier transform, they cannot be applied to structures of complex, irregular cross section.

It is in such circumstances that time-domain methods, such as the TLM method, deploy their full potential and become invaluable tools for the microwave and millimeter-wave engineer.

2. HISTORICAL BACKGROUND

Two distinct models describing the phenomenon of light were developed in the seventeenth century: the corpuscular model by Isaac Newton and the wave model by Christian Huygens [1]. At the time of their conception, these models were considered incompatible. However, modern quantum theory has demonstrated that light in particular, and electromagnetic radiation in general, possess both granular (photons) and wave properties. These aspects are complementary, and one or the other usually dominates, depending on the phenomenon under study.

At microwave and millimeter-wave frequencies, the granular nature of electromagnetic radiation is not very evident, manifesting itself only in certain interactions with matter, while the wave aspect predominates in all situations involving propagation and scattering. This suggests that the model proposed by Huygens, and later refined by Fresnel, could form the basis for a general method of treating microwave propagation and scattering problems, particularly in view of its considerable success in predicting the diffraction and interference of light waves.

Indeed, Johns and Beurle [2] described in 1971 a novel numerical technique for solving two-dimensional scattering problems, which was based on Huygens's model of wave propagation. Inspired by earlier network simulation techniques [3–5], this method employed a Cartesian mesh of open two-wire transmission lines to simulate two-dimensional propagation of delta function impulses. Subsequent papers by Johns and Akhtarzad [6–16] extended the method to three dimensions and included the effect of dielectric loading and losses. Building on the groundwork laid by these original authors, other researchers [17–35] have added various features and improvements such as variable mesh size, condensed nodes, error correction techniques, and extension to anisotropic media.

The following section briefly describes the discretized version of Huygens's wave model, which is suitable for implementation on a digital computer and forms the algorithm of the TLM method. A detailed description of this model can be found in a very interesting paper by Johns [9].

3. HUYGENS'S PRINCIPLE AND ITS DISCRETIZATION

According to Huygens [1], a wave front (Fig. 1) consists of a number of secondary radiators (b) that give rise to spherical wavelets (d). The envelope of these wavelets forms a new wave front, which in turn gives rise to a new generation of spherical wavelets, and so on. In spite of certain difficulties in the mathematical formulation of this mechanism, its application nevertheless leads to an accurate description of wave propagation and scattering, as will be shown below.

In order to implement Huygens's model on a digital computer, one must

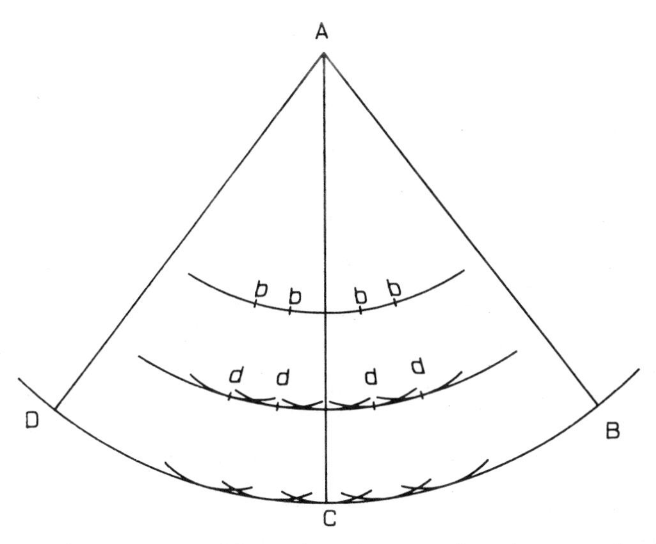

Fig. 1 Huygens's principle and formation of a wave front by secondary wavelets.

formulate it in discretized form. To this end, both space and time are represented in terms of finite elementary units, Δl and Δt, which are related by the velocity of light—such that

$$\Delta t = \Delta l / c \tag{1}$$

Accordingly, two-dimensional space is modeled by a Cartesian matrix of points or nodes, separated by the mesh parameter Δl (see Fig. 2*a*). The unit time Δt is then the time required for an electromagnetic pulse to travel from one node to the next.

Assume that a delta function impulse is incident upon one of the nodes from the negative y direction. The energy in the impulse is unity. In accordance with Huygens's principle, this energy is scattered isotropically in all four directions, each radiated impulse carrying one-fourth of the incident energy. The corresponding field quantities must then be 1/2 in magnitude. Furthermore, the reflection coefficient "seen" by the incident impulse must be negative to ensure field continuity at the node (Fig. 2*b*).

This two-dimensional model has a network analogue in the form of a mesh of orthogonal transmission lines, or a transmission line matrix (Figs. 2*c* and *d*), forming a Cartesian array of shunt nodes that have the same scattering properties as the nodes in Figs. 2*a* and *b*. Figure 2*e* shows the scheme for numbering the node branches in the formulation of the scattering algorithm.

It will be shown in the next section that there is a direct equivalence between the voltages and currents on the line mesh and the electric and magnetic fields of Maxwell's equations [2]. Such a mesh can thus model

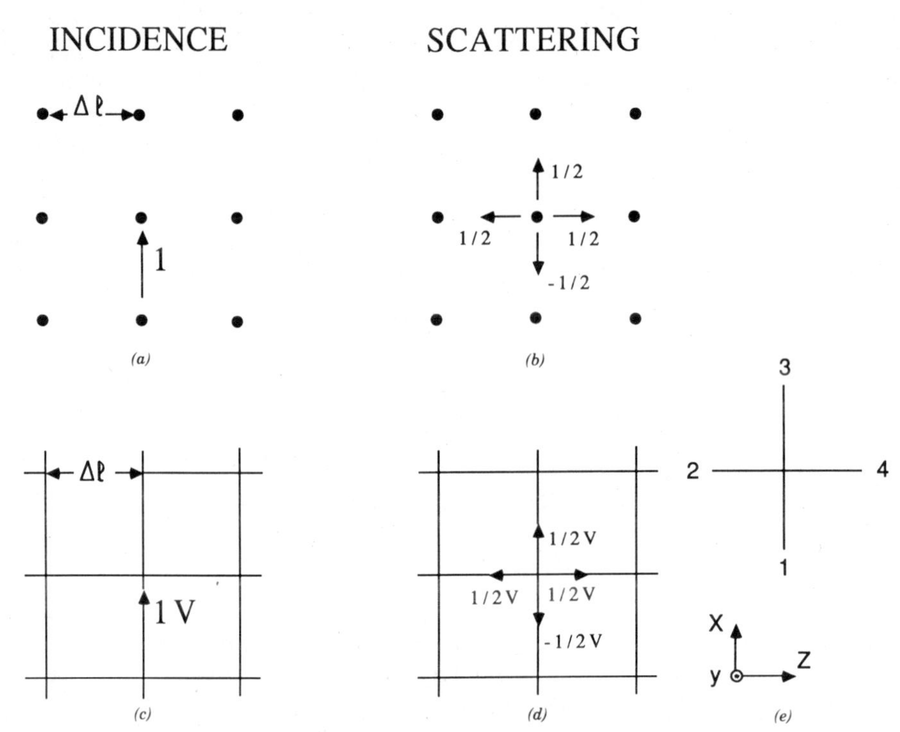

Fig. 2 The discretized Huygens's wave model in two-dimensional space (*a* and *b*), and the equivalent transmission line model (*c*, *d*, and *e*). (*a*) Incidence of a Dirac impulse at a space point. (*b*) Scattering of the Dirac impulse. (*c*) Incidence of a voltage impulse at a node in the equivalent Cartesian mesh of transmission lines. (*d*) Scattering of the voltage impulse at the node. (*e*) Coordinates and numbering scheme for node branches. (After Hoefer [36].) Copyright © 1985 IEEE.

two-dimensional wave problems by the procedure called the two-dimensional TLM method.

4. THE TWO-DIMENSIONAL TLM METHOD

A considerable number of guided wave problems are two-dimensional. Propagation of TE_{n0} modes in rectangular homogeneous waveguides, including scattering at discontinuities of constant dimensions in the direction of the **E** field, involves only two spatial directions. Examples are inductive strips, irises, and filters made thereof, as well as T junctions, bends, and *n* furcations in the *H* plane. Furthermore, the cutoff frequencies of all modes in inhomogeneously filled cylindrical waveguides of arbitrary cross-sectional geometry can be found using two-dimensional analysis. This includes all planar, quasi-planar, and dielectric waveguiding structures.

The two-dimensional TLM method is therefore of considerable practical importance. Furthermore, the simplicity of the two-dimensional TLM network makes the understanding of the method, its implementation and its limitations relatively easy. This section describes the wave properties of the two-dimensional TLM network and shows how it can model the propagation, reflection, refraction, and attenuation of fields in a two-dimensional structure.

It must be kept in mind that Huygens's principle is a continuous model of wave propagation. The sequence of scattering events occurs in infinitesimally small steps and is thus not frequency- or space-dispersive. In other words, all frequency components in the spectrum of the scattered Dirac impulses travel at the same speed in all directions. In the discrete TLM model, however, this is only true as long as the mesh parameter Δl is small with respect to the wavelength, as will be discussed. It is thus important to study the wave properties of the discrete TLM network in order to assess the limitations of the model and, eventually, to determine and correct errors due to the finite length of the mesh parameter Δl.

4.1. Wave Propagation in Shunt-Connected TLM Networks

4.1.1. Scattering of Dirac Impulses at Mesh Nodes

It was pointed out in Section 3 that the discretized form of Huygens's principle can be represented by the scattering of voltage impulses in a mesh of orthogonal transmission lines. This can easily be verified by considering the local reflection coefficient Γ_i "seen" by an impulse traveling toward a shunt node, as shown in Fig. 2*c*. The three outgoing lines appear in parallel, terminating the incoming line in a normalized impedance of 1/3. Hence

$$\Gamma_i = \frac{1/3 - 1}{1/3 + 1} = -\frac{1}{2} \tag{2}$$

and the transmission coefficient for each outgoing line is

$$T_i = 1 + \Gamma_i = +\frac{1}{2} \tag{3}$$

Thus, if a unit Dirac voltage impulse is incident on a node in the TLM mesh of Fig. 2*c*, it will be scattered in the form of a reflected impulse of $-\frac{1}{2}$ V and three transmitted impulses of $+\frac{1}{2}$ V, as shown in Fig. 2*d*.

The more general case of four impulses being incident on the four branches of a node can be obtained by superposition from the previous case. Hence, if at time $t = k\,\Delta t$, voltage impulses designated by ${}_kV^i_1$, ${}_kV^i_2$, ${}_kV^i_3$, and ${}_kV^i_4$ are incident on lines 1–4, respectively, on any junction node, then the total voltage impulse reflected along line n at time $(k+1)\,\Delta t$ will be

$$
{}_{k+1}V_n^r = \frac{1}{2}\left(\sum_{m=1}^{4} {}_kV_m^i\right) - {}_kV_n^i \tag{4}
$$

The positive integer variable k represents the number of time steps Δt that have passed since the beginning of the computation. It is called the number of iterations.

This situation is conveniently described by a scattering matrix equation [7] relating the reflected voltages at time $(k+1)\,\Delta t$ to the incident voltages at the previous time step $k\,\Delta t$

$$
{}_{k+1}\begin{bmatrix} V_1 \\ V_2 \\ V_3 \\ V_4 \end{bmatrix}^r = \frac{1}{2}\begin{bmatrix} -1 & 1 & 1 & 1 \\ 1 & -1 & 1 & 1 \\ 1 & 1 & -1 & 1 \\ 1 & 1 & 1 & -1 \end{bmatrix} \cdot {}_k\begin{bmatrix} V_1 \\ V_2 \\ V_3 \\ V_4 \end{bmatrix}^i \tag{5}
$$

Furthermore, any impulse emerging from a node at position (z, x) (reflected impulse) automatically becomes an incident impulse on the neighboring node. (See Fig. 3.) Hence

$$
\begin{aligned}
{}_{k+1}V_1^i(z,x) &= {}_{k+1}V_3^r(z,x-1) \qquad & {}_{k+1}V_2^i(z,x) &= {}_{k+1}V_4^r(z-1,x) \\
{}_{k+1}V_3^i(z,x) &= {}_{k+1}V_1^r(z,x+1) \qquad & {}_{k+1}V_4^i(z,x) &= {}_{k+1}V_2^r(z+1,x)
\end{aligned} \tag{6}
$$

where coordinates have been normalized to Δl.

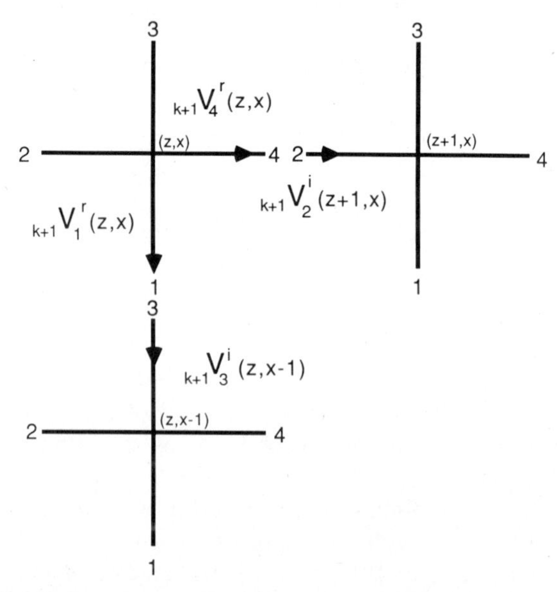

Fig. 3 Transfer of impulses between neighboring nodes.

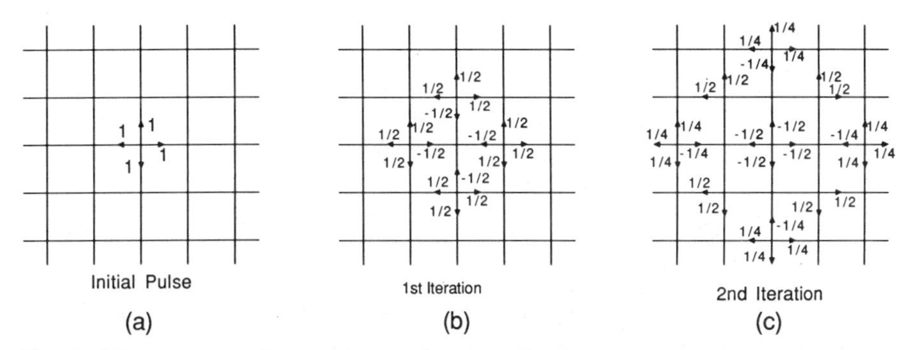

Fig. 4 Three consecutive scatterings in a two-dimensional TLM network excited by a Dirac impulse. (*a*) Initial impulse; (*b*) first iteration; (*c*) second iteration. (After Akhtarzad [13].)

Consequently, if the magnitudes, positions, and directions of all impulses are known at time $k\,\Delta t$, the corresponding values at time $(k+1)\,\Delta t$ can be obtained by operating (5) and (6) on each node in the network. The impulse response of the network is then found by initially fixing the magnitudes, directions, and positions of all impulses at $t = 0$ and then calculating the state of the network at successive time intervals.

The scattering process described above forms the basic algorithm of the TLM method. Three consecutive scatterings are shown in Fig. 4, visualizing the spreading of the injected voltage across the two-dimensional network.

This sequence of events closely resembles the disturbance of a pond due to a falling drop of water. However, there is one obvious difference: the discrete nature of the TLM mesh, which causes dispersion of the velocity of the wave front. In other words, the velocity of a signal component in the mesh depends on its direction of propagation and its frequency. To appreciate the importance of this dispersion, note that the process in Fig. 4 depicts a short episode of the response of the TLM network to a single impulse containing all frequencies. Thus, harmonic solutions to a problem can be obtained from the impulse response via the Fourier transform. Solutions are accurate only at frequencies for which the dispersion can be neglected. This will be discussed next.

4.1.2. Wave Properties of Infinitesimally Fine Meshes

The building block, or unit cell, of the two-dimensional TLM network is the shunt node shown in Fig. 5*a*. It can be approximated by the lumped element model shown in Fig. 5*b*. L and C are the inductance and capacitance per unit length for an individual line. Note that the node capacitance is twice that of an individual line section due to the parallel connection at the node.

As long as the mesh parameter Δl is very small with respect to the shortest wavelength of interest, the size of each mesh element can be

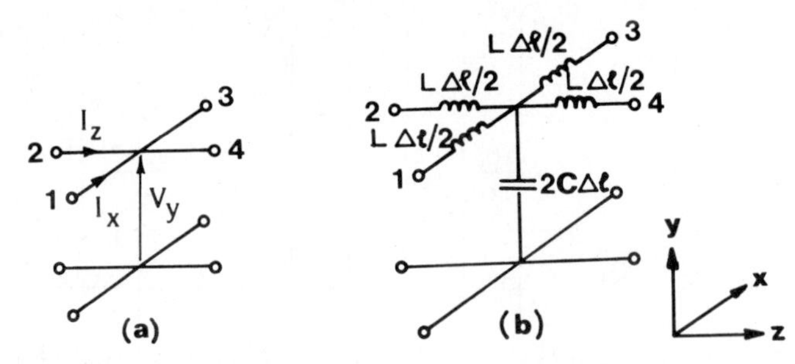

Fig. 5 The building block of the two-dimensional TLM network. (*a*) Shunt node; (*b*) equivalent lumped element model. (After Akhtarzad [13].)

considered infinitesimal, and the voltage and current changes in the x and z directions from one node to the next are

$$\frac{\partial V_y}{\partial x} = -L\frac{\partial I_x}{\partial t} \qquad \frac{\partial V_y}{\partial z} = -L\frac{\partial I_z}{\partial t} \qquad \frac{\partial I_z}{\partial z} + \frac{\partial I_x}{\partial x} = -2C\frac{\partial V_y}{\partial t} \tag{7}$$

These expressions can be combined to yield the two-dimensional wave equation

$$\frac{\partial^2 V_y}{\partial x^2} + \frac{\partial^2 V_y}{\partial z^2} = 2LC\frac{\partial^2 V_y}{\partial t^2} \tag{8}$$

Hence, the two-dimensional TLM network variables can simulate any phenomenon governed by the general two-dimensional wave equation in Cartesian coordinates u, v,

$$\frac{\partial^2 \Phi}{\partial u^2} + \frac{\partial^2 \Phi}{\partial v^2} = \mu\varepsilon\frac{\partial^2 \Phi}{\partial t^2} \tag{9}$$

Comparing, for instance, these relations with the expansions of Maxwell's equations for $\partial/\partial y = 0$, and $E_x = E_z = H_y = 0$ (which describe the TE_{n0} modes in a rectangular waveguide, z being the longitudinal direction),

$$\frac{\partial E_y}{\partial x} = -\mu\frac{\partial H_z}{\partial t} \qquad \frac{\partial E_y}{\partial z} = +\mu\frac{\partial H_x}{\partial t} \qquad \frac{\partial H_x}{\partial z} - \frac{\partial H_z}{\partial x} = +\varepsilon\frac{\partial E_y}{\partial t} \tag{10}$$

and

$$\frac{\partial^2 E_y}{\partial x^2} + \frac{\partial^2 E_y}{\partial z^2} = \mu\varepsilon\frac{\partial^2 E_y}{\partial t^2} \tag{11}$$

the following equivalences between field and TLM mesh parameters are seen to exist:

$$E_y \equiv V_y \qquad H_z \equiv I_x \qquad H_x \equiv -I_z \qquad \mu \equiv L \qquad \varepsilon \equiv 2C \tag{12}$$

For elementary transmission lines in the TLM network, and for $\mu_r = \varepsilon_r = 1$, the inductance and capacitance per unit length are related by

$$1/\sqrt{LC} = 1/\sqrt{\mu_0 \varepsilon_0} = c \tag{13}$$

where c is the velocity of light in free space.

However, the velocity term in the wave equation (8) for the two-dimensional TLM mesh is $1/\sqrt{2LC} = c/\sqrt{2}$. Hence, if voltage and current waves on each transmission line component travel at the speed of light, the complete network of intersecting transmission lines represents a medium of relative permittivity twice that of free space. This means that as long as the equivalent circuit in Fig. 5 is valid, the propagation velocity in the TLM mesh is $1/\sqrt{2}$ times the velocity of light.

The field equations (10) and (11) also govern all TM modes at cutoff in uniform waveguides of arbitrary cross section or in homogeneous subregions thereof. The TLM network voltage then describes the longitudinal electric field, while the network currents model the transverse magnetic field. In addition, the dual nature of electric and magnetic fields allows us to simulate with the same network the behavior of TE modes at cutoff. In this case, the network voltage represents the longitudinal magnetic field, while the network currents correspond to the transverse electric field components. However, care must be taken to extend the duality to boundary conditions and dielectric properties, as will be discussed in Section 4.2.

4.1.3. Wave Properties of Coarse Meshes

The analogy between field and mesh parameters is perfect as long as the mesh is extremely fine compared with the wavelength. However, the use of very small Δl values leads to prohibitive memory and CPU time requirements in a computer simulation. On the other hand, if the cell size is increased and approaches the order of a wavelength, the TLM mesh can no longer be considered as a continuum; it must be treated as an anisotropic, periodic structure. It is thus important to evaluate the slow-wave properties of the mesh in order to assess its limitations as a continuous model.

First, consider the *diagonal* propagation of a plane wave front at 45° with respect to the mesh axes (Fig. 6*a*). The constituent waves on the branches have identical amplitudes and phases in both axial directions. Two such identical waves converging toward a network node always "see" a match, whatever their wavelength. Therefore, the propagation velocity in the mesh is independent of frequency and equal to $1/\sqrt{2}$ of the phase velocity on the mesh lines.

The *axial* propagation of a plane wave front in the x or z direction, however, is frequency-dispersive. For the purpose of analysis, assume that waves are traveling in the positive z direction and are invariant in the x and

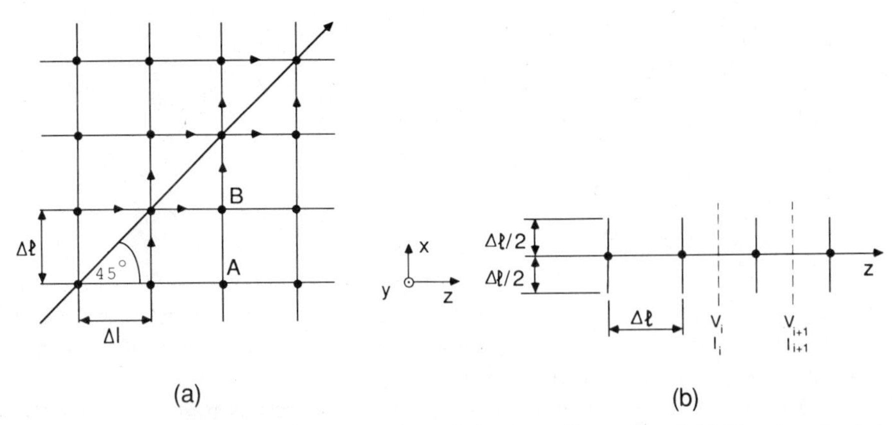

Fig. 6 The slow wave characteristics of the two-dimensional TLM network. (*a*) Diagonal propagation of a plane wave front; (*b*) axial propagation of a plane wave front. (After Akhtarzad [13].)

y directions. Under these conditions, any wave component traveling transversely along a mesh line from A to B must, by symmetry, be accompanied by a similar wave traveling from B to A. It follows that a wave traveling over the matrix can be represented by the passage of a wave down a transmission line having open-circuited stubs of length $\Delta l/2$, as shown in Fig. 6*b*.

To analyze such a periodic structure, we divide it into individual cells and express the voltage and current at the output of one cell in terms of the voltage and current at the input. For a given ratio $\Delta l/\lambda = \theta/2\pi$, where λ is the wavelength on the mesh lines, Johns and Beurle [2] calculate the following transmission equation for one cell:

$$\begin{bmatrix} V_i \\ I_i \end{bmatrix} = \begin{bmatrix} \cos\theta/2 & j\sin\theta/2 \\ j\sin\theta/2 & \cos\theta/2 \end{bmatrix} \begin{bmatrix} 1 & 0 \\ 2j\tan\theta/2 & 1 \end{bmatrix} \cdot \begin{bmatrix} \cos\theta/2 & j\sin\theta/2 \\ j\sin\theta/2 & \cos\theta/2 \end{bmatrix} \begin{bmatrix} V_{i+1} \\ I_{i+1} \end{bmatrix} \tag{14}$$

If the waves on the periodic structure have a propagation constant $\gamma_n = \alpha_n + j\beta_n$, then

$$\begin{bmatrix} V_i \\ I_i \end{bmatrix} = \begin{bmatrix} e^{\gamma_n \Delta l} & 0 \\ 0 & e^{\gamma_n \Delta l} \end{bmatrix} \begin{bmatrix} V_{i+1} \\ I_{i+1} \end{bmatrix} \tag{15}$$

The solution of eqs. (14) and (15) gives

$$\cosh\gamma_n \Delta l = \cos\theta - \tan\frac{\theta}{2}\sin\theta \tag{16}$$

For the lowest frequency propagation region, $\gamma_n = j\beta_n$, and eq. (16) reduces to

$$\sin\frac{\beta_n\,\Delta l}{2} = \sqrt{2}\sin\frac{\omega\,\Delta l}{2c} = \sqrt{2}\sin\frac{\beta\,\Delta l}{2} \tag{17a}$$

which can also be written in the following form:

$$\frac{\beta}{\beta_n} = \frac{\pi\,\Delta l/\lambda}{\sin^{-1}[\sqrt{2}\sin(\pi\,\Delta l/\lambda)]} \tag{17b}$$

where β and β_n are the propagation constants on the mesh lines and in the network, respectively. The resulting ratio of velocities $v_n/c = \omega/(\beta_n c) = \beta/\beta_n$, is plotted in Fig. 7 as a function of the mesh parameter Δl, normalized to the wavelength on the mesh lines ($\Delta l/\lambda$). Note that λ is equal to the free-space wavelength as long as eq. (13) applies. It appears that the first cutoff occurs at $\Delta l/\lambda = \frac{1}{4}$. At very low frequencies, i.e., small $\Delta l/\lambda$ values, however, the propagation velocity in the mesh approximates to $c/\sqrt{2}$, as shown earlier.

For *arbitrary* directions of propagation in the mesh, the velocity ratio lies somewhere between the curves for axial and diagonal propagation. To

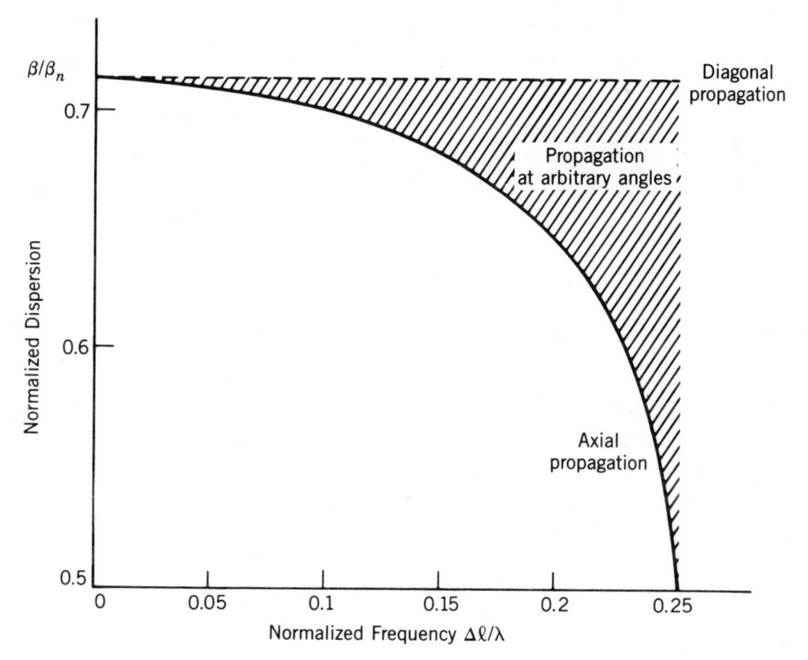

Fig. 7 Dispersion of the velocity of waves in a two-dimensional TLM network. (After Akhtarzad [13].)

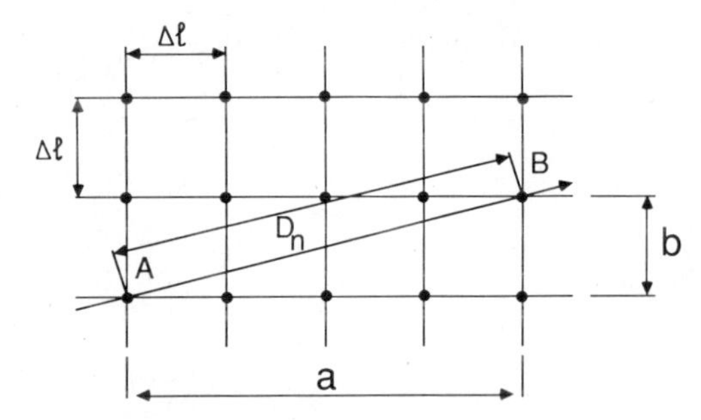

Fig. 8 Scheme for estimating the velocity of a plane wave in arbitrary direction in the two-dimensional TLM network. (After Saguet [17].)

estimate v_n/c for arbitrary propagation angles, Saguet [17] has suggested the following modification of eq. (17) (see Fig. 8).

If D_n is the distance covered by the wave front going directly from A to B, and $D = a + b$ is the minimal distance between these points along mesh lines, the following modified dispersion relation applies:

$$\frac{\beta}{\beta_n} = \frac{d_n \pi \, \Delta l/\lambda}{\sin^{-1}[d_n\sqrt{2}\sin(\pi \, \Delta l/\lambda)]} \tag{18}$$

where $d_n = D_n/D$. It can be easily verified that eq. (18) includes the limiting cases of axial ($d_n = 1$) and diagonal ($d_n = 1/\sqrt{2}$) propagation.

In conclusion, the TLM network simulates isotropic propagation only as long as all frequencies are well below the network cutoff frequency, in which case the network propagation velocity may be considered equal to $c/\sqrt{2}$ in all directions. However, when the propagation is predominantly in one direction, eqs. (17) and (18) can be taken into account to reduce errors due to velocity dispersion, thus extending the method to higher frequencies or reducing the mesh density and hence computational expenditure.

4.2. Wave Propagation in Series-Connected TLM Networks

In accordance with Babinet's principle, which is based on the dual nature of the electric and magnetic fields, voltage and current, and impedance and admittance, the same wave properties can be modeled by a series-connected mesh of transmission lines as shown in Fig. 9.

Using the same ideas and definitions as in Section 4.1.1, one obtains the following voltage scattering matrix for a series node:

$${}_{k+1}\begin{bmatrix} V_1 \\ V_2 \\ V_3 \\ V_4 \end{bmatrix}^r = \frac{1}{2}\begin{bmatrix} 1 & 1 & 1 & -1 \\ 1 & 1 & -1 & 1 \\ 1 & -1 & 1 & 1 \\ -1 & 1 & 1 & 1 \end{bmatrix} \cdot {}_{k}\begin{bmatrix} V_1 \\ V_2 \\ V_3 \\ V_4 \end{bmatrix}^i \tag{19}$$

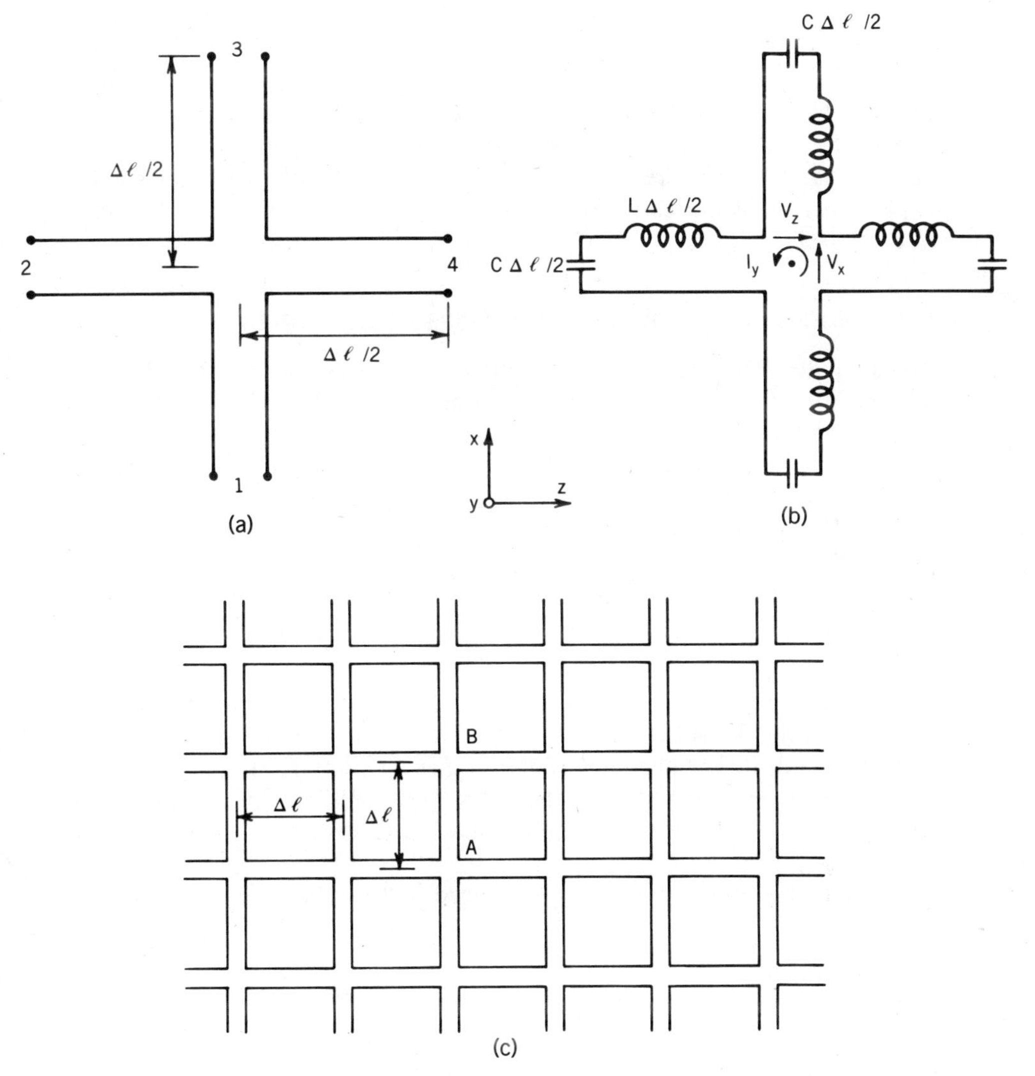

Fig. 9 The series node and the configuration of a two-dimensional TLM series network. (*a*) Series node topology; (*b*) equivalent node network; (*c*) two-dimensional series mesh. (After Akhtarzad [13].)

The transmission line equations are in this case

$$\frac{\partial I_y}{\partial x} = -C\,\frac{\partial V_z}{\partial t} \qquad \frac{\partial I_y}{\partial z} = -C\,\frac{\partial V_x}{\partial t} \qquad \frac{\partial V_x}{\partial z} + \frac{\partial V_z}{\partial x} = -2L\,\frac{\partial I_y}{\partial t} \tag{20}$$

If voltages in the TLM mesh represent E fields in a medium, then the field equations solved by the series node are

$$\frac{\partial H_y}{\partial x} = +\varepsilon \frac{\partial E_z}{\partial t} \qquad \frac{\partial H_y}{\partial z} = -\varepsilon \frac{\partial E_x}{\partial t} \qquad \frac{\partial E_x}{\partial z} - \frac{\partial E_z}{\partial x} = -\mu \frac{\partial H_y}{\partial t} \tag{21}$$

and here the permeability of the simulated space is twice that of free space.

For the solution of two-dimensional problems, only one of the two network types is required. However, the treatment of three-dimensional fields calls for a combination of series and shunt nodes, as will be described in Section 5.

4.3. Modeling of Lossless and Lossy Boundaries

To simulate electromagnetic structures, one must model not only the propagation of fields, but also their total or partial reflections at boundaries. The equivalence between fields and TLM mesh voltages or currents suggests that this could be achieved by introducing suitable reflection coefficients in the mesh. The value of these reflection coefficients depends on the nature of both the boundaries and the simulated field component.

4.3.1. *Lossless Boundaries*

Consider a shunt-connected two-dimensional mesh in which the voltage V_y simulates an electric field. In this case, electric and magnetic walls are represented by short and open circuits, respectively, at the appropriate positions in the TLM mesh (Fig. 10*a*). However, if the voltage simulates a magnetic field, then a magnetic wall becomes a short circuit, and an electric wall becomes an open circuit in the mesh. Generally speaking, the physical nature of the reflection must be the same in the structure and in the TLM model.

In all cases, a boundary must be placed halfway between two nodes so

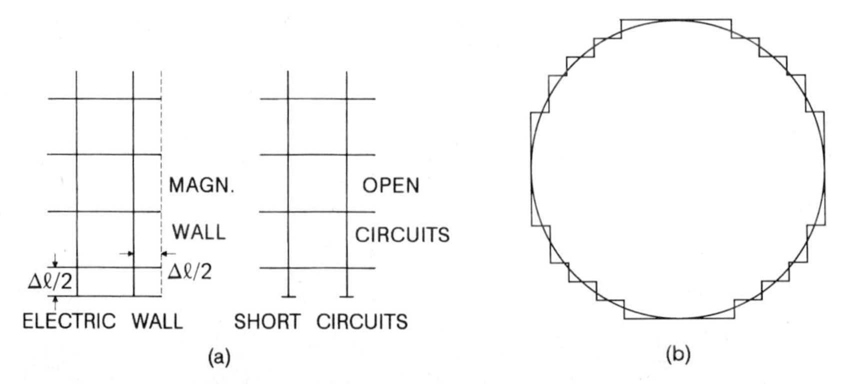

Fig. 10 Representation of boundaries in the TLM mesh. (*a*) Electric and magnetic walls in a shunt-type mesh in which V_y simulates E_y. (*b*) Approximation of a curved boundary by a piecewise straight boundary.

that reflected impulses reach the boundary nodes in synchronism with the other impulses in the network. In practice, this is achieved by making the mesh parameter Δl an integer fraction of the structure dimensions. Curved walls can be approximated by piecewise straight boundaries as shown in Fig. 10*b*.

In the computation, the lossless reflection of an impulse is achieved by returning it, after one unit time step Δt, with equal (open circuit) or opposite sign (short circuit) to its boundary node of origin. Figure 11 shows three boundaries A, B, C in a two-dimensional shunt-connected TLM network. Note that the mesh extends to one node beyond the boundaries, so that for each boundary we have a row of inner and outer boundary nodes.

Let boundary A be a perfect magnetic wall. The reflection coefficient $+1$ is then modeled by making the computation

$$ {}_kV_3^i(i, j-1) = {}_kV_1^r(i, j) = {}_kV_3^r(i, j-1) \tag{22} $$

for each external node ($z = i$, $x = j$). This ensures that at each iteration the impulses transmitted toward the boundary at all nodes on the line $x = j - 1$ are returned with equal magnitude and phase. Similarly, if boundary B is an electric wall, we write for the external node at ($z = l$, $x = m$):

$$ {}_kV_1^i(l, m+1) = {}_kV_3^r(l, m) = -{}_kV_1^r(l, m+1) \tag{23} $$

thus simulating a reflection coefficient of -1.

Ideal electric and magnetic walls are especially useful in reducing computational expenditure when the structure under study possesses some symmetry. The cross sections of many transmission line structures can be

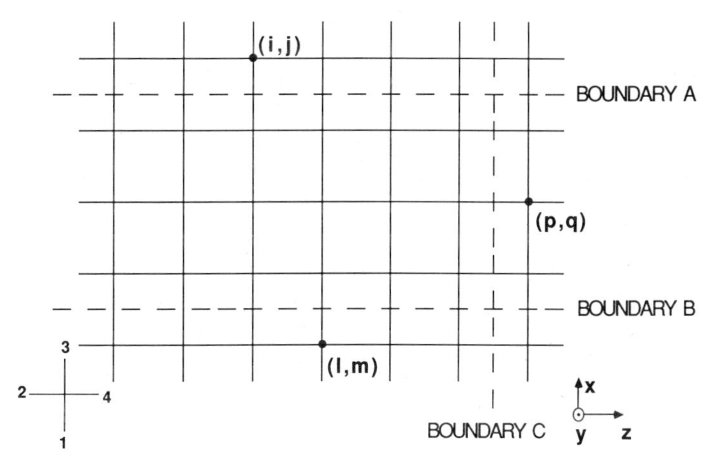

Fig. 11 Modeling of general boundaries in a two-dimensional shunt-connected TLM mesh. (After Akhtarzad [13].)

divided into two or even four symmetrical subregions by appropriate walls. However, by solving the field problem in a single subregion, only modes with the corresponding symmetry will be found.

4.3.2. Lossy Boundaries

Let C be a general boundary of surface impedance Z_c. The boundary reflection coefficient is then

$$\rho = \frac{Z_c - Z_0}{Z_c + Z_0} \tag{24}$$

where $Z_0 = \sqrt{\mu_0/\varepsilon_0}$ is the characteristic impedance of the mesh lines. For the boundary nodes such as $(z = p, x = q)$ we write

$$_kV_4^i(p-1, q) = {_kV_2^r}(p, q) = \rho \, {_kV_4^r}(p-1, q) \tag{25}$$

This formalism allows us to represent a variety of boundary conditions as long as the reflection coefficient ρ can be considered real. The resulting complex reflection coefficient would alter the shape of the Dirac impulses, which cannot be accounted for in the TLM method. Some typical examples of real impedance walls follow.

Free-Space Discontinuities. Assume that boundaries A and B are perfect electric walls. The mesh then simulates TE_{m0} modes in a rectangular waveguide. If we terminate the TLM mesh lines at boundary C with $Z_c = Z_0 = \sqrt{\mu_0/\varepsilon_0}$, making $\rho = 0$ in (25), we effectively simulate a free-space discontinuity (open waveguide radiating into an infinite parallel waveguide) at all frequencies.

Narrowband Matched Load. If in the same structure we terminate the mesh lines at C with $Z_c = Z_0\lambda_g/\lambda$ (the dispersive wave impedance in the guide for a given TE_{m0} mode), we simulate a matched load for the appropriate mode and frequency. A broadband match for all modes can be simulated only by a continuation of the mesh beyond C, in which losses are gradually introduced as in a real waveguide termination. Introduction of losses in the mesh will be described below.

Lossy Waveguide Walls. The wave impedance inside an imperfect conductor is

$$Z_c = \sqrt{\frac{\mu\omega}{2\sigma}}\,(1+j) \tag{26}$$

μ and σ are the permeability and conductivity of the medium that can be modeled by a lossy mesh (see next section). If the mesh parameter Δl is small enough, the penetration of the fields into the conductor and the resulting complex reflection are properly described. In practice, however, waveguide walls are highly conductive; the penetration depth is thus very small, and it is more convenient to represent the boundary by a surface of impedance Z_c. Fortunately, Z_c is usually much smaller than Z_0, and the imaginary part of ρ can be neglected. Hence

$$\rho \sim -1 + 2\sqrt{\frac{\mu\omega\varepsilon_0}{2\sigma\mu_0}} \tag{27}$$

Note that since ρ depends on the frequency ω, the loss calculations are accurate only for that frequency that has been selected in determining ρ. The field solution at that frequency can be extracted from the impulse response through Fourier transform, as will be discussed in Section 4.5.

4.4. Modeling of Lossless and Lossy Materials

So far it has been assumed that the propagation medium was lossless, homogeneous space. However, most practical structures, such as planar and quasi-planar transmission media, contain lossy dielectric and/or magnetic materials. This section describes how lossless and lossy inhomogeneous structures can be represented in a two-dimensional mesh.

4.4.1. Lossless Homogeneous Materials

Any lossless, homogeneous, and isotropic material can be modeled without any modification of the TLM algorithm described above. Simply remember that the phase velocities in the TLM network (v_n) and in the material (v_m) have the ratio

$$\frac{v_n}{v_m} = \frac{v_n}{c}\sqrt{\varepsilon_r\mu_r} \tag{28}$$

where v_n/c is given in Fig. 8 and has a low-frequency value of $1/\sqrt{2}$. ε_r and μ_r are the relative permittivity and permeability of the material modeled by the TLM mesh.

4.4.2. Lossy Homogeneous Materials

Lossy materials can be simulated by introducing losses into the TLM mesh. This can be done in two ways. Either the losses are distributed continuously along the mesh lines or they are introduced at the nodes in the form of a lumped resistive element or a stub terminated in its characteristic impedance (loss stub), with the lines themselves being lossless.

In the first case, the amplitude of impulses traveling from one node to the next is reduced by a factor $e^{-\alpha\,\Delta l}$, where α is the attenuation constant of the lossy mesh lines. In the second case, energy is extracted at the mesh nodes at each scattering event.

The first method is particularly suitable for simulating losses in homogeneous media because the impulse response can then be obtained simply by modifying the response obtained under loss-free conditions.

The second method is better suited for inhomogeneous structures; it will be described in Section 4.4.3.

To relate the attenuation constant α of the mesh lines to the loss tangent of the simulated material, we rewrite the third line of eqs. (7) and (10) in time-harmonic form and include a loss term. For the lossy mesh we have

$$\frac{\partial I_z}{\partial z} + \frac{\partial I_x}{\partial x} = -j\omega 2\left(C + \frac{G}{j\omega}\right)V_y \tag{29}$$

In the lossy medium, the corresponding expression is

$$\frac{\partial H_x}{\partial z} - \frac{\partial H_z}{\partial x} = j\omega\left(\varepsilon + \frac{\sigma}{j\omega}\right)E_y \tag{30}$$

It appears that a conductance G per unit length of the mesh lines represents a conductivity of $\sigma = 2G$ in the medium. Hence, if the complex propagation constant on the individual mesh lines is

$$\gamma = j\omega\left[LC\left(1 + \frac{G}{j\omega C}\right)\right]^{1/2} \tag{31}$$

and the propagation constant of TEM waves in the medium is

$$\gamma_n = j\omega\left[\mu\varepsilon\left(1 + \frac{\sigma}{j\omega\varepsilon}\right)\right]^{1/2} \tag{32}$$

then the equivalence of eqs. (29) and (30) implies that

$$\gamma_n = \sqrt{2}\gamma \tag{33}$$

This equality holds only as long as the mesh size is much smaller than the wavelength. For higher frequencies, we must analyze the wave properties of the mesh exactly as in Section 4.1.2., with γ and γ_n being complex this time. Since the TLM method cannot account for complex line impedances that would distort the Dirac impulses, we must restrict our analysis to cases for which α, $\alpha_n \ll 1$. We then obtain for axial propagation

$$\frac{\beta}{\beta_n} = \frac{\pi\,\Delta l/\lambda}{\sin^{-1}[\sqrt{2}\sin(\pi\,\Delta l/\lambda)]} \tag{34}$$

$$\frac{\alpha}{\alpha_n} = \frac{[\cos(2\pi\ \Delta l/\lambda)]^{1/2}}{\sqrt{2}\cos(\pi\ \Delta l/\lambda)} \tag{35}$$

where $\beta = 2\pi/\lambda$.

The ratio of propagation constants (34) is the same as in the lossless case [eq. (17) and Fig. 7]. Both α/α_n and β/β_n are plotted in Fig. 12 versus the normalized frequency $\Delta l/\lambda$. The velocity dispersion error is larger for the attenuation constant ratio. Both ratios depend on the direction of propagation, becoming worst in the axial and least in the diagonal direction. Clearly, as long as the wavelength is large compared with Δl, eq. (33) holds.

This computation of lossy structures thus differs from the lossless case in two respects. First, the amplitude of the impulses is reduced at each iteration to account for the losses in the mesh lines. Second, the time interval between iterations must be increased slightly, because the phase constant increases due to the attenuation. Consider the following expressions for the real and imaginary parts of the propagation constant on the individual mesh lines [eq. (31)]:

$$\alpha = \omega\sqrt{\frac{LC}{2}}\left[-1 + \sqrt{1 + \left(\frac{G}{\omega C}\right)^2}\right]^{1/2} \tag{36}$$

$$\beta = \omega\sqrt{\frac{LC}{2}}\left[1 + \sqrt{1 + \left(\frac{G}{\omega C}\right)^2}\right]^{1/2} \tag{37}$$

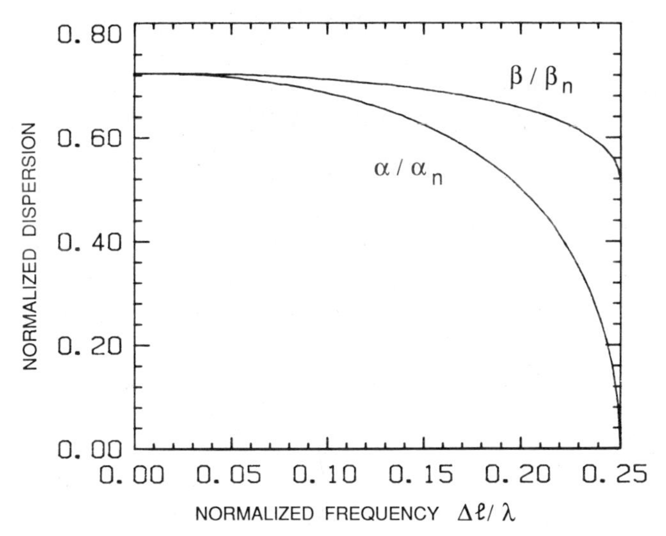

Fig. 12 Dispersion of the attenuation and phase constants of plane waves traveling along a main axis in a lossy two-dimensional TLM mesh. (After Akhtarzad [13].)

During their transit from one node to the next, the impulses are thus attenuated by a factor $e^{-\alpha \Delta l}$, where

$$\alpha \Delta l = \sqrt{2}\, \pi \frac{\Delta l}{\lambda} [-1 + \sqrt{1 + (\tan \delta)^2}]^{1/2} \tag{38}$$

and $\tan \delta = G/\omega C$ is the loss factor. Hence, if the attenuation constant is the same everywhere in the mesh (homogeneous lossy material), the output impulse response value $({}_kA_i')$ for electric or magnetic fields at any node and at any instant $k\,\Delta t$ is related to the value $({}_kA_i)$ in the lossless case as follows:

$${}_kA_i' = {}_kA_i\, e^{-k\alpha \Delta l} \tag{39}$$

Hence, no modification of the main TLM program is required except for the adjustment of the output function, and many different loss conditions can be covered with a single simulation by merely recalculating the impulse response using different attenuation constants α. At the same time, the phase constant β is increased by a factor

$$r = \left[\frac{1 + \sqrt{1 + (\tan \delta)^2}}{2}\right]^{1/2} \tag{40}$$

which increases the transit time for pulses to $t = r\,\Delta l/c$. This time delay must be accounted for when translating the TLM impulse response into the frequency domain by Fourier transform. That operation will be explained in Section 4.5.

While these modifications of the lossless procedure are easily implemented for homogeneous dielectrics, difficulties arise when the dielectric is inhomogeneous. Because of the complex nature of the wave impedance in lossy media, the reflection coefficient at interfaces becomes complex. This distorts the shape of impulses, which cannot be accounted for in the TLM method. However, if losses are small, the imaginary component can be neglected, and inhomogeneous problems can be treated with reasonable accuracy. Nevertheless, the alternative procedure described in the following section is more flexible and easier to implement in the case of inhomogeneous lossy media.

4.4.3. Lossy Inhomogeneous Materials

This is the most general and most realistic case encountered in micro- and millimeter-wave engineering: Both the permittivity (or permeability) and the loss tangent vary across the structure. To model such a situation the TLM mesh can be loaded at the nodes situated inside the material with additional reactive and dissipative elements; their values are directly related to the local permittivity (permeability) and loss tangent, respectively.

Introduction of Shunt Stubs. Consider, for example, a lossy dielectric with a complex permittivity

$$\varepsilon = \varepsilon_0 \varepsilon_r + \frac{\sigma}{j\omega} = \varepsilon_0 \varepsilon_r (1 - j \tan \delta) \quad . \tag{41}$$

Such a medium can be modeled by a shunt-connected mesh to which the following elements have been added at each node (see Fig. 13):

1. An open-ended shunt stub of length $\Delta l/2$ and normalized characteristic admittance y_0, called a *permittivity stub.*
2. A lumped normalized shunt conductance g_0 or, alternatively, a matched shunt stub of normalized characteristic admittance g_0, called a *loss stub*.

The normalizing admittance is the characteristic admittance of the main mesh lines.

The length of the permittivity stubs is fixed at $\Delta l/2$ to ensure synchronism of impulses on all mesh elements, but their characteristic admittance y_0 is variable to describe any value of dielectric constant. At low frequencies, these stubs add a lumped capacitance of $Cy_0 \, \Delta l/2$ at each node, where C is the capacitance per unit length of the main mesh lines. The total shunt capacitance at each node thus becomes $2C \, \Delta l(1 + y_0/4)$. The real conductance $g_0 Cc$ appears in parallel with that capacitance; c is the velocity of light in free space.

Impulse Scattering at Stub-Loaded Shunt Nodes. A Dirac impulse incident on a stub-loaded node is scattered into six lines, but only five of these impulses will return to the node. The sixth is absorbed by the loss stub and must not be accounted for. The reflection coefficient "seen" by an impulse incident on one of the four mesh line branches is

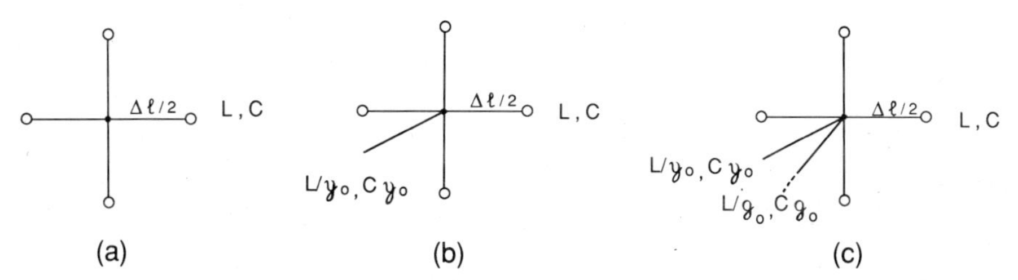

Fig. 13 Simulation of permittivity and losses in a shunt-connected TLM mesh using reactive and dissipative stubs. (*a*) Node without stubs. (*b*) Node with a permittivity stub. (*c*) Node with permittivity and loss stubs. (After Akhtarzad [13].)

$$S_{ii} = -\frac{y-2}{y}, \qquad i = 1, \ldots, 4 \tag{42}$$

and the transmission coefficient for each outgoing line, including the stubs, is

$$S_{ki} = 1 + S_{ii} = \frac{2}{y} \qquad k \neq i \tag{43}$$

where $y = 4 + y_0 + g_0$. Impulses entering the permittivity stubs are reflected at the open end and become incident impulses on the node:

$${}_{k+1}V_5^i(z, x) = {}_{k+1}V_5^r(z, x) \tag{44}$$

where they are subject to a reflection coefficient

$$S_{55} = \frac{2y_0 - y}{y} \tag{45}$$

and transmission coefficients

$$S_{i5} = 1 + S_{55} = \frac{2y_0}{y}, \qquad i \neq 5 \tag{46}$$

The impulse scattering matrix for a stub-loaded shunt node thus becomes, with the additional impulse V_5 on the permittivity stub,

$${}_{k+1}\begin{bmatrix} V_1 \\ V_2 \\ V_3 \\ V_4 \\ V_5 \end{bmatrix}^r = \frac{1}{y}\begin{bmatrix} -(y-2) & 2 & 2 & 2 & 2y_0 \\ 2 & -(y-2) & 2 & 2 & 2y_0 \\ 2 & 2 & -(y-2) & 2 & 2y_0 \\ 2 & 2 & 2 & -(y-2) & 2y_0 \\ 2 & 2 & 2 & 2 & 2y_0 - y \end{bmatrix} \cdot {}_k\begin{bmatrix} V_1 \\ V_2 \\ V_3 \\ V_4 \\ V_5 \end{bmatrix}^i \tag{47}$$

Again, $y = 4 + y_0 + g_0$. The subscripts 1–4 correspond to the coordinate directions specified in Figs. 2 and 3. As mentioned previously, the impulsive excitation of the TLM mesh contains all frequencies in the spectrum. It is therefore important to study the wave properties of the network at all frequencies in order to assess the validity of the TLM model.

Wave Properties of Stub-Loaded Shunt Meshes. As long as the mesh parameter Δl is small compared with the wavelength, the voltage and current changes in the x and z directions of the shunt-connected mesh are

$$\begin{gathered} \frac{\partial V_y}{\partial x} = -L\frac{\partial I_x}{\partial t} \qquad \frac{\partial V_y}{\partial z} = -L\frac{\partial I_z}{\partial t} \\ \frac{\partial I_z}{\partial z} + \frac{\partial I_x}{\partial x} = -2C\left(1 + \frac{y_0}{4}\right)\frac{\partial V_y}{\partial t} - \frac{g_0 Cc}{\Delta l}V_y \end{gathered} \tag{48}$$

Comparing, for instance, these relations with the expansions of Maxwell's equations for $\partial/\partial y = 0$, and $E_x = E_z = H_y = 0$ (which describe the TE_{n0} modes in a rectangular waveguide),

$$\frac{\partial E_y}{\partial x} = -\mu \frac{\partial H_z}{\partial t} \qquad \frac{\partial E_y}{\partial z} = +\mu \frac{\partial H_x}{\partial t}$$
$$\frac{\partial H_x}{\partial z} - \frac{\partial H_z}{\partial x} = +\varepsilon_0 \varepsilon_r \frac{\partial E_y}{\partial t} + \sigma E_y \tag{49}$$

the following equivalences between field and TLM mesh parameters can be established:

$$E_y \equiv V_y \qquad H_z \equiv I_x \qquad H_x \equiv -I_z$$
$$\mu_0 \equiv L \qquad \varepsilon_0 \equiv 2C \qquad \varepsilon_r \equiv 1 + y_0/4 \qquad \sigma \equiv g_0 Cc/\Delta l \tag{50}$$

For elementary transmission lines in the TLM network, the inductance and capacitance per unit length are related by

$$\frac{1}{\sqrt{LC}} = \frac{1}{\sqrt{\mu_0 \varepsilon_0}} = c \qquad \sqrt{\frac{L}{C}} = \sqrt{\frac{\mu_0}{\varepsilon_0}} \tag{51}$$

where c is the velocity of light in free space.

The frequency-dependent propagation constant of TEM waves in the stub-loaded mesh is found with the procedure outlined in Section 4.1.3 for the transmission equations of a stub-loaded cell.

For *axial* propagation (propagation in the direction of the mesh axes), the equation is [7]

$$\begin{bmatrix} V_i \\ I_i \end{bmatrix} = \begin{bmatrix} \cos\theta/2 & j\sin\theta/2 \\ j\sin\theta/2 & \cos\theta/2 \end{bmatrix} \begin{bmatrix} 1 & 0 \\ g_0 + j(2+y_0)\tan\theta/2 & 1 \end{bmatrix} \cdot \begin{bmatrix} \cos\theta/2 & j\sin\theta/2 \\ j\sin\theta/2 & \cos\theta/2 \end{bmatrix} \begin{bmatrix} V_{i+1} \\ I_{i+1} \end{bmatrix} \tag{52}$$

where $\theta = 2\pi\,\Delta l/\lambda$. If the waves on the periodic structure have a propagation constant $\gamma_n = \alpha_n + j\beta_n$, then

$$\begin{bmatrix} V_i \\ I_i \end{bmatrix} = \begin{bmatrix} e^{\gamma_n \Delta l} & 0 \\ 0 & e^{\gamma_n \Delta l} \end{bmatrix} \begin{bmatrix} V_{i+1} \\ I_{i+1} \end{bmatrix} \tag{53}$$

Combining eqs. (52) and (53), we obtain for $\alpha\,\Delta l$ and $\alpha_n\,\Delta l \ll 1$

$$\frac{\beta}{\beta_n} = \frac{\pi\,\Delta l/\lambda}{\sin^{-1}[\sqrt{2(1+y_0/4)}\,\sin(\pi\,\Delta l/\lambda)]} \tag{54}$$

$$\frac{\alpha}{\alpha_n} = \frac{[1 - 2(1 + y_0/4)\sin^2(\pi\,\Delta l/\lambda)]^{1/2}}{\sqrt{2(1 + y_0/4)}\cos(\pi\,\Delta l/\lambda)} \tag{55}$$

where

$$\alpha = \frac{g_0}{4\,\Delta l(1 + y_0/4)} \qquad \beta = \frac{2\pi}{\lambda}$$

The first network cutoff frequency is obtained from (54), when the argument of the $\sin^{-1}$ function in the denominator is unity:

$$\left(\frac{\Delta l}{\lambda}\right)_{\text{cutoff}} = \frac{1}{\pi}\sin^{-1}\frac{1}{\sqrt{2(1 + y_0/4)}} \tag{56}$$

Figures 14 and 15 show the phase constant and attenuation constant ratios as a function of the normalized frequency for various values of the normalized characteristic admittance y_0. The normalizing quantities α and β represent equivalent propagation characteristics of the mesh transmission lines, and they have been chosen such that at low frequencies we can write

$$\gamma_n = \sqrt{2(1 + y_0/4)}\,\gamma \tag{57}$$

where $\gamma = \alpha + j\beta$. Note that with increasing y_0 the useful frequency range is reduced.

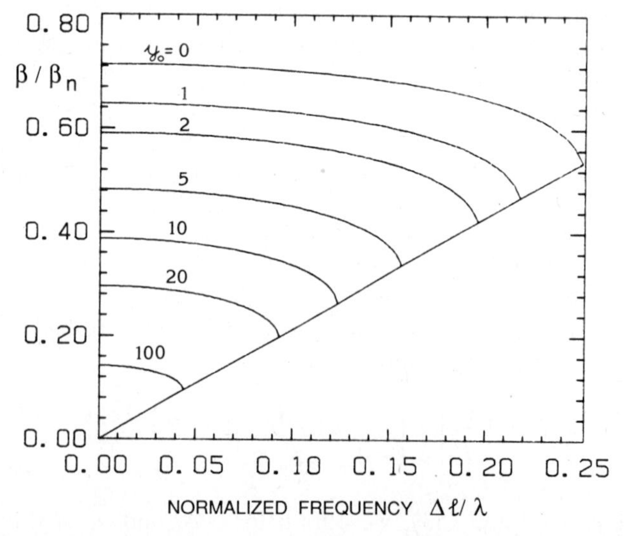

Fig. 14 Phase characteristic for plane waves traveling along a main axis on a two-dimensional stub-loaded shunt TLM network. (After Akhtarzad [13].)

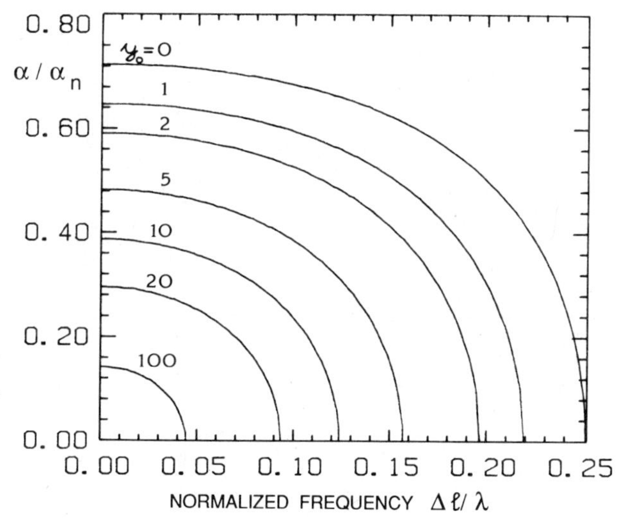

Fig. 15 Attenuation characteristic for plane waves traveling along a main axis on a two-dimensional stub-loaded shunt TLM network. (After Akhtarzad [13].)

For *diagonal* propagation, the velocity is essentially independent of frequency [7] as shown in Section 4.1.3. The largest velocity error thus occurs in the direction of the mesh axes. For propagation in an *arbitrary* direction, Saguet [17] proposes the following approximate formula:

$$\frac{\beta}{\beta_n} = \frac{d_n \pi \, \Delta l/\lambda}{\sin^{-1}[d_n \sqrt{2(1 + y_0/4)} \sin(\pi \, \Delta l/\lambda)]} \tag{58}$$

where $d_n = D_n/D$, and $D = a + b$, as defined in Fig. 8.

Introduction of Series Stubs. Identical wave properties can be achieved with series-connected meshes, which, by duality, can be loaded with reactive and resistive series stubs. If voltages in the series mesh represent E-field components in the medium, then the stubs describe the permeability and magnetic losses, respectively. A node of such a network is shown in Fig. 16. The additional series elements are

1. A short-circuited series stub of length $\Delta l/2$ and normalized characteristic impedance z_0, called the *permeability stub*.
2. *A lumped normalized series resistance* r_0 or, alternatively, a matched series stub of normalized characteristic impedance r_0, called the *loss stub*.

The normalizing impedance is the characteristic impedance of the main mesh lines. The loss stub is usually not included in the series mesh since

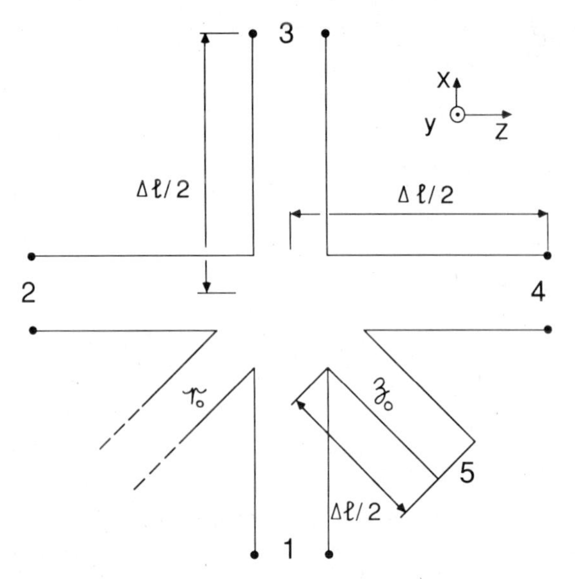

Fig. 16 A series node with a permeability stub (z_0) and a loss stub (r_0). (After Akhtarzad [13].)

there is no corresponding term in Maxwell's equations. Nevertheless, it can be introduced to represent the imaginary part of a complex permeability. At low frequencies, the permeability stub adds a lumped series inductance of Lth $z_0\,\Delta l/2$ at each node, bringing the total inductance to $2L\,\Delta l(1 + z_0/4)$. The real resistance $r_0 Lc$ appears in series with that inductance. L is the inductance per unit length of the main mesh lines, and c is the velocity of light in free space.

Impulse Scattering at Stub-Loaded Series Nodes. Following the same procedure as for shunt nodes, and taking into account the signs of voltages and currents in Fig. 16, we obtain this scattering matrix for a stub-loaded series node:

$$
{}_{k+1}\begin{bmatrix} V_1 \\ V_2 \\ V_3 \\ V_4 \\ V_5 \end{bmatrix}^r = \frac{1}{z}\begin{bmatrix} z-2 & 2 & 2 & -2 & -2 \\ 2 & z-2 & -2 & 2 & 2 \\ 2 & -2 & z-2 & 2 & 2 \\ -2 & 2 & 2 & z-2 & -2 \\ -2z_0 & 2z_0 & 2z_0 & -2z_0 & z-2z_0 \end{bmatrix} \cdot {}_{k}\begin{bmatrix} V_1 \\ V_2 \\ V_3 \\ V_4 \\ V_5 \end{bmatrix}^i \tag{59}
$$

where $z = 4 + z_0 + r_0$.

Wave Properties of Stub-Loaded Series Meshes. For low frequencies, the voltage and current changes in the x and z directions of the stub-loaded series mesh are in time-harmonic form:

$$\frac{\partial I_y}{\partial x} = -j\omega C V_z \qquad \frac{\partial I_y}{\partial z} = -j\omega C V_x$$
$$\frac{\partial V_x}{\partial z} + \frac{\partial V_z}{\partial x} = -j\omega 2L\left(1 + \frac{z_0}{4}\right) I_y - \frac{r_0 Lc}{\Delta l} I_y \tag{60}$$

If voltages in the TLM mesh represent E fields in a medium, then the field equations solved by the series node are

$$\frac{\partial H_y}{\partial x} = +j\omega\varepsilon E_z \qquad \frac{\partial H_y}{\partial z} = -j\omega\varepsilon E_x$$
$$\frac{\partial E_x}{\partial z} - \frac{\partial E_z}{\partial x} = -j\omega\mu_0\mu_r(1 - j\tan\delta) H_y \tag{61}$$

and the following equivalences can be established between field and TLM mesh parameters:

$$H_y \equiv I_y \qquad E_z \equiv -V_z \qquad E_x \equiv +V_x$$
$$\mu_0 \equiv 2L \qquad \mu_r \equiv 1 + z_0/4 \qquad \varepsilon \equiv C \tag{62}$$
$$\tan\delta = \frac{\mu_r''}{\mu_r'} \equiv \frac{r_0 c}{2\omega\,\Delta l(1 + z_0/4)}$$

The propagation and phase constant ratios are

$$\frac{\beta}{\beta_n} = \frac{\pi\,\Delta l/\lambda}{\sin^{-1}[\sqrt{2(1 + z_0/4)}\sin(\pi\,\Delta l/\lambda)]} \tag{63}$$

$$\frac{\alpha}{\alpha_n} = \frac{[1 - 2(1 + z_0/4)\sin^2(\pi\,\Delta l/\lambda)]^{1/2}}{[2(1 + z_0/4)]^{1/2}\cos(\pi\,\Delta l/\lambda)} \tag{64}$$

where

$$\alpha = \frac{r_0}{4\,\Delta l(1 + z_0/4)} \qquad \beta = \frac{2\pi}{\lambda}$$

From (64) the first network cutoff frequency is obtained:

$$\left(\frac{\Delta l}{\lambda}\right)_{\text{cutoff}} = \frac{1}{\pi}\sin^{-1}\frac{1}{\sqrt{2(1 + z_0/4)}} \tag{65}$$

Hence, Figs. 14 and 15 describe the wave properties of series networks as well, provided that y_0 is replaced by z_0 of the permeability stub.

Note also that the introduction of losses in localized form at the nodes, unlike the method described in Section 4.4.2, does not delay the impulses on the mesh lines, nor does it introduce complex mesh line impedances.

Dielectric Interface Conditions. We have seen in the previous section that the introduction of reactive stubs slows down a wave propagating on the TLM mesh. However, this is not sufficient for the simulation of a structure with inhomogeneous dielectric; we must also ensure that the interface conditions at boundaries between dielectrics are satisfied as well.

Consider a region filled partially with air and partially with a dielectric of relative permittivity ε_r (see Fig. 17*a*). The electric field tangential to the interface can be modeled by the voltage in a shunt network, and we have the equivalences

$$\varepsilon_r = 1 + \frac{y_0}{4} \qquad v_2 = \frac{v_1}{\sqrt{1 + y_0/4}} \qquad Z_2 = \frac{Z_1}{\sqrt{1 + y_0/4}}$$

where v_1, v_2 and Z_1, Z_2 are the velocities and intrinsic impedances in regions 1 and 2, respectively. Similarly, for the magnetic material in Fig. 17*b*, where the magnetic field tangential to the interface is modeled by the current in a series network, we have

$$\mu_r = 1 + \frac{z_0}{4} \qquad v_2 = \frac{v_1}{\sqrt{1 + z_0/4}} \qquad Z_2 = Z_1\sqrt{1 + z_0/4}$$

In both cases, the velocities as well as the intrinsic impedances, and hence the interface conditions, are properly represented. Note that the interface lies halfway between the nodes situated in the respective media. However,

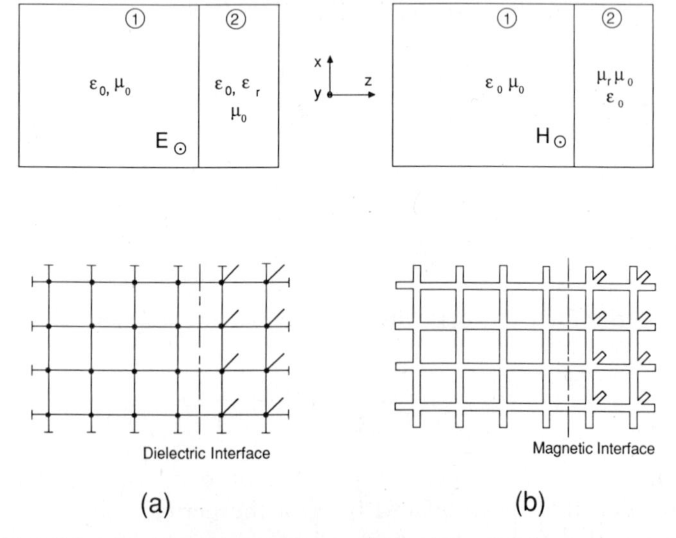

Fig. 17 Modeling of inhomogeneous regions by stub-loaded meshes. (*a*) Inhomogeneous dielectric modeled by a shunt-connected mesh. (*b*) Inhomogeneous magnetic material modeled by a series-connected mesh.

when the network in Fig. 17*a* is used to model the field problem in Fig. 17*b*, and vice versa, we have in the first case

$$\mu_r = 1 + \frac{y_0}{4} \qquad v_2 = \frac{v_1}{\sqrt{1 + y_0/4}} \qquad Z_2 = \frac{Z_1}{\sqrt{1 + y_0/4}}$$

and in the second case

$$\varepsilon_r = 1 + \frac{z_0}{4} \qquad v_2 = \frac{v_1}{\sqrt{1 + z_0/4}} \qquad Z_2 = Z_1\sqrt{1 + z_0/4}$$

While the velocities are represented correctly in both cases, the impedances are not, and corrective boundary reflection and transmission coefficients [7] must be introduced to achieve a realistic model. If

$$r = \frac{\text{intrinsic impedance of medium 1}}{\text{intrinsic impedance of medium 2}}$$

we have for the impulses traveling on the mesh lines across the interface:

$$\Gamma_{11} = \frac{1 - r}{1 + r} \qquad \Gamma_{22} = \frac{r - 1}{r + 1} \qquad T_{12} = \frac{2}{1 + r} \qquad T_{21} = \frac{2r}{1 + r} \tag{66}$$

The application of these coefficients is demonstrated in Fig. 18. Note that the network voltage or current can model any field component in the structure, provided that the boundary and interface conditions are properly represented in the TLM mesh. Thus, the voltage in the shunt network could model, for example, the magnetic field component H_z, which is perpendicular to the air–dielectric interface. In this case, the electric walls parallel to the z axis must be represented by open circuits, and those parallel to the x

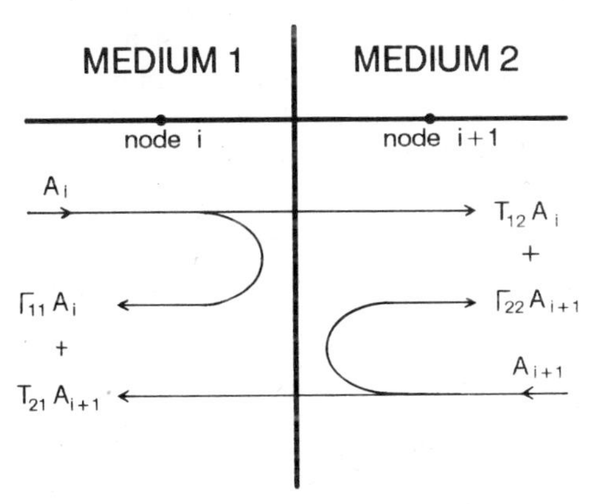

Fig. 18 Application of boundary reflection and transmission coefficients at a dielectric interface. (After Saguet [17].)

axis by short circuits in the shunt mesh. If ε_r changes from region 1 to region 2, the stubs in the dielectric properly describe the velocity and, since the normal magnetic field is continuous across the boundary, the interface conditions as well. However, if μ_r is different in the two regions, corrective interface reflection and transmission coefficients must be introduced to account for the discontinuity in H_z across the interface.

4.5. Computation of the Frequency Response of a TLM Structure

The TLM method provides a means of obtaining the output impulse function at any observation point in a space in which wave propagation is taking place. This is achieved in the computer by storing the amplitude of pulses entering each node in a transmission line matrix. In this section the computation of the network response to sinusoidal and to arbitrary excitation will be described.

4.5.1. *Impulsive Excitation of the TLM Mesh*

In most cases, the amplitudes of impulses throughout the mesh are initially set to zero. The network is then excited at selected source or input points with delta function impulses. As time progresses, impulses travel from one node to the next along the transmission lines and are scattered at each node. Each iteration in the computer represents a time interval of $\Delta l/c$, and the new values of the incident impulse amplitudes for each node are calculated for each iteration. The network therefore becomes filled with impulses as waves spread out from the source points and are reflected at the boundaries.

This type of excitation is best when the network is to be excited at many frequencies simultaneously. However, it is also possible to inject impulses simultaneously at all nodes and to select the amplitudes of these impulses in a way to approximate the field distribution of a specific mode. The steady-state solution for that mode will then be reached with a much smaller number of iterations. In any case, the output impulse function at a particular point in the mesh is simply obtained by observing the stream of pulses as they pass through the point in question. It can be written as

$$F(t) = \sum_{k=1}^{\infty} {}_kA\, \delta(t - kt_0) \tag{67}$$

The solutions for all frequencies within the passband of the network are now simultaneously available in this impulse response; the response to any arbitrary excitation function can be extracted from the impulse response $F(t)$ by convolving one with the other.

Of particular interest is the response to a sinusoidal excitation that is obtained by taking the Fourier transform of the impulse response. Since $F(t)$

is a series of delta functions, the Fourier integral becomes a summation, and the real and imaginary parts of the output spectrum are

$$\mathrm{Re}\left[F\left(\frac{\Delta l}{\lambda}\right)\right] = \sum_{k=1}^{N} {}_kA \cos\left(2\pi k \frac{\Delta l}{\lambda}\right) \tag{68}$$

$$\mathrm{Im}\left[F\left(\frac{\Delta l}{\lambda}\right)\right] = \sum_{k=1}^{N} {}_kA \sin\left(2\pi k \frac{\Delta l}{\lambda}\right) \tag{69}$$

where $F(\Delta l/\lambda)$ is the frequency response, ${}_kA$ is the value of the output impulse response at time $t = k\,\Delta l/\lambda$, and N is the total number of iterations. Note that in a practical computation, N is always finite, which results in a truncation of the impulse response.

The value for ${}_kA$ at a given node is usually the total node voltage (shunt mesh) or node current (series mesh) at the kth scattering event. For example, at a stub-loaded shunt node [see Fig. 13c] the value for ${}_kA$ is

$${}_kA = \frac{2}{y}\left(\sum_{m=1}^{4} {}_kV_m^i + {}_kV_5^i y_0\right) \tag{70}$$

where $y = 4 + y_0 + g_0$.

In the case of a closed structure, the frequency response (68) and (69) represents its mode spectrum. A typical example is given in Fig. 19a, which shows the cutoff spectrum of TE modes in a WR(90) waveguide. The cutoff frequencies correspond to the $\Delta l/\lambda$ values for which the modulus $|F| = (\mathrm{Re}^2\{F\} + \mathrm{Im}^2\{F\})^{1/2}$ has a peak, i.e.,

$$f_c = \left(\frac{c}{\Delta l}\right)\left(\frac{\Delta l}{\lambda}\right)_{\text{peak}} \tag{71}$$

Figure 19a shows clearly that the spectral response is not a line spectrum but a superposition of $(\sin x)/x$ functions (Gibbs phenomenon). This is due to the inevitable truncation of the impulse response. If the number of iterations is too small, the lobes become quite wide and interfere with each other. This results in a displacement of their maxima and gives rise to the so-called truncation error. In Section 5.6 we will see how this error can be estimated and reduced considerably by using a Hanning window in the Fourier transform (Fig. 19b) [36].

Note that, as in a real field measurement, the position of input and output points as well as the nature of the field component will affect the magnitudes of the spectral lines. If, for example, the input and output nodes are situated close to a minimum of a particular mode field, the corresponding eigenfrequency will not appear in the frequency response. This feature can be used to either suppress or enhance certain modes.

It was pointed out in Section 4.4.2 that the impulse transit time is increased when lossy mesh lines are used to simulate losses in a homoge-

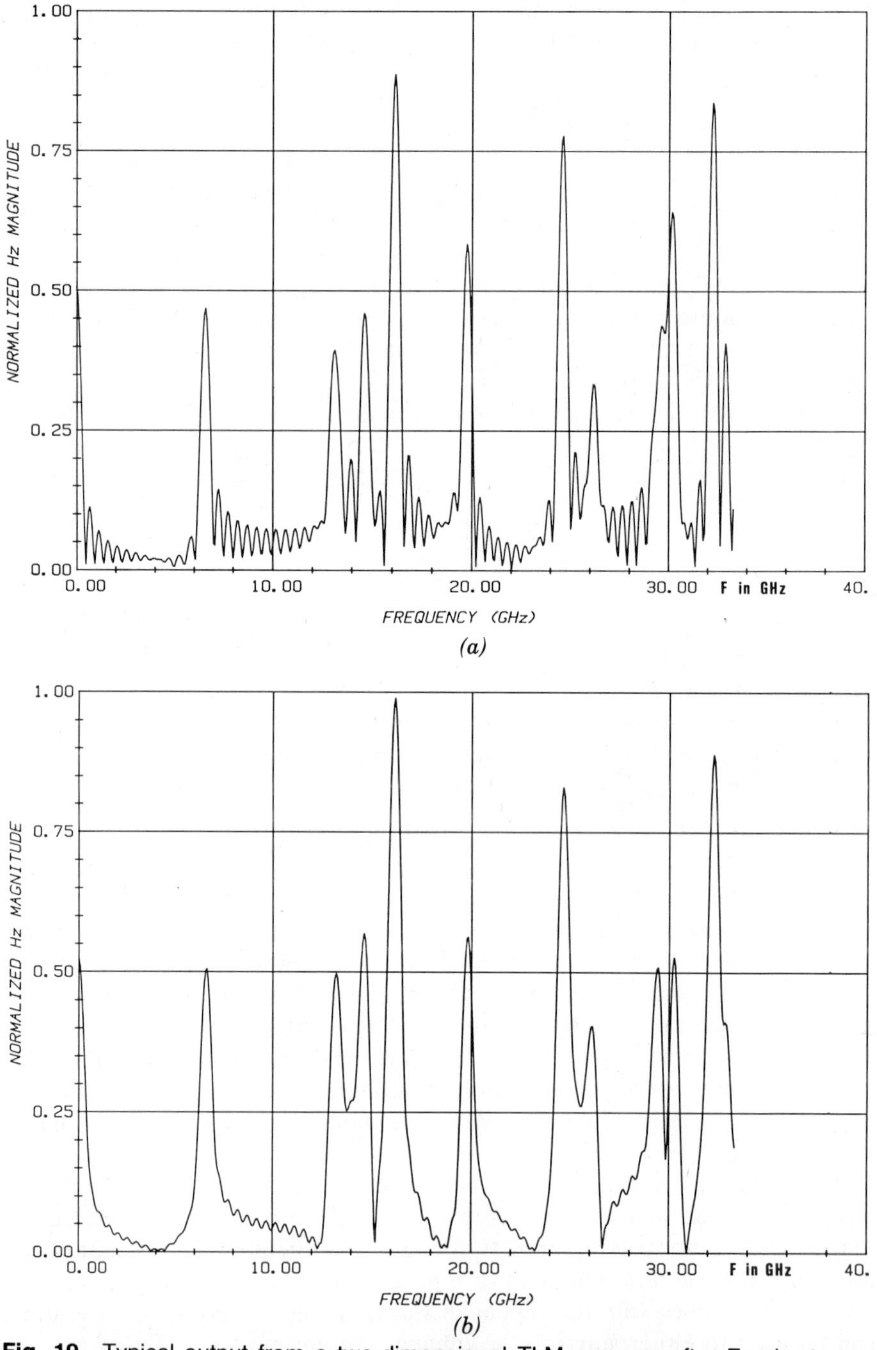

Fig. 19 Typical output from a two-dimensional TLM program after Fourier transform. (*a*) Cutoff spectrum of a WR-90 waveguide (42×92 mesh nodes, 3500 iterations). (*b*) The same spectrum after convolution of the output impulse function with a Hanning window. (After Hoefer [36].) Copyright © 1985 IEEE.

neous structure. This delay can be taken into account by computing the Fourier transform of the impulse response as follows:

$$\mathrm{Re}\left[F\left(\frac{\Delta l}{\lambda}\right)\right] = \sum_{k=1}^{N} {}_kA\, e^{k\alpha\,\Delta l} \cos\left(r2\pi k \frac{\Delta l}{\lambda}\right) \tag{72}$$

$$\mathrm{Im}\left[F\left(\frac{\Delta l}{\lambda}\right)\right] = \sum_{k=1}^{N} {}_kA\, e^{k\alpha\,\Delta l} \sin\left(r2\pi k \frac{\Delta l}{\lambda}\right) \tag{73}$$

where $\alpha\,\Delta l$ and r are given by eqs. (38) and (40), respectively.

4.5.2. Continuous Excitation of the TLM Mesh

It is also possible to simulate an excitation by a continuous waveform: A stream of impulse functions with the exciting waveform as an envelope is injected at the appropriate input nodes. These impulses are added at each iteration to the impulses already present in the mesh, thus continuously injecting energy into the system. Provided that the first cutoff frequency is well above the highest frequency component of the envelope, the sampling theorem by Shannon applies.

The response of the mesh to the sampled waveform is thus computed directly in the time domain. This procedure has certain advantages when modeling scattering in open or lossy structures. It is also advantageous when treating excitation problems involving heating effects. An example in point is the computation of the temperature rise in lossy objects exposed to electromagnetic fields.

4.6. Computation of Fields and Impedances

4.6.1. Computation of Fields

Since the network voltages and currents are directly proportional to field quantities in the simulated structure, the TLM method also yields the field distribution. The correspondence between field and mesh parameters depends on the type of mesh chosen for a given problem.

For a *shunt-connected* mesh in which the voltage simulates an electric field, we have, in accordance with Fig. 13,

$${}_kE_y \equiv {}_kV_y = \frac{2}{y}\left(\sum_{m=1}^{4} {}_kV^i_m + {}_kV^i_5 y_0\right) \tag{74}$$

$$-{}_kH_x \equiv {}_kI_z = ({}_kV^i_2 - {}_kV^i_4)/Z_0 \tag{75}$$

$${}_kH_z \equiv {}_kI_x = ({}_kV^i_1 - {}_kV^i_3)/Z_0 \tag{76}$$

where $y = 4 + y_0 + g_0$, and $Z_0 = \sqrt{L/C} = \sqrt{\mu_0/\varepsilon_0}$ is the characteristic impedance of the mesh lines. [See equivalences in eq. (12).]

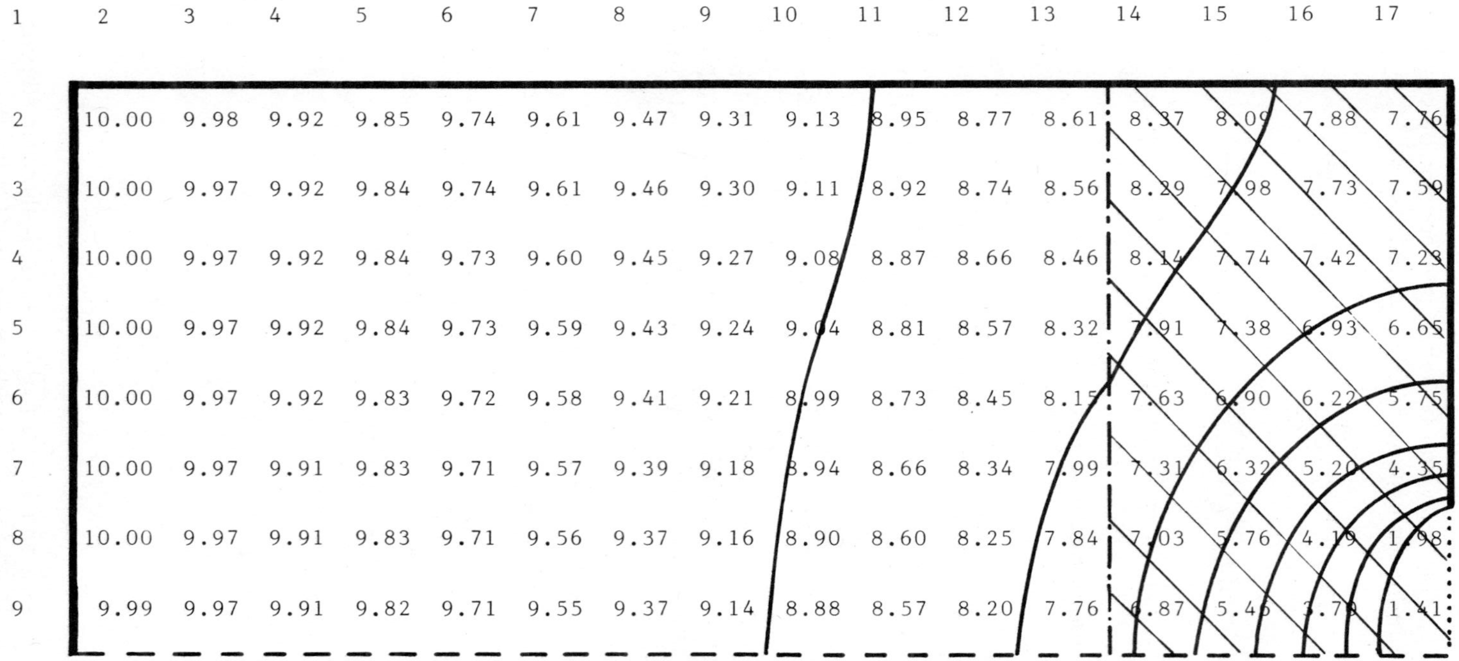

1	2	3	4	5	6	7	8	9	10	11	12	13	14	15	16	17
2	10.00	9.98	9.92	9.85	9.74	9.61	9.47	9.31	9.13	8.95	8.77	8.61	8.37	8.09	7.88	7.76
3	10.00	9.97	9.92	9.84	9.74	9.61	9.46	9.30	9.11	8.92	8.74	8.56	8.29	7.98	7.73	7.59
4	10.00	9.97	9.92	9.84	9.73	9.60	9.45	9.27	9.08	8.87	8.66	8.46	8.14	7.74	7.42	7.23
5	10.00	9.97	9.92	9.84	9.73	9.59	9.43	9.24	9.04	8.81	8.57	8.32	7.91	7.38	6.93	6.65
6	10.00	9.97	9.92	9.83	9.72	9.58	9.41	9.21	8.99	8.73	8.45	8.15	7.63	6.90	6.22	5.75
7	10.00	9.97	9.91	9.83	9.71	9.57	9.39	9.18	8.94	8.66	8.34	7.99	7.31	6.32	5.20	4.35
8	10.00	9.97	9.91	9.83	9.71	9.56	9.37	9.16	8.90	8.60	8.25	7.84	7.03	5.76	4.19	1.98
9	9.99	9.97	9.91	9.82	9.71	9.55	9.37	9.14	8.88	8.57	8.20	7.76	6.87	5.46	3.79	1.41

Fig. 20 Transverse electric field lines of the dominant mode at cutoff in an insulated finline, computed with a two-dimensional TLM program. Only the upper half of the cross section is shown. (After Hoefer and Shih [23].) Copyright © 1980 IEEE.

For a *series-connected* mesh in which the node current simulates a magnetic field, we write, in accordance with Fig. 16 and eqs. (20) and (21),

$$ {}_kH_y \equiv {}_kI_y = \frac{2}{z}\left(\sum_{m=1}^{4} {}_kV^i_m + {}_kV^i_5 z_0\right)\frac{1}{Z_0} \tag{77} $$

$$ {}_kE_x \equiv {}_kV_x = {}_kV^i_2 - {}_kV^i_4 \tag{78} $$

$$ -{}_kE_z \equiv {}_kV_z = {}_kV^i_1 - {}_kV^i_3 \tag{79} $$

Analogous expressions are easily derived for other relationships between mesh and field parameters.

The impulse response at each node contains information on the field distribution at all frequencies within the passband of the mesh. This information can be extracted by Fourier transform [eqs. (68) and (69)]. To obtain the field configuration of a particular mode in a closed structure, its cutoff frequency must be determined first. Then the Fourier transform of the network variable ${}_kA$ [see eq. (70)] representing the desired field component is computed at each node during a second run, with $\Delta l/\lambda$ corresponding to the eigenfrequency of the mode. The field solution is built up at each node on a cumulative basis after each iteration. The impulse response function for a particular node thus need not to be stored during field computations. The field between nodes can be obtained by interpolation techniques (splines). Figure 20 shows a typical field picture for the dominant mode in an insulated finline, obtained in this manner. (Only one-fourth of the cross section is shown in view of symmetry.)

4.6.2. Computation of Impedances

Impedances can, in turn, be obtained from the field quantities. The local field impedance can be found directly as the ratio of voltages and currents at a node (local field impedance), taking into account the equivalence between mesh and field parameters. Note that the mesh impedance depends on frequency in the same way as the phase velocity, and the dispersion characteristics of the phase constant of a particular mesh apply to the mesh impedance as well. For example, the field impedance in the z direction of a shunt mesh, E_y/H_x, is the ratio of (74) and (75). The intrinsic impedance of such a mesh is

$$ Z_n = Z_0 \frac{\beta}{\beta_n} \tag{80} $$

where Z_0 is the intrinsic impedance of free space and β/β_n is given by eq. (54). In a stub-loaded series mesh (but also in a stub-loaded shunt mesh in which the voltage simulates a magnetic field), the intrinsic network impedance is

$$Z_n = \frac{Z_0}{b/b_n} \tag{81}$$

Impedances defined on the basis of specific field integrals (such as the voltage–power, voltage–current, or power–current impedance in a non-TEM waveguide) are computed by stepwise integration of the discrete field values. This procedure is identical to that used in finite element and finite difference methods of analysis. Typical results for impedances computed with the two-dimensional TLM method can be found in References 2 and 7.

5. THE THREE-DIMENSIONAL TLM METHOD

The extension of the TLM procedure to three space dimensions requires a network that can model the propagation of three electric and three magnetic field components. It has been demonstrated in Sections 4.1 and 4.2 that a shunt-connected two-dimensional TLM mesh can simulate, for example, the components E_y, H_x, and H_z, while a series-connected mesh can describe E_x, E_z, and H_y. The corresponding field equations are eqs. (10) and (21). Hence, an orthogonal combination of both network types should be able to handle general hybrid three-dimensional field problems.

Indeed, Akhtarzad and Johns [10] presented such a network in 1974. In subsequent publications [11–16] they examined its wave properties and applied it to various waveguide and microstrip problems. In the following, this TLM network will be described and analyzed in detail.

5.1. The Structure of the Three-Dimensional TLM Network

The three-dimensional TLM network by Akhtarzad and Johns [10] is constructed by interweaving shunt and series nodes in all three coordinate directions. To understand this procedure, consider first a shunt node in the *xz* plane connected to two series nodes, one in the *yz* and one in the *xy* plane, as shown in Fig. 21. At the shunt node, the voltage is common to both line pairs. If the line voltage represents the E_y component, and the line currents represent H_x and H_z components, we have

$$\frac{\partial H_x}{\partial z} - \frac{\partial H_z}{\partial x} = \varepsilon \frac{\partial E_y}{\partial t} \tag{82}$$

as in eq. (10). In the series nodes, the current is common to both connected lines, and we have for the node in the *yz* plane

$$\frac{\partial E_z}{\partial y} - \frac{\partial E_y}{\partial z} = -\mu \frac{\partial H_x}{\partial t} \tag{83}$$

and similarly for the series node in the *xy* plane

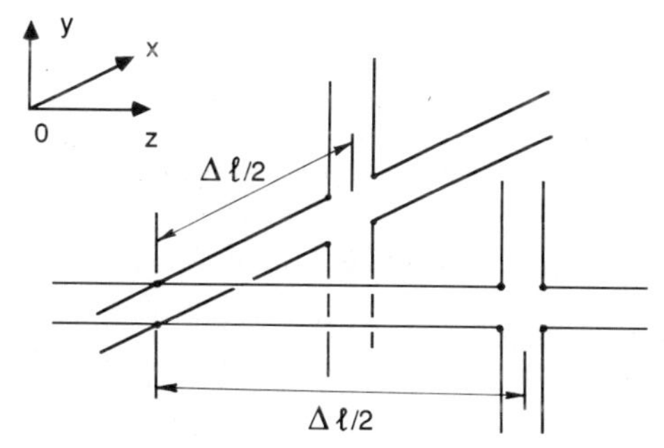

Fig. 21 Part of a three-dimensional node containing one shunt and two series nodes. (After Akhtarzad [13].)

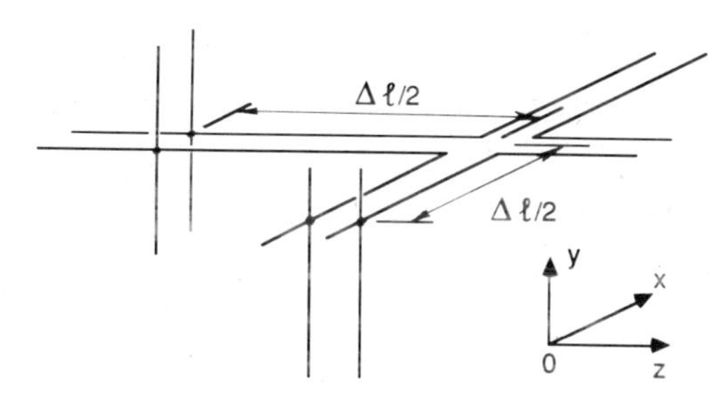

Fig. 22 Part of a three-dimensional node containing one series and two shunt nodes. (After Akhtarzad [13].)

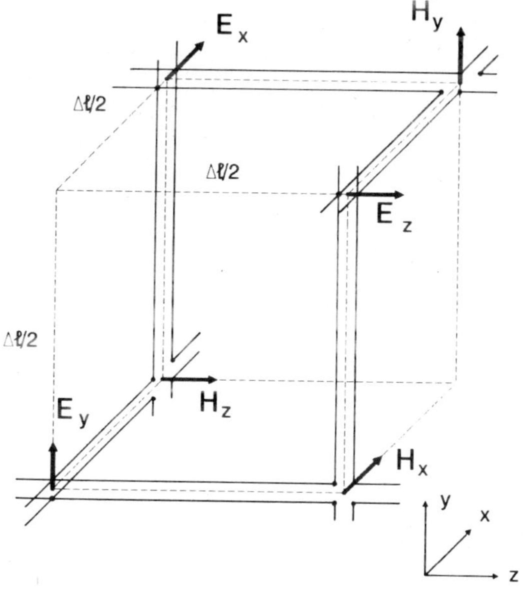

Fig. 23 The complete three-dimensional TLM cell featuring three series and three shunt nodes. (After Akhtarzad [13].)

$$\frac{\partial E_y}{\partial x} - \frac{\partial E_x}{\partial y} = -\mu \frac{\partial H_z}{\partial t} \tag{84}$$

These three expressions represent half of Maxwell's equations in Cartesian coordinates. The other half are obtained by connecting a series node in the *xz* plane with shunt nodes in the *yz* and *xy* planes, as shown in Fig. 22.

The related field equations are

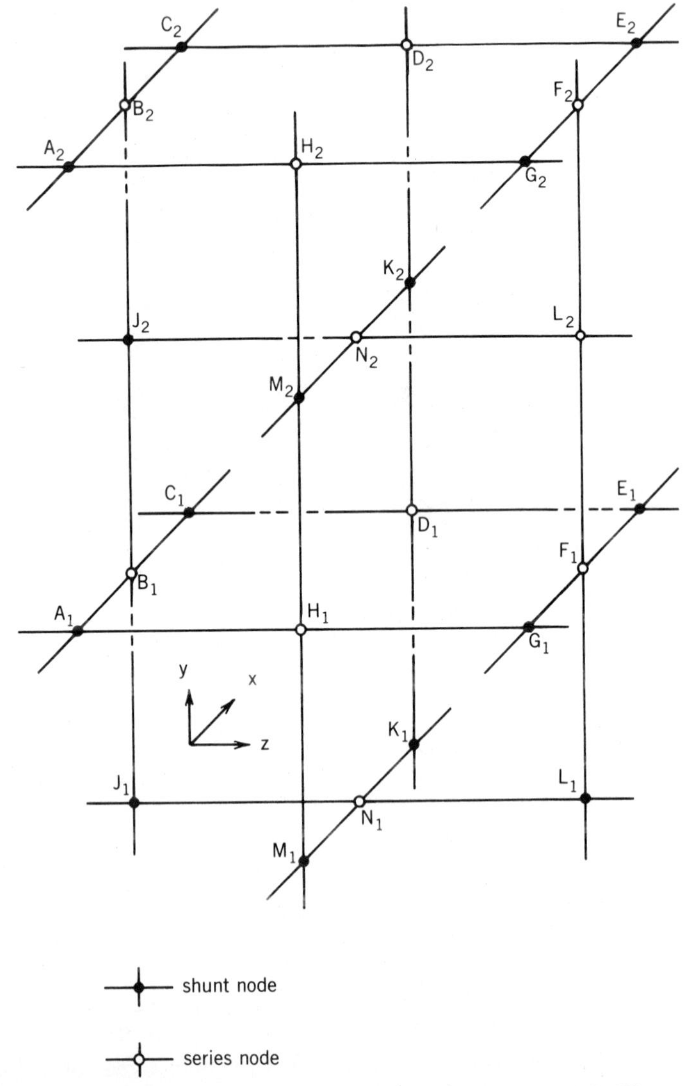

Fig. 24 Three-dimensional network model. (After Akhtarzad [13].)

$$\frac{\partial E_x}{\partial z} - \frac{\partial E_z}{\partial x} = -\mu \frac{\partial H_y}{\partial t} \tag{85}$$

$$\frac{\partial H_z}{\partial y} - \frac{\partial H_y}{\partial z} = \varepsilon \frac{\partial E_x}{\partial t} \tag{86}$$

$$\frac{\partial H_y}{\partial x} - \frac{\partial H_x}{\partial y} = \varepsilon \frac{\partial E_z}{\partial t} \tag{87}$$

Combining both circuits yields all six Maxwell's equations. The resulting configuration (Fig. 23) forms the unit cell of a three-dimensional TLM network, which is illustrated schematically in Fig. 24. Here, a single line represents a pair of wires.

Since the six field components are defined at points separated by a distance $\Delta l/2$, the network is termed a "distributed node mesh."

5.2. Wave Propagation in the Three-Dimensional TLM Network

5.2.1. *Scattering of Dirac Impulses at Mesh Nodes*

Since the three-dimensional mesh is made up of two-dimensional shunt and series nodes, the scattering of Dirac impulses at the nodes, and the corresponding numerical procedures, are exactly the same as those described in Sections 4.1 and 4.2. However, the wave properties of the three-dimensional mesh differ from the two-dimensional case, as can be shown by considering the propagation of plane waves along the main axes of the network.

5.2.2. *Wave Properties of the Three-Dimensional TLM Network*

Suppose the network in Fig. 24 supports a uniform plane wave traveling in the positive z direction, with electric and magnetic field vectors pointing in the positive y and negative x direction, respectively. They will appear as voltages and currents at shunt nodes in planes like $A_1C_1E_1G_1$. Because of the uniformity of the plane wave, identical voltages will travel in all planes $A_nC_nE_nG_n$ parallel to the former. A current impulse proceeding from H_1 to H_2 will encounter at M_2 an identical current impulse traveling from H_2 to H_1, and thus no impulses are injected into the plane $J_2K_2L_2M_2N_2$. This confirms that there are no E_x, E_z, and H_y components. Similarly, voltage impulses traveling from A_n to C_n are identical to those traveling from C_n to A_n. Consequently, an impulse propagating in the z direction along a line through A_1 and G_1 "sees" a pair of open shunt stubs at A_1 and G_1 and a pair of short-circuited series stubs at H_1, as shown in Fig. 25.

This equivalent network may be analyzed, as in the two-dimensional case (Section 4.1.3), by dividing it into elementary cells and expressing the voltage and current at the output of one cell in terms of the voltage and

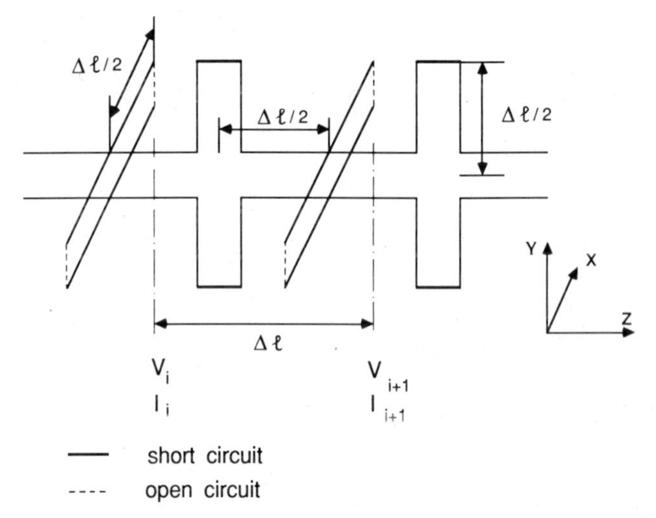

Fig. 25 Equivalent network for plane waves traveling in the direction of a principal mesh axis (z axis) in a three-dimensional TLM mesh. (After Akhtarzad [13].)

current at its input. For a given ratio $\Delta l/\lambda = \theta/2\pi$, where λ is the wavelength on the mesh lines, Akhtarzad and Johns [15] calculate the following transmission equation for one cell:

$$\begin{bmatrix} V_i \\ I_i \end{bmatrix} = \boldsymbol{T} \cdot \begin{bmatrix} 1 & 2j \tan \theta/2 \\ 0 & 1 \end{bmatrix} \cdot \boldsymbol{T} \cdot \boldsymbol{T} \cdot \begin{bmatrix} 1 & 0 \\ 2j \tan \theta/2 & 1 \end{bmatrix} \cdot \boldsymbol{T} \cdot \begin{bmatrix} V_{i+1} \\ I_{i+1} \end{bmatrix} \tag{88}$$

where the matrix $\boldsymbol{T}$ is given by

$$\boldsymbol{T} = \begin{bmatrix} \cos(\theta/4) & j \sin(\theta/4) \\ j \sin(\theta/4) & \cos(\theta/4) \end{bmatrix} \tag{89}$$

If the waves on the periodic structure have a propagation constant $\gamma_n = \alpha_n + j\beta_n$, then

$$\begin{bmatrix} V_i \\ I_i \end{bmatrix} = \begin{bmatrix} e^{\gamma_n \Delta l} & 0 \\ 0 & e^{\gamma_n \Delta l} \end{bmatrix} \begin{bmatrix} V_{i+1} \\ I_{i+1} \end{bmatrix} \tag{90}$$

The solution of eqs. (88) and (90) gives

$$\cosh \gamma_n \Delta l = 1 - 8 \sin^2(\theta/2) \tag{91}$$

For the lowest frequency propagation region, and for $\gamma_n = j\beta_n$, eq. (91) reduces to

$$\sin \frac{\beta_n \, \Delta l}{2} = 2 \sin \frac{\omega \, \Delta l}{2c} = 2 \sin \frac{\beta \, \Delta l}{2} \tag{92a}$$

which can also be written as

$$\frac{\beta}{\beta_n} = \frac{\pi \, \Delta l/\lambda}{\sin^{-1}[2 \sin(\pi \, \Delta l/\lambda)]} \tag{92b}$$

where β and β_n are the propagation constants on the mesh lines and in the network, respectively. One can see immediately that for very small Δl values, β/β_n tends toward 1/2. This means that along the main coordinate axes the low-frequency wave velocity on the network is half the phase velocity on the individual mesh lines, which is consistent with the observation that the shunt stubs double the capacitance and the series stubs double the inductance per unit length.

The wave velocities along the surface and space diagonals of a TLM unit cell (Fig. 23) also tend toward the same value (see Reference 15 for a detailed demonstration). Thus it is reasonable to assume that for low frequencies ($\Delta l/\lambda \ll 1$) the wave velocity on the TLM network is the same in all directions. If on the mesh lines, signals travel at the velocity of light, we have

$$v_n = \frac{1}{\sqrt{2L2C}} = \frac{c}{2} \qquad \text{for} \qquad \frac{\Delta l}{\lambda} \ll 1 \tag{93}$$

5.3. Modeling of Lossless and Lossy Boundaries

Since in the three-dimensional TLM network, voltages represent E-field components and currents represent H-field components, there is no need for dual representations of boundaries. Thus, electric walls are always represented by short circuits, and magnetic walls always by open circuits in the mesh.

5.3.1. Positioning of Electric Walls

In the two-dimensional TLM network, short circuits were placed halfway between nodes, as shown in Section 4.3. In the three-dimensional case, however, short circuits can be placed either between two nodes by imposing a -1 impulse reflection coefficient, or directly through shunt nodes by adding, for example, a stub of infinite characteristic admittance. As Akhtarzad [13] has pointed out, the latter procedure presents certain advantages from the programming point of view and is implemented as follows. Referring to Fig. 24, an electric wall parallel to the xy plane is introduced by short-circuiting the shunt nodes that lie in planes at right angles to it, such as nodes A, C, and J, or E, G, and L. Short circuits across nodes J, K, L, and M simulate a conducting wall in the xz plane. Finally, short circuits across nodes A, G, and M represent an electric wall in the yz plane.

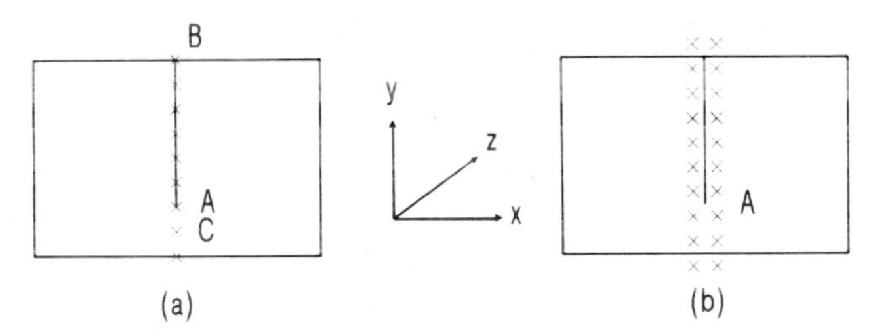

Fig. 26 Positioning of a thin electric fin in a three-dimensional TLM mesh. (*a*) Positioning across shunt nodes; (*b*) positioning halfway between shunt nodes. (After Saguet [17].)

Saguet [17] has studied both positioning alternatives and their effect on the accuracy when modeling electromagnetic cavities and waveguides. While both methods are equivalent for the representation of the exterior walls, they lead to different results in the description of metallic strips or discontinuities. Consider the case of a finned rectangular waveguide shown in Fig. 26. In Fig. 26*a*, the metal fin cuts across shunt nodes, making the tangential components E_y and E_z equal to zero from node B to and including node A. The metallic edge will thus appear to be situated somewhere between A and C. However, if the fin is placed halfway between nodes (Fig. 26*b*), E_y is not zero at node A, and the position of the edge is defined twice as accurately. Furthermore, since more nodes are situated in the highly nonuniform edge field, a better resolution and hence higher accuracy (usually twice as high, according to Saguet [17]) can be obtained.

The positioning of electric walls between nodes is also advantageous when dealing with planar structures where dielectric and conducting interfaces lie in the same plane. This aspect will be discussed in Section 5.4. Nevertheless, the advantages in modeling accuracy are obtained at the expense of more complex programming.

5.3.2. Positioning of Walls of Symmetry (Magnetic Walls)

In certain situations, the computational effort can be reduced considerably by the introduction of walls of symmetry, or magnetic walls. In order to preserve the symmetry of the three-dimensional TLM network, such walls must necessarily run across series nodes (see Fig. 27). From a modeling point of view, this can be achieved by adding a series stub of infinite characteristic impedance to the boundary nodes that lie in planes perpendicular to these boundaries. Again referring to Fig. 24, a magnetic wall parallel to the xy plane is created by open-circuiting nodes D, H, and N. Open circuits in nodes B, D, F, and H simulate a magnetic wall parallel to the xz plane, etc.

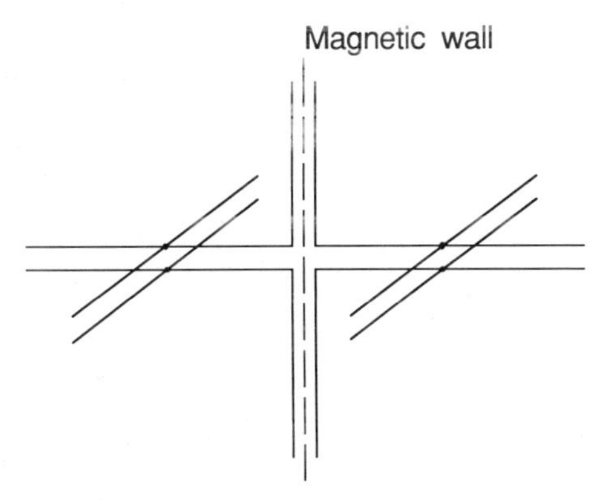

Fig. 27 Positioning of a magnetic wall in a three-dimensional TLM mesh. (After Saguet [17].)

5.3.3. Lossy Boundaries

The boundaries may be made lossy by introducing imperfect reflection coefficients. If boundaries are introduced halfway between two nodes, the procedure is exactly the same as in the two-dimensional case (Section 4.3.2). If the walls are placed across nodes, losses can be introduced at the nodes by adding loss stubs to them. Stub loading of the three-dimensional TLM mesh will be discussed in the following section. In all cases it must be remembered that the value of the imperfect boundary reflection coefficient depends on the frequency so that the simulation is strictly accurate only for the frequency at which it has been determined.

Also note that the intrinsic impedance of the three-dimensional TLM mesh is the same as that of the constituent transmission lines, since both the apparent permeability and permittivity are twice that of the lines. Hence

$$Z_{0n} = \sqrt{2L/2C} = \sqrt{\mu_0/\varepsilon_0} \tag{94}$$

Consequently, in order to simulate a wall of arbitrary real impedance, the lines perpendicular to that wall must be terminated in the same impedance. Of course, the three-dimensional TLM method cannot account for complex reflection coefficients for the reasons already discussed in the two-dimensional case.

5.4. Modeling of Lossless and Lossy Materials

5.4.1. Isotropic Materials

The method becomes really useful for microwave and millimeter-wave circuit analysis only when it can handle inhomogeneous media with losses.

This is accomplished by introducing reactive and dissipative stubs at the nodes. The effect of these stubs on the impulse scattering matrix of nodes has been described in Section 4.4.3, and it is identical for the three-dimensional case. However, since, in the three-dimensional network, one can have inductive, capacitive, and loss stubs at the same time, its wave propagation characteristics incorporate all of them. In the following, expressions for the real and imaginary parts of the network propagation constant will be derived.

The Stub-Loaded Three-Dimensional TLM Network. In analogy to the two-dimensional case, material properties are simulated in the following way:

1. Electrical permittivity is increased by adding an open-ended shunt stub of length $\Delta l/2$ and normalized characteristic admittance y_0 at shunt nodes (permittivity stub).

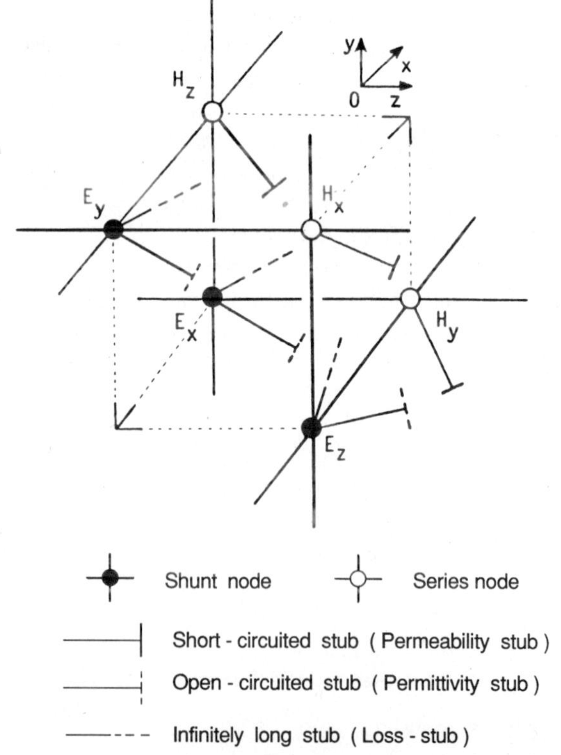

Fig. 28 A three-dimensional node equipped with reactive and dissipative stubs for the modeling of permittivity, permeability, and losses. (Two-dimensional node separation and reactive stub length = $\Delta l/2$). (After Akhtarzad [13].)

2. Magnetic permeability is increased by adding a short-circuited series stub of length $\Delta l/2$ and normalized characteristic impedance z_0 at series nodes (permeability stub).
3. Losses are introduced by adding an infinitely long shunt stub of normalized characteristic admittance g_0 at shunt nodes (loss stub). In principle, losses can also be introduced in the form of series stubs at series nodes, but in most cases it is convenient and economical to embody all losses in the shunt stubs.

In all cases, the normalizing immittance is that of the main mesh lines, also called "link lines."

Figure 28 shows schematically a three-dimensional unit cell with completely equipped nodes. The relationship between the parameters of the stub-loaded network and the constitutive parameters of the simulated space have already been derived in detail for the two-dimensional case (Section 4.4) and can simply be summarized as follows:

1. To increase the permittivity of the simulated space by a factor ε_r, add permittivity stubs with a normalized admittance $y_0 = 4(\varepsilon_r - 1)$ at the shunt nodes.
2. To increase the permeability of the simulated space by a factor μ_r, add permeability stubs with a normalized impedance $z_0 = 4(\mu_r - 1)$ at the series nodes.
3. To introduce a conductivity σ, add loss stubs of normalized admittance $g_0 = \sigma\, \Delta l \sqrt{L/C}$ at the shunt nodes.

Wave Properties of the Stub-Loaded Three-Dimensional TLM Network. Following the procedure outlined in Section 5.1, the equivalent network for plane wave propagation in the axial direction is obtained as shown in Fig. 29.

The transmission equation for one cell is [15]

$$\begin{bmatrix} V_i \\ I_i \end{bmatrix} = \boldsymbol{T} \cdot \begin{bmatrix} 1 & j(2+z_0)\tan(\theta/2) \\ 0 & 1 \end{bmatrix} \cdot \boldsymbol{T} \cdot \boldsymbol{T} \cdot \begin{bmatrix} 1 & 0 \\ g_0 + j(2+y_0)\tan(\theta/2) & 1 \end{bmatrix} \cdot \boldsymbol{T} \cdot \begin{bmatrix} V_{i+1} \\ I_{i+1} \end{bmatrix} \tag{95}$$

where the matrix $\boldsymbol{T}$ is given by

$$\boldsymbol{T} = \begin{bmatrix} \cos(\theta/4) & j\sin(\theta/4) \\ j\sin(\theta/4) & \cos(\theta/4) \end{bmatrix} \tag{96}$$

and $\theta = 2\pi\, \Delta l/\lambda$. If the waves on the mesh have a propagation constant $\gamma_n = \alpha_n + j\beta_n$, then

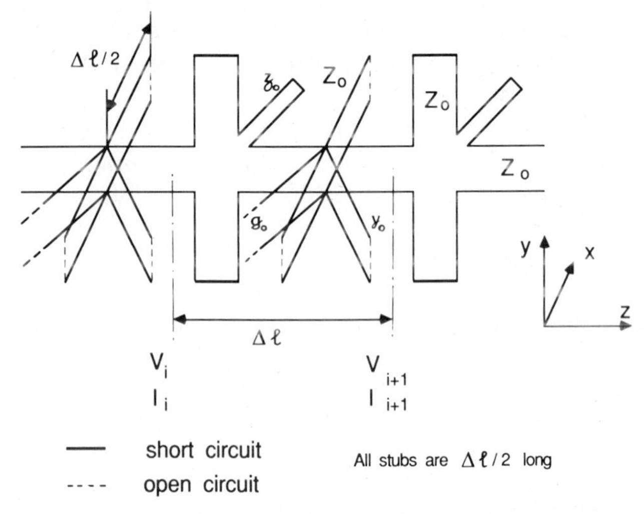

Fig. 29 Equivalent network for plane waves traveling in the direction of a principal mesh axis (*z* axis) in a stub-loaded three-dimensional TLM mesh. (After Akhtarzad [13].)

$$\begin{bmatrix} V_i \\ I_i \end{bmatrix} = \begin{bmatrix} e^{\gamma_n \Delta l} & 0 \\ 0 & e^{\gamma_n \Delta l} \end{bmatrix} \begin{bmatrix} V_{i+1} \\ I_{i+1} \end{bmatrix} \tag{97}$$

Assuming small losses ($\alpha \, \Delta l$ and $\alpha_n \, \Delta l \ll 1$), the solution of eqs. (95) and (97) gives

$$\cos(\beta_n \, \Delta l) = 1 - 8\left(1 + \frac{y_0}{4}\right)\left(1 + \frac{z_0}{4}\right) \sin^2 \frac{\theta}{2} \tag{98}$$

$$\alpha_n \, \Delta l \sin(\beta_n \, \Delta l) = \frac{g_0}{2} \, (4 + z_0) \sin \frac{\theta}{2} \cos \frac{\theta}{2} \tag{99}$$

which can also be written

$$\frac{\beta}{\beta_n} = \frac{\pi \, \Delta l/\lambda}{\sin^{-1}\{2[(1 + y_0/4)(1 + z_0/4)]^{1/2} \sin(\pi \, \Delta l/\lambda)\}} \tag{100}$$

and

$$\frac{\alpha}{\alpha_n} = \frac{1}{2} \left(\frac{1 + z_0/4}{1 + y_0/4}\right)^{1/2} \frac{[1 - 4(1 + y_0/4)(1 + z_0/4) \sin^2(\pi \, \Delta l/\lambda)]^{1/2}}{\cos(\pi \, \Delta l/\lambda)} \tag{101}$$

where

$$\alpha = \frac{g_0}{4\,\Delta l}\left(\frac{1 + z_0/4}{1 + y_0/4}\right) \qquad \beta = \frac{2\pi}{\lambda}$$

The normalizing quantities α and β represent equivalent propagation characteristics of the mesh transmission lines, and they have been chosen such that at low frequencies we can write

$$\gamma_n = 2[(1 + y_0/4)(1 + z_0/4)]^{1/2}\gamma \tag{102}$$

The first network cutoff frequency is obtained from (100) as

$$\left(\frac{\Delta l}{\lambda}\right)_{\text{cutoff}} = \frac{1}{\pi}\sin^{-1}\frac{1}{2[(1 + y_0/4)(1 + z_0/4)]^{1/2}} \tag{103}$$

Figures 30 and 31 show the phase constant and attenuation constant ratios in the axial direction as a function of the normalized frequency $\Delta l/\lambda$. The value of z_0 has been set to zero ($\mu_r = 1$), and y_0 is the sole parameter. The larger the shunt stub admittance y_0, the smaller the useful frequency range. As in the two-dimensional TLM network, the dispersion of the wave velocity is worst in the direction of the main mesh axes and zero in the diagonal direction. For propagation in an *arbitrary* direction, the approximate correction procedure proposed by Saguet [17] leads to the expression

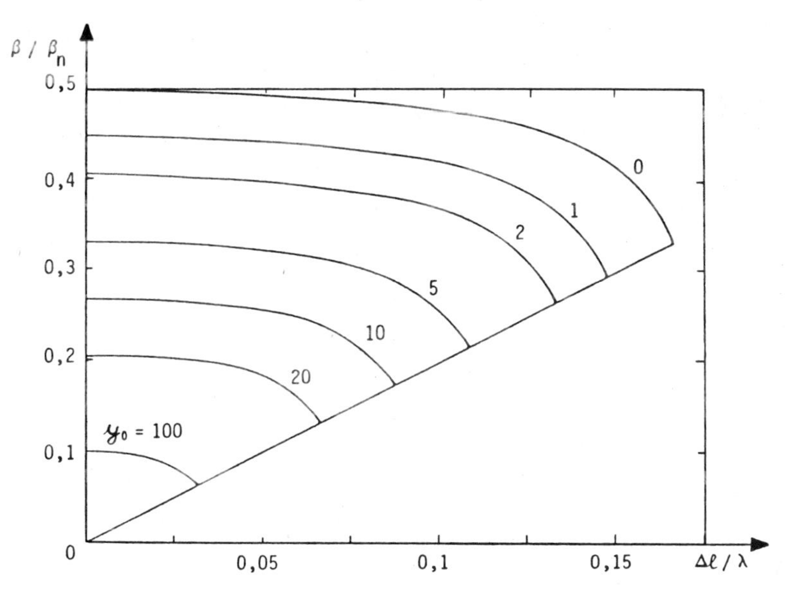

Fig. 30 Phase characteristic for plane waves traveling along a main axis on a three-dimensional stub-loaded TLM network. (After Akhtarzad [13].)

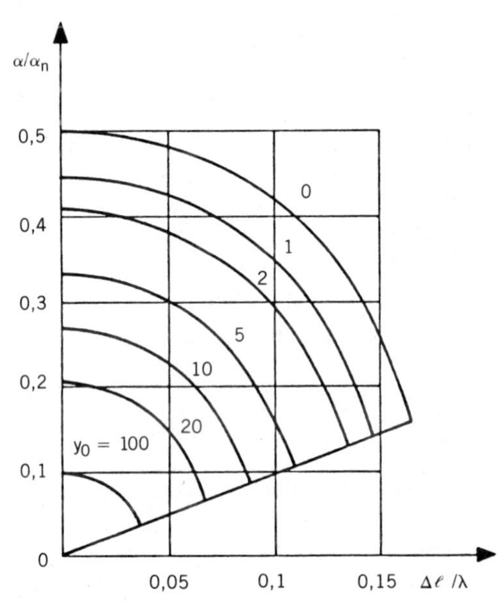

Fig. 31 Attenuation characteristic for plane waves traveling along a main axis on a three-dimensional stub-loaded TLM network. (After Akhtarzad [13].)

$$\frac{\beta}{\beta_n} = \frac{d_n \pi \, \Delta l/\lambda}{\sin^{-1}\{2d_n[(1+y_0/4)(1+z_0/4)]^{1/2} \sin(\pi \, \Delta l/\lambda)\}} \tag{104}$$

where $d_n = D_n/D$. D_n is the distance covered by the wave front going directly from one point A to another point B in the mesh, while D is the shortest distance from A to B *along mesh lines*.

Saguet suggests that eq. (104) can even be used in inhomogeneous structures such as partially dielectric filled and planar waveguides by using a value for the permittivity stubs such that

$$1 + \frac{y_0}{4} = \varepsilon_{\text{eff}} = \left(\frac{\Delta l/\lambda_{\varepsilon=1}}{\Delta l/\lambda_{\varepsilon_{\text{inhomogeneous}}}} \right)^2 \tag{105}$$

In conclusion, the three-dimensional TLM network simulates isotropic propagation as long as all frequencies are much smaller than the network cutoff frequency given by eq. (103). In this case, the network propagation velocity is virtually the same in all directions. However, when the wave fronts in the network propagate predominantly in one direction, the dispersion of the wave velocity can be corrected using eq. (104).

5.4.2. Anisotropic Materials

An extension of the three-dimensional TLM method to anisotropic materials has been developed by Mariki and Yeh [18, 37] and applied to the analysis of microstrip lines on anisotropic substrates. Provided that the properties of a material are characterized by diagonal tensors of rank 2, it can be modeled by making the immittance of reactive and dissipative stubs different in the three coordinate directions. To this end, the isotropic case is simply modified in the following manner (refer to Fig. 28 for identifying the nodes that model the six field components).

To model a material with anisotropic *permittivity* characterized by a tensor

$$\bar{\bar{\boldsymbol{\varepsilon}}} = \varepsilon_0 \begin{bmatrix} \varepsilon_{xx} & 0 & 0 \\ 0 & \varepsilon_{yy} & 0 \\ 0 & 0 & \varepsilon_{zz} \end{bmatrix} \tag{106}$$

add the following *permittivity stubs* to the *shunt* nodes in the TLM network:

To the shunt node modeling	Add a permittivity stub of normalized characteristic admittance
E_x	$y_{xx} = 4(\varepsilon_{xx} + 1)$
E_y	$y_{yy} = 4(\varepsilon_{yy} + 1)$
E_z	$y_{zz} = 4(\varepsilon_{zz} + 1)$

To model a material with anisotropic *permeability* characterized by a tensor

$$\bar{\bar{\boldsymbol{\mu}}} = \mu_0 \begin{bmatrix} \mu_{xx} & 0 & 0 \\ 0 & \mu_{yy} & 0 \\ 0 & 0 & \mu_{zz} \end{bmatrix} \tag{107}$$

add the following *permeability stubs* to the *series* nodes in the TLM network:

To the series node modeling	Add a permeability stub of normalized characteristic impedance
H_x	$z_{xx} = 4(\mu_{xx} + 1)$
H_y	$z_{yy} = 4(\mu_{yy} + 1)$
H_z	$z_{zz} = 4(\mu_{zz} + 1)$

Finally, to model a material with anisotropic *conductivity* characterized by a tensor

$$\bar{\bar{\boldsymbol{\sigma}}} = \begin{bmatrix} \sigma_{xx} & 0 & 0 \\ 0 & \sigma_{yy} & 0 \\ 0 & 0 & \sigma_{zz} \end{bmatrix} \tag{108}$$

add the following *loss stubs* to the *shunt* nodes in the TLM network:

To the shunt node modeling	Add a loss stub of normalized characteristic susceptance
E_x	$g_{xx} = \sigma_{xx} Z_0 \, \Delta l$
E_y	$g_{yy} = \sigma_{yy} Z_0 \, \Delta l$
E_z	$g_{zz} = \sigma_{zz} Z_0 \, \Delta l$

where Z_0 is the characteristic impedance of the main mesh lines.

All three types of stubs can be present simultaneously in a single network. Figure 32 shows a unit cell equipped with a complete set of stubs to simulate $\bar{\bar{\boldsymbol{\varepsilon}}}$, $\bar{\bar{\boldsymbol{\mu}}}$, and $\bar{\bar{\boldsymbol{\sigma}}}$ in tensor form.

Boundaries and dielectric interfaces are modeled in the same way as in the isotropic case. All formulas for the dispersion characteristics, node scattering matrices, impedances, and impulse response apply to the anisotropic case as well, provided that the appropriate stub immittances are introduced in these formulas. Thus, the treatment of anisotropic materials requires only minor modifications of the regular three-dimensional TLM program.

5.4.3. Positioning of Dielectric Interfaces

If two adjacent regions in an electromagnetic structure are simulated by reactance stubs of different immittances, the question arises as to where the

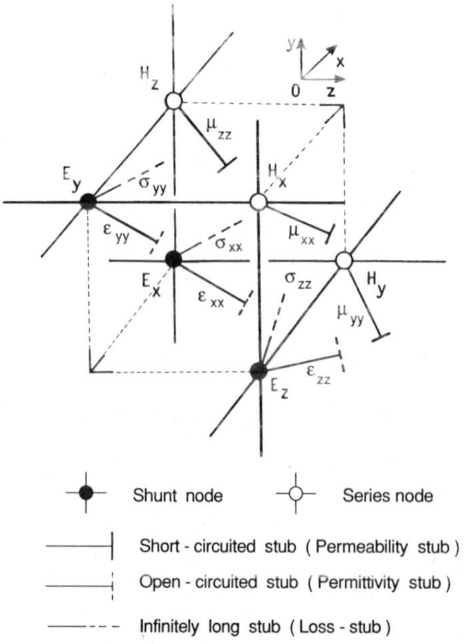

Fig. 32 A three-dimensional node equipped with reactive and dissipative stubs for the modeling of anisotropic permittivity, permeability, and losses. (Two-dimensional node separation and reactive stub length = $\Delta l/2$.)

interface between the two media is actually situated. It is reasonable to assume that the effect of a stub extends $\Delta l/2$ from the node to which it is attached. For example, in Fig. 24, the permittivity stubs (not shown) on the shunt nodes K_1, L_1, M_1, and J_1 describe the permittivity of the medium in the positive y direction, as far as a plane through the shunt nodes A_1, C_1, E_1, and G_1. Therefore, the dielectric interface, say in the xz plane, always cuts through the shunt nodes K_1, L_1, M_1, and J_1. All permittivity stubs below this plane will have a value of, say, y_0 and a value of y_0' above the plane. Now the question is, Do we assign dielectric stub values of y_0 or y_0' to the shunt nodes lying on the dielectric interface? Or should we rather assume an average value of $(y_0 + y_0')/2$, which seems to be a good compromise? Studies by both Akhtarzad [13] and Saguet [17] showed that the latter, although best suited, is undesirable from the computing point of view. This is because the computer program will have to handle more than one value of permittivity within a three-dimensional node, which would increase the storage requirements considerably. However, both authors found that the error in assigning a value of y_0 or y_0' to the stub on the boundary is relatively small and perfectly acceptable in cases where the difference in permittivity or permeability is not extreme (less than a ratio of 10).

If the remaining ambiguity in the position of the dielectric interface is of minor consequence, particularly when the dielectric region is several Δl thick, a more serious problem arises when a conducting strip is placed on a dielectric (as in microstrips or finlines). If conducting walls are placed across shunt nodes as suggested by Akhtarzad [13], and dielectric boundaries appear between shunt nodes, it is impossible to place a metallic and a dielectric boundary into the same plane. This results in a misalignment, causing either a recess or a protrusion of dielectric as shown in Fig. 33. This effect can be neglected only in a high-resolution mesh. Otherwise, corrections can be made by averaging results as discussed in Section 6.4. If

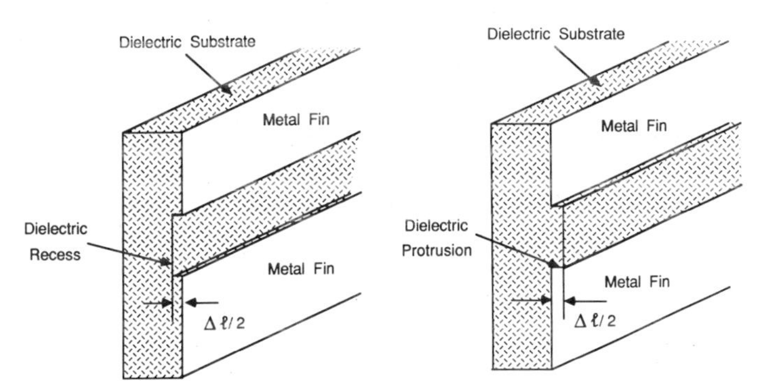

Fig. 33 Misalignment of conducting boundaries and dielectric interfaces in the three-dimensional TLM simulation of planar structures. (After Shih [27].)

metallic boundaries are placed between nodes—a measure resulting in more complicated programming—the problem of misalignment does not occur.

5.5. Excitation and Output

Since the three-dimensional TLM mesh is essentially a combination of two-dimensional nodes, excitation and output procedures are exactly the same as in the two-dimensional case. Each shunt and series node represents one field component, which is obtained in the manner described in Section 4.6. However, it must be kept in mind that the three-dimensional mesh represents a medium with a characteristic impedance equal to that of the individual mesh lines, i.e., 377 Ω. The slow-wave propagation velocity is only half that on the lines. Hence, when the resonant frequencies of an air-filled structure are computed with the three-dimensional TLM method, the $\Delta l/\lambda$ values generated by the program must be doubled to yield the free-space wavelengths corresponding to these frequencies. On the other hand, if three-dimensional mesh is to be excited by a stream of impulses modulated by a sinusoidal waveform of given frequency, the excitation frequency must be halved prior to convolution with the input impulse stream.

6. ERRORS AND THEIR CORRECTION

Like all other numerical techniques, the TLM method is subject to various sources of error and must be applied with caution in order to yield reliable and accurate results. The main sources of error are due to the following circumstances:

1. The impulse response must be truncated in time.
2. The propagation velocity in the TLM mesh depends on the direction of propagation and on the frequency.
3. The spatial resolution is limited by the finite mesh size.
4. Boundaries and dielectric interfaces are not aligned in the three-dimensional TLM model when electric walls are placed across shunt nodes.

The resulting errors will be discussed below, and ways of eliminating or at least significantly reducing these errors will be described. These measures apply equally to the two-dimensional and three-dimensional TLM methods unless indicated otherwise.

6.1. Truncation Error

The practical requirement to truncate the output impulse function leads to the so-called truncation error: Due to the finite duration of the impulse response, its Fourier transform is not a line spectrum but rather a superposition of $(\sin x)/x$ functions (Gibbs's phenomenon) that may interfere with each other such that their maxima are slightly shifted. The resulting error in the eigenfrequencies of a structure, the truncation error [6], is given by

$$E_T \leq \frac{\Delta S}{\Delta l/\lambda_c} = \pm \frac{3\lambda_c}{SN^2\pi^2 \Delta l} \tag{109}$$

where N is the number of iterations and S is the distance in the frequency domain between two neighboring spectral peaks. (See Fig. 34.) This expression shows that the truncation error decreases with increasing separation S and increasing number of iterations N. It is thus desirable to suppress all unwanted modes close to the desired mode by choosing appropriate input and output points in the TLM network. The choice of input and output points is governed by the same considerations as the selection of input and output probes in cavities or guides for the selection or suppression of specific modes. This is a direct consequence of the fact that the TLM method uses a physical model of the structure and establishes a one-to-one relationship between fields and network parameters.

Another technique for reducing the truncation error was proposed by Saguet and Pic [21]. They used a Hanning window in the Fourier transform, resulting in a considerable attenuation of the side lobes. In this process, the output impulse response is first convolved with the Hanning profile:

$$f_h(k) = 0.5\left(1 + \cos\frac{\pi k}{N}\right), \qquad k = 1, 2, 3, \ldots, N \tag{110}$$

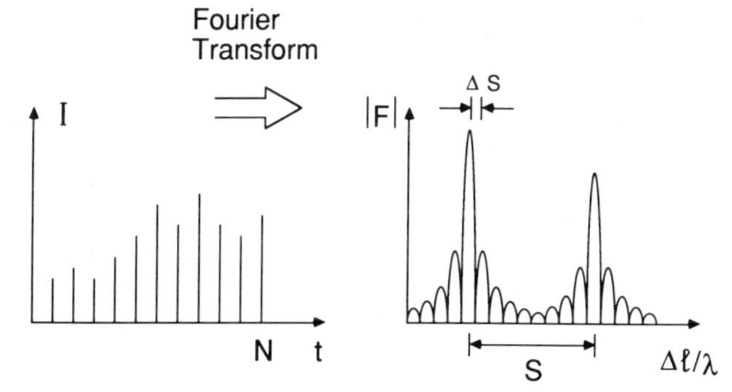

Fig. 34 (*a*) Truncated output impulse response; (*b*) resulting truncation error in the frequency domain. (After Hoefer [36].) Copyright © 1985 IEEE.

where k is the iteration variable or counter and N is the total number of iterations. The filtered impulse response is then Fourier transformed. The resulting improvement can be appreciated by comparing Figs. 19*a* and *b*.

Finally, the number of iterations may be made very large, but this leads to increased CPU time. As a general rule, it is recommended to choose the number of iterations such that the truncation error given by (109) is reduced to a fraction of a percent and can be neglected.

6.2. Velocity Error

If the wavelength in the TLM network is large compared with the network parameter Δl, it can be assumed that the fields propagate with the same frequency-independent velocity in all directions. However, when the wavelength decreases, the velocity becomes dispersive and depends on the direction of propagation (see Figs. 7, 12, 14, 30, 41, and 46). At first glance, the resulting velocity error can be reduced only by choosing a very dense mesh ($\Delta l/\lambda \ll 1$), so that one stays in the linear part of the dispersion curves. However, if wave fronts in the structure propagate essentially in a single direction (for example, computation of the dominant cutoff frequency in a rectangular waveguide), the velocity error can be corrected directly using the appropriate dispersion relations for β/β_n: (18), (58), (104), or (147). The computed value for $\Delta l/\lambda$ is then corrected by multiplying it with the correction factor K_v as follows:

$$\left(\frac{\Delta l}{\lambda}\right)_{\text{corrected}} = K_v \left(\frac{\Delta l}{\lambda}\right)_{\text{computed}} \tag{111}$$

where

$$K_v = \frac{(\beta/\beta_n)_{\Delta l=0}}{(\beta/\beta_n)_{\Delta l}} \tag{112}$$

Note that for wave fronts propagating in the axial direction, $(\beta/\beta_n)_{\Delta l=0}$ is $1/\sqrt{2}$ in the homogeneous two-dimensional case, and $1/2$ in the homogeneous three-dimensional case. Of course, the frequency values obtained from the TLM program after the velocity error corrections are characteristic of a nondispersive TLM mesh. To find the corresponding frequencies of the commensurate air-filled structure, the corrected values must be multiplied by $\sqrt{2}$ and 2, respectively.

Fortunately, the velocity error is also reduced whenever the coarseness error (which will be described next) is corrected.

6.3. Coarseness Error

Coarseness error occurs when the TLM mesh is too coarse to resolve highly nonuniform fields such as can be found at corners and wedges. This error is

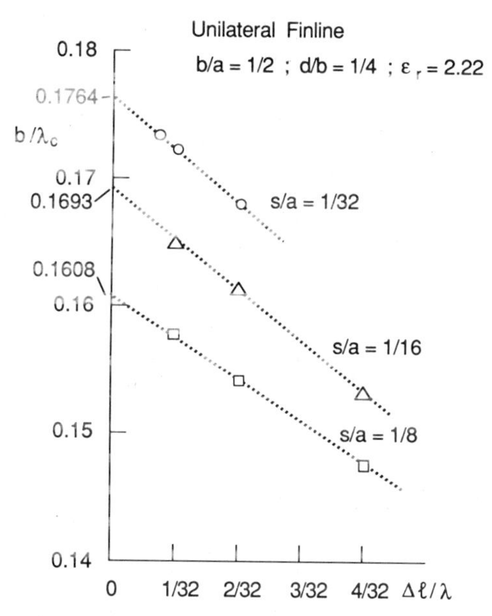

Fig. 35 Elimination of coarseness error by linear extrapolation of results obtained with TLM meshes of different parameters $\Delta l/b$, for a unilateral finline. (After Shih and Hoefer [26].) Copyright © 1980 IEEE.

the dominant source of inaccuracies when analyzing planar structures that contain such regions. A possible but impractical measure would be to choose a very fine mesh. However, this would lead to large memory requirements, particularly for three-dimensional problems. A better remedy is to introduce a graded mesh (see Section 7) to provide higher resolution in the nonuniform field region [28–30]. This approach is described in the next section; however, it requires more complicated programming. Yet another approach, proposed by Shih and Hoefer [26], is to compute the structure several times using coarse meshes of different mesh parameter Δl, and then to extrapolate the obtained results for $\Delta l = 0$ as shown in Fig. 35. Both measures effectively reduce the error by one order of magnitude and simultaneously correct the velocity error.

6.4. Misalignment of Dielectric Interfaces and Reflecting Boundaries in Three-Dimensional Structures

In a three-dimensional TLM network, magnetic walls are placed across nodes, and so are electric walls in most cases, while dielectric interfaces appear halfway between nodes. This can be a problem when simulating planar structures such as microstrips or fin lines. In the TLM model, the dielectric either protrudes or is undercut by $\Delta l/2$ as shown in Fig. 33. The resulting error is acceptable in most cases in which the dielectric region itself contains at least three or four nodes. If a graded mesh is used, a fine mesh size should be deployed in the vicinity of dielectric-backed metal edges. To further reduce the error due to boundary misalignment, one must make two computations, one with recessed and one with protruding dielectric, and take the average of the results [27]. The problem does not occur in variations of the three-dimensional TLM method involving condensed node configurations proposed by Saguet and Pic [32, 33], and Johns [38–40], which will be described in the next section.

7. VARIATIONS OF THE TLM METHOD

Over the years, a number of researchers have modified the original TLM procedure with the aim to reduce errors, memory requirements, and CPU time. Some effort has also been directed toward improving the efficiency of programming techniques [25]. In the following, the most interesting and significant innovations are discussed.

7.1. Two-Dimensional TLM Networks with Variable Mesh Size

To ensure synchronism of scattering at all nodes, the conventional TLM network features square or cubic mesh elements with a uniform mesh parameter throughout. This can lead to considerable numerical expenditure if the structure contains sharp corners or fins producing highly nonuniform fields and thus demands a high-density mesh. Saguet and Pic [29] and Al-Mukthar and Sitch [30] have independently proposed ways to implement irregularly graded TLM meshes that, as in the finite element method, allow the density of the network to be adapted to the local nonuniformity of the fields.

Figure 36 shows a typical graded mesh used by Saguet and Pic [29] for the

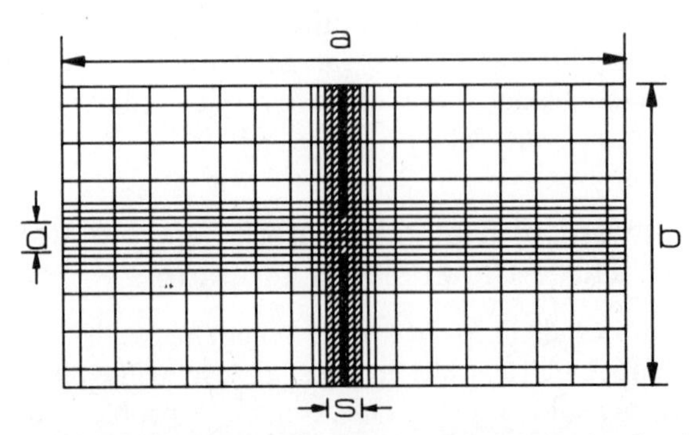

Fig. 36 Two-dimensional TLM network with variable mesh size for the computation of cutoff frequencies of finlines. (After Saguet and Pic [29].)

computation of cutoff frequencies in a finline. A generic node in the rectangular mesh section is shown in Fig. 37. The inductance and capacitance per unit length as well as the length of the branches are now different in the two directions, and the following equations apply:

$$\Delta l_1 \frac{\partial I_x}{\partial x} + \Delta l_2 \frac{\partial I_z}{\partial z} = -(C_1 \Delta l_1 + C_2 \Delta l_2) \frac{\partial V_y}{\partial t} \tag{113}$$

$$\frac{\partial V_y}{\partial x} = -L_1 \frac{\partial I_x}{\partial t} \tag{114}$$

$$\frac{\partial V_y}{\partial z} = -L_2 \frac{\partial I_z}{\partial t} \tag{115}$$

If length Δl_2 of the cell is N times as large as its width Δl_1, $L_2 = NL_2$, and $C_2 = C_1/N$, these equations become

$$\frac{\partial I_x}{\partial x} + \frac{\partial NI_z}{\partial z} = -2C_1 \frac{\partial V_y}{\partial t} \tag{116}$$

$$\frac{\partial V_y}{\partial x} = -L_1 \frac{\partial I_x}{\partial t} \tag{117}$$

$$\frac{\partial V_y}{\partial z} = -L_1 \frac{\partial NI_z}{\partial t} \tag{118}$$

One thus obtains the same system of equations as for the square mesh cell with the current I_z being replaced by NI_z.

The velocity of propagation is $1/\sqrt{2L_1C_1}$ in both main directions since

$$\sqrt{L_1C_1} = \sqrt{\frac{L_2}{N} C_2 N} = \sqrt{L_2C_2} \tag{119}$$

However, the characteristic impedances of the two lines are different:

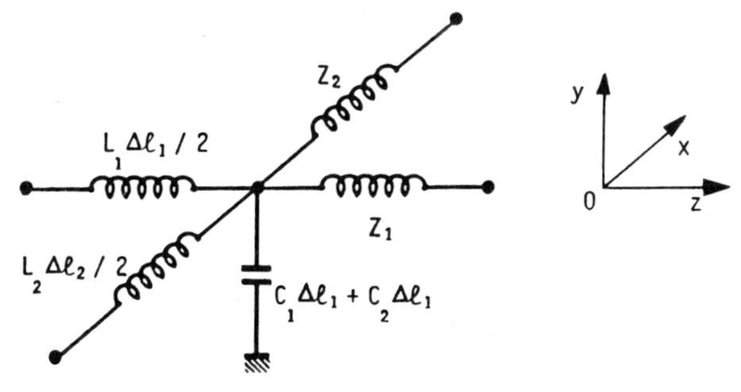

Fig. 37 Equivalent circuit of a node in a non-square two-dimensional TLM mesh. (After Saguet and Pic [29].)

$$\sqrt{\frac{L_1}{C_1}} = \frac{1}{N}\sqrt{\frac{L_2}{C_2}} \qquad \text{or} \qquad Z_1 = \frac{Z_2}{N} \tag{120}$$

Consequently the impulse scattering matrix of the nodes must be modified as follows:

$$V_n^r = \frac{2\sum_m Y_m Y_m^i}{\sum_m Y_m} - V_n^i \tag{121}$$

where Y_m is the characteristic admittance of the mth branch, V_n^r is the reflected impulse voltage on branch n, and V_m^i is the incident impulse voltage on branch m. Branch m can also represent a permittivity or a conductivity stub in the case of inhomogeneous and lossy media simulation.

Note that the size of the mesh cells is not arbitrary as in the case of finite elements; the length of each side is an *odd integer multiple* N of the smallest cell length in the network. To preserve synchronism, impulses traveling on longer branches are kept in store for N iterations before being reinjected at the next node. This increases the memory requirement by approximately a factor N; however, the CPU time is reduced proportionally to the number of nodes saved by the employment of a graded mesh.

For the configuration shown in Fig. 36, Saguet and Pic found that computing time was reduced between 3.5 and 5 times over a uniform mesh, depending on the relative size of the larger cells.

A different approach has been proposed by Al-Mukhtar and Sitch [28, 30]. They describe two possible ways to modify the characteristics of mesh

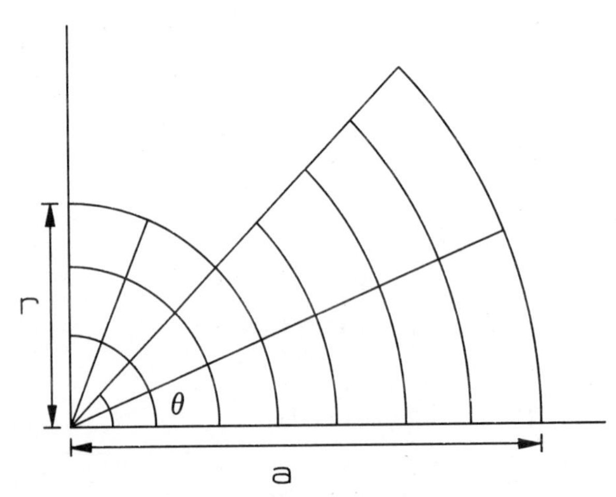

Fig. 38 A radial TLM mesh for the treatment of circular ridged waveguides. (After Al-Mukhtar [28].)

elements in order to ensure synchronism—one involving the insertion of series stubs between nodes and loading of nodes by shunt stubs, and the other involving modification of inductivity and capacity per unit length in such a way that propagation velocity in a branch becomes proportional to its length. The work by Al-Mukhtar and Sitch also covers the representation of radial meshes (see Fig. 38) as well as three-dimensional inhomogeneous structures. They report an economy of 45% in computer expenditure for a two-dimensional ridged waveguide problem, and a 40% reduction in storage and 80% in run time for a three-dimensional finline problem thanks to mesh grading. For further details the reader is directed to Reference 30.

7.2. Three-Dimensional TLM Networks with Condensed Nodes and Variable Mesh Size

We have seen that in the original three-dimensional TLM network, the mesh lines are interconnected at two-dimensional shunt and series nodes that are half a mesh parameter apart (see Fig. 23). For this reason the classical TLM network is termed an expanded-node network. The spatial separation of the six field components can introduce errors into the description of boundaries and dielectric interfaces. This inconvenience has stimulated the development of condensed node schemes by Saguet, Pic, and Tedjini [17, 32–34], Amer [35], and Johns [38–40]. While the first four references describe an asymmetrical node, those of Johns introduce a symmetrical node, a new concept that breaks away from the traditional lumped-element representation of the unit cell by deriving the scattering matrix of the node directly from the field equations.

All condensed node schemes lead to considerable savings in computer resources, particularly when they are combined with a graded mesh technique in which the density of the mesh can be adapted to the degree of field nonuniformity.

7.2.1. *Properties of the Condensed Asymmetrical Node*

The condensed asymmetrical node has been described in detail by Saguet et al. [17–34] and by Amer [35]. When an elementary transmission line cell is represented by a half T instead of a full T (see Fig. 39), the series and the shunt nodes can be connected at one point A, resulting in a single three-dimensional node with 12 branches. The network itself then becomes a three-dimensional Cartesian mesh with two lines, corresponding to two polarizations, in each branch.

As in the distributed node mesh, a dielectric can be simulated by adding open-circuited $\Delta l/2$ stubs at shunt nodes. Assuming for convenience that the link lines have a unity characteristic admittance, the stub characteristic admittance is $Y_0 = 4(\varepsilon_r - 1)$. Short-circuited $\Delta l/2$ series stubs at the series nodes simulate magnetic permeability. Their characteristic impedance is

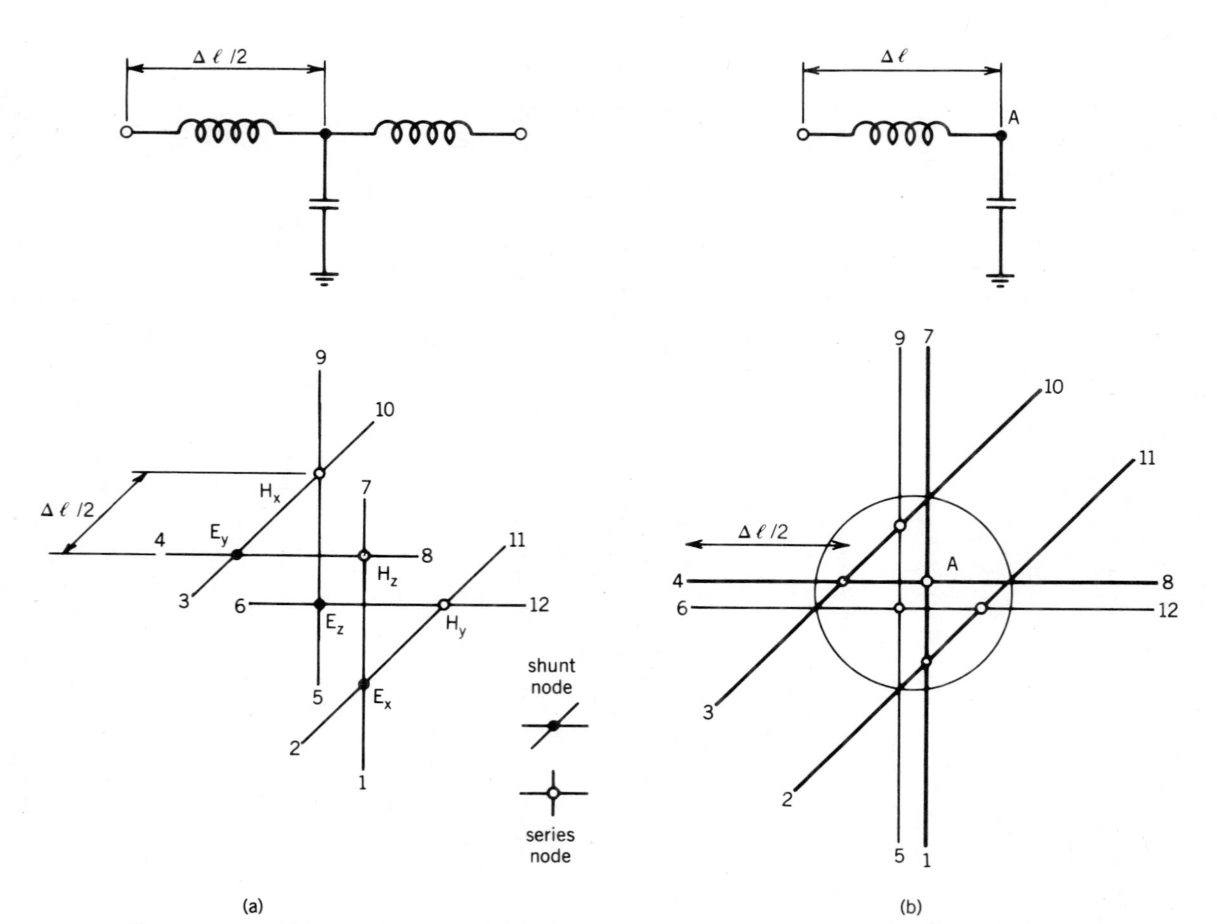

Fig. 39 Topologies of (*a*) a distributed TLM node and (*b*) a nonsymmetrical condensed node. The equivalent *LC* circuits show the structure of the generic shunt nodes. (After Saguet [17].)

$Z_0 = 4(\mu_r - 1)$. Loss stubs of characteristic admittance $G_0 = \sigma Z_{\mathrm{air}}\,\Delta l$ at the shunt nodes simulate finite conductivity.

A completely equipped condensed node has 21 branches (six less than the distributed node). Its equivalent circuit is shown in Fig. 40. Each branch is numbered and is represented by its immittance. An impulse incident on any given branch "sees" the driving immittance of the network presented to that branch. The scattering matrix of the asymmetrical node can thus be obtained from Fig. 40 [see (122) on page 558].

This matrix relates the reflected to the incident impulse voltages:

$$\begin{bmatrix} V_1 \\ V_2 \\ \cdot \\ \cdot \\ \cdot \\ V_{17} \\ V_{18} \end{bmatrix}^{\mathrm{refl}} = \boldsymbol{S} \cdot \begin{bmatrix} V_1 \\ V_2 \\ \cdot \\ \cdot \\ \cdot \\ V_{17} \\ V_{18} \end{bmatrix}^{\mathrm{inc}} \tag{123}$$

V_{13}, V_{14}, and V_{15} are the impulse voltages on the three permittivity stubs, and V_{16}, V_{17}, and V_{18} those on the permeability stubs. The impulses on the loss stubs are not included in the scattering matrix, even though the stub conductance enters into the expressions for the matrix elements:

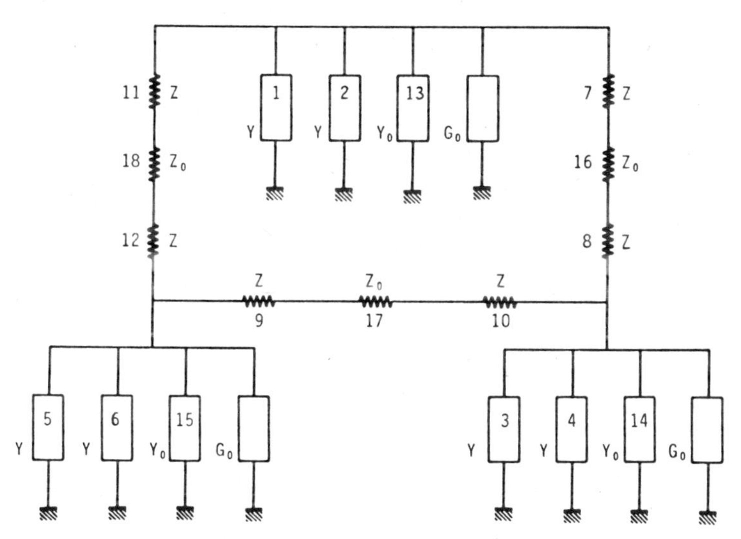

Fig. 40 Equivalent circuit of a three-dimensional condensed asymmetrical node equipped with reactive and lossy stubs. (After Saguet [17].)

$$
S = \left[\begin{array}{cccccccccccccccccc}
A-1 & A & B & B & B & B & C & -C & 0 & 0 & C & -C & AY_0 & BY_0 & BY_0 & CZ_0 & 0 & CZ_0 \\
A & A-1 & B & B & B & B & C & -C & 0 & 0 & C & -C & AY_0 & BY_0 & BY_0 & CZ_0 & 0 & CZ_0 \\
B & B & A-1 & A & B & B & -C & C & -C & C & 0 & 0 & BY_0 & AY_0 & BY_0 & -CZ_0 & -CZ_0 & 0 \\
B & B & A & A-1 & B & B & -C & C & -C & C & 0 & 0 & BY_0 & AY_0 & BY_0 & -CZ_0 & -CZ_0 & 0 \\
B & B & B & B & A-1 & A & 0 & 0 & C & -C & -C & C & BY_0 & BY_0 & AY_0 & 0 & CZ_0 & CZ_0 \\
B & B & B & B & A & A-1 & 0 & 0 & C & -C & -C & C & BY_0 & BY_0 & AY_0 & 0 & CZ_0 & CZ_0 \\
C & C & -C & -C & 0 & 0 & 1-D & D & E & -E & E & -E & CY_0 & -CY_0 & 0 & DZ_0 & EZ_0 & EZ_0 \\
-C & -C & C & C & 0 & 0 & D & 1-D & -E & E & -E & E & -CY_0 & CY_0 & 0 & -DZ_0 & -EZ_0 & -EZ_0 \\
0 & 0 & -C & -C & C & C & E & -E & 1-D & D & -E & E & 0 & -CY_0 & CY_0 & -EZ_0 & -DZ_0 & -EZ_0 \\
0 & 0 & C & C & -C & -C & -E & E & D & 1-D & E & -E & 0 & CY_0 & -CY_0 & EZ_0 & DZ_0 & EZ_0 \\
C & C & 0 & 0 & -C & -C & E & -E & -E & E & 1-D & D & CY_0 & 0 & -CY_0 & -EZ_0 & -EZ_0 & -DZ_0 \\
-C & -C & 0 & 0 & C & C & -E & E & E & -E & D & 1-D & -CY_0 & 0 & CY_0 & EZ_0 & EZ_0 & DZ_0 \\
A & A & B & B & B & B & C & -C & 0 & 0 & C & -C & AY_0-1 & BY_0 & BY_0 & CZ_0 & 0 & CZ_0 \\
B & B & A & A & B & B & -C & C & -C & C & 0 & 0 & BY_0 & AY_0-1 & BY_0 & -CZ_0 & -CZ_0 & 0 \\
B & B & B & B & A & A & 0 & 0 & C & -C & -C & C & BY_0 & BY_0 & AY_0-1 & 0 & CZ_0 & CZ_0 \\
CZ_0 & CZ_0 & -CZ_0 & -CZ_0 & 0 & 0 & DZ_0 & -DZ_0 & EZ_0 & -EZ_0 & EZ_0 & -EZ_0 & CY_0Z_0 & -CY_0Z_0 & 0 & 1-DZ_0 & EZ_0^2 & EZ_0^2 \\
0 & 0 & -CZ_0 & -CZ_0 & CZ_0 & CZ_0 & EZ_0 & -EZ_0 & -DZ_0 & DZ_0 & -EZ_0 & EZ_0 & 0 & -CY_0Z_0 & CY_0Z_0 & -EZ_0^2 & 1-DZ_0 & -EZ_0^2 \\
CZ_0 & CZ_0 & 0 & 0 & -CZ_0 & -CZ_0 & EZ_0 & -EZ_0 & -EZ_0 & EZ_0 & -DZ_0 & DZ_0 & CY_0Z_0 & 0 & -CY_0Z_0 & -EZ_0^2 & -EZ_0^2 & 1-DZ_0
\end{array}\right] \tag{122}
$$

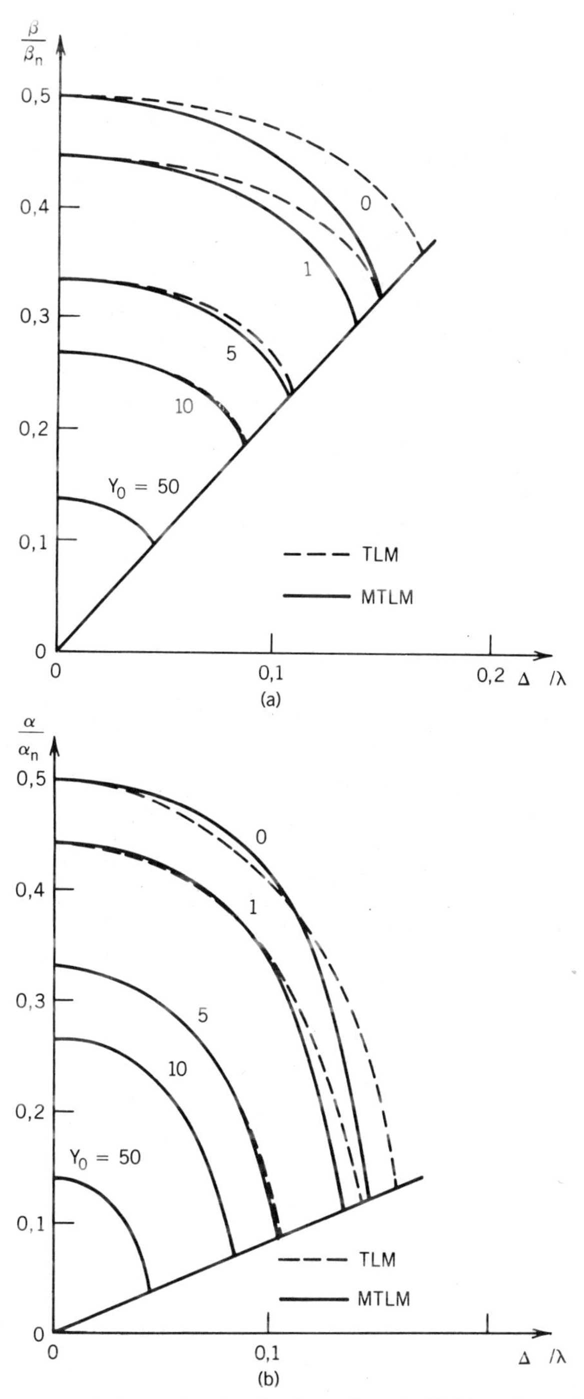

Fig. 41 Comparison of dispersion in distributed node TLM and condensed asymmetrical node TLM networks. (*a*) Phase constant; (*b*) attenuation constant. (After Saguet [17].)

$$
\begin{aligned}
A &= \frac{2[1+(2+Z_0)(2+Y_0+G_0)]}{(2+Y_0+G_0)[3+(2+Z_0)(2+Y_0+G_0)]} \\
B &= \frac{2}{(2+Y_0+G_0)[3+(2+Z_0)(2+Y_0+G_0)]} \\
C &= \frac{2}{3+(2+Z_0)(2+Y_0+G_0)} \\
D &= \frac{2[1+(2+Y_0)(2+Y_0+G_0)]}{(2+Z_0)[3+(2+Z_0)(2+Y_0+G_0)]} \\
E &= \frac{2}{(2+Z_0)[3+(2+Z_0)(2+Y_0+G_0)]}
\end{aligned}
\tag{124}
$$

At first glance the computation of this matrix seems to be quite arduous. However, many of its elements being identical, it is rather easy to program and requires 30% less computing time than the regular TLM procedure. The wave propagation characteristics of the condensed node mesh are slightly different from those of the regular mesh, as the comparison in Fig. 41 reveals. However, the resulting velocity error is virtually the same in both methods and can be corrected as discussed in Section 6.2. The error resulting from the asymmetry of the node structure is negligible, according to Amer [35]. In addition, the orientation of the nodes can be alternated or inverted within a structure so that symmetry can be achieved at a larger scale [17].

7.2.2. *The Graded Mesh Technique in Three Dimensions Using Condensed Asymmetrical Nodes*

The deployment of a graded mesh is essential to the efficient modeling of structures containing highly concentrated nonuniform field regions. In the following, the theory of the graded mesh, as developed by Saguet [17], will be described briefly.

Consider the elementary condensed series and shunt nodes in Fig. 42. $N_x\,\Delta l$, $N_y\,\Delta l$, $N_z\,\Delta l$ represent the length of their branches in the x, y, and z directions, respectively, Δl is the basic mesh parameter, and N_x, N_y, N_z are integers.

The node equation for the shunt node (Fig. 42*a*) is

$$
N_x \frac{\partial I_x}{\partial x} + N_z \frac{\partial I_z}{\partial z} = -(C_1 N_x + C_3 N_z)\frac{\partial V_y}{\partial t} \tag{125}
$$

The related field equation is

$$
\frac{\partial H_z}{\partial x} - \frac{\partial H_x}{\partial z} = -\varepsilon \frac{\partial E_y}{\partial t} \tag{126}
$$

Provided that

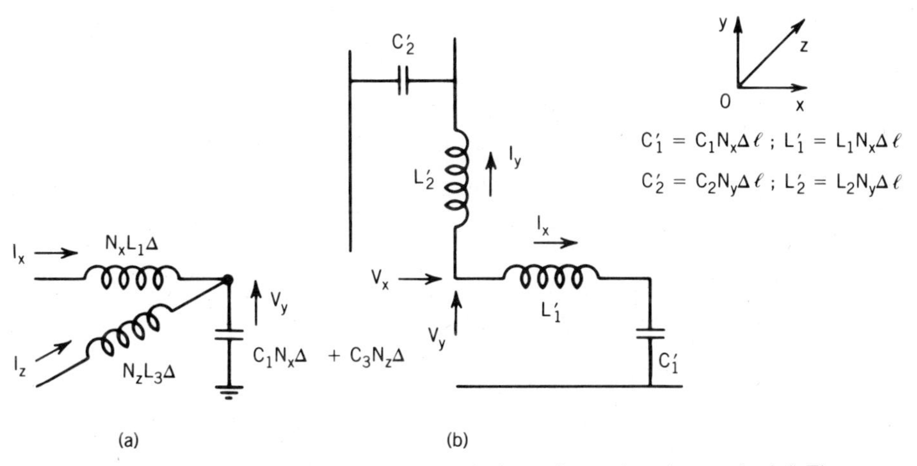

Fig. 42 Structure of basic nodes in a graded condensed node mesh. (*a*) Elementary shunt node; (*b*) elementary series node. (After Saguet [17].)

$$C_1 N_x = C_3 N_z = C \tag{127}$$

the following equivalences between network and field quantities can be established:

$$N_z I_z \equiv -H_x \qquad N_x I_x \equiv H_z \qquad V_y \equiv E_y \qquad \varepsilon \equiv 2C \tag{128}$$

The phase velocity must be the same in all directions. Hence

$$\sqrt{L_1 C_1} = \sqrt{L_2 C_2} = \sqrt{L_3 C_3} = \sqrt{LC} \tag{129}$$

The inductance values then become, in view of (127),

$$L_1 = LN_x \qquad L_2 = LN_y \qquad L_3 = LN_z \tag{130}$$

Consequently, the characteristic admittances of the mesh lines in the three coordinate directions are

$$Y_x = \sqrt{\frac{C_1}{L_1}} = \frac{1}{N_x}\sqrt{\frac{C}{L}} \qquad Y_y = \sqrt{\frac{C_2}{L_2}} = \frac{1}{N_y}\sqrt{\frac{C}{L}} \qquad Y_z = \sqrt{\frac{C_3}{L_3}} = \frac{1}{N_z}\sqrt{\frac{C}{L}} \tag{131}$$

The equations for the series node (Fig. 42*b*) can be obtained in a similar manner:

$$\frac{\partial N_x V_y}{\partial x} - \frac{\partial N_y V_x}{\partial y} = -(L_2 N_y + L_1 N_x)\frac{\partial I}{\partial t} \tag{132}$$

where

$$I = I_y = -I_x \tag{133}$$

The related field equation is

$$\frac{\partial E_y}{\partial x} - \frac{\partial E_x}{\partial y} = -\mu \frac{\partial H_z}{\partial t} \tag{134}$$

If we write

$$L_2 N_y = L_1 N_x = L \tag{135}$$

we obtain the following equivalences between network and field quantities:

$$I \equiv H_z \qquad N_x V_y \equiv E_y \qquad N_y V_x \equiv E_x \qquad \mu \equiv 2L \tag{136}$$

For equal phase velocity in all directions we must have

$$C_1 = CN_x \qquad C_2 = CN_y \qquad C_3 = CN_z \tag{137}$$

which yields the characteristic impedances of the three branches:

$$Z_x = \sqrt{\frac{L_1}{C_1}} = \frac{1}{N_x}\sqrt{\frac{L}{C}} \qquad Z_y = \sqrt{\frac{L_2}{C_2}} = \frac{1}{N_y}\sqrt{\frac{L}{C}} \qquad Z_z = \sqrt{\frac{L_3}{C_3}} = \frac{1}{N_z}\sqrt{\frac{L}{C}} \tag{138}$$

For convenience, the reference admittance $Y = \sqrt{L/C}$ is set equal to unity in all computations. Note that for a given direction, the characteristic impedance ratio of the mesh lines is N^2 (N_x^2, N_y^2, N_z^2, respectively, depending on the nature of the node to which they are connected). On the other hand, the field magnitudes must be identical at all elementary nodes. However, that condition is not satisfied by the equivalence relations (128) and (136). It thus becomes necessary to place ideal transformers between series and shunt nodes to establish the correct field values and to match the line impedances.

The equivalent circuit of a three-dimensional condensed node with three different branch lengths is shown in Fig. 43. For simplicity, reactive and dissipative stubs have been omitted. Z_i and Y_j are the characteristic branch immittances of the series and shunt nodes, respectively.

For field modeling, the structure under study is divided into subregions in which the mesh size, permittivity, and permeability are constant, so that all nodes in one subregion havea the same impulse scattering matrix. This matrix is computed only once for each subregion. During the iteration process, impulses traveling on branches longer than the elementary length Δl are stored in memory during a number of iterations corresponding to N_x, N_y, or N_z, as the case may be, before they are reinjected into the

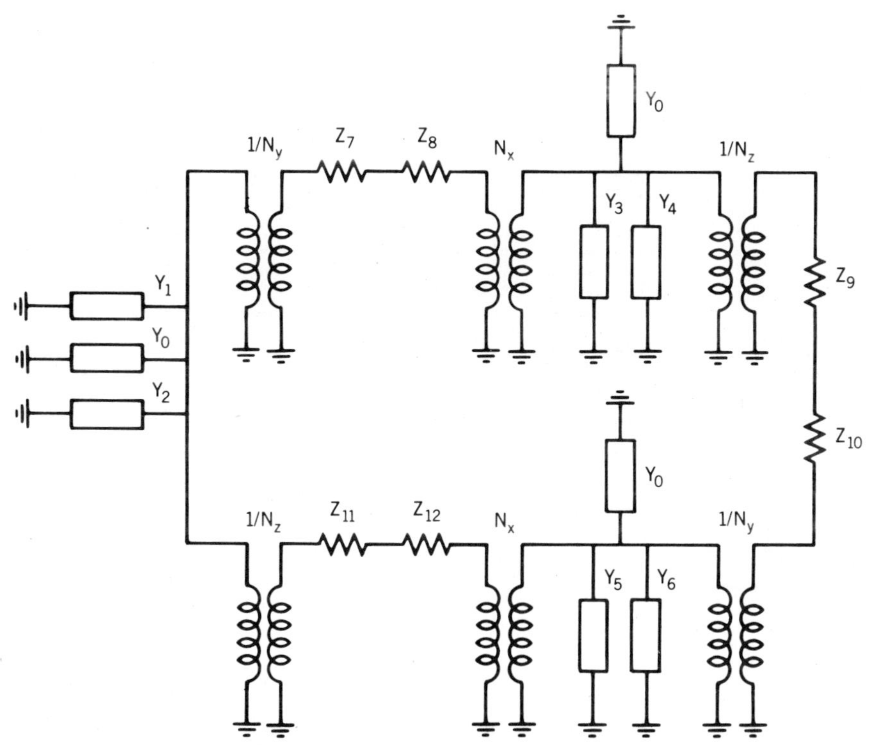

Fig. 43 Equivalent circuit of a three-dimensional condensed asymmetrical node in a graded TLM mesh. Stubs are not included. (After Saguet [17].)

neighboring node. To ensure synchronism of scattering events, N_x, N_y, and N_z must be integer numbers. All further processing is the same as in the conventional TLM procedure.

7.2.3. Properties of the Condensed Symmetrical Node

A symmetrical three-dimensional condensed node has been developed by Johns [38–40]. In contrast to the node structures described previously, this node cannot be represented by an equivalent lumped element network (which makes it more difficult to study the general wave properties of the corresponding network). Thus, its scattering matrix must be derived directly from the behavior of the fields.

Figure 44 depicts the symmetrical condensed node. Each of the six branches consists of two uncoupled transmission lines arranged in space quadrature, representing two polarizations. All 12 lines have the same characteristic impedance equal to the free-space value. The ports are numbered and oriented according to the voltages shown in Fig. 44.

Associating, for example, with a voltage impulse V_1^i incident at port 1,

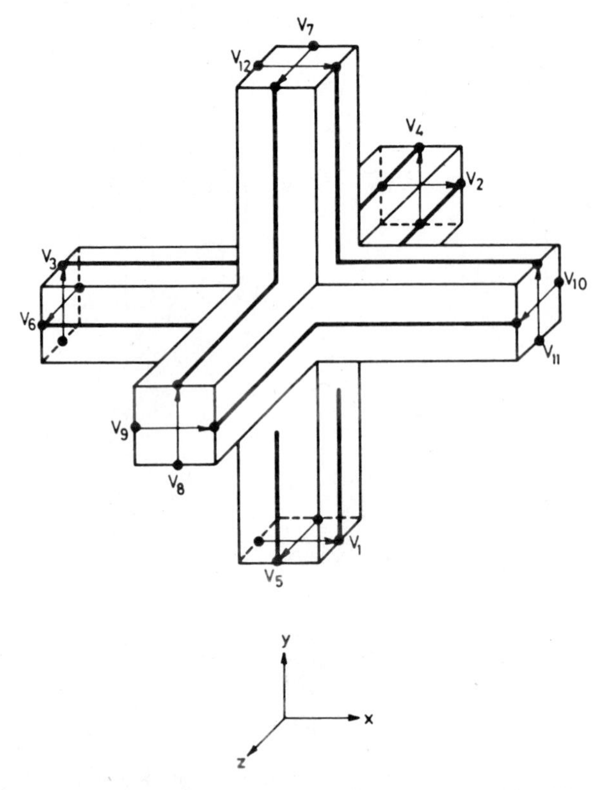

Fig. 44 The symmetrical condensed three-dimensional node. (After Johns [40].) Copyright © 1987 IEEE.

the field quantities E_x and H_z, and invoking the field equations that involve coupling between these two quantities, namely

$$\frac{\partial H_z}{\partial y} - \frac{\partial H_y}{\partial z} = \varepsilon \frac{\partial E_x}{\partial t} \tag{139}$$

and

$$\frac{\partial E_y}{\partial x} - \frac{\partial E_x}{\partial y} = -\mu \frac{\partial H_z}{\partial t} \tag{140}$$

Johns obtains scattered impulses in ports 1, 2, 3, 9, 11, and 12. If the same procedure is applied to all other ports, and the results are combined while respecting the field continuity and energy conservation conditions, the following impulse scattering matrix for the symmetrical condensed node results:

$$
S = \frac{1}{2}\begin{bmatrix}
 & 1 & 1 & & & & & & 1 & & -1 & \\
1 & & & & & 1 & & & & -1 & & 1 \\
1 & & & 1 & & & & 1 & & & & -1 \\
 & & 1 & & 1 & & -1 & & & & 1 & \\
 & & & 1 & & 1 & & -1 & & 1 & & \\
 & 1 & & & 1 & & 1 & & -1 & & & \\
 & & & -1 & & 1 & & 1 & & 1 & & \\
 & & 1 & & -1 & & 1 & & & & 1 & \\
1 & & & & & -1 & & & & 1 & & 1 \\
 & -1 & & & 1 & & 1 & & 1 & & & \\
-1 & & & 1 & & & & 1 & & & & 1 \\
 & 1 & -1 & & & & & & 1 & & 1 &
\end{bmatrix} \tag{141}
$$

Reactive $\Delta l/2$ stubs can be added to the node to simulate dielectric permittivity and magnetic permeability, as in other TLM networks. A fully equipped node is characterized by a scattering matrix of size 12×12, which is derived and presented in Reference 40.

The excitation of the network and the determination of fields and impedances is similar to the procedure employed in the other mesh schemes. For example, if only the field $E_x = 1$ is to be excited at a node with a permittivity stub (port 13, not shown in Fig. 44), the following incident impulses are required:

$$
V_1^1 = \tfrac{1}{2} \qquad V_2^i = \tfrac{1}{2} \qquad V_9^i = \tfrac{1}{2} \qquad V_{12}^i = \tfrac{1}{2} \qquad V_{13}^i = \tfrac{1}{2} \tag{142}
$$

The low-frequency velocity of waves on the symmetrical condensed-node mesh is half that on the mesh lines, i.e., $c/2$. A full propagation analysis of the symmetrical condensed node has not yet been presented. However, Johns [40] has determined that at 45° with respect to the main axes the dispersion in the symmetrical condensed node mesh is the same as that in the expanded node mesh, whereas in the direction of the main axes, where dispersion is at its worst in the expanded node mesh, there is no dispersion at all in the symmetrical condensed node mesh.

The validity of the latter has been tested extensively by Allen et al. [41], who affirm that the symmetrical condensed node is more accurate than the other mesh schemes.

7.3. The Scalar TLM Method

In those cases where electromagnetic fields can be decomposed into TE and TM modes (or LSE and LSM modes) it is only necessary to solve the scalar wave equation. Choi and Hoefer [42] have described a scalar TLM network to simulate a single field component or a Hertzian potential in three-dimensional space. The scalar TLM mesh can be thought of as a two-

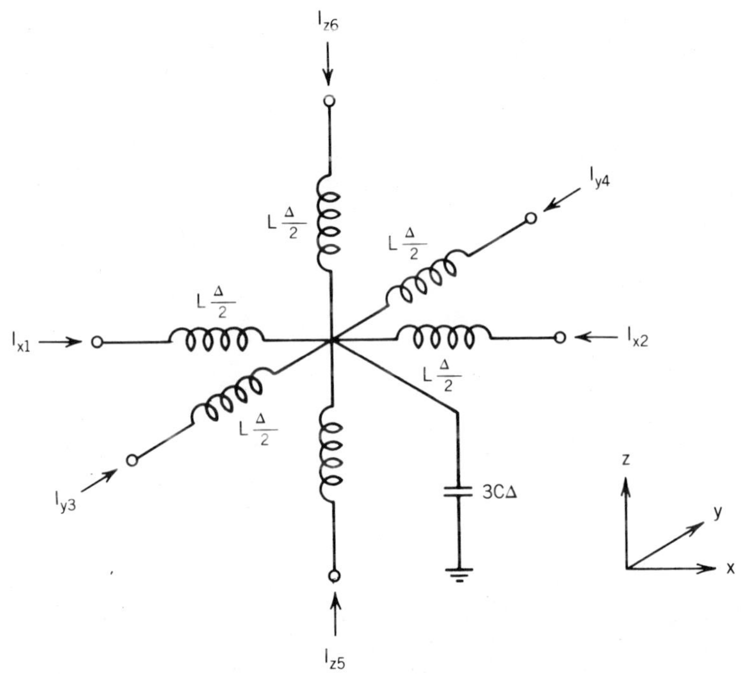

Fig. 45 Basic node of a three-dimensional scalar TLM network. (After Choi and Hoefer [42].) Copyright © 1984 IEEE.

dimensional network to which additional transmission lines are connected orthogonally at each node as shown in Fig. 45. Such a structure could be realized in the form of a three-dimensional grid of coaxial lines.

At low frequencies ($\Delta l/\lambda \ll 1$), the differential equations governing the voltages and currents at the scalar node are

$$\left(\nabla^2 - 3LC \frac{\partial^2}{\partial t^2}\right) V = 0 \tag{143}$$

and

$$\left[\nabla^2 - (\nabla \times \nabla) - 3LC \frac{\partial^2}{\partial t^2}\right] \mathbf{I} = 0 \tag{144}$$

where $\mathbf{I} = I_x \hat{\mathbf{x}} + I_y \hat{\mathbf{y}} + I_z \hat{\mathbf{z}}$.

Note that although eq. (144) may not look like a wave equation, the second term involving the double curl of $\mathbf{I}$ is zero. Thus, V and each component of $\mathbf{I}$ satisfy the scalar wave equation

$$\left(\nabla^2 - 3LC \frac{\partial^2}{\partial t^2}\right) \psi(\mathbf{u}, t) = 0 \tag{145}$$

The impulse scattering matrix of the scalar node is found from Fig. 45:

$$S = \frac{1}{3}\begin{bmatrix} -2 & 1 & 1 & 1 & 1 & 1 \\ 1 & -2 & 1 & 1 & 1 & 1 \\ 1 & 1 & -2 & 1 & 1 & 1 \\ 1 & 1 & 1 & -2 & 1 & 1 \\ 1 & 1 & 1 & 1 & -2 & 1 \\ 1 & 1 & 1 & 1 & 1 & -2 \end{bmatrix} \tag{146}$$

The voltage impulses traveling across such a network represent the scalar variable to be simulated. Boundary reflection coefficients depend on both the nature of the boundary and that of the quantity to be simulated. For example, impulses will be subject to a reflection coefficient of −1 at a lossless electric wall if they represent either a tangential electric or a normal magnetic field component. A normal electric or a tangential magnetic field will be reflected with a coefficient of +1 in the same circumstances.

The slow-wave velocity in the three-dimensional scalar mesh is $c/\sqrt{3}$ as opposed to $c/2$ in the conventional TLM network. For higher frequencies, the velocity dispersion characteristics have been computed by Choi [42]. They are:

Along the Main Axes

$$\frac{\beta}{\beta_n} = \frac{\pi\,\Delta l/\lambda}{\sin^{-1}[\sqrt{3}\sin(\pi\,\Delta l/\lambda)]} \tag{147}$$

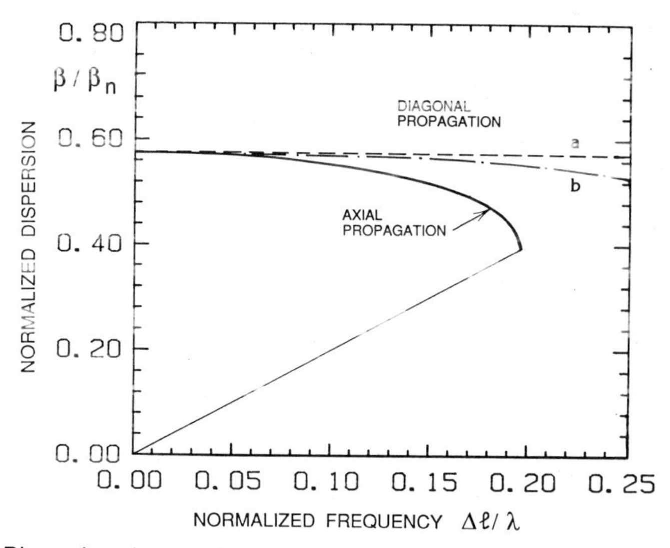

Fig. 46 Dispersion characteristics of the three-dimensional scalar TLM network for plane wave propagation in the direction of (*a*) a space diagonal, (*b*) a face diagonal, and a principal axis.

Along the Diagonal of a Unit Square Plane

$$\frac{\beta}{\beta_n} = \frac{\pi \, \Delta l/\lambda}{\sqrt{2} \sin^{-1}[\sqrt{3/2} \sin(\pi \, \Delta l/\lambda)]} \tag{148}$$

Along the Diagonal of a Unit Cube

$$\frac{\beta}{\beta_n} = 1/\sqrt{3} \tag{149}$$

These dispersion characteristics are plotted in Fig. 46.

Dielectric or magnetic materials as well as losses may be simulated using reactive and dissipative stubs, and the extension of the above expressions to include them is straightforward.

The scalar method requires only one-fourth as much memory space and is seven times faster than the conventional method for a commensurate problem. However, its application to electromagnetic problems is severely restricted, as it can be applied to scalar wave problems only.

7.4. Alternative Networks for Modeling Maxwell's Equations

Yoshida and co-workers [20, 24, 31, 43] have described a network similar to the TLM mesh, differing only in the way the basic cell element has been modeled. Instead of series and shunt nodes, this network contains so-called electric and magnetic nodes that are both "shunt-type nodes": While at the electric node the voltage variable represents an electric field, at the magnetic node it symbolizes a magnetic field. The resulting ambivalence in the nature of the network voltage and current must be removed by inserting gyrators between the two types of nodes as shown in Fig. 47. The wave properties of this network are identical with that of the conventional TLM mesh. Errors and limitations are the same, and so are the possibilities of introducing losses in isotropic as well as anisotropic dielectric and magnetic materials.

8. APPLICATIONS OF THE TLM METHOD

In the previous sections, the flexibility, versatility, and generality of the transmission line matrix method and its variants have been demonstrated. In the following, an overview of potential applications of the method will be given, and references describing specific applications will be indicated. This list is not exhaustive, and many more applications can be found, not only in electromagnetism but also in other fields dealing with wave phenomena, such as optics and acoustics.

For completeness, it should be mentioned that the TLM procedure is by no means restricted to wave-related problems but can also be used to model

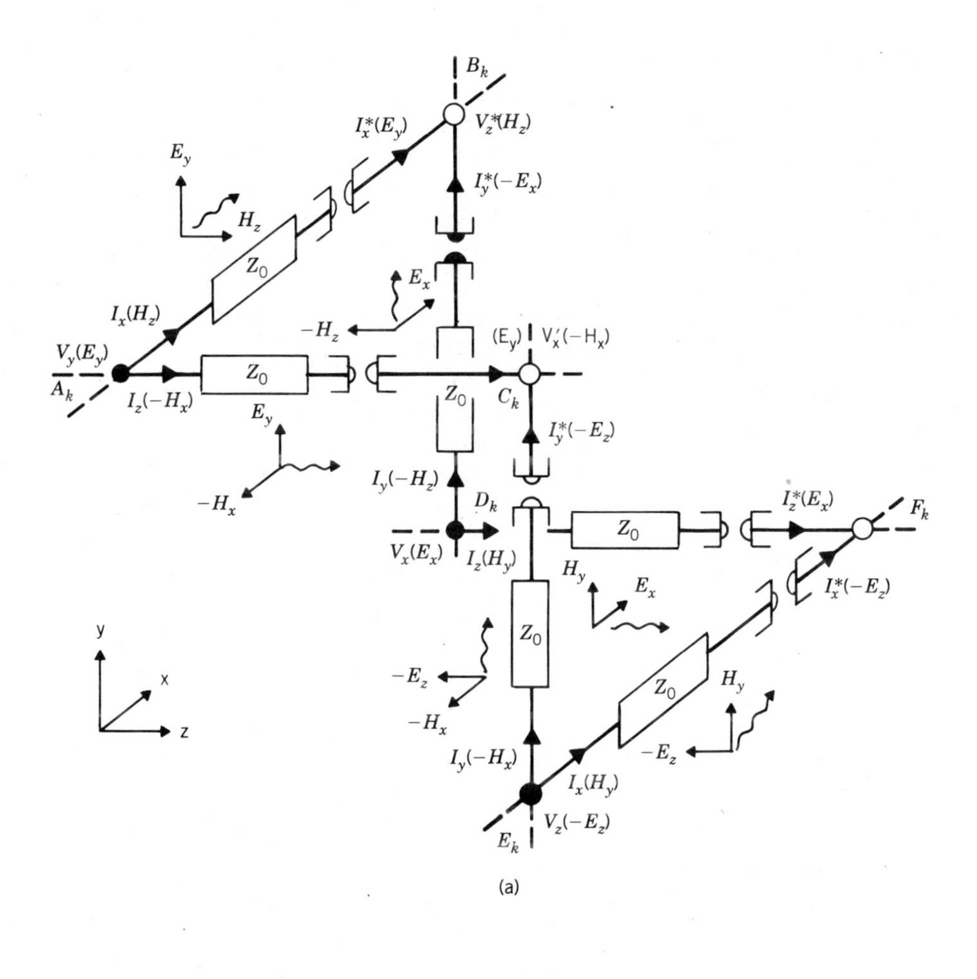

(a)

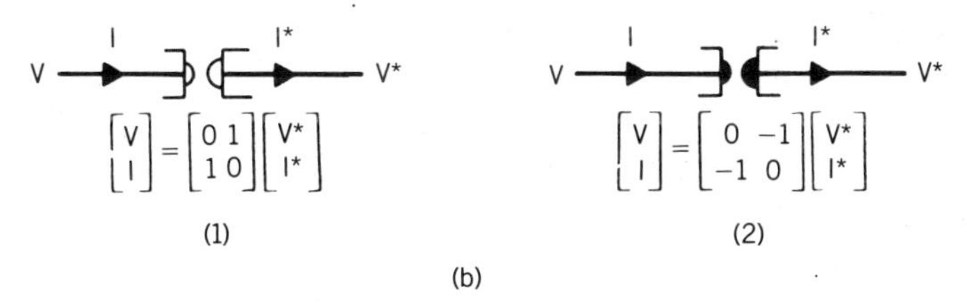

(b)

Fig. 47 Alternative network for three-dimensional TLM analysis proposed by Yoshida and Fukai [43]. (*a*) Equivalent circuit of an alternative three-dimensional TLM cell. (*b*) Definition of gyrators in (*a*): (1) positive gyrator, (2) negative gyrator.

and solve linear and nonlinear lumped networks [44–46] and diffusion problems [47, 48]. Readers with a special interest in these applications should consult these references for more details.

Wave problems can be simulated in unbounded and bounded space, either in the time domain or, via Fourier analysis, in the frequency domain. Arbitrary homogeneous or inhomogeneous nonlinear structures with anisotropic space- and time-dependent electrical properties, including losses, can be simulated in two and three dimensions. This can be accomplished by making the characteristic immittances of the reactive and absorbing stubs dependent on the instantaneous values of the local field components.

The following are some typical application examples.

1. Two-dimensional scattering problems in rectangular waveguides (field distribution of propagating and evanescent modes, wave impedance, scattering parameters of discontinuities)

   - Open-circuited rectangular waveguide (TE_{10}) [2]
   - Bifurcation in rectangular waveguide (TE_{10}) [2]
   - Scattering on arbitrarily shaped two-dimensional discontinuities in rectangular waveguide (TE_{10}) including losses

2. Two-dimensional eigenvalue problems (Eigenfrequencies, mode fields)

   - Cutoff frequencies and mode fields in homogeneous waveguides of arbitrary cross section, such as ridged waveguides [6, 8, 13, 26, 27]
   - Cutoff frequencies and mode fields in inhomogeneous waveguides of arbitrary cross section, such as dielectric loaded waveguides, finlines, image lines [7, 13, 16, 17, 19, 22, 23, 26, 27, 29]

3. Three-dimensional eigenvalue and hybrid field problems (dispersion characteristics of planar transmission lines, wave impedances, losses, eigenfrequencies, mode fields, Q factors of resonators, modeling of discontinuities)

   - Characteristics of dielectric-loaded cavities [10, 13–15, 17, 19, 27, 28, 32–34, 42, 49]
   - Dispersion characteristics and scattering in inhomogeneous planar transmission line structures, including anisotropic substrate [11–14, 17–19, 27, 31–34, 37, 42, 43, 49–52]
   - Modeling of uniaxial and multiaxial discontinuities in inhomogeneous planar transmission line structures (microstrips, finlines) [12, 13, 17, 19, 27, 50, 53]
   - Transient analysis of transmission line structures [20, 24, 31, 43]

4. Other applications, such as

   - Imaging, free-space scattering, and EMP response of aircraft [41, 54, 55, 55a]
   - Lumped network analysis [35, 44–46, 56, 57]
   - Diffusion problems [47, 48]
   - Underwater acoustic scattering [58]
   - Computation of magnetic fields [59]

To give the reader an opportunity to study the TLM programming technique and to solve some typical waveguide field problems, a simple two-dimensional TLM program is reproduced and explained in the appendix. It can treat two-dimensional inhomogeneous wave problems with arbitrary geometry. The program is written in Turbo-Pascal and runs on a personal computer.

General-purpose two- and three-dimensional programs in Fortran can be found in Akhtarzad's thesis [13] or are available from Johns [60] or the author. They can be adapted to most of the applications described above. If the various improvements and modifications described in Section 7 are implemented in these programs, versatile and powerful numerical tools for the solution of complicated field problems are indeed obtained.

9. DISCUSSION AND CONCLUSION

In this chapter the physical principles, the formulation, and the implementation of the transmission line matrix method of analysis have been described. Numerous features and applications of the method have been discussed, in particular the principal sources of error and their correction, the inclusion of losses, inhomogeneous and anisotropic properties of materials, and the capability to analyze transient as well as steady-state wave phenomena.

The method is limited only by the amount of memory storage required, which depends on the complexity of the structure and the nonuniformity of fields set up in it. In general, the smallest feature in the structure should contain at least three nodes for good resolution. The total storage requirement for a given computation can be found by considering that each two-dimensional node requires five real number storage places, and an additional number equal to the number of iterations are needed to store the output impulse function. A basic three-dimensional node requires 12 number locations; if it is completely equipped with permittivity, permeability, and loss stubs, the required number of stores goes up to 26. Again, one real number must be stored per output function and per iteration. The number of iterations required varies between several hundred and several thousand, depending on the size and complexity of the TLM mesh.

With respect to computational expenditure, the TLM method compares favorably with finite element and finite difference methods. Its accuracy is even slightly better by virtue of the Fourier transform, which ensures that the field function between nodes is automatically circular rather than linear as in the two other methods. There exist a number of similarities between the TLM method and the finite difference–time domain method, which have been discussed in a short paper by Johns [61].

The main advantage of the TLM method, however, is the ease with which even the most complicated structures can be evaluated. The great flexibility and versatility of the method reside in the fact that the TLM network incorporates all the properties of the electromagnetic fields and their interaction with the boundaries and materials. Hence, it is not necessary to reformulate the electromagnetic problem for every new structure; its parameters are simply entered into a general-purpose program in the form of codes for boundaries, losses, permeability and permittivity, and excitation of the fields. Furthermore, by solving the problem in an iterative fashion through simulation of wave propagation in the time domain, the solution of large numbers of simultaneous equations is avoided. There are no problems with convergence, stability, or spurious solutions.

Another advantage of the TLM method resides in the large amount of information generated in one single computation. Not only is the impulse response of a structure obtained, yielding, in turn, its response to any excitation, but also the characteristics of the dominant and higher-order modes are accessible in the frequency domain through the Fourier transform.

To increase numerical efficiency and reduce the various errors associated with the method, more programming effort must be invested. Such an effort may be worthwhile when faced with the problem of modeling a three-dimensional discontinuity in an inhomogeneous transmission medium or when studying the overall electromagnetic properties of a monolithic circuit.

Finally, the TLM method can be applied to problems in other areas such as thermodynamics, optics, and acoustics. Not only is it a very powerful and versatile numerical tool, but because of its affinity with the mechanism of wave propagation, it can provide new insights into the physical nature and behavior of waves.

ACKNOWLEDGMENTS

I wish to thank my colleagues, Professor P. Saguet, of the Institut National Polytechnique de Grenoble, and Professor A. Beyer, of the University of Duisburg, for numerous discussions and helpful suggestions during the preparation of the manuscript.

APPENDIX. A TWO-DIMENSIONAL INHOMOGENEOUS TLM PROGRAM FOR THE PERSONAL COMPUTER

This appendix contains:

- The listing of a two-dimensional TLM program (TLM_INHO.PAS) written in Pascal for compilation with Turbo-Pascal
- A typical input file (TLM_INHO.INP) with detailed explanations of its various features
- The output file (TLM_INHO.OUT) created by the program when the above input file is processed

As an example, the dominant cutoff frequency of a rectangular waveguide partially filled with a dielectric slab is computed. By working through the example, the reader will learn how to input other structures and interpret the results.

For speedy execution, it is recommended that the program TLM_INHO.PAS be compiled with Turbo-87, provided that a 7078 math coprocessor is installed in the computer.

The size of the arrays used in the computations can be increased to cover a larger grid and more iterations. However, this will increase memory size and CPU time, and it may be preferable to implement the program on a mainframe computer. Nevertheless, in its present form the program is excellent for experimenting and learning about the TLM method.

```
- - - - - - - - - - - - - -  top of file  - - - - - - - - - - - - - -

                              {FILE TLM_INHO.PAS}
program tlm_inho(input,output,datin,datout);

 {                                CREATED BY

        THE LABORATORY FOR ELECTROMAGNETICS AND MICROWAVES,
                DEPARTMENT OF ELECTRICAL ENGINEERING,
                        UNIVERSITY OF OTTAWA,
                OTTAWA, ONTARIO, CANADA  K1N 6N5

                                featuring

 a simplified 2-dimensional TLM program for inhomogeneous structures.
 The program accepts input from file "TLM_INHO.inp" and returns results
 in "TLM_INHO.out". This version contains an optional correction for
 the velocity error in axial direction and a polynomial curve-fit
 routine for the accurate positioning of the spectral peak.}

type
   title=string[80];            {title lines used in data input file}
var
   nx,ny:           integer;    {number of nodes in mesh}
   io,it,jo,ni:     integer;    {output point (io,jo), output type & number
                                 of iterations}
```

```
  kb,kc,kd,ke:   integer;   {number of boundaries,computational boxes,
                             dielectric boundaries & excitation points
                             or lines}

  v:  array[1..5,1..12,1..12] of real;  {working space }
  r:        array [1..12] of real;
  rc,rd: array [1..10] of real ;    {reflection coefficients,relative
                                     permittivity of dielectric}
  va: array [1..6] of real;         {initial excitation values}
  ehre,ehim,d:  real;               {field magnitudes (real&imaginary)
                                     and normalized frequencies}
  eh:    array [1..300] of real;    {storage for field results}
  ib:    array [1..12,1..8] of integer;
  ibd: array [1..10,1..8] of integer; {waveguide & air-dielectric
                                       boundaries and codes}
  ie: array [1..5,1..7] of integer;   {excitation points and code
                                       (115)}
  ia: array[1..8,1..4] of integer;    {computational boxes = working
                                       areas with homogeneous
                                       permittivity}

  datin,datout:   text;               {input & output files for
                                       data & results}

  header:            title;
  l,jd,j,m,i,ic,
  pt,ptp,ptm,nn: integer;             {iteration counters}
  pcf,cf,d1,d2,ds: real;              {normalized frequencies & step
                                          size}

  peak,a,cs,max,yo: real;
  npt:               integer;
  data: array [1..101,1..2] of real;
  out: array [1..101,1..70] of char;

procedure readnxny;
begin
   readln(datin,header);writeln(datout,header);
   readln(datin,header);writeln(datout,header);
   readln(datin,nx,ny);
   writeln(datout,nx:4,ny:4);

end;

procedure readbound;
begin
   readln(datin,header);writeln(datout,header);
   readln(datin,header);writeln(datout,header);
   kb:=0;
   repeat
      kb:=kb+1;
      for m:=1 to 8
      do begin
         read(datin,ib[kb,m]);
         write(datout,ib[kb,m]:4);
         if m=4 then write(datout,'    ');
      end;
      readln(datin,r[kb],it);
      writeln(datout,'      ',r[kb]:10:6,'           ',it);
   until it=0;
end;

procedure readdielbound;
begin
   kd:=0;
   readln(datin,header);writeln(datout,header);
   readln(datin,header);writeln(datout,header);
```

```
   repeat
      kd:=kd+1;
      for m:=1 to 8
      do begin
         read(datin,ibd[kd,m]);
         write(datout,ibd[kd,m]:4);
         if m=4 then write(datout,'    ');
      end;
      readln(datin,rc[kd],it);
      writeln(datout,'             ',rc[kd]:8:6,'             ',it);
   until it=0;
end;

procedure readcompbox;
begin
   readln(datin,header);writeln(datout,header);
   readln(datin,header);writeln(datout,header);
   kc:=0;
   repeat
      kc:=kc+1;
      for m:=1 to 4
      do begin
         read(datin,ia[kc,m]);
         write(datout,ia[kc,m]:4);
      end;
      readln(datin,rd[kc],it);
      writeln(datout,'                  ',rd[kc]:8:4,
      '                              ',it);
   until it=0;
end;

procedure readexcitation;
begin
   readln(datin,header);writeln(datout,header);
   readln(datin,header);writeln(datout,header);
   ke:=0;
   repeat
      ke:=ke+1;
      for m:=1 to 7
      do begin
         read(datin,ie[ke,m]);
         write(datout,ie[ke,m]:4);
         if m=4 then write(datout,'  ');
      end;
      readln(datin,va[ke],it);
      writeln(datout,'       ',va[ke]:10:6,'                         ',it);
   until it=0;
end;

procedure readfreq;
begin
    readln(datin,header);
    readln(datin,header);
    readln(datin,d1,d2,ds);
    readln(datin,header);
    readln(datin,header);
    readln(datin,io,jo,l,ni,yo);
    writeln(datout,'Output point is (',io:4,',',jo:4,')');
    writeln(datout,'Number of iterations is ',ni:4);
    writeln(datout,'Permittivity stub admittance is  ',yo:6:4);
    writeln(datout,'   D1           D2           Step Size');
    writeln(datout,d1:8:6,'    ',d2:8:6,'    ',ds:8:4);
    writeln('Finished reading input. Working hard on TLM now.');
    writeln('Please be patient!');end;
```

```
procedure iterate;
var
  a,vx,vy,vxy:   real;

begin
   { CLEAR WORKING SPACE }
   for j:=1 to ny
   do begin
      for i:= 1 to nx
      do begin
         for m:= 1 to 5
         do begin
            v[m,i,j]:=0.0;
         end;
      end;
   end;

  { INITIALIZE EXCITATION POINTS }
   for nn:=1 to ke
   do begin
      for j:=ie[nn,3] to ie[nn,4]
      do begin
         for i:=ie[nn,1] to ie[nn,2]
         do begin
            m:=ie[nn,5];
            while(m<=ie[nn,7])
            do begin
               v[m,i,j]:=va[nn];
               m:=m+ie[nn,6];
            end;
            v[5,i,j]:=va[nn];
          end;
      end;
   end;

   for ic:=1 to ni
   do begin

   { SET UP BOUNDARY CONDITIONS }
     for nn:=1 to kb
     do begin
        for j:= ib[nn,3] to ib[nn,4]
        do begin
           for i:= ib[nn,1] to ib[nn,2]
           do begin
              vxy:= v[ib[nn,6],i,j];
              v[ib[nn,6],i,j]:=r[nn]*v[ib[nn,5],i+ib[nn,8],j+ib[nn,7]];
              v[ib[nn,5],i+ib[nn,8],j+ib[nn,7]]:=r[nn]*vxy;
           end;
        end;
     end;

  { PERFORM IMPEDANCE MODIFICATIONS AT AIR-DIELECTRIC BOUNDARIES }
    if ibd[1,1]<>0
    then begin
     for nn:=1 to kd
     do begin
        for j:= ibd[nn,3] to ibd[nn,4]
        do begin
           for i:=ibd[nn,1] to ibd[nn,2]
           do begin
              vx:=v[ibd[nn,6],i,j];
              vy:=v[ibd[nn,5],i+ibd[nn,8],j+ibd[nn,7]];
              v[ibd[nn,6],i,j]:=-rc[nn]*vy+(1.0+rc[nn])*vx;
```

```
              v[ibd[nn,5],i+ibd[nn,8],j+ibd[nn,7]]:=rc[nn]*vx+
              (1.0-rc[nn])*vy;
            end;
          end;
        end;
      end;

 { BOXES COMPUTATION - BASED ON APPROPRIATE VOLTAGE SCATTERING MATRIX
   FOR SHUNT INHOMOGENEOUS NODE }
      for nn:=1 to kc
      do begin
         for j:=ia[nn,3] to ia[nn,4]
         do begin
            for i:=ia[nn,1] to ia[nn,2]
            do begin
               a:=(v[1,i,j+1]+v[1,i,j]+v[2,i,j]+v[2,i+1,j]
                  +v[5,i,j]*rd[nn])*2.0/(rd[nn]+4.0);
               v[5,i,j]:=a-v[5,i,j];
               v[1,i,j]:=a-v[1,i,j];
               v[2,i,j]:=a-v[2,i,j];
               vy:=a-v[1,i,j+1];
               vx:=a-v[2,i+1,j];
               v[2,i+1,j]:=v[4,i,j];
               v[1,i,j+1]:=v[3,i,j];
               v[3,i,j]:=vy;
               v[4,i,j]:=vx;
            end;
         end;
      end;
      case 1 of
         3:eh[ic]:=0.5*(v[1,io,jo]+v[2,io,jo]+v[3,io,jo]+v[4,io,jo]);
         2:eh[ic]:=v[3,io,jo]-v[1,io,jo];
         1:eh[ic]:=v[4,io,jo]-v[2,io,jo];
      end;
   end;
end;

procedure fourier;
var
  r,ra,rb,t:  real;
  cs,u,uk:    real;
  ehre,ehim,ehmod: real;
  d:          real;
begin
   npt:=0;
   t:=0.0;
   max:=0;
   r:=0.5*sqrt(1.0+t*t);
   if t<> 0.0
   then ra:=6.283184*sqrt(-0.5*r)
   else ra:=0.0;
   rb:=6.283184*sqrt(0.5+r);
   d:=d1;
   while d<=d2
   do begin
      ehre:=0.0;
      ehim:=0.0;
      uk:=exp(-d*ra);
      u:=uk;
      for ic:=1 to ni
      do begin
         cs:=ic*rb*d;
         ehre:=ehre+(eh[ic]*cos(cs)*uk);
         ehim:=ehim-(eh[ic]*sin(cs)*uk);
```

```
         uk:=uk*u;
      end;
      ehmod:=sqrt(ehre*ehre+ehim*ehim);
      npt:=npt+1;
      data[npt,1]:=d;
      data[npt,2]:=ehmod;
      d:=d+ds;
   end;
   for j:=1 to npt
   do begin
      if data[j,2]>max
      then begin
         max:=data[j,2];
         peak:=data[j,1];
         pt:=j;
      end;
   end;
end;

procedure results;
begin
   writeln(datout);
   writeln(datout,'         RESULTS');
   writeln(datout,'         -------' );
   writeln(datout,npt:4,' points plotted.');
   writeln(datout,'  Maximum plotted field magnitude:');
   writeln(datout,'      ',max:10:6);writeln(datout);
   cf:=1/cf;
   writeln(datout,'  The velocity correction factor (Do/D) is ',cf:8:6);
   writeln(datout);
   writeln(datout,'  MESHSIZE/MESH WAVELENGTH     MESHSIZE/LINK LINE ',
'   FIELD MAGNITUDE');
   writeln(datout,'           (D)                    (Do)    WAVELENGTH',
'      (EHMOD)');
   writeln(datout,'  ------------------------   -----------------------',
'   --------------');
   writeln(datout);
   for j:=1 to npt
   do begin
      writeln(datout,'           ', data[j,1]:8:6,'              '
            , (data[j,1]*cf):8:6,'             ',data[j,2]:8:6);
   end;
   writeln(datout);
   writeln(datout,'Maximum field value at D  = ',peak:8:6);
   writeln(datout,'       corresponding to Do = ',pcf:8:6,
'    <=== FINAL RESULT');
   writeln(datout);
end;

procedure correct;
var
   a,p,ga,gb: real;
   ve:        char;
{ VELOCITY ERROR CORRECTION IN AXIAL DIRECTION }
begin
   WRITE('Velocity error correction? (y/n)  ');readln(ve);
     if (ve='y') or (ve='Y')
     then begin
   p:=3.141592653*peak;
   ga:=sqrt(2)*sin(p);
   gb:=sqrt(1-ga*ga);
   cf:=p/Arctan(ga/gb);
   pcf:=peak/cf;
     end;
```

```
      if (ve='n') or (ve='N')
      then begin
    cf:=1/sqrt(2);
    pcf:=peak/cf;
    end;
end;

procedure curvefit;
var
   ai,bmax:  real;
begin
   if (pt<>1) and (pt<>jd)
   then begin
      ptp:=pt+1;
      ptm:=pt-1;
      ai:=((data[ptm,2]-max)*(data[ptm,1]-data[ptp,1])-
           (data[ptm,2]-data[ptp,2])*(data[ptm,1]-data[pt,1]));
      ai:=ai/((data[ptm,1]-data[pt,1])*
              (data[pt,1]-data[ptp,1])*(data[ptm,1]-data[ptp,1]));
      bmax:=(data[ptm,2]-max)/
         (data[ptm,1]-data[pt,1])-ai*(data[ptm,1]+data[pt,1]);
      peak:=-bmax/(2*ai);
   end;
end;

procedure output;
   var u: real;

   begin
      writeln(datout,'                                    Graph of EHMOD vs. D');
      writeln(datout,'   D                                         EHMOD');
     u:=max/70;
     if u=0
     then writeln('No plot generated - all values equal zero.')
     else begin
        writeln(datout,'          ', 7*u:7:1,14*u:7:1,21*u:7:1,28*u:7:1,
              35*u:7:1,42*u:7:1,49*u:7:1,56*u:7:1,63*u:7:1,max:7:1);
        for j:=1 to npt
        do begin
           data[j,2]:=data[j,2]/max;
           data[j,2]:=trunc(70*data[j,2]);
           for i:=1 to 70
           do begin
              out[j,i]:=' ';
              if data[j,2]=i
              then out[j,i]:='*';
           end;
           write(datout,' ',data[j,1]:6:4,'|');
           for i:=1 to 70
           do begin
              write(datout,out[j,i]);
           end;
           writeln(datout,' ');
        end;
     end;
end;

                    {MAIN PROGRAM BLOCK}
begin
   textmode(bw80);
   assign(datin,'tlm_inho.inp');assign(datout,'tlm_inho.out');
   reset (datin);
   rewrite (datout);
   textmode(bw80);
```

```
   readnxny;
   readbound;
   readdielbound;
   readcompbox;
   readexcitation;
   readfreq;
   iterate;
   fourier;
   curvefit;
   correct;
   results;
   output;
   close(datout);
end.

- - - - - - - - - - - - - -  end of file  - - - - - - - - - - - - -
```

A.1. Input File

The following is a typical TLM_INHO.INP input file in the prescribed format. It describes the specific example given below.

```
--------------------- top of input file------------------------
                          INPUT DATA
Mesh size
    12          12
Waveguide boundaries
xmin xmax ymin ymax              code              reflection coeff.
  2    11    1    1            1 3 1 0                1.0              1
 11    11    2   11            2 4 0 1               -1.0              1
  2    11   11   11            1 3 1 0                1.0              1
  1     1    2   11            2 4 0 1                1.0              0
Air-dielectric boundaries
xmin xmax ymin ymax              code              reflection coeff.
  10   10    2    7          2 4 0 1                0.500000           1
  11   11    7    7          1 3 1 0               -0.500000           0
Computational boxes
xmin xmax ymin ymax         permittivity stub value (Yo)
  1    10    1   11                   0.0                              1
  11   11    1    7                   8.00                             1
  11   11    8   11                   0.0                              0
Excitation point or lines
xmin xmax ymin ymax     code            initial value
  2     3    2   11      1  1  4              1.0                      0
Frequency range & step size
     D1           D2            Ds
  0.005         0.03          0.0005

Output point,output type, # of iterations and Yo at output point
 2  3               3              300                  0.0
----------------------- end of input file ---------------------
```

A.2. Instructions for the Preparation and Interpretation of the Input File

The sample input file TLM_INHO.INP has been configured to compute the dominant cutoff frequency of a rectangular waveguide that is partially filled with a dielectric slab. The geometry of its cross section is shown in Fig. 48.

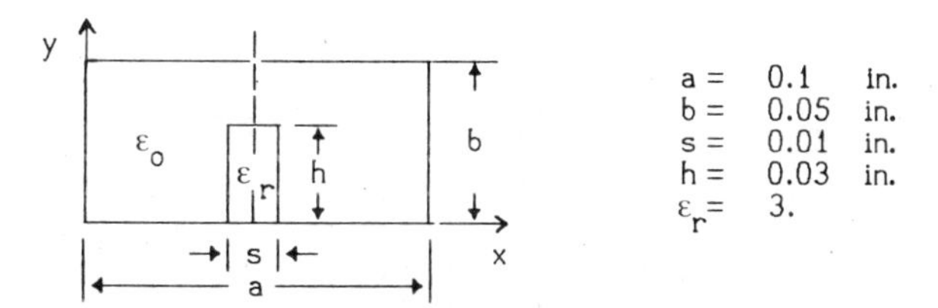

Fig. 48 Cross section and dimensions of a rectangular waveguide partially filled with a dielectric slab.

A.3. Considerations of Symmetry

The structure is symmetrical about the axis $x = a/2$. Hence, the cutoff frequencies of all modes with

- even symmetry of the magnetic field tangential to $x = a/2$ (or odd symmetry of the electric field tangential to $x = a/2$) can be obtained by placing an electric wall at $x = a/2$, and those of all modes with

- odd symmetry of the magnetic field tangential to $x = a/2$ (or even symmetry of the electric field tangential to $x = a/2$) can be obtained by placing a magnetic wall at $x = a/2$.

Note. The dominant mode in the structure is the distorted TE_{10} mode in the unloaded rectangular waveguide. It has an even symmetry of the tangential electric field at $x = a/2$. It can thus be computed by placing a magnetic wall at $x = a/2$. By taking advantage of symmetry, a better resolution of the field space is achieved by extending only a fraction of the structure over the complete TLM mesh, which in this program has been limited to 12×12 nodes.

A.4. The Layout of the TLM Mesh

The boundaries of the waveguide are placed halfway between nodes. The layout shown in Fig. 49 makes the best possible use of the available 12×12 mesh. There are three electric walls, one magnetic wall (wall of symmetry), and two air–dielectric boundaries. The dashed lines across nodes delimit so-called computational boxes.

The position of the boundaries, their reflection coefficients, and the constituent parameters of the dielectric subregions must now be specified in the input file. Since we want to compute the cutoff frequency of a TE-type mode, we must simulate the magnetic field in the z direction by the voltage in the TLM mesh. This determines the nature of the reflection coefficients at the boundaries. In the following section the various entries in the input file will be explained in detail.

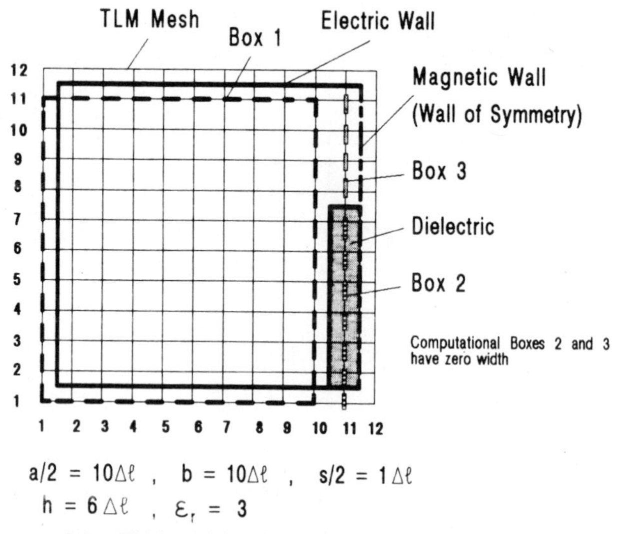

Fig. 49 Layout of the TLM grid for the preparation of the input file to compute the dominant cutoff frequency in the structure of Fig. 48.

A.5. Explanation of Input File Entries

A.5.1. Mesh Size

The mesh size specifies the number of nodes in the x and y directions respectively. A maximum of 12×12 nodes can be handled by the program.

A.5.2. Waveguide Boundaries

Boundaries can be either electric or magnetic walls. All boundaries must be situated halfway between two nodes. Twelve boundaries are accepted by the program. Their position and nature are specified by x, y coordinates, a code, and a reflection coefficient as follows:

x, y Coordinates

- Horizontal boundaries extend from $\Delta l/2$ to the left of $x$min to $\Delta l/2$ to the right of $x$max, and lie $\Delta l/2$ above $y$min (= $y$max).
- Vertical boundaries extend from $\Delta l/2$ below $y$min to $\Delta l/2$ above $y$max, and lie $\Delta l/2$ to the right of $x$min (= $x$max).

Code for Boundaries

1310	2401
horizontal boundary	vertical boundary

Reflection Coefficient. Since the program simulates the magnetic field perpendicular to the xy plane,

- Electric walls have a reflection coefficient of 1.0.
- Magnetic walls have a reflection coefficient of −1.0.

A.5.3. Air–Dielectric Boundaries

All dielectric interfaces must, as the waveguide boundaries, be situated halfway between nodes. Their position is coded in the same way. A maximum of 10 such boundaries are accepted by the program. The reflection coefficient (RC) is obtained as follows:

- For vertical air–dielectric boundaries:

$$\mathrm{RC} = \frac{\varepsilon_{rr} - \varepsilon_{rl}}{\varepsilon_{rr} + \varepsilon_{rl}}$$

  where ε_{rr} is the relative dielectric constant of the medium to the right of the boundary, and ε_{rl} that of the medium to the left of the boundary.
- For horizontal air–dielectric boundaries:

$$\mathrm{RC} = \frac{\varepsilon_{ra} - \varepsilon_{rb}}{\varepsilon_{ra} + \varepsilon_{rb}}$$

  where ε_{ra} is the relative dielectric constant of the medium above the boundary, and ε_{rb} that of the medium below the boundary.

Note. All data lines must terminate in a 1 except the last line in each data block, which must end with a 0.

A.5.4. Computational Boxes

Computational boxes are homogeneous rectangular subsections of the waveguide cross section (they contain only one kind of dielectric). A maximum of eight computational boxes are accepted by the program. The limits of these boxes are situated on the nodes. All nodes in a computational box, including the nodes on all four sides of the box, therefore have the same permittivity stub value. Computational boxes situated at the left and the lower waveguide walls must include the nodes immediately outside these walls. Boxes are defined by the coordinates of their walls. They can be of zero width (the box then becomes a line).

Permittivity Stub Value (Y_0). The permittivity stub value is calculated from the relative dielectric constant of the dielectric in the corresponding computational box as follows:

$$Y_0 = 4(\varepsilon_r - 1)$$

A.5.5. Excitation Points or Lines

They define the nodes at which the initial impulses are injected. A maximum of five data lines in the input file are accepted by the program. Excitation points are defined by the node coordinates.

Excitation Code

114 Impulses are launched on all four branches of the excitation nodes.

123 Impulses are launched on the vertical branches of the excitation nodes only.

224 Impulses are launched on the horizontal branches of the excitation nodes only.

The initial value is the magnitude of the excitation impulses and is arbitrary.

A.5.6. Frequency Range and Step Size

This input determines the relative frequency range over which the Fourier transform of the impulse response is computed. The lower and upper limits of the range (D1 and D2) as well as the step size Ds are given in terms of Meshsize/TLM mesh wavelength.

A.5.7. Output Point, Type, Number of Iterations and Y_0 at Output Point

The output point is given by its coordinates. The type determines the nature of the field that is output. Hence

- Output type 1 yields the y component of the electric field.
- Output type 2 yields the x component of the electric field.
- Output type 3 yields the z component of the magnetic field.

The maximum number of iterations accepted by the program is 300. Y_0 is computed as in Section A5.4.

A.6. Output File

This is the output file TLM_INHO.OUT created by the .COM file obtained when compiling TLM_INHO.PAS. The input data are reproduced for verification and documentation. The table and the * graph represent the Fourier-transformed impulse response of the TLM mesh. The peak value corresponds to the cutoff wavelength $\Delta l/\lambda$, which must be multiplied by the velocity correction factor D_0/D. Velocity error correction has been performed in the computation. If no velocity error correction is made, D_0/D is automatically set to $\sqrt{2}$ by the program.

```
-------------------- top of output file -----------------------
                          INPUT DATA
Mesh size
  12  12
Waveguide boundaries
xmin xmax ymin ymax          code              reflection coeff.
   2  11   1   1       1   3   1   0          1.000000            1
  11  11   2  11       2   4   0   1         -1.000000            1
   2  11  11  11       1   3   1   0          1.000000            1
   1   1   2  11       2   4   0   1          1.000000            0
Air-dielectric boundaries
xmin xmax ymin ymax          code              reflection coeff.
  10  10   2   7       2   4   0   1            0.500000            1
  11  11   7   7       1   3   1   0           -0.500000             0
Computational boxes
xmin xmax ymin ymax      permittivity stub value (Yo)
   1  10   1  11                0.0000                            1
  11  11   1   7                8.0000                            1
  11  11   8  11                0.0000                            0
Excitation point or lines
xmin xmax ymin ymax    code          initial value
   2   3   2  11     1   1   4        1.000000                    0
Output point is (   2,   3)
Number of iterations is  300
Permittivity stub admittance is  0.0000
   D1           D2          Step Size
0.005000     0.030000        0.0005


      RESULTS
      -------
51 points plotted.
Maximum plotted field magnitude:
   105.132702

The velocity correction factor (Do/D) is 1.414814

MESHSIZE/MESH WAVELENGTH      MESHSIZE/LINK LINE    FIELD MAGNITUDE
         (D)                    (Do)     WAVELENGTH       (EHMOD)
------------------------   ----------------------   --------------

        0.005000                 0.007074              9.691107
        0.005500                 0.007781              7.302061
        0.006000                 0.008489              7.266252
        0.006500                 0.009196              10.480938
        0.007000                 0.009904              14.046049
        0.007500                 0.010611              15.956953
        0.008000                 0.011319              15.229763
        0.008500                 0.012026              11.707402
        0.009000                 0.012733              6.792179
        0.009500                 0.013441              7.665348
        0.010000                 0.014148              14.828026
        0.010500                 0.014856              21.540565
        0.011000                 0.015563              25.188545
        0.011500                 0.016270              24.204131
        0.012000                 0.016978              17.811897
        0.012500                 0.017685              6.977647
        0.013000                 0.018393              13.074145
        0.013500                 0.019100              32.700531
        0.014000                 0.019807              53.858716
        0.014500                 0.020515              73.790773
        0.015000                 0.021222              90.134029
```

```
          0.015500                     0.021930             100.984975
          0.016000                     0.022637             105.132702
          0.016500                     0.023344             102.207449
          0.017000                     0.024052             92.705070
          0.017500                     0.024759             77.885828
          0.018000                     0.025467             59.569651
          0.018500                     0.026174             39.868318
          0.019000                     0.026881             20.919260
          0.019500                     0.027589             5.110563
          0.020000                     0.028296             9.050598
          0.020500                     0.029004             16.960261
          0.021000                     0.029711             20.403022
          0.021500                     0.030418             19.709172
          0.022000                     0.031126             15.783717
          0.022500                     0.031833             9.840018
          0.023000                     0.032541             3.360481
          0.023500                     0.033248             3.735540
          0.024000                     0.033956             8.431405
          0.024500                     0.034663             11.247953
          0.025000                     0.035370             11.798373
          0.025500                     0.036078             10.204510
          0.026000                     0.036785             6.935280
          0.026500                     0.037493             2.699259
          0.027000                     0.038200             1.686376
          0.027500                     0.038907             5.432719
          0.028000                     0.039615             7.915454
          0.028500                     0.040322             8.771430
          0.029000                     0.041030             7.955381
          0.029500                     0.041737             5.747885
          0.030000                     0.042444             2.799343
Maximum field value at D  = 0.016043
      corresponding to Do = 0.022698    <=== FINAL RESULT

                         Graph of EHMOD vs. D
  D                                EHMOD
           10.5   21.0    31.5   42.1   52.6    63.1   73.6   84.1    94.6  105
0.0050|      *
0.0055|    *
0.0060|    *
0.0065|      *
0.0070|         *
0.0075|          *
0.0080|          *
0.0085|       *
0.0090|    *
0.0095|     *
0.0100|         *
0.0105|              *
0.0110|                *
0.0115|                *
0.0120|           *
0.0125|    *
0.0130|        *
0.0135|                      *
0.0140|                                    *
0.0145|                                                   *
0.0150|                                                              *
0.0155|                                                                      *
0.0160|
0.0165|                                                                       *
0.0170|                                                               *
0.0175|                                                     *
0.0180|                                        *
0.0185|                           *
0.0190|             *
```

```
0.0195|  *
0.0200|     *
0.0205|          *
0.0210|            *
0.0215|            *
0.0220|         *
0.0225|     *
0.0230| *
0.0235| *
0.0240|    *
0.0245|      *
0.0250|      *
0.0255|     *
0.0260|   *
0.0265|*
0.0270|*
0.0275|  *
0.0280|    *
0.0285|    *
0.0290|    *
0.0295|  *
0.0300|*
----------------------  end of output file  --------------------
```

REFERENCES

1. C. Huygens, *Traité de la Lumière*, Leiden, 1690.
2. P. B. Johns and R. L. Beurle, "Numerical solution of 2-dimensional scattering problems using a transmission-line matrix," *Proc. Inst. Electr. Eng.*, vol. 118, pp. 1203–1208, Sept. 1971.
3. J. R. Whinnery and S. Ramo, "A new approach to the solution of high-frequency field problems," *Proc. IRE*, vol. 32, pp. 284–288, May 1944.
4. G. Kron, "Equivalent circuit of the field equations of Maxwell I," *Proc. IRE*, vol. 32, pp. 289–299, May 1944.
5. J. R. Whinnery, C. Concordia, W. Ridgway, and G. Kron, "Network analyzer studies of electromagnetic cavity resonators," *Proc. IRE*, vol. 32, pp. 360–367, June 1944.
6. P. B. Johns, "Application of the transmission-line matrix method to homogeneous waveguides of arbitrary cross-section," *Proc. Inst. Electr. Eng.*, vol. 119, pp. 1086–1091, Aug. 1972.
7. P. B. Johns, "The solution of inhomogeneous waveguide problems using a transmission-line matrix," *IEEE Trans. Microwave Theory Tech.*, vol. MTT-22, pp. 209–215, Mar. 1974.
8. S. Akhtarzad and P. B. Johns, "Numerical solution of lossy waveguides: T.L.M. computer program," *Electron. Lett.*, vol. 10, pp. 309–311, July 25, 1974.
9. P. B. Johns, "A new mathematical model to describe the physics of propagation," *Radio Electron. Eng.*, vol. 44, pp. 657–666, Dec. 1974.

10. S. Akhtarzad and P. B. Johns, "Solution of 6-component electromagnetic fields in three space dimensions and time by the T.L.M. method," *Electron. Lett.*, vol. 10, pp. 535–537, Dec. 12, 1974.
11. S. Akhtarzad and P. B. Johns, "T.L.M. analysis of the dispersion characteristics of microstrip lines on magnetic substrates using 3-dimensional resonators," *Electron. Lett.*, vol. 11, pp. 130–131, Mar. 20, 1975.
12. S. Akhtarzad and P. B. Johns, "Dispersion characteristic of a microstrip line with a step discontinuity," *Electron. Lett.*, vol. 11, pp. 310–311, July 10, 1975.
13. S. Akhtarzad, "Analysis of lossy microwave structures and microstrip resonators by the TLM method," Ph.D. Dissertation, University of Nottingham, England, July 1975.
14. S. Akhtarzad and P. B. Johns, "Three-dimensional transmission-line matrix computer analysis of microstrip resonators," *IEEE Trans. Microwave Theory Tech.*, vol. MTT-23, pp. 990–997, Dec. 1975.
15. S. Akhtarzad and P. B. Johns, "Solution of Maxwell's equations in three space dimensions and time by the T.L.M. method of analysis," *Proc. Inst. Electr. Eng.*, vol. 122, pp. 1344–1348, Dec. 1975.
16. S. Akhtarzad and P. B. Johns, "Generalised elements for T.L.M. method of numerical analysis," *Proc. Inst. Electr. Eng.*, vol. 122, pp. 1349–1352, Dec. 1975.
17. P. Saguet, "Analyse des milieux guidés—la méthode MTLM," Doctoral Thesis, Inst. Natl. Polytech., Grenoble, 1985.
18. G. E. Mariki, "Analysis of microstrip lines on inhomogeneous anisotropic substrates by the TLM numerical technique," Ph.D. Thesis, University of California, Los Angeles, June 1978.
19. W. J. R. Hoefer and A. Ros, "Fin line parameters calculated with the TLM method," *IEEE-MTT Int. Microwave Symp. Dig.*, pp. 341–343, Apr.–May 1979.
20. N. Yoshida, I. Fukai, and J. Fukuoka, "Transient analysis of two-dimensional Maxwell's equations by Bergeron's method," *Trans. Inst. Electron. Commun. Eng. Jpn.*, vol. J62B, pp. 511–518, June 1979.
21. P. Saguet and E. Pic, "An improvement for the TLM method," *Electron. Lett.*, vol. 16, pp. 247–248, Mar. 27, 1980.
22. Y. C. Shih, W. J. R. Hoefer, and A. Ros, "Cutoff frequencies in fin lines calculated with a two-dimensional TLM-program," *IEEE-MTT Int. Microwave Symp. Dig.*, pp. 261–263, June 1980.
23. W. J. R. Hoefer and Y.-C. Shih, "Field Configuration of Fundamental and Higher Order Modes in Fin Lines obtained with the TLM Method," presented at URSI and Int. *IEEE-AP Symp.*, June 1980.
24. N. Yoshida, I. Fukai, and J. Fukuoka, "Transient analysis of three-dimensional electromagnetic fields by nodal equations," *Trans. Inst. Electron. Commun. Eng. Jpn.*, vol. J63B, pp. 876–883, Sept. 1980.
25. A. Ros, Y.-C. Shih, and W.J.R. Hoefer, "Application of an accelerated TLM method to microwave systems," *Conf. Proc.—10th Eur. Microwave Conf. Dig.*, pp. 382–388, Sept. 1980.

26. Y.-C. Shih and W. J. R. Hoefer, "Dominant and second-order mode cutoff frequencies in fin lines calculated with a two-dimensional TLM program," *IEEE Trans. Microwave Theory Tech.*, vol. MTT-28, pp. 1443–1448, Dec. 1980.
27. Y.-C. Shih, "The analysis of fin lines using transmission line matrix and transverse resonance methods," M.A.Sc. Thesis, University of Ottawa, Canada, 1980.
28. D. Al-Mukhtar, "A transmission line matrix with irregularly graded space," Ph.D. Thesis, University of Sheffield, England, Aug. 1980.
29. P. Saguet and E. Pic, "Le maillage rectangulaire et le changement de maille dans la méthode TLM en deux dimensions," *Electron. Lett.*, vol. 17, pp. 277–278, Apr. 1981.
30. D. A. Al-Mukhtar and J. E. Sitch, "Transmission-line matrix method with irregularly graded space," *IEE Proc., Part H: Microwaves, Opt. Antennas*, vol. 128, pp. 299–305, Dec. 1981.
31. N. Yoshida, I. Fukai, and J. Fukuoka, "Application of Bergeron's method to anisotropic media," *Trans. Inst. Electron. Commun. Eng. Jpn.*, vol. J64B, pp. 1242–1249, Nov. 1981.
32. P. Saguet and E. Pic, "Utilisation d'un nouveau type de noeud dans la méthode TLM en 3 dimensions," *Electron. Lett.*, vol. 18, pp. 478–480, May 1982.
33. P. Saguet, "Le maillage parallelépipédique et le changement de maille dans la méthode TLM en trois dimensions," *Electron. Lett.*, vol. 20, pp. 222–224, Mar. 1984.
34. P. Saguet and S. Tedjini, "Méthode des lignes de transmission en trois dimensions: Modification du processus de simulation," *Ann. Télécommun.*, vol. 40, pp. 145–152, 1985.
35. A. Amer, "The condensed node TLM method and its application to transmission in power systems," Ph.D. Thesis, University of Nottingham, 1980.
36. W. J. R. Hoefer, "The transmission-line matrix method—theory and application," *IEEE Trans. Microwave Theory Tech.*, vol. MTT-33, pp. 882–893, Oct. 1985.
37. G. E. Mariki and C. Yeh, "Dynamic three-dimensional TLM analysis of microstrip lines on anisotropic substrates," *IEEE Trans. Microwave Theory Tech.*, vol. MTT-33, pp. 789–799, Sept. 1985.
38. P. B. Johns, "New symmetrical condensed node for three-dimensional solution of electromagnetic-wave problems by TLM," *Electron. Lett.*, vol. 22, pp. 162–164, Jan. 1986.
39. P. B. Johns, "Use of condensed and symmetrical TLM nodes in computer-aided electromagnetic design," *IEE Proc., Part H: Microwaves, Opt. Antennas*, vol. 133, Oct. 1986.
40. P. B. Johns, "A symmetrical condensed node for the TLM method," *IEEE Trans. Microwave Theory Tech.*, vol. MTT-35, pp. 370–377, Apr. 1987.
41. R. Allen, A. Mallik, and P. B. Johns, "Numerical results for the symmetrical condensed TLM node," *IEEE Trans. Microwave Theory Tech.*, vol. MTT-35, pp. 378–382, Apr. 1987.
42. D. H. Choi and W. J. R. Hoefer, "The simulation of three-dimensional wave

propagation by a scalar TLM model," *IEEE-MTT Int. Microwave Symp. Dig.*, pp. 70–71, May 1984.

43. N. Yoshida and I. Fukai, "Transient analysis of a stripline having a corner in three-dimensional space," *IEEE Trans. Microwave Theory Tech.*, vol. MTT-32, pp. 491–498, May 1984.
44. J. W. Bandler, P. B. Johns, and M. R. M. Rizk, "Transmission-line modeling and sensitivity evaluation for lumped network simulation and design in the time domain," *J. Franklin Inst.*, vol. 304, pp. 15–23, 1977.
45. P.B. Johns and M. O'Brien, "Use of the transmission-line modelling (T.L.M.) method to solve non-linear lumped networks," *Radio Electron. Eng.*, vol. 50, pp. 59–70, Jan./Feb. 1980.
46. C. R. Brewitt-Taylor and P. B. Johns, "On the construction and numerical solution of transmission-line and lumped network models of Maxwell's equations," *Int. J. Numer. Methods Eng.*, vol. 15, pp. 13–30, 1980.
47. P. B. Johns, "A simple explicit and unconditionally stable numerical routine for the solution of the diffusion equation," *Int. J. Numer. Methods Eng.*, vol. 11, pp. 1307–1328, 1977.
48. P. B. Johns and G. Butler, "The consistency and accuracy of the TLM method for diffusion and its relationship to existing methods," *Int. J. Numer. Methods Eng.*, vol. 19, pp. 1549–1554, 1983.
49. S. Akhtarzad and P. B. Johns, "Analysis of a wide range of microwave resonators: TLM method," *Electron. Lett.*, vol. 11, pp. 599–600, Nov. 1975.
50. P. Saguet, S. Tedjini, and W. J. R. Hoefer, "TLM analysis of finline T-junctions for computer-aided design," *17th Eur. Microwave Conf. Dig.*, pp. 653–658, Sept. 1987.
51. N. G. Alexopoulos, "Integrated-circuit structures on anisotropic substrates," *IEEE Trans. Microwave Theory Tech.*, vol. MTT-33, pp. 847–881, Oct. 1985.
52. E. M. El-Sayed and M. N. Morsy, "Use of transmission-line matrix method in determining the resonant frequencies of loaded microwave ovens," *J. Microwave Power*, vol. 19, pp. 65–71, 1984.
53. J. E. Sitch and P. B. Johns, "Transmission-line matrix analysis of continuous waveguiding structures using stepped-impedance cavities," *IEE J. Microwaves, Opt. Acoust.*, vol. 1, pp. 181–184, Sept. 1977.
54. P. B. Johns and W. N. R. Stevens, "Imaging by numerical analysis of wave scattering in the time domain using transmission-line modelling, signal processing and optimisation," *IEE Proc., Part A: Phys. Sci., Meas. Instrum., Manage. Educ., Rev.*, vol. 129, May 1982.
55. P. B. Johns and A. Mallik, "EMP response of aircraft using TLM," presented at *6th Symp. Tech. Exhib. Electromag. Compat.*, Mar. 1985.

55a P. B. Johns and S. Akhtarzad, "The use of time domain diakoptics in time discrete models of fields," *Int. J. Numer. Methods Eng.*, vol. 17, pp. 1–14, 1981.

56. P. B. Johns, "Numerical modelling by the TLM method," in *Large Engineering Systems* (A. Wexler, ed.), Pergamon, Oxford, 1977.
57. P. Naylor, C. Christopoulos, and P. B. Johns, "Analysis of the coupling of electromagnetic radiation into wires using transmission line modelling," *5th Int. Conf. EMC, IERE Publ.*, 71, pp. 129–135, 1986.

58. E. A. Orme, P. B. Johns, and J. M. Arnold, "A hybrid computational method for underwater acoustic scattering," *Scattering Phenomena Underwater Acoust. Conf., Proc. Inst. Acoust.*, vol. 7, Part 3, April 1985.
59. P. B. Johns, A. Wright, and J. E. Sitch, "Determination of magnetic fields in power transformers," *Proc. Compumag. Conf. Comput. Magn. Fields*, pp. 261–268, 1976.
60. S. Akhtarzad and P. B. Johns, "TLMRES—the TLM computer program for the analysis of microstrip resonators," *IEEE Trans. Microwave Theory Tech.*, vol. MTT-24, p. 675, Oct. 1976.
61. P. B. Johns, "On the relationship between TLM and finite difference methods for Maxwell's equations," *IEEE Trans. Microwave Theory Tech.*, vol. MTT-35, pp. 60–61, Jan. 1987.

9

The Mode-Matching Method

Y. C. Shih*
Technical Consultant
Torrance, California

1. INTRODUCTION

Mode matching is one of the most frequently used methods for formulating boundary-value problems. Generally speaking, this technique is useful when the geometry of the structure can be identified as a junction of two or more regions, each belonging to a separable coordinate system. In other words, in each region there exists a set of well-defined solutions of Maxwell's equations that satisfies all the boundary conditions except at the junction. When the solutions are orthonormal, they are referred to as the normal modes.

The first step in the mode-matching procedure entails the expansion of unknown fields in the individual regions in terms of their respective normal modes. Since the functional form of the normal modes is known, the problem reduces to that of determining the set of modal coefficients associated with the field expansions in various regions. This procedure, in conjunction with the orthogonality property of the normal modes, eventually leads to an infinite set of linear simultaneous equations for the unknown modal coefficients. In general, it is not possible to extract an exact solution of this infinite system of equations, and one is forced to resort to approximation techniques, such as truncation or iteration. The accuracy of the approximated results should be verified carefully because of the relative convergence problem found in the evaluation of the mode-matching equations.

The mode-matching method has been applied to solve for the scattering problem due to various discontinuities in waveguides [1–4], finlines [5, 6], and microstrip lines [7–9]. It has also been extended to analyze composite structures such as E-plane filters [10, 11], direct-coupled cavity filters, waveguide impedance transformers [12], power dividers [13], and microstrip

*Present address: Microwave Products Division, Hughes Aircraft Company, Torrance, California.

filters [14]. These are closed-region scattering problems where the set of normal modes is discrete. A generalization of this technique can be extended to the case of a continuous mode spectrum.

In addition to the scattering problem, the mode-matching method is useful in solving eigenvalue problems. It can be formulated to obtain the resonant frequency of a cavity, the cutoff frequency of a waveguide, or the propagation constant of a transmission line. It is especially suitable for the analysis of planar transmission lines, such as finlines and microstrip lines, with finite metal thickness [15, 16].

The mode-matching method has been described in detail by Mittra and Lee [17], in an effort to obtain exact analytic solutions. However, only a small class of problems can have the luxury of having exact solutions. Besides, the recent advances in computer technology make it very desirable to develop this method into a numerical procedure that can handle more generalized problems. Therefore, this chapter examines the method again with a focus on the development of an efficient but accurate numerical procedure.

The method will be demonstrated by a set of progressive examples. The structures are chosen to be simple so that the concepts can be illustrated without tedious derivation yet cover wide enough range for general applications. Section 2 presents formulations for a family of waveguide discontinuities. We start out with a scattering problem due to an infinitely thin septum. The structure is then progressively modified to form other types of scattering problems by a step discontinuity and a finite-length septum. Finally, the structure is modified into an eigenvalue problem of solving the resonant frequency of a cavity with a tuning septum. Section 3 discusses the relative convergence problem associated with the numerical mode-matching procedure. In Section 4 two numcrical examples are presented in detail to provide a better understanding. It is also demonstrated that certain guidelines can be derived from the numerical results to avoid the relative convergence problem and to achieve an efficient procedure.

2. FORMULATION

2.1. Bifurcated Waveguide

The problem of scattering by an infinitely thin and semiinfinitely long septum in a waveguide will be treated first. The geometry of this bifurcated waveguide is shown in Fig. 1. The waveguide walls and the septum are assumed to be perfect electric or magnetic walls (i.e., lossless). For convenience of discussion, we also assume that there is no field or structural variation in the y direction. Therefore, the y dependent functions are eliminated in our derivation.

The mode-matching procedure begins with expanding the tangential components of electric and magnetic fields at the junction in terms of the

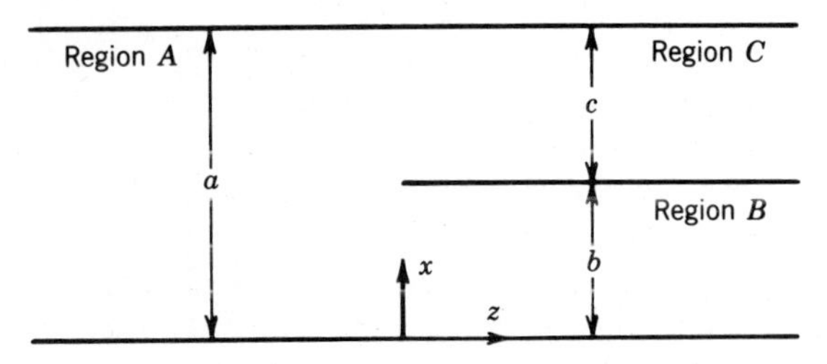

Fig. 1 Bifurcated waveguide.

normal modes in each region. For TE_{n0} ($n = 1, 2, \ldots$) excitation, we write down the tangential electric and magnetic fields as follows:

For Region A ($z < 0$)

$$E_y = \sum_{n=1}^{M} [A_n^+ \phi_{an}(x)\, e^{-\gamma_{an} z} + A_n^- \phi_{an}(x)\, e^{\gamma_{an} z}] \tag{1a}$$

$$H_x = \sum_{n=1}^{M} [A_n^+ Y_{an} \phi_{an}(x)\, e^{-\gamma_{an} z} - A_n^- Y_{an} \phi_{an}(x)\, e^{\gamma_{an} z}] \tag{1b}$$

For Region B ($z > 0, 0 < x < b$)

$$E_y = \sum_{n=1}^{K} [B_n^+ \phi_{bn}(x)\, e^{-\gamma_{bn} z} + B_n^- \phi_{bn}(x)\, e^{\gamma_{bn} z}] \tag{2a}$$

$$H_x = \sum_{n=1}^{K} [B_n^+ Y_{bn} \phi_{bn}(x)\, e^{-\gamma_{bn} z} - B_n^- Y_{bn} \phi_{bn}(x)\, e^{\gamma_{bn} z}] \tag{2b}$$

For Region C ($z > 0, b < x < a$)

$$E_y = \sum_{n=1}^{L} [C_n^+ \phi_{cn}(x)\, e^{-\gamma_{cn} z} + C_n^- \phi_{cn}(x)\, e^{\gamma_{cn} z}] \tag{3a}$$

$$H_x = \sum_{n=1}^{L} [C_n^+ Y_{cn} \phi_{cn}(x)\, e^{-\gamma_{cn} z} - C_n^- Y_{cn} \phi_{cn}(x)\, e^{\gamma_{cn} z}] \tag{3b}$$

where ϕ_{an}, ϕ_{bn}, and ϕ_{cn} are normal modes in regions A, B, and C, with propagation constants γ_{an}, γ_{bn}, and γ_{cn}, respectively. The normal modes satisfy the orthogonality relation defined by

$$\int \phi_{im}(x) \phi_{in}(x)\, dx = \delta_{mn}\,, \qquad i = a, b, c \tag{4}$$

where δ_{mn} is the Kronecker delta. A_n^+, B_n^-, and C_n^- are the given incident field coefficients from regions A, B, and C, while A_n^-, B_n^+, and C_n^+ are the unknown excited field coefficients in regions A, B, and C, respectively. The wave admittance is defined by

$$Y_{in} = \gamma_{in}/j\omega\mu , \qquad i = a, b, c \tag{5}$$

Applying the continuity condition at the junction ($z = 0$), we obtain an equation for the electric field as

$$\sum_{n=1}^{M} (A_n^+ + A_n^-)\phi_{an}(x) = \begin{cases} \sum_{n=1}^{K} (B_n^+ + B_n^-)\phi_{bn}(x) & 0 < x < b \\ \sum_{n=1}^{L} (C_n^+ + C_n^-)\phi_{cn}(x) & b < x < a \end{cases} \tag{6}$$

and a corresponding one for the magnetic field,

$$\sum_{n=1}^{M} (A_n^+ - A_n^-)Y_{an}\phi_{an}(x) = \begin{cases} \sum_{n=1}^{K} (B_n^+ - B_n^-)Y_{bn}\phi_{bn}(x) & 0 < x < b \\ \sum_{n=1}^{L} (C_n^+ - C_n^-)Y_{cn}\phi_{cn}(x) & b < x < a \end{cases} \tag{7}$$

From (6) and (7) we can derive a set of equations involving the unknown coefficients only, by making use of the property of mode orthogonality. To this end, (6) and (7) are multiplied by $\phi_{am}(x)$ and integrated with respect to x from 0 to a. This yields

$$A_m^+ + A_m^- = \sum_{n=1}^{K} H_{mn}(B_n^+ + B_n^-) + \sum_{n=1}^{L} \bar{H}_{mn}(C_n^+ + C_n^-) ,$$
$$m = 1, 2, \ldots , M \tag{8a}$$

$$Y_{am}(A_m^+ - A_m^-) = \sum_{n=1}^{K} Y_{bn}H_{mn}(B_n^+ - B_n^-) + \sum_{n=1}^{L} Y_{cn}\bar{H}_{mn}(C_n^+ - C_n^-) ,$$
$$m = 1, 2, \ldots , M \tag{8b}$$

where

$$H_{mn} = \int_0^b \phi_{am}(x)\phi_{bn}(x)\, dx \qquad \text{and} \qquad \bar{H}_{mn} = \int_b^a \phi_{am}(x)\phi_{cn}(x)\, dx$$

A second set of equations can be obtained by multiplying (6) and (7) by $\phi_{bm}(x)$ and $\phi_{cm}(x)$, respectively, and integrating with respect to x from 0 to b and from b to a, respectively. This yields

$$\sum_{n=1}^{M} H_{nm}(A_n^+ + A_n^-)B_m^+ + B_m^- , \qquad m = 1, 2, \ldots, K \quad \text{(9a)}$$

$$\sum_{n=1}^{M} H_{nm}Y_{an}(A_n^+ - A_n^-) = Y_{bm}(B_m^+ - B_m^-) , \qquad m = 1, 2, \ldots, K \quad \text{(9b)}$$

$$\sum_{n=1}^{M} \bar{H}_{nm}(A_n^+ + A_n^-) = C_m^+ + C_n^- , \qquad m = 1, 2, \ldots, L \quad \text{(10a)}$$

$$\sum_{n=1}^{M} \bar{H}_{nm}Y_{an}(A_n^+ - A_n^-) = Y_{cm}(C_m^+ - C_m^-) , \qquad m = 1, 2, \ldots, L \quad \text{(10b)}$$

It is easier to handle the simultaneous equations (8)–(10) in matrix form. Before doing that, let us further simplify (8b), (9b), and (10b) by dividing them by Y_{am}, Y_{bm}, and Y_{cm}, respectively and define $\boldsymbol{Z} = \boldsymbol{Y}^{-1}$. The matrix form of the above equations would now read as

$$\mathbf{a}^+ + \mathbf{a}^- = \boldsymbol{H}(\mathbf{b}^+ + \mathbf{b}^-) + \bar{\boldsymbol{H}}(\mathbf{c}^+ + \mathbf{c}^-) \quad \text{(11a)}$$

$$\mathbf{a}^+ - \mathbf{a}^- = \boldsymbol{Z}_a\boldsymbol{H}\boldsymbol{Y}_b(\mathbf{b}^+ - \mathbf{b}^-) + \boldsymbol{Z}_a\bar{\boldsymbol{H}}\boldsymbol{Y}_c(\mathbf{c}^+ - \mathbf{c}^-) \quad \text{(11b)}$$

$$\boldsymbol{H}^{\mathrm{t}}(\mathbf{a}^+ + \mathbf{a}^-) = \mathbf{b}^+ + \mathbf{b}^- \quad \text{(12a)}$$

$$\boldsymbol{Z}_b\boldsymbol{H}^{\mathrm{t}}(\mathbf{a}^+ - \mathbf{a}^-) = \mathbf{b}^+ - \mathbf{b}^- \quad \text{(12b)}$$

$$\bar{\boldsymbol{H}}^{\mathrm{t}}(\mathbf{a}^+ + \mathbf{a}^-) = \mathbf{c}^+ + \mathbf{c}^- \quad \text{(13a)}$$

$$\boldsymbol{Z}_c\bar{\boldsymbol{H}}^{\mathrm{t}}(\mathbf{a}^+ - \mathbf{a}^-) = \mathbf{c}^+ - \mathbf{c}^- \quad \text{(13b)}$$

where $\boldsymbol{H}$ is a matrix of size $M \times K$ with generic element H_{mn} as defined before, while $\bar{\boldsymbol{H}}$ is a matrix of size $M \times L$ with generic element $\bar{H}_{mn}$. The superscript t denotes the transpose operation, and Y_i and Z_i ($i = a, b, c$) are diagonal matrices with diagonal elements Y_{in} and Z_{in} ($i = a, b, c$). The coefficient vectors are defined as

$$\mathbf{a}^+ = \begin{bmatrix} A_1^+ \\ A_2^+ \\ \cdot \\ \cdot \\ \cdot \\ A_M^+ \end{bmatrix} \qquad \mathbf{b}^- = \begin{bmatrix} B_1^- \\ B_2^- \\ \cdot \\ \cdot \\ B_K^- \end{bmatrix} \qquad \mathbf{c}^- = \begin{bmatrix} C_1^- \\ C_2^- \\ \cdot \\ \cdot \\ \cdot \\ C_L^- \end{bmatrix}$$

$$\boldsymbol{a}^- = \begin{bmatrix} A_1^- \\ A_2^- \\ \cdot \\ \cdot \\ \cdot \\ A_M^- \end{bmatrix} \qquad \mathbf{b}^+ = \begin{bmatrix} B_1^+ \\ B_2^+ \\ \cdot \\ \cdot \\ \cdot \\ B_K^+ \end{bmatrix} \qquad \mathbf{c}^+ = \begin{bmatrix} C_1^+ \\ C_2^+ \\ \cdot \\ \cdot \\ \cdot \\ C_L^+ \end{bmatrix}$$

$\mathbf{a}^+$, $\mathbf{b}^-$, and $\mathbf{c}^-$ are column vectors of the excitation terms and $\mathbf{a}^-$, $\mathbf{b}^+$, and $\mathbf{c}^+$ are column vectors of unknown modal coefficients.

Note that in eqs. (11)–(13), the matrices are of different sizes and some of them are not even square. This makes the matrix operation more difficult. To alleviate this problem, let us assume $K + L = M$ and define the following composite matrices and vectors:

$$\boldsymbol{G} = [\boldsymbol{H} \,\vdots\, \bar{\boldsymbol{H}}], \qquad \boldsymbol{Y}_d = \left[\begin{array}{c:c} \boldsymbol{Y}_b & \boldsymbol{0} \\ \hdashline \boldsymbol{0} & \boldsymbol{Y}_c \end{array}\right], \qquad \boldsymbol{Z}_d = \left[\begin{array}{c:c} \boldsymbol{Z}_b & \boldsymbol{0} \\ \hdashline \boldsymbol{0} & \boldsymbol{Z}_c \end{array}\right]$$

$$\mathbf{d}^- = \begin{bmatrix} \mathbf{b}^- \\ \mathbf{c}^- \end{bmatrix} \qquad \mathbf{d}^+ = \begin{bmatrix} \mathbf{b}^+ \\ \mathbf{c}^+ \end{bmatrix}$$

Equations (11)–(13) now become

$$\mathbf{a}^+ + \mathbf{a}^- = \boldsymbol{G}(\mathbf{d}^+ + \mathbf{d}^-) \tag{14a}$$

$$\mathbf{a}^+ - \mathbf{a}^- = \boldsymbol{Z}_a \boldsymbol{G} \boldsymbol{Y}_d(\mathbf{d}^+ - \mathbf{d}^-) \tag{14b}$$

$$\boldsymbol{G}^{\mathrm{t}}(\mathbf{a}^+ + \mathbf{a}^-) = \mathbf{d}^+ + \mathbf{d}^- \tag{14c}$$

$$\boldsymbol{Z}_d \boldsymbol{G}^{\mathrm{t}} \boldsymbol{Y}_a(\mathbf{a}^+ - \mathbf{a}^-) = \mathbf{d}^+ - \mathbf{d}^- \tag{14d}$$

In fact, this treatment makes more sense from a physical point of view. Let us now refer back to Fig. 1. To the left of the junction, we have region A supporting M electric ports (if we treat every normal mode as an individual electric port). To the right of the junction, we have regions B and C, which should also support M electric ports altogether. This consideration is also tied to the relative convergence phenomenon, which will be discussed later.

When $M \to \infty$, we can show that $\boldsymbol{G}^{-1} \equiv \boldsymbol{G}^{\mathrm{t}}$. Therefore, (14a) and (14b) are equivalent to (14c) and (14d). Two independent vectors are required to solve for two unknown vectors. Hence, for four pairs of equations [i.e., (14a) and (14b), (14b) and (14c), (14c) and (14d), and (14d) and (14a)], substituting one equation into the other in the same pair, we have eight ways of obtaining a solution for $\mathbf{a}^+$ and $\mathbf{d}^-$. They are defined graphically in Fig. 2 for clarity. The approaches indicated by a solid arrow are classified as the formulations of the first kind, and those indicated by a dashed arrow are of the second kind. The final solutions are expressed by scattering parameters defined by

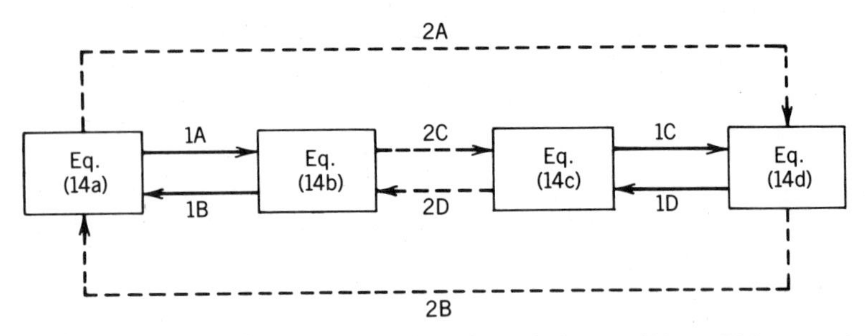

Fig. 2 Classification of mode-matching formulations. (After Shih et al. [24] copyright © 1985 by IEEE.)

$$\mathbf{a}^- = \boldsymbol{S}_{11}\mathbf{a}^+ + \boldsymbol{S}_{12}\mathbf{d}^- \tag{15a}$$

$$\mathbf{d}^+ = \boldsymbol{S}_{21}\mathbf{a}^+ + \boldsymbol{S}_{22}\mathbf{d}^- \tag{15b}$$

where $\boldsymbol{S}_{ij}$ $(i = 1, 2; j = 1, 2)$ are the generalized scattering parameters representing the amplitude of the scattered field at port i due to the unit incident field at port j. The detailed expressions for scattering parameters are included in Appendix A. Although the eight ways of solution are theoretically equivalent, their numerical behavior is somewhat different, because of the matrix truncation and the numerical errors associated with the computers.

2.2. Step Discontinuity

The problem of scattering by a step discontinuity in a waveguide is treated next. Figure 3a shows the geometry of a waveguide step discontinuity formed by joining two waveguides of different widths. The heights of the waveguides are assumed to be the same so that there is no field variation in the y direction. For convenience of analysis, an auxiliary structure is introduced in Fig. 3b, where the transverse wall at the junction is recessed

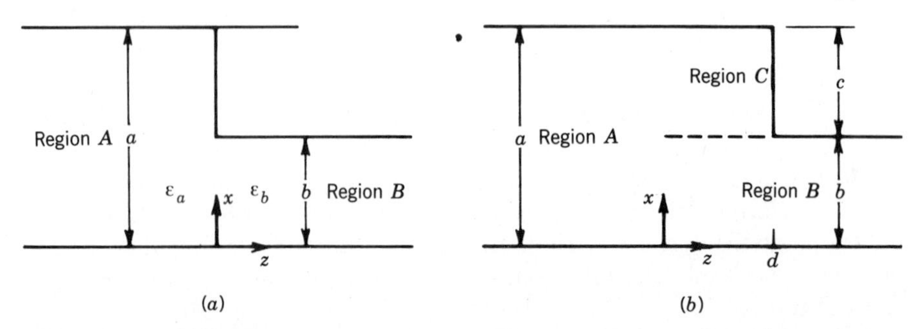

Fig. 3 (a) Waveguide step discontinuity; (b) an auxiliary structure for analysis.

to create a new region, C, of depth d. The original structure is recovered by letting $d = 0$.

Compared to the bifurcated waveguide structure in Fig. 1, the structure in Fig. 3b differs in that in region C a perfect boundary condition (e.g., an electric wall or a magnetic wall), instead of a radiation condition, is introduced and imposes a known relationship between the incident fields and the scattered fields. The additional boundary condition can easily be treated using a generalized scattering matrix technique as will be discussed in Chapter 10. Here we will look at an alternative approach that considers the boundary condition from the beginning when we set up the mode-matching equations.

Because of the transverse boundary, the normal modes in region C are in the form of standing waves in the z direction. We can also visualize the situation as the scattered fields being reflected at the boundary and turning into the incident fields. In this way, the mode-matching procedure described in the previous section can be followed step by step by replacing $C_n^- = \rho_{cn} C_n^+$, $n = 1, 2, \ldots$, where $\rho_{cn} = \rho_c \exp(-2\gamma_{cn} d)$ and ρ_c is the electric field reflection coefficient at the boundary ($\rho_c = -1$ for an electric wall and $\rho_c = 1$ for a magnetic wall).

The tangential electric and magnetic fields at the junction are first expanded in terms of the normal modes on both sides of the junction. The equations are then set up by enforcing the continuity condition for the tangential electric and magnetic fields. The modal orthogonality property is applied to obtain the linear simultaneous equations for the unknown scattered model coefficients. Written in matrix form, the equations are

$$\mathbf{a}^+ + \mathbf{a}^- = \boldsymbol{G}(\bar{\boldsymbol{R}}_p \mathbf{d}^+ + \mathbf{d}^-) \tag{16a}$$

$$\mathbf{a}^+ - \mathbf{a}^- = \boldsymbol{Z}_a \boldsymbol{G} \boldsymbol{Y}_d (\bar{\boldsymbol{R}}_m \mathbf{d}^+ - \mathbf{d}^-) \tag{16b}$$

$$\boldsymbol{G}^{\mathrm{t}}(\mathbf{a}^+ + \mathbf{a}^-) = \bar{\boldsymbol{R}}_p \mathbf{d}^+ + \mathbf{d}^- \tag{16c}$$

$$\boldsymbol{Z}_d \boldsymbol{G}^{\mathrm{t}} \boldsymbol{Y}_a (\mathbf{a}^+ - \mathbf{a}^-) = \bar{\boldsymbol{R}}_m \mathbf{d}^+ - \mathbf{d}^- \tag{16d}$$

where

$$\bar{\boldsymbol{R}}_p = \left[\begin{array}{c|c} \boldsymbol{I} & \boldsymbol{0} \\ \hline \boldsymbol{0} & \boldsymbol{R}_p'' \end{array}\right] \qquad \bar{\boldsymbol{R}}_m = \left[\begin{array}{c|c} \boldsymbol{I} & \boldsymbol{0} \\ \hline \boldsymbol{0} & \boldsymbol{R}_m'' \end{array}\right]$$

$$\boldsymbol{R}_p'' = \begin{bmatrix} 1+\rho_{c1} & 0 & & & \\ 0 & 1+\rho_{c2} & & & \\ & & \cdot & & \\ & & & \cdot & \\ & & & & \cdot \\ & & & & 1+\rho_{cL} \end{bmatrix}$$

$$R''_m = \begin{bmatrix} 1-\rho_{c1} & 0 & & & \\ 0 & 1-\rho_{c2} & & & \\ & & \cdot & & \\ & & & \cdot & \\ & & & & \cdot \\ & & & & 1-\rho_{cL} \end{bmatrix}$$

and

$$\mathbf{d}^- = \begin{bmatrix} B_1^- \\ B_2^- \\ \cdot \\ \cdot \\ \cdot \\ B_K^- \\ \hline 0 \\ \cdot \\ \cdot \\ \cdot \end{bmatrix}$$

Adopting the same classification defined in Fig. 2 for eq. (16), there are eight ways to obtain the scattering matrix for the step discontinuity. The detailed formulas are included in Appendix B. For the general case of $d \neq 0$, all the formulations require a matrix inversion of size $M \times M$. For the limiting case of having a magnetic wall at the upper half of the junction (i.e., $d = 0$ and $\rho_{c_n} = 1$, $n = 1, 2, \ldots, L$), special modifications must be made for some cases due to numerical considerations. Specifically, formulations 1D and 2B need to invert a $(M + L) \times (M + L)$ matrix and 2C needs to invert a $K \times K$ matrix. A more detailed explanation is given in Appendix C. Hence, 2C is most attractive in this example because of its potential advantage in numerical efficiency.

2.3. Finite-Length Septum

The next case to be considered is shown in Fig. 4 where a bifurcated waveguide is terminated in both regions B and C by perfect electric or magnetic walls. This is a single-port scattering problem with the incident field coming from region A only. The solutions are useful for obtaining the scattering parameters of some waveguide discontinuities. For instance, letting $d_1 = d_2 = t/2$ and assuming the same terminating boundary conditions, the structure is equivalent to a finite-length septum of length t. Letting $d_1 = 0$ and $d_2 = t/2$, the structure is then equivalent to an inductive iris of thickness t, as shown in Fig. 5. These waveguide discontinuities are useful

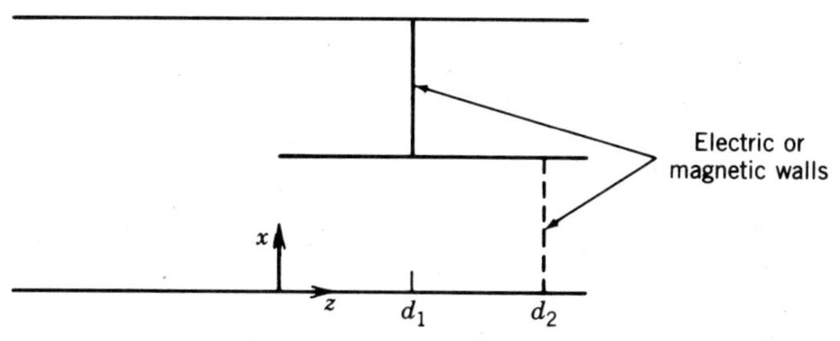

Fig. 4 Bifurcated waveguide terminated in regions *B* and *C* by perfect electric or magnetic walls.

geometries for building *E*-plane filters [18], direct-coupled cavity filters [19], and other waveguide circuitries [20].

Similar to the step discontinuity case, the mode-matching procedure is carried out with the following modifications:

$$C_n^- = \rho_{cn} C_n^+ , \qquad n = 1, 2, \ldots, L$$

$$B_n^- = \rho_{bn} B_n^+ , \qquad n = 1, 2, \ldots, K$$

where

$$\rho_{cn} = \rho_c \exp(-2\gamma_{cn} d_1) \qquad \rho_{bn} = \rho_b \exp(-2\gamma_{bn} d_2)$$

and ρ_c and ρ_b are the electric field reflection coefficients at the corresponding boundary. The final equations are

$$\mathbf{a}^+ + \mathbf{a}^- = \boldsymbol{G}\bar{\boldsymbol{R}}_p \mathbf{d}^+ \tag{17a}$$

$$\mathbf{a}^+ - \mathbf{a}^- = \boldsymbol{Z}_a \boldsymbol{G} \boldsymbol{Y}_d \bar{\boldsymbol{R}}_m \mathbf{d}^+ \tag{17b}$$

$$\boldsymbol{G}^{\mathrm{t}}(\mathbf{a}^+ + \mathbf{a}^-) = \bar{\boldsymbol{R}}_p \mathbf{d}^+ \tag{17c}$$

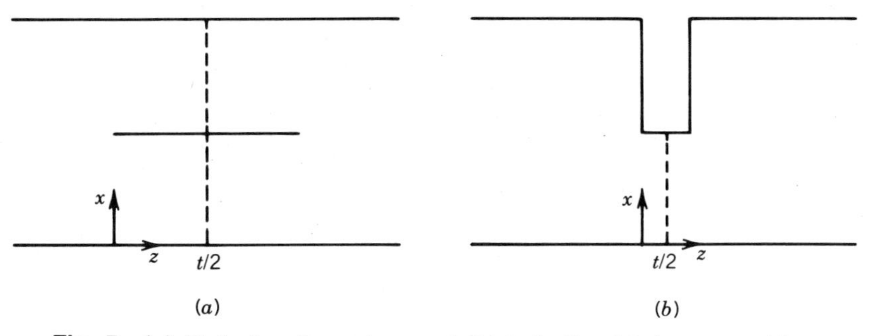

Fig. 5 (*a*) Finite-length septum and (*b*) inductive iris in waveguides.

$$Z_d G^t Y_a(\mathbf{a}^+ - \mathbf{a}^-) = \bar{R}_m \mathbf{d}^+ \tag{17d}$$

where

$$\bar{R}_p = \left[\begin{array}{c|c} R'_p & 0 \\ \hline 0 & R''_p \end{array}\right] \qquad \bar{R}_m = \left[\begin{array}{c|c} R'_m & 0 \\ \hline 0 & R''_m \end{array}\right]$$

and

$$R'_p = \begin{bmatrix} 1+\rho_{b1} & 0 & & \\ 0 & 1+\rho_{b2} & & \\ & & \ddots & \\ & & & 1+\rho_{bK} \end{bmatrix}$$

$$R'_m = \begin{bmatrix} 1-\rho_{b1} & 0 & & \\ 0 & 1-\rho_{b2} & & \\ & & \ddots & \\ & & & 1-\rho_{bK} \end{bmatrix}$$

In most of the applications, we are only interested in the solution for the reflection coefficients in region A. Therefore, using any pair of independent equations, we can eliminate $\mathbf{d}^+$ and obtain an expression for $\mathbf{a}^-$ in terms of $\mathbf{a}^+$. For example, substituting (17c) into (17b), we obtain

$$\mathbf{a}^- = S_{11}\mathbf{a}^+ = [I + Z_a G Y_d \bar{R}_m \bar{R}_p^{-1} G^t]^{-1} [I - Z_a G Y_d \bar{R}_m \bar{R}_p^{-1} G^t]\mathbf{a}^+ \tag{18}$$

2.4. Cavity with Tuning Septum

Finally, we consider the geometry shown in Fig. 6. In this case, regions A, B, and C of the bifurcated waveguide are terminated with perfect boundaries. The enclosed structure is essentially a cavity with a tuning septum. Since there is no incident field, the problem becomes an eigenvalue problem. The eigenvalue is the resonant frequency of the cavity.

Following a similar line of discussion, we can apply the mode-matching procedure here by further replacing $A_n^+ = \rho_{an} A_n^-$, $n = 1, 2, \ldots, M$ with $\rho_{an} = \rho_a \exp(-2\gamma_{an} d_3)$, where ρ_a is the electric field reflection coefficient at the boundary in region A. The final equations are

$$R_p \mathbf{a}^- = G\bar{R}_p \mathrm{d}^+ \tag{19a}$$

$$R_m \mathbf{a}^- = -Z_a G Y_d \bar{R}_m \mathbf{d}^+ \tag{19b}$$

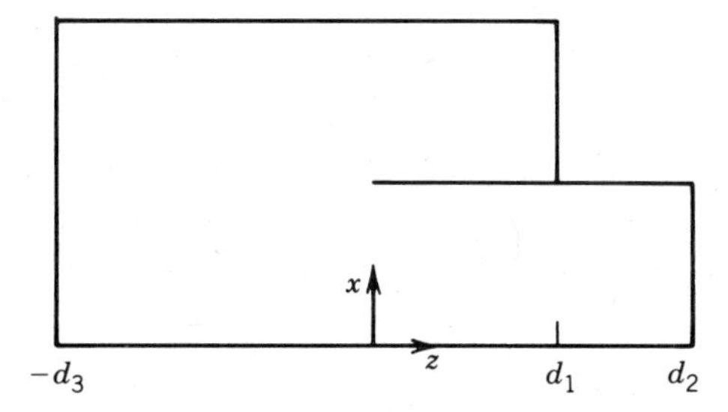

Fig. 6 Cavity with a tuning septum.

$$\mathbf{G}^{t}\boldsymbol{R}_{p}\mathbf{a}^{-} = \bar{\boldsymbol{R}}_{p}\mathbf{d}^{+} \tag{19c}$$

$$\boldsymbol{Z}_{d}\boldsymbol{G}^{t}\boldsymbol{Y}_{a}\boldsymbol{R}_{m}\mathbf{a}^{-} = -\bar{\boldsymbol{R}}_{m}\mathbf{d}^{+} \tag{19d}$$

where

$$\boldsymbol{R}_{p} = \begin{bmatrix} 1+\rho_{a1} & 0 & & \\ 0 & 1+\rho_{a2} & & \\ & & \ddots & \\ & & & 1+\rho_{aM} \end{bmatrix}$$

$$\boldsymbol{R}_{m} = \begin{bmatrix} 1-\rho_{a1} & 0 & & \\ 0 & 1-\rho_{a2} & & \\ & & \ddots & \\ & & & 1-\rho_{aM} \end{bmatrix}$$

Using a pair of independent equations, we can eliminate either $\mathbf{a}^{-}$ or $\mathbf{d}^{+}$ to form a set of linear homogeneous equations. The resonant frequency is the frequency that makes the determinant of the coefficient matrix equal to zero.

This approach to eigenvalue problems has been used for obtaining the propagation constant of finlines and microstrip lines. It is especially suitable for the cases when the conductor has finite thickness [15, 16].

3. RELATIVE CONVERGENCE PROBLEM

The basis of the mode-matching method is the expansion of an electromagnetic field in terms of an infinite series of normal modes. Because a computer's capacity for numerical calculation is finite, we have to truncate the infinite series to obtain a numerical solution. This approach is more useful and more accurate than most analytical approximations. Unfortunately, a digital computer always gives a solution even if the problem to be solved is not well-posed or has no unique solution. Therefore, it is important to ensure the validity of a numerical solution.

A series can be truncated only if it is convergent. All the researchers who have encountered these problems have tried to show the convergence of the numerical results. A natural way is to plot the numerical values of some desired parameters versus the number of terms retained. The truncation is considered appropriate when the change in the parameters is smaller than certain criteria. This method presents some limitations in the mode-matching technique where we need to truncate two or more infinite series simultaneously. Indeed, we have observed that the numerical results converge to different values depending on the way we truncate the series. This phenomenon, called relative convergence, was first studied on a bifurcated parallel-plate waveguide problem through a mode-matching technique [21]. This finding throws doubt on the validity of many numerical computations and on modal analysis.

The relative convergence problem has been studies in detail to find the cause of the problem and ways to avoid it. It has been found that relative convergence is related to the violation of field distributions at the edge of a conductor at the boundary [17] and is also related to the ill-conditioned situation of the linear system in the computation process [22]. Therefore, either the edge condition or the condition number of the linear system can be used as a criterion to ensure the validity of modal analyses. Another commonly used criterion is to plot the field distributions on both sides of the boundary and observe their matching conditions. Based on these criteria, simple rules have been drawn for the truncation of the infinite series. In the following numerical examples, we discuss the relative convergence problem in more detail.

4. NUMERICAL EXAMPLES

In this section, two numerical examples are chosen to demonstrate the application of the mode-matching method. A quantitative comparison will be made between the various formulations with respect to their numerical behavior. The relative convergence problem will be addressed in more detail using the numerical results. We will also show that there exists a simple rule for truncating the infinite series to ensure an accurate result.

The first example is a step-discontinuity problem between two parallel-plate waveguides of different widths. Both sides of the waveguide are open to the free space. In the present analysis the waveguide sides are assumed to be bounded by perfect magnetic walls. This assumption remains valid as long as the radiation loss is negligible. The structure has commonly been used in the analysis of the microstrip discontinuity problems with an appropriate waveguide model for the microstrip line [23]. Therefore, the present formulation can be useful for analyses of microstrip discontinuities.

In the second example we deal with a waveguide double-step discontinuity problem. This example is used to show that the formulation is in fact very general and is applicable to complex structures. Although both TE and TM modal fields are excited and are functions of x and y coordinates, the same formulation applies. The derivation of some parameters is more laborious, and the numerical calculations are more involved. However, the numerical behavior is similar, and similar rules can be used to avoid the relative convergence problem.

4.1. Step Discontinuity

The first example is the microstrip step discontinuity problem based on the waveguide model [7, 8, 24]. The range of validity of this model for microstrip discontinuity analysis has been discussed in the literature [25]; therefore, we will limit ourselves to the solution of the structure in Fig. 7. The parallel-plate waveguides are idealized with magnetic side walls and are filled with dielectrics of permittivities ε_a and ε_b in regions A and B, respectively.

The formulas derived in Section 2.1 are directly applied here with the following parameters:

$$\phi_{an}(x) = \sqrt{\varepsilon_{n0}/a}\cos(k_{an}x)\,, \qquad k_{an} = [(n-1)\pi/a]$$
$$\phi_{bn}(x) = \sqrt{\varepsilon_{n0}/b}\cos(k_{bn}x)\,, \qquad k_{bn} = [(n-1)\pi/b]$$
$$\phi_{cn}(x) = \sqrt{\varepsilon_{n0}/c}\cos[k_{cn}(a-x)]\,, \qquad k_{cn} = [(n-1)\pi/c]$$

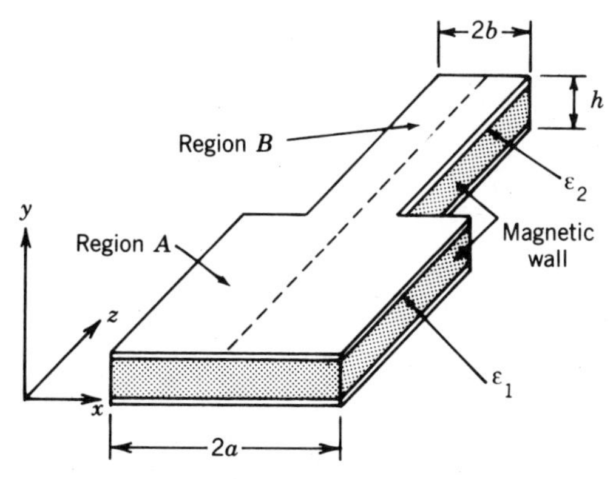

Fig. 7 Waveguide model for microstrip step discontinuity. (After Shih et al. [24] copyright © 1988 by IEEE.)

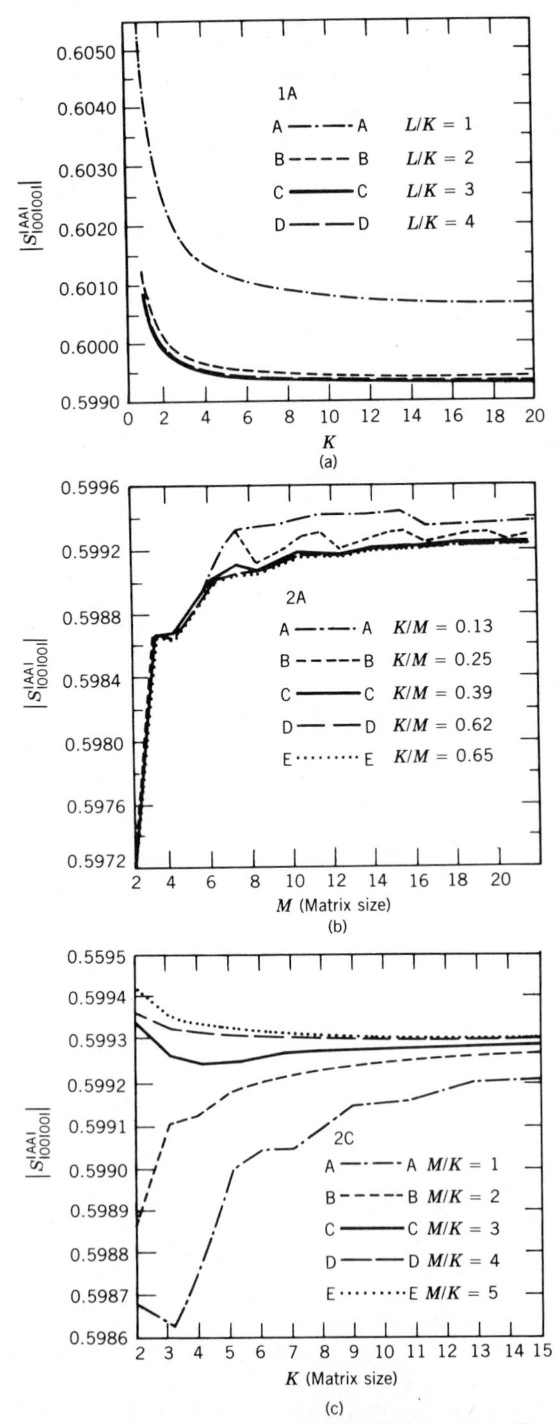

Fig. 8 Convergence study for various formulations. (*a*) Formulation 1A; (*b*) formulation 2A; (*c*) formulation 2C. (After Shih et al. [24] copyright © 1988 by IEEE.)

$$\varepsilon_{n0} = \begin{cases} 1 & n = 1 \\ 2 & n > 1 \end{cases}$$

$$\gamma_{an} = \sqrt{k_{an}^2 - \varepsilon_a k_0^2} \qquad \gamma_{bn} = \sqrt{k_{bn}^2 - \varepsilon_b k_0^2} \qquad \gamma_{cn} = \sqrt{k_{cn}^2 - \varepsilon_c k_0^2}$$

$$k_0^2 = \omega^2 \mu_0 \varepsilon_0 \qquad \rho_c = 1$$

$$H_{mn} = \sqrt{\frac{\varepsilon_{m0}\varepsilon_{n0}}{ab}} \frac{(-1)^n k_{am} \sin(k_{am} b)}{k_{am}^2 - k_{bn}^2}$$

$$\bar{H}_{mn} = \sqrt{\frac{\varepsilon_{m0}\varepsilon_{n0}}{ac}} \cdot \frac{(-1)^{n+1} k_{am} \sin(k_{am} b)}{k_{am}^2 - k_{cn}^2}$$

where ω is the angular frequency, and μ_0 and ε_0 are the free-space permeability and permittivity, respectively.

For numerical computations we have chosen the structural parameters as $a = 100$, $b = 26.1$, $\varepsilon_a = 2.2$, $\varepsilon_b = 2.1$. The dominant mode (TEM) reflection and transmission coefficients at the junction are calculated by varying the matrix size for different K/M ratios.

After extensive studies, we have found that formulations 1A, 1B, and 1C are numerically identical. Similarly, 2A and 2D are numerically identical. Since 1D and 2B have an apparent disadvantage in numerical calculations, they are not considered here. Therefore, only three sets of data, corresponding to 1A, 2A, and 2C, are shown in Fig. 8. In each formulation, the indices L, K, and M are involved. The numerical results are affected by the ratios among these indices, i.e., the relative convergence phenomenon. It is observed that 2A and 2C suffer very little from the relative convergence problem. The problem is more serious in 1A, as can be seen in Fig. 8*a*. With different L/K ratios, curves A and C converge to different values. To identify the correct value, we have to examine the field behavior at the junction. Figure 9 is a plot of the transverse magnetic field, H_x, for various

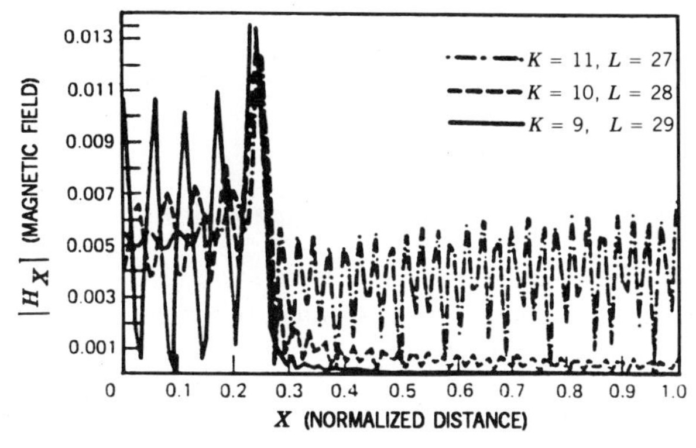

Fig. 9 Relative convergence problem of formulation 1A demonstrated by field plots. (After Shih et al. [24] copyright © 1988 by IEEE.)

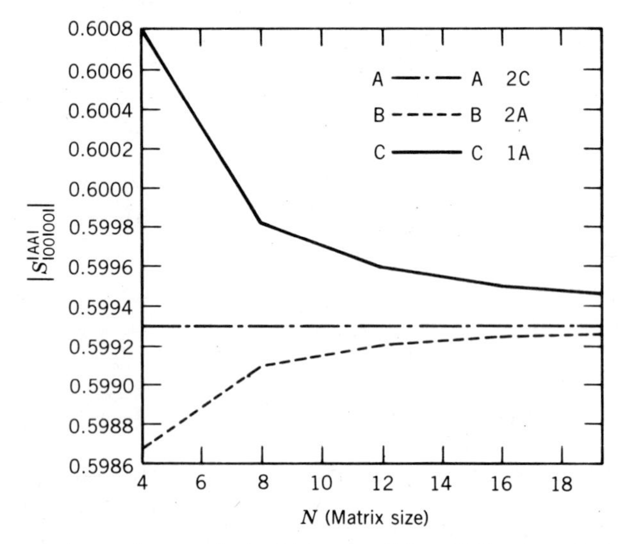

Fig. 10 Comparison of numerical efficiency. (After Shih et al. [24] copyright © 1988 by IEEE.)

L/K ratios using formulation 1A. Three ratios, i.e., $L/K = 29/9$, $28/10$, and $27/11$ are chosen as examples. It has been found that the field behaves reasonably well across the junction for $L/K = 28/10$, which is very close to the ratio c/b. It is obvious that the field behavior becomes more unreasonable when the L/K ratio deviates from $L/K = c/b$. This observation agrees with the study conducted by Lee et al. [26] and Mittra et al. [27] on waveguide inductive iris windows. We therefore concluded that the correct ratio of $L/K = c/b$ should be maintained to obtain a correct numerical value.

A comparative study on the numerical efficiency for different approaches has also been done. In this case, $L/K = 3$, which is close to c/b, is chosen. The results of the dominant mode reflection coefficient are evaluated as a function of the matrix size required and are shown in Fig. 10. It is now obvious that 2C has a definite advantage over other approaches.

4.2. Waveguide Double-Step Discontinuity

The problem is a transverse junction between two rectangular waveguides as shown in Fig. 11. The junction is formed by joining two waveguides of different sizes with concentric axes. For such a double-step junction, an incident TE wave would excite all the TE_{mn} and TM_{mn} normal modes. If we truncate the index m at M1 and the index n at N1 for numerical calculations, we will be handling matrix operations of dimension $2 \times (M1 \times N1)$. For a

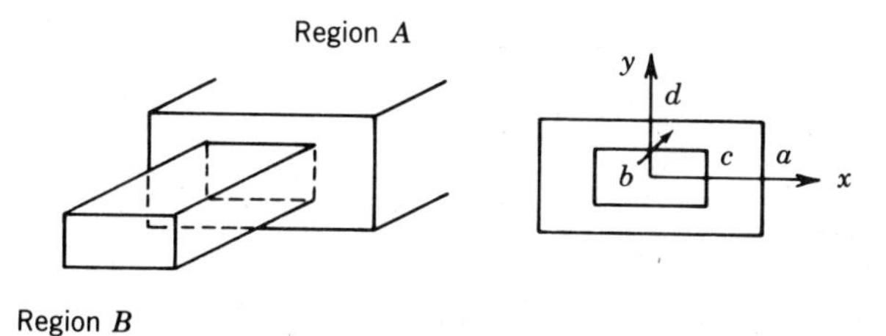

Fig. 11 Waveguide double-step discontinuity. (After Shih and Gray [4] copyright © 1983 by IEEE.)

scattering problem relating to the fundamental TE_{10} mode, it is more appropriate to represent the field by TE_{mn}-to-x modes, because in this manner the matrices to be handled are of the dimension $M1 \times N1$, half of that of the previous case. Still, for a moderate value of $M1 = N1 = 20$, the matrix size (400) becomes so large that the computation task is difficult and expensive. Therefore, it is essential that guidelines be available for use in obtaining rapid convergence.

The derivation in Section 2.2 is directly applicable if an appropriate arrangement of a single index is used for the normal modes with double indices. Let us define the index i for the normal modes in region A and the index j in region B. The sequence of the normal modes is based on their respective values of propagation constant. The corresponding parameters are

$$\phi_{ai} = \sqrt{\frac{2\varepsilon_{n0}}{ab}} \cos k_{axm}x \cos k_{ayn}y\ ;$$

$$k_{axm} = \frac{(2m-1)\pi}{a}\ , \quad k_{ayn} = \frac{2n\pi}{b}$$

$$\phi_{bj} = \sqrt{\frac{2\varepsilon q_0}{cd}} \cos k_{bxp}x \cos k_{byq}y\ ;$$

$$k_{bxp} = \frac{(2p-1)\pi}{c}\ , \quad k_{byq} = \frac{2q\pi}{d}$$

$$\gamma_{amn} = \sqrt{k_{axm}^2 + k_{ayn}^2 - k_0^2} \qquad \gamma_{bpq} = \sqrt{k_{bxp}^2 + k_{byq}^2 - k_0^2}$$

$$Y_{ai} = \frac{k_{axm}^2 - k_0^2}{j\omega\mu_0\gamma_{amn}} \qquad Y_{bj} = \frac{k_{bxp}^2 - k_0^2}{j\omega\mu_0\gamma_{bpq}} \qquad \rho_c = -1$$

$$H_{ij} = \int_0^a \int_0^b \phi_i \phi_j\, dy\, dx$$

Case 1: $n \neq 0,\ q \neq 0$

$$H_{ij} = \sqrt{\frac{16}{abcd}} \frac{2(-1)^p k_{bxp} \cos(k_{axm} c/2)}{k_{axm}^2 - k_{bxp}^2} \times \frac{2(-1)^q k_{ayn} \sin(k_{ayn} d/2)}{k_{ayn}^2 - k_{byq}^2}$$

Case 2: $n = 0,\ q \neq 0$

$$H_{ij} = 0$$

Case 3: $n \neq 0,\ q = 0$

$$H_{ij} = \sqrt{\frac{8}{abcd}} \frac{2(-1)^p k_{bxp} \cos(k_{axm} c/2)}{k_{axm}^2 - k_{bxp}^2} \frac{2 \sin(k_{ayn} d/2)}{k_{ayn}}$$

Case 4: $n = 0,\ q = 0$

$$H_{ij} = \sqrt{\frac{4}{abcd}} \frac{2(-1)^p k_{bxp} \cos(k_{axm} c/2)}{k_{axm}^2 - k_{bxp}^2} d$$

where $(m = 1, 2, \ldots, \mathrm{M1};\ n = 1, 2, \ldots, \mathrm{N1})$ and $(p = 1, 2, \ldots, \mathrm{P1};\ q = 1, 2, \ldots, \mathrm{Q1})$ are the double indices for the normal mode in regions A and B, respectively.

Since the step discontinuity is formed by a perfect electric wall at the junction between regions A and C, there is one formulation that allows us to obtain the solution by inverting a smaller matrix, similar to the formulation 2C discussed in the previous section. In the present case, $\rho_{cn} = -1$ $(n = 1, 2, \ldots, L)$. Therefore, $\boldsymbol{R}''_p = \boldsymbol{0}$ and $\boldsymbol{R}''_m = 2\boldsymbol{I}$, where $\boldsymbol{0}$ and $\boldsymbol{I}$ are the zero matrix and identity matrix, respectively. Substituting the submatrices into eqs. (16a) and (16b), we obtain the following simplified equations:

$$\mathbf{a}^+ + \mathbf{a}^- = \boldsymbol{H}(\mathbf{b}^+ + \mathbf{b}^-) \tag{20}$$

$$\boldsymbol{Z}_b \boldsymbol{H}^{\mathrm{t}} \boldsymbol{Y}_a (\mathbf{a}^+ - \mathbf{a}^-) = \mathbf{b}^+ - \mathbf{b}^- \tag{21a}$$

$$\boldsymbol{Z}_c \bar{\boldsymbol{H}}^{\mathrm{t}} \boldsymbol{Y}_a (\mathbf{a}^+ - \mathbf{a}^-) = 2\mathbf{c}^+ \tag{21b}$$

Following the procedure of formulation 2A, we use eq. (20) to obtain an expression for $\mathbf{a}^-$ and substitute it into (21a) to obtain an expression for $\mathbf{b}^+$ as

$$\mathbf{b}^+ = (\boldsymbol{I} + \boldsymbol{Z}_b \boldsymbol{H}^{\mathrm{t}} \boldsymbol{Y}_a \boldsymbol{H})^{-1} [(\boldsymbol{I} - \boldsymbol{Z}_b \boldsymbol{H}^{\mathrm{t}} \boldsymbol{Y}_a \boldsymbol{H}) \mathbf{b}^- + 2\boldsymbol{Z}_b \boldsymbol{H}^{\mathrm{t}} \boldsymbol{Y}_a \mathbf{a}^+] \tag{22}$$

The scattering matrices are given as

$$S_{22} = (I + Z_b H^t Y_a H)^{-1}(I - Z_b H^t Y_a H) \tag{23a}$$

$$S_{21} = 2(I + Z_b H^t Y_a H)^{-1} Z_b H^t Y_a \tag{23b}$$

$$S_{12} = H(I + S_{22}) \tag{23c}$$

$$S_{11} = H S_{21} - I \tag{23d}$$

where I is the identity matrix; Y_a $(M \times M)$ and Z_b $(L \times L)$ are diagonal matrices whose diagonal elements are the characteristic wave admittance and impedance of the corresponding normal modes; $H(M \times L)$ and its transpose H^t $(L \times M)$ are the transformation matrices defined above. The scattering parameters S_{11}, S_{21}, S_{12}, and S_{22} are matrices of dimensions $M \times M$, $L \times M$, $M \times L$, and $L \times L$, respectively. It is interesting to point out that the expressions in eq. (23) are identical to those obtained in Reference 2 using a conservation of complex power technique. As a result, the solutions of the modal analysis always satisfy the condition of power conservation. This condition, therefore, can be used to help debug the computer code, but not as an indication of the validity of the numerical results.

For numerical calculation, let us consider first a special case of $b = d$, i.e., two waveguides of equal height. This reduces the structure to an H-plane step junction, very similar to the previous example, except that the side walls are now electric walls instead of magnetic walls. In this case, any TE_{n0} incident waves excite only TE_{n0} waves with $n = 1, 2, \ldots$.

In previous publications [28, 29], the TE_{10} characteristics of the H-plane step junction have been analyzed by means of various methods. Although the published data are not exact, they serve as a good reference for the convergence study. As an example, we consider the case of a single incident TE_{10} wave from the larger waveguide. The parameters used for calculations are shown in Fig. 12. Since the narrow guide is below cutoff at the chosen frequency, the normalized input admittance looking from the larger guide is purely susceptive, i.e., $Y_{in} = jB$. In Fig. 12 the susceptance B is plotted versus M for fixed ratios $K/M = 1/3$, $2/3$, and $3/3$. Note that as M increases, all three curves converge to values within 0.5% of each other. Furthermore, these values agree with the *Waveguide Handbook* data [28] within its specified 1% accuracy. It is also observed that the values of B converge with respect to M to the asymptotic value most rapidly when $K/M = c/a = 1/3$. This observation agrees very well with the previous example.

In Fig. 13 the susceptance B is calculated with a fixed M and a varying K. It is noticed that B approaches the asymptotic value before K reaches the value $(c/a)M$; beyond that point it drops and converges to a smaller value.

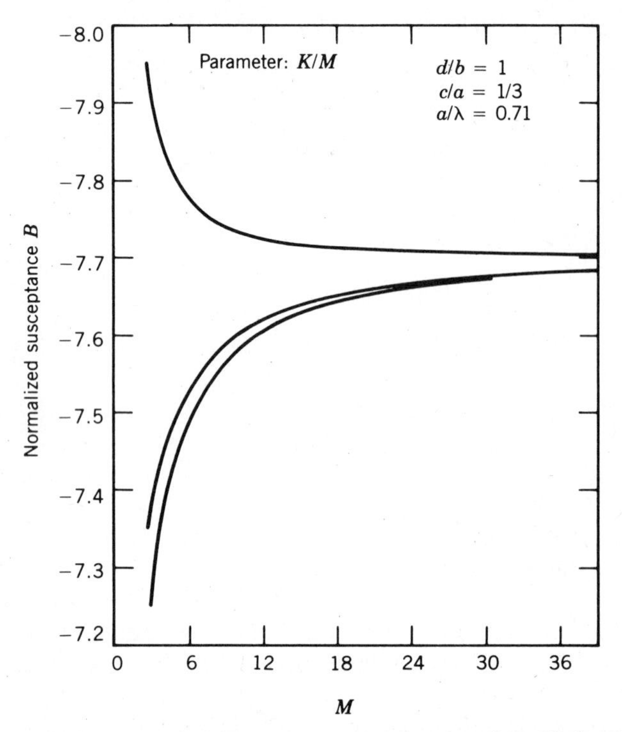

Fig. 12 Convergence study of TE_{10}-mode susceptance of an *H*-plane step discontinuity with fixed *K/M*. (After Shih and Gray [4] copyright © 1983 by IEEE.)

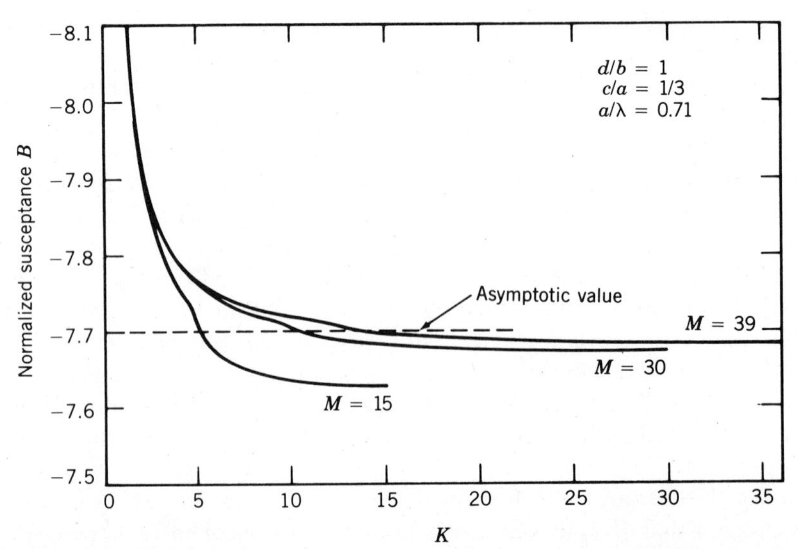

Fig. 13 Convergence study of TE_{10}-mode susceptance of an *H*-plane step discontinuity with fixed *M*. (After Shih and Gray [4] copyright © 1983 by IEEE.)

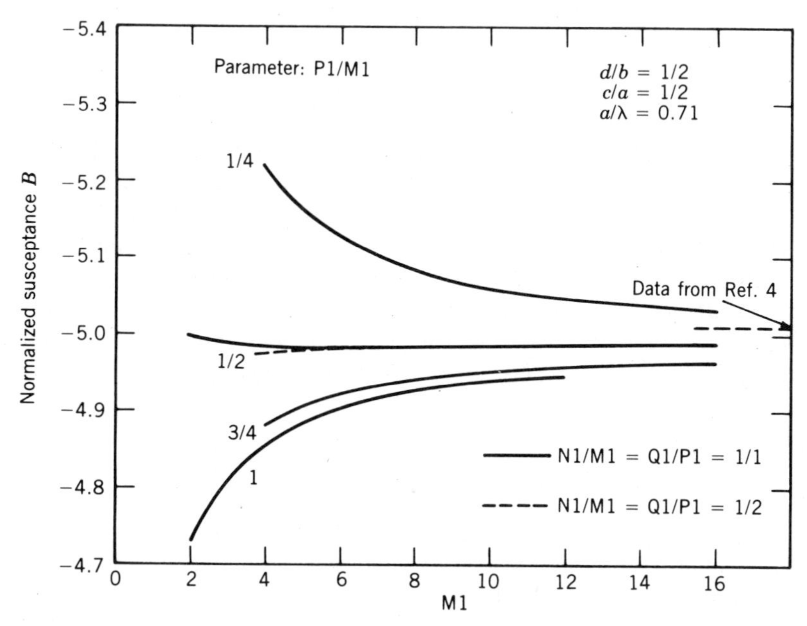

Fig. 14 Convergence study of TE_{10}-mode susceptance of a double-step discontinuity between two rectangular waveguides with fixed P1/M1. (After Shih and Gray [4] copyright © 1983 by IEEE.)

However, the error is not very large. From this we conclude that as long as both M and K are large, the K/M ratio does not affect the solution accuracy significantly. However, for efficient numerical computations one should keep the ratio K/M equal to c/a. This criterion becomes more important for the case of a double-step junction where the numerical computation is much more involved.

For a double-step junction a number of numerical calculations were made, and it was noticed that the convergence behavior was similar to the case of the H-plane step junction. In particular, we observed that the solutions approach the correct value as long as the values of M1, N1, P1, and Q1 are large. Furthermore, accurate and efficient solutions can be obtained by maintaining the ratios P1/M1 = c/a; Q1/N1 = d/b; N1/M1 = b/a.

This behavior is demonstrated by a typical example in Fig. 14. Note that the change of N1/M1 from 1/1 to 1/2 does not affect the convergence rate very much, but it does reduce the computation efforts significantly.

5. CONCLUSION

The mode-matching method is a very general tool for the numerical computation of electromagnetic field problems. It is a rigorous full-wave analysis suitable for the treatment of two- or three-dimensional field problems, including both scattering and eigenvalue problems. When combined with the generalized scattering matrix method, they together become a powerful tool for analyzing many composite waveguide structures commonly found in practical situations. Furthermore, the straightforward formulation will lend itself to common acceptance by personal computer users. However, compared to those methods that are problem-oriented and optimized, the efficiency of this method is not very high. In the given examples, we have always emphasized achieving an accurate and efficient numerical procedure. More recently, many researchers have proposed ways to improve the efficiency of the mode-matching method. For example, Mansour and Macphie [30] have proposed an improved transmission matrix formulation for cascaded discontinuities to reduce the number of matrix-inversion operations. Similarly, Alessandi et al. [31] have proposed the use of an admittance matrix formulation for certain boundary enlargement/reduction cascaded discontinuities. Bogelsack and Wolff [32] have applied a projection method to a mode-matching solution for microstrip lines with finite metallization thickness with significant improvement in efficiency. In conclusion, the numerical aspect of the classical mode-matching method is still awaiting exploration.

APPENDIX A. SCATTERING PARAMETERS FOR BIFURCATED WAVEGUIDES

Formulation 1A

$$\boldsymbol{S}_{22} = (\boldsymbol{Z}_a\boldsymbol{G}\boldsymbol{Y}_d + \boldsymbol{G})^{-1}(\boldsymbol{Z}_a\boldsymbol{G}\boldsymbol{Y}_d - \boldsymbol{G})$$

$$\boldsymbol{S}_{21} = 2(\boldsymbol{Z}_a\boldsymbol{G}\boldsymbol{Y}_d + \boldsymbol{G})^{-1}$$

$$\boldsymbol{S}_{12} = \boldsymbol{G}(\boldsymbol{I} + \boldsymbol{S}_{22})$$

$$\boldsymbol{S}_{11} = \boldsymbol{G}\boldsymbol{S}_{21} - \boldsymbol{I}$$

Formulation 1B

$$\boldsymbol{S}_{22} = (\boldsymbol{Z}_a\boldsymbol{G}\boldsymbol{Y}_d + \boldsymbol{G})^{-1}(\boldsymbol{Z}_a\boldsymbol{G}\boldsymbol{Y}_d - \boldsymbol{G})$$

$$\boldsymbol{S}_{21} = 2(\boldsymbol{Z}_a\boldsymbol{G}\boldsymbol{Y}_d + \boldsymbol{G})^{-1}$$

$$\boldsymbol{S}_{12} = \boldsymbol{Z}_a\boldsymbol{G}\boldsymbol{Y}_d(\boldsymbol{I} - \boldsymbol{S}_{22})$$

$$\boldsymbol{S}_{11} = \boldsymbol{I} - \boldsymbol{Z}_a\boldsymbol{G}\boldsymbol{Y}_d\boldsymbol{S}_{21}$$

Formulation 1C

$$S_{11} = (Z_d G^t Y_a + G^t)^{-1}(Z_d G^t Y_a - G^t)$$
$$S_{12} = 2(Z_d G^t Y_a + G^t)^{-1}$$
$$S_{21} = G^t(I + S_{11})$$
$$S_{22} = G^t S_{12} - I$$

Formulation 1D

$$S_{11} = (Z_d G^t Y_a + G^t)^{-1}(Z_d G^t Y_a - G^t)$$
$$S_{12} = 2(Z_d G^t Y_a + G^t)^{-1}$$
$$S_{21} = Z_d G^t Y_a(I - S_{11})$$
$$S_{22} = I - Z_d G^t Y_a S_{12}$$

Formulation 2A

$$S_{22} = (I + Z_d G^t Y_a G)^{-1}(I - Z_d G^t Y_a G)$$
$$S_{21} = 2(I + Z_d G^t Y_a G)^{-1} Z_d G^t Y_a$$
$$S_{12} = G(I + S_{22})$$
$$S_{11} = G S_{21} - I$$

Formulation 2B

$$S_{11} = (G Z_d G^t Y_a + I)^{-1}(G Z_d G^t Y_a - I)$$
$$S_{12} = 2(G Z_d G^t Y_a + I)^{-1} G$$
$$S_{21} = Z_d G^t Y_a(I - S_{11})$$
$$S_{22} = I - Z_d G^t Y_a S_{12}$$

Formulation 2C

$$S_{22} = (G^t Z_a G Y_d + I)^{-1}(G^t Z_a G Y_d - I)$$
$$S_{21} = 2(G^t Z_a G Y_d + I)^{-1} G^t$$
$$S_{12} = Z_a G Y_d(I - S_{22})$$
$$S_{11} = I - Z_a G Y_d S_{21}$$

Formulation 2D

$$S_{11} = (I + Z_a G Y_d G^t)^{-1}(I - Z_a G Y_d G^t)$$

$$S_{12} = 2(I + Z_a G Y_d G^t)^{-1} Z_a G Y_d$$

$$S_{21} = G^t(I + S_{11})$$

$$S_{22} = G^t S_{21} - I$$

APPENDIX B. SCATTERING PARAMETERS FOR STEP DISCONTINUITY

Formulation 1A

$$S_{22} = (Z_a G Y_d \bar{R}_m + G\bar{R}_p)^{-1}(Z_a G Y_d - G)$$

$$S_{21} = 2(Z_a G Y_d \bar{R}_m + G\bar{R}_p)^{-1}$$

$$S_{12} = G(\bar{R}_p S_{22} + I)$$

$$S_{22} = G\bar{R}_p S_{21} - I$$

Formulation 1B

$$S_{22} = (Z_a G Y_d \bar{R}_m + G\bar{R}_p)^{-1}(Z_a G Y_d - G)$$

$$S_{21} = 2(Z_a G Y_a \bar{R}_m + G\bar{R}_p)^{-1}$$

$$S_{12} = Z_a G Y_d (I - \bar{R}_m S_{22})$$

$$S_{22} = I - Z_a G Y_d \bar{R}_m S_{21}$$

Formulation 1C

$$S_{11} = (Z_d G^t Y_a + \bar{R}_m \bar{R}_p^{-1} G^t)^{-1}(Z_d G^t Y_a - \bar{R}_m \bar{R}_p^{-1} G^t)$$

$$S_{12} = (Z_d G^t Y_a + \bar{R}_m \bar{R}_p^{-1} G^t)^{-1}(\bar{R}_m \bar{R}_p^{-1} + I)$$

$$S_{21} = \bar{R}_p^{-1} G^t(I + S_{11})$$

$$S_{22} = \bar{R}_p^{-1}(G^t S_{12} - I)$$

Formulation 1D

$$S_{11} = (\bar{R}_p \bar{R}_m^{-1} Z_d G^t Y_a + G^t)^{-1} (\bar{R}_p \bar{R}_m^{-1} Z_d G^t Y_a - G^t)$$

$$S_{12} = (R_p \bar{R}_m^{-1} Z_d G^t Y_a + G^t)^{-1} (\bar{R}_p \bar{R}_m^{-1} + I)$$

$$S_{21} = \bar{R}_m^{-1} Z_d G^t Y_a (I - S_{11})$$

$$S_{22} = \bar{R}_m^{-1} (I - Z_d G^t Y_a S_{12})$$

Formulation 2A

$$S_{22} = (\bar{R}_m + Z_d G^t Y_a G \bar{R}_p)^{-1} (I - Z_d G^t Y_a G)$$

$$S_{21} = 2(\bar{R}_m + Z_d G^t Y_a G \bar{R}_p)^{-1} Z_d G^t Y_a$$

$$S_{12} = G(\bar{R}_p S_{22} + I)$$

$$S_{11} = G \bar{R}_p S_{21} - I$$

Formulation 2B

$$S_{11} = (G \bar{R}_p \bar{R}_m^{-1} Z_d G^t Y_a + I)^{-1} (G \bar{R}_p \bar{R}_m^{-1} Z_d G^t Y_a - I)$$

$$S_{12} = (G \bar{R}_p \bar{R}_m^{-1} Z_d G^t Y_a + I)^{-1} G (\bar{R}_p \bar{P}_m^{-1} + I)$$

$$S_{21} = \bar{R}_m^{-1} Z_d G^t Y_a (I - S_{11})$$

$$S_{22} = \bar{R}_m^{-1} (I - Z_d G^t Y_a S_{12})$$

Formulation 2C

$$S_{22} = (\bar{R}_p + G^t Z_a G Y_d \bar{R}_m)^{-1} (G^t Z_a G Y_d - I)$$

$$S_{21} = 2(\bar{R}_p + G^t Z_a G Y_d \bar{R}_m)^{-1} G^t$$

$$S_{12} = Z_a G Y_d (I - \bar{R}_m S_{22})$$

$$S_{11} = I - Z_a G Y_d \bar{R}_m S_{21}$$

Formulation 2D

$$S_{11} = (I + Z_a G Y_d \bar{R}_m \bar{R}_p^{-1} G^t)^{-1} (I - Z_a G Y_d \bar{R}_m \bar{R}_p^{-1} G)$$

$$S_{12} = (I + Z_a G Y_d \bar{R}_m \bar{R}_p^{-1} G^t)^{-1} Z_a G Y_d (\bar{R}_m \bar{R}_p^{-1} + I)$$

$$S_{21} = \bar{R}_p^{-1} G^t (I + S_{11})$$

$$S_{22} = \bar{R}_p^{-1} (G^t S_{12} - I)$$

APPENDIX C. COMPARISON OF DIFFERENT FORMULATIONS

When evaluating the various formulations of the step discontinuity, we encounter numerical problems in some examples. For instance, in the limiting case of having a magnetic wall at the junction of region C ($d = 0$, $\rho_{cn} = 1$, $n = 1, 2, \ldots, L$), the determinant of the matrix $\bar{\boldsymbol{R}}_m$ is zero. Therefore, formulations 1D and 2B cannot be evaluated by a computer because of the matrix inversion $\bar{\boldsymbol{R}}_m^{-1}$. If one insists on using these procedures, the problem can be solved by rearranging the matrix equations. For example, let us consider the case of 1D and rewrite eq. (16d) as follows:

$$\boldsymbol{Z}_d \boldsymbol{G} \boldsymbol{Y}_a (\mathbf{a}^+ - \mathbf{a}^-) = \bar{\boldsymbol{R}}_m \mathbf{d}^+ - \mathbf{d}^- \tag{24}$$

Substituting the submatrices into (24), we obtain

$$\left[\begin{array}{c|c} \boldsymbol{Z}_b & \boldsymbol{0} \\ \hline \boldsymbol{0} & \boldsymbol{Z}_c \end{array}\right] \left[\begin{array}{c} \boldsymbol{H}^{\mathrm{t}} \\ \hline \bar{\boldsymbol{H}}^{\mathrm{t}} \end{array}\right] \boldsymbol{Y}_a (\mathbf{a}^+ - \mathbf{a}^-) = \left[\begin{array}{c|c} \boldsymbol{I} & \boldsymbol{0} \\ \hline \boldsymbol{0} & \boldsymbol{0} \end{array}\right] \left[\begin{array}{c} \mathbf{b}^+ \\ \hline \mathbf{c}^+ \end{array}\right] - \left[\begin{array}{c} \mathbf{b}^- \\ \hline \boldsymbol{0} \end{array}\right] \tag{25}$$

After some algebraic manipulations, the equation is broken down as

$$\boldsymbol{Z}_b \boldsymbol{H}^{\mathrm{t}} \boldsymbol{Y}_a (\mathbf{a}^+ - \mathbf{a}^-) = \mathbf{b}^+ - \mathbf{b}^- \tag{26}$$

$$\boldsymbol{Z}_c \bar{\boldsymbol{H}}^{\mathrm{t}} \boldsymbol{Y}_a (\mathbf{a}^+ - \mathbf{a}^-) = \boldsymbol{0} \tag{27}$$

where (26) contains K equations and (27) contains L equations. Now, (26) can be used to solve for $\mathbf{b}^+$. If we substitute the results back into (16c), (16c) will then contain M equations with $M + L$ unknowns (i.e., the unknown coefficients $\mathbf{a}^-$ and $\mathbf{c}^+$). Therefore, we have to solve (16c) and (27) simultaneously; this requires the inversion of a matrix of size $(M + L) \times (M + L)$.

Similarly, the solution of formulation 2B requires a matrix inversion of size $(M + L) \times (M + L)$.

On the other hand, formulation 2C is an interesting approach because of the fact that $\boldsymbol{R}''_m = [\boldsymbol{0}]$ allows a simplified solution. Let us rewrite eqs. (16b) and (16c) here:

$$\mathbf{a}^+ - \mathbf{a}^- = \boldsymbol{Z}_a \boldsymbol{G} \boldsymbol{Y}_d (\bar{\boldsymbol{R}}_m \mathbf{d}^+ - \mathbf{d}^-) \tag{28}$$

$$\boldsymbol{G}^{\mathrm{t}} (\mathbf{a}^+ + \mathbf{a}^-) = \bar{\boldsymbol{R}}_p \mathbf{d}^+ + \mathbf{d}^- \tag{29}$$

Substituting the submatrices and performing some simplifications, we obtain

$$\mathbf{a}^+ - \mathbf{a}^- = \boldsymbol{Z}_a \boldsymbol{H} \boldsymbol{Y}_b (\mathbf{b}^+ - \mathbf{b}^-) \tag{30}$$

$$\boldsymbol{H}^{\mathrm{t}} (\mathbf{a}^+ + \mathbf{a}^-) = \mathbf{b}^+ + \mathbf{b}^- \tag{31a}$$

$$\bar{H}^{t}(\mathbf{a}^{+} + \mathbf{a}^{-}) = 2\mathbf{c}^{+} \tag{31b}$$

Using (30) to obtain an expression for $\mathbf{a}^{-}$ and substituting into (31a), we obtain an expression for $\mathbf{b}^{+}$ as

$$\mathbf{b}^{+} = (H^{t}Z_{a}HY_{b} + I)^{-1}[(H^{t}Z_{a}HY_{b} - I)\mathbf{b}^{-} + 2H^{t}\mathbf{a}^{+}] \tag{32}$$

The scattering parameters of the junction are then obtained:

$$S_{22} = (H^{t}Z_{a}HY_{b} + I)^{-1}(H^{t}Z_{a}HY_{b} - I) \tag{33a}$$

$$S_{21} = 2(H^{t}Z_{a}HY_{b} + I)^{-1}H^{t} \tag{33b}$$

$$S_{12} = Z_{a}HY_{b}(I - S_{22}) \tag{33c}$$

$$S_{11} = I - Z_{a}HY_{b}S_{21} \tag{33d}$$

Notice that in this solution only one inversion of matrix of size $K \times K$ is required.

REFERENCES

1. A. Wexler, "Solution of waveguide discontinuities by modal analysis," *IEEE Trans. Microwave Theory Tech.*, vol. MTT-15, pp. 508–517, Sept. 1967.
2. R. Safavi-Naini and R. H. Macphie, "On solving waveguide junction scattering problems by conservation of complex power technique," *IEEE Trans. Microwave Theory Tech.*, vol. MTT-29, pp. 337–343, Apr. 1981.
3. R. R. Mansour and R. H. Macphie, "Scattering at an *N*-furcated parallel-plate waveguide junction," *IEEE Trans. Microwave Theory Tech.*, vol. MTT-33, pp. 830–835, Sept. 1985.
4. Y. C. Shih and K. Gray, "Convergence of numerical solutions of step-type waveguide discontinuity problems by modal analysis," *1983 IEEE MTT-S Int. Microwave Symp. Dig.*, pp. 233–235, 1983.
5. A. S. Omar and K. Schunemann, "Transmission matrix representation of finline discontinuity," *IEEE Trans. Microwave Theory Tech.*, vol. MTT-33, pp. 765–770, Sept. 1985.
6. R. Vahldieck and W. J. R. Hoefer, "Finline and metal insert filters with improved passband separation and increased stopband attenuation," *IEEE Trans. Microwave Theory Tech.*, vol. MTT-33, pp. 1333–1339, Dec. 1985.
7. I. Wolff, G. Kompa, and R. Mehran, "Calculation method for microstrip discontinuities and T-junctions," *Electron. Lett.*, vol. 8, pp. 177–179, Apr. 1972.
8. W. Menzel and I. Wolff, "A method for calculating the frequency-dependent properties of microstrip discontinuities," *IEEE Trans. Microwave Theory Tech.*, vol. MTT-25, pp. 107–112, Feb. 1977.

9. T. S. Chu, T. Itoh, and Y. C. Shih, "Comparative study of mode-matching formulations for microstrip discontinuity problems," *IEEE Trans. Microwave Theory Tech.*, vol. MTT-33, pp. 1018–1023, Oct. 1985.
10. Y. C. Shih, "Design of waveguide E-plane filters with all metal insert," *IEEE Trans. Microwave Theory Tech.*, vol. MTT-32, pp. 695–704, July 1984.
11. F. Arndt et al., "E-plane integrated filters with improved stopband attenuation," *IEEE Trans. Microwave Theory Tech.*, vol. MTT-32, pp. 1391–1394, Oct. 1984.
12. F. Arndt et al., "Computed-optimized multisection transformers between rectangular waveguides of adjacent frequency bands," *IEEE Trans. Microwave Theory Tech.*, vol. MTT-32, pp. 1479–1484, Nov. 1984.
13. F. Arndt et al., "Optimized E-plane T-junction series power divider," *IEEE Trans. Microwave Theory Tech.*, vol. MTT-35, pp. 1052–1059, Nov. 1987.
14. R. Mehran, "Computer-aided design of microstrip filters considering dispersion, loss, and discontinuity effects," *IEEE Trans. Microwave Theory Tech.*, vol. MTT-27, pp. 239–245, Mar. 1979.
15. A. Beyer and I. Wolff, "A solution of the earthed fin line with finite metalization thickness," *1980 IEEE Int. Microwave Symp. Dig.*, pp. 258–260, 1980.
16. G. Kowalski and R. Pregla, "Dispersion characteristic of shielded microstrips with finite thickness," *Arch. Elektron. Ubertragungstech.*, vol. 25, pp. 193–196, 1971.
17. R. Mittra and W. W. Lee, *Analytical Techniques in the Theory of Guided Waves*, Macmillan, New York, 1971.
18. Y. C. Shih, T. Itoh, and L. Q. Bui, "Computer-aided design of millimeter-wave E-plane filters," *IEEE Trans. Microwave Theory Tech.*, vol. MTT-31, pp. 135–141, Feb. 1983.
19. S. C. Kashyap and M. A. K. Hamid, "Frequency response of waveguide filters with thick diaphragms," *Int. J. Electron.*, vol. 32, pp. 169–180, 1972.
20. C. P. Jethwa and R. L. Gunshor, "An analytical equivalent circuit representation for waveguide-mounted Gunn oscillators," *IEEE Trans. Microwave Theory Tech.*, vol. MTT-20, pp. 565–572, Sept. 1972.
21. R. Mittra, "Relative convergence of the solution of a doubly infinite set of equations," *J. Res. Natl. Bur. Stand., Sect. D*, vol. 67, pp. 245–254, Mar.–Apr. 1963.
22. M. Leroy, "On the convergence of numerical results in modal analysis," *IEEE Trans. Antennas Propag.*, vol. AP-31, pp. 655–659, July 1983.
23. G. Kompa, "S-Matrix computation of microstrip discontinuities with a planar waveguide model," *Arch. Elektron. Ubertragungstech.*, vol. 30, pp. 58–64, 1975.
24. Y. C. Shih, T. S. Chu, and T. Itoh, "Comparative study of mode-matching formulations for microstrip discontinuity problems," *1985 IEEE MTT-S. Int. Microwave Symp. Dig.*, pp. 435–438, 1985.
25. R.H. Jansen, "Improved microstrip model for the analysis of MIC components," *Arch. Elektron. Ubertragungstech.*, vol. 30, pp. 502–504, 1976.
26. S. W. Lee, W. R. Jones, and J. J. Campbell, "Convergence of numerical solutions of iris-type discontinuity problems," *IEEE Trans. Microwave Theory Tech.*, vol. MTT-19, pp. 528–536, June 1971.

27. R. Mittra, T. Itoh, and T. S. Li, "Analytical and numerical studies of the relative convergence phenomenon arising in the solution of an integral equation by the moment method," *IEEE Trans. Microwave Theory Tech.*, vol. MTT-20, pp. 96–104, Feb. 1972.
28. N. Marcuvitz, *Waveguide Handbook*, M.I.T. Radiat. Lab. Ser., vol. 10, McGraw-Hill, New York, 1951.
29. L. Lewin, *Theory of Waveguides*, Wiley, New York, 1975.
30. R. R. Mansour and R. H. Macphie, "An improved transmission matrix formulation of cascaded discontinuities and its application to E-plane circuits," *IEEE Trans. Microwave Theory Tech.*, vol. MTT-34, pp. 1490–1498, Dec. 1986.
31. F. Alessandi, G. Bartolucci, and R. Sorrentino, "Admittance matrix formulation of waveguide discontinuity problems: Computer-aided design of branch-guide directional couplers," *IEEE Trans. Microwave Theory Tech.*, submitted.
32. F. Bogelsack and I. Wolff, "Application of a projection method to a mode-matching solution for microstrip lines with finite metalization thickness," *IEEE Trans. Microwave Theory Tech.*, vol. MTT-35, pp. 918–921, Oct. 1987.

10

Generalized Scattering Matrix Technique

Tatsuo Itoh
Department of Electrical and Computer Engineering
The University of Texas at Austin
Austin, Texas

1. INTRODUCTION

It is well known that only a very limited number of waveguide discontinuity problems can be solved exactly. Therefore, most discontinuity problems are handled numerically. It is often difficult to assess the accuracy of solutions obtained purely numerically. There are, however, certain classes of discontinuity problems that can be reduced or decomposed to much simpler problems.

The generalized scattering matrix technique treats these classes of problems. It is an extension of the conventional scattering matrix in a single-moded transmission line system. In the conventional form, the *S* parameters describe the reflection and transfer characteristics of a junction. The generalized version takes into account the scattering phenomena of the dominant and all of the higher-order modes including evanescent ones.

When there are two cascaded junctions, we can exactly describe the overall scattering phenomena including the interactions between the two junctions by using the genralized *S* parameters of each junction. Since all the higher-order modes are included, the interaction between junctions can be correctly described even if the distance between them is infinitesimally small. Therefore, the scattering characteristics of a rather complicated discontinuity can be described if we can find the generalized *S* parameters of each of the simpler junctions created by decomposition of the original junction.

In this chapter, we first describe the general idea and a format of calculation. We then demonstrate the usefulness of this technique by way of several practical examples.

2. DEFINITION OF GENERALIZED SCATTERING MATRIX

The generalized scattering matrix (GSM) technique, introduced by Mittra and Pace [1], is an extension of the conventional scattering matrix technique very familiar to all microwave engineers. The conventional scattering matrix technique keeps track of the signal flow from a junction in a single-moded transmission line system. The effect of the higher-order modes is included in the expression of the S parameter. This is an extremely useful tool for microwave circuit design.

There are at least two instances for which the conventional S parameter method fails to provide satisfactory solutions. One is when the transmission line system is multimoded, and another is when two discontinuities are located in an extreme proximity. In the first case, all of the propagating modes interact via discontinuity, and S parameters can be defined for each of these modes. On the other hand, in the second case, interactions between two discontinuities via all higher-order modes can no longer be neglected even if these modes are evanescent.

The generalized scattering matrix (GSM) technique is a somewhat more "field theory" type of concept than the circuit-oriented conventional S parameters. The GSM consists of scattered modal coefficients from a discontinuity. The definition can be best illustrated by way of an example. Figure 1 shows a well-known waveguide bifurcation problem. It is assumed that this is a two-dimensional problem with the TE_{n0} mode entering the discontinuity from the left, region I. The fields are reflected (or scattered) back to region I and transmitted (or scattered) into regions II and III. Since these fields can be written in terms of the modes (in this case TE_{m0}'s only) in each region, the modal coefficients indicate how strongly and in what phase each mode is excited. How one can obtain such information has been a primary theme of scholars and engineers with training in the so-called boundary value problems. Here, we simply assume that such information is somehow available. We also assume that the amplitude of the incident mode

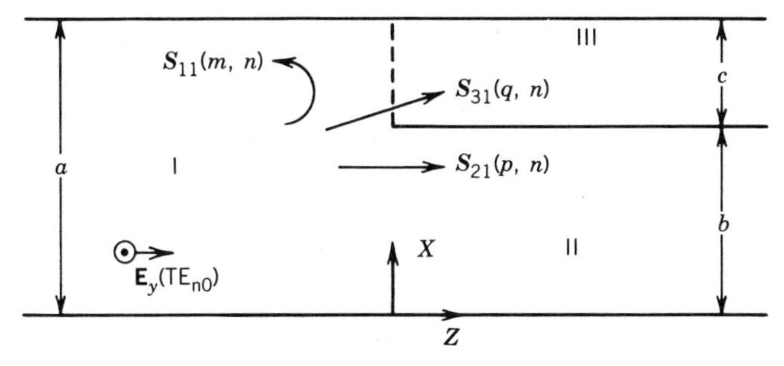

Fig. 1 Waveguide bifurcation.

is 1. Then the complex amplitude of the mth scattered mode reflected (not necessarily propagating) in region I is

$$S_{11}(m, n)$$

whereas, the pth and qth scattered modes transmitted (not necessarily propagating) in regions II and III are

$$S_{21}(p, n) \qquad \text{and} \qquad S_{31}(q, n)$$

Other GSM coefficients can be similarly defined. Note that S_{11}, S_{21}, S_{31}, etc., are in general infinite-dimensional matrices.

3. SIMPLE USE OF THE GENERALIZED SCATTERING MATRIX

The generalized scattering matrix (GSM) method was initially introduced to solve problems related to the auxiliary problem for which the solution is easily or exactly available [2]. In fact, the structure in Fig. 1 is exactly solvable either by the Wiener–Hopt technique or by the residue calculus technique. Consider Fig. 2. A standard way of solving this problem would be to expand the field in terms of modal expansion in each region, followed by a mode matching at the junction $z = 0$. The GSM can handle this problem much more efficiently and accurately. The auxiliary structure in Fig. 3 provides a clue. It is seen that if the solution for Fig. 3 is obtained, the one for Fig. 2 is recovered by letting $\delta \rightarrow 0$. Also notice that Fig. 3 has two junctions. Junction A is exactly the same as the one in Fig. 1, whereas junction B is a simple transition for which no mode coupling takes place.

We now identify the GSM associated with junctions A and B by means of superscripts A and B. For instance, $S_{12}^{A}(m, n)$ is the amplitude of the mth mode transmitted in regions I when the nth mode enters junction A from the right in region II. The GSMs at junction B are easily found. For instance,

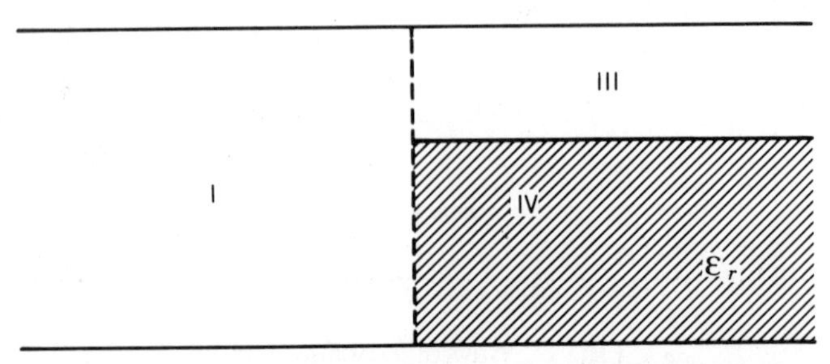

Fig. 2 Singly nonhomogeneous waveguide junction.

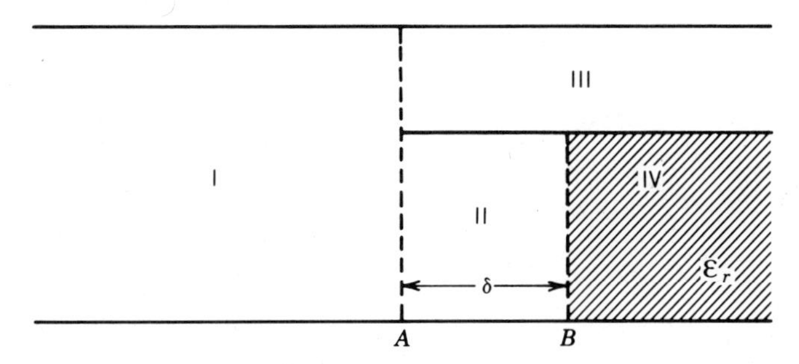

Fig. 3 Auxiliary geometry for Fig. 2.

$$S_{22}^{B}(m, n) = \frac{\beta_n - \bar{\beta}_n}{\beta_n + \bar{\beta}_n}\, \delta_m^n \tag{1}$$

$$S_{42}^{B}(m, n) = \frac{2\beta_n}{\beta_n + \bar{\beta}_n}\, \delta_m^n \tag{2}$$

where

$$\beta_n = \left[k^2 - \left(\frac{n\pi}{b}\right)^2\right]^{1/2} = -j\left[\left(\frac{n\pi}{b}\right)^2 - k^2\right]^{1/2} \tag{3}$$

$$\bar{\beta}_n = \left[\varepsilon_r k^2 - \left(\frac{n\pi}{b}\right)^2\right]^{1/2} = -j\left[\left(\frac{n\pi}{b}\right)^2 - \varepsilon_r k^2\right]^{1/2} \tag{4}$$

where δ_m^n is the Kronecker delta.

Now, let us consider what will happen when a TE wave is incident upon junction A from regions I in the case $\delta \rightarrow 0$. Since we can describe this wave once all the modal coefficients are known, the wave can be represented by a column vector $\bar{\boldsymbol{\phi}}$ such that ϕ_i is the coefficient of the ith mode in the incident field. We now refer to Fig. 4 and observe the following:

1. At junction A, fields are reflected back into region I and transmitted into regions II and III. The mode vector for the field reflected into I is $S_{11}^{A}\bar{\boldsymbol{\phi}}$. The one into III is $S_{31}^{A}\bar{\boldsymbol{\phi}}$. The wave transmitted into II is $S_{21}^{A}\bar{\boldsymbol{\phi}}$.
2. Part of $S_{21}^{A}\bar{\boldsymbol{\phi}}$ is reflected back toward junction A as $S_{22}^{B}(S_{21}^{A}\bar{\boldsymbol{\phi}})$, and part is transmitted into region IV as $S_{42}^{B}(S_{21}^{A}\bar{\boldsymbol{\phi}})$. Notice that $\delta \rightarrow 0$, and therefore no wave propagation phenomena between A and B need be considered. (The case of nonzero δ will be treated later.)
3. Multiple reflection phenomena as depicted in Fig. 4 take place. The fields in regions I and IV are the sums of the fields transmitted or reflected successively.

The field in region IV can be expressed by the mode vector $\bar{\boldsymbol{\psi}}$:

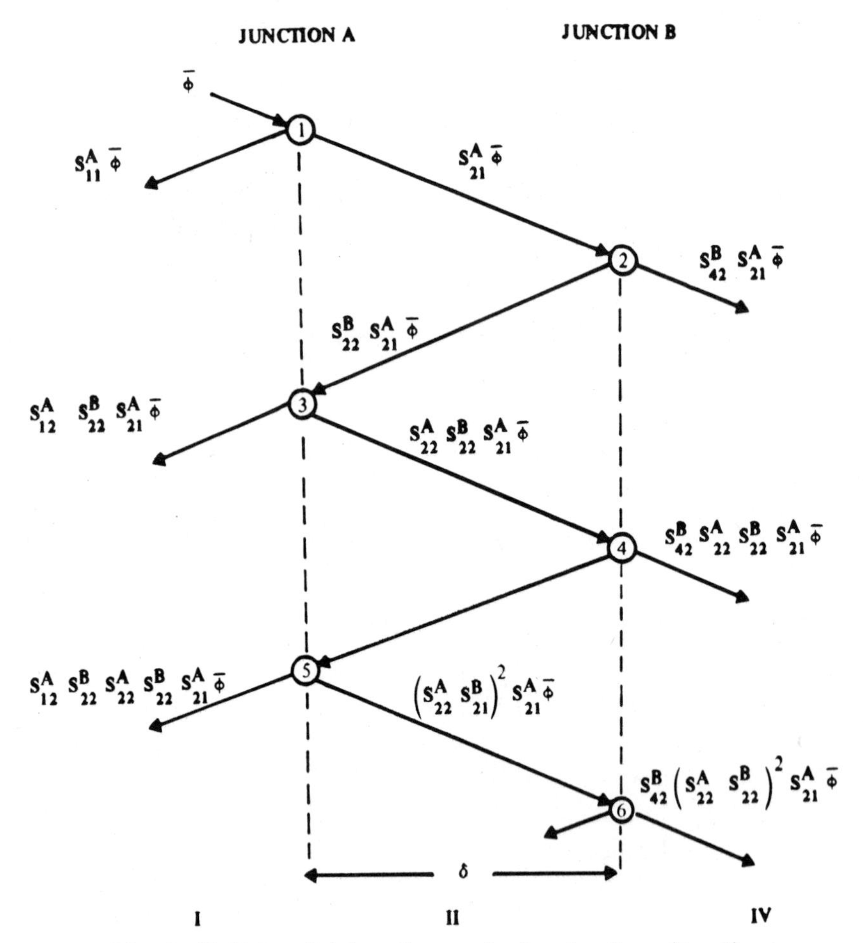

Fig. 4 Multiple reflection diagram for two junctions ($\delta \to 0$).

$$\bar{\psi} = S_{42}^{B} S_{21}^{A} \bar{\phi} + S_{42}^{B} S_{22}^{A} S_{22}^{B} S_{21}^{A} \bar{\phi} + S_{42}^{B} (S_{22}^{A} S_{22}^{B})^{2} S_{21}^{A} \bar{\phi} + \cdots$$

$$= \sum_{n=0}^{\infty} S_{42}^{B} (S_{22}^{A} S_{22}^{B})^{n} S_{21}^{A} \bar{\phi} = S_{42}^{B} (I - S_{22}^{A} S_{22}^{B})^{-1} S_{21}^{A} \bar{\phi} \tag{5}$$

where I is the identity matrix.

The summation is of a Neuman series form and can be expressed as in the last line. Proof of the convergence was given by Pace [3]. Since junctions A and B join, we can define the composite GSM for the combined junction. Hence, we can write

$$\bar{\psi} = S_{41} \bar{\phi} \tag{6}$$

where

$$S_{41} = S^{B}_{42}(I - S^{A}_{22}S^{B}_{22})^{-1}S^{A}_{21} \tag{7}$$

The other GSMs for the composite junction can be found in a similar manner.

The process described above is mathematically exact because of the use of matrices of infinite size. However, in practice, the size of the matrices must be finite, and approximate results are obtained from these truncated matrices. In many applications, surprisingly accurate results can be obtained by a relatively small matrix size.

4. EXAMPLES FOR CASCADED JUNCTIONS

In the preceding section, the generalized scattering matrix method as applied to a boundary value problem has been discussed. In this section, the GSM method is applied to a more "circuit"-oriented problem. In many microwave circuits, it is a common scene that there are several discontinuities appearing in succession. A good example is an iris-coupled waveguide filter. As long as these discontinuities are far apart with respect to the *guide* wavelength, the conventional *S*-parameter technique works quite well. However, in many cases this is not the case, and interaction by way of higher-order modes needs to be taken into account. Rozzi introduced the concept of "accessible modes" to account for the interaction by higher-order modes [4]. Use of the GSM methods, however, automatically includes all of the higher-order mode interactions in principle and can retain as many modes as necessary for desired accuracy.

The application will be illustrated here by way of the *E*-plane septum shown in Fig. 5. Such a septum is an essential element of increasingly popular *E*-plane filters [5]. The dielectric layer is inserted at the center of the waveguide parallel to the *E* plane. Since in practice only the TE_{10} mode is the propagating one and the structure is symmetric with respect to the midplane, only the odd-order modes are excited at the junction. Therefore, a magnetic wall can be placed at the midplane and only half of the structure as given at the top of Fig. 6 needs to be analyzed. This structure can be identified with two identical junctions *A* and *B* placed back to back at a distance *d*. The junction *A* (or *B*) is again one of exactly solvable structure and is therefore characterized by a three-port GSM that has nine elements,

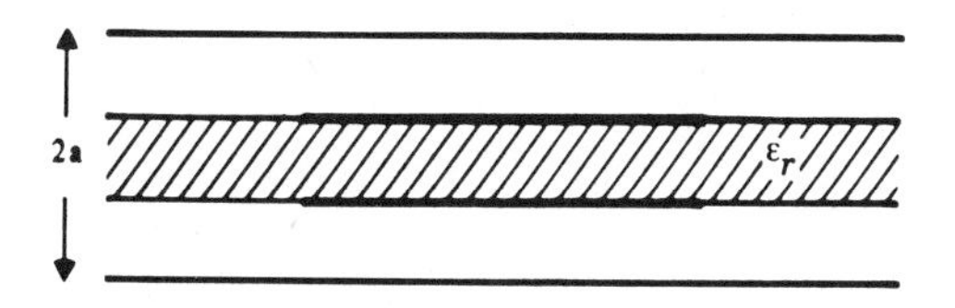

Fig. 5 Top view of bilateral septum.

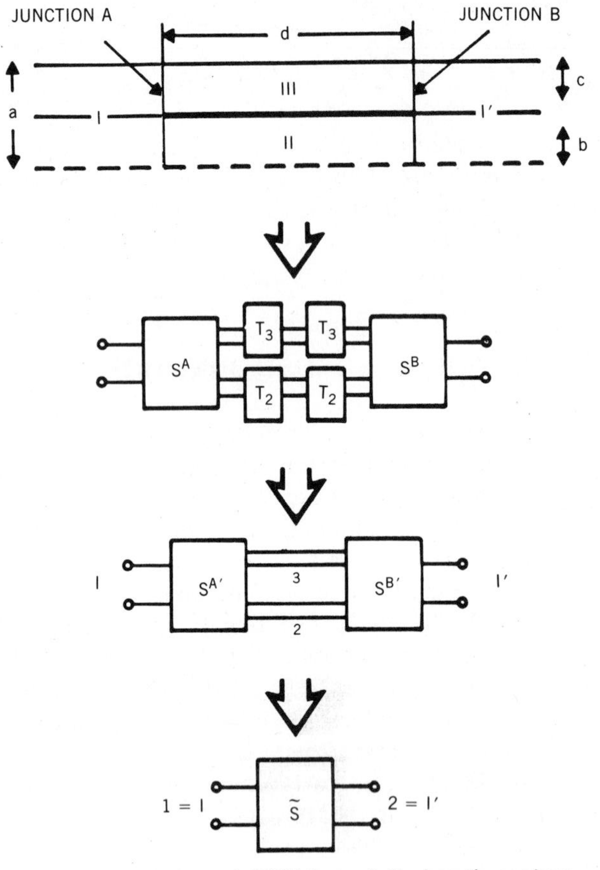

Fig. 6 Derivation of GSM for a finite-length septum.

each of which is of infinite dimensions. Hence, S^A takes the following form:

$$S^A = \begin{bmatrix} S_{11}^A & S_{12}^A & S_{13}^A \\ S_{21}^A & S_{22}^A & S_{23}^A \\ S_{31}^A & S_{32}^A & S_{33}^A \end{bmatrix} \tag{8}$$

The elements S_{11}^A, etc., are infinite size matrices. The specific form of the matrix elements is given in Reference 5.

We now cascade S^A and S^B by way of regions II and III. To take advantage of symmetry, we first combine S^A and S^B with one-half of the septum. Notice that regions II and III are smaller waveguides of length d. Half-lengths of these waveguides may be characterized by the transmission matrices T_2 and T_3, which are diagonal matrices of infinite size. The ith diagonal elements of T_2 and T_3 are

$$T_{2ii} = e^{-\gamma_{2i} d/2} \tag{9}$$

$$T_{3ii} = e^{-\gamma_{3i} d/2} \tag{10}$$

where γ_{2i} and γ_{3i} are the propagation constants of the ith mode in regions II and III. The GSM for junction A plus one-half length of septum is designated $\boldsymbol{S}^{A'}$. It is given by

$$\boldsymbol{S}^{A'} = \boldsymbol{T}\boldsymbol{S}^{A}\boldsymbol{T} \tag{11}$$

Similarly,

$$\boldsymbol{S}^{B'} = \boldsymbol{T}\boldsymbol{S}^{B}\boldsymbol{T} \tag{12}$$

where the transmission matrix $\boldsymbol{T}$ is

$$\boldsymbol{T} = \begin{bmatrix} \boldsymbol{I} & \boldsymbol{0} & \boldsymbol{0} \\ \boldsymbol{0} & \boldsymbol{T}_2 & \boldsymbol{0} \\ \boldsymbol{0} & \boldsymbol{0} & \boldsymbol{T}_3 \end{bmatrix} \tag{13}$$

$\boldsymbol{I}$ and $\boldsymbol{0}$ are the identity matrix and the zero matrix, respectively. Equation (11) implies that the reference planes for junction A are moved by $d/2$ in regions II and III and by zero distance in region I.

Although the multiple reflection technique in Fig. 4 may be extended to the process of combining $\boldsymbol{S}^{A'}$ and $\boldsymbol{S}^{B'}$, we can find the composite matrix for a septum S by the following matrix manipulation. In reference to Fig. 6, we have

$$\boldsymbol{S}^{A'} = \begin{bmatrix} \boldsymbol{S}_{\mathrm{II}}^{A'} & \boldsymbol{S}_{\mathrm{I}2}^{A'} & \boldsymbol{S}_{\mathrm{I}3}^{A'} \\ \boldsymbol{S}_{2\mathrm{I}}^{A'} & \boldsymbol{S}_{22}^{A'} & \boldsymbol{S}_{23}^{A'} \\ \boldsymbol{S}_{3\mathrm{I}}^{A'} & \boldsymbol{S}_{32}^{A'} & \boldsymbol{S}_{33}^{A'} \end{bmatrix} \tag{14}$$

$$\boldsymbol{S}^{B'} = \begin{bmatrix} \boldsymbol{S}_{\mathrm{I'I'}}^{B'} & \boldsymbol{S}_{\mathrm{I'}2}^{B'} & \boldsymbol{S}_{\mathrm{I'}3}^{B'} \\ \boldsymbol{S}_{2\mathrm{I'}}^{B'} & \boldsymbol{S}_{22}^{B'} & \boldsymbol{S}_{23}^{B'} \\ \boldsymbol{S}_{3\mathrm{I'}}^{B'} & \boldsymbol{S}_{32}^{B'} & \boldsymbol{S}_{33}^{B'} \end{bmatrix} \tag{15}$$

Ports I, I′, 2, and 3 are identified in the figure. Let us trace the wave vector at ports 2 and 3 connected to $\boldsymbol{S}^{A'}$ and $S^{B'}$. The incident and reflected wave vectors at each port for each matrix are related. For instance,

$$\boldsymbol{b}_2^{A'} = \boldsymbol{S}_{2\mathrm{I}}^{A'}\boldsymbol{a}_{\mathrm{I}}^{A'} + \boldsymbol{S}_{22}^{A'}\boldsymbol{a}_2^{A'} + \boldsymbol{S}_{23}^{A'}\boldsymbol{a}_3^{A'}$$

where $\boldsymbol{b}_2^{A'}$ denotes the wave vector exiting port 2 of $\boldsymbol{S}^{A}$ and $\boldsymbol{a}_i$ $(i = \mathrm{I}, 2, 3)$ is the incident wave vector at each port. Similar relations are given for all ports for $\boldsymbol{S}^{A'}$ and $\boldsymbol{S}^{B'}$. Since ports 2 and 3 are common to $\boldsymbol{S}^{A'}$ and $\boldsymbol{S}^{B'}$, the following boundary conditions must be satisfied:

$$\boldsymbol{b}_2^{A'} = \boldsymbol{a}_2^{B'} \qquad \boldsymbol{a}_2^{A'} = \boldsymbol{b}_2^{B'} \tag{16a}$$

$$\boldsymbol{b}_3^{A'} = \boldsymbol{a}_3^{B'} \qquad \boldsymbol{a}_3^{A'} = \boldsymbol{b}_3^{B'} \tag{16b}$$

When these conditions are imposed, we can find the relations between $\boldsymbol{a}_{\mathrm{I}}^{A'}$, $\boldsymbol{b}_{\mathrm{I}}^{A'}$, $\boldsymbol{a}_{\mathrm{I}}^{B'}$, and $\boldsymbol{b}_{\mathrm{I}}^{B'}$ since these relations provide the generalized scattering matrix between port I and port I′. In the present case, it is readily seen that $\boldsymbol{S}^{A'} = \boldsymbol{S}^{B'}$ because the physical configurations of junctions A and B are identical. Finally, ports I and I′ are renamed as ports 1 and 2 of the composite structure, that is, of the septum. The resultant GSM of the septum $\tilde{\boldsymbol{S}}$ is

$$\tilde{\boldsymbol{S}} = \begin{bmatrix} \tilde{\boldsymbol{S}}_{11} & \tilde{\boldsymbol{S}}_{12} \\ \tilde{\boldsymbol{S}}_{21} & \tilde{\boldsymbol{S}}_{22} \end{bmatrix} = \begin{bmatrix} \tilde{\boldsymbol{S}}_{12} & \tilde{\boldsymbol{S}}_{13} \\ \tilde{\boldsymbol{S}}'_{12} & \boldsymbol{S}'_{13} \end{bmatrix} \begin{bmatrix} \boldsymbol{I} - \tilde{\boldsymbol{S}}_{22} & -\tilde{\boldsymbol{S}}_{23} \\ -\tilde{\boldsymbol{S}}_{32} & \boldsymbol{I} - \tilde{\boldsymbol{S}}_{33} \end{bmatrix}^{-1} \begin{bmatrix} \boldsymbol{S}'_{21} & \tilde{\boldsymbol{S}}_{21} \\ \tilde{\boldsymbol{S}}'_{31} & \tilde{\boldsymbol{S}}_{31} \end{bmatrix} + \begin{bmatrix} \boldsymbol{S}'_{11} & \tilde{\boldsymbol{S}}_{11} \\ \tilde{\boldsymbol{0}} & \boldsymbol{S}'_{11} \end{bmatrix} \tag{17}$$

where

$$\tilde{\boldsymbol{S}}_{ij} = \sum_{k=2}^{3} \boldsymbol{S}'_{ik} \boldsymbol{S}'_{kj} \tag{18}$$

and $\boldsymbol{S}'_{ik}$, etc., are the (i, k) minor of $\boldsymbol{S}^{A'}$.

Notice that $\boldsymbol{S}$ is of infinite dimensions. In circuit analysis, only $\tilde{S}_{11}(1, 1)$, $\tilde{S}_{12}(1, 1)$, $\tilde{S}_{21}(1, 1)$, and $\tilde{S}_{22}(1, 1)$ are usually required if the septum is isolated. In conventional notation they are written as $\tilde{S}_{11}$, $\tilde{S}_{12}$, $\tilde{S}_{21}$, and $\tilde{S}_{22}$. On the other hand, if the septum is cascaded with another one at a relatively short distance w, the GSM method must be used.

The process of cascading two septa is relatively simple. As shown in Fig. 7, let the GSM of septum A be $\tilde{\boldsymbol{S}}_a$ and that of septum B be $\tilde{\boldsymbol{S}}_b$. Note that both $\tilde{\boldsymbol{S}}_a$ and $\tilde{\boldsymbol{S}}_b$ are made of 2×2 infinite-dimensioned minors. The septum A represented by $\tilde{\boldsymbol{S}}_a$ is cascaded first with the waveguide section of length w represented by the transmission matrix $\boldsymbol{T}_1$:

$$\boldsymbol{T}_1 = \begin{bmatrix} \boldsymbol{I} & \boldsymbol{0} \\ \boldsymbol{0} & \boldsymbol{T} \end{bmatrix} \qquad T_{ii} = \exp(-\gamma_{1i} w)$$

where γ_{1i} is the propagation constant of the $i$the mode in the waveguide section of length w. The result of cascading $\tilde{\boldsymbol{S}}_a$ and $\boldsymbol{T}_1$ results in

$$\tilde{\boldsymbol{S}}'_a = \boldsymbol{T}_1 \tilde{\boldsymbol{S}}_a \boldsymbol{T}_1 \tag{19}$$

The next step is the cascading of $\tilde{\boldsymbol{S}}'_a$ and $\tilde{\boldsymbol{S}}_b$ to obtain the GSM of the

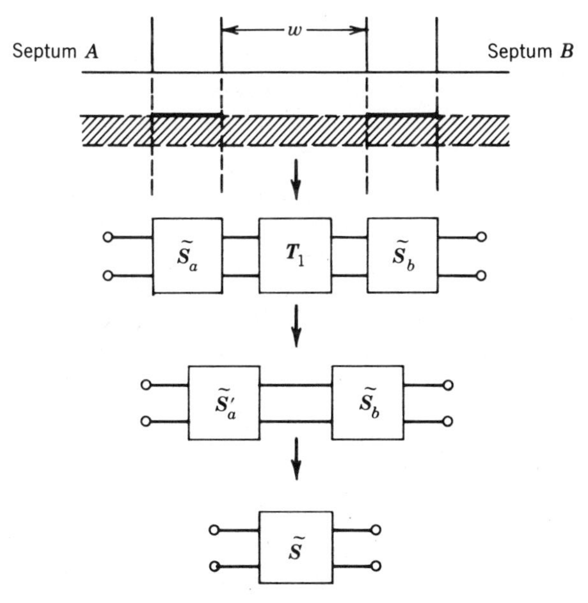

Fig. 7 Derivation of S parameters for cascading septa.

entire structure. The process is similar to, but simpler than, that for (17). The result is

$$\tilde{S} = \begin{bmatrix} \tilde{S}'_{a12}\tilde{S}_{b11} \\ \tilde{S}_{b21} \end{bmatrix} [I - \tilde{S}_{a22}\tilde{S}_{b11}]^{-1} [\tilde{S}'_{a21} \qquad \tilde{S}'_{a22}\tilde{S}_{b12}] + \begin{bmatrix} \tilde{S}_{a11} & (\tilde{S}'_{a12}\tilde{S}_{b12}) \\ \mathbf{0} & \tilde{S}_{b22} \end{bmatrix} \tag{20}$$

5. CONCLUSION

In this chapter, the concept of the generalized scattering matrix (GSM) method has been introduced and its algorithm presented. The method is useful for (1) analyzing a complicated junction that can be decomposed to several simpler junctions and (2) characterizing cascaded junctions that are in close proximity in terms of electrical length. The second feature is particularly useful in many microwave passive components. The method is mathematically exact provided that *all* the matrices of infinite order are available. Naturally, in actual computations, the matrix has to be kept to a reasonable size.

APPENDIX. COMPUTER PROGRAM DESCRIPTION

Subroutine STEP2 calculates eq. (17) for a given scattering matrix SS for the single junction of a septum. SS is the input, and ST11, ST21, ST12, ST22 are the output, or $\tilde{S}_{11}$, etc., in (17). MSIZE, NMAX, PI, ZJ in the COMMON block must be declared from the calling program. MSIZE is the size of the truncated minor matrices such as $\tilde{S}_{11}$, $\tilde{S}_{21}$, and S_{11}. NMAX is not used in the subroutine. PI is π, and ZJ is $\sqrt{-1}$.

Subroutine CASCADE provides the results of eq. (20). The inputs $\tilde{S}_a$ and $\tilde{S}_b$ are placed by COMMON/S2/ into ST11 through ST22 and SD11 through SD22, respectively. The output $\bar{S}$ in (20) is placed in ST11 through ST22.

Both of these subroutines require several subroutines for matrix manipulations. They are CMSUM, CMSUB, CMDOT2, CMMUL, and COPY. The functions these subroutines provided are easily seen.

```
      SUBROUTINE STEP2(D,ST11,ST21,ST12,ST22,NS,SS,NSS,SA,SB,SG)
      IMPLICIT COMPLEX(S,Z)
      DIMENSION SS(NSS,1),SA(1),SB(1),SG(1)
      COMMON/CONS/ZJ,MSIZE,NMAX,PI,CFACT
      DIMENSION S11(5,5),S21(5,5),S31(5,5),S12(5,5),S22(5,5),S32(5,5)
     1          ,S13(5,5),S23(5,5),S33(5,5)
      DIMENSION ZT2(5,5),ZT3(5,5),ZTP(5,5)
      DIMENSION SS11(5,5),SS21(5,5),SS31(5,5),SS12(5,5),SS22(5,5)
     1          ,SS32(5,5),SS13(5,5),SS23(5,5),SS33(5,5)
      DIMENSION ST11(NS,1),ST21(NS,1),ST12(NS,1),ST22(NS,1)
      DIMENSION Z11(5,5),Z21(5,5),Z31(5,5),Z12(5,5),Z22(5,5),Z32(5,5)
     1          ,Z13(5,5),Z23(5,5),Z33(5,5),ZZ1(10,10),ZZ2(10,10)
      DIMENSION WA(20)
C
C     CONVERT MATRIX SS TO S11 S21 S31 .......
C
      DO 3 I=1,MSIZE
      DO 3 J=1,MSIZE
      S11(I,J)=SS(I,J)
      S21(I,J)=SS(I,J+MSIZE)
      S31(I,J)=SS(I,J+2*MSIZE)
      S12(I,J)=SS(I+MSIZE,J)
      S22(I,J)=SS(I+MSIZE,J+MSIZE)
      S32(I,J)=SS(I+MSIZE,J+2*MSIZE)
      S13(I,J)=SS(I+2*MSIZE,J)
      S23(I,J)=SS(I+2*MSIZE,J+MSIZE)
      S33(I,J)=SS(I+2*MSIZE,J+2*MSIZE)
    3 CONTINUE
C
C     DEFINE TRANSMISSION MATRIX ZT2,ZT3
C
      LDS=5
      LDTT=5
      DO 10 I=1,MSIZE
      DO 10 J=1,MSIZE
      IF(I.EQ.J)GO TO 5
      ZT2(I,J)=(0.,0.)
      ZT3(I,J)=(0.,0.)
      GO TO 10
    5 ZT2(I,J)=CEXP(-ZJ*SB(I)*D)
      ZT3(I,J)=CEXP(-ZJ*SG(I)*D)
   10 CONTINUE
C
C     TST OPERATION
C
C     S11 DOES NOT CHANGE
```

```
C
      CALL COPY(S11,SS11,LDS,LDS,MSIZE)
C
C     S12
C
      CALL CMMUL(S12,ZT2,SS12,LDS,LDTT,LDS,MSIZE)
C
C     S13
C
      CALL CMMUL(S13,ZT3,SS13,LDS,LDTT,LDS,MSIZE)
C
C     S21
C
      CALL CMMUL(ZT2,S21,SS21,LDTT,LDS,LDS,MSIZE)
C
C     S31
C
      CALL CMMUL(ZT3,S31,SS31,LDTT,LDS,LDS,MSIZE)
C
C     S22
C
      CALL CMMUL(S22,ZT2,ZTP,LDS,LDTT,LDTT,MSIZE)
      CALL CMMUL(ZT2,ZTP,SS22,LDTT,LDTT,LDS,MSIZE)
C
C     S23
C
      CALL CMMUL(S23,ZT3,ZTP,LDS,LDTT,LDTT,MSIZE)
      CALL CMMUL(ZT2,ZTP,SS23,LDTT,LDTT,LDS,MSIZE)
C
C     S32
C
      CALL CMMUL(S32,ZT2,ZTP,LDS,LDTT,LDTT,MSIZE)
      CALL CMMUL(ZT3,ZTP,SS32,LDTT,LDTT,LDS,MSIZE)
C
C     S33
C
      CALL CMMUL(S33,ZT3,ZTP,LDS,LDTT,LDTT,MSIZE)
      CALL CMMUL(ZT3,ZTP,SS33,LDTT,LDTT,LDS,MSIZE)
C
C     EQUATION (18)
C
      LDZ=5
      LDST=5
      CALL CMDOT2(SS12,SS13,LDS,SS21,SS31,LDS,Z11,LDZ,MSIZE)
      CALL CMDOT2(SS12,SS13,LDS,SS22,SS32,LDS,Z12,LDZ,MSIZE)
      CALL CMDOT2(SS12,SS13,LDS,SS23,SS33,LDS,Z13,LDZ,MSIZE)
      CALL CMDOT2(SS22,SS23,LDS,SS21,SS31,LDS,Z21,LDZ,MSIZE)
      CALL CMDOT2(SS22,SS23,LDS,SS22,SS32,LDS,Z22,LDZ,MSIZE)
      CALL CMDOT2(SS22,SS23,LDS,SS23,SS33,LDS,Z23,LDZ,MSIZE)
      CALL CMDOT2(SS32,SS33,LDS,SS21,SS31,LDS,Z31,LDZ,MSIZE)
      CALL CMDOT2(SS32,SS33,LDS,SS22,SS32,LDS,Z32,LDZ,MSIZE)
      CALL CMDOT2(SS32,SS33,LDS,SS23,SS33,LDS,Z33,LDZ,MSIZE)
C
C     BUILD COMPOSITE MATRIX; SIZE 2*MSIZE BY 2*MSIZE
C
      DO 50 I=1,MSIZE
      DO 50 J=1,MSIZE
      ZZ2(I,J)=Z22(I,J)
      ZZ2(I+MSIZE,J)=Z32(I,J)
      ZZ2(I,J+MSIZE)=Z23(I,J)
      ZZ2(I+MSIZE,J+MSIZE)=Z33(I,J)
      ZZ1(I,J+MSIZE)=(0.,0.)
      ZZ1(I+MSIZE,J)=(0.,0.)
```

```
         IF(I.EQ.J)GO TO 45
         ZZ1(I,J)=(0.,0.)
         ZZ1(I+MSIZE,J+MSIZE)=(0.,0.)
         GO TO 50
      45 ZZ1(I,J)=(1.,0.)
         ZZ1(I+MSIZE,J+MSIZE)=(1.,0.)
      50 CONTINUE
         LDZZ=10
         NN=2*MSIZE
         CALL CMSUB(ZZ1,ZZ2,ZZ2,LDZZ,LDZZ,LDZZ,NN)
         IJOB=0
C
C     INVERSE MATRIX IN EQUATION (17)
C     NEED IMSL LINK
C
      CALL LEQT1C(ZZ2,NN,LDZZ,ZZ1,NN,LDZZ,IJOB,WA,IER)
      IF(IER.NE.0)WRITE(6,1001)IER
 1001 FORMAT(//,'RETURN CODE FROM THE MATRIX INVERSION ROUTINE',I5)
C
C     DECOMPOSITE INTO SUBMATRICES
C
      DO 80 I=1,MSIZE
      DO 80 J=1,MSIZE
      ST11(I,J)=ZZ1(I,J)
      ST12(I,J)=ZZ1(I,J+MSIZE)
      ST21(I,J)=ZZ1(I+MSIZE,J)
      ST22(I,J)=ZZ1(I+MSIZE,J+MSIZE)
   80 CONTINUE
C
C     GET FINAL SCATTERING MATRIX
C
      CALL CMDOT2(ST11,ST12,LDST,SS21,SS31,LDS,Z22,LDZ,MSIZE)
      CALL CMDOT2(ST11,ST12,LDST,Z21,Z31,LDZ,Z23,LDZ,MSIZE)
      CALL CMDOT2(ST21,ST22,LDST,SS21,SS31,LDS,Z32,LDZ,MSIZE)
      CALL CMDOT2(ST21,ST22,LDST,Z21,Z31,LDZ,Z33,LDZ,MSIZE)
      CALL CMDOT2(Z12,Z13,LDZ,Z22,Z32,LDZ,Z21,LDZ,MSIZE)
      CALL CMDOT2(Z12,Z13,LDZ,Z23,Z33,LDZ,Z31,LDZ,MSIZE)
      CALL CMDOT2(SS12,SS13,LDS,Z22,Z32,LDZ,ST21,LDST,MSIZE)
      CALL CMDOT2(SS12,SS13,LDS,Z23,Z33,LDZ,Z13,LDZ,MSIZE)
      CALL CMSUM(Z21,SS11,ST11,LDZ,LDS,LDST,MSIZE)
      CALL CMSUM(Z31,Z11,ST12,LDZ,LDZ,LDST,MSIZE)
      CALL CMSUM(Z13,SS11,ST22,LDZ,LDS,LDST,MSIZE)
      RETURN
      END
      SUBROUTINE CASCAD(ALFA,W)
C
C      OUTPUTING S-MATRIN IS PLACED IN ST, AFTER CALCULATION,
C      THE CONTENT IN SD IS DESTROYED
C
      IMPLICIT COMPLEX(S,Z)
      COMMON/CONS/ZJ,MSIZE,NMAX,PI,CFACT
      COMMON/S2/ST11,ST21,ST12,ST22,SD11,SD21,SD12,SD22
      COMPLEX ALFA(1)
      DIMENSION ST11(5,5),ST21(5,5),ST12(5,5),ST22(5,5)
     1          ,SD11(5,5),SD21(5,5),SD12(5,5),SD22(5,5)
     2          ,ZT1(5,5),Z12(5,5),Z21(5,5),ZTP(5,5),ZU(5,5)
      DIMENSION WA(20)
C
C     DEFINE TRANSMISSION MATRIX IN EQUATION (15)
C
      DO 10 I=1,MSIZE
      DO 10 J=1,MSIZE
        IF(I.EQ.J) GO TO 5
```

```
          ZT1(I,J)=(0.,0.)
          ZU(I,J)=(0.,0.)
          GO TO 10
    5     ZT1(I,J)=CEXP(-ZJ*ALFA(I)*W)
          ZU(I,J)=(1.,0.)
   10 CONTINUE
      LDS=5
      LDZ=5
C
C     EQUATION (19)
C
C     Z11 = ST11
C
C     Z12 =
C
      CALL CMMUL(ST12,ZT1,Z12,LDS,LDZ,LDZ,MSIZE)
C
C     Z21 =
C
      CALL CMMUL(ZT1,ST21,Z21,LDZ,LDS,LDZ,MSIZE)
C
C     Z22 = ST22
C
      CALL CMMUL(ST22,ZT1,ZTP,LDS,LDZ,LDZ,MSIZE)
      CALL CMMUL(ZT1,ZTP,ST22,LDZ,LDZ,LDS,MSIZE)
C
C     COMPUTE EQUATION (20) TO OBTAIN THE
C     COMPOSITE S-MATRIX FOR TWO CASCADING UNIT
C
      CALL CMMUL(ST22,SD11,ST21,LDS,LDS,LDS,MSIZE)
      CALL CMSUB(ZU,ST21,ST21,LDZ,LDS,LDS,MSIZE)
      IJOB=0
      CALL LEQT1C(ST21,MSIZE,LDS,ZU,MSIZE,LDZ,IJOB,WA,IER)
      CALL CMMUL(Z12,SD11,ZTP,LDZ,LDS,LDZ,MSIZE)
      CALL CMMUL(ZTP,ZU,SD11,LDZ,LDZ,LDS,MSIZE)
      CALL CMMUL(SD21,ZU,ZTP,LDS,LDZ,LDZ,MSIZE)
      CALL CMMUL(ST22,SD12,SD21,LDS,LDS,LDS,MSIZE)
      CALL CMMUL(SD11,Z21,ST12,LDS,LDZ,LDS,MSIZE)
      CALL CMSUM(ST12,ST11,ST11,LDS,LDS,LDS,MSIZE)
      CALL CMMUL(ZTP,Z21,ST21,LDZ,LDZ,LDS,MSIZE)
      CALL CMMUL(SD11,SD21,Z21,LDS,LDS,LDZ,MSIZE)
      CALL CMMUL(Z12,SD12,ST12,LDS,LDZ,LDS,MSIZE)
      CALL CMSUM(Z21,ST12,ST12,LDZ,LDS,LDS,MSIZE)
      CALL CMMUL(ZTP,SD21,ST22,LDZ,LDS,LDS,MSIZE)
      CALL CMSUM(ST22,SD22,ST22,LDS,LDS,LDS,MSIZE)
      RETURN
      END
      SUBROUTINE CMSUM(A,B,C,LDA,LDB,LDC,N)
C
C     C=A+B
C
      COMPLEX A(LDA,1),B(LDB,1),C(LDC,1)
      DO 10 I=1,N
      DO 10 J=1,N
   10 C(I,J)=A(I,J)+B(I,J)
      RETURN
      END
      SUBROUTINE CMSUB(A,B,C,LDA,LDB,LDC,N)
C
C     C=A-B
C
      COMPLEX A(LDA,1),B(LDB,1),C(LDC,1)
      DO 10 I=1,N
      DO 10 J=1,N
```

```
   10 C(I,J)=A(I,J)-B(I,J)
      RETURN
      END
      SUBROUTINE CMDOT2(A1,A2,LDA,B1,B2,LDB,C,LDC,N)
C
C     C=(A1,A2)*(B1,B2)=A1B1+A2B2
C
      COMPLEX A1(LDA,1),A2(LDA,1),B1(LDB,1),B2(LDB,1),C(LDC,1)
      COMPLEX T1(10,10)
      CALL CMMUL(A1,B1,T1,LDA,LDB,10,N)
      CALL CMMUL(A2,B2,C,LDA,LDB,LDC,N)
      CALL CMSUM(T1,C,C,10,LDC,LDC,N)
      RETURN
      END
      SUBROUTINE CMMUL(A,B,C,LDA,LDB,LDC,N)
      COMPLEX A(LDA,1),B(LDB,1),C(LDC,1)
C
C     C=A*B
C
      DO 10 I=1,N
      DO 10 J=1,N
      C(I,J)=0.0
      DO 10 K=1,N
   10 C(I,J)=C(I,J)+A(I,K)*B(K,J)
      RETURN
      END
      SUBROUTINE COPY(A,B,LDA,LDB,N)
C
C     MATRIX A IS COPIED TO MATRIX B
C
      COMPLEX A(LDA,1),B(LDB,1)
      DO 10 I=1,N
      DO 10 J=1,N
   10 B(I,J)=A(I,J)
      RETURN
      END
```

REFERENCES

1. R. Mittra and J. Pace, *A New Technique for Solving a Class of Boundary Value Problems*, Rep. 72, Antenna Laboratory, University of Illinois, Urbana, 1963.
2. G. F. VanBlaricum, Jr. and R. Mittra, "A modified residue-calculus technique for solving a class of boundary value problems Part II. Waveguide phased arrays, modulated surfaces, and diffraction gratings," *IEEE Trans. Microwave Theory Tech.*, vol. MTT-17, pp. 310–319, June 1969.
3. J. Pace, *The Generalized Scattering Matrix Analysis of Waveguide Discontinuity Problems*, Rep. 1, Antenna Laboratory, University of Illinois, Urbana, 1964.
4. T. E. Rozzi and W. F. G. Mecklenbrauker, "Wide-band network modeling of interacting inductive irises and steps," *IEEE Trans. Microwave Theory Tech.*, vol. MTT-23, pp. 235–244, Feb. 1975.
5. Y.-C. Shih, T. Itoh and L. Q. Bui, "Computer-aided design of millimeter-wave E-plane filters, "*IEEE Trans. Microwave Theory Tech.*, vol. MTT-31, pp. 135–142, Feb. 1983.

11

Transverse Resonance Technique

R. Sorrentino
Department of Electronic Engineering
Università di Roma Tor Vergata
Rome, Italy

1. INTRODUCTION

The transverse resonance technique (TRT) originated as an application of the microwave circuit formalism in the direction perpendicular to the actual power flow in a cylindrical waveguide.

Over the years it has evolved from a technique to evaluate the dispersion relation for the dominant mode of certain types of waveguides, such as ridged waveguides, to more general and analytically elaborate techniques to evaluate the full propagation characteristics of these and other homogeneous and inhomogeneous structures. Recent developments include the modeling of discontinuities.

TRT typically applies to structures derived from conventional waveguides, for which simple analytical solutions are available, with the addition of some discontinuities placed across a transverse direction, i.e., a direction orthogonal to the waveguide axis. We will refer to them as "transversely discontinuous waveguides." In the simplest case, discontinuities are represented by abrupt changes of the dielectric filling material (dielectric slab-loaded waveguides). More generally, in addition to dielectric inhomogeneity, metallic obstacles and septa may be placed across a transverse coordinate.

In the conventional formulation of the transverse resonance technique, a suitable transverse equivalent network is established to compute the cutoff frequencies and possibly some additional characteristics of the structure.

In a more elaborate and rigorous formulation, a full wave analysis is developed using a representation of the electromagnetic field in the structure in terms of sets of modes properly chosen to simplify the boundary value problem. In terms of the microwave network formalism, this corresponds to establishing a generalized transverse equivalent network. In other words, a rigorous formulation of the boundary value problem is achieved by combining the transverse resonance concept with the generalized matrix characterization of the transverse discontinuity (or discontinuities). This formulation will therefore be referred to as the generalized transverse resonance technique. This terminology is used to stress the connection of this technique with the microwave network formalism. Other terms, such as transverse modal analysis [1] or transverse resonance diffraction [2], have also been used.

It should be stressed that the transverse resonance concept is also commonly used in connection with other numerical techniques, such as the spectral domain immittance approach [3] or the method of lines [4]. Some similarities with these and other methods are therefore to be expected.

TRT can be applied to structures where all longitudinal discontinuities are orthogonal to the same transverse coordinate. It is not restricted to rectangular geometries but can also be applied, for example, in a cylindrical coordinate system. However, it cannot be applied to arbitrary geometries, for which more numerically oriented, thus more flexible, techniques, such as the finite difference technique, must be used.

The basic idea in the transverse resonance approach is that the boundary value problem in certain types of uniform waveguides can more easily be formulated in terms of a set of modes differing from the usual TE and TM modes with respect to the longitudinal axis. The simplification is particularly evident in the class of dielectric slab-loaded rectangular waveguides and corresponding problems in cylindrical coordinates.

In the usual approach, the solution of Maxwell's equations in uniform and homogeneous waveguides is obtained by separation of the longitudinal and transverse dependence of the field components. This leads to the decomposition of the general field solution into TE and TM modes with respect to the axial direction. The solution of Maxwell's equations is reduced to that of two scalar eigenvalue problems. (For waveguides with multiply connected cross sections, TEM mode(s) must also be taken into account.) This is the universally adopted approach to the analysis of homogeneous waveguides with cross sections of separable geometries.

It is easily recognized, however, that in some instances the above is not the most convenient approach for EM field computation. Alternative formulations, based on different modal sets, can lead to more expeditious and simpler solutions. These modal sets are obtained through a different separation of Maxwell's equations. In the conventional formulation of TE and TM modes, these are implicitly assumed as transverse to the axial direction, which is also the direction of propagation. Different choices of what is assumed as the axial direction or the direction of propagation lead to different modal sets.

To avoid confusion between different sets of modes, we will from now on use the notation $\mathrm{TE}^{(\xi)}$ and $\mathrm{TM}^{(\xi)}$ to indicate transverse electric and magnetic fields with respect to the generic ξ direction, which does not necessarily coincide with the waveguide axis.

1.1. Transverse Resonance Analysis of Rectangular Waveguides

As an introduction to the TR concept and to become familiar with the different modal representations, we will refer to the simple case of a rectangular waveguide (RW). This example is useful to demonstrate the most elementary application of the TR concept and at the same time to show how this technique can be applied to the different modal representations.

One way to evaluate the rectangular waveguide modes is to consider the parallel-plate waveguide (PPW) first (Fig. 1*a*). The rectangular waveguide structure is recovered by adding two metallic plates orthogonal to the previous ones (Fig. 1*b*). The PPW modes will be reflected back and forth in the transverse x direction so that a standing wave along x is established. The boundary conditions on the conducting planes at $x = 0, a$ make the field resonate between these planes. In this manner, it is seen that the rectangular waveguide modes can be derived from PPW modes imposing the condition of resonance in the transverse x direction. (It is indeed well known that the normal modes of the rectangular waveguide result from the multiple reflections of plane waves by the four waveguide walls.)

To evaluate the RW modes we can start from different types of PPW modes. The conventional $\mathrm{TM}^{(z)}$ or $\mathrm{TE}^{(z)}$ modes of the rectangular waveguide can be obtained by superimposing two corresponding $\mathrm{TM}^{(z)}$ and $\mathrm{TE}^{(z)}$ modes of the parallel-plate waveguide with opposite components of the phase velocity in the x direction. The resultant standing wave pattern must comply with the conditions imposed by the lateral walls, i.e., that the E-field y and z components have nulls at $x = 0, a$.

Another set of modes can be derived for the rectangular waveguide that have no H or E components along the transverse x direction. These modes have been called LSM (longitudinal section magnetic) and LSE (longitudinal section electric) modes [5]. For uniformity, we will use instead the notation $\mathrm{TM}^{(x)}$ and $\mathrm{TE}^{(x)}$. These modes are obtained using an x-directed, instead of a z-directed, Hertzian potential (see the appendix). This corresponds to assuming the x direction to be the direction of propagation.

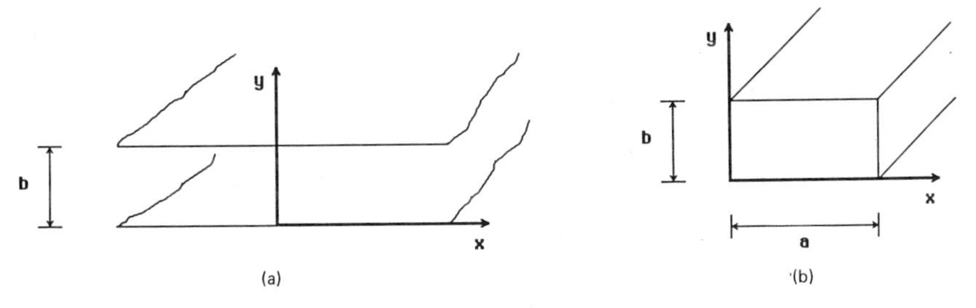

Fig. 1 (*a*) Parallel-plate waveguide (PPW) and (*b*) rectangular waveguide (RW) obtained by addition of metallic plates at $x = 0, a$.

Let us consider the case of $\mathrm{TE}^{(x)}$ modes. For the parallel-plate waveguide, the potential ψ, which is a function of the y and z coordinates, is expressed by, apart from a normalization coefficient,

$$\psi_n = \exp(-j\beta z)\cos(n\pi y/b) \tag{1}$$

The order of the mode, $n = 1, 2, \ldots$, is the number of maxima of the field along the y direction. The "axial" field dependence can be assumed to be that of a propagating wave, $\exp(-jk_x x)$.

Contrary to the usual formulation of $\mathrm{TM}^{(z)}$ and $\mathrm{TE}^{(z)}$ waveguide modes, the potential is now represented by a complex function of the coordinates. Actually, the z dependence is in the form of a propagating wave. Since the assumed x dependence is of the same form, it results that the physical nature of the solution we are searching for corresponds to a traveling wave in a direction between the x and z directions. It is worth specifying that in order to compute the modal characteristics it is not strictly required that we assume a propagating wave along z. A standing wave may also be assumed. This is obtained by superimposing two waves (1) with opposite β values so that the exponential function is replaced by a trigonometric function. Both formulations will be used throughout this chapter. The latter leads to a resonant behavior of the EM field in the axial direction also and suggests further developments of the TRT toward the analysis of discontinuities, as discussed in Section 5.

Using the formulas given in the appendix after a proper axis rotation ($x \to y$, $y \to z$, $z \to x$), the following EM field components are derived from (1):

$$\begin{aligned}
\mathbf{E}_{tn} &= \left(j\beta_n \cos\frac{n\pi y}{b}\,\hat{\mathbf{y}}_0 - \frac{n\pi}{b}\sin\frac{n\pi y}{b}\,\hat{\mathbf{z}}_0 \right)\exp(-j\beta_n z) \\
\mathbf{H}_{tn} &= Y_0\hat{\mathbf{x}}_0 \times \mathbf{E}_{tn} = Y_0\left(\frac{n\pi}{b}\sin\frac{n\pi y}{b}\,\hat{\mathbf{y}}_0 + j\beta_n\cos\frac{n\pi y}{b}\,\hat{\mathbf{z}}_0 \right)\exp(-j\beta_n z) \\
H_{xn} &= \frac{k_{cn}^2}{j\omega\mu}\cos\frac{n\pi y}{b}\exp(-j\beta_n z)
\end{aligned} \tag{2}$$

where

$$Y_0 = \frac{k_x}{\omega\mu} \qquad k_{cn}^2 = \left(\frac{n\pi}{b}\right)^2 + \beta_n^2 \tag{3}$$

The x dependence $\exp(-jk_x x)$ has been omitted in these expressions.

We may now use the transmission line formalism to express the EM field propagating along the positive or negative x direction. The addition of the two side walls to recover the RW structure simply corresponds to placing two short circuits on the equivalent transmission line. This is represented by the *transverse* equivalent circuit of Fig. 2. The condition for nonzero voltage

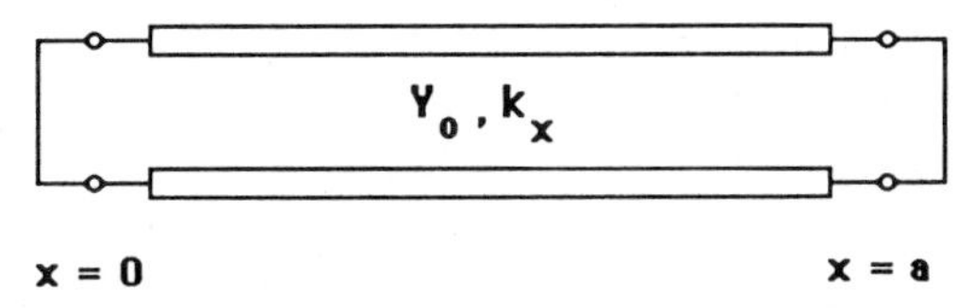

Fig. 2 Transverse equivalent circuit of the RW of Fig. 1*b*. Transmission line may represent either $TE^{(x)}$ modes [formulas (1)–(4)] or $TE^{(z)}$ modes [formulas (7)–(9)].

and current in the absence of any source, thus the resonance condition, is easily found to be

$$Z_0 \tan k_x a = 0 \tag{4}$$

with $Z_0 = 1/Y_0$.

The resonant condition (4) leads to the allowed values for the transverse phase constant k_x:

$$k_{xm} = m\pi/a \qquad m = 0, 1, 2, \ldots \tag{5}$$

The propagation constant β in the z direction is then found, as usual, from the condition of separability,

$$\beta_{mn}^2 = k_0^2 - (m\pi/a)^2 - (n\pi/b)^2 \tag{6}$$

where $k_0^2 = \omega^2 \mu_0 \varepsilon_0$.

It is noted that these $TE_{mn}^{(x)}$ modes have the same propagation constant as the ordinary $TE_{mn}^{(z)}$ modes, but different field components. It can be easily demonstrated that the set of $TE_{mn}^{(x)}$ and $TM_{mn}^{(x)}$ modes can be derived from the set of $TE_{mn}^{(z)}$ and $TM_{mn}^{(z)}$ modes and vice versa by linear combination.

We show next that the transverse resonance approach is not necessarily related to a formulation in terms of $TE^{(x)}$ and $TM^{(x)}$ modes but, as already mentioned, can also be applied using other sets of modes. The transverse resonance concept, in fact, is simply that of considering the propagation in a transverse (say x) instead of longitudinal direction, without necessarily assuming modal sets transverse to that direction. Depending on the boundary value problem to be solved, one set can be more convenient than another for applying the TRT.

We consider the case of $TE^{(z)}$ modes. Apart from a normalization coefficient, the appropriate potential for the PPW is given by

$$\psi_n = \exp(-jk_x x)\cos(n\pi y/b) \tag{7}$$

The corresponding field components are evaluated as before using the formulas in the appendix. They can also be obtained from (2) by a proper rotation of the coordinate system. Using the subscript t to indicate components transverse to x, we may write, omitting for simplicity both the x dependence $\exp(-jk_x x)$ and the z dependence $\exp(-j\beta_n z)$,

$$\mathbf{E}_{tn} = -jk_x \cos\frac{n\pi y}{b}\,\hat{\mathbf{y}}_0$$

$$\mathbf{H}_{tn} = \frac{n\pi}{b}\frac{\beta_n}{\omega\mu}\sin\frac{n\pi y}{b}\,\hat{\mathbf{y}}_0 + \frac{k_x^2 + (n\pi/b)^2}{j\omega\mu}\cos\frac{n\pi y}{b}\,\hat{\mathbf{z}}_0 \tag{8}$$

$$E_{xn} = \frac{n\pi}{b}\sin\frac{n\pi y}{b} \qquad H_{xn} = jk_x\frac{\beta_n}{\omega\mu}\cos\frac{n\pi y}{b}$$

The transmission line representation of the above fields is formulated in terms of the transverse propagation constant k_x and the characteristic admittance

$$Y_{0n} = \frac{k_x^2 + (n\pi/b)^2}{\omega\mu k_x} \tag{9}$$

relating the E- and H-field components along y and z, respectively.

To compute the propagation constant of the waveguide, we can now apply the same transverse equivalent network of Fig. 2 with characteristic admittance given by (9). For such a simple circuit, the resonant condition (4) is solved regardless of the characteristic admittance so that the same k_x values (5) as for the $\mathrm{TE}_{mn}^{(x)}$ are obtained. The superposition of two fields (8) with opposite transverse wavenumbers (5) yield the usual $\mathrm{TE}_{mn}^{(z)}$ modes of a rectangular waveguide. Modes represented by eqs. (7) and (8) can be viewed as $\mathrm{TE}^{(z)}$ propagating in the transverse x direction and are therefore somewhat different from the usual $\mathrm{TE}^{(z)}$ modes of the rectangular waveguide. The terminology used in Reference 6 is that of *H-type* modes.

With this elementary example we have shown that the transverse resonance technique can be applied to different sets of modes defined in the same guiding structure. These alternative sets differ in that they have no E or H components along different assumed directions. In the simple case of the rectangular waveguide, no advantage is gained by using $\mathrm{TM}^{(x)}$ and $\mathrm{TE}^{(x)}$ instead of $\mathrm{TM}^{(z)}$ and $\mathrm{TE}^{(z)}$ modal formulations. In the case of more complicated boundary value problems, on the other hand, one type of representation may be much more convenient than another. The possibility of simplifying the boundary value problems of guiding structures as well as of the relevant discontinuities using suitable TRT formulations will be the subject of the next sections.

1.2. Organization of the Chapter

Transverse resonance finds a rigorous and relatively simple application in the analysis of inhomogeneous structures uniform along a transverse coordinate, such as dielectric slab filled waveguides. This classical formulation is presented and discussed in Section 2.

The boundary value problem becomes more complicated in cases when, looking into the transverse direction, a discontinuity other than a simple change of dielectric material is seen in the waveguide cross section. For this class of problems, for which the ridged waveguide will be taken as the representative example, we have to deal with transversely discontinuous structures. The conventional transverse resonance formulation involves an approximate evaluation of the discontinuities in terms of equivalent lumped elements. This formulation is generally capable of providing a satisfactory approximation of the characteristics of the waveguide, leading to a dramatic reduction in computational effort. This is discussed in Section 3.

Though the conventional transverse resonance technique can be applied also to transversely discontinuous inhomogeneous structures, such as finlines, only first approximation results are obtainable in this way. There are some cases when either the approximate formulation does not provide sufficiently accurate results or some additional information is required that cannot be extracted from it. A rigorous formulation obtained in terms of a generalized transverse resonance technique is illustrated in Section 4. This approach typically applies to a variety of configurations for MICs and millimeter-wave structures. The coplanar waveguide on a metal–insulator–semiconductor substrate is used as a representative case.

Section 5 presents the application of the generalized TRT to the characterization of discontinuities. The basic ideas of the conventional TRT, as applied to two-dimensional problems, are extended to the general case of three-dimensional boundary value problems for analysis of discontinuities, junctions, and, in general, N-port microwave circuits.

Simple examples of computer programs to illustrate the practical implementation of the different formulations of TRT are given in Section 6.

2. INHOMOGENEOUS WAVEGUIDES UNIFORM ALONG A TRANSVERSE COORDINATE

The transverse resonance concept introduced in the previous section has its most obvious application in the analysis of rectangular waveguides loaded with one or more slabs of dielectric material(s) parallel to one side of the rectangular cross section.

Considerable attention has been given to E-plane dielectric-loaded waveguides in the past decades because of their theoretical advantages of higher power-handling capacity and broader bandwidth over conventional waveguides [7–11]. In practice, however, it was shown that the increase in bandwidth is very modest and power handling is drastically compromised by possible air gaps between the dielectric and the guide wall [12]. The dielectric-loaded waveguide is nonetheless of notable interest, not only because it finds application in several waveguide components, such as phase changers, but also because it is representative of a number of other guiding structures. The basic methodology for the analysis of dielectric-loaded waveguides can easily be extended to other inhomogeneous structures, as will be shown later.

Consider first a rectangular waveguide loaded with one dielectric slab placed against one waveguide wall, so that the cross section consists of two homogeneous regions, as illustrated in Fig. 3. This problem is treated in classical textbooks such as those of References 5 and 13.

It is well known that, with the exception of $TE^{(z)}$ modes having no y dependence ($TE^{(z)}_{m0}$ modes), neither $TM^{(z)}$ nor $TE^{(z)}$ alone can exist in such a waveguide. The continuity conditions for the electromagnetic field at the interface $x = a_1$, in fact, corresponds to three scalar equations (continuity of E_y, H_y, and E_z or H_z depending on whether a $TM^{(z)}$ or $TE^{(z)}$ mode is being considered). These conditions cannot be satisfied assuming the existence of two $TM^{(z)}$ (or $TE^{(z)}$) waves, one in region 1 ($0 < x < a_1$) and one in region 2 ($a_1 < x < a_2$). The continuity conditions at the interface can, on the other hand, be satisfied if a hybrid wave, i.e., a combination of one $TM^{(z)}$ and one $TE^{(z)}$ wave, is assumed in each region. In this manner one obtains a homogeneous set of four equations, corresponding to the continuity of E_y, H_y, E_z, and H_z. Unknowns are the four amplitudes of $TM^{(z)}$ and $TE^{(z)}$ waves in regions 1 and 2. The condition for nontrivial solutions yields the characteristic equation for the slab-loaded waveguide, from which the propagation constants of the hybrid modes are computed.

It is evident that the above procedure, though conceptually simple, is nevertheless quite cumbersome and becomes extremely complicated when a greater number of dielectric slabs (thus of dielectric interfaces) are present. This is due to the fact that the boundary conditions at the dielectric

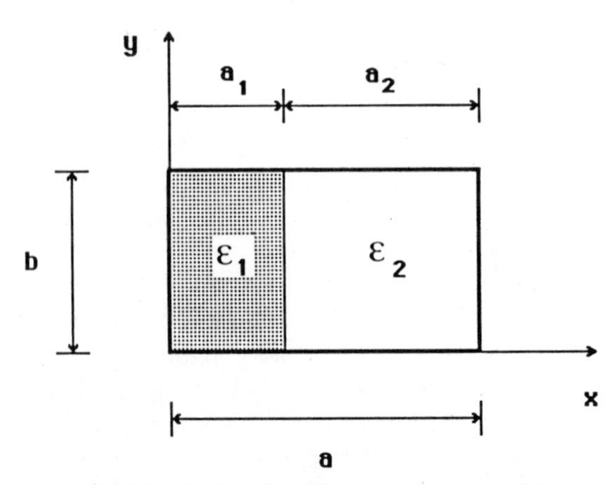

Fig. 3 Dielectric-slab-loaded waveguide.

interfaces cannot be satisfied by $TM^{(z)}$ or $TE^{(z)}$ fields alone, but a combination of them is required.

It is immediately found, on the contrary, that the boundary conditions can be satisfied independently by $TM^{(x)}$ and $TE^{(x)}$ fields. This is evident when one considers the simple discontinuity due to an abrupt change (in the axial z direction) of the dielectric filling in a waveguide. The cross-sectional distribution of the electromagnetic field is not altered by the change of the dielectric, only its wave impedance is. As a consequence, the continuity conditions across the dielectric interface are satisfied by assuming the same $TM^{(z)}$ or $TE^{(z)}$ mode on both sides.

Let us refer for simplicity to $TE^{(x)}$ modes, the treatment being essentially the same for $TM^{(x)}$ modes. The field components can be derived from an x-directed vector potential, which is a function of the y and z coordinates. Because of the required continuity at the interface $x = a_1$, the same potential $\psi(y, z)$ must be assumed in both regions. As already seen in the previous section, an appropriate expression is

$$\psi_n = \exp(-j\beta z)\cos(n\pi y/b) \tag{10}$$

Any normalization coefficient has been omitted as unnecessary in the present analysis.

Note that the longitudinal propagation constant β is still to be determined. The $TE^{(x)}$ field components are derived from (10) using the formulas given in the appendix. Because of expression (10) for the transverse potential, the boundary conditions at $y = 0, b$ are already satisfied. Additional boundary conditions are still to be imposed at $x = 0, a$ and $x = a_1$. Adopting the transmission line formalism to represent the field propagation along the x axis, these boundary conditions are represented in a straightforward manner and lead to the transverse equivalent circuit of Fig. 4.

Two transmission line sections correspond to the $TE^{(x)}$ mode in the two regions $x < a_1$ and $x > a_1$. The lateral walls at $x = 0, a$ correspond to short circuits. The field continuity at $x = a_1$ implies the continuity of both the voltage and current, and thus is represented by the direct connection of the two line sections.

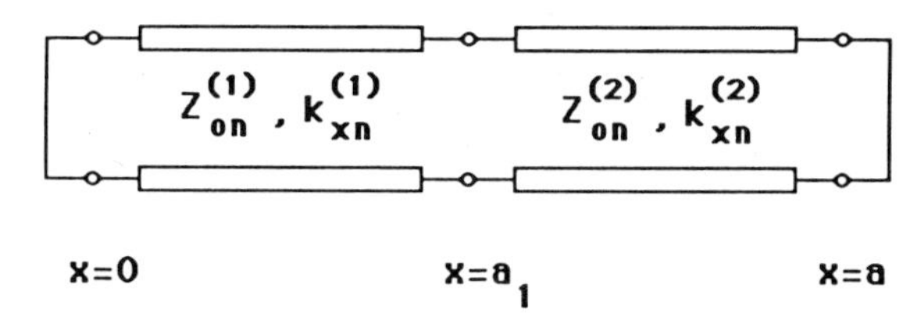

Fig. 4 Transverse equivalent network of the slab-loaded waveguide of Fig. 3.

For nonzero voltages and currents in the transverse equivalent circuit, the resonance condition must be satisfied. This is easily found to be

$$Z_0^{(1)} \tan(k_x^{(1)} a_1) + Z_0^{(2)} \tan(k_x^{(2)} a_2) = 0 \tag{11}$$

where the values of the characteristic impedances and transverse propagation constants will depend on the type of mode being considered. The same equivalent circuit of Fig. 4, in fact, holds for both $\mathrm{TE}^{(x)}$ and $\mathrm{TM}^{(x)}$ modes. For $\mathrm{TE}^{(x)}$ modes the characteristic impedance is given by

$$Z_0^{(i)} = \omega\mu / k_x^{(i)}\,, \qquad i = 1, 2 \tag{12a}$$

while, for $\mathrm{TM}^{(x)}$ modes,

$$Z_0^{(i)} = k_x^{(i)} / \omega\varepsilon\,, \qquad i = 1, 2 \tag{12b}$$

The transverse propagation constants $k_x^{(i)}$ in the two dielectric materials can be expressed in terms of the longitudinal propagation constant β using the separability condition:

$$k_x^{(i)2} = k_0^2 \varepsilon_i - \beta^2 - (n\pi/b)^2\,, \qquad i = 1, 2 \tag{13}$$

Inserting (13), (12a), or (12b), into (11), we obtain the dispersion relation for the slab-loaded rectangular waveguide. This is a transcendental equation of the general form of a complex function of β and ω equated to zero:

$$f(\beta, \omega) = 0 \tag{14}$$

The cutoff frequencies are obtained solving (14) with $\beta = 0$.

It is important to note that because of the dielectric inhomogeneity of the waveguide, the frequency behavior of β as it results from the dispersion relation (14) is different from that for a homogeneous waveguide. The latter is completely determined once the cutoff frequency is known. For the inhomogeneous waveguide, on the other hand, the dispersion relation must be solved at each frequency. This is a general feature of inhomogeneously filled waveguides.

The above procedure can be generalized fairly easily to waveguides inhomogeneously filled with any number of dielectric slabs, provided, of course, that all interfaces are parallel to the same transverse coordinate. In this way $\mathrm{TE}^{(x)}$ and $\mathrm{TM}^{(x)}$ can exist independently in the structure. The generalization may also include the case when the side walls at $x = 0, a$ are replaced by some surface reactances. In this manner the same general formulation is used for a number of different guiding structures, such as the H-guide with single [14] or laminated dielectric [15], the nonradiative dielectric waveguide [16], etc.

The structure under consideration is schematically represented in Fig. 5 together with its transverse equivalent network. The side walls at $x = 0, a$ may represent any surface reactance, obviously including the usual case of an ideal short circuit. Such reactances are indicated as X_1, X_2 in the transverse equivalent circuit.

When the upper and lower plates at $y = 0$, b are removed, the structure of Fig. 5 with metallic side walls at $x = 0$ and $x = a$ may represent a partially filled parallel-plate waveguide [17] or, with one side wall removed, a surface wave structure [17, 18].

When no metallic side wall exists at $x = 0$ and $x = a$, the structure of Fig. 5 represents an H-guide. In this case jX_1, jX_2 are the transverse wave impedances of the $TE^{(x)}$ or $TM^{(x)}$ mode under consideration. In order for such a mode to propagate along z with no attenuation, it must be decaying outside the central region $0 < x < a$. In other words, the transverse wavenumber k_x as well as the transverse wave impedances must be imaginary for $x < 0$ and $x > a$.

The analysis of the general structure of Fig. 5 being presented here on the basis of the transverse resonance technique is essentially the same as that of References 6 and 11. The dispersion relation is obtained, as already stated, by the resonant condition of the transverse equivalent network shown in Fig. 5.

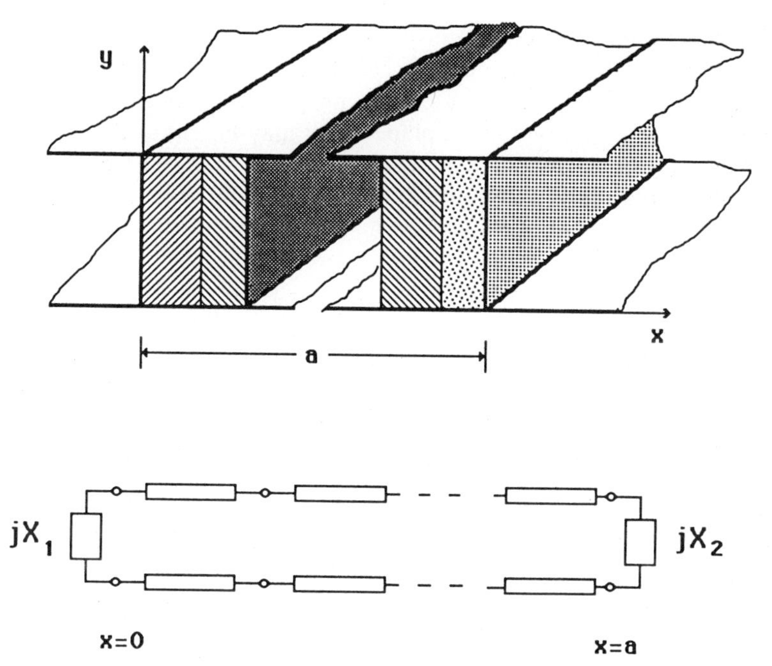

Fig. 5 Waveguide loaded with laminated dielectric and equivalent transverse network.

Let $Z_0^{(i)} = 1/Y_0^{(i)}$ be the characteristic impedance of the ith line section and θ_i the corresponding electrical length. Values of $Z_0^{(i)}$ are given by eq. (12a) or (12b) and depend on the type and order n [see eq. (13)] of the mode being considered. Voltages and currents at its ends are related by the chain matrix

$$[\boldsymbol{T}_i] = \begin{bmatrix} \cos\theta_i & jZ_0^{(i)}\sin\theta_i \\ jY_0^{(i)}\sin\theta_i & \cos\theta_i \end{bmatrix} \tag{15}$$

The overall chain matrix

$$[\boldsymbol{T}] = \prod_{i=1}^{N} [\boldsymbol{T}_i] = \begin{bmatrix} t_{11} & jt_{12} \\ jt_{21} & t_{22} \end{bmatrix} \tag{16}$$

relates voltages and currents at the two terminating reactances X_1, X_2. The resonant condition of the network is easily found to be

$$X_1 = -\frac{t_{11}X_2 + t_{12}}{t_{21}X_2 + t_{22}} \tag{17}$$

For the rectangular waveguide, $X_1 = X_2 = 0$, and (17) reduces to

$$t_{12} = 0 \tag{18}$$

For illustration, a simple computer program for the analysis of a rectangular waveguide loaded with N dielectric slabs is given in Section 6.

As a further example, consider the nonradiating dielectric waveguide (NRD) of Fig. 6. The symmetry plane $x = 0$ may be either an electric wall (odd symmetry) or a magnetic wall (even symmetry). We consider $\mathrm{TM}^{(x)}$ modes. The transverse equivalent circuit is obtained from the general one by putting

$$X_1 = 0 \quad \text{(odd modes)} \qquad 1/X_1 = 0 \quad \text{(even modes)}$$

$$X_2 = \frac{\alpha_{xn}}{\omega\varepsilon_0} \qquad Z_{0n} = \frac{k_{xn}}{\omega\varepsilon_0\varepsilon_r}$$

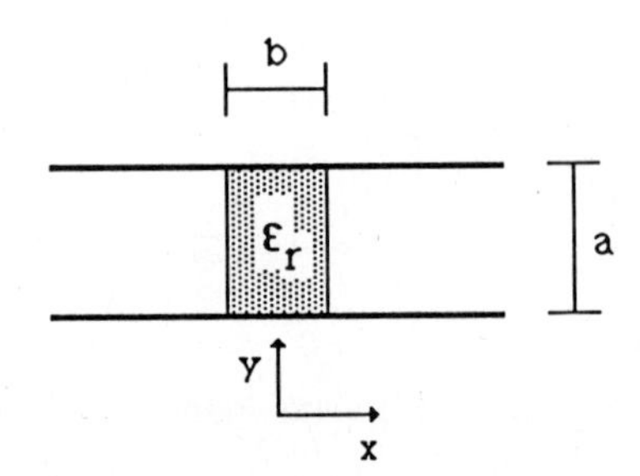

Fig. 6 The nonradiating dielectric (NRD) waveguide.

where k_{xn} and α_{xn} are the transverse phase constant of the mode in the dielectric filled region and its transverse attenuation constant in the empty region, respectively.

$$k_{xn} = \left[k_0^2 \varepsilon_r - \beta^2 - \left(\frac{n\pi}{a} \right)^2 \right]^{1/2} \qquad \alpha_{xn} = \left[-k_0^2 + \beta^2 + \left(\frac{n\pi}{a} \right)^2 \right]^{1/2}$$

Using the above expressions in the resonant condition (17) with the chain matrix elements given by (15), one easily obtains, for odd modes ($X_1 = 0$),

$$\alpha_{xn} - \frac{k_{xn}}{\varepsilon_r} \tan \theta = 0$$

and for even modes ($1/X_1 = 0$),

$$\alpha_{xn} + \frac{k_{xn}}{\varepsilon_r} \cot \theta = 0$$

where the electrical length is given by $\theta = k_{xn} b/2$. These equations are equivalent to those given in Reference 16.

A procedure similar to that presented here for the dielectric-slab-loaded waveguide can be applied to a class of anisotropic-slab-loaded waveguides, provided that the anisotropic behavior is not in the y direction [19].

Transverse resonance techniques can be applied to the analysis of circular waveguides and coaxial cables loaded with dielectric cylinders. In such cases the transverse resonance condition is imposed in the radial direction using the concept of radially propagating waves [20, 21].

3. CONVENTIONAL TRANSVERSE RESONANCE TECHNIQUE FOR TRANSVERSELY DISCONTINUOUS WAVEGUIDES

The class of inhomogeneous structures of the previous section has been shown to be susceptible to exact and relatively simple analytical solution in terms of $\text{TM}^{(x)}$ and $\text{TE}^{(x)}$ modes, the propagation constants being computed from the resonance condition of an exact equivalent network in the transverse direction. The existence of simple $\text{TM}^{(x)}$ and $\text{TE}^{(x)}$ solutions was possible because of the uniformity of the structure along the other transverse coordinate (y). In particular, the cross section had a simple separable geometry. The presence of any discontinuity in the transverse direction would produce a coupling between $\text{TM}^{(x)}$ and $\text{TE}^{(x)}$ modes, so that such elementary solutions can no longer be obtained. While a plane dielectric interface normal to the transverse x direction does not couple $\text{TM}^{(x)}$ with $\text{TE}^{(x)}$ fields, different types of obstacles in the transverse direction will.

The presence of obstacles normal to the transverse direction produces reactive energy storage in their proximities. This can be taken into account in the transverse equivalent network by a proper lumped reactance. The resonance condition of such networks constitutes the characteristic equation for the structure. The application of the transverse resonance method is therefore based on the equivalent circuit characterization of the longitudinal discontinuities contained in the guide. Fortunately, the solution for many practical discontinuity problems has been made available in the *Waveguide Handbook* [20] or in some classical paper (see Reference 22).

When the equivalent network characterization of the discontinuities is known with good approximation, the TRT is extremely useful in providing the propagation properties of the guide with minimum computational effort.

In the case of homogeneously filled waveguides, the general field problem can still be resolved into TM and TE modes in the longitudinal z direction ($TM^{(z)}$ and $TE^{(z)}$ modes). The cross-sectional field distribution is independent of the frequency, and knowledge of the cutoff frequency is sufficient to determine the propagation constant at any frequency. Examples of this class of structures are the single- and double-ridge waveguides, the grooved guide, etc.

When the waveguide is inhomogeneously filled, the cross-sectional field distribution is a function of frequency, and the dispersion behavior is different from the homogeneous case. In such cases the conventional TRT can provide simple and useful results in conjunction with some approximate dispersion relation, provided that the structural inhomogeneity is weak. This approach has been used to obtain very simple characterizations of finlines, and an example of computer analysis is provided in Section 6.

It can be observed that the conventional application of the TRT suffers from some limitations. First, longitudinal discontinuities must be noninteracting, both with each other and with the waveguide walls, so that they can be modeled as purely lumped elements. Second, the characterization of the discontinuity is generally available only for given incident field distributions, such as the principal TEM mode of the parallel-plate waveguide or the dominant TE_{10} mode of the rectangular guide. As a consequence, the method does not permit evaluation of the complete modal spectrum of the guide. Finally, local field distribution in the proximity of the discontinuities cannot be computed. These limitations will be removed by the generalized transverse resonance technique described in the next section.

The ridged waveguide will be taken as a representative example of transversely discontinuous waveguides to be solved by conventional TRT. Originally suggested by Ramo and Whinnery in 1944 [23], the transverse resonance concept was applied by Cohn [24] in 1947 to determine the cutoff frequency and impedance of a ridged waveguide. The same basic technique was employed later by a number of authors to improve and extend Cohn's work [25–27]. Modern computing capabilities make it possible to implement the necessary equations on a pocket calculator [28].

Figure 7 shows the cross-sectional geometries of single- and double-ridge waveguides. The dominant mode is a $TE^{(z)}$ mode with even symmetry with

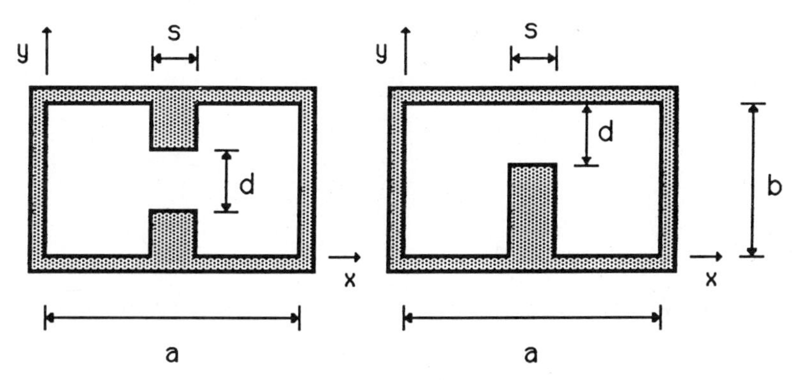

Fig. 7 Cross-sectional geometries of (left) double- and (right) single-ridge waveguides.

respect to the plane $x = a/2$. In contrast with both the ordinary rectangular waveguide and E-plane dielectric-loaded waveguides, a y dependence of the field is produced by the presence of the ridges. In addition, an x component of the E field is necessary to satisfy the boundary conditions at the ridge edges. These alterations, however, are localized in the proximities of the ridges. A rigorous analysis of the ridged waveguide is apparently quite involved.

Looking in the transverse x direction, however, the ridged waveguide is seen as a composite structure consisting of waveguides of different heights. Transverse step discontinuities produce the excitation of both $\text{TE}^{(x)}$ and $\text{TM}^{(x)}$ higher-order modes. In the usual conditions, these are well below cutoff, so they contribute with some reactive energy stored in the proximities of the ridge edges. If the characterization of these step discontinuities is available, the transverse resonance approach would provide a very simple and straightforward way to compute the characteristic equation of the structure. This is actually the case, as will be shown next.

For clarity of explanation, let us reduce the problem to a more familiar structure. Suppose that a standing wave regime in the longitudinal z direction is established in the ridged waveguide by superimposing two dominant $\text{TE}^{(z)}$ modes with equal amplitudes traveling in opposite z directions. We may now insert two perfectly conducting planes transverse to z at consecutive nulls of the transverse electric field. These additional planes are spaced $\lambda_g/2$ apart and do not alter the standing wave field distribution in the ridged waveguide. This is actually transformed into a resonator of length $l = \lambda_g/2$ in the z direction. Looking in the x direction, the same resonator is seen as a structure composed of one rectangular waveguide with reduced height d inserted between two rectangular waveguides with the same width l and higher b. The equivalent circuit looking in the x direction is that of Fig. 8. The shunt capacitances account for the reactive energy associated with stray field at the junctions. The line sections correspond to $\text{TE}_{10}^{(x)}$ modes in

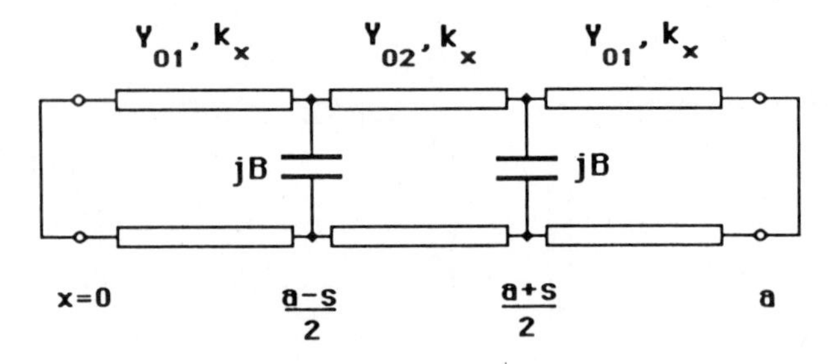

Fig. 8 Transverse equivalent network of ridged waveguides.

the three waveguide sections and are therefore characterized by the same transverse propagation constant

$$k_x = (k_0^2 - \beta^2)^{1/2} \tag{19}$$

where β is the longitudinal propagation constant. The characteristic impedances are chosen to be proportional to the corresponding waveguide heights [22]. Because of symmetry considerations, the resonance of the transverse equivalent circuit occurs when an open- or short-circuit condition ($I = 0$ or $V = 0$, respectively) is established at $x = a/2$. The former condition is that of the lowest resonance and corresponds to the dominant mode $TE_{10}^{(z)}$ of the ridged waveguide. This mode possesses an even symmetry with respect to the center of the waveguide. With reference to the symbols of Fig. 8, the resonance condition is obtained by equating to zero the sum of the three admittances seen at the connection between the lines:

$$-jY_{01} \cot[k_x(a - s)/2] + jB + jY_{02} \tan(k_x s/2) = 0$$

Rearranging this equation and using $Y_{02}/Y_{01} = b/d$, we obtain

$$\frac{b}{d} \tan \frac{k_x s}{2} - \cot \frac{k_x(a-s)}{2} + \frac{B}{Y_{01}} = 0 \tag{20}$$

The normalized susceptance of the waveguide step B/Y_{01} is also a function of the transverse propagation constant $k_x = 2\pi/\lambda_x$. Equation (20) can therefore be regarded as a purely geometrical relationship between the transverse wavelength λ_x and the waveguide geometrical parameters. Once the former quantity has been computed, the longitudinal phase constant β at any frequency is then evaluated through (19).

It is worth specifying that the above analysis could be carried out at cutoff condition, $\beta = 0$, instead of in the longitudinal resonance condition ($\beta = 2\pi/\lambda_g = \pi/l$). (Actually, as already specified, k_x is independent of β.) In this manner, the problem would be that of a parallel-plate guide (PPW) instead of a rectangular guide (RW) discontinuity problem. However, the same transverse resonance condition [eq. (20)] would be obtained, since the equivalent circuit for the PPW is the same as for the RW on replacement of λ with λ_g [29]. It must be stressed, however, that this result holds because of the dielectric homogeneity of the waveguide, so that k_x is independent of the frequency.

If the ridge is very narrow ($k_x s \ll 1$), the shunt susceptance may be approximated with one-half of that of an infinitely thin diaphragm [28, 30]:

$$\frac{B}{Y_0} = \frac{2b}{\lambda_x}\left[\ln(\csc\theta) + \frac{Q_2 \cos^4\theta}{1 + Q_2 \sin^4\theta} + \left(\frac{b}{4\lambda_x}\right)^2 (1 - 3\sin^2\theta)^2 \cos^4\theta\right] \tag{21}$$

where

$$Q_2 = \frac{1}{[1 - (b/\lambda_x)^2]^{1/2}} - 1 \qquad \theta = \frac{\pi d}{2b}$$

For finite s, the more complicated expressions for the change in height of a rectangular guide given in Reference 31 must be adopted.

The transverse equivalent circuit of Fig. 8 and consequently the characteristic equation (20) are valid provided that the step discontinuities are sufficiently far both from each other and from the terminal walls ($x = 0, a$) that no higher-order mode interaction can occur. In particular, only the dominant $TE_{10}^{(x)}$ mode must be propagating, all higher-order modes being below cutoff. In the general case, specifically when the complete spectrum of the ridged waveguide must be evaluated, a rigorous general analysis must be carried out [32].

The main features of the above-described formulation of the TRT can be summarized as follows.

The technique is applied to a homogeneous metallic waveguide in the presence of longitudinal discontinuities. A transverse equivalent network is first determined, where the transverse discontinuities are modeled as lumped elements inserted at the connections of cascaded transmission lines. In rectangular geometry, these correspond to the dominant $TE^{(x)}$ modes in the uniform transverse sections of the waveguide. In contrast with the transverse equivalent circuit for the inhomogeneous waveguides of the previous section, a lumped reactive element is now necessary to account for the energy stored by $TE^{(x)}$ and $TM^{(x)}$ higher-order modes excited at each longitudinal discontinuity. The validity of such an equivalent circuit is subject to the higher-order modes in the transverse direction being below cutoff so that no interaction between discontinuities occurs.

The feasibility of this technique is subject to the availability of the equivalent network parameters of the discontinuities involved. Such parameters are known for a number of practical discontinuities with good approximation. In some cases it is possible to adapt known results to slightly different configurations. When a weak inhomogeneity is present, as in the case of the finline, where the dielectric substrate is very thin and has low permittivity, an approximate transverse equivalent network can be used to compute the cutoff frequency of the dominant mode ($\beta = 0$) [33, 34]. Dispersion is then approximated assuming

$$\beta = k\sqrt{\kappa_e}[1 - (f_c/f)^2]^{1/2} \tag{22}$$

where k is the free-space wavenumber and κ_e is the effective dielectric constant as defined by Meier [35]. A finline computer analysis based on this approximation is given in Section 6.

The final step is to impose the resonance condition of the transverse equivalent network. For this type of circuit, i.e., when lumped reactances are inserted at the connections between uniform line sections, the resonance condition is derived in a straightforward way by slightly modifying the procedure derived for the general circuit of Fig. 5. The resultant final equation can be solved numerically by known methods such as the Newton–Raphson or the secant method. The lowest root for the transverse propagation constant is used to compute the longitudinal propagation constant of the dominant mode.

The above-described formulation of the TRT is also applicable, with proper modifications, to semi-open guiding structures, such as slitted rectangular and circular waveguides [6, 21], open groove guides [36], or other structures susceptible of application as leaky wave antennas. In such cases the transverse equivalent network comprises a lumped conductance to account for the power leakage. A discussion of antenna applications is outside the scope of this chapter. The interested reader is directed to Reference 37.

As already mentioned, the above-illustrated conventional formulation of TRT is not only approximate (the approximation being essentially due to the characterization of the discontinuities) but, more important, cannot be used for determining the complete spectrum of the waveguide or the exact field distribution of even the dominant mode. For this, a complete analysis of the waveguide is needed. A rigorous analytical technique based on the generalized transverse resonance approach is presented in the next section.

4. GENERALIZED TRANSVERSE RESONANCE TECHNIQUE FOR TRANSVERSELY DISCONTINUOUS INHOMOGENEOUS WAVEGUIDES

An extensive variety of transmission structures used for microwave integrated circuits up to the millimeter-wave range consist basically of a dielectric layer with a metallized pattern deposited on one or both sides. The dielectric substrate may be either in an open environment or inserted into a rectangular waveguide enclosure. Striplines, microstrip lines, coplanar strips, coplanar waveguides, finlines, etc., all belong to this category. The most popular approach to the analysis of these and other planar structures is the spectral domain technique. The main limitation of the spectral domain technique is that, at least in its conventional formulation, it cannot take into account the finite thickness of the metallization. While the zero thickness approximation is good enough for most practical problems, it has been pointed out that, particularly in the millimeter-wave range, the effects of finite metallization thickness may lead to noticeable errors. An alternative approach that does not suffer this type of limitation is the TRT. In contrast with the formulation of Section 3, a rigorous formulation of the transverse resonance concept is necessary to achieve the required accuracy. This type of formulation is presented and discussed in this section.

In 1968, S. B. Cohn proposed the slot line as a candidate structure for microwave integrated circuits, in some respects complementary to the microstrip line [38]. Cohn also presented a method of analysis that consisted, first, of the conversion of the slot-line configuration into a rectangular waveguide discontinuity problem [39]. This was then solved by a transverse resonance technique. Some approximations were made in order to simplify and accelerate computations. The method was then used not only for an extensive characterization of the slot line [40], but also for the approximate analysis of integrated finlines [41].

It should be stressed that the availability of modern computer facilities has a strong impact on the use of numerical methods. Numerical analyses, once requiring prohibitively long computer times, have become easily affordable. Rigorous and elaborate numerical procedures can run on PCs with no or very little need of any analytical preprocessing. By simple modifications, the analytical technique developed by Cohn for the slot line can easily be formulated in a rigorous yet easily affordable manner.

Actually, the same basic ideas as those of Cohn's work were later applied to a number of configurations, from the microstrip line [42] to coplanar waveguides [43–45], and to closed integrated structures such as finlines and, in general, quasi-planar structures [1, 46–51].

TRT can also be applied to open dielectric waveguides. In a very illuminating paper on the use of such a technique [52], Peng and Oliner have first applied TRT, in the simple form of Section 2, to determine the modal spectrum of layered dielectric structures. A generalized TRT was then developed to characterize the dielectric strip waveguide, viewed as the connection of two dielectric steps in the transverse direction.

In this section we will concentrate on the exact analysis of transmission lines for microwave and millimeter-wave integrated circuits using a generalized TRT.

By suitably modifying Cohn's method [39], a full-wave transverse resonance analysis of planar transmission lines can be developed that accounts for lossy substrates as well as for finite thickness of the metallization. The representative structure for illustration of the analytical technique is chosen here as the metal–insulator–semiconductor coplanar waveguide (MISCPW) of References 43–45. The structure geometry is sketched in Fig. 9 along with the reference frame. Note that the z axis has been assumed orthogonal to the substrate.

The structure can be considered a coplanar waveguide (CPW) fabricated on an inhomogeneous substrate consisting of L isotropic and, generally, lossy layers. Unlike the ridged structure discussed in Section 3, in the present case purely standing waves along the longitudinal x axis are prevented by the structure losses. This is not a difficulty, however, since the existence of longitudinal standing waves is not strictly necessary to the performance of transverse resonance analysis of uniform transmission lines.

To apply TRT, the configuration of Fig. 9 is converted into a parallel-plate waveguide (PPW) discontinuity problem by inserting two longitudinal electric or magnetic walls perpendicular to the substrate. These auxiliary walls should be placed sufficiently far from the slots as not to perturb the electromagnetic field, which is essentially confined to the proximity of the slots. Strictly speaking, the modal spectrum of the transmission line is modified by the presence of these fictitious walls. In the case of lossy structures, the effect of these walls can be made negligibly small. In the case of lossless structures, on the other hand, due care should be exercised before the results for the closed structure are extended to the open structure. As far as guided CPW modes are concerned, like the dominant quasi-TEM, they propagate with no attenuation and are confined to the central region. The addition of the lateral walls does not actually alter the field distribution. It is nevertheless evident that the radiation properties will be altered as any power leakage toward the transverse y direction is eliminated. On the other hand, additional transverse resonances are intro-

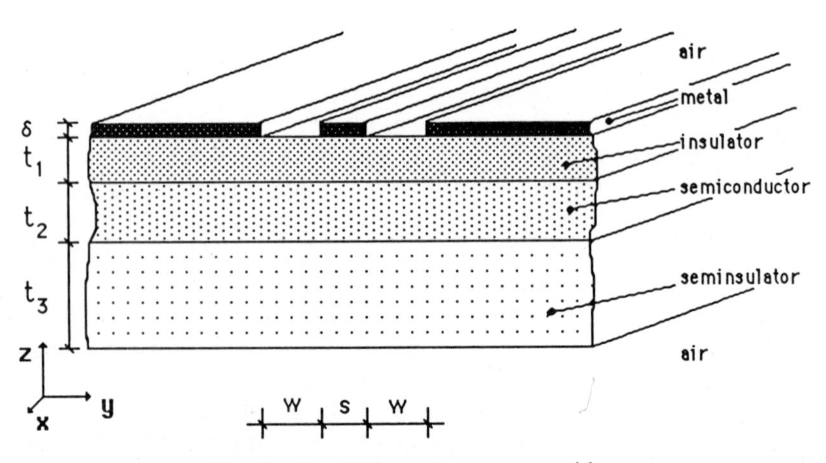

Fig. 9 The MIS coplanar waveguide.

duced by the lateral walls so as to noticeably affect the modal spectrum of the line [44].

Symmetry of the CPW, which we assume for simplicity, permits the analysis to be reduced to one half of the structure by replacing the symmetry plane with a magnetic or electric wall for even or odd modes, respectively. The original CPW is thus converted into the structure depicted in Fig. 10.

Looking in the (transverse) z direction, this structure is seen as a parallel-plate waveguide with magnetic plates, loaded with a number $(L-2)$ of lossy slabs and with a metallic iris of finite thickness.

Because of the finite metallization thickness, the slot region too may be regarded as a PPW section with electric plates. The entire structure therefore appears as the cascade of a number $(L+1)$ of PPW sections (including the semi-infinite terminating sections). The section between the planes $z = z_1 = 0$ and $z = z_2 = \delta$ has a reduced height $s = s_1$, while all the other sections have $s = s_2 > s_1$. Each PPW section can support the propagation of $\mathrm{TE}^{(z)}$ and $\mathrm{TM}^{(z)}$ modes. The modes are uncoupled and thus propagate independently, except at the two step discontinuities between the iris region and the outer PPW ($z = z_1, z_2$). The metallic discontinuity actually makes this structure differ from that of Section 2. In particular, it produces the coupling of $\mathrm{TE}^{(z)}$ and $\mathrm{TM}^{(z)}$ modes. The rigorous technique presented here is a generalization of the one studied in the previous sections. Basically, it consists of a mode-matching technique applied in the transverse direction, making use of $\mathrm{TE}^{(z)}$ and $\mathrm{TM}^{(z)}$ mode expansions.

We will briefly summarize the field theoretical analysis of Fig. 10 and then show that the solution can be sought on the basis of a generalized transverse equivalent circuit. This contains an infinite number of transmission lines corresponding to the modes (propagating or evanescent) excited in the various PPW sections.

According to the transverse resonance approach, the electromagnetic

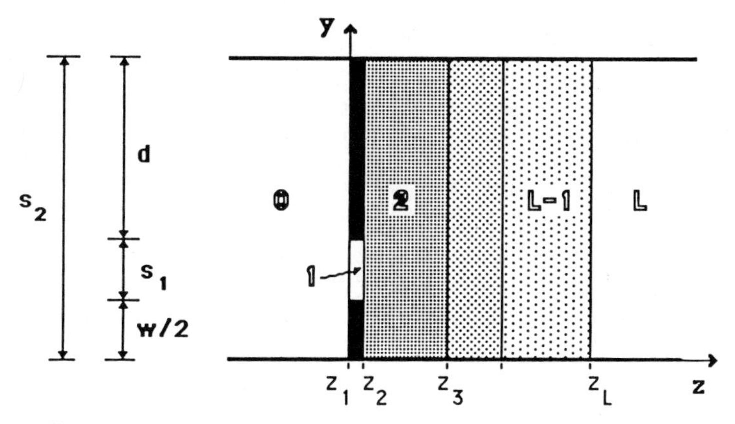

Fig. 10 Reduced MISCPW structure for transverse analysis.

field inside each PPW section of Fig. 10 is expressed in the form of a series of $TE^{(z)}$ and $TM^{(z)}$ modes. The appropriate potentials in the transverse xy plane are, apart from a normalization coefficient,

$$\psi_m = \sin \frac{m\pi y}{s_2} \exp(-k_x x) \quad \text{and} \quad \phi_m = \cos \frac{m\pi y}{s_2} \exp(-k_x x) \tag{23}$$

respectively. These expressions are valid for all sections except Section 1. For this, s_2 must be replaced by s_1 and y by $y - w/2$. The EM field components in each section are then derived as usual (see the appendix). It can be noted that for each value of m in (23), except for $m = 0$, there are four unknown amplitudes associated with four EM waves: two $TE^{(z)}$ and two $TM^{(z)}$ waves traveling in the $+z$ and $-z$ directions, respectively. For $m = 0$, only two $TM^{(z)}$ waves are present. (They are actually TEM waves with respect to a direction between the x and z axes.)

The transverse equivalent circuit representation is shown in Fig. 11. Such a network generalizes those already discussed in connection with the dielectric-slab-loaded waveguide (Fig. 5) and with the ridged waveguide (Fig. 8). Each mode of the PPW is represented by an equivalent transmission line. Mode coupling, which occurs at each step discontinuity, is represented by a generalized network with an infinite number of ports. Each transmission line on the left-hand side and on the far right-hand side is terminated by its characteristic impedance, because the structure of Fig. 10 is supposed to be infinite in both the positive and negative z directions. Different terminations can be assumed to generalize the problem to include, for instance, the presence of a back metallization of the substrate [44, 53].

In the normal operating conditions of the MISCPW, the EM field is confined to the dielectric substrate and is exponentially decaying in the air regions ($I = 0$ and $I = L$). This means that the PPW modes in air are below

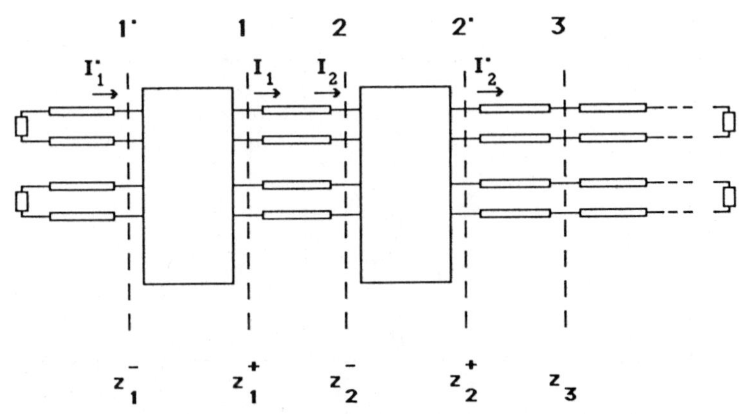

Fig. 11 Generalized equivalent transverse network of Fig. 10.

cutoff and the corresponding equivalent lines have imaginary characteristic impedances.

To completely define the transverse equivalent circuit of Fig. 11 it is necessary to identify the generalized equivalent network of the step discontinuity. This can be done by a mode-matching technique.

Let us refer to the discontinuity between regions 1 and 2 (Fig. 10). The EM field in the two regions is expanded in terms of $\mathrm{TE}^{(z)}$ and $\mathrm{TM}^{(z)}$ modes of the corresponding PPW. Note that for region 1 the plates are metallic, while for the other regions they may be either metallic or magnetic. In any case, at $z = z_1$ we may write (see the appendix)

$$\begin{aligned} \mathbf{E}_t^{(1)} &= \sum_m V_m^{(1)} \mathbf{e}_m^{(1)}(x, y) \qquad & \mathbf{E}_t^{(2)} &= \sum_n V_n^{(2)} \mathbf{e}_n^{(2)}(x, y) \\ \mathbf{H}_t^{(1)} &= \sum_m I_m^{(1)} \mathbf{h}_m^{(1)}(x, y) \qquad & \mathbf{H}_t^{(2)} &= \sum_n I_n^{(2)} \mathbf{h}_n^{(2)}(x, y) \end{aligned} \tag{24}$$

Explicit expressions for the eigenvectors $\mathbf{e}$ and $\mathbf{h}$ are obtained from the TE and TM potentials (23) using the formula given in the appendix. The running index m (or n) in the above series (24), however, is to be interpreted as an index for numbering the modes (TE as well as TM modes), not as the spatial frequency index used in (23). In fact, apart from the case $m = 0$, for each spatial frequency [m in (23)] there are two terms in (24), corresponding to one TE mode and one TM mode.

The boundary conditions at the interface require

$$\begin{aligned} \mathbf{E}_t^{(2)} &= \begin{cases} \mathbf{0} & \text{on } S_2 - S_1 \\ \mathbf{E}_t^{(1)} & \text{on } S_1 \end{cases} \\ \mathbf{H}_t^{(2)} &= \mathbf{H}_t^{(1)} \qquad \text{on } S_1 \end{aligned} \tag{25}$$

S_1 is the aperture region $(w/2 < y < w/2 + s_1)$ corresponding to the cross section of region 1, S_2 is the cross section corresponding to region 2 $(0 < y < s_2)$. Inserting (24) into (25) and using the orthogonal properties of the eigenfunctions $\mathbf{e}_m$ and $\mathbf{h}_m = \mathbf{z}_0 \times \mathbf{e}_m$, we obtain

$$V_n^{(2)} = \sum_m g_{nm} V_m^{(1)} \qquad I_m^{(1)} = \sum_n g_{mn} I_n^{(2)} \tag{26}$$

where

$$g_{mn} = \int_{S_1} \mathbf{e}_m^{(1)} \cdot \mathbf{e}_n^{(2)} \, dS \tag{27}$$

Equations (26) provide the required characterization of the step. In the practical computation the series will be truncated to $\bar{M}$ and $\bar{N}$ for regions 1 (narrow) and 2 (wide), respectively. This corresponds to a representation of the step as an $(\bar{M} + \bar{N})$-port network. Effects related to this truncation will be discussed briefly later. The above equations can be put in the matrix form

$$\mathbf{V}^{(2)} = \boldsymbol{G}\mathbf{V}^{(1)} \qquad \mathbf{I}^{(1)} = \boldsymbol{G}^{\mathrm{t}}\mathbf{I}^{(2)} \tag{28}$$

with obvious meaning of the symbols.

We can now specialize the above result to the two networks of Fig. 11 by writing

$$\begin{aligned} \mathbf{V}'^{(1)} &= \boldsymbol{G}\mathbf{V}^{(1)} \qquad \mathbf{I}^{(1)} = \boldsymbol{G}^{\mathrm{t}}\mathbf{I}'^{(1)} \\ \mathbf{V}'^{(2)} &= \boldsymbol{G}\mathbf{V}^{(2)} \qquad \mathbf{I}^{(2)} = \boldsymbol{G}^{\mathrm{t}}\mathbf{I}'^{(2)} \end{aligned} \tag{29}$$

where primed quantities refer to external ports, as indicated in Fig. 11. Matrix $\boldsymbol{G}$ depends only on the geometry of the step and thus is the same for both discontinuities.

Voltages and currents at ports 1′ and 2′ are related by the admittances seen looking to the left ($\boldsymbol{Y}_1'$) and to the right ($\boldsymbol{Y}_2'$), respectively. These admittances can be calculated in an elementary way using standard transmission line theory. We may write

$$\mathbf{I}'^{(1)} = -\boldsymbol{Y}_1'\mathbf{V}'^{(1)} \qquad \mathbf{I}'^{(2)} = \boldsymbol{Y}_2'\mathbf{V}'^{(2)} \tag{30}$$

where $\boldsymbol{Y}_1'$ and $\boldsymbol{Y}_2'$ are diagonal matrices. These relations permit the quantities relative to external ports 1′ and 2′ to be eliminated from (29):

$$\mathbf{I}^{(1)} = -\boldsymbol{G}^{\mathrm{t}}\boldsymbol{Y}_1'\boldsymbol{G}\mathbf{V}^{(1)} \qquad \mathbf{I}^{(2)} = \boldsymbol{G}^{\mathrm{t}}\boldsymbol{Y}_2'\boldsymbol{G}\mathbf{V}^{(2)}$$

or, synthetically,

$$\mathbf{I}^{(1)} = -\boldsymbol{Y}_1\boldsymbol{V}^{(1)} \qquad \mathbf{I}^{(2)} = \boldsymbol{Y}_2\mathbf{V}^{(2)} \tag{31}$$

with

$$\boldsymbol{Y}_i = \boldsymbol{G}^{\mathrm{t}}\boldsymbol{Y}_i'\boldsymbol{G}$$

These are $2\bar{M}$ equations relating the $2\bar{M}$ currents to the $2\bar{M}$ voltages at the ends of the $\bar{M}$ transmission lines inserted between the generalized equivalent networks of the step discontinuities.

So we have reduced the problem to that of determining the resonance condition of such lines that appear terminated on the *admittances* $\boldsymbol{Y}_1$ and $\boldsymbol{Y}_2$. Note that these are not diagonal matrices, as a consequence of the modes being coupled by the step discontinuity. The transmission line equations for such line sections provide $2\bar{M}$ additional relations. In this manner we can obtain a homogeneous system of equations in an equal number of unknowns. The condition for nontrivial solution, i.e., the resonance condition, constitutes the characteristic equation for the structure.

Depending on the matrix representation used for the $\bar{M}$ line sections, different computational procedures can be developed. One possibility is to characterize each line section in terms of a transmission (or $ABCD$) matrix. From the transmission matrix of a line section,

$$\begin{bmatrix} A & B \\ C & D \end{bmatrix} = \begin{bmatrix} \cosh \gamma l & Z_0 \sinh \gamma l \\ Y_0 \sinh \gamma l & \cosh \gamma l \end{bmatrix}$$

we easily obtain

$$\mathbf{V}^{(1)} = \boldsymbol{C}\mathbf{V}^{(2)} + \boldsymbol{Z}_0\boldsymbol{S}\mathbf{I}^{(2)} \qquad \mathbf{I}^{(1)} = \boldsymbol{Y}_0\boldsymbol{S}\mathbf{V}^{(2)} + \boldsymbol{C}\mathbf{I}^{(2)} \tag{32}$$

where $\boldsymbol{C}$, $\boldsymbol{S}$, $\boldsymbol{Y}_0$, $\boldsymbol{Z}_0$ are diagonal matrices

$$\begin{aligned} \boldsymbol{C} &= \mathbf{diag}[\cosh \gamma_m \delta] \\ \boldsymbol{S} &= \mathbf{diag}[\sinh \gamma_m \delta] \\ \boldsymbol{Z}_0 = \boldsymbol{Y}_0^{-1} &= \mathbf{diag}[\boldsymbol{Z}_{0m}] \end{aligned} \tag{33}$$

γ_m is the propagation constant and Z_{0m} the characteristic impedance of the mth line; $\delta = z_1$ is the metallization thickness and thus the line length.

Combining (31) and (32), it is possible to express the current vectors $\mathbf{I}^{(1)}$ and $\mathbf{I}^{(2)}$ as well as the voltage vector $\mathbf{V}^{(1)}$ in terms of $\mathbf{V}^{(2)}$ only. The resonant condition is obtained from the determinant of the resulting $\bar{M} \times \bar{M}$ coefficient matrix equated to zero. By simple algebraic manipulations, we get

$$\|\boldsymbol{Y}_0\boldsymbol{S} + \boldsymbol{Y}_1\boldsymbol{C} + \boldsymbol{C}\boldsymbol{Y}_2 + \boldsymbol{Y}_1\boldsymbol{Z}_0\boldsymbol{S}\boldsymbol{Y}_2\| = 0 \tag{34}$$

According to expressions (33), in the limit of zero fin thickness ($\delta = 0$), $\boldsymbol{S}$ tends to the null matrix, and $\boldsymbol{C}$ tends to the identity matrix, so that (34) simply reduces to

$$\|\boldsymbol{Y}_1 + \boldsymbol{Y}_2\| = 0 \tag{35}$$

This equation also follows directly from (31) since, in the case of zero metallization thickness ($a_2 = z_1$) the continuity conditions are simply that $\mathbf{V}^{(1)} = \mathbf{V}^{(2)}$ and $\mathbf{I}^{(1)} = -\mathbf{I}^{(2)}$.

When finite metal thickness may be assumed, slot region 1 reduces to a mere interface, and no propagation effect across it has to be taken into account. Different sets of basis functions over the aperture region can be chosen. In particular, the singular behavior of the field at metallic edges can be incorporated into the basis functions in order to accelerate the convergence rate of the solution. This method has been used commonly in conjunction with the spectral domain technique [54] and leads to a very small matrix size [eq. (35)]. Actually, in the case of zero metal thickness, the generalized transverse resonance technique, as discussed here, becomes substantially equivalent to the spectral domain method.

In the numerical computations, the number of terms used in the expansions (23) must be truncated to finite values. It is known that the numerical solution of discontinuities with sharp edges by mode-matching techniques is affected by the relative convergence phenomenon [55–58]. To ensure the proper convergent behavior of the solution and at the same time take advantage of the increased convergence rate, the highest order of modes M and N used for the field expansions in the narrow and wide PPW sections, respectively, must be chosen in a proper way. The mode ratio $R = N/M$ should be as close as possible to the height ratio

$$N/M = s_2/s_1 \tag{36}$$

Such a condition implies that the highest spatial frequency must be approximately the same in adjacent regions. In practical cases M values to a few units are sufficient to guarantee a good approximation. Since the lateral walls at $y = \pm s_2$ must be far away from the slot region, N, on the other hand, may assume much higher values.

It must be mentioned that, according to some investigations [59, 60], the fastest convergence is achieved with a mode ratio 1.5 times the height ratio. This conclusion has been drawn in the case of zero thickness of the metallization.

This is illustrated in Fig. 12*a* for the case of a unilateral finline structure. The convergence behavior of the computed phase constant is shown as a function of the mode ratio $R = N/M$ for different highest orders (M) of modes in the aperture region. The dimensional ratio is equal to 8. It is observed that convergence is faster for higher M values. It is also noticed that, for high mode ratios, all curves are very close and approximately intersect for $R = 8 \times 1.5 = 12$. For this critical value of the mode ratio, no appreciable variation of the results is detected on increasing the highest mode order from $M = 2$ to $M = 8$.

The case of finite metal thickness is illustrated in Fig. 12*b*. Apart from slightly different β values, also the convergence behavior is somewhat altered. It is seen, in particular, that convergence is reached faster because of the weaker singularity of the field at the edges. Intersection of the curves occurs approximately at $R = 10$.

From the above arguments, we see the advantage of reducing all unknowns to those of the slot region. A final matrix equation of relatively small size (M) is obtained, while a much higher ($\approx Ms_2/s_1$) number of modal terms are used for the field description in the wider regions.

The technique described can be extended fairly easily to other planar or quasi-planar configurations. When the printed circuit is inserted into a waveguide housing, for instance, the only modification to be made is to replace the admittances [eq. (30)] seen at the slot region edges by the appropriate values.

A somewhat different formulation of the generalized transverse reso-

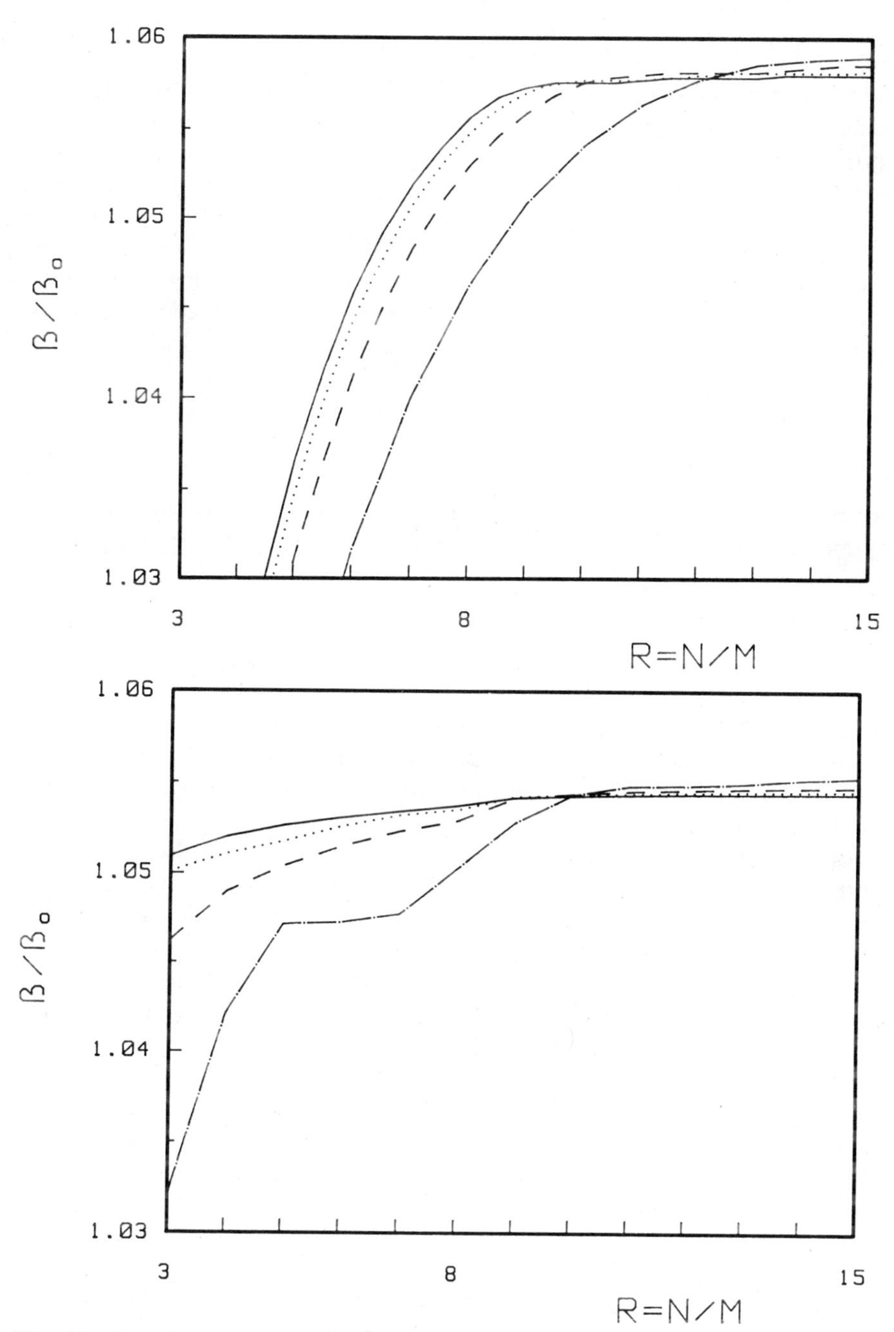

Fig. 12 Convergence behavior of the solution for a unilateral finline with metallization thickness ($\delta = 35\ \mu$m). WR28 waveguide housing, frequency = 34 GHz, ε_r = 2.22, substrate thickness $d = 0.254$ mm, slot width $w = 0.445$ mm. (– · –) $M = 2$; (– – –) $M = 4$, (· · ·) $M = 6$, (———) $M = 8$.

nance technique has been developed by Vahldieck and Bornemann [46, 48–50] for the analysis of finlines and quasi-planar transmission lines. A generalized transmission line matrix description of the discontinuities, including both iris-type and step discontinuities to account for holding grooves, is developed. In this manner, the transverse resonance condition is obtained after simple multiplication of the transmission line matrices of the various sections. This procedure generalizes that described in Section 2 in connection with the slab-loaded waveguide [eqs. (16)–(18)]. The only drawback with this procedure is that the transmission line matrix formulation requires the same number of modal terms to be used in each section, so that condition (36) is not satisfied. However, specific algorithms can be developed to implement the transmission line formulation using different numbers of field expansion terms on the two sides of the discontinuities [61].

5. ANALYSIS OF DISCONTINUITIES AND JUNCTIONS BY THE GENERALIZED TRANSVERSE RESONANCE TECHNIQUE

5.1. Introduction

In the preceding sections transverse resonance techniques for the analysis of uniform guiding structures have been discussed. The assumption of a given longitudinal dependence of the electromagnetic field as an exponentially propagating or a standing wave has reduced the problem to a two-dimensional one, i.e., a problem in the transverse coordinates. Two basic formulations have been presented.

The conventional formulation is based on the characterization of the longitudinal discontinuities in terms of equivalent lumped circuits, so that the transverse resonance condition can be expressed in an elementary way.

In the generalized formulation, longitudinal discontinuities are modeled in terms of a generalized matrix representation on the basis of a mode-matching technique applied in the transverse direction. The idea behind this approach is that the relevant boundary value problem is formulated more easily in the transverse than in the axial direction. The same idea can be applied also to the analysis of three-dimensional problems, as in the modeling of transverse discontinuities. Needless to say, the characterization of discontinuities is of fundamental importance in the design of any microwave circuit, but particularly for microwave and millimeter-wave integrated circuits where tolerances are very strict and tuning is very difficult or even impossible.

The rigorous generalized formulation of the transverse resonance technique can be extended further to provide a useful tool for the characterization of a large variety of discontinuity problems in planar and quasi-planar configurations, such as microstrips, striplines, and finlines. Actually this technique has been used to characterize one-port terminations, as in the shorted finline [62], as well as two-port discontinuities [47, 63], up to four-port junctions, as in cross-coupled striplines [64]. In addition to N-port discontinuities and junctions, the generalized TRT is applicable to transitions between different guiding media, such as microstrip-to-slot line transitions [65]. The case of two-port junctions or discontinuities is considered in this section to illustrate the method.

The analytical method consists schematically of the following steps:

1. A resonant cavity is created by enclosing the discontinuity or junction by auxiliary reactive walls sufficiently apart from it.
2. A field analysis based on TRT is performed to compute the resonance frequencies and, possibly, the corresponding field distributions.
3. The network matrix representation (scattering matrix, impedance matrix, etc.) of the discontinuity is evaluated by means of the resonance frequencies and/or field distributions.

The determination of the equivalent circuit parameters of a two-port network via evaluation of the resonance frequencies of the network inserted between two reactive terminations is equivalent to the experimental technique known as the tangent or Weissfloch method [66, 67]. Analytical formulations have been presented by Collin in conjunction with a variational approach [68] and by Jansen in conjunction with a spectral domain approach [69]. The implementation of this method in conjunction with generalized transverse resonance analysis has been proposed by Sorrentino and Itoh [47]. The analytical technique proposed there for finline discontinuities is reported here in a slightly modified from for generality of presentation.

The representative structure used here to illustrate the method is that of Fig. 13. An arbitrarily shaped discontinuity between the reference planes p_1 and p_2 is connected to two uniform finlines. The analysis is carried out under the following hypotheses:

1. Only the dominant modes can be propagated in each of the two finline sections. The line lengths connected to the discontinuity must be long enough that no higher-order mode interaction can occur with other external circuit elements. This allows the discontinuity region to be modeled as a two-port network. Fig. 13*b* shows the longitudinal equivalent circuit of the discontinuity enclosed in a cavity with end walls at t_1, t_2.
2. The structure is lossless and reciprocal. With this assumption three real quantities are needed to characterize the two-port network. Generalizations to include losses as well as nonreciprocity, however, are possible.
3. The fins are infinitesimally thin. This simplification is not always allowed, particularly at higher frequencies. The way to account for

finite conductor thickness, however, has already been illustrated in the previous section. For the sake of simplicity, we therefore omit this aspect of the problem.

For the purpose of illustrating the method while avoiding unnecessary complications, we consider a bilateral finline. This simplifies the transverse problem, as the symmetry plane at $z = 0$ can be replaced by an open-circuit plane (magnetic wall). A short-circuit plane (electric wall) would be used for odd higher-order excitations. Unilateral and antipodal finline structures can be treated in a similar manner, except that the analytical formulation is more complicated because of the reduced or absent degree of symmetry.

For clarity of presentation the treatment is divided into two parts. The first step is to create a resonant structure in the longitudinal direction. The parameters of the two-port equivalent network of the discontinuity are evaluated via the resonant condition. The second part is the transverse resonance field analysis to compute the resonances of the cavity. The field analysis is, in principle, the same as in the previous section, except that

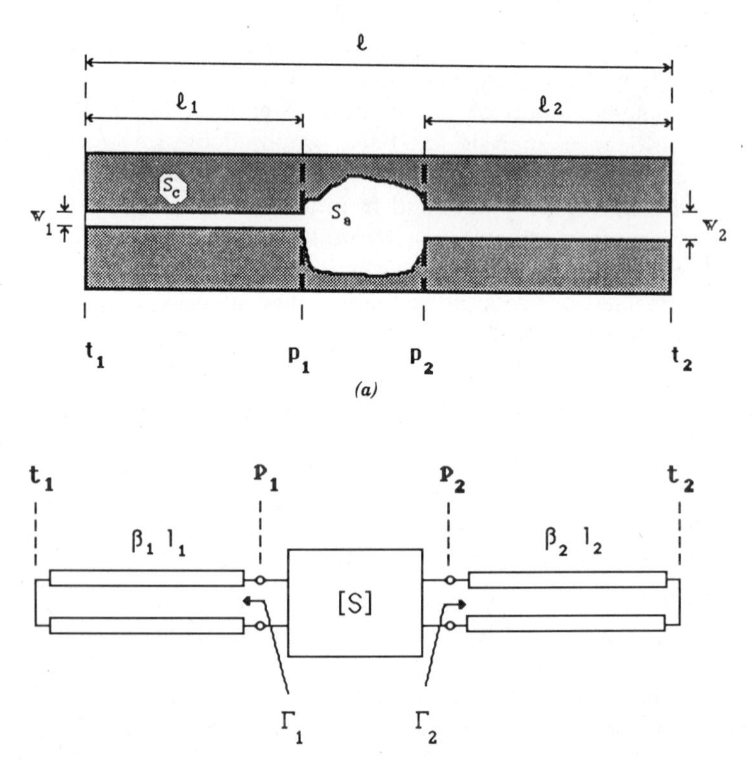

Fig. 13 (*a*) A two-port finline discontinuity (longitudinal section), and (*b*) longitudinal equivalent circuit.

because of the presence of a discontinuity a more complicated field variation in the longitudinal direction must be considered instead of a simple exponential dependence.

5.2. Computation of Equivalent Circuit Parameters

In Cohn's method of analysis of the slot line [38] a standing wave was assumed along the longitudinal axis so that the analysis could be reduced to a rectangular waveguide discontinuity problem. This procedure was used in Section 3 in connection with the ridged waveguide analysis. As already mentioned, however, in such cases the analysis can also be performed assuming a propagating wave.

For discontinuous structures, on the other hand, the assumption of a standing wave is particularly useful in order to apply a transverse resonance analysis.

In a uniform transmission line, a standing wave is created by simply superimposing two waves of equal amplitudes traveling in opposite directions. In an N-port junction, waves of proper amplitudes and phases must impinge on the outputs of the junction in such a way that the superposition with the reflected and transmitted waves cancel to zero at some locations along the feeding lines. This is equivalent to inserting perfect electric (or magnetic) planes at voltage (or current) nulls so as to obtain a resonant structure containing the junction under consideration. The locations of these nulls, and thus the cavity size, are related to the parameters of the junction so that such parameters can be evaluated in terms of the null locations.

Consider first a one-port reactive termination such as the shorted finline of Fig. 14a. In this case, because of the lossless character of the structure, a standing wave is automatically created by any incident wave. Voltage null locations are easily determined in terms of the equivalent reactance X of the short end. Let l be the distance of a voltage null from the edge of the shorting septum. With reference to Fig. 14, such a distance is related to the equivalent reactance of the short end by

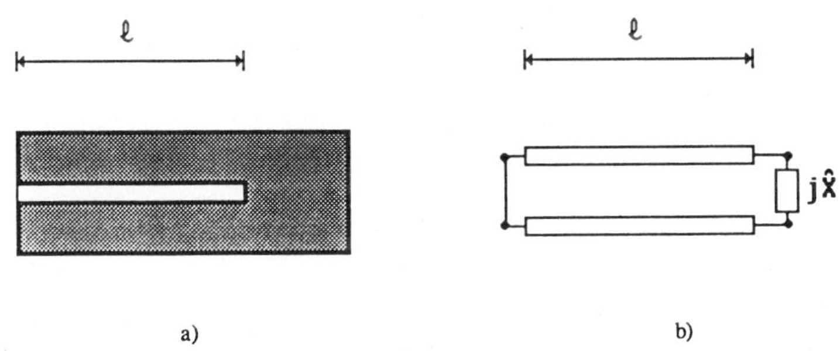

Fig. 14 (a) Shorted finline, and (b) equivalent circuit.

$$\tan\frac{2\pi l}{\lambda} + \hat{X} = 0 \tag{37}$$

where ^ indicates normalization with respect to the finline characteristic impedance. In terms of the reflection coefficient of the septum Γ, (37) is equivalent to

$$\Gamma = -\exp(2j\theta) \tag{38}$$

where $\theta = 2\pi l/\lambda$ is the electrical length. Conversely, the knowledge of the locations of the nulls permits, through the above relation, the computation of the equivalent reactance. Equation (37) or (38) is nothing but the resonant condition of the transmission line length l loaded with the reactance $\hat{X}$. The latter is therefore evaluated at any given frequency through the resonant line length l. It must be specified that the electric planes enclosing the resonant cavity must be located sufficiently apart from the strip edge so as not to perturb the reactive field (i.e., evanescent higher-order modes) confined to its proximity. We will return to this point later.

Consider now the two-port discontinuity problem of Fig. 13. Let s_{ij} ($i, j = 1, 2$) be the scattering parameters evaluated at the reference planes $p_1 p_2$ of the two-port network representative of the discontinuity region. Let a_i and b_i be the incident and reflected waves, respectively, at the ith port. Suppose a standing wave regime is established so that voltage nulls are located at the planes $t_1 t_2$. Insertion of electric plates at $t_1 t_2$, thus creating a resonant cavity, does not perturb the field distribution. This assumption is valid as long as evanescent higher-order modes have negligibly small amplitudes at those planes. This condition can always be met by shifting the terminal planes by one (or more) half-wavelengths away from the discontinuity.

Incident and reflected waves at the reference planes p_1, p_2 are related by the scattering matrix $\boldsymbol{S}$

$$\mathbf{b} = \boldsymbol{S}\mathbf{a} \tag{39}$$

If voltage nulls are located at distances l_1, l_2, incident and reflected wave amplitudes are related by

$$\mathbf{a} = \mathbf{diag}[\rho_i]\mathbf{b} \tag{40}$$

where

$$\rho_i = -\exp(j2\theta_i) = -\exp(j2\beta_i l_i) \tag{41}$$

is the reflection coefficient seen from the ith port of the junction and $\beta_i = 2\pi/\lambda_i$ is the phase constant of the ith finline.

Eliminating **b** from (39) and (40), we obtain a homogeneous system in the amplitudes of the incident waves:

$$(S - \mathbf{diag}[1/\rho_i)]\mathbf{a} = \mathbf{0} \tag{42}$$

For nontrivial solutions to exist, the determinant of the coefficient matrix must vanish:

$$\|S - \mathbf{diag}[1/\rho_i])\| = 0 \tag{43}$$

In explicit form, (43) is written as

$$(s_{11} - 1/\rho_1)(s_{22} - 1/\rho_2) - s_{12}^2 = 0 \tag{44}$$

where reciprocity has been assumed, so that $s_{12} = s_{21}$.

Equation (44) is the resonant condition of the cavity obtained by inserting two perfectly conducting plates at t_1 and t_2. If the scattering parameters are given, (44) provides a relation between ρ_1 and ρ_2, thus, through (41), between l_1 and l_2. We can arbitrarily choose one line length, e.g., l_1. The resonant condition determines the other distance, apart from multiples of half-wavelengths.

Conversely, if the distances l_1, l_2, and thus the reflection coefficients (41), are known, the scattering parameters can be computed through the resonance condition (43). It suffices to determine three sets of resonant lengths l_1 and l_2 for the same frequency. Let (l_{1a}, l_{2a}), (l_{1b}, l_{2b}), (l_{1c}, l_{2c}) be three such pairs of resonant lengths and

$$\Gamma_{1a} = -1/\rho_{1a} = \exp(-2j\theta_{1a}), \quad \text{etc.}$$

Using (44), we obtain

$$s_{11} = -(A_1\Gamma_{1a} + B_1\Gamma_{1b} + C_1\Gamma_{1c})/\Delta$$

$$s_{22} = -(A_2\Gamma_{2a} + B_2\Gamma_{2b} + C_2\Gamma_{2c})/\Delta$$

where

$$A_1 = \Gamma_{2a}(\Gamma_{1b} - \Gamma_{1c}) \qquad A_2 = \Gamma_{1a}(\Gamma_{2c} - \Gamma_{2b})$$

$$B_1 = \Gamma_{2b}(\Gamma_{1c} - \Gamma_{1a}) \qquad B_2 = \Gamma_{1b}(\Gamma_{2a} - \Gamma_{2c})$$

$$C_1 = \Gamma_{2c}(\Gamma_{1a} - \Gamma_{1b}) \qquad C_2 = \Gamma_{1c}(\Gamma_{2b} - \Gamma_{2a})$$

$$\Delta = A_1 + B_1 + C_1 = A_2 + B_2 + C_2$$

s_{12} can be evaluated, apart from a phase uncertainty of π, through (44).

The resonance condition (43) can be written in terms of any two-port matrix representation such as the impedance matrix. In this case, (44) is replaced by

$$(Z_{11} + Z_1)(Z_{22} + Z_2) - Z_{12}Z_{21} = 0$$

Z_i being the impedance seen from the ith port of the equivalent network. Considering normalized impedances only, we have

$$Z_i = j \tan(\beta_i l_i)$$

Another equivalent network representation consists of an ideal transformer n:1 inserted between two transmission lines of lengths ψ_1 and ψ_2. In this case the resonant condition is just the tangent relation [66, 67] between the two added line lengths θ_1 and θ_2. This forms the basis for the Weissfloch method for the characterization of waveguide discontinuities. This method is based on the experimental determination of the relation between l_1 and l_2. From the plot of l_1 vs. l_2 it is possible to compute the three unknown parameters of the two-port junction or discontinuity.

The generalized TRT being illustrated here is based on the same concept, except that it uses a transverse resonance field analysis to compute the distances l_1 and l_2.

The procedure to compute the parameters of the junction from the resonant lengths can be extended to a higher number of ports. In Reference 64 it has been applied to symmetrical four-port transitions.

It is observed that no field quantity computation is required to evaluate the unknown parameters of the junction. This is due to the resonance condition (43) being dependent only on the electrical lengths. In particular, no characteristic impedance definition is involved. It is required, however, to perform a number of resonance experiments, thus computer field analyses, equal to the number of unknown parameters.

The number of field analyses required to characterize an N-port junction is reduced using the procedure suggested by Jansen [69]. For each resonance experiment, however, it is required, in addition to the resonant lengths, to compute the field quantities in order to evaluate the eigensolutions $\mathbf{a}$ of (42). In this manner only N resonances, thus N field analyses, are sufficient to compute the scattering matrix of the junction:

$$S = [\mathbf{b}_1, \mathbf{b}_2, \ldots, \mathbf{b}_N][\mathbf{a}_1, \mathbf{a}_2, \ldots, \mathbf{a}_N]^{-1}$$

where $\mathbf{a}_i$ and $\mathbf{b}_i$ are the incident and reflected wave vectors at the ith resonance experiment.

5.3. Transverse Resonance Field Analysis

To compute the resonant lengths, a transverse resonance analysis is developed. The addition of two terminal reactive walls has converted the problem, as seen in the transverse direction, into a discontinuity problem in a rectangular waveguide instead of that of a finline discontinuity. Looking at the structure in the transverse direction, the longitudinal cross-sectional pattern of Fig. 13 is viewed as an irregularly shaped iris discontinuity placed in a rectangular waveguide of size $l \times b$.

As already mentioned, to simplify the analysis we consider the case of a bilateral finline. The symmetry plane at the center of the dielectric substrate is replaced by a perfect magnetic conductor. The reduced cross section for analysis is shown in Fig. 15*a*. The transverse resonance analysis of such a structure can be carried out along similar lines as the uniform line problem of the previous section. The differences are that (1) we are now not considering the finite fin thickness and (2) a resonance is established in both the transverse and longitudinal directions. The assumption of zero thickness simplifies the generalized transverse equivalent network. Instead of two steps we have an infinitely thin metallic iris that is represented by only one generalized network, as shown in Fig. 15*b*. Unlike the uniform finline problem, the longitudinal resonance involves an infinite number of modal terms, instead of just one, in the z direction, so that a doubly infinite series is required for the modal field representation in the resonator.

In each region shown in the cross-sectional view of Fig. 15, the EM field is expanded in terms of $\mathrm{TE}^{(x)}$ and $\mathrm{TM}^{(x)}$ modes, for which the appropriate potentials are

$$\psi_{mn} = P_{mn} \cos \frac{m\pi z}{l} \cos \frac{n\pi y}{b} \qquad \phi_{mn} = P_{mn} \sin \frac{m\pi z}{l} \sin \frac{n\pi y}{b} \tag{45}$$

where P_{mn} are normalization coefficients.

The EM field components in the plane transverse to x are then expressed as

$$\mathbf{E}_t^{(i)} = \sum_{mn} V_{mn}^{(i)}(x)\mathbf{e}_{mn}(y, z) \tag{46a}$$

$$\mathbf{H}_t^{(i)} = \sum_{mn} I_{mn}^{(i)}(x)\mathbf{h}_{mn}(y, z) \qquad i = 1, 2 \tag{46b}$$

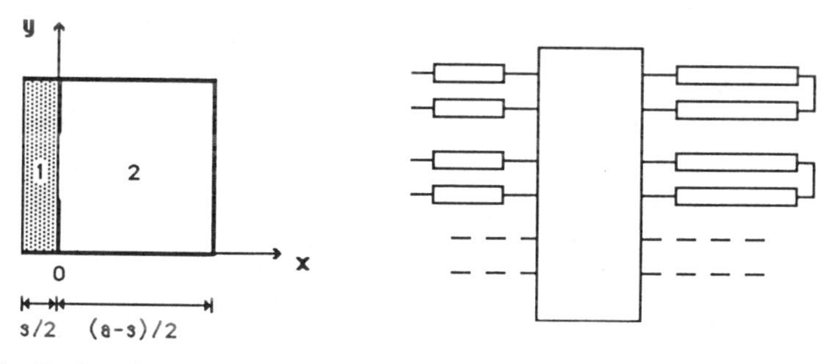

Fig. 15 Reduced cross section of the bilateral finline discontinuity of Fig. 13 and generalized equivalent transverse network.

The $\mathbf{e}_{mn}$ eigenvectors are obtained from (45) using the expressions (73), (74), and (76) of the appendix. An appropriate rotation ($x \rightarrow z$, $z \rightarrow -x$) of the reference frame, however, must be made, as we are now dealing with transverse-to-x instead of transverse-to-z modes. Each modal component of the above expansion can be represented by an equivalent transmission line section. With reference to the generalized transverse equivalent circuit of Fig. 14b, these line sections are terminated by open circuits on the left side (region 1) or by a short circuit on the right side (region 2). At the plane of the fins ($x = 0$), equivalent voltages and currents are related by the admittances seen looking to the two sides. In matrix form we may write

$$\mathbf{I}^{(i)} = \mathbf{diag}[\mathbf{Y}_{0n}^{(i)}]\mathbf{V}^{(i)} \tag{47}$$

with

$$\mathbf{Y}_n^{(1)} = -j\mathbf{Y}_{0n}^{(1)} \tan\left(\beta_{mn}^{(1)} \frac{s}{2}\right) \qquad \mathbf{Y}_n^{(2)} = j\mathbf{Y}_{0n}^{(2)} \cot\left(\beta_{mn}^{(2)} \frac{a-s}{2}\right) \tag{48}$$

$$Y_{0n}^{(i)} = \begin{cases} \beta_{mn}^{(i)}/\omega\mu & \text{for TE modes} \\ \omega\varepsilon_i/\beta_{mn}^{(i)} & \text{for TM modes} \end{cases} \tag{49}$$

β_{mn} are the phase constants of the transverse equivalent transmission lines on the two sides of the fins

$$\beta_{mn}^{(i)} = (k_i^2 - \kappa_{mn}^2)^{1/2}, \qquad k_i^2 = \omega^2\mu\varepsilon_i$$

$$\kappa_{mn}^2 = \left(\frac{m\pi}{l}\right)^2 + \left(\frac{n\pi}{b}\right)^2$$

The boundary conditions at $x = -s/2$ and $x = (a-s)/2$ are satisfied, as they are incorporated into eqs. (46). We must now impose the boundary conditions on the plane of the fins $x = 0$.

In the case of infinitesimally thin fins, the boundary conditions can be formulated either in terms of the slot E-field representation or in terms of the current density representation on the metallic fins. Hofman [57] has shown that one formulation is preferable to the other depending on the relative surface of the slot region S_a and metal region S_c. In some structures, both slot-type and strip-type boundary value problems are present, as for the microstrip-to-slot transition, so that both types of formulations have to be adopted [63].

On the aperture plane $x = 0$, electromagnetic field quantities must obey the following conditions:

$$\mathbf{E}_t^{(1)} = \mathbf{E}_t^{(2)} = \begin{cases} \mathbf{E}_0 & \text{on } S_a \quad (50a) \\ \mathbf{0} & \text{on } S_c \quad (50b) \end{cases}$$

$$\hat{\mathbf{x}}_0 \times (\mathbf{H}_t^{(1)} - \mathbf{H}_t^{(2)}) = \begin{cases} \mathbf{0} & \text{on } S_a \quad (51a) \\ \mathbf{J}_0 & \text{on } S_c \quad (51b) \end{cases}$$

where, as indicated in Fig. 13, S_a is the aperture region and $S_c = S - S_a$ is the metallic portion of the cross section. As is known, the above equations are redundant. Either $\mathbf{E}_0$ or $\mathbf{J}_0$ is sufficient to determine the EM field in the structure. In the slot field formulation, the aperture field $\mathbf{E}_0$ is expanded in terms of a set of orthonormal vector functions defined over S_a, while in the current density formulation $\mathbf{J}_0$ is to be expanded in orthonormal vector functions on the metal surface S_m [1].

Following the first approach, we write

$$\mathbf{E}_0 = \sum_p U_p \mathbf{e}_p^{(0)} \tag{52}$$

We first insert (52) and (46a) into (50), then scalar multiply by $\mathbf{e}_{mn}$, and finally integrate over the cross section S to get

$$V_{mn}^{(1)} = V_{mn}^{(2)} = \sum_p U_p g_{mnp} \tag{53}$$

where

$$g_{mnp} = \int_{S_a} \mathbf{e}_{mn} \cdot \mathbf{e}_p^{(0)} \, dS \tag{54}$$

In matrix form,

$$\mathbf{V}^{(1)} = \mathbf{V}^{(2)} = \boldsymbol{G}\mathbf{U} \tag{55}$$

We now proceed in a similar way inserting the H-field expansion (46b) into (51a). We note that the H-field continuity is now imposed on the slot S_a, which is a portion of the entire section S. We use the set $\mathbf{e}_p$ to test the resultant equation on S_a and obtain

$$\sum_{mn} g_{mnp}(I_{mn}^{(1)} - I_{mn}^{(2)}) = 0 \tag{56a}$$

In matrix form,

$$\boldsymbol{G}^{\mathrm{t}}(\mathbf{I}^{(1)} - \mathbf{I}^{(2)}) = \mathbf{0} \tag{56b}$$

where the superscript t stands for the transpose of the matrix. We now use the E–H field relations, thus eq. (47), to combine (55) and (56) and obtain a homogeneous system of equations in the U expansion coefficients

$$\boldsymbol{G}^{\mathrm{t}}(\boldsymbol{Y}^{(1)} + \boldsymbol{Y}^{(2)})\boldsymbol{G}\mathbf{U} = \mathbf{0} \tag{57}$$

The resonance condition is finally obtained by equating to zero the determinant of coefficient matrix:

$$\|\boldsymbol{G}^{\mathrm{t}}(\boldsymbol{Y}^{(1)} + \boldsymbol{Y}^{(2)})\boldsymbol{G}\| = 0 \tag{58}$$

The alternative procedure consists of expanding the current density on the fins $\mathbf{J}_0$ instead of the electric field on the slot:

$$\mathbf{J}_0 = \sum_q J_q j_q^{(0)} \tag{59}$$

Inserting (46b) and (59) into (51), taking the scalar product with $\mathbf{j}_q$, and integrating over the entire cross section S, we get

$$I_{mn}^{(1)} - I_{mn}^{(2)} = \sum_q f_{mnq} J_q \tag{60}$$

with

$$f_{mnq} = \int_{S_c} \mathbf{e}_{mn} \cdot \mathbf{j}_q^{(0)} \, dS \tag{61}$$

In matrix form we write

$$\mathbf{I}^{(1)} - \mathbf{I}^{(2)} = \boldsymbol{F}\mathbf{J} \tag{62}$$

The continuity of the electric field between regions 1 and 2 across S [eqn. (50a)] implies, as before,

$$V_{mn}^{(1)} = V_{mn}^{(2)} \tag{63}$$

The last boundary condition is (50b), i.e., the vanishing of the tangential electric field over S_c. Using $\mathbf{j}_q$ as a testing function, we obtain

$$\sum_{mn} f_{mnq} V_{mn}^{(1)} = 0 \tag{64}$$

Equations (63) and (64) are also written

$$\mathbf{V}^{(1)} = \mathbf{V}^{(2)} \tag{63$'$}$$

$$\boldsymbol{F}^{t}\mathbf{V}^{(1)} = 0 \tag{64$'$}$$

Similar to the previous formulation, we can now get a homogeneous system in the expansion coefficients for the current density on the metallic fins. Inserting (47) into (62) and using (63′) and (64′) to eliminate the voltages, we obtain

$$\boldsymbol{F}^{t}(\mathbf{Z}^{(1)} + \mathbf{Z}^{(2)})\boldsymbol{F}\mathbf{J} = \mathbf{0} \tag{65}$$

where

$$\mathbf{Z}^{(i)} = [\boldsymbol{Y}^{(i)}]^{-1} \tag{66}$$

Instead of (58) the resonance condition is now

$$\|F^{t}(Z^{(1)} + Z^{(2)})F\| = 0 \tag{67}$$

The resonance condition, in either form (58) or (67), can be regarded as a transcendental function of ω, l_1, l_2 equated to zero. For any given frequency, this equation is solved for three different pairs of l_1, l_2 from which the parameters of the discontinuity are computed.

5.4. Computation of Basis Functions

The procedure developed so far is quite general. It covers, in particular, the case of uniform finlines, where the knowledge of the resonance frequencies immediately provides the phase constant of the modes. Depending on the metallization pattern, the choice of the basis functions $\mathbf{e}_p$ or $\mathbf{j}_q$ must be made. Actually the feasibility of the method resides in the availability of a set of functions suitable to represent the field or the current density on the plane of the discontinuity. In many cases, the metallization pattern has a simple geometry, so that a complete set of orthogonal functions is that of the TE and TM functions for a waveguide with an identical cross section. The complete set of the normal modes of a cylindrical waveguide, in fact, can be used to represent either the E field or the current density on the plane of the discontinuity. The idea of relating the characterization of a discontinuity to the analysis of an irregularly shaped waveguide was used by Muilwyk and Davies [70].

Figure 16 illustrates some typical finline discontinuities that can be analyzed using waveguide basis functions for the aperture fields. It is interesting to note that if the metallized portion S_c is interchanged with the aperture S_a, the unilateral finline configuration is converted into a suspended microstrip. Thus the analysis of a finline discontinuity is easily transformed into that of a suspended microstrip discontinuity by simply replacing (59) with (66).

For the slot configurations of Fig. 16*a–l*, basis functions can be expressed in a closed elementary form. For the step discontinuity of Fig. 16*m*, as well as for the other composite discontinuities, the basis functions can be computed approximately.

Let us refer to the step discontinuity. For simplicity we consider the symmetrical case, so that analysis can be reduced to one half of the structure. Extension to the general case is straightforward. With reference to Fig. 17, the symmetry plane $y = 0$ is chosen as a perfectly conducting plane. The problem is that of determining the modal functions of a homogeneous uniform waveguide with such a stepped cross section. When symmetry is taken into consideration, it is recognized that this problem is the same as that of the ridged or grooved guide and can be solved equivalently by a field-matching or a generalized transverse resonance technique.

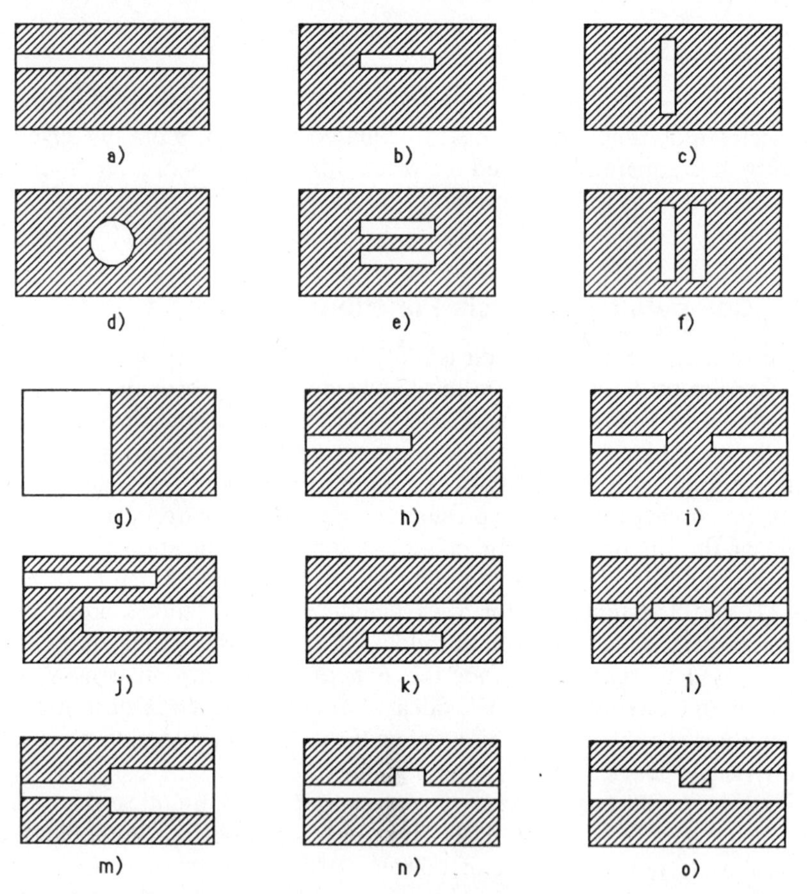

Fig. 16 Examples of *E*-plane circuit discontinuity problems that can be attacked by generalized transverse resonance technique. (*a*) Uniform finline; (*b*–*f*) resonators; (*g*–*h*) shorting septa; (*i*) inductive strip; (*j*–*l*) edge- and end-coupled lines; (*m*) step; (*n*) inductive notch; (*o*) capacitive strip.

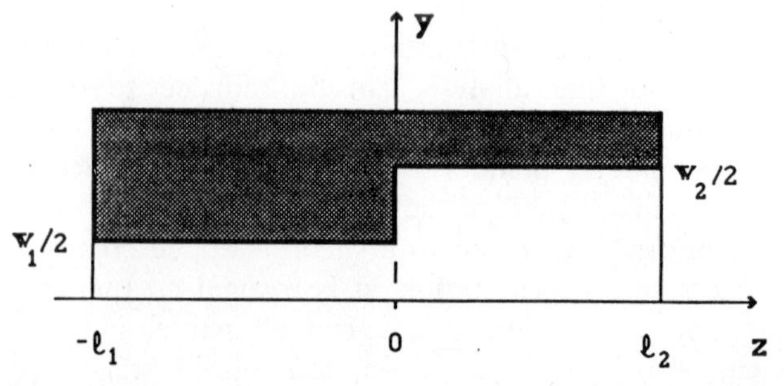

Fig. 17 Reduced geometry of a finline step discontinuity.

Following Reference 36, the potential for a $\mathrm{TE}^{(z)}$ mode is expressed as

$$\psi = \begin{cases} \psi_1 = \sum_r A_r \psi_r^{(1)} & z < 0 \\ \psi_2 = \sum_s B_s \psi_s^{(2)} & z > 0 \end{cases} \tag{68}$$

where

$$\psi_r^{(1)} = \cos k_{1r}(z + l_1) \cos \frac{2\pi r y}{w_1} \tag{69a}$$

$$\psi_s^{(2)} = \cos k_{2s}(z - l_2) \cos \frac{2\pi s y}{w_2} \tag{69b}$$

k_{1r} and k_{2s} are related to the eigenvalue

$$k_c^2 = k_{1r}^2 + \left(\frac{2r\pi}{w_1}\right)^2 = k_{2s}^2 + \left(\frac{2s\pi}{w_2}\right)^2, \qquad r, s = 0, 1, 2, \ldots \tag{70}$$

Using the orthogonality of the above functions, the continuity conditions for the EM field at $z = 0$ lead to the following homogeneous system in the expansion coefficients A_r, B_r:

$$\sum_r A_r f_{rs} \cos k_{1r} l_1 - \frac{d_2}{2\delta_s} B_s \cos k_{2s} l_2 = 0, \qquad s = 0, 1, 2, \ldots \tag{71a}$$

$$\frac{d_1}{2\delta_r} A_r k_{1r} l_1 + \sum_s B_s f_{rs} k_{2s} \sin k_{2s} l_2 = 0, \qquad r = 0, 1, 2, \ldots \tag{71b}$$

where $\delta_r = 1$ for $r = 0$, $\delta_r = 2$ for $r \neq 0$, and

$$f_{rs} = \int_0^{w_2/2} \cos \frac{2r\pi y}{w_1} \cos \frac{2s\pi y}{w_2} \, dy$$

The computational procedure follows the same lines as in similar cases already illustrated. For numerical evaluation, the series are truncated to a finite number of terms. The condition for nontrivial solution of homogeneous system (71) leads to a transcendental equation for k_c^2. For each eigenvalue, the corresponding eigenfunction ψ is evaluated solving system (71). The vector basis function to be used in (52) are finally obtained by the usual expressions, which are given in the appendix. A similar procedure is applied to compute TM eigenvectors.

The technique described above to evaluate the set of basis functions is applicable either for zero or finite fin thickness. When zero thickness may be assumed, as discussed at the end of Section 4, convergence behavior of the numerical solution can be improved by a proper choice of the basis functions. Instead of the eigensolutions for waveguide modes, it is possible to select basis functions approximating the actual field behavior. In particular, they should incorporate the singular behavior at the edges. In addition, basis functions can be constructed on the basis of the field distribution computed for the uniform line case, in combination with suitable perturbation terms in the proximity of the discontinuity [69]. With this technique, even complicated printed circuit configurations can be attacked, interaction and coupling effects among the various circuit elements being included in the overall characterization of the circuit.

6. EXAMPLES OF COMPUTER PROGRAMS*

Three computer program samples are presented in this section to provide the reader with examples of the application of the techniques described in this chapter. They are not intended to provide general and complete characterizations of the structures involved. These have been chosen simple enough that the reader can easily understand how the program is constructed and how it works and implement all additional modifications and extensions.

The three programs require the evaluation of the roots of a real function, and one of them requires the computation of the determinant of a real matrix. Routines to carry out such computations are available in any computer library and are therefore omitted from the listing.

Program No. 1. SLABGUIDE

The first example concerns the analysis of a rectangular waveguide loaded with N slabs of lossless dielectric materials. The analytical procedure is that described in Section 2. The characteristic equation is obtained by multiplying the chain matrices of the various dielectric slabs and then equating to zero the term t_{12}, as in eq. (18). Input variables are frequency, number of slabs ($N < 10$) and dielectric constants, and geometry of the structure. Mode type (TE or TM) and order in the y direction must also be specified. The cutoff frequency is first computed by solving the characteristic equation for $\beta = 0$. The program then computes, at the given frequency, the phase constant and the effective permittivity. The latter quantity is defined as $(\beta/\beta_0)^2$.

*Computer programs 1 and 2 have been developed by G. Schiavon, and program 3 by G. Schiavon and P. Tognolatti.

```
C*********************************************************************
C
C                         PROGRAM SLABGUIDE
C
C       COMPUTES CUTOFF FREQUENCY, PROPAGATION CONSTANT "BETA" AND
C       EFFECTIVE PERMITTIVITY OF A RECTANGULAR GUIDE LOADED WITH
C       M DIELECTRIC SLABS.
C
C--------------------- PARAMETERS -------------------------------------
C
        PARAMETER MAXSLABSNUMBER=10
C
C------------- TYPES AND VARIABLES DECLARATION -----------------------
C
        IMPLICIT DOUBLE PRECISION (A-H,O-Z)
        DOUBLE PRECISION MU0,K0,K02,KV2,KV
C
        DIMENSION S(MAXSLABSNUMBER),EPSR(MAXSLABSNUMBER),T(2,2),TT(2,2)
C
        EXTERNAL CHARFUNCT
C
C------------- COMMON BLOCK ------------------------------------------
C
        COMMON PI,FREQ,B,M,S,EPS0,EPSR,MU0,N,ITYPE,IVAR
C
C------------- CONSTANTS ---------------------------------------------
C
        PI=4.*DATAN(1.D+0)
        EPS0=8.854187818D-012
        MU0=1.256637061D-006
C
C------------- INPUT DATA --------------------------------------------
C
        PRINT *,'GUIDE HEIGHT ? (MM)'
        READ  *,B
        B=B*1.E-3
        PRINT *,'MODE TYPE ? (0:TE 1:TM)'
        READ  *,ITYPE
        PRINT *,'MODE ORDER ?'
        READ  *,N
        PRINT *,'NUMBER OF SLABS ? '
        READ  *,M
        C=0.
        DO 1 I=1,M
          PRINT 100,I
100       FORMAT (1X,'RELATIVE DIELECTRIC CONSTANT OF SLAB #',I2)
          READ  *,EPSR(I)
          PRINT 101,I
101       FORMAT (1X,'THICKNESS OF SLAB #',I2,' (MM)')
          READ  *,S(I)
          S(I)=S(I)*1.E-3
          C=C+S(I)
1       CONTINUE
C
C--------------------------------------------------------------------
C
        PRINT *
        PRINT *,'GUIDE WIDTH IS ',C*1.E+3,' (MM)'
        PRINT *
C

        IVAR=0
C
```

```
C       CALL ROOT(CHARFUNCT, ... )
C
C       THIS ROUTINE (NOT PROVIDED IN THIS LISTING) SHOULD
C       COMPUTE THE ROOT OF THE CHARACTERISTIC EQUATION
C       (ZERO OF "CHARFUNCT"). THE ROOT COMPUTED IS THE
C       CUTOFF FREQUENCY.
C
        PRINT *
        PRINT 102,FREQ*1.D-9
102     FORMAT (1X,'THE CUTOFF FREQUENCY IS :',F8.3,' GHZ')
        PRINT *
C
        PRINT *,'FREQUENCY OF INTEREST ? (GHZ)'
        PRINT *,'(MUST BE GREATER THAN CUTOFF)'
        READ  *,FREQ
        FREQ=FREQ*1.E+9
C
        IVAR=1
C
C       CALL ROOT(CHARFUNCT, ... )
C
C       IN THIS CASE (IVAR = 1) THE ROOT COMPUTED IS THE
C       PROPAGATION CONSTANT "BETA" AT THE GIVEN FREQUENCY.
C
        K02=4.*PI*PI*FREQ*FREQ*MU0*EPS0
        K0=DSQRT(K02)
        EPSEFF=BETA*BETA/K02
C
        PRINT *
        PRINT *,'PROPAGATION CONSTANT (BETA) :',BETA,' ( 1/M )'
        PRINT *,'BETA-0/BETA RATIO :',K0/BETA
        PRINT *,'EFFECTIVE DIELECTRIC CONSTANT :',EPSEFF
        PRINT *
C
        STOP
        END
C
C*******************************************************************


C*******************************************************************
C
        DOUBLE PRECISION FUNCTION CHARFUNCT (X)
C
C       THIS FUNCTION COMPUTES THE CHARACTERISTIC FUNCTION
C
C-------------------- PARAMETERS -----------------------------------
C
        PARAMETER MAXSLABSNUMBER=10
C
C------------ TYPES AND VARIABLES DECLARATION ----------------------
C
        IMPLICIT DOUBLE PRECISION (A-H,O-Z)
        DOUBLE PRECISION MU0,K0D2,KD2,KD
C
        DIMENSION S(MAXSLABSNUMBER),EPSR(MAXSLABSNUMBER)
        DIMENSION T(2,2),TT(2,2)
C
C------------ COMMON BLOCK -----------------------------------------
C
        COMMON PI,FREQ,B,M,S,EPS0,EPSR,MU0,N,ITYPE,IVAR
C
C-------------------------------------------------------------------
```

```
C
      IF (IVAR .EQ. 0) THEN
        FREQ=X
        BETA=0.D+0
      ELSE
        BETA=X
      END IF
C
      TT(1,1)=1.D+0
      TT(1,2)=1.D+0
      TT(2,1)=1.D+0
      TT(2,2)=1.D+0
C
      OMEGA=2*PI*FREQ
      ONE=0.D+0
      DO 2 I=1,M
        KOD2=OMEGA*OMEGA*MUO*EPSO*EPSR(I)
        W=BETA**2+(N*PI/B)**2
        KD2=KOD2-W
        IF (KD2 .GE. 0.) THEN
              KD=DSQRT(KD2)
              IUCTFF=0
        ELSE
              KD=DSQRT(-KD2)
              IUCTFF=1
        END IF
        THETA=KD*S(I)
        IF (ITYPE .EQ. 0) THEN
              Z0=OMEGA*MU0/KD
        ELSE
              Z0=KD/OMEGA/EPS0/EPSR(I)
        END IF
        Y0=1/Z0
        IF (IUCTFF .EQ. 0) THEN
              T(1,1)=DCOS(THETA)
              T(2,2)=T(1,1)
              SINTHETA=DSIN(THETA)
              T(1,2)=Z0*SINTHETA
              T(2,1)=Y0*SINTHETA
        ELSE
              T(1,1)=DCOSH(THETA)
              T(2,2)=T(1,1)
              SINHTHETA=DSINH(THETA)
              T(1,2)=(-1)**ITYPE*Z0*SINHTHETA
              T(2,1)=-((-1)**ITYPE)*Y0*SINHTHETA
        END IF
        T11=TT(1,1)*T(1,1)-ONE*TT(1,2)*T(2,1)
        T12=TT(1,1)*T(1,2)+ONE*TT(1,2)*T(2,2)
        T21=ONE*TT(2,1)*T(1,1)+TT(2,2)*T(2,1)
        T22=TT(2,2)*T(2,2)-ONE*TT(2,1)*T(1,2)
        TT(1,1)=T11
        TT(1,2)=T12
        TT(2,1)=T21
        TT(2,2)=T22
        ONE=1.
2     CONTINUE
C
      CHARFUNCT=TT(1,2)
      RETURN
C
      END
C
C*********************************************************************
```

Program No. 2. CTRUNI

The second computer sample is an application of the conventional TRT described in Section 3 for computing the phase constant of the dominant mode of a unilateral finline. In the equivalent transverse network, the fins and dielectric substrate are represented by a shunt capacitance. From the resonance condition the program determines the cutoff frequencies of the structure with and without the dielectric substrate (thus using $\varepsilon_r = 1$). The cutoff frequency ratio approximately determines the effective dielectric constant. The phase constant at any frequency is then computed by formula (22). The evaluation of the equivalent susceptance of the fins in the presence and in the absence of the dielectric is made according to Reference 33 on the basis of the results reported in the *Waveguide Handbook*.

Input variables are geometrical parameters, dielectric permittivity, and frequency. Output quantities are the phase constant and effective permittivity.

```
C***********************************************************************
C
C                            PROGRAM CTRUNI
C
C        THIS PROGRAM ANALYZES A UNILATERAL FINLINE STRUCTURE BY
C        CONVENTIONAL TRT METHOD (REFERENCE : SCHIEBLICH ET AL.,IEEE
C        TRANS. MTT-32, NO. 12, DEC. 1984). CUTOFF FREQUENCY,
C        PROPAGATION CONSTANT "BETA" AND EFFECTIVE DIELECTRIC CONSTANT
C        ARE COMPUTED.
C
C------------ TYPES AND VARIABLES DECLARATION ------------------------
C
         IMPLICIT DOUBLE PRECISION (A-H,O-Z)
         DOUBLE PRECISION MUO,KO,KX,KXO,KO2,KX2,KE
C
         EXTERNAL CHARFUNCTO
         EXTERNAL CHARFUNCT
C
C------------ COMMON BLOCK ------------------------------------------
C
         COMMON PI,EPSR,B,C1,C3,PW,PD,PB
C
C------------ CONSTANTS ---------------------------------------------
C
         PI=4.*DATAN(1.D+0)
         EPSO=8.854187818D-012
         MUO=1.256637061D-006
C
C------------ INPUT DATA --------------------------------------------
C
         PRINT *,'GUIDE HEIGHT ? (MM)'
         READ  *, B
         B=B*1.D-3
         PRINT *,'GUIDE WIDTH ? (MM)'
         READ  *, C
         C=C*1.D-3
         PRINT *,'SLOT WIDTH ? (MM)'
         READ  *, D
         D=D*1.D-3
         PRINT *,'SLOT POSITION (DISTANCE OF ITS CENTRE FROM LOWER',
     #          'GUIDE WALL) ? (MM)'
```

```
      READ  *, COF
      COF=COF*1.D-3
      PRINT *,'FIN POSITION (METALLIZATION DISTANCE FROM SIDE',
     #         'GUIDE WALL) ? (MM)'
      READ  *, C1
      C1=C1*1.D-3
      PRINT *,'DIELECTRIC THICKNESS ? (MM)'
      READ  *, C2
      C2=C2*1.D-3
      C3=C-C1-C2
      PRINT *,'RELATIVE DIELECTRIC CONSTANT ?'
      READ  *, EPSR
C
C-----------------------------------------------------------------------
C
      ALFAW=PI*D/2./B
      BETAW=PI/2.*(1.-(2*COF/B-1.))
      RD=C2/D
      RB=C2/B
C
      PW=-DLOG(DSIN(ALFAW)*DSIN(BETAW))
      PD=RD*DATAN(1./RD)+DLOG(DSQRT(1.+RD*RD))
      PB=RB*DATAN(1./RB)+DLOG(DSQRT(1.+RB*RB))
C
C     CALL ROOT (CHARFUNCTO, ... )
C     CALL ROOT (CHARFUNCT, ... )
C
C     THE ROUTINE ROOT (NOT PROVIDED IN THIS LISTING) SHOULD
C     COMPUTE THE ROOTS OF THE CHARACTERISTIC EQUATIONS (ZEROS
C     OF "CHARFUNCTO" AND "CHARFUNCT"). THE ROOTS COMPUTED ARE
C     THE CUTOFF WAVENUMBERS OF THE AIR-FILLED RIDGED WAVEGUIDE
C     AND OF THE UNILATERAL FINLINE AT THE GIVEN FREQUENCY.
C
      FC0=KX0/2./PI/DSQRT(MU0*EPS0)
      FC=KX/2./PI/DSQRT(MU0*EPS0)
C
      PRINT *
      PRINT 100,FC*1.D-9
100   FORMAT (1X,'THE CUTOFF FREQUENCY IS :',F8.3,' (GHZ)')
      PRINT *
C
      PRINT *,'FREQUENCY OF INTEREST ? (GHZ)'
      PRINT *,'(MUST BE GREATER THAN CUTOFF)'
      READ  *, FREQ
      FREQ=FREQ*1.D+9
C
      OMEGA=2*PI*FREQ
      K02=OMEGA*OMEGA*MU0*EPS0
      K0=DSQRT(K02)
C
      KE=FC0*FC0/FC/FC
      BETA=2.*PI*FREQ*DSQRT(MU0*EPS0)*DSQRT(KE)*
     #        DSQRT(1.-FC*FC/FREQ/FREQ)
      EPSEFF=BETA*BETA/K02
C
      PRINT *
      PRINT *,'PROPAGATION CONSTANT (BETA) :',BETA,' ( 1/M )'
      PRINT *,'BETA-0/BETA RATIO :',K0/BETA
      PRINT *,'EFFECTIVE DIELECTRIC CONSTANT :',EPSEFF
      PRINT *
C
      STOP
      END
C
C***********************************************************************
```

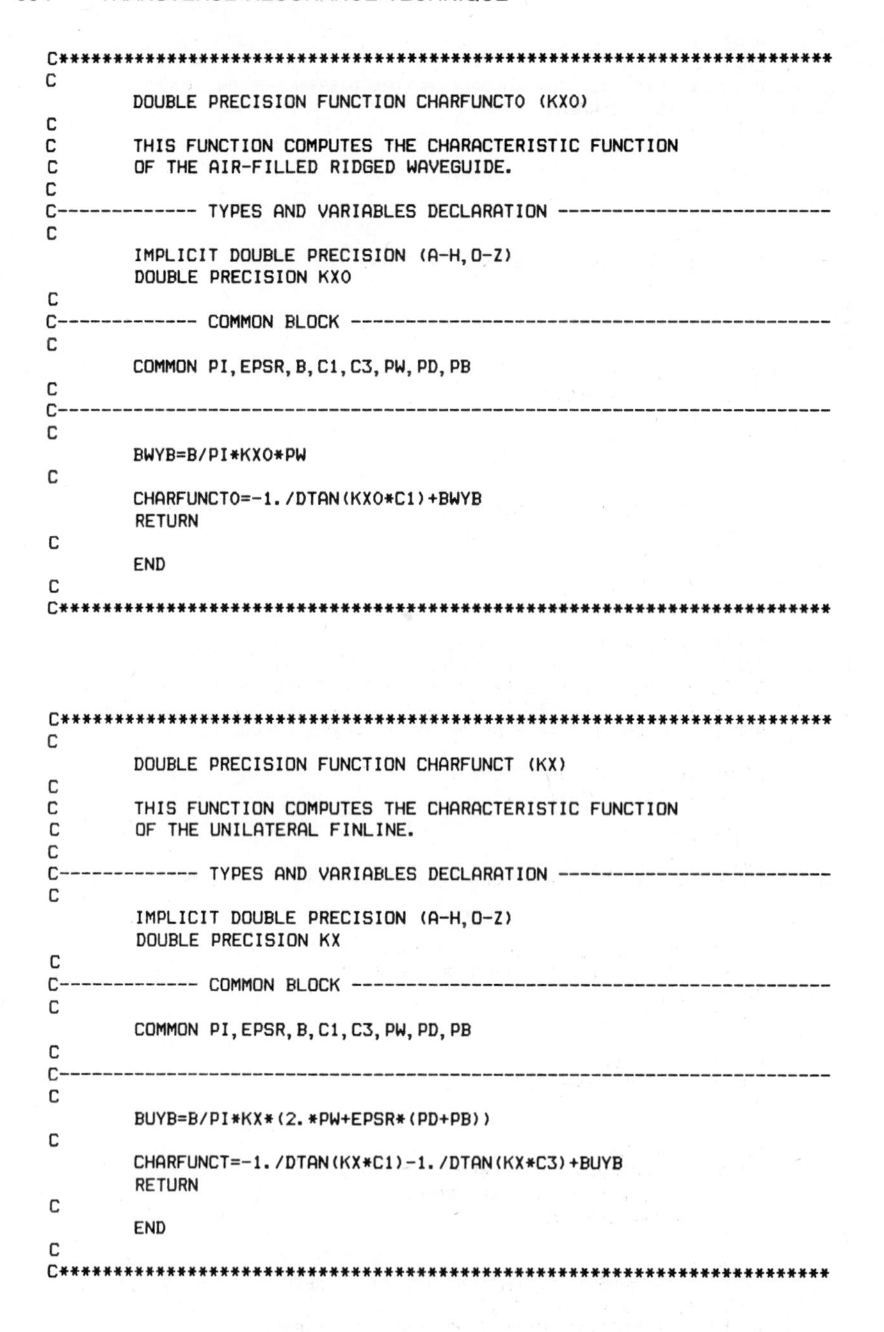

```
C***********************************************************************
C
      DOUBLE PRECISION FUNCTION CHARFUNCTO (KXO)
C
C     THIS FUNCTION COMPUTES THE CHARACTERISTIC FUNCTION
C     OF THE AIR-FILLED RIDGED WAVEGUIDE.
C
C------------ TYPES AND VARIABLES DECLARATION -----------------------
C
      IMPLICIT DOUBLE PRECISION (A-H,O-Z)
      DOUBLE PRECISION KXO
C
C------------ COMMON BLOCK ------------------------------------------
C
      COMMON PI,EPSR,B,C1,C3,PW,PD,PB
C
C-------------------------------------------------------------------
C
      BWYB=B/PI*KXO*PW
C
      CHARFUNCTO=-1./DTAN(KXO*C1)+BWYB
      RETURN
C
      END
C
C***********************************************************************


C***********************************************************************
C
      DOUBLE PRECISION FUNCTION CHARFUNCT (KX)
C
C     THIS FUNCTION COMPUTES THE CHARACTERISTIC FUNCTION
C     OF THE UNILATERAL FINLINE.
C
C------------ TYPES AND VARIABLES DECLARATION -----------------------
C
      IMPLICIT DOUBLE PRECISION (A-H,O-Z)
      DOUBLE PRECISION KX
C
C------------ COMMON BLOCK ------------------------------------------
C
      COMMON PI,EPSR,B,C1,C3,PW,PD,PB
C
C-------------------------------------------------------------------
C
      BUYB=B/PI*KX*(2.*PW+EPSR*(PD+PB))
C
      CHARFUNCT=-1./DTAN(KX*C1)-1./DTAN(KX*C3)+BUYB
      RETURN
C
      END
C
C***********************************************************************
```

Program No. 3. GTRUNI

The unilateral finline is now analyzed by a generalized TRT. The analytical technique is the one described in Section 4, except that the metal thickness is assumed to be zero. Under this simplifying hypothesis, the whole spectrum of the finline can in principle be evaluated by this program. The maximum order M of the basis functions in the slot must be given among the input variables. Correspondingly, the maximum order in the waveguide region (N) is chosen as $1.5M$ times the dimensional ratio.

```
C*********************************************************************
C
C                        PROGRAM GTRUNI
C
C       COMPUTES PROPAGATION CONSTANT "BETA" AND EFFECTIVE PERMITTIVITY
C       OF A UNILATERAL FINLINE BY GENERALIZED TRT.
C
C------------- TYPES AND VARIABLES DECLARATION -------------------------
C
        IMPLICIT DOUBLE PRECISION (A-H,O-Z)
        INTEGER NN,QQ
        DOUBLE PRECISION MUO,KO,KO2,L
C
        EXTERNAL CHARFUNCT
C
C------------- COMMON BLOCKS ------------------------------------------
C
        COMMON PI,L,B,COFS,D
        COMMON /CF/ EPSO,EPSR,MUO,FREQ,NN,QQ,A1,A2,A3
C
C------------- CONSTANTS ----------------------------------------------
C
        PI=4.*DATAN(1.D+0)
        EPSO=8.854187818D-012
        MUO=1.256637061D-006
C
C------------- INPUT DATA ---------------------------------------------
C
        PRINT *,'GUIDE HEIGHT ? (MM)'
        READ  *, B
        B=B*1.D-3
        PRINT *,'GUIDE WIDTH ? (MM)'
        READ  *, A
        A=A*1.D-3
        PRINT *,'SLOT WIDTH ? (MM)'
        READ  *, D
        D=D*1.D-3
        PRINT *,'SLOT POSITION (DISTANCE OF ITS CENTRE FROM LOWER ',
     #          'GUIDE WALL) ? (MM)'
        READ  *, COF
        COF=COF*1.D-3
        COFS=COF-D/2.
        PRINT *,'FIN POSITION (METALIZATION DISTANCE FROM SIDE',
     #          'GUIDE WALL) ? (MM)'
        READ  *, A1
        A1=A1*1.D-3
        PRINT *,'DIELECTRIC THICKNESS ? (MM)'
        READ  *, A2
        A2=A2*1.D-3
        A3=A-A1-A2
        PRINT *,'RELATIVE DIELECTRIC CONSTANT ?'
        READ  *, EPSR
        PRINT *,'MAXIMUM ORDER OF GUIDE MODES ?'
        READ  *, QQ
        NN=(QQ*D/B)/1.5
```

```
C
C       QQ : MAXIMUM ORDER OF WAVE-GUIDE MODES
C       NN : MAXIMUM ORDER OF SLOT BASIS FUNCTIONS
C
        PRINT *,'FREQUENCY OF INTEREST ? (GHZ)'
        READ  *, FREQ
        FREQ=FREQ*1.D+9
C
C----------------------------------------------------------------------
C
C       CALL ROOT(CHARFUNCT, ... )
C
C       THIS ROUTINE (NOT PROVIDED IN THIS LISTING) SHOULD
C       COMPUTE THE ROOT OF THE CHARACTERISTIC EQUATION
C       (ZERO OF "CHARFUNCT"). THE ROOT COMPUTED IS THE
C       PROPAGATION CONSTANT "BETA" AT THE GIVEN FREQUENCY.
C
        OMEGA=2*PI*FREQ
        K02=OMEGA*OMEGA*MUO*EPSO
        KO=DSQRT(K02)
        EPSEFF=BETA*BETA/K02
C
        PRINT *
        PRINT *,'PROPAGATION CONSTANT (BETA) :',BETA,' ( 1/M )'
        PRINT *,'BETA-0/BETA RATIO :',KO/BETA
        PRINT *,'EFFECTIVE DIELECTRIC CONSTANT :',EPSEFF
        PRINT *
C
        STOP
        END
C
C**********************************************************************


C**********************************************************************
C
        DOUBLE PRECISION FUNCTION CHARFUNCT (BETA)
C
C       THIS FUNCTION COMPUTES THE CHARACTERISTIC FUNCTION
C       OF THE UNILATERAL FINLINE
C
C------------------------- PARAMETERS ---------------------------------
C
        PARAMETER MAXGUIDETERMS=280
        PARAMETER MAXSLOTSTERMS=40
C
C------------- TYPES AND VARIABLES DECLARATION ------------------------
C
        IMPLICIT DOUBLE PRECISION (A-H,O-Z)
        DIMENSION T(1:MAXGUIDETERMS,1:MAXSLOTSTERMS)
        DIMENSION TTJBT(1:MAXSLOTSTERMS,1:MAXSLOTSTERMS)
        DIMENSION BUFF(1:MAXGUIDETERMS,1:MAXSLOTSTERMS)
        DOUBLE PRECISION JBL(1:MAXGUIDETERMS),JBR(1:MAXGUIDETERMS)
        DOUBLE PRECISION KV,KV2,KD,KD2,K02,KOD2,MUO,L
        INTEGER DIME,DIMV,N,NN,Q,QQ
C
C------------- COMMON BLOCKS ------------------------------------------
C
        COMMON PI,L,B,COFS,D
        COMMON /CF/ EPSO,EPSR,MUO,FREQ,NN,QQ,A1,A2,A3
C
C----------------------------------------------------------------------
C
```

```
C       COMPUTES THE TRANSFORM MATRIX [ T ] FROM THE SET OF BASIS
C       FUNCTIONS TO THE SET OF WAVE-GUIDE MODES.
C
        L=PI/BETA
C
        DIME=2*QQ+1
        DIMV=2*NN+1
        Q=0
        DO 1 I=1,DIME
          IF (Q .EQ. QQ+1) Q=1
          N=0
          DO 2 J=1,DIMV
                IF (N .EQ. NN+1) N=1
                IF (J .GT. NN+1) THEN
                  IF (I .GT. QQ+1) THEN
                        T(I,J)=FF(N,Q)
                  ELSE
                        T(I,J)=0.
                  END IF
                ELSE
                  IF (I .GT. QQ+1) THEN
                        T(I,J)=PF(N,Q)
                  ELSE
                        T(I,J)=PP(N,Q)
                  END IF
                END IF
                N=N+1
2         CONTINUE
          Q=Q+1
1       CONTINUE
C
C        COMPUTES THE ADMITTANCE MATRICES [ JBR ] AND [ JBL ]
C        SEEN FROM THE SLOT.
C
         OMEGA=2*PI*FREQ
         K02=OMEGA*OMEGA*MU0*EPS0
         K0D2=K02*EPSR
         Q=0
         DO 3 I=1,DIME
           IF (Q .EQ. QQ+1) Q=1
           GAMMA2=BETA**2+(Q*PI/B)**2
           KV2=K02-GAMMA2
           KD2=K0D2-GAMMA2
           ISIGNV=1
           ISIGND=1
           ICASE=1
           IF (KV2 .LT. 0.D+0) THEN
                 ISIGNV=-ISIGNV
                 ICASE=ICASE+1
           END IF
           IF (KD2 .LT. 0.D+0) THEN
                 ISIGND=-ISIGND
                 ICASE=ICASE+1
           END IF
           KV=DSQRT(ISIGNV*KV2)
           KD=DSQRT(ISIGND*KD2)
           IF (I .GT. QQ+1) THEN
                 Y0=ISIGNV*OMEGA*EPS0/KV
                 Y0D=ISIGND*OMEGA*EPS0*EPSR/KD
           ELSE
                 Y0=KV/(OMEGA*MU0)
                 Y0D=KD/(OMEGA*MU0)
```

```
          END IF
          IF (ICASE .EQ. 1) THEN
                TAN1=DTAN(KV*A1)
                TAN2=DTAN(KD*A2)
                TAN3=DTAN(KV*A3)
          ELSE IF (ICASE .EQ. 2) THEN
                TAN1=DTANH(KV*A1)
                TAN2=DTAN(KD*A2)
                TAN3=DTANH(KV*A3)
          ELSE IF (ICASE .EQ. 3) THEN
                TAN1=DTANH(KV*A1)
                TAN2=DTANH(KD*A2)
                TAN3=DTANH(KV*A3)
          END IF
          JBR(I)=-Y0/TAN1
          JBL(I)=(ISIGND*Y0D*TAN2-Y0/TAN3)/(1+Y0/Y0D*TAN2/TAN3)
          Q=Q+1
3       CONTINUE
C
        DO 4 I=1,DIME
          DO 5 J=1,DIMV
                BUFF(I,J)=(JBL(I)+JBR(I))*T(I,J)
5         CONTINUE
4       CONTINUE
C
        DO 6 I=1,DIMV
          DO 7 J=1,DIMV
            TTJBT(I,J)=0.
            DO 8 III=1,DIME
                TTJBT(I,J)=TTJBT(I,J)+T(III,I)*BUFF(III,J)
8           CONTINUE
7         CONTINUE
6       CONTINUE
C
C       CALL DETERM (TTJBT,DET, ... )
C
C       THIS ROUTINE (NOT PROVIDED IN THIS LISTING) SHOULD
C       COMPUTE THE DETEMINANT "DET" OF THE MATRIX :
C       [ TTJBT ] = TRAS [ T ] * ([ JBR ] + [ JBL ]) * [ T ]
C
        CHARFUNCT=DET
        RETURN
C
      END
C
C*********************************************************************


        DOUBLE PRECISION FUNCTION FF(N,Q,PCOEG,PCOES)
C
        IMPLICIT DOUBLE PRECISION (A-H,O-Z)
        INTEGER N,Q
        DOUBLE PRECISION L
        COMMON PI,L,B,COFS,D
C
        CB=Q*COFS/B
        DB=Q*D/B
        IF (N .EQ. 0 .AND. Q .EQ. 0) THEN
          FACT=0.
```

```
      ELSE IF (N .EQ. DB) THEN
        FACT=PI*DCOS(CB*PI)
      ELSE
        FACT=(2.*N/(N*N-DB*DB))*(DSIN(CB*PI)-(-1)**N*DSIN((CB+DB)*PI))
      END IF
      GAMMAG2=(PI/L)**2+(Q*PI/B)**2
      PG=PCOEGUIDE(Q)
      PS=PCOESLOT(N)
      FF=GAMMAG2*PS*PG*D*L*PI*FACT/(4.*PI*PI)
      RETURN
      END



      DOUBLE PRECISION FUNCTION PP(N,Q,PCOEG,PCOES)
C
      IMPLICIT DOUBLE PRECISION (A-H,O-Z)
      INTEGER N,Q
      DOUBLE PRECISION L
      COMMON PI,L,B,COFS,D
C
      CB=Q*COFS/B
      DB=Q*D/B
      IF (N .EQ. O .AND. Q .EQ. O) THEN
        FACT=2.*PI
      ELSE IF (N .EQ. DB) THEN
        FACT=PI*DCOS(CB*PI)
      ELSE
        FACT=(2.*DB/(N*N-DB*DB))*(DSIN(CB*PI)-
     #          (-1)**N*DSIN((CB+DB)*PI))
      END IF
      GAMMAS2=(PI/L)**2+(N*PI/D)**2
      PG=PCOEGUIDE(Q)
      PS=PCOESLOT(N)
      PP=GAMMAS2*PS*PG*D*L*PI*FACT/(4.*PI*PI)
      RETURN
      END



      DOUBLE PRECISION FUNCTION PF(N,Q,PCOEG,PCOES)
C
      IMPLICIT DOUBLE PRECISION (A-H,O-Z)
      INTEGER N,Q
      DOUBLE PRECISION L
      COMMON PI,L,B,COFS,D
C
      CB=Q*COFS/B
      DB=Q*D/B
      PG=PCOEGUIDE(Q)
      PS=PCOESLOT(N)
      PF=PS*PG*PI/2.*(DSIN(CB*PI)-(-1)**N*DSIN((CB+DB)*PI))
      RETURN
      END
```

```
      DOUBLE PRECISION FUNCTION PCOEGUIDE (IQ)
C
C     NORMALIZATION COEFFICIENTS OF THE MODES
C
      IMPLICIT DOUBLE PRECISION (A-H,O-Z)
      DOUBLE PRECISION L
      COMMON PI,L,B,COFS,D
C
      DQ=2.
      IF (IQ.EQ.0) DQ=1.
      GAMMAG2=(PI/L)**2+(IQ*PI/B)**2
      PCOEGUIDE=DSQRT(2.*DQ/L/B/GAMMAG2)
      RETURN
      END
```

```
      DOUBLE PRECISION FUNCTION PCOESLOT (IN)
C
C     NORMALIZATION COEFFICIENTS OF THE BASIS FUNCTIONS
C
      IMPLICIT DOUBLE PRECISION (A-H,O-Z)
      DOUBLE PRECISION L
      COMMON PI,L,B,COFS,D
C
      DN=2.
      IF (IN.EQ.0) DN=1.
      GAMMAS2=(PI/L)**2+(IN*PI/D)**2
      PCOESLOT=DSQRT(2.*DN/L/D/GAMMAS2)
      RETURN
      END
```

APPENDIX. FIELD EXPANSION IN WAVEGUIDES

We briefly recall the modal formalism to represent the general field distribution in a uniform cylindrical waveguide filled with a homogeneous isotropic dielectric medium with constants ε and μ.

It is known that the most general electromagnetic field in a waveguide can be obtained as the superposition of an infinite number of elementary solutions (normal modes) that belong to one out of three groups: TE, TM, or TEM modes.

Each modal component of the electromagnetic field of a $\mathrm{TM}^{(z)}$ (or $\mathrm{TE}^{(z)}$) mode can be derived from a z-directed electric (or magnetic) vector potential that is separated into a longitudinal and a transverse component. It can be demonstrated that any well-behaved (i.e., piecewise-continuous and square-integrable) vector function defined over the cross-sectional domain S of the waveguide can be expressed as the superposition of the modal eigenfunctions. For a complete proof of this statement, the reader is referred to the excellent book by Kurokawa [71].

In what follows we assume the z axis as the reference longitudinal direction. It does not necessarily coincide with the waveguide longitudinal axis. A reference frame rotation might be necessary in some instances before applying the present notation.

Following the notation of Marcuvitz's *Waveguide Handbook* [71], the electric and magnetic field components transverse to the z direction are expressed as

$$\mathbf{E}_t(x, y, z) = \sum_m V_m(z)\mathbf{e}_m(x, y)$$

$$\mathbf{H}_t(x, y, z) = \sum_m I_m(z)\mathbf{h}_m(x, y) \tag{72}$$

The transverse dependence can be entirely derived from the electric field modal vectors $\mathbf{e}_m$,

$$\mathbf{h}_m = \mathbf{z}_0 \times \mathbf{e}_m \tag{73}$$

$\mathbf{z}_0$ being the unit vector of the z axis. In (72) the summation includes all three groups of modal functions. They are obtained as follows.

For TE Modes:

$$\mathbf{e}_m = \mathbf{z}_0 \times \nabla_t \psi_m \tag{74}$$

where ψ_m is a solution of the eigenvalue problem,

$$\nabla_t^2 \psi_m + k_{cm}^2 \psi_m = 0 \qquad \text{in } S$$

$$\frac{\partial \psi_m}{\partial n} = 0 \quad \text{on } C \qquad \psi_m = 0 \quad \text{on } C' \tag{75}$$

where n is the outwardly directed normal to the surface S, and C and C' are the portions of electric and magnetic conductors of the contour of S in the xy plane, respectively.

For TM Modes:

$$\mathbf{e}_m = -\nabla_t \phi_m \tag{76}$$

where ϕ_m is a solution of the eigenvalue problem

$$\nabla_t^2 \phi_m + k_{cm}^2 \phi_m = 0 \qquad \text{in } S$$

$$\phi_m = 0 \quad \text{on } C \qquad \frac{\partial \phi_m}{\partial n} = 0 \quad \text{on } C' \tag{77}$$

TEM modes can exist only for multiply connected cross sections, as for the coaxial cable. A TEM mode can be obtained as a special case of TM modes with zero eigenvalue k_c^2:

$$\mathbf{e}_m = -\nabla_t \phi_m \tag{78}$$

where ϕ_m is a solution of the Laplace equation

$$\begin{aligned} &\nabla_t^2 \phi_m = 0 \qquad \text{in } S \\ &\hat{\mathbf{n}} \times \nabla_t \phi_m = 0 \quad \text{on } C \qquad \hat{\mathbf{n}} \cdot \nabla \phi_m = 0 \quad \text{on } C' \end{aligned} \tag{79}$$

The longitudinal components of the electric and magnetic fields are expressed by series containing only TM or TE terms, respectively,

$$\begin{aligned} E_z &= \sum_m I_m(z) e_{zm} \\ H_z &= \sum_m V_m(z) h_{zm} \end{aligned} \tag{80}$$

where

$$e_{zm} = \frac{k_{cm}^2}{j\omega\varepsilon} \phi_m \qquad \text{and} \qquad h_{zm} = \frac{k_{cm}^2}{j\omega\mu} \psi_m$$

In the absence of any sources of the electromagnetic field inside the waveguide, from homogeneous Maxwell equations the longitudinal dependence through the functions V_m and I_m is found to be

$$\begin{aligned} V_m(z) &= V_m^+ \exp(-j\beta_m z) + V_m^- \exp(j\beta_m z) \\ I_m(z) &= [V_m^+ \exp(-j\beta_m z) - V_m^- \exp(j\beta_m z)] / Z_{0m} \end{aligned}$$

with

$$\beta_m = (\omega^2 \mu\varepsilon - k_{cm}^2)^{1/2}$$

and

$$Z_{0m} = \begin{cases} \dfrac{\omega\mu}{\beta_m} & \text{for TE modes} \\[2ex] \dfrac{\beta_m}{\omega\varepsilon} & \text{for TM modes} \end{cases}$$

For TEM modes ($k_c = 0$), both of the above expressions could be used. It is customary, however, to assume the static field definitions of voltage and current to define the characteristic impedance for TEM modes.

REFERENCES

1. H.-Y. Yee, "Transverse modal analysis for printed circuit transmission lines," *IEEE Trans. Microwave Theory Tech.*, vol. MTT-33, pp. 808–816, Sept. 1985.
2. C. A. Olley and T. E. Rozzi, "Systematic characterization of the spectrum of unilateral finline," *IEEE Trans. Microwave Theory Tech.*, vol. MTT-34, pp. 1147–1156, Nov. 1986.
3. T. Itoh, "Spectral domain immittance approach for dispersion characteristics of generalized printed transmission lines," *IEEE Trans. Microwave Theory Tech.*, vol. MTT-28, pp. 733–736, July 1980.
4. S. B. Worm and R. Pregla, "Hybrid-mode analysis of arbitrarily shaped planar microwave structures by the method of lines," *IEEE Trans. Microwave Theory Tech.*, vol. MTT-32, pp. 191–196, Feb. 1984.
5. R. E. Collin, *Field Theory of Guided Waves*, Chapter 6, McGraw-Hill, New York, 1960.
6. L. O. Goldstone and A. A. Oliner, "Leaky-wave antennas. I. Rectangular waveguides," *IRE Trans. Antennas Propag.*, vol. AP-7, pp. 307–319, Oct. 1959.
7. P. H. Vartanian, W. P. Ayres, and A. L. Helgesson, "Propagation in dielectric slab loaded rectangular waveguide," *IRE Trans. Microwave Theory Tech.*, vol. MTT-6, pp. 215–222, Apr. 1958.
8. N. Eberhardt, "Propagation in the off center E-plane dielectrically loaded waveguide," *IEEE Trans. Microwave Theory Tech.*, vol. MTT-15, pp. 282–289, May 1967.
9. R. Seckelmann, "Propagation of TE modes in dielectric loaded waveguides," *IEEE Trans. Microwave Theory Tech.*, vol. MTT-14, pp. 518–527, Nov. 1966.
10. F. E. Gardiol, "Higher order modes in dielectrically loaded rectangular waveguides," *IEEE Trans. Microwave Theory Tech.*, vol. MTT-16, pp. 919–924, Nov. 1968.
11. T. K. Findakly and H. M. Haskal, "On the design of dielectric loaded waveguides," *IEEE Trans. Microwave Theory Tech.*, vol. MTT-24, pp. 39–43, Jan. 1976.
12. F. Gardiol, "Comments on 'The design of dielectric loaded waveguides,'" *IEEE Trans. Microwave Theory Tech.*, vol. MTT-25, pp. 624–625, July 1977.
13. R. F. Harrington, *Time-Harmonic Electromagnetic Fields*, McGraw-Hill, New York, 1961.
14. F. J. Tischer, "A waveguide structure with low loss," *Arch. Elektron. Ubertragungstech.*, vol. 7, pp. 592–596, Dec. 1953.
15. F. J. Tischer, "H-Guide with laminated dielectric slab," *IEEE Trans. Microwave Theory Tech.*, vol. MTT-18, pp. 9–15, Jan. 1970.
16. T. Yoneyama and S. Nishida, "Nonradiative dielectric waveguide for millimeter-wave integrated circuits," *IEEE Trans. Microwave Theory Tech.*, vol. MTT-29, pp. 1188–1192, Nov. 1981.
17. A. Hessel, "General characteristics of traveling-wave antennas," in *Antenna Theory* (R. E. Collin and F. J. Zucker, Eds.), Chapter 19, McGraw-Hill, New York, 1969.

18. R. E. Collin, *Field Theory of Guided Waves*, Chapter 11, McGraw-Hill, New York, 1960.
19. F. E. Gardiol, "Anisotropic slabs in rectangular waveguides," *IEEE Trans. Microwave Theory Tech.*, vol. MTT-18, 461–467, Aug. 1970.
20. N. Marcuvitz, *Waveguide Handbook*, Sect. 8, McGraw-Hill, New York, 1951.
21. L. O. Goldstone and A. A. Oliner, "Leaky-wave antennas. II. Circular waveguides," *IRE Trans. Antennas Propag.*, vol. AP-9, pp. 280–290, May 1961.
22. J. R. Whinnery and H. W. Jamieson, "Equivalent circuits for discontinuities in transmission lines," *Proc. IRE*, vol. 32, pp. 98–116, Feb. 1944.
23. S. Ramo and J. Whinnery, *Fields and Waves in Modern Radio*, Wiley, New York, 1944.
24. S. B. Cohn, "Properties of ridge wave guide," *Proc. IRE*, vol. 35, pp. 783–788, Aug. 1947.
25. T. G. Mihran, "Closed- and open-ridge waveguide," *Proc. IRE*, vol. 37, pp. 640–644, June 1949.
26. S. Hopfer, "The design of ridged waveguides," *IRE Trans. Microwave Theory Tech.*, vol. MTT-3, pp. 20–29, Oct. 1955.
27. J. R. Pile, "The cutoff wavelength of the TE_{10} mode in ridged rectangular waveguide of any aspect ratio," *IEEE Trans. Microwave Theory Tech.*, vol. MTT-14, pp. 175–183, Apr. 1966.
28. W. J. R. Hoefer, "Quickly now, where does waveguide cutoff occur?," *Microwaves*, pp. 70–74, Dec. 1979.
29. N. Marcuvitz, *Waveguide Handbook*, p. 161, McGraw-Hill, New York, 1951.
30. N. Marcuvitz, *Waveguide Handbook*, p. 218, McGraw-Hill, New York, 1951.
31. N. Marcuvitz, *Waveguide Handbook*, p. 307, McGraw-Hill, New York, 1951.
32. J. P. Montgomery, "On the complete eigenvalue solution of ridged waveguide," *IEEE Trans. Microwave Theory Tech.*, vol. MTT-19, pp. 547–555, June 1971.
33. J. Piotrowski, "Dispersion in fin-lines," *Arch. Elektron. Ubertragungstech.*, vol. 38, pp. 278–280, Apr. 1984.
34. C. Schieblich, J. K. Piotrowski, and J. H. Hinken, "Synthesis of optimum finline tapers using dispersion formulas for arbitrary slot widths and locations," *IEEE Trans. Microwave Theory Tech.*, vol. MTT-32, pp. 1638–1645, Dec. 1984.
35. P. J. Meier, "Integrated fin-line millimeter components," *IEEE Trans. Microwave Theory Tech.*, vol. MTT-22, pp. 1209–1216, Dec. 1974.
36. A. A. Oliner and P. Lampariello, "The dominant mode properties of open groove guide: An improved solution," *IEEE Trans. Microwave Theory Tech.*, vol. MTT-33, pp. 755–764, Sept. 1985.
37. T. Tamir, "Leaky-wave antennas," in *Antenna Theory* (R. E. Collin and F. J. Zucker, Eds.), Chapter 20, pp. 280 sgg., McGraw-Hill, New York, 1969.
38. S. B. Cohn, "Slot line—An alternative transmission medium for integrated circuits," *1968 IEEE Trans. G-MTT Int. Microwave Symp. Dig.*, pp. 104–109, 1968.
39. S. B. Cohn, "Slot line on a dielectric substrate," *IEEE Trans. Microwave Theory Tech.*, vol. MTT-17, pp. 768–778, Oct. 1969.
40. E. A. Mariani, C. P. Heinzman, J. P. Agrios, and S. B. Cohn, "Slot line

characteristic," *IEEE Trans. Microwave Theory Tech.*, vol. MTT-17, pp. 1091–1096, Dec. 1969.

41. R. N. Simons and M. Tech, "Analysis of millimetre-wave integrated fin line," *IEE Proc., Part H: Microwaves, Opt. Antennas*, vol. 130, pp. 166–169, Mar. 1983.
42. F. Arndt and G. U. Paul, "The reflection definition of the characteristic impedance of microstrips," *IEEE Trans. Microwave Theory Tech.*, vol. MTT-27, pp. 724–731, Aug. 1979.
43. R. Sorrentino and G. Leuzzi, "Full-wave analysis of integrated transmission lines on layered lossy media," *Electron. Lett.*, vol. 18, pp. 607–609, July 1982.
44. G. Leuzzi, A. Silbermann, and R. Sorrentino, "Mode propagation in laterally bounded conductor-backed coplanar waveguides," *1983 IEEE MTT-S Int. Microwave Symp. Dig.*, pp. 393–395, June 1983.
45. R. Sorrentino, G. Leuzzi, and A. Silbermann, "Characteristics of metal–insulator–semiconductor coplanar waveguides for monolithic microwave circuits," *IEEE Trans. Microwave Theory Tech.*, vol. MTT-32, pp. 410–416, Apr. 1984.
46. R. Vahldieck, "Accurate hybrid mode analysis of various finline configurations including multilayered dielectrics, finite metallization thickness and substrate holding grooves," *IEEE Trans. Microwave Theory Tech.*, vol. MTT-32, pp. 1454–1460, Nov. 1984.
47. R. Sorrentino and T. Itoh, "Transverse resonance analysis of finline discontinuities," *IEEE Trans. Microwave Theory Tech.*, vol. MTT-32, pp. 1633–1638, Dec. 1984.
48. J. Bornemann, "Rigorous field theory analysis of quasiplanar waveguides," *IEE Proc., Part H: Microwaves, Opt. Antennas*, vol. 132, pp. 1–6, Feb. 1985.
49. R. Vahldieck and J. Bornemann, "A modified mode-matching technique and its application to a class of quasi-planar transmission lines," *IEEE Trans. Microwave Theory Tech.*, vol. MTT-33, pp. 916–926, Oct. 1985.
50. J. Bornemann and F. Arndt, "Calculating the characteristic impedance of finlines by transverse resonance method," *IEEE Trans. Microwave Theory Tech.*, vol. MTT-34, pp. 85–92, Jan. 1986.
51. H.-Y. Yee and K. Wu, "Printed circuit transmission line characteristic impedance by transverse modal analysis," *IEEE Trans. Microwave Theory Tech.*, vol. MTT-34, pp. 1157–1163, Nov. 1986.
52. S.-T. Peng and A. A. Oliner, "Guidance and leakage properties of a class of open dielectric waveguides: Part I—Mathematical formulations," *IEEE Trans. Microwave Theory Tech.*, vol. MTT-29, pp. 843–855, Sept. 1981.
53. Y. C. Shih and T. Itoh, "Analysis of conductor backed coplanar waveguide," *Electron. Lett.*, vol. 17, pp. 538–540, June 1982.
54. L.-P. Schmidt and T. Itoh, "Spectral domain analysis of dominant and higher order modes in fin-lines," *IEEE Trans. Microwave Theory Tech.*, vol. MTT-28, pp. 981–985, Sept. 1980.
55. R. Mittra and S. W. Lee, *Analytical Techniques in the Theory of Guided Waves*, Macmillan, New York, 1971.
56. S. W. Lee, W. R. Jones, and J. J. Campbell, "Convergence of numerical

solutions of iris-type discontinuity problems," *IEEE Trans. Microwave Theory Tech.*, vol. MTT-19, pp. 528–536, June 1971.

57. R. Mittra, T. Itoh, and T.-S. Li, "Analytical and numerical studies of the relative convergence phenomenon arising in the solution of an integral equation by the moment method," *IEEE Trans. Microwave Theory Tech.*, vol. MTT-20, pp. 96–104, Feb. 1972.
58. Y. C. Shih and K. G. Gray, "Convergence of numerical solutions of step-type waveguide discontinuity problems by modal analysis," *1983 IEEE MTT-S Int. Symp. Dig.*, pp. 233–235, 1983.
59. H. Hofman, "Dispersion of planar waveguides for millimeter-wave application," *Arch. Elektron. Ubertragungstech.*, vol. 31, pp. 40–44, Jan. 1977.
60. G. Schiavon, R. Sorrentino, and P. Tognolatti, "Characterization of coupled finlines by generalized transverse resonance method," *Int. J. Numer. Model.*, vol. 1, pp. 45–59, March 1988.
61. R. R. Mansour and R. H. MacPhie, "An improved transmission matrix formulation of cascaded discontinuties and its application to E-plane circuits," *IEEE Trans. Microwave Theory Tech.*, vol. MTT-34, pp. 1490–1498, Dec. 1986.
62. G. Bartolucci and R. Sorrentino, "Analysis of end-effect in short-circuited finlines," *Proc. 8th Coll. Microwave Comm.* (*MICROCOLL*), pp. 55–56, Aug. 1986.
63. G. Bartolucci and R. Sorrentino, "Equivalent circuit modeling of finline end-coupling and short-end by generalized transverse resonance analysis," *J. Electromagn. Waves Appl.*, vol. 2, no. 1, pp. 63–76, 1988.
64. T. Uwano, R. Sorrentino, and T. Itoh, "Characteristic of stripline crossing by transverse resonance analysis," *IEEE Trans. Microwave Theory Tech.*, vol. MTT-35, pp. 1369–1376, Dec. 1987.
65. T. Uwano, R. Sorrentino, and T. Itoh, "Characterization of microstrip-to-slotline transition discontinuities by transverse resonance analysis," *Proc. 17th Eur. Microwave Conf.*, pp. 317–322, Sept. 1987.
66. R. E. Collin, *Field Theory of Guided Waves*, Sect. 5.8, McGraw-Hill, New York, 1960.
67. N. Marcuvitz, *Waveguide Handbook*, Sect. 3.4, McGraw-Hill, New York, 1951.
68. R. E. Collin, *Field Theory of Guided Waves*, Sect. 8.1, McGraw-Hill, New York, 1960.
69. R. H. Jansen, "Hybrid mode analysis of end effects of planar microwave and millimetrewave transmission lines," *IEE Proc., Part H: Microwaves, Opt. Antennas*, vol. 128, pp. 77–86, Apr. 1981.
70. C. A. Muilwyk and J. B. Davies, "The numerical solution of rectangular waveguide junctions and discontinuities of arbitrary cross section," *IEEE Trans. Microwave Theory Tech.*, vol. MTT-15, pp. 450–455, Aug. 1967.
71. K. Kurokawa, *An Introduction to the Theory of Microwave Circuits*, Chapter 3, Academic Press, New York, 1969.
72. N. Marcuvitz, *Waveguide Handbook*, Sect. 8, McGraw-Hill, New York, 1951.

INDEX

Accessible modes, 627
Admissible function, 45, 59
Air–dielectric boundaries, 581, 583
Air–dielectric interface, 525
Akhtarzad, 503, 504, 506, 507, 509, 511, 515, 517, 520–522, 532–534, 536, 537, 540–544, 547, 571
Al-Mukthar, 552, 554, 555
Allen, 565
Alternative networks, 568, 569
Amer, 555
Analysis, 35
Anisotropic:
  conductivity, 545
  dielectric substrate, 410
  materials, 50, 544, 546, 568
  permeability, 545
  permittivity, 545, 546
  slab loaded waveguides, 649
  structures, 505
  substrates, 98, 545, 570
Antennas, 86
Antennas, microstrip:
  broadband, 303
  circularly polarized, 300
  multiport network model, 307
  multiport patches, 307
    two-port circular, 309
    two-port rectangular, 307
Arbitrarily sparse matrices, 112
Arbitrary cross section, 505
Arbitrary direction, 507, 521, 543
Area coordinates, 65, 66
Asymmetrical node, 555, 557, 559, 560, 563
Asymmetric couplers, 123
Attenuation characteristic, 521, 544
Attenuation constant, 513, 514, 515, 516, 520, 543, 559
Axial propagation, 505, 506, 514, 519

Babinet's principle, 508
Back-substitution, 114
Back-transformation, 484
Backward difference formula, 103
Band matrices, 112, 114
Basis coefficients, 63
Basis function, 37, 39, 56, 81, 341, 661, 675, 677
Basic nodes, 561
Bessel function, 87, 90
Beurle, 498
Beyer, 572
Bidiagonal matrix, 18
Bifurcated waveguides, 593, 614
Boundaries:
  air-dielectric, 581, 583
  code for, 582
  conducting, 547
  dielectric, 547, 548
  general, 511
  horizontal, 582, 583
  lossless, 510, 537
  lossy, 510, 512, 537, 539
  reflecting, 551
  vertical, 582, 583
  waveguide, 581, 582
Boundary conditions, 75, 137, 383, 435, 463, 512
  Dirichlet, 384, 385
  inhomogeneous, 411, 422
  Neumann, 384, 385
  periodic, 412, 441
Boundary element method, 86
Boundary reflection coefficient, 512, 525, 567, 582, 583
Broadband match, 512

Capacitance, 123
Capacitive stubs, 540
Cascaded discontinuities, 19
Cascaded junctions, 627
Catastrophe, 111
Cauchy's principal value, 157
Central difference formula, 103
Chain matrix, 152
Change in height, 653
Channel directional coupler, 97
Channel waveguide, 101, 102

Characteristic equation, 660
Characteristic impedance, 49, 343, 349, 423, 442, 453, 512, 553, 652
Choi, 565–567
Choleski, 114
Circuits:
 open, 525
 planar, 496
 quasi-planar, 496
 short, 526
Circular ridged waveguides, 554
Circular waveguides, 649, 654
Coarseness error, 550, 551
Coaxial cables loaded with dielectric cylinders, 649
Coaxial probes, 194
Complementary principles, 86
Complete polynomial, 64, 65
Complex plane, 155
Computational boxes, 581, 583
Computational expenditure, 572
Computation of magnetic fields, 571
Computer storage, 83, 85
Computing time, 83, 85, 114
Condensed node, 552, 555–557, 559–565
Condition, 108, 109
Condition number, 108, 110, 116, 604
Conducting boundaries, 547
Conducting wall, 537
Conductivity stub, 554
Conjugate gradient algorithm, 115
Conjugate gradient method, 115
Conservation of complex power, 611
Continuity conditions, 392
Continuous excitation, 529
Conventions and notations, 135
Convergence, 119, 398, 423, 481, 572, 613
Coplanar guides, 45
Coplanar waveguide electric fields, basis functions, 353
Coplanar waveguides (CPW), 123, 353, 655–657
Correction of errors, 548
Coupled integral equations, 336
Coupled-line filter, 428
Coupling integrals:
 crossing, 476
 impedance step, 470
Cross-coupled striplines, 665
Crossing:
 electromagnetic field, 474
 microstrip, 472
 mode matching technique, 476
 reference planes, 472
 scattering parameters, 488
 subregions, 477
Crout, 114
Current distribution, 207, 433
Curved boundary, 510, 511
Cutoff frequency, 53, 452, 520, 553, 570
Cutoff spectrum, 528
Cutoff wavelength, 48
Cutoff wavenumber, 49

Dense, 83, 112
Density, 73
Desegmentation, 27, 29
Design, 35
Deterministic problem, 36
Diagonal direction, 543
Diagonal propagation, 505, 506, 521
Dielectric:
 boundary, 547
 homogeneous, 516
 inhomogeneous, 516
 interface, 524, 525, 546–548, 551
 lossy, 517
 slab, 581
 waveguides, 95, 97, 383, 500
Dielectric-loaded waveguides, 570
Dielectric slab-loaded waveguide, 643, 649, 658
Difference operator, 386, 397, 407, 413, 418
Diffracted field, 137
Diffusion problems, 570, 571
Dirac delta function, 11, 38, 95, 104, 188, 195
Dirac impulses, 501, 517
 scattering, 501, 535
Direct methods, 113
Dirichlet boundary conditions, 75–77
Discontinuities, 416, 422, 428, 500, 538, 570
 microstrips, in:
  characterization, 273
  compensation, 276
 step discontinuity, 428
 waveguides, in characterization, 270
Discrete Fourier transformation, 431
Discretization, 384, 385
 of dielectric constants, 407, 440
 non-equidistant, 396
 two-dimensional, 416
Dispersion, 101, 457, 503, 559, 654
Dispersion characteristics, 531, 546, 567, 568, 570
Dispersion curves, 425
Dispersion relation, 508, 550
Dispersive wave impedance, 512

Dissipative stubs, 517, 540, 546
Distributed node, 555, 556, 559
Double-layer microstrip, 153, 160
Double-ridge waveguides, 650
Duality principles, 86
Dyadics, 135, 138, 142

Edge condition, 604
Effective dielectric constant, 25, 424, 453, 654
Effective refraction index, 100
Effective scattering area, 86
Effective widths, 26, 453
Eigenfrequencies, 570
Eigenfunctions, 659
Eigenvalue problems, 570
Eigenvalues, 91, 345, 434
  indirect system, 395
  for periodic boundary conditions, 441
Eigenvector, 343, 395, 434, 441
Electric field integral equation (EFIE), 91, 92, 93, 94, 139
Electric fin, 538
Electric surface current, 137, 207
Electric vector potential, 155, 176
Electric wall, 510, 511, 512, 537, 581
Electromagnetic field, 474
Electromagnetic resonator, 50
Electrostatic problem, 36
Element matrices, 71
Elements, 59
EMP response of aircraft, 571
Enclosed structure, 355
Energy levels, 50
Energy-type expression, 41
Entire domain, 189, 204
E-plane filters, 627
E-plane septum, 627
Equivalence theorems, 139
Equivalent circuit:
  de-embedding technique, 206
  of a discontinuity, 206
  of a resonator, 203
Equivalent transmission lines, 24
Error residual, 37, 115
Errors, 548, 568
Essential boundary conditions, 51–53, 76
Euclidian norm, 109
Euler equation, 47, 48, 50–52, 85
Euler-Lagrange equation, 47
Excitation, 548
  code, 584
  field, 137
  lines, 584
  points, 584
Expanded node network, 555

Face diagonal, 567
Fast inversion, 421
Fermat's principle, 42–44, 47
Ferrite substrates, 410
Field components, 384, 388, 408, 414, 419
Field distribution, 529, 531
  magnetic waveguide model, 450
  microstrip, 450
Field expansion in waveguides, 690
Field-matching method, 13
Field profile, 102
Fields of an elementary source, 146
Finite conductivity ground plane, 160
Finite difference, 102, 104–107, 120
Finite difference method, 2, 33
Finite element method, 4, 33, 59
Finite element program, 97, 123
Finite elements, 106, 107, 123
Finite-length septum, 593, 600
Finite metallization thickness, 400, 424, 657
Finline discontinuity, 14, 670
Finlines, 351, 547, 551–553, 555, 570, 650, 654, 655, 664
Finned rectangular waveguide, 538
Fin thickness, 671
First-order elements, 62
First-order shape functions, 66
First variation, 46
Five-point formula, 105
Forward difference formula, 103
Forward elimination, 113
Fourier transform, 6, 9, 22, 23, 108, 135, 187, 334–336, 497, 503, 513, 516, 526–529, 531, 549, 572, 584
  boundary conditions, 339
  field equations, 337
  Helmholtz equation, 337
  Poynting power flow, 344
  scalar potentials, 337
Free nodal values, 78
Free-space discontinuities, 512
Free variables, 77
Frequency range, 584
Frequency response, 526, 527
Fresnel, 498
Frontal method, 115
Fukai, 569
Full matrices, 112
Full wave analysis, 383
Functional, 43

Galerkin method, 10, 11, 23, 36, 82, 83, 85, 95, 185, 189, 195, 341
Gaussian beam, 34
Gauss method, 113, 114, 121, 122
Gauss–Seidel algorithm, 115, 116, 117, 122
Generalized Galerkin method, 38
Generalized matrix representation, 664
Generalized scattering matrix, 19, 622
Generalized scattering parameters, 598
Gibbs phenomenon, 527, 549
Global coordinates, 69, 71
Global matrices, 72, 73, 75, 77
Graded mesh, 551, 552, 554, 555, 560, 561, 563
Green's function, 8, 9, 28, 36, 84, 85, 87, 90, 143, 153, 162, 187, 210
Green's functions:
  approximate values, 164
  dyadic for the fields, 138, 154
  dyadic for the vector potential, 142, 154
  interpolation among numerical values, 191
  near and far field values, 164, 165
  partial image representation, 163
  scalar for the scalar potential, 145, 154
  singularities, 144, 167, 190
  symmetry properties, 141
Green's functions for planar components:
  evaluation of, 224–227
    expansion into eigenfunctions, 226
    method of images, 225
  mixed boundary, 237–247
    circular sectors, 247
    rectangles, 237–240
    triangles, 240–247
  open boundary, 227–231
    annular ring, 230
    annular sectors, 231
    circle, 230
    circular sectors, 230
    rectangle, 227
    triangles, 227, 229
  shorted boundary, 231–237
    circle, 236
    circular sectors, 236, 237
    rectangles, 231, 232
    triangles, 234
Green's theorem, 46, 48, 49
Grooved guide, 650, 675
Ground plane:
  finite conductivity, 160
  as terminal plane, 137
Guide wavelength, 460
Gyrators, 569
Gyrotropic material, 50

H-guide, 646, 647
$H^{(1)}_0(kr)$, 90
Hadamard, 109
Hankel form, 87
Hankel function, 27
Hankel integral transform, 136
Hanning window, 527, 528, 549
Heating effects, 529
Helmholtz equation, 9, 22, 27, 48, 50, 383, 419, 422, 431
Hertz–Debye potentials, 142
Hertzian potential, 565
Higher-order elements, 66
Hoefer, 500, 528, 530, 549, 551, 565, 566
Hollow conducting waveguide, 40, 90, 95, 104
Hollow waveguide problem, 90
Homogeneous, 505, 550
Homogeneous system of equations, 342
Huygens, 498
  principle, 498, 499, 501
  wave model, 498, 500
Hybrid modes, 98, 450
Hybrid problems, 532, 570

Image guide, 425
Immittance approach, 345
  Poynting power flow, 366
Impedances, 525, 529, 531, 546
  power-current, 532
  voltage-current, 532
  voltage-power, 532
Impedance matrices for planar segments, 313–326
  annular ring, 319, 320
  annular sector, 320–322
  circle:
    double summation expression, 316–317
    single summation expression, 269
  circular sectors, 317–319
  rectangle:
    double summation expression, 313
    single summation expression, 314
  triangles, 322
    equilateral, 322–324
    90°–60°–30°, 325, 326
    right-angled isosceles, 324–325
Impedance matrix, 28, 95, 197, 208
Impedance step:
  coupling integrals, 470
  mode matching technique, 466
  potential functions, 466
  reference planes, 466
Impedance wall, 137

Imperfect conductor, 512
Impulse response, 503, 514, 516, 526, 527, 529, 531, 546, 548, 549, 572
Impulse scattering, 517, 518, 554, 562, 564, 567
Impulsive excitation, 526
Indirect methods, 113
Inductive stubs, 540
Infinite elements, 73, 74
Infinitely long shunt stub, 541
Infinitely thin septum, 593
Infinite parallel waveguide, 512
Infinite series, 163
Infinitesimally thin fins, 672
Inhomogeneous:
  dielectric, 405, 524
  magnetic, 524
  regions, 524
  structures, 555
  substrate, 656
  waveguides, 570
Inhomogeneously filled waveguides, 646, 650
Inner product, 10
Input file, 580
  explanation, 582
  interpretation, 580
  preparation, 580
Input impedance, 196, 197, 198, 199
Input points, 549
Insulated finline, 530, 531
Integral equation, 10, 27, 33, 82, 88, 89, 107, 162, 163, 203
Integral equation method, 22
Integral equation of the second kind, 89
Integral equations for layered media:
  associated boundary conditions, 137
  electric field integral equation, 139
  magnetic field integral equation, 139
  mixed potential integral equation, 145
  quasi-static case, 163, 203
  static case, 162, 203
Integral formulation, 80
Integral operation, 37
Integrated optical structures, 97
Integrated optics, 34
Interface conditions, 524, 526
Interfaces, 516
Interior problem, 90
Interpolation, 191
Interpolation polynomial, 63, 65, 66, 68
Intrinsic impedances, 524, 531, 539
Inward radiating wave, 90
Isotropic materials, 568
Isotropic propagation, 508, 544
Iterations, number of, 584
Iterative algorithms, 115
Iterative methods, 113, 115

Jacobi method, 115, 116, 117
Johns, 498, 532, 555, 563, 564, 571

Kronecker product, 418

Lagrange interpolation polynomial, 66
Lagrangian function, 41
Laplace equation, 2, 4, 63, 120
Layered media, 146, 152
Layout, 581
Leaky wave antennas, 654
Least squares residual method, 38
$LiNbO_3$, 50, 98, 99, 101, 102
Link lines, 541, 555
Local coordinates, 65–67, 69
Loss factor, 516
Lossless, 510, 513, 516
Loss stub, 513, 517, 521, 522, 540, 541, 545, 557
Loss tangent, 514, 516
Loss term, 514
Losses, 512, 516, 540, 546, 570
  dielectric, 156
  ohmic (surface impedance), 137, 160
  radiation, 208
Lossy, 358, 510, 512, 513, 514
Lossy mesh, 512, 514
Lossy waveguide walls, 512
Lower bound, 86
Low-frequency velocity, 565
LSE mode, 565
L-shaped patch, 198
LSM mode, 565
Lumped networks, 570, 571

Magnetic field integral equation (MFIE), 91, 92, 93, 94, 139
Magnetic vector potential:
  integral formulation, 155
  numerical values, 176
Magnetic walls, 510, 511, 538, 539
Mariki, 544
Matched load, 512
Materials:
  anisotropic, 544, 546, 568
  isotropic, 539, 568
  lossless, 513, 515, 539
    homogeneous, 513
    inhomogeneous, 514
  lossy, 513, 515, 516, 539
    homogeneous, 516
    inhomogeneous, 516, 539, 554

Matrix, 108
Matrix computations, 108, 112
Matrix formulation, 147
Matrix methods, 113
Matrix norms, 109
Maxwell's equation, 107, 497, 499, 504, 519, 522, 534, 535
Meander line, 428
Mesh nodes, 535
Mesh points, 2, 3
Mesh size, 582
Metal-insulator-semiconductor coplanar waveguide (MISCPW), 656, 658
Metallic strips, 538
Metallic waveguide, 653
Metallization thickness, 655, 661
Method of lines, 16, 638
Method of moments, 36, 38, 210
  entire domain basis functions, 189, 204
  Galerkin testing, 185, 189
  point-matching, 188
  razor testing along segments, 187
  rooftop basis functions, 184, 185, 204
Method of subsections, 39
Microstrip discontinuities, 447
Microstrip line, 9, 206, 449, 655
  shielded microstrip lines, 335
Microstrip problems, 532
Microstrip resonator, 8, 18
Microstrips, 33, 45, 86, 123, 423, 433, 545, 547, 551, 570, 664
  characteristic impedance, 453
  coupled, 428
  crossing, 472
  dispersion, 457
  effective dielectric constant, 453
  effective width, 453
  field distribution, 450
  impedance step, 466
  phase velocity, 453
  T-junction, 481
Microstrip structures:
  coupled patches, 203
  discontinuities, 206
  double layer, 153, 181
  L-shaped resonator, 198
  practical excitations, 193
  single layer, 176
Microstrip-to-slot line, 665
Microwave image guide, 97
Microwave planar guides, 33
Microwave structures, 102, 152, 153, 166, 176, 181, 193, 198, 203, 206
Minimized, 47
Misalignment, 547, 548, 551, 552
Mixed potential integral equation (MPIE), 145, 185, 188, 198, 210
Modes, even or odd, 657
Mode fields, 570
Mode-matching method, 12, 13, 14, 16, 19, 432, 592, 593, 599, 601–603, 613, 664
  crossing, 476
  impedance step, 466
Mode spectrum, 527
Moment methods, 9
Multiaxial discontinuities, 570
Multilayer structure, 335, 348
Multiport analysis, 196
Mutual coupling, 204

Narrowband matched load, 512
Natural boundary conditions, 50, 51, 52, 53, 76
Network cutoff frequency, 508
Network propagation velocity, 508
Neumann boundary conditions, 75, 76, 77
Neumann coefficients, 463
Newton, 498
Newton–Raphson algorithm, 157
Nodal values, 62, 63, 64, 77, 85
Node scattering matrices, 546
Nonlinear problems, 497
Nonradiating dielectric waveguide, 646, 648
Non-reciprocal components:
  circulators, wideband, 263, 296
  planar analysis, 263–270
  power dividers and combiners, 263, 296
Nonsymmetrical node, 556
Nonuniform fields, 550, 551
Normal components, 140
Normal modes, 609, 611
Numerical efficiency, 572
Numerical integration algorithm:
  for multiple integrals, 191
  for unbounded oscillating functions, 170
  on the spectral real axis, 167
Numerical integration techniques, 161, 167, 169, 170, 191

Ohmic losses, 160
One-layer microstrip, 176
Open circuits, 525
Open groove guides, 654
Open structure, 338
Open waveguide, 512
Optical channel guide, 97, 100, 101
Optical components, 50
Optical fiber, 97
Optical planar guides, 98

Optical rib guide, 97
Optical structures, 102
Optical waveguides, 97
Optimization, 35, 115
Orthogonality, 13, 14, 463
Orthogonality relation, 594
Oscillating functions, 170
Output, 548
file, 584
impulse function, 526
points, 549, 584
Outward radiating wave, 90

Parallel-plate waveguide (PPW), 639, 656–659
partially filled, 647
Parameter, 398
Parseval's relation, 24
Parseval's theorem, 341, 356, 359
Partial image representation, 163
Pascal's triangle, 64, 65
Periodic structures, 416, 431
Permeability stubs, 521, 522, 523, 540, 541, 545, 546, 557
Permittivity stubs, 517, 518, 540, 541, 545, 546, 554, 557
Permittivity stub value, 583
Perturbation formula, 52, 53, 55
Perturbation theory, 53
Phase characteristic, 520, 543
Phase constant, 516, 520, 523, 531, 543, 559
Phase velocity, 123, 453, 513, 531, 537
Pic, 549, 552–555
Planar, 500
Planar circuit, 26
analysis, 216, 220
contour integral approach, 247–256
desegmentation method, 256, 260–263
Green's function approach, 221–224
multiple sub-ports, 224
segmentation method, 256, 257–260
anisotropic spacing media, with, 263–270
contour integral approach, 264, 265
eigenfunction expansion approach, 265, 266
mode matching approach, 266–270
applications, 270
antennas, microstrip, 298
circulators, wideband, 296, 298
discontinuities characterization, 270–278
filters, 289–296
non-reciprocal dividers and combiners, 296, 298
power dividers, reciprocal, 278–289
boundary conditions, 217–219
electric wall, 219
impedance boundary, 219
magnetic wall, 218
mixed, 237–247
open circuit, 218
short circui, 218
excitation methods, 219–220
impedance matrix, elements of, 223. *See also* Impedance matrices for planar segments
Planar structures, 547, 551
Planar transmission, 334
Plane wave, 34, 90
Pocklington's integral equation, 94
Point-matching method, 10, 11
Poisson equation, 63
Pole location, 157
Polynomial shape functions, 64
Positive definite, 112, 119
Potential functions:
impedance step, 466
waveguide model, 463
Potentials, 155, 176
Hertz-Debye, 142
normal fields, 140
possible choices, 140
Sommerfeld potentials, 143
transverse potentials, 144
Power dividers:
branch-line, shape optimization, 287, 288
circular disc, 282–284
disc hybrid, 279
sector shaped, 333
Practical substrates, 152
Preconditioned Conjugate Gradient (PCCG) algorithm, 116
Preconditioning, 116
Prescribed nodal values, 78
Principal axis, 567
Principal boundary conditions, 51
Principal value, 169
Principle of superposition, 497
Program, 33, 63, 97, 122–124
Projection, 36, 38
Propagation constant, 395, 506, 507, 514, 515, 519, 523, 536, 537
Propagation velocity, 505, 507, 548, 553

Q factors, 570
Quasi-planar, 500
Quasi-planar structures, 655
Quasi-static case, 163
Quasi-static solution, 33, 123
Quasi-static values, 203
Quasi-TEM, 9, 45, 86
Quasi-TEM mode, 450

Radar cross section, 86
Radiation, 86
Radiation condition, 74
Rayleigh–Ritz, 5, 56–59, 63, 79, 85, 97, 99, 106
Razor testing, 187
Reactive stubs, 517, 524, 540, 546
Real axis, 167
Reciprocity, 95
Reciprocity theorem, 195
Rectangular elements, 64
Rectangular waveguide, 53, 54, 504, 519, 550, 570, 581, 639, 648, 651
  bifurcation, 570
  open-circuited, 570
  scattering on arbitrarily shaped discontinuities, 570
  TE_{n0} modes, 504, 519
Reference planes:
  back-transformation, 484
  crossing, 472
  impedance step, 466
  waveguide model, 466
Reflection coefficient, 516, 582, 583
Relative convergence, 593, 604, 607, 662
Relaxation, 116
Residual error, 82
Residue calculus, 624
Resonance condition, 652, 660, 669, 675
Resonant frequency, 49, 50, 56
Resonators, 416, 425
Ridged waveguide, 555, 650–653, 658, 675
  double-ridge, 650
  single-ridge, 650
Ritz parameters, 56
Rooftop functions, 204

Saguet, 508, 521, 525, 538, 543, 544, 549, 552–557, 559–561, 572
Sampling theorem, 529
Scalar field approximation, 101
Scalar Green's function, 145
Scalar potential, 91, 153, 181, 190
  associated surface waves, 166, 180
  definition, 145, 155
  numerical values, 178, 182
Scalar TLM method, 565
Scalar wave problems, 568
Scattering, 86
Scattering matrix, 207, 501, 508, 546, 622
Scattering parameters, 488
  crossing, 481
  of a discontinuity, 207
  of a set of coupled resonators, 204
  T-junction, 488
Scattering problems, 570
Schrödinger equation, 50
Second-order shape functions, 66
Segmentation, 27, 28
Semiconductor, 34
Semi-open guiding structures, 654
Series-connected, 508, 521, 524, 532
Series network, 509, 527, 531
Series node, 508, 509, 522, 523, 532, 534, 535, 545, 555, 561
Series resistance, 521
Series stubs, 521, 537, 541, 555
Shannon, 529
Shape functions, 63, 64, 68, 70
Shielded microstrip, 8
Shih, 530, 551
Short circuit, 526
Short circuited series, 541
Shunt-connected, 501, 510, 511, 517, 524, 529, 532
Shunt mesh, 527, 531
Shunt nodes, 499, 503, 504, 517, 518, 532, 534, 535, 538, 545, 555, 561
Shunt stubs, 517, 537, 540, 541
Simultaneous displacement method, 115, 116, 117, 119
Single-layer microstrip, 152, 155
Singular behavior at the edges, 678
Singularities, 144, 167
Sitch, 552, 555
Slotline electric fields, basis functions, 352
Slotlines, 351, 655, 667
Slow wave characteristics, 506
Slow-wave structure, 358
Slow-wave velocity, 567
Snell's law, 43, 44, 47
Sommerfeld integral, 165, 170, 171, 172, 176
  Cauchy's principal value, 157
  infinite series expansion, 163
  numerical evaluation, 166
  relation to Green's functions, 136
  zero frequency values, 164
Sommerfeld potentials, 143, 144, 153
Space diagonals, 537, 567
Space domain, 108
S parameters, 622
Sparse, 112
Sparsity, 72, 73
Spatial resolution, 548
Special band, 114
Spectral counterparts, 135
Spectral domain, 9, 22, 23, 136
  complex spectral plane, 155

pole location, 157
spectral domain analysis, 139
spectral variables, 135
technique, 655, 661
Spectral domain approach (SDA), 334, 335
basis functions, 357
discrete Fourier transform, 356
Green's function, 336
immittance approach, 638
Spectral response, 527
Spectral variable, 136
Spurious modes, 101
Spurious solutions, 572
Stability, 572
Static case, 162
Stationariness, 47
Stationary, 49
Steepest descent algorithm, 115
Steepest-gradient method, 115
Step discontinuity, 8, 598, 605, 608, 610, 616, 618, 651, 653, 657–659, 675
Step size, 584
Stored energy, 100
Stratified media:
chain matrix formalism, 152
fields of embedded sources, 146
practical substrates, 152
Stripline currents, basis functions, 342
Striplines, 75, 81, 119, 664
Stronger coupling, 204
Structure functions, 459
Stub-loaded, 519, 522, 524, 531, 540–544
Sturm–Liouville differential equation, 405
Subdomains, 59
Subregions, 39, 477
Subsectional basis functions, 184
Subsections, 39
Successive displacement algorithm, 115, 116, 117, 119, 122
Successive overrelaxation method, 3
Surface current, 185, 190
Surface current density, 199
Surface diagonals, 537
Surface impedance, 137, 512
Surface waves, 166, 180, 208
for a double layer substrate, 160
numerical techniques, 157, 161
poles in the complex spectral plane, 155, 157, 161, 167
for a single layer substrate, 155
structure, 647
Suspended substrate, 123
Symmetrical node, 555, 563–565
Symmetry, 75, 77, 84, 581
Symmetry properties, 141
Synthesis, 35
System equation:
in the spatial domain, 394, 421
in the transform domain, 392

Tangent method, 665
Taylor expansions, 2
TE, 48
TE_{m0} modes, 512
TE_{n0} modes, 504
Tedjini, 555
TE modes, 49, 80, 96, 505, 527, 565
TE polarization, 89
TE wave, 88
TEM mode, 75, 86
TEM-mode stripline, 85
TEM structures, 49
TEM waveguide, 532
Tensor, 99
Terminal plane, 137
Test functions, 10, 37
Thin-wire approximation, 94
Third-order element, 67
Thompson's theorem, 45
Time domain, 107, 108
T-junction:
microstrip, 481
scattering parameters, 488
TLM algorithm, 513
TLM analysis, 569
TLM grid, 582
TLM mesh:
as an anisotropic, periodic structure, 505
arbitrary direction, 508, 543
continuous excitation of, 529
discrete, 501
dispersion of velocity, 503
fields, 529, 531
impedances, 529, 531, 539, 546
impulsive excitation, 526
layout, 581
nondispersive, 550
nonsquare, 553
permeability stub, 540, 541, 545
permittivity stub, 518, 540, 541, 545, 554
phase velocities, 513, 531
radial, 554
scalar, 565–567
series-connected, 508, 521, 524, 531, 532
series node, 523, 527, 533, 535, 545
shunt-connected, 501, 510, 511, 517, 524, 525, 529, 531, 532
shunt node, 518, 527, 533, 535, 538, 545

TLM mesh (*Continued*)
  slow wave characteristics, 506
  storage requirement, 571
  stub-loaded, 519–521, 524, 531, 540, 542–544
  three-dimensional, 531, 532, 534, 535, 538–544, 547, 551, 552, 555, 557, 562–564, 566, 567, 569
  two-dimensional, 501, 503–505, 508–511, 515, 520, 521, 528, 530, 532, 533, 535, 536, 543, 552, 553, 555
  velocity, 508
  wave propagation, 501
TLM method, 496, 497, 500, 503, 526, 572
  algorithm of, 498
  applications, 568
  computation expenditure, 572
  historical background, 498
  scalar, 565
  three-dimensional, 548
  two-dimensional, 500, 501, 503, 532, 548
  variations of, 552
TLM model, 501, 510, 518, 551
TLM network, *see* TLM mesh
TLM procedure, 568
TLM program, 516, 528, 546, 571, 573
TM, 48
TM modes, 80, 90, 96, 505, 565
TM plane wave, 87
TM polarization, 89, 90
TM surface wave, 166
TM wave, 88
TM waveguide mode, 90
Toeplitz matrices, 421, 431
Transformation:
  inverse, 394, 403, 421
  matrices, 402, 407, 414, 418, 434, 441
  orthogonal, 397
  to principal axes, 387
Transient analysis, 570
Transient time, 103
Transmission coefficient, 525
Transmission equation, 506, 519, 541
Transmitting antenna, 89
Transverse equivalent circuit, 645, 647
Transverse modal analysis, 638
Transverse potentials, 144
Transverse resonance diffraction, 637, 638
Transverse resonance technique, 14
Triangles, 63
Triangular elements, 61, 62, 64, 67
Triangulation, 113
Tridiagonal matrix, 17
Truncation error, 527, 549, 550
Two-layer microstrip, 181

Underwater acoustic scattering, 571
Uniaxial discontinuities, 570
Uniform conducting cylinder, 87, 90
Unilateral finline, 551, 662, 682, 684
Uniqueness, 92
Unit cell, 541
Upper bound, 86

Van der Pol, 174
Variable band, 114
Variable band matrices, 112
Variable mesh, 123, 552
Variational approach, 85, 86
Variational expression, 4, 41, 42, 49, 50, 51, 52, 53, 56, 82, 99
Variational form, 99, 104
Variational formulation, 85
Variational method, 33, 38, 40
Vector norms, 109
Vector potentials, 91, 176
Vectors, 112
Velocities, 524, 525
Velocity dispersion, 503, 508, 515
Velocity error, 521, 550, 551
Vibrating string, 50, 56, 59

Walls of symmetry, 538
Wave amplitudes, 478
Wave equation, two-dimensional, 504
Waveguide discontinuities:
  arbitrary cross-sectional geometry, 500
  bends, 500
  cylindrical, 500
  filters, 500
  irises, 500
  inductive strips, 500
  inhomogeneously filled, 500
  n furcations in the *H* plane, 500
  T-junctions, 500
Waveguide model, 449
  boundary conditions, 463
  cutoff frequencies, 452
  guide wavelength, 460
  microstrip discontinuities, 447
  Neumann coefficients, 463
  orthogonality, 463
  potential functions, 463
  reference planes, 466
  structure functions, 459
Waveguides, 25, 95, 500, 505, 512, 532, 570, 582
  circular, 649, 654
  dielectric, 383, 405, 425
  inhomogeneously filled, 646, 650

and lossless dielectric materials, 678
metallic, 653
optical, 405
quasiplanar, 382
rectangular, 639, 648, 651
ridged, 650–653, 658, 675
single-ridge, 650
slab-loaded, 664
slitted rectangular, 654
Wave impedance, 512, 516
Wavelength, 550
Wave propagation, 501, 508, 560
Wave properties, 518, 568
of coarse meshes, 505
of infinitesimally fine meshes, 503
of series meshes, 522, 523
of shunt meshes, 518
slow, 505
of stub-loaded meshes, 522
of three-dimensional TLM network, 535, 541
Wave velocities, 537, 567
Weighted-averages algorithm, 172
Weighted-residual approach, 115
Weighted residuals, 33, 35, 36, 38, 63, 83, 108
Weight functions, 37, 38
Weighting functions, 82
Weissfloch method, 665, 670
Well-conditioned problem, 109
Well-posed problem, 109
Wiener-hopt technique, 624
Wire-grid modeling, 94, 107
WR(90) waveguide, 527, 538

Yeh, 544
Yoshida, 568, 569

Zero thickness, 677